Understanding Social Problems

9th edition

Understanding Social Problems

9th edition

Linda A. Mooney

East Carolina University

David Knox

East Carolina University

Caroline Schacht

East Carolina University

CENGAGE
Learning·

Australia • Brazil • Japan • Korea • Mexico • Singapore • Spain • United Kingdom • United States

CENGAGE
Learning®

Understanding Social Problems,
Ninth Edition
Linda A. Mooney, David Knox,
and Caroline Schacht

Product Director: Marta Lee-Perriard

Product Manager: Seth Dobrin

Content Developer: Nicole Bridge

Content Coordinator: Coco Bator

Product Assistant: Jessica Alderman

Media Developer: John Chell

Marketing Manager: Kara Kindstrom

Content Project Manager: Cheri Palmer

Art Director: Caryl Gorska

Manufacturing Planner: Judy Inouye

Rights Acquisitions Specialist: Don Schlotman

Production Service: Jill Traut, MPS Limited

Photo Researcher: PreMedia Global

Text Researcher: PreMedia Global

Copy Editor: Heather McElwain

Illustration and Composition: MPS Limited

Text Designer: Lisa Buckley

Cover Designer: Lee Friedman

Cover Image: Mario Tama/Getty Images

Design Element Credits: Nicemonkey/Alamy;
   Kirsty Pargeter/Alamy

Library of Congress Control Number: 2013952121

ISBN-13: 978-1-285-74650-0

ISBN-10: 1-285-74650-0

Cengage Learning
200 First Stamford Place, 4th Floor
Stamford, CT 06902
USA

Cengage Learning is a leading provider of customized learning solutions with office locations around the globe, including Singapore, the United Kingdom, Australia, Mexico, Brazil, and Japan. Locate your local office at **www.cengage.com/global.**

Cengage Learning products are represented in Canada by Nelson Education, Ltd.

To learn more about Cengage Learning Solutions, visit **www.cengage.com.**

Purchase any of our products at your local college store or at our preferred online store **www.cengagebrain.com.**

Printed in the United States of America
3 4 5 6 7 18 17 16 15

It always seems impossible until it's done.

—Nelson Mandela, July 18, 1918–December 5, 2013, Civil rights activist, Nobel Peace Prize winner, first black president of South Africa

Brief Contents

Contents

PART 3 Problems of Inequality

PART 4 Problems of Globalization

Features

Preface

Understanding Social Problems is intended for use in a college-level sociology course. We recognize that many students enrolled in undergraduate sociology classes are not sociology majors. Thus, we have designed our text with the aim of inspiring students—no matter what their academic major or future life path may be—to care about the social problems affecting people throughout the world. In addition to providing a sound theoretical and research basis for sociology majors, *Understanding Social Problems* also speaks to students who are headed for careers in business, psychology, health care, social work, criminal justice, and the nonprofit sector, as well as to those pursuing degrees in education, fine arts and the humanities, or to those who are "undecided." Social problems, after all, affect each and every one of us, directly or indirectly. And everyone—whether a leader in business or politics, a stay-at-home parent, or a student—can become more mindful of how his or her actions (or inactions) perpetuate or alleviate social problems. We hope that *Understanding Social Problems* not only informs but also inspires, planting seeds of social awareness that will grow no matter what academic, occupational, and life path students choose.

New to This Edition

The ninth edition of *Understanding Social Problems* features an increased focus on wealth and economic inequality, and a renaming of Chapter 6 from "Poverty and Economic Inequality" to "Economic Inequality, Wealth, and Poverty." Pedagogical features that students and professors have found useful have been retained, including a running glossary, list of key terms, chapter reviews, and *Test Yourself* sections. Most of the opening vignettes in the ninth edition are new, as are many of the *What Do You Think?* sections, which are designed to engage students in critical thinking. Many of the boxed chapter features (*The Human Side, Social Problems Research Up Close,* and *Self and Society*) have been updated or replaced with new content. This revised edition also includes a new photo essay on "Globesity," and a new *Animals and Society* essay on the role that companion animals play in the lives of homeless individuals. Finally, the ninth edition has updated research, data, tables, figures, and new photos in each chapter, as well as new and revised material, detailed as follows.

Chapter 1 ("Thinking about Social Problems") examines the use of cell phones and distracted driving as a recent social problem. This revised chapter features a new *Self and Society* feature that will help readers assess personal values, and a new section on web-based surveys. The *Photo Essay* on "Students Making a Difference" includes new material on student protests arising from rapes in India. A new *What Do You Think?* section asks students to consider the new mandate in many universities that requires students to take courses in global issues.

Chapter 2 ("Physical and Mental Health and Health Care") begins with a new opening vignette that conveys the emotional and financial difficulty of paying for medical care. This revised chapter has expanded the focus on the problems of overweight and obesity, and includes a new photo essay on "Globesity." A new section "Are Americans the Healthiest Population in the World?" discusses a U.S. National Research Council and Institute of Medicine report that found that, despite the fact that the United States spends more on health care per person than any other industrialized country, Americans die sooner and have higher rates of disease or injury. This new section explores the social and economic factors that contribute to the poor U.S. health. We added a new section on "Mental Illness among College Students," along with a new table on "Top Reasons for Mental Health Diagnosis Disclosure or Nondisclosure among College Students." The section that addresses ways to improve mental health care includes new Associated Press guidelines for how editors and reporters report about mental illness and initiatives in the Affordable Care Act that are designed to improve access to mental health care. This revised chapter also features a new section on "Complementary and Alternative Medicine (CAM)," including a new table on "States with Legalized Medical Marijuana," and a new *The Human Side* feature, "Testimony from Medical Marijuana Patients." Other changes to this chapter include a new discussion of the growing medical tourism industry; updated information about the widespread use of antibiotics and the rise of antibiotic-resistant "superbugs"; an expanded discussion of "Socioeconomic Status and Health"; a new section on "Life Expectancy and Mortality in Low, Middle, and High Income Countries"; a new section on "Improving Health in Low-, Middle-, and High-Income Countries"; an expanded and updated section on "The High Cost of Health Care"; and new sections on "Regulation of Food Marketing to Youth," "Local and State Antiobesity Policies," and "Workplace Wellness Programs." New research on the health of U.S. Hispanic populations was added. A new *Social Problems Research Up Close* feature describes research on "Texting Healthy Lifestyle Messages to Teens." Finally, although the rollout of the Affordable Care Act (ACA) is still in process as this text goes to press, we have updated information about the ACA.

Chapter 3 ("Alcohol and Other Drugs") has been reorganized in this edition, and now features a "Synthetic Drugs" section, as well as new sections on "Gender and the War on Drugs," "Race, Ethnicity, and the War on Drugs," the "Cost of the War on Drugs," and the impact of the war on drugs on Latin American countries. New topics discussed include the impact of social media sites on alcohol and drug use among adolescents, tobacco companies' sale of small, inexpensive, flavored cigars to avoid federal regulations, the prescription drug epidemic among youth, and parental attitudes toward psychotherapeutic drugs. Also new to this chapter is the impact of drugs in schools on students and on the learning environment, and changes in drug use over time. Turning to drug abuse prevention, research in the ninth edition compares the cost of drug treatment versus incarceration of drug addicts, discusses variables that predict drug court successes, and reviews the use of monetary incentives in smoking cessation programs. There is also updated information on recent court decisions in tobacco and alcohol litigation and on U.S. drug policies. New *What Do You Think?* questions, designed to start a lecture and teach students critical thinking skills include (1) whether or not electronic cigarettes should be under the control of the U.S. Food and Drug Administration, (2) should employees who use marijuana, even if it is after work and legal in the state, be able to be fired because such use remains illegal under federal law, (3) why is binge drinking increasing among women and girls, and (4) should women who are pregnant be arrested for chemical endangerment of a child, even if the child is born healthy? There are also new statistics, tables, and figures from the National Survey on Drug Use and Health, the World Health Organization, the World Drug Report, European Monitoring Centre for Drugs and Drug Addiction, Monitoring the Future, the Center for Disease Control and Prevention, Alcoholics Anonymous, National Institute on Drug Abuse, Office of National Drug Control Policy, among other sources. Finally, there is a new *The Human Side* entitled "Real Stores, Real People," which enlightens students on the consequences of drug abuse for users, family, and friends.

New to **Chapter 4** ("Crime and Social Control") are the most recent statistics, tables, and figures from the Federal Bureau of Investigation, the U.S. Department of Justice, the National Crime Victimization Survey, the Bureau of Justice Statistics, the National Gang Center, National Center for Victims of Crime, Internet Crime Complaint Center, the Innocence Project, the National Security Council (NSC), and the United Nations Office on Drugs and Crime. There is a new opening vignette, and a greater emphasis on transnational crime including human trafficking, child pornography, and organized crime. This revised chapter is grounded in theory and discusses research on the effects of legalizing prostitution on sex trafficking, causes of increased violence against women and girls, the prevalence of children's exposure to violence, and how media reports of crime and violence impact fear of crime. Among other crime-related topics new to this edition are the Sandy Hook Elementary School killings, the renewed debate over gun control, and the resulting pressure from the National Rifle Association to limit gun regulations. There are expanded sections on white-collar crime, computer crime, and organized crime, and new discussions on why people fail to report violent victimization, litigation against technology companies, data breaches, the demographics of youth gangs, theories of why criminal behavior declines with age, corporate pollution and mortality rates, the *punitivity ratio,* the problems associated with lethal injection, and global gun ownership and firearm murders. Combining the latest topics with sociologically meaningful questions, this chapter also includes many new *What Do You Think?* sections. Students are asked to think about whether or not "stand your ground laws" are necessary to protect innocent people, whether or not toy guns that look realistic should be banned, the problems associated with the privatization of prisons, why the number of serial killings has changed so dramatically over the last 50 years, whether Edward Snowden is a traitor or a hero, the problems associated with violent victimization, and at what age, if any, parents should be allowed to give their child a gun.

In response to reviewer feedback, **Chapter 5** ("Family Problems") was reorganized so that the topic of divorce precedes the topic of abuse. This revised chapter includes new data on changing patterns and trends in U.S. families, including a new section on three-generation households in the United States. A new section on "Arranged Marriages versus Self-Choice Marriages" has also been added. This revised chapter also includes updated information about the legal status of same-sex couples, an expanded and updated section on "Parental Alienation," new research and data on intimate partner violence and child abuse, an updated discussion of gender differences in the perpetration of intimate partner violence, and new research on relationship "churning" (the pattern of breaking up and reconciling) and its link to physical and verbal abuse. A new *What Do You Think?* section asks why sibling bullying—perhaps the most common form of abuse in families—is accepted as normal and expected behavior.

Chapter 6 features an increased focus on wealth and economic inequality, and a renaming of the chapter from "Poverty and Economic Inequality" to "Economic Inequality, Wealth, and Poverty" A new opening vignette illustrates how economic inequality leads to different life outcomes. This revised chapter includes new data on inequality in the global distribution of household wealth, inequality in the United States, and updated Census data on poverty and poverty thresholds. A new section looks at "The 'One Percent': Wealth in the United States." There is also a new discussion of political inequality and political alienation among the poor, and new sections on "Legal Inequality," "Region and Poverty," and "Reducing U.S. Economic Inequality." This revised chapter expands the discussion of the Occupy Wall Street movement, updates material on "Natural Disasters, Economic Inequality, and Poverty" (e.g., earthquakes in Haiti and Chile, and Hurricane Sandy), and includes new coverage of meritocracy and, progressive taxes. This chapter contains a new *Animals and Society* feature that presents research on the role that companion animals play in the lives of homeless individuals, and a new *Social Problems Research Up Close* feature: "Patchwork: Poor Women's Stories of Resewing the Shredded Safety Net." A new "What Do You Think?" section asks readers how their eating patterns would change if they depended on SNAP benefits, which averaged $4.50 per recipient per day in 2012, for their food. Finally, this chapter presents new and updated material on corporate welfare, international poverty reduction efforts, and myths about welfare.

Chapter 7 ("Work and Unemployment") includes a revised and expanded discussion of socialism and capitalism, new material on *Citizens United v. Federal Election Commission,* and a new section on "Worker Cooperatives: An Alternative to Capitalism." Other new topics in this ninth edition include 3-D printing as an example of automation in the workplace, the toxic workplace, and workers' self-directed enterprises. This chapter also includes an updated table on "Employer-Based Work–Life Benefits and Policies in the United States." A new *What Do You Think?* section asks readers if Americans who critically examine capitalism are "un-American" or disloyal to the United States. A new *The Human Side* feature describes working conditions in the poultry processing industry.

Students will be particularly interested in the updated and expanded discussion of employment concerns of recent college graduates.

Chapter 8 ("Problems in Education") has been significantly revised with a new opening vignette, a new *Self and Society* feature on student ethics, and a new *The Human Side* on "The Death of Dylan Hockley," one of the 26 victims at Sandy Hook Elementary School. There are new statistics, tables, and figures from the National Center for Educational Statistics, Organization for Economic Cooperation and Development (OECD), *Programme for International Student Assessment* (PISA), Children's Defense Fund, UNESCO NAEP, Assessment of Educational Progress (NAEP), Trends in International Mathematics and Science Study (TIMSS), World Literacy Foundation, AFE, and the American Federation of Teachers. The Higher Education section has been reorganized and now includes sections on the cost of higher education, higher education and race and ethnicity, and community colleges. This heavily revised chapter also has new sections on socioeconomic integration, teacher effectiveness, the lack of financial support for schools, educational policy across the states, common core standards, testing and accountability, and advocacy and grassroots movements in education. There are a range of new topics and discussions on global literacy rates, the trend toward increased school segregation, *earnings premiums,* the societal costs of high school dropouts, the consequences of school budget cuts, the unintended consequences of the GED, the relationship between zero-tolerance school policies and the school-to-prison pipeline, the Tuition Equality Act, and the Student Loan Certainty Act. New terms include, although are not limited to, the "rug rat race" and "parent trigger laws." New *What Do You Think?* sections provide questions that include: (1) Should corporations be able to advertise in schools? (2) Should undocumented graduates of a states' high schools be able to pay in-state tuition? (3) Should teachers be required to take cultural sensitivity courses? (4) Should a teacher's salary be determined by how well their students perform? and, finally, (5) Should parents who homeschool their children receive financial compensation for doing so?

Chapter 9 ("Race, Ethnicity, and Immigration") opens with a new vignette about the Trayvon Martin/George Zimmerman case. This chapter features a new *Self and Society* feature: "How Do You Explain White Racial Advantage?" New data on U.S. intermarriage rates and perceptions of race and ethnic relations in the United States have been added. This updated chapter also includes a revised discussion of the history of U.S. immigration, adding the 1965 Hart-Celler Act and an updated discussion of state and federal immigration policies. The chapter presents new journalistic rules prohibiting the use of the terms *illegal immigrant* and *illegal alien.* We have added a new discussion of how U.S. civil rights legislation was enacted, in part, as a response to the Cold War. A provocative new *What Do You Think?* section examines the use of the "N-word." Other new topics include institutional racism, color-blind racism, and "racial grammar." Finally, this chapter contains updated information about affirmative action, a discussion of changes to the Voting Rights Act, and new data on the incidence of hate crimes and hate crimes on campus.

Chapter 10 ("Gender Inequality") begins with a new opening vignette, new *Social Problems Research Up Close* on "Overdoing Gender," and a new *The Human Side* on sex discrimination in the workplace. There are new statistics, tables, and figures from the World Economic Forum, World Health Organization, U.S. Census Bureau, Center for American Women and Politics, Amnesty International, American Association of University Women, Equal Employment Opportunity Commission, International Labour Organization, United Nations, Institute for Women's Policy Research, and the United Nations Educational, Scientific and Cultural Organization. New discussion lecture starters and questions include: Should transgender college students be permitted to compete in sports? Does gender role socialization explain the overrepresentation of men in politics? Should girls who "make the team" be able to play football with middle school boys? Should tradition or gender equality rule at a holy prayer site? and finally, Is there a "war against women"? Sections new to this edition include (1) the persistence of occupational sex segregation, (2) why the gender pay gap still exists, (3) gender and school policies and programs, (4) and the cult of thinness. New topics of discussion include Ridgeway's *Framed by Gender,* labor policies that discourage women from working full time, *gender* *deviance,* academic performance of students in single-sex versus coed schools, gender representation in top grossing family films, PSAs by "Jane," ordination of female bishops in the Anglican Church, "soft" versus "tough" countries and masculine gender norms, body image and femininity and masculinity, WWII comfort stations, gender identity threat, and the *masculine overcompensation* thesis.

Chapter 11 ("Sexual Orientation and the Struggle for Equality") begins with a new opening vignette on the repeal of the Defense of Marriage Act (DOMA). There is also a new *Social Problems Research Up Close* on suicide and lesbian, gay, and bisexual youth, and a new *The Human Side* entitled "An Open Letter to the Church from My Generation." New *What Do You Think?* questions include: (1) What should take precedence when human rights clash with a segment of the populations religious beliefs? (2) Will Pope Francis's comment that he would not judge gay and lesbian people have an impact on peoples' attitudes toward those who are same-sex attracted? (3) Should the Boy Scouts of America ban gay scout leaders? (4) If the denial of marriage equality to lesbians, gays, and bisexuals is based on religious ideology, is that a violation of the Second Amendment? (5) Do boycotts of businesses or organizations with antigay positions have any impact in the broader fight for LGBT equality? and, finally, (6) Does repealing antigay laws attack religious freedom of those who are morally opposed to same-sex relations? There are also new table, figures, and statistics from the Human Rights Campaign, the National College Health Association, the Palm Center, Exodus International, the Williams Institute, the Gallup Poll, the Gay, Lesbian and Straight Education Network, the International Gay and Lesbian Human Rights Commission, the International Lesbian, Gay, Bisexual, Trans, and Intersex Association, the Pew Research Center, and the National Coalition of Anti-Violence Programs.

Chapter 12 ("Population Growth and Aging") has been extensively updated to include the most recent population statistics available from the United Nations and includes updated figures on population distribution and growth and population aging. The chapter includes a new discussion of the "club sandwich generation," which consists of adults who are taking care of three generations (children, parents, and grandparents, or grandchildren, children, and parents). The discussion about Social Security and retirement security has been updated, and a new *What Do You Think?* question pertains to whether mandatory retirement is ever justified, or if it is age discrimination. The *Animals and Society* feature on "Pet Overpopulation" has been revised to include a new discussion of the "no kill" movement in animal shelters.

Chapter 13 ("Environmental Problems") presents several new key terms: *Earth Overshoot Day; perceived obsolescence; biomass;* and *climate denial machine.* Another new key term is *pinkwashing,* which is the practice of using the color pink or pink ribbons and other marketing strategies that suggest that a company is helping to fight breast cancer when the company may be manufacturing or selling chemicals linked to cancer. Along with updated information about global warming and climate change, we have added new topics: the effects of climate change on recreation and the growing problem of plastics contaminating the oceans and harming marine life. This revised chapter includes an updated discussion of how industries use their power and wealth to influence politicians' environmental and energy policies as well as the public's beliefs about environmental issues. This chapter also includes a new discussion of the Environmental Literacy Improvement Act—a bill proposed by the American Legislative Exchange Council (ALEC) that aims to weaken teachers' ability to teach K–12 students the science of climate change. New research findings on the effects of fracking are presented. This revised chapter includes updated statistics and information from a variety of sources, including the Global Footprint Network, the United Nations Development Programme, United Nations Environment Program, and the Environmental Protection Agency. We have also added mention of Rachel Carson's contribution to environmental activism.

Chapter 14 ("Science and Technology") begins with a new opening vignette on Facebook and a case of mistaken identity, a new *Self and Society,* and a new *The Human Side.* New figures, tables, and statistics come from the National Science Foundation, the World Economic Forum, the United Nations, the OpenNet Initiative, the Digest of Education Statistics, the U.S. Citizenship and Immigration Services, the Entertainment Software Association, the Federal Trade Commission, the Information Technology and Innovation Foundation, the

International Federation of Robotics, the National Institute of Health, the President's Council of Advisors on Science and Technology, among other sources. New topics of research and discussion include protests of Monsanto and genetically modified organisms; proposed state fetus personhood laws; state-level restrictions on abortion; Internet recommender systems; the X Keyscore computer program; technology's impact on manufacturing and service-sector jobs; the relationship between technology, declining incomes, and greater economic inequality; maintaining large genetic databases on citizens; the demographics of computers, the Internet, and smartphone users; technology and health care cost; the Marketplace Fairness Act; the book *The Shallows: What the Internet is Doing to Our Brains;* racism in online and off-line role-playing games; sexism in video game advertising; *vicarious management;* games with roots in white supremacy; the networked readiness index; the E-government Development Index; Stem Cell Research Advancement Act; and the National Security Administration (NSA) surveillance program. In addition to these topics, new *What Do You Think?* questions ask students to think about why abortion rates are higher in countries where abortion is illegal, whether or not technology leads to loneliness and isolation, and the consequences of home genetic testing kits.

Chapter 15 ("Conflict, War, and Terrorism") contains new statistics from the Department of Defense, the U.S. State Department, the Stockholm International Peace Research Institute, the Pew Research Center, the National Counterterrorism Center, the United Nations, the National Center for Veterans Analysis and Statistics, the Office of Weapons Removal and Abatement, the White House, the Institute for Economics & Peace, the U.S. Central Command, the Congressional Research Service, the Defense Manpower Data Center, the International Campaign to Ban Landmines, the National Counterterrorism Center, and the Office of Management and Budget. There are several new *What Do You Think?* section questions, including (1) Does the use of drones prevent violence by killing terrorist leaders, or spur more violence by inspiring others to carry out terrorist attacks? (2) Under what conditions do you think the United States should send arms to support fighters in other countries? (3) In what ways do you think inequality contributes to a "culture of rape," both in the military and in civilian society? (4) Are the actions of Manning and Snowden those of traitors or heroes? (5) Under what conditions do you think the United States should withdraw foreign aid from countries it has supported for decades? (6) Does setting preconditions make it more or less likely that the United States will change Iran's policies? There is also a new *Social Problems Research Up Close* on "Combat, Mental Illness, and Military Suicides." New and updated topics include the Arab Spring, the Syrian Civil War, the Egyptian coup, the use of chemical weapons, the Westgate Mall attacks in Nairobi, the Boston Marathon bombings, Bradley Manning and Edward Snowden controversies, the democratic peace theory, military policy regarding gay marriages and benefits, women in combat, the military sexual assault crisis, security and ethical concerns surrounding drone strikes, the drawdowns from Iraq and Afghanistan, the costs of the Iraq and Afghan wars, military contractors and the MPRI, U.S.–Chinese relations, Iran nuclear negotiations, North Korea missile tests, and the Ground Combat Exclusion Policy.

Features and Pedagogical Aids

We have integrated a number of features and pedagogical aids into the text to help students learn to think about social problems from a sociological perspective. Our mission is to help students think critically about social problems and their implications, and to increase their awareness of how social problems relate to their personal lives.

Boxed Features

Animals and Society. Several chapters contain a feature called *Animals and Society,* which examines issues, problems, policies, and/or programs concerning animals within the context of the social problem discussed in that chapter. For example, Chapter 5 ("Family Problems") includes an *Animals and Society* feature that examines "Pets and

Domestic Violence," and in Chapter 14 ("Science and Technology"), the *Animals and Society* feature discusses "The Use of Animals in Scientific Research."

Self and Society. Each chapter includes a social survey designed to help students assess their own attitudes, beliefs, knowledge, or behaviors regarding some aspect of the social problem under discussion. In Chapter 5 ("Family Problems"), for example, the "Abusive Behavior Inventory" invites students to assess the frequency of various abusive behaviors in their own relationships. The *Self and Society* feature in Chapter 3 ("Alcohol and Other Drugs") allows students to measure the consequences of their own drinking behavior and compare it to respondents in a national sample, and students can assess their knowledge of nuclear weapons in Chapter 15 ("Conflict, War, and Terrorism").

The Human Side. Each chapter includes a boxed feature that makes the social problems under discussion more salient by describing personal experiences of individuals who have been affected by them. *The Human Side* feature in Chapter 4 ("Crime and Social Control"), for example, describes the horrific consequences of being a victim of rape, and *The Human Side* feature in Chapter 9 ("Race, Ethnicity, and Immigration") describes the experiences of an immigrant day laborer who was victimized by a violent hate crime. Further, in Chapter 11 ("Sexual Orientation and the Struggle for Equality"), *The Human Side* feature details Dannika Nash's blog entry inspired by Macklemore's marriage equality anthem "Same Love."

Social Problems Research Up Close. This feature, found in every chapter, presents examples of social science research, summarizing the sampling and methods involved in data collection, and presenting findings and conclusions of the research study. The *Social Problems Research Up Close* topics include bullying, job loss in midlife, computer hacking, the changing nature of marriage in the United States, and two-faced racism.

Photo Essay. Chapter 1 ("Thinking about Social Problems") features a photo essay on "Students Making a Difference." Chapter 2 ("Physical and Mental Health and Health Care") includes a new photo essay titled "Globesity." "Prison Programs that Work" is in Chapter 4 ("Crime and Social Control"). In Chapter 6 ("Economic Inequality, Wealth, and Poverty"), a photo essay covers the topic "Lack of Clean Water and Sanitation among the Poor." Chapter 7 ("Work and Unemployment") includes a photo essay on "Child Labor in U.S. Agriculture," and a photo essay in Chapter 10 ("Gender Inequality") is titled, "The Gender Continuum." Lastly, Chapter 13 ("Environmental Problems") depicts the horrors of the "Deepwater Horizon Oil Rig Explosion and Fukushima Nuclear Power Plant Accident."

In-Text Learning Aids

Vignettes. Each chapter begins with a vignette designed to engage students and draw them into the chapter by illustrating the current relevance of the topic under discussion. For example, Chapter 2 ("Physical and Mental Health and Health Care"), begins with an account of a woman who was devastated by the high medical bills incurred by her husband. Chapter 9 ("Race, Ethnicity, and Immigration") begins with the story of the Trayvon Martin and George Zimmerman case, and Chapter 15 ("Conflict, War, and Terrorism") opens with a portrayal of a young veteran with PTSD (post-traumatic stress disorder).

Key Terms. Important terms and concepts are highlighted in the text where they first appear. To reemphasize the importance of these words, they are listed at the end of every chapter and are included in the glossary at the end of the text.

Running Glossary. This ninth edition continues the running glossary that highlights the key terms in every chapter by putting the key terms and their definitions in the text margins.

What Do You Think? Sections. Each chapter contains multiple sections called *What Do You Think?* These sections invite students to use critical thinking skills to answer questions about issues related to the chapter content. For example, one *What Do You*

Think? feature in Chapter 3 ("Alcohol and Other Drugs") asks students "Did the Colorado Court of Appeals that decided workers can be fired for using marijuana after work hours—even if it's prescribed for a medical condition make the right choice?" A *What Do You Think?* Feature in Chapter 11 ("Sexual Orientation and the Struggle for Equality") asks if the repeal of antigay laws attacks religious freedom and denies free speech to those who are morally opposed to homosexuality. In Chapter 12 ("Population Growth and Aging"), a *What Do You Think?* feature asks readers if birthday cards and jokes that make fun of aging are a form of ageism.

Glossary. All key terms are defined in the page margins as well as in the end-of-text glossary.

Understanding [Specific Social Problem] Sections. All too often, students, faced with contradictory theories and study results walk away from social problems courses without any real understanding of their causes and consequences. To address this problem, chapter sections titled "Understanding [specific social problem]" cap the body of each chapter just before the chapter summaries. Unlike the chapter summaries, these sections synthesize the material presented in the chapter, summing up the present state of knowledge and theory on the chapter topic.

Supplements

The ninth edition of *Understanding Social Problems* comes with a full complement of supplements designed with both faculty and students in mind.

Supplements for the Instructor

Online Instructor's Resource Manual with Test Bank. This supplement offers instructors learning objectives, key terms, lecture outlines, student projects, classroom activities, Internet exercises, and video suggestions. Test items include multiple-choice and true-false questions with answers and text references, as well as short-answer and essay questions for each chapter.

Cengage Learning Testing Powered by Cognero. The Test Bank is also available through Cognero, a flexible, online system that allows you to author, edit, and manage test bank content as well as create multiple test versions in an instant. You can deliver tests from your school's learning management system, your classroom, or wherever you want.

Online PowerPoints. These vibrant, Microsoft® PowerPoint® lecture slides for each chapter assist you with your lecture, by providing concept coverage using images, figures, and tables directly from the textbook!

AIDS in Africa DVD. Southern Africa has been overcome by a pandemic of unparalleled proportions. This documentary series focuses on Namibia, a new democracy, and the many actions that are being taken to control HIV/AIDS there. Included in this series are four documentary films created by the Project Pericles scholars at Elon University.

The Wadsworth Sociology Video Library Vol. I–IV. These DVDs drive home the relevance of course topics through short, provocative clips of current and historical events. Perfect for enriching lectures and engaging students in discussion, many of the segments on this volume have been gathered from the BBC Motion Gallery. Ask your Cengage Learning representative for a list of contents.

Supplements for the Student

Aplia for *Understanding Social Problems* helps students learn to use their sociological imagination through compelling content and thought-provoking questions. Students complete interactive activities that encourage them to think critically by practicing and applying course concepts. These valuable critical thinking skills help students become thoughtful and engaged members of society. Aplia for *Understanding Social Problems* includes the following features:

- Auto-assigned, auto-graded homework holds students accountable for the material before they come to class, increasing their effort and preparation.
- Immediate, detailed explanations for every answer enhance student comprehension of sociological theories and concepts.
- Grades are automatically recorded in the instructor's Aplia Gradebook.
- Gradebook Analytics allow instructors to monitor and address performance on a student-by-student and topic-by-topic basis.

Go to login.cengagebrain.com to access Aplia for *Understanding Social Problems*.

CourseReader for Sociology. CourseReader for Sociology allows you to create a fully customized online reader in minutes. Access a rich collection of thousands of primary and secondary sources, readings, and audio and video selections from multiple disciplines. Each selection includes a descriptive introduction that puts it into context, and every selection is further supported by both critical thinking and multiple-choice questions designed to reinforce key points. This easy-to-use solution allows you to select exactly the content you need for your courses, and is loaded with convenient pedagogical features like highlighting, printing, note taking, and downloadable MP3 audio files for each reading. You have the freedom to assign and customize individualized content at an affordable price.

Acknowledgments

This text reflects the work of many people. We would like to thank the following for their contributions to the development of this text: Seth Dobrin, Product Manager; Nicole Bridge, Content Developer; John Chell, Media Developer; Cheri Palmer, Senior Content Project Manager; Jill Traut, Project Manager at MPS Limited; Caryl Gorska, Senior Art Director; Sachin Das, Associate Project Manager-Text Permissions; and Don Schlotman, Rights Acquisitions Editor. We would also like to acknowledge the support and assistance of Carol L. Jenkins, Emily Schacht, Molly Clever, Marieke Van Willigen, James and Mabelle Miller, Don and Jean Fowler, and Megan Allen. To each, we send our heartfelt thanks. Special thanks also to George Glann, whose valuable contributions have assisted in achieving the book's high standard of quality from edition to edition.

Additionally, we are indebted to those who read the manuscript in its various drafts and provided valuable insights and suggestions, many of which have been incorporated into the final manuscript:

Aimee Burgdorf, Southwest Tennessee Community College

Dr. Rebecca Fahrlander, Metropolitan Community College

Sharon Hardesty, Eastern Kentucky University

Robert Hollenbaugh, Irvine Valley College

Monica Johnson, Northwest Mississippi Community College

Judith Kirwan Kelley, Curry College

Steve McGlamery, Radford University

Nancy Sonleitner, University of Tennessee at Martin

Warren Waren, University of Central Florida

Steve Willis, National Park Community College Hot Springs

We are also grateful to reviewers of the previous editions: Maria D. Cuevas, Yakima Valley Community College; Kim Gilbert, Iowa Lakes Community College; Heather S. Kindell, Morehead State University; John J. Leiker, Utah State University; Sharon A. Nazarchuk, Lackawanna College; Jewrell Rivers, Abraham Baldwin Agricultural College; Michelle Willms, Itawamba Community College; Sally Vyain, Ivy Tech Community College of Indiana; David Allen, University of New Orleans; Patricia Atchison, Colorado

State University; Wendy Beck, Eastern Washington University; Walter Carroll, Bridgewater State College; Deanna Chang, Indiana University of Pennsylvania; Roland Chilton, University of Massachusetts; Verghese Chirayath, John Carroll University; Margaret Choka, Pellissippi State Technical Community College; Kimberly Clark, DeKalb College-Central Campus; Anna M. Cognetto, Dutchess Community College; Robert R. Cordell, West Virginia University at Parkersburg; Barbara Costello, Mississippi State University; William Cross, Illinois College; Kim Davies, Blinn College; Jane Ely, State University of New York-Stony Brook; William Feigelman, Nassau Community College; Joan Ferrante, Northern Kentucky University; Robert Gliner, San Jose State University; Roberta Goldberg, Trinity Washington University; Roger Guy, Texas Lutheran University; Julia Hall, Drexel University; Millie Harmon, Chemeketa Community College; Madonna Harrington-Meyer, University of Illinois; Sylvia Jones, Jefferson Community College; Nancy Kleniewski, University of Massachusetts, Lowell; Daniel Klenow, North Dakota State University; Sandra Krell-Andre, Southeastern Community College; Pui-Yan Lam, Eastern Washington University; Mary Ann Lamanna, University of Nebraska; Phyllis Langton, George Washington University; Cooper Lansing, Erie Community College; Linda Kaye Larrabee, Texas Tech University; Tunga Lergo, Santa Fe Community College, Main Campus; Dale Lund, University of Utah; Lionel Maldonado, California State University, San Marcos; J. Meredith Martin, University of New Mexico; Judith Mayo, Arizona State University; Peter Meiksins, Cleveland State University; JoAnn Miller, Purdue University; Clifford Mottaz, University of Wisconsin-River Falls; Lynda D. Nyce, Bluffton University; Frank J. Page, University of Utah; James Peacock, University of North Carolina; Barbara Perry, Northern Arizona University; Ed Ponczek, William Rainey Harper College; Donna Provenza, California State University at Sacramento; Cynthia Reynaud, Louisiana State University; Carl Marie Rider, Longwood University; Jeffery W. Riemer, Tennessee Technological University; Cherylon Robinson, University of Texas at San Antonio; Rita Sakitt, Suffolk County Community College; Mareleyn Schneider, Yeshiva University; Paula Snyder, Columbus State Community College; Lawrence Stern, Collin County Community College; John Stratton, University of Iowa; D. Paul Sullins, The Catholic University of America; Vickie Holland Taylor, Danville Community College; Joseph Trumino, St. Vincent's College of St. John's University; Robert Turley, Crafton Hills College; Alice Van Ommeren, San Joaquin Delta College; Joseph Vielbig, Arizona Western University; Harry L. Vogel, Kansas State University; Jay Watterworth, University of Colorado at Boulder; Robert Weaver, Youngstown State University; Jason Wenzel, Valencia Community College; Rose Weitz, Arizona State University; Bob Weyer, County College of Morris; Oscar Williams, Diablo Valley College; Mark Winton, University of Central Florida; Diane Zablotsky, University of North Carolina; Joan Brehm, Illinois State University; Doug Degher, Northern Arizona University; Heather Griffiths, Fayetteville State University; Amy Holzgang, Cerritos College; Janet Hund, Long Beach City College; Kathrin Parks, University of New Mexico; Craig Robertson, University of North Alabama; Matthew Sanderson, University of Utah; Jacqueline Steingold, Wayne State University; and William J. Tinney, Jr., Black Hills State University.

Finally, we are interested in ways to improve the text and invite your feedback and suggestions for new ideas and material to be included in subsequent editions. You can contact us at mooneyl@ecu.edu, knoxd@ecu.edu, or cschacht@suddenlink.net.

Andrew Rich/Getty Images

1

Thinking about Social Problems

"Unless someone like you cares a whole awful lot, nothing is going to get better. It's not."

—Dr. Seuss, The Lorax

included economic issues (i.e., wages, corporate corruption, the gap between the rich and poor, etc.) and were the clear majority of responses, and noneconomic issues such as distrust of government, health care, immigration, gun control, education, and poverty (Gallup 2013a). Moreover, a 2013 survey indicated that just 30 percent of Americans were satisfied "with the way things are going in the United States" at that time. Although this number has increased over the last few years (for example, it was 21 percent at the same time in 2012), it is significantly lower than a decade ago when 60 percent of Americans were satisfied with the direction of the country (Gallup 2013b).

President Obama is keenly aware of the challenges facing the nation and the long journey ahead. In his second Inaugural Address he stated that:

> . . . our journey is not complete until our wives, our mothers, and daughters can earn a living equal to their efforts . . . until our gay brothers and sisters are treated like anyone else under the law . . . until no citizen is forced to wait for hours to exercise the right to vote . . . until we find a better way to welcome the striving, hopeful immigrants who still see America as a land of opportunity . . . until all our children, from the streets of Detroit to the hills of Appalachia to the quiet lanes of Newtown, know that they are cared for and cherished and always safe from harm. That is our generation's task, to make these words, these rights, these values of life and liberty and the pursuit of happiness real for every American. (Obama 2013)

After the economic turndown of 2008, the U.S. Congress passed and President Obama signed into law the American Recovery and Reinvestment Act of 2009. The stimulus package was designed to help failing industries, create jobs, promote consumer spending, rescue the failed housing market, and encourage energy-related investments. To date, the distribution of stimulus funds amounts to over $793 billion (Recovery.gov 2013).

IN AN APRIL 2013 Gallup Poll, a random sample of Americans were asked, "What do you think is the most important problem facing this country today?" Leading problems

Problems related to poverty, inadequate education, crime and violence, oppression of minorities, environmental destruction, and war and terrorism as well as many other social issues are both national and international concerns. Such problems present both a threat and a challenge to our national and global society. The primary goal of this textbook is to facilitate increased awareness and understanding of problematic social conditions in U.S. society and throughout the world.

Although the topics covered in this book vary widely, all chapters share common objectives: to explain how social problems are created and maintained; to indicate how they affect individuals, social groups, and societies as a whole; and to examine programs and policies for change. We begin by looking at the nature of social problems.

What Is a Social Problem?

There is no universal, constant, or absolute definition of what constitutes a social problem. Rather, social problems are defined by a combination of objective and subjective criteria that vary across societies, among individuals and groups within a society, and across historical time periods.

Objective and Subjective Elements of Social Problems

objective element of a social problem Awareness of social conditions through one's own life experiences and through reports in the media.

Although social problems take many forms, they all share two important elements: an objective social condition and a subjective interpretation of that social condition. The **objective element of a social problem** refers to the existence of a social condition. We become aware of social conditions through our own life experience, through the media, and through education. We see the homeless, hear gunfire in the streets, and see battered women in hospital emergency rooms. We read about employees losing their jobs as businesses downsize and factories close. In television news reports, we see the anguished faces of parents whose children have been killed by violent youths.

What Do You Think? For a condition to be defined as a social problem, there must be public awareness of the condition. How do you think the widespread use of communication technology—such as smartphones, Facebook, Twitter, and YouTube—have affected public awareness of problematic social conditions? Can you think of social problems that you became aware of through communication technology that you probably would not have been aware of if such technology were not accessible?

The **subjective element of a social problem** refers to the belief that a particular social condition is harmful to society or to a segment of society and that it should and can be changed. We know that crime, drug addiction, poverty, racism, violence, and pollution exist. These social conditions are not considered social problems, however, unless at least a segment of society believes that these conditions diminish the quality of human life.

By combining these objective and subjective elements, we arrive at the following definition: A **social problem** is a social condition that a segment of society views as harmful to members of society and in need of remedy.

Variability in Definitions of Social Problems

Individuals and groups frequently disagree about what constitutes a social problem. For example, some Americans view gun control as a necessary means of reducing gun violence whereas others believe that gun control is a threat to civil rights and individual liberties. Similarly, some Americans view the availability of abortion as a social problem, whereas others view restrictions on abortion as a social problem.

Definitions of social problems vary not only within societies but also across societies and historical time periods. For example, before the 19th century, a husband's legal right and marital obligation was to discipline and control his wife through the use of physical force. Today, the use of physical force is regarded as a social problem rather than a marital right.

> . . . some Americans view gun control as a necessary means of reducing gun violence whereas others believe that gun control is a threat to civil rights and individual liberties.

Social problems change over time not only because *definitions* of conditions change, as in the example of the use of force in marriage, but also because the *conditions* themselves change. The use of cell phones while driving was not considered a social problem in the 1990s, as cell phone technology was just beginning to become popular. Now, with most U.S. adults having a cell phone, the problem of "distracted driving" has become a national problem. According to the National Highway Traffic Safety Administration (2013), in 2011, 21,000 vehicle crashes in the United States involved cell phone use. Almost half of drivers say they answer incoming calls and about a quarter are willing to make calls while driving. Fourteen percent of drivers in 2012 said they sent text messages while driving (National Highway Traffic Safety Administration 2013).

What Do You Think? Many drivers see using mobile phones while driving as risky when other drivers do it, but view their own mobile phone use while driving as safe (National Highway Traffic Safety Administration 2013). Why do you think this is so?

subjective element of a social problem The belief that a particular social condition is harmful to society, or to a segment of society, and that it should and can be changed.

social problem A social condition that a segment of society views as harmful to members of society and in need of remedy.

The massacre at Sandy Hook Elementary School in Newtown, Connecticut, where 20 children and 6 staff members were killed, sparked a national discussion on gun control. To date, gun control and all that it entails (e.g., background checks, ban on automatic weapons) is one of the most hotly debated and divisive social issues in the United States.

Because social problems can be highly complex, it is helpful to have a framework within which to view them. Sociology provides such a framework. Using a sociological perspective to examine social problems requires knowledge of the basic concepts and tools of sociology. In the remainder of this chapter, we discuss some of these concepts and tools: social structure, culture, the "sociological imagination," major theoretical perspectives, and types of research methods.

Elements of Social Structure and Culture

Although society surrounds us and permeates our lives, it is difficult to "see" society. By thinking of society in terms of a picture or image, however, we can visualize society and therefore better understand it. Imagine that society is a coin with two sides: On one side is the structure of society and on the other is the culture of society. Although each side is distinct, both are inseparable from the whole. By looking at the various elements of social structure and culture, we can better understand the root causes of social problems.

Elements of Social Structure

The **structure** of a society refers to the way society is organized. Society is organized into different parts: institutions, social groups, statuses, and roles.

Institutions. An **institution** is an established and enduring pattern of social relationships. The five traditional institutions are family, religion, politics, economics, and education, but some sociologists argue that other social institutions, such as science and technology, mass media, medicine, sports, and the military, also play important roles in modern society. Many social problems are generated by inadequacies in various institutions. For example, unemployment may be influenced by the educational institution's failure to prepare individuals for the job market and by alterations in the structure of the economic institution.

Social Groups. Institutions are made up of social groups. A **social group** is defined as two or more people who have a common identity, interact, and form a social relationship. For example, the family in which you were reared is a social group that is part of the family institution. The religious association to which you may belong is a social group that is part of the religious institution.

Social groups can be categorized as primary or secondary. **Primary groups**, which tend to involve small numbers of individuals, are characterized by intimate and informal interaction. Families and friends are examples of primary groups. **Secondary groups**, which may involve small or large numbers of individuals, are task oriented and are characterized by impersonal and formal interaction. Examples of secondary groups include employers and their employees and clerks and their customers.

structure The way society is organized including institutions, social groups, statuses, and roles.

institution An established and enduring pattern of social relationships.

social group Two or more people who have a common identity, interact, and form a social relationship.

primary groups Usually small numbers of individuals characterized by intimate and informal interaction.

secondary groups Involving small or large numbers of individuals, groups that are task-oriented and are characterized by impersonal and formal interaction.

Statuses. Just as institutions consist of social groups, social groups consist of statuses. A **status** is a position that a person occupies within a social group. The statuses we occupy largely define our social identity. The statuses in a family may consist of mother, father, stepmother, stepfather, wife, husband, partner, child, and so on. Statuses can be either ascribed or achieved. An **ascribed status** is one that society assigns to an individual on the basis of factors over which the individual has no control. For example, we have no control over the sex, race, ethnic background, and socioeconomic status into which we are born. Similarly, we are assigned the status of child, teenager, adult, or senior citizen on the basis of our age—something we do not choose or control.

An **achieved status** is assigned on the basis of some characteristic or behavior over which the individual has some control. Whether you achieve the status of college graduate, spouse, parent, bank president, or prison inmate depends largely on your own efforts, behavior, and choices. One's ascribed statuses may affect the likelihood of achieving other statuses, however. For example, if you are born into a poor socioeconomic status, you may find it more difficult to achieve the status of college graduate because of the high cost of a college education.

Every individual has numerous statuses simultaneously. You may be a student, parent, tutor, volunteer fund-raiser, female, and Hispanic. A person's *master status* is the status that is considered the most significant in a person's social identity. In the United States, a person's occupational status is typically regarded as a master status. If you are a full-time student, your master status is likely to be student.

Roles. Every status is associated with many **roles**, or the set of rights, obligations, and expectations associated with a status. Roles guide our behavior and allow us to predict the behavior of others. As students, you are expected to attend class, listen and take notes, study for tests, and complete assignments. Because you know what the role of teacher involves, you can predict that your teachers will lecture, give exams, and assign grades based on your performance on tests.

A single status involves more than one role. For example, the status of prison inmate includes one role for interacting with prison guards and another role for interacting with other prison inmates. Similarly, the status of nurse involves different roles for interacting with physicians and with patients.

Elements of Culture

Whereas the social structure refers to the organization of society, the **culture** refers to the meanings and ways of life that characterize a society. The elements of culture include beliefs, values, norms, sanctions, and symbols.

Whereas the social structure refers to the organization of society, the culture refers to the meanings and ways of life that characterize a society.

Beliefs. **Beliefs** refer to definitions and explanations about what is assumed to be true. The beliefs of an individual or group influence whether that individual or group views a particular social condition as a social problem. Does secondhand smoke harm nonsmokers? Are nuclear power plants safe? Does violence in movies and on television lead to increased aggression in children? Our beliefs regarding these issues influence whether we view the issues as social problems. Beliefs influence not only how a social condition is interpreted but also the existence of the condition itself.

Values. **Values** are social agreements about what is considered good and bad, right and wrong, desirable and undesirable. Frequently, social conditions are viewed as social problems when the conditions are incompatible with or contradict closely held values. For example, poverty and homelessness violate the value of human welfare; crime contradicts the values of honesty, private property, and nonviolence; racism, sexism, and heterosexism violate the values of equality and fairness.

status A position that a person occupies within a social group.

ascribed status A status that society assigns to an individual on the basis of factors over which the individual has no control.

achieved status A status that society assigns to an individual on the basis of factors over which the individual has some control.

roles The set of rights, obligations, and expectations associated with a status.

culture The meanings and ways of life that characterize a society, including beliefs, values, norms, sanctions, and symbols.

beliefs Definitions and explanations about what is assumed to be true.

values Social agreements about what is considered good and bad, right and wrong, desirable and undesirable.

Which of the following do you consider to be "essential" or "very important" in your life? (Indicate with a check mark after the item.)

1. Raising a family _____
2. Becoming an authority in my field _____
3. Influencing social values _____
4. Being very well off financially _____
5. Helping others who are in difficulty _____
6. Becoming successful in a business of my own _____
7. Adopting "green" practices to protect the environment _____
8. Helping to promote racial understanding _____
9. Improving my understanding of other countries and cultures _____
10. Developing a meaningful philosophy of life _____
11. Keeping up to date with political affairs _____

Percentage of First-Year College Students Identifying Various Personal Values as Being "Essential" or "Very Important"*

1. Raising a family: 74%
2. Becoming an authority in my field: 60%
3. Influencing social values: 42%
4. Being very well off financially: 81%
5. Helping others who are in difficulty: 72%
6. Becoming successful in a business of my own: 41%
7. Adopting "green" practices to protect the environment: 40%
8. Helping to promote racial understanding: 35%
9. Improving my understanding of other countries and cultures: 51%
10. Developing a meaningful philosophy of life: 46%
11. Keeping up to date with political affairs: 35%

*Percentages are rounded.

Source: Pryor et al. 2012.

Values play an important role not only in the interpretation of a condition as a social problem but also in the development of the social condition itself. For example, most Americans view capitalism, characterized by free enterprise and the private accumulation of wealth, positively (Newport 2012). Nonetheless, a capitalist system, in part, is responsible for the inequality in American society as people compete for limited resources. This chapter's *Self and Society* feature allows you to identify values that are important to you and to compare your values to a national sample of first-year college students.

Norms and Sanctions. **Norms** are socially defined rules of behavior. Norms serve as guidelines for our behavior and for our expectations of the behavior of others.

There are three types of norms: folkways, laws, and mores. *Folkways* refer to the customs, habits, and manners of society—the ways of life that characterize a group or society. In many segments of our society, it is customary to shake hands when being introduced to a new acquaintance, to say "excuse me" after sneezing, and to give presents to family and friends on their birthdays. Although no laws require us to do these things, we are expected to do them because they are part of the cultural tradition, or folkways, of the society in which we live.

Laws are norms that are formalized and backed by political authority. It is normative for a Muslim woman to wear a veil. However, in the United States, failure to remove the

norms Socially defined rules of behavior, including folkways, laws, and mores.

veil for a driver's license photo is grounds for revoking the permit. Such is the case of a Florida woman who brought suit against the state, claiming that her religious rights were being violated because she was required to remove her veil for the driver's license photo (Canedy 2002). She appealed the decision to Florida's District Court of Appeal and lost. The Court recognized, however, "the tension created as a result of choosing between following the dictates of one's religion and the mandates of secular law" (Associated Press 2006).

Mores are norms with a moral basis. Both littering and child sexual abuse are violations of law, but child sexual abuse is also a violation of our mores because we view such behavior as immoral.

All norms are associated with **sanctions**, or social consequences for conforming to or violating norms. When we conform to a social norm, we may be rewarded by a positive sanction. These may range from an approving smile to a public ceremony in our honor. When we violate a social norm, we may be punished by a negative sanction, which may range from a disapproving look to the death penalty or life in prison. Most sanctions are spontaneous expressions of approval or disapproval by groups or individuals—these are referred to as informal sanctions. Sanctions that are carried out according to some recognized or formal procedure are referred to as formal sanctions. Types of sanctions, then, include positive informal sanctions, positive formal sanctions, negative informal sanctions, and negative formal sanctions (see Table 1.1).

TABLE 1.1 Types and Examples of Sanctions		
	Positive	Negative
Informal	Being praised by one's neighbors for organizing a neighborhood recycling program	Being criticized by one's neighbors for refusing to participate in the neighborhood recycling program
Formal	Being granted a citizen's award for organizing a neighborhood recycling program	Being fined by the city for failing to dispose of trash properly

© Cengage Learning

Symbols. A **symbol** is something that represents something else. Without symbols, we could not communicate with one another or live as social beings.

The symbols of a culture include language, gestures, and objects whose meanings the members of a society commonly understand. In our society, a red ribbon tied around a car antenna symbolizes Mothers Against Drunk Driving; a peace sign symbolizes the value of nonviolence; and a white-hooded robe symbolizes the Ku Klux Klan. Sometimes people attach different meanings to the same symbol. The Confederate flag is a symbol of southern pride to some and a symbol of racial bigotry to others.

The elements of the social structure and culture just discussed play a central role in the creation, maintenance, and social response to various social problems. One of the goals of taking a course in social problems is to develop an awareness of how the elements of social structure and culture contribute to social problems. Sociologists refer to this awareness as the "sociological imagination."

The Sociological Imagination

The **sociological imagination**, a term C. Wright Mills (1959) developed, refers to the ability to see the connections between our personal lives and the social world in which we live. When we use our sociological imagination, we are able to distinguish between "private troubles" and "public issues" and to see connections between the events and conditions of our lives and the social and historical context in which we live.

For example, that one person is unemployed constitutes a private trouble. That millions of people are unemployed in the United States constitutes a public issue. Once we understand that other segments of society share personal troubles such as intimate partner abuse, drug addiction, criminal victimization, and poverty, we can look for the elements of social structure and culture that contribute to these public issues and private troubles. If the various elements of social structure and culture contribute to private troubles and public issues, then society's social structure and culture must be changed if these concerns are to be resolved.

sanctions Social consequences for conforming to or violating norms.

symbol Something that represents something else.

sociological imagination The ability to see the connections between our personal lives and the social world in which we live.

When we use our sociological imagination, we are able to distinguish between "private troubles" and "public issues" and to see connections between the events and conditions of our lives and the social and historical context in which we live.

Rather than viewing the private trouble of obesity and all of its attending health concerns as a result of an individual's faulty character, lack of self-discipline, or poor choices regarding food and exercise, we may understand the obesity epidemic as a public issue that results from various social and cultural forces, including government policies that make high-calorie foods more affordable than healthier, fresh produce; powerful food lobbies that fight against proposals to restrict food advertising to children; and technological developments that have eliminated many types of manual labor and replaced it with sedentary "desk jobs"(see Chapter 2).

Theoretical Perspectives

Theories in sociology provide us with different perspectives with which to view our social world. A perspective is simply a way of looking at the world. A **theory** is a set of interrelated propositions or principles designed to answer a question or explain a particular phenomenon; it provides us with a perspective. Sociological theories help us to explain and predict the social world in which we live.

Sociology includes three major theoretical perspectives: the structural-functionalist perspective, the conflict perspective, and the symbolic interactionist perspective. Each perspective offers a variety of explanations about the causes of and possible solutions to social problems.

Structural-Functionalist Perspective

The structural-functionalist perspective is based largely on the works of Herbert Spencer, Emile Durkheim, Talcott Parsons, and Robert Merton. According to structural functionalism, society is a system of interconnected parts that work together in harmony to maintain a state of balance and social equilibrium for the whole. For example, each of the social institutions contributes important functions for society: Family provides a context for reproducing, nurturing, and socializing children; education offers a way to transmit a society's skills, knowledge, and culture to its youth; politics provides a means of governing members of society; economics provides for the production, distribution, and consumption of goods and services; and religion provides moral guidance and an outlet for worship of a higher power.

The structural-functionalist perspective emphasizes the interconnectedness of society by focusing on how each part influences and is influenced by other parts.

The structural-functionalist perspective emphasizes the interconnectedness of society by focusing on how each part influences and is influenced by other parts. For example, the increase in single-parent and dual-earner families has contributed to the number of children who are failing in school because parents have become less available to supervise their children's homework. As a result of changes in technology, colleges are offering more technical programs, and many adults are returning to school to learn new skills that are required in the workplace. The increasing number of women in the workforce has contributed to the formulation of policies against sexual harassment and job discrimination.

Structural functionalists use the terms *functional* and *dysfunctional* to describe the effects of social elements on society. Elements of society are functional if they contribute to social stability and dysfunctional if they disrupt social stability. Some aspects of society can be both functional and dysfunctional. For example, crime is dysfunctional in that it is associated with physical violence, loss of property, and fear. But according to Durkheim and other functionalists, crime is also functional for society because it leads to heightened awareness of shared moral bonds and increased social cohesion.

theory A set of interrelated propositions or principles designed to answer a question or explain a particular phenomenon.

Sociologists have identified two types of functions: manifest and latent (Merton 1968). **Manifest functions** are consequences that are intended and commonly recognized. **Latent functions** are consequences that are unintended and often hidden. For example, the manifest function of education is to transmit knowledge and skills to society's youth. But public elementary schools also serve as babysitters for employed parents, and colleges offer a place for young adults to meet potential mates. The babysitting and mate-selection functions are not the intended or commonly recognized functions of education; hence, they are latent functions.

What Do You Think? In viewing society as a set of interrelated parts, structural functionalists argue that proposed solutions to social problems may lead to other social problems. For example, urban renewal projects displace residents and break up community cohesion. Racial imbalance in schools led to forced integration, which in turn generated violence and increased hostility between the races. What are some other "solutions" that have led to social problems? Do all solutions come with a price to pay? Can you think of a solution to a social problem that has no negative consequences?

Structural-Functionalist Theories of Social Problems

Two dominant theories of social problems grew out of the structural-functionalist perspective: social pathology and social disorganization.

Social Pathology. According to the social pathology model, social problems result from some "sickness" in society. Just as the human body becomes ill when our systems, organs, and cells do not function normally, society becomes "ill" when its parts (i.e., elements of the structure and culture) no longer perform properly. For example, problems such as crime, violence, poverty, and juvenile delinquency are often attributed to the breakdown of the family institution; the decline of the religious institution; and inadequacies in our economic, educational, and political institutions.

Social "illness" also results when members of a society are not adequately socialized to adopt its norms and values. People who do not value honesty, for example, are prone to dishonesties of all sorts. Early theorists attributed the failure in socialization to "sick" people who could not be socialized. Later theorists recognized that failure in the socialization process stemmed from "sick" social conditions, not "sick" people. To prevent or solve social problems, members of society must receive proper socialization and moral education, which may be accomplished in the family, schools, churches, or workplace and/or through the media.

Social Disorganization. According to the social disorganization view of social problems, rapid social change (e.g., the cultural revolution of the 1960s) disrupts the norms in a society. When norms become weak or are in conflict with each other, society is in a state of **anomie**, or *normlessness*. Hence, people may steal, physically abuse their spouses or children, abuse drugs, commit rape, or engage in other deviant behavior because the norms regarding these behaviors are weak or conflicting. According to this view, the solution to social problems lies in slowing the pace of social change and strengthening social norms. For example, although the use of alcohol by teenagers is considered a violation of a social norm in our society, this norm is weak. The media portray young people drinking alcohol, teenagers teach each other to drink alcohol and buy fake identification cards (IDs) to purchase alcohol, and parents model drinking behavior by having a few drinks after work or at a social event. Solutions to teenage drinking may involve strengthening norms against it through public education, restricting media depictions of youth and alcohol, imposing stronger sanctions against the use of fake IDs to purchase alcohol, and educating parents to model moderate and responsible drinking behavior.

manifest functions Consequences that are intended and commonly recognized.

latent functions Consequences that are unintended and often hidden.

anomie A state of normlessness in which norms and values are weak or unclear.

Conflict Perspective

Contrary to the structural-functionalism perspective, the conflict perspective views society as composed of different groups and interests competing for power and resources. The conflict perspective explains various aspects of our social world by looking at which groups have power and benefit from a particular social arrangement. For example, feminist theory argues that we live in a patriarchal society—a hierarchical system of organization controlled by men. Although there are many varieties of feminist theory, most would hold that feminism "demands that existing economic, political, and social structures be changed" (Weir & Faulkner 2004, p. xii).

The origins of the conflict perspective can be traced to the classic works of Karl Marx. Marx suggested that all societies go through stages of economic development. As societies evolve from agricultural to industrial, concern over meeting survival needs is replaced by concern over making a profit, the hallmark of a capitalist system. Industrialization leads to the development of two classes of people: the bourgeoisie, or the owners of the means of production (e.g., factories, farms, businesses), and the proletariat, or the workers who earn wages.

The division of society into two broad classes of people—the "haves" and the "have-nots"—is beneficial to the owners of the means of production. The workers, who may earn only subsistence wages, are denied access to the many resources available to the wealthy owners. According to Marx, the bourgeoisie use their power to control the institutions of society to their advantage. For example, Marx suggested that religion serves as an "opiate of the masses" in that it soothes the distress and suffering associated with the working-class lifestyle and focuses the workers' attention on spirituality, God, and the afterlife rather than on worldly concerns such as living conditions. In essence, religion diverts the workers so that they concentrate on being rewarded in heaven for living a moral life rather than on questioning their exploitation.

Conflict Theories of Social Problems

There are two general types of conflict theories of social problems: Marxist and non-Marxist. Marxist theories focus on social conflict that results from economic inequalities; non-Marxist theories focus on social conflict that results from competing values and interests among social groups.

Marxist Conflict Theories. According to contemporary Marxist theorists, social problems result from class inequality inherent in a capitalistic system. A system of haves and have-nots may be beneficial to the haves but often translates into poverty for the have-nots. As we will explore later in this textbook, many social problems, including physical and mental illness, low educational achievement, and crime, are linked to poverty.

In addition to creating an impoverished class of people, capitalism also encourages "corporate violence." *Corporate violence* can be defined as actual harm and/or risk of harm inflicted on consumers, workers, and the general public as a result of decisions by corporate executives or managers. Corporate violence can also result from corporate negligence; the quest for profits at any cost; and willful violations of health, safety, and environmental laws (Reiman & Leighton 2013). Our profit-motivated economy encourages individuals who are otherwise good, kind, and law abiding to knowingly participate in the manufacturing and marketing of defective products, such as brakes on American jets, fuel tanks on automobiles, and salmonella-contaminated peanut butter.

In 2010, a British Petroleum (BP) oil well off the coast of Louisiana ruptured, killing 11 people and spewing millions of gallons of oil into the Gulf of Mexico (see Chapter 13). Evidence suggests that BP officials knew of the unstable cement seals on the rigs long before what has been called the worst offshore disaster in U.S. history (Pope 2011). As of 2013, BP had paid over $1.87 billion in damages (Finn 2013).

Marxist conflict theories also focus on the problem of **alienation**, or powerlessness and meaninglessness in people's lives. In industrialized societies, workers often have little power or control over their jobs, a condition that fosters in them a sense of powerlessness in their lives. The specialized nature of work requires employees to perform limited and

alienation A sense of powerlessness and meaninglessness in people's lives.

repetitive tasks; as a result, workers may come to feel that their lives are meaningless.

Alienation is bred not only in the workplace but also in the classroom. Students have little power over their education and often find that the curriculum is not meaningful to their lives. Like poverty, alienation is linked to other social problems, such as low educational achievement, violence, and suicide.

Marxist explanations of social problems imply that the solution lies in eliminating inequality among classes of people by creating a classless society. The nature of work must also change to avoid alienation. Finally, stronger controls must be applied to corporations to ensure that corporate decisions and practices are based on safety rather than on profit considerations.

Non-Marxist Conflict Theories. Non-Marxist conflict theorists, such as Ralf Dahrendorf, are concerned with conflict that arises when groups have opposing values and interests. For example, antiabortion activists value the life of unborn embryos and fetuses; pro-choice activists value the right of women to control their own bodies and reproductive decisions. These different value positions reflect different subjective interpretations of what constitutes a social problem. For antiabortionists, the availability of abortion is the social problem; for pro-choice advocates, the restrictions on abortion are the social problem. Sometimes the social problem is not the conflict itself but rather the way that conflict is expressed. Even most pro-life advocates agree that shooting doctors who perform abortions and blowing up abortion clinics constitute unnecessary violence and lack of respect for life. Value conflicts may occur between diverse categories of people, including nonwhites versus whites, heterosexuals versus homosexuals, young versus old, Democrats versus Republicans, and environmentalists versus industrialists.

Solving the problems that are generated by competing values may involve ensuring that conflicting groups understand each other's views, resolving differences through negotiation or mediation, or agreeing to disagree. Ideally, solutions should be win-win, with both conflicting groups satisfied with the solution. However, outcomes of value conflicts are often influenced by power; the group with the most power may use its position to influence the outcome of value conflicts. For example, when Congress could not get all states to voluntarily increase the legal drinking age to 21, it threatened to withdraw federal highway funds from those that would not comply.

Mark Wilson/Getty Images

Preschooler Jacob Hurley, who became seriously ill after eating peanut butter manufactured by the Peanut Corporation of America, is shown sitting with his father Peter Hurley, who is testifying before a House Energy and Commerce Committee hearing on Capitol Hill in Washington, DC, in January 2009. Nine deaths and over 700 illnesses resulted from the salmonella-tainted peanuts and, in 2013, former officials of the company were indicted on over 76 criminal counts (Schoenberg & Mattingly 2013).

Symbolic Interactionist Perspective

Both the structural-functionalist and the conflict perspectives are concerned with how broad aspects of society, such as institutions and large social groups, influence the social world. This level of sociological analysis is called *macrosociology:* It looks at the big picture of society and suggests how social problems are affected at the institutional level.

Microsociology, another level of sociological analysis, is concerned with the social-psychological dynamics of individuals interacting in small groups. Symbolic interactionism reflects the microsociological perspective and was largely influenced by the work of early sociologists and philosophers such as Max Weber, Georg Simmel, Charles Horton Cooley, G. H. Mead, W. I. Thomas, Erving Goffman, and Howard Becker. Symbolic interactionism emphasizes that human behavior is influenced by definitions and meanings that are created and maintained through symbolic interaction with others.

We develop our self-concept by observing how others interact with us and label us. By observing how others view us, we see a reflection of ourselves that Cooley called the "looking-glass self."

Sociologist W. I. Thomas (1931/1966) emphasized the importance of definitions and meanings in social behavior and its consequences. He suggested that humans respond to their definition of a situation rather than to the objective situation itself. Hence, Thomas noted that situations that we define as real become real in their consequences.

Symbolic interactionism also suggests that social interaction shapes our identity or sense of self. We develop our self-concept by observing how others interact with us and label us. By observing how others view us, we see a reflection of ourselves that Cooley calls the "looking-glass self."

Last, the symbolic interactionist perspective has important implications for how social scientists conduct research. German sociologist Max Weber argued that, to understand individual and group behavior, social scientists must see the world through the eyes of that individual or group. Weber called this approach *verstehen*, which in German means "to understand." *Verstehen* implies that, in conducting research, social scientists must try to understand others' views of reality and the subjective aspects of their experiences, including their symbols, values, attitudes, and beliefs.

Symbolic Interactionist Theories of Social Problems

A basic premise of symbolic interactionist theories of social problems is that a condition must be *defined or recognized* as a social problem for it to *be* a social problem. Three symbolic interactionist theories of social problems are based on this general premise.

Blumer's Stages of a Social Problem. Herbert Blumer (1971) suggested that social problems develop in stages. First, social problems pass through the stage of *societal recognition*—the process by which a social problem, for example, drunk driving, is "born." Drunk driving wasn't illegal until 1939, when Indiana passed the first state law regulating alcohol consumption and driving (Indiana State Government 2013). Second, *social legitimation* takes place when the social problem achieves recognition by the larger community, including the media, schools, and churches. As the visibility of traffic fatalities associated with alcohol increased, so did the legitimation of drunk driving as a social problem. The next stage in the development of a social problem involves *mobilization for action,* which occurs when individuals and groups, such as Mothers Against Drunk Driving, become concerned about how to respond to the social condition. This mobilization leads to the *development and implementation of an official plan* for dealing with the problem, involving, for example, highway checkpoints, lower legal blood-alcohol levels, and tougher regulations for driving drunk.

Blumer's stage-development view of social problems is helpful in tracing the development of social problems. For example, although sexual harassment and date rape occurred throughout the 20th century, these issues did not begin to receive recognition as social problems until the 1970s. Social legitimation of these problems was achieved when high schools, colleges, churches, employers, and the media recognized their existence. Organized social groups mobilized to develop and implement plans to deal with these problems. Groups successfully lobbied for the enactment of laws against sexual harassment and the enforcement of sanctions against violators of these laws. Groups also mobilized to provide educational seminars on date rape for high school and college students and to offer support services to victims of date rape.

Some disagree with the symbolic interactionist view that social problems exist only if they are recognized. According to this view, individuals who were victims of date rape in the 1960s may be considered victims of a problem, even though date rape was not recognized as a social problem at that time.

Labeling Theory. Labeling theory, a major symbolic interactionist theory of social problems, suggests that a social condition or group is viewed as problematic if it is labeled as such. According to labeling theory, resolving social problems sometimes involves changing the meanings and definitions that are attributed to people and situations. For example, so long as teenagers define drinking alcohol as "cool" and "fun," they will continue to abuse alcohol. So long as our society defines providing sex education and contraceptives to teenagers as inappropriate or immoral, the teenage pregnancy rate in the United States will continue to be higher than that in other industrialized nations. Individuals who label their own cell phone use while driving as safe will continue to use their cells phones as they drive, endangering their own lives and the lives of others.

Social Constructionism. Social constructionism is another symbolic interactionist theory of social problems. Similar to labeling theorists and symbolic interactionism in general, social constructionists argue that individuals who interpret the social world around them socially construct reality. Society, therefore, is a social creation rather than an objective given. As such, social constructionists often question the origin and evolution of social problems. For example, social constructionist theory has been used to ". . . analyze the history of the temperance and prohibition movements[,] . . . the rise of alcoholism as a disease movement in the post-prohibition era[,] . . . and the crusade against drinking and driving in the 1980s in the United States. . . . These studies [each] analyzed the shifts in social meanings attributed to alcohol beverage use and to problems within the changing landscapes of social, economic, and political power relationships in American society" (Herd 2011). Central to this idea of the social construction of social problems are the media, universities, research institutes, and government agencies, which are often responsible for the public's initial "take" on the problem under discussion.

Table 1.2 summarizes and compares the major theoretical perspectives, their criticisms, and social policy recommendations as they relate to social problems. The study of

TABLE 1.2 Comparison of Theoretical Perspectives

	Structural Functionalism	Conflict Theory	Symbolic Interactionism
Representative theorists	Emile Durkheim Talcott Parsons Robert Merton	Karl Marx Ralf Dahrendorf	George H. Mead Charles Cooley Erving Goffman
Society	Society is a set of interrelated parts; cultural consensus exists and leads to social order; natural state of society—balance and harmony.	Society is marked by power struggles over scarce resources; inequities result in conflict; social change is inevitable; natural state of society—imbalance.	Society is a network of interlocking roles; social order is constructed through interaction as individuals, through shared meaning, making sense out of their social world.
Individuals	Individuals are socialized by society's institutions; socialization is the process by which social control is exerted; people need society and its institutions.	People are inherently good but are corrupted by society and its economic structure; institutions are controlled by groups with power; "order" is part of the illusion.	Humans are interpretive and interactive; they are constantly changing as their "social beings" emerge and are molded by changing circumstances.
Cause of social problems?	Rapid social change; social disorganization that disrupts the harmony and balance; inadequate socialization and/or weak institutions.	Inequality; the dominance of groups of people over other groups of people; oppression and exploitation; competition between groups.	Different interpretations of roles; labeling of individuals, groups, or behaviors as deviant; definition of an objective condition as a social problem.
Social policy/ solutions	Repair weak institutions; assure proper socialization; cultivate a strong collective sense of right and wrong.	Minimize competition; create an equitable system for the distribution of resources.	Reduce impact of labeling and associated stigmatization; alter definitions of what is defined as a social problem.
Criticisms	Called "sunshine sociology"; supports the maintenance of the status quo; needs to ask "functional for whom?"; does not deal with issues of power and conflict; incorrectly assumes a consensus.	Utopian model; Marxist states have failed; denies existence of cooperation and equitable exchange; cannot explain cohesion and harmony.	Concentrates on micro issues only; fails to link micro issues to macro-level concerns; too psychological in its approach; assumes label amplifies problem.

© Cengage Learning

Each chapter in this book contains a *Social Problems Research Up Close* box that describes a research study that examines some aspect of a social problem, and is presented in a report, book, or journal. Academic sociologists, those teaching at community colleges, colleges, or universities, as well as other social scientists, primarily rely on journal articles as the means to exchange ideas and information. Some examples of the more prestigious journals in sociology include the *American Sociological Review*, the *American Journal of Sociology*, and *Social Forces*. Most journal articles begin with *an introduction and review of the literature*. Here, the investigator examines previous research on the topic, identifies specific research areas, and otherwise "sets the stage" for the reader. Often in this section, research hypotheses are set forth, if applicable. A researcher, for example, might hypothesize that the sexual behavior of adolescents has changed over the years as a consequence of increased fear of sexually transmitted diseases and that such changes vary on the basis of sex.

The next major section of a journal article is *sample and methods*. In this section, an investigator describes how the research sample was selected, the characteristics of the research sample, the details of how the research was conducted, and how the data were analyzed (see Appendix). Using the sample research question, a sociologist might obtain data from the Youth Risk Behavior Surveillance Survey collected by the Centers for Disease Control and Prevention.

This self-administered questionnaire is distributed biennially to more than 10,000 high school students across the United States.

The final section of a journal article includes the *findings and conclusions*. The findings of a study describe the results, that is, what the researcher found as a result of the investigation. Findings are then discussed within the context of the hypotheses and the conclusions that can be drawn. Often, research results are presented in tabular form. Reading tables carefully is an important part of drawing accurate conclusions about the research hypotheses. In reading a table, you should follow the steps listed here (see the table within this box):

1. *Read the title of the table and make sure that you understand what the table contains.* The title of the table indicates the unit of analysis (high school students), the dependent variable (sexual risk behaviors), the independent variables (sex and year), and what the numbers represent (percentages).
2. *Read the information contained at the bottom of the table, including the source and any other explanatory information.* For example, the information at the bottom of this table indicates that the data are from the Centers for Disease Control and Prevention, that "sexually active" was defined as having intercourse in the last three months, and that data on condom use were only from those students who were defined as being currently sexually active.

3. *Examine the row and column headings.* This table looks at the percentage of males and females, over four years, who reported ever having sexual intercourse, having four or more sex partners in a lifetime, being currently sexually active, and using condoms during the last sexual intercourse.
4. *Thoroughly and carefully examine the data in the table, looking for patterns between variables.* As indicated in the table, the percentage of males engaging in "risky" sexual behavior has gone up between 2005 and 2011 for two of the four categories and are the highest or near highest recorded over the time period for (1) ever having had sexual intercourse, and (2) having four or more partners during their lifetime. The percentage of males using protection during sex has decreased over the years and, in 2011, was at its lowest recorded level with only 67.0 percent of high school males reporting condom use. Females, as with males, report an increase in four or more sex partners during their lifetime and a decrease in condom use. However, between 2005 and 2011, there is very little difference in the percent of females reporting ever having sexual intercourse, and the percent reporting currently being sexually active has decreased over time. The difference between male and female sexually risky behaviors should also be noted. Contrary to "commonsense" beliefs, in 2011, males were less likely to be sexually active and more likely to have used a condom during last intercourse.

social problems is based on research as well as on theory, however. Indeed, research and theory are intricately related. As Wilson (1983) stated:

> Most of us think of theorizing as quite divorced from the business of gathering facts. It seems to require an abstractness of thought remote from the practical activity of empirical research. But theory building is not a separate activity within sociology. Without theory, the empirical researcher would find it impossible to decide what to observe, how to observe it, or what to make of the observations. (p. 1)

Social Problems Research

Most students taking a course in social problems will not become researchers or conduct research on social problems. Nevertheless, we are all consumers of research that is reported in the media. Politicians, social activist groups, and organizations attempt to justify their decisions, actions, and positions by citing research results. As consumers of research, we need to understand that our personal experiences and casual observations are less reliable than generalizations based on systematic research. One strength of scientific research is that

5. *Use the information you have gathered in Step 4 to address the hypotheses.* Clearly, sexual practices, as hypothesized, have changed over time. For example, contrary to expectations, both males and females, when comparing data from 2005 to 2011, report a general decrease in condom use during sexual intercourse. Further, the percentage of males and females reporting four or more sex partners has also increased during the same time period. Look at the table and see what patterns you detect, and how these patterns address the hypothesis.

6. *Draw conclusions consistent with the information presented.* From the table, can we conclude that sexual practices have changed over time? The answer is probably yes, although the limitations of the survey, the sample, and the measurement techniques used always should be considered. Can we conclude that the observed changes are a consequence of the fear of sexually transmitted diseases? The answer is *no,* and not just because of the results. Having no measure of fear of sexually transmitted diseases over the time period studied, we are unable to come to such a conclusion. More information, from a variety of sources, is needed. The use of multiple methods and approaches to study a social phenomenon is called *triangulation.*

Percentages of High School Students Reporting Sexually Risky Behaviors, by Sex and Survey Year

Survey Year	Ever Had Sexual Intercourse	Four Or More Sex Partners During Lifetime	Currently Sexually Active*	Condom Used During Last Intercourse†
Male				
2005	47.9	16.5	33.3	70.0
2007	49.8	17.9	34.3	68.5
2009	46.1	16.2	32.6	70.4
2011	49.2	17.8	34.2	67.0
Females				
2005	45.7	12.0	34.6	55.9
2007	45.9	11.8	35.6	54.9
2009	45.7	11.2	35.7	57.0
2011	45.6	12.6	34.2	53.6

*Sexual intercourse during the three months preceding the survey
†Among currently sexually active students
Source: Centers for Disease Control and Prevention 2008, 2010, 2012.

it is subjected to critical examination by other researchers (see this chapter's *Social Problems Research Up Close* feature). The more you understand how research is done, the better able you will be to critically examine and question research rather than to passively consume research findings. In the remainder of this section, we discuss the stages of conducting a research study and the various methods of research that sociologists use.

> The more you understand how research is done, the better able you will be to critically examine and question research rather than to passively consume research findings.

Stages of Conducting a Research Study

Sociologists progress through various stages in conducting research on a social problem. In this section, we describe the first four stages: (1) formulating a research question, (2) reviewing the literature, (3) defining variables, and (4) formulating a hypothesis.

Formulating a Research Question. A research study usually begins with a research question. Where do research questions originate? How does a particular researcher come to ask a particular research question? In some cases, researchers have a personal interest in a specific topic because of their own life experiences. For example, a researcher who has experienced spouse abuse may wish to do research on such questions as "What factors are associated with domestic violence?" and "How helpful are battered women's shelters in helping abused women break the cycle of abuse in their lives?" Other researchers may ask a particular research question because of their personal values—their concern for humanity and the desire to improve human life. Researchers may also want to test a particular sociological theory, or some aspect of it, to establish its validity or conduct studies to evaluate the effect of a social policy or program. Research questions may also be formulated by the concerns of community groups and social activist organizations in collaboration with academic researchers. Government and industry also hire researchers to answer questions such as "How many vehicle crashes are caused by 'distracted driving' involving the use of cell phones?" and "What types of cell phone technologies can prevent the use of cell phones while driving?"

Reviewing the Literature. After a research question is formulated, researchers review the published material on the topic to find out what is already known about it. Reviewing the literature also provides researchers with ideas about how to conduct their research and helps them formulate new research questions. A literature review serves as an evaluation tool, allowing a comparison of research findings and other sources of information, such as expert opinions, political claims, and journalistic reports.

What Do You Think? In a free society, there must be freedom of information. That is why the U.S. Constitution and, more specifically, the First Amendment protect journalists' sources. If journalists are compelled to reveal their sources, their sources may be unwilling to share information, and this would jeopardize the public's right to know. A journalist cannot reveal information given in confidence without permission from the source or a court order. Do you think sociologists should be granted the same protections as journalists? If a reporter at your school newspaper uncovered a scandal at your university, should he or she be protected by the First Amendment?

Defining Variables. A **variable** is any measurable event, characteristic, or property that varies or is subject to change. Researchers must operationally define the variables they study. An *operational definition* specifies how a variable is to be measured. For example, an operational definition of the variable "religiosity" might be the number of times the respondent reports going to church or synagogue. Another operational definition of "religiosity" might be the respondent's answer to the question "How important is religion in your life?" (for example, 1 is not important; 2 is somewhat important; 3 is very important).

Operational definitions are particularly important for defining variables that cannot be directly observed. For example, researchers cannot directly observe concepts such as "mental illness," "sexual harassment," "child neglect," "job satisfaction," and "drug abuse." Nor can researchers directly observe perceptions, values, and attitudes.

Formulating a Hypothesis. After defining the research variables, researchers may formulate a **hypothesis**, which is a prediction or educated guess about how one variable is related to another variable. The **dependent variable** is the variable that researchers want to explain; that is, it is the variable of interest. The **independent variable** is the variable that is expected to explain change in the dependent variable. In formulating a hypothesis, researchers predict how the independent variable affects the dependent variable. For example, Kmec (2003) investigated the impact of segregated work environments on minority wages, concluding that "minority concentration in different jobs, occupations,

variable Any measurable event, characteristic, or property that varies or is subject to change.

hypothesis A prediction or educated guess about how one variable is related to another variable.

dependent variable The variable that the researcher wants to explain; the variable of interest.

independent variable The variable that is expected to explain change in the dependent variable.

and establishments is a considerable social problem because it perpetuates racial wage inequality" (p. 55). In this example, the independent variable is workplace segregation, and the dependent variable is wages.

Methods of Data Collection

After identifying a research topic, reviewing the literature, defining the variables, and developing hypotheses, researchers decide which method of data collection to use. Alternatives include experiments, surveys, field research, and secondary data.

From the film Obedience copyright 1968 by Stanley Milgram, copyright renewed 1993 by Alexandra Milgram, and distributed by Alexander Street Press

Experiments. **Experiments** involve manipulating the independent variable to determine how it affects the dependent variable. Experiments require one or more experimental groups that are exposed to the experimental treatment(s) and a control group that is not exposed. After a researcher randomly assigns participants to either an experimental group or a control group, the researcher measures the dependent variable. After the experimental groups are exposed to the treatment, the researcher measures the dependent variable again. If participants have been randomly assigned to the different groups, the researcher may conclude that any difference in the dependent variable among the groups is due to the effect of the independent variable.

An example of a "social problems" experiment on poverty would be to provide welfare payments to one group of unemployed single mothers (experimental group) and no such payments to another group of unemployed single mothers (control group). The independent variable would be welfare payments; the dependent variable would be employment. The researcher's hypothesis would be that mothers in the experimental group would be less likely to have a job after 12 months than mothers in the control group.

The major strength of the experimental method is that it provides evidence for causal relationships, that is, how one variable affects another. A primary weakness is that experiments are often conducted on small samples, often in artificial laboratory settings; thus, the findings may not be generalized to other people in natural settings.

Surveys. **Survey research** involves eliciting information from respondents through questions. An important part of survey research is selecting a sample of those to be questioned. A **sample** is a portion of the population, selected to be representative so that the information from the sample can be generalized to a larger population. For example, instead of asking all middle school children about their delinquent activity, the researcher would ask a representative sample of them and assume that those who were not questioned would give similar responses. After selecting a representative sample, survey researchers either interview people, ask them to complete written questionnaires, or elicit responses to research questions through computers. After selecting a representative sample, survey researchers either interview people, ask them to complete written questionnaires, or elicit responses to research questions through web-based surveys.

Interviews. In interview survey research, trained interviewers ask respondents a series of questions and make written notes about or tape-record the respondents' answers. Interviews may be conducted over the phone or face-to-face.

One advantage of interview research is that researchers are able to clarify questions for respondents and follow up on answers to particular questions. Researchers often conduct face-to-face interviews with groups of individuals who might otherwise be inaccessible. For example, some AIDS-related research attempts to assess the degree to which individuals engage in behavior that places them at high risk for transmitting or contracting HIV. Street youth and intravenous drug users, both high-risk groups for HIV infection, may not have a telephone or address because of their transient lifestyle. These groups may be accessible, however, if the researcher locates their hangouts and conducts face-to-face interviews.

In one of the most famous experiments in the social sciences, Stanley Milgram found that 65 percent of a sample of ordinary citizens were willing to use harmful electric shocks—up to 450 volts—on an elderly man with a heart condition simply because the experimenter instructed them to do so. It was later revealed that the man was not really receiving the shocks and that he had been part of the experimental manipulation. The experiment, although providing valuable information, raised many questions on the ethics of scientific research.

experiments Research methods that involve manipulating the independent variable to determine how it affects the dependent variable.

survey research A research method that involves eliciting information from respondents through questions.

sample A portion of the population, selected to be representative so that the information from the sample can be generalized to a larger population.

The most serious disadvantages of interview research are cost and the lack of privacy and anonymity. Respondents may feel embarrassed or threatened when asked questions that relate to personal issues such as drug use, domestic violence, and sexual behavior. As a result, some respondents may choose not to participate in interview research on sensitive topics. Those who do participate may conceal or alter information or give socially desirable answers to the interviewer's questions (e.g., "No, I do not use drugs" or "No, I do not text while driving.").

Questionnaires.   Instead of conducting personal or phone interviews, researchers may develop questionnaires that they either mail, post online, or give to a sample of respondents. Questionnaire research offers the advantages of being less expensive and less time-consuming than face-to-face or telephone surveys. Questionnaire research also provides privacy and anonymity to the research participants, thus increasing the likelihood that respondents will provide truthful answers.

The major disadvantage of mail or online questionnaires is that it is difficult to obtain an adequate response rate. Many people do not want to take the time or make the effort to complete a questionnaire. Others may be unable to read and understand the questionnaire.

Web-based surveys.   In recent years, technological know-how and the expansion of the Internet have facilitated the use of online surveys. Web-based surveys, although still less common than interviews and questionnaires, are growing in popularity and are thought by some to reduce many of the problems associated with traditional survey research (Farrell & Petersen 2010). For example, the response rate of telephone surveys has been declining as potential respondents have caller ID, unlisted telephone numbers, answering machines, or no home (i.e., landline) telephone (Farrell & Petersen 2010). On the other hand, the use of and access to the Internet continues to grow. In 2011, the number of Americans connected to the Internet was the highest in all previous years (File 2013).

Field Research. **Field research** involves observing and studying social behavior in settings in which it occurs naturally. Two types of field research are participant observation and nonparticipant observation.

In participant observation research, researchers participate in the phenomenon being studied so as to obtain an insider's perspective on the people and/or behavior being observed. Palacios and Fenwick (2003), two criminologists, attended dozens of raves over a 15-month period to investigate the South Florida drug culture. In nonparticipant observation research, researchers observe the phenomenon being studied without actively participating in the group or the activity. For example, Simi and Futrell (2009) studied white power activists by observing and talking to organizational members but did not participate in any of their organized activities.

Sometimes sociologists conduct in-depth detailed analyses or case studies of an individual, group, or event. For example, Fleming (2003) conducted a case study of young auto thieves in British Columbia. He found that, unlike professional thieves, the teenagers' behavior was primarily motivated by thrill seeking—driving fast, the rush of a possible police pursuit, and the prospect of getting caught.

The main advantage of field research on social problems is that it provides detailed information about the values, rituals, norms, behaviors, symbols, beliefs, and emotions of those being studied. A potential problem with field research is that the researcher's observations may be biased (e.g., the researcher becomes too involved in the group to be objective). In addition, because field research is usually based on small samples, the findings may not be generalizable.

Secondary Data Research. Sometimes researchers analyze secondary data, which are data that other researchers or government agencies have already collected or that exist in forms such as historical documents, police reports, school records, and official records of marriages, births, and deaths. A major advantage of using secondary data in studying social problems is that the data are readily accessible, so researchers avoid

field research Research that involves observing and studying social behavior in settings in which it occurs naturally.

the time and expense of collecting their own data. Secondary data are also often based on large representative samples. The disadvantage of secondary data is that researchers are limited to the data already collected.

Ten Good Reasons to Read This Book

Most students reading this book are not majoring in sociology and do not plan to pursue sociology as a profession. So, why should students take a course on social problems? How can reading this textbook about social problems benefit you?

1. *Understanding that the social world is too complex to be explained by just one theory will expand your thinking about how the world operates.* For example, juvenile delinquency doesn't have just one cause—it is linked to (1) an increased number of youths living in inner-city neighborhoods with little or no parental supervision (social disorganization theory); (2) young people having no legitimate means of acquiring material wealth (anomie theory); (3) youths being angry and frustrated at the inequality and racism in our society (conflict theory); and (4) teachers regarding youths as "no good" and treating them accordingly (labeling theory).

2. *Developing a sociological imagination will help you see the link between your personal life and the social world in which you live.* In a society that values personal responsibility, there is a tendency to define failure and success as a consequence of individual free will. The *sociological imagination* enables us to understand how social forces influence our personal misfortunes and failures, and contribute to personal successes and achievements.

3. *Understanding globalization can help you become a safe, successful, and productive world citizen.* Social problems cross national boundaries. Problems such as obesity, war, climate change, human trafficking, and overpopulation are global problems. Problems that originate in one part of the world may affect other parts of the world, and may be caused by social policies in other nations. Thus, understanding social problems requires consideration of the global interconnectedness of the world. And solving today's social problems requires collective action among citizens across the globe. To better prepare students for a globalized world, many colleges and universities have made changes to the curriculum such as adding new general education or core curriculum courses on global concerns and perspectives, revamping existing courses to increase emphasis on global issues, and offering a "global certificate" that students can earn by completing a certain number of courses with an international focus (Wilhelm 2012).

What Do You Think? Some colleges and universities have instituted policies that require students to take one or more global courses—courses with a global or international focus—in order to graduate. Do you think colleges and universities should require some minimum number of global courses for undergraduates? Why or why not?

4. *Understanding the difficulty involved in "fixing" social problems will help you make decisions about your own actions, for example, who you vote for or what charity you donate money to.* It is important to recognize that "fixing" social problems is a very difficult and complex enterprise. One source of this difficulty is that we don't all agree on what the problems are. We also don't agree on what the root causes are of social problems. Is the problem of gun violence in the United States a problem caused by gun availability? Violence in the media? A broken mental health care system? Masculine gender norms? If we socialized boys to be more nurturing and gentle, rather than aggressive and competitive, we might reduce gun violence, but we would also potentially create a generation of boys who would not want to sign up for combat duties in the military, and our armed forces would not have

Some of us know early in life exactly what we want to be, or at least what we think we want to be, when we "grow up." Starting in high school and continuing in college, students are often frantic about what to "do." I see it all the time as a sociology student adviser. "I really love sociology but what can I do with a sociology degree?" Actually, I don't find that question particularly surprising. Few of us grow up hearing about sociology or knowing what it is. Even I didn't consciously choose to be a sociologist as one might choose a career in law or nursing, or increasingly, in business and computer science.

There is a theory called "drift theory," which argues that delinquency is a "relatively inarticulate oral tradition" (Matza 1990, p. 52), in which youths drift back and forth between conforming and non-conforming behavior.

Although believing strongly and hoping fervently that sociology is more than a "relatively inarticulate oral tradition," I think the concept of drift applied to me. I had no vision, no calling, no great quest to be a sociologist. I don't even remember, although this was many years ago, the first time I was aware of the fact that such creatures existed.

But as a child of 13, I left the safety and security of suburbia to take a train to downtown Cleveland, Ohio, to an area called Hough to investigate why some people, who were very, very different than me, were burning down a town. It was, as you may know, part of the urban riots and civil rights protests of the 1960s. I suppose, looking back, it was my first field research.

Throughout high school, I consciously drifted in and out of various cliques, fascinated by each, and at Kent State University, where I did my undergraduate work, I was faced with the reality of anti–Viet Nam war demonstrations not unlike the many student demonstrations of today (e.g., Occupy Wall Street). There was one difference though. At Kent State, four students were killed and nine injured by Ohio National Guardsmen called in to quiet the protests.

I think these events, as various events you could isolate in your own lives, were instrumental in molding me as a person and, eventually, as a sociologist. Although, as I said, I had no burning desire to be a sociologist growing up, today I am possessed by a ". . . very special kind of passion . . ." (Berger 1963, p.12), driven by a demon, perhaps the one that led me to Hough so many years ago.

I feel privileged and honored to be a sociologist, and to be able to do what I want to do—and to get paid for it.

The beauty of sociology, among other things, is that it provides a framework, a lens if you will, to problem-solve within a variety of venues—social service departments; consulting firms; hospitals; federal, county, or local government agencies; nursing homes and rehabilitation facilities; law offices; and so forth. Or, as I did, you might want to become an academic sociologist.

Not only will you find sociologists working in almost every imaginable location, the list of what they do is endless. As you may know, you can find sociology courses on race, class, and gender; social movements; family; criminology; sexuality; mass media; religion; the environment; education; health; social psychology; aging; immigration; . . . and yes, social problems. We have such diversity of topics because we do such a variety of jobs. Would you like to be a Foreign Affairs Officer and work in the State Department? Among other degrees listed for this entry-level position is a bachelor's degree in sociology (U.S. Department of State 2013).

But then there's that other pesky problem—what about pay? I know you're all smart enough to know that income in itself is not predictive of job satisfaction. You could earn $249,999 a year, a lot of money to you and me, but if you never saw your spouse/partner, developed ulcers from the stress, worked in a life-threatening environment, and had little job security, I hope most of you would run the other way. So let's take a more balanced approach.

As discussed in the *Wall Street Journal* (2013) and *Forbes* (Smith 2013), an annual listing of the 200 "Best and Worst Jobs" indicate that sociologists are near the top, checking in at number 19. Imagine that you are number 19 out of 200 students graduating from your high school class. Not bad. And the best part is that these calculations were based on official data (e.g., from the U.S. Department of Labor) in *five areas*—(1) environmental factors (e.g., stamina required, competitiveness), (2) income, (3) outlook (e.g., expected employment growth), (4) physical demands (e.g., requires lifting), and (5) stress (e.g., deadlines, travel) (Career Cast 2013).

I admit it would be nice to be number one but right now, I couldn't be happier with number 19.

Source: Mooney 2015.

enough recruits. Thus, solving one social problem (gun violence) may create another social problem (too few recruits). It should also be noted that although some would see low military recruitment as a problem, others would see it as a positive step toward a less militaristic society.

5. *Although this is a social problems book, it may actually make you more, rather than less, optimistic.* Yes, all the problems discussed in the book are real, and they may seem insurmountable, but they aren't. You'll read about positive social change (for example, the number of people who smoke cigarettes in the United States has dramatically dropped, as have rates of homophobia, racism, and sexism). Life expectancy has increased, and more people go to college than ever before. Change for the better can and does happen.

6. *Knowledge is empowering.* Social problems can be frightening, in part, because most people know very little about them beyond what they hear on the news or from their friends. Misinformation can make problems seem worse than they are. The more accurate the information you have, the more you will realize that we, as a society, have the power to solve the problems, and the less alienated you will feel.

7. *The* Self and Society *exercises increase self-awareness and allow you to position yourself within the social landscape.* For example, earlier in this chapter, you had the opportunity to assess your personal values and to compare your responses to a national sample of first-year college students.

8. The Human Side *features make you a more empathetic and compassionate human being by personalizing the topic at hand.* The study of social problems is always about the quality of life of individuals. By conveying the private pain and personal triumphs associated with social problems, we hope to elicit a level of understanding that may not be attained through the academic study of social problems alone. The Human Side in this chapter highlights one of the author's paths to becoming a sociologist.

9. *The* Social Problems Research Up Close *features teach you the basics of scientific inquiry, making you a smarter consumer of "pop" sociology, psychology, anthropology, and the like.* These boxes demonstrate the scientific enterprise, from theory and data collection to findings and conclusions. Examples of research topics featured in later chapters of this book include bullying in schools, juvenile delinquency, how marriage in the United States is changing, greenwashing, responses to masculinity threats, and military suicides.

10. *Learning about social problems and their structural and cultural origins helps you—individually or collectively—make a difference in the world.* Individuals can make a difference in society through the choices they make. You may choose to vote for one candidate over another, demand the right to reproductive choice or protest government policies that permit it, drive drunk or stop a friend from driving drunk, repeat a homophobic or racist joke or chastise the person who tells it, and practice safe sex or risk the transmission of sexually transmitted diseases.

   Collective social action is another, often more powerful way to make a difference. You may choose to create change by participating in a **social movement**—an organized group of individuals with a common purpose of promoting or resisting social change through collective action. Some people believe that, to promote social change, one must be in a position of political power and/or have large financial resources. However, the most important prerequisite for becoming actively involved in improving levels of social well-being may be genuine concern and dedication to a social "cause." This chapter's photo essay visually portrays students acting collectively to change the world.

Understanding Social Problems

At the end of each chapter, we offer a section with a title that begins with "Understanding . . . ," in which we reemphasize the social origin of the problem being discussed, the consequences, and the alternative social solutions. Our hope is that readers will end each chapter with a "sociological imagination" view of the problem and with an idea of how, as a society, we might approach a solution.

Sociologists have been studying social problems since the Industrial Revolution. Industrialization brought about massive social changes: The influence of religion declined, and families became smaller and moved from traditional, rural communities to urban settings. These and other changes have been associated with increases in crime, pollution, divorce, and juvenile delinquency. As these social problems became more widespread, the need to understand their origins and possible solutions became more urgent. The field of sociology developed in response to this urgency. Social problems provided the initial impetus for the development of the field of sociology and continue to be a major focus of sociology.

There is no single agreed-on definition of what constitutes a social problem. Most sociologists agree, however, that all social problems share two important elements: an objective social condition and a subjective interpretation of that condition. Each of the three major theoretical perspectives in sociology—structural-functionalist, conflict, and symbolic interactionist—has its own notion of the causes, consequences, and solutions of social problems.

social movement An organized group of individuals with a common purpose to either promote or resist social change through collective action.

Photo Essay

Student activism is not new nor is it unique to the United States. In the 1930s, the American Youth Congress (AYC) protested racial injustice, educational inequality, and the looming involvement of the United States in WWII. Called the "student brain of the New Deal" by some, the political power of the AYC would not be felt again until the student demonstrations of the 1960s (The Eleanor Roosevelt Papers 2008). Today, however, there is a new activism as students all over the world protest perceived injustices (Rifkind 2009). Aided by new technologies, social networking sites such as Facebook and Twitter allow for "virtual activism" as hundreds of thousands of students join online causes. This chapter's photo essay highlights some of the most prominent examples of student activism, past and present. Although the faces have changed over time, the passion and dedication with which students voice their concerns has not.

AP Images/Jeff Widener, File

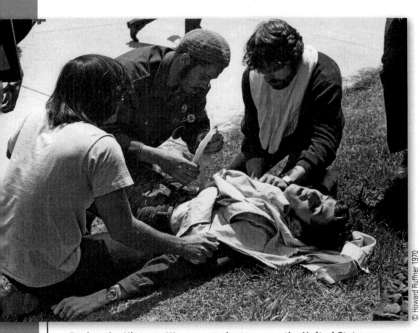

© Howard Ruffner 1970

▲ During the Vietnam War era, students across the United States were vocal about their opposition to America's involvement in the war. In 1970, at Kent State University, the Ohio National Guard opened fire on unarmed student demonstrators, resulting in four deaths and nine injuries and leading singer/songwriter Neil Young to compose "Ohio" ("Tin soldiers and Nixon coming, we're finally on our own. This summer I hear the drumming, four dead in Ohio . . ."). The "Kent State Massacre" sparked nationwide campus protests and the only nationwide student strike in U.S. history. A government report on antiwar demonstrations concluded that the shootings of students by the Ohio National Guard were unjustified (The Scranton Report 1971). No criminal charges were ever filed. Here, an injured student lies on the ground as onlookers stare in disbelief.

Dibyangshu Sarkar/AFP/Getty Images

▲ Rape is a universal crime, although the extent to which the laws are enforced and women are, in fact, protected varies dramatically. Recently, there have been national protests in India sparked by the brutal gang rape of a young college student who later died, and the repeated rape and sexual abuse of a 5-year-old girl who was found semiconsciousness three days after being abducted (Park 2013). Many of the protests are led by students from Nehru University and University of Delhi (Gottipati, Trivedi, & Rai 2013). The protests reflect not only the most recent rash of crimes against women but also the failure of the Indian justice system to prosecute such cases in a timely manner, if at all. As one activist stated, "We have been screaming ourselves hoarse demanding greater security for women and girls. But the government, the police, and others responsible for public security have ignored the daily violence that women face" (George 2012, p. 1).

◄ In 1989, thousands of students from universities across China sat peacefully in Tiananmen Square protesting for democratic reforms and social justice. On June 3rd, tanks entered the square and opened fired on the unarmed students, killing or injuring hundreds, perhaps thousands. There is no official tally of the casualties due to the Chinese government's subsequent clampdown on media and the reporting of any dissident activities, a policy that continues today. Here an unknown man brings to a halt the People's Liberation Army as they advanced to disburse peaceful student demonstrations.

Jack Moebes/Historical/Corbis

▲ On February 1, 1960, four African American students entered a Greensboro Woolworth's store to buy school supplies (Sykes 1960; Schlosser 2000). If their money was good enough to buy school supplies, why not a cup of coffee, they reasoned? At 4:30 p.m., they sat at the "whites only" lunch counter, intending to place an order. The four young men sat at the counter until closing but were never served. The next day, more students sat at the counter—they too were never served. As news of the "sit-in" spread, students returned to the Greensboro Woolworth's and to other lunch counters across the South. White and Black American students alike from New York to San Francisco began picketing Woolworth's in support of the "Greensboro Four." This one act by four students was the pivotal step in propelling forward what became known as the American civil rights movement (Schlosser 2000).

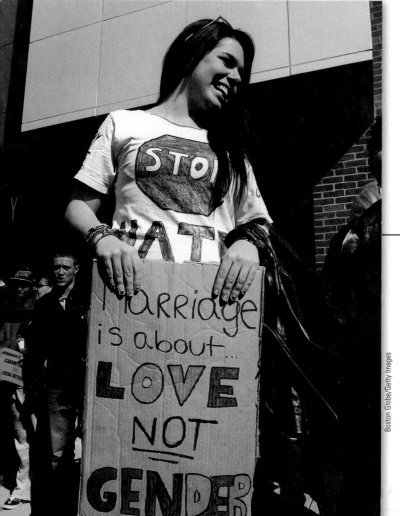

Boston Globe/Getty Images

◄ The term *marriage equality* is fairly new but has become the rallying call of many college and high school students alike. Eliza Byard, the Executive Director of the Gay, Lesbian and Straight Education Network (GLSEN), recently commented that the real heroes of the marriage equality movement are students. Citing a *Washington Post*-ABC News Poll, Byard (2013) noted that fully 81 percent of 18- to 29-year-olds support marriage equality. Student groups that support marriage equality, in addition to GLSEN, include the Gay-Straight Alliance Network, the American Medical Student Association, the National Youth Advocacy Coalition, the hundreds of LGBT centers at colleges across the United States, and Campus Pride, which—among other initiatives—developed a *Campus Pride Index* that rates colleges on their LGBT friendliness.

23

- **What is a social problem?**
  Social problems are defined by a combination of objective and subjective criteria. The objective element of a social problem refers to the existence of a social condition; the subjective element of a social problem refers to the belief that a particular social condition is harmful to society or to a segment of society and that it should and can be changed. By combining these objective and subjective elements, we arrive at the following definition: A social problem is a social condition that a segment of society views as harmful to members of society and in need of remedy.

- **What is meant by the structure of society?**
  The structure of a society refers to the way society is organized.

- **What are the components of the structure of society?**
  The components are institutions, social groups, statuses, and roles. Institutions are an established and enduring pattern of social relationships and include family, religion, politics, economics, and education. Social groups are defined as two or more people who have a common identity, interact, and form a social relationship. A status is a position that a person occupies within a social group and that can be achieved or ascribed. Every status is associated with many roles, or the set of rights, obligations, and expectations associated with a status.

- **What is meant by the culture of society?**
  Whereas social structure refers to the organization of society, culture refers to the meanings and ways of life that characterize a society.

- **What are the components of the culture of society?**
  The components are beliefs, values, norms, and symbols. Beliefs refer to definitions and explanations about what is assumed to be true. Values are social agreements about what is considered good and bad, right and wrong, desirable and undesirable. Norms are socially defined rules of behavior. Norms serve as guidelines for our behavior and for our expectations of the behavior of others. Finally, a symbol is something that represents something else.

- **What is the sociological imagination, and why is it important?**
  The sociological imagination, a term that C. Wright Mills (1959) developed, refers to the ability to see the connections between our personal lives and the social world in which we live. It is important because, when we use our sociological imagination, we are able to distinguish between "private troubles" and "public issues" and to see connections between the events and conditions of our lives and the social and historical context in which we live.

- **What are the differences between the three sociological perspectives?**
  According to structural functionalism, society is a system of interconnected parts that work together in harmony to maintain a state of balance and social equilibrium for the whole. The conflict perspective views society as composed of different groups and interests competing for power and resources. Symbolic interactionism reflects the microsociological perspective and emphasizes that human behavior is influenced by definitions and meanings that are created and maintained through symbolic interaction with others.

- **What are the first four stages of a research study?**
  The first four stages of a research study are formulating a research question, reviewing the literature, defining variables, and formulating a hypothesis.

- **How do the various research methods differ from one another?**
  Experiments involve manipulating the independent variable to determine how it affects the dependent variable. Survey research involves eliciting information from respondents through questions. Field research involves observing and studying social behavior in settings in which it occurs naturally. Secondary data are data that other researchers or government agencies have already collected or that exist in forms such as historical documents, police reports, school records, and official records of marriages, births, and deaths.

- **What is a social movement?**
  Social movements are one means by which social change is realized. A social movement is an organized group of individuals with a common purpose to either promote or resist social change through collective action.

TEST YOURSELF

1. Definitions of social problems are clear and unambiguous.
   a. True
   b. False
2. The social structure of society contains
   a. statuses and roles
   b. institutions and norms
   c. sanctions and social groups
   d. values and beliefs
3. The culture of society refers to its meaning and the ways of life of its members.
   a. True
   b. False
4. Alienation
   a. refers to a sense of normlessness
   b. is focused on by symbolic interactionists
   c. can be defined as the powerlessness and meaninglessness in people's lives
   d. is a manifest function of society
5. Blumer's stages of social problems begin with
   a. mobilization for action
   b. societal recognition
   c. social legitimation
   d. development and implementation of a plan

6. The independent variable comes first in time; i.e., it precedes the dependent variable.
   a. True
   b. False
7. The third stage in defining a research study is
   a. formulating a hypothesis
   b. reviewing the literature
   c. defining the variables
   d. formulating a research question
8. A sample is a subgroup of the population—the group to whom you actually give the questionnaire.
   a. True
   b. False

9. Studying police behavior by riding along with patrol officers would be an example of
   a. participant observation
   b. nonparticipant observation
   c. field research
   d. both a and c
10. Students benefit from reading this book because it
    a. provides global coverage of social problems
    b. highlights social problems research
    c. encourages students to take pro-social action
    d. all of the above

Answers: 1. B; 2. A; 3. A; 4. C; 5. B; 6. A; 7. C; 8. A; 9. D; 10. D.

KEY TERMS

achieved status 5
alienation 10
anomie 9
ascribed status 5
beliefs 5
culture 5
dependent variable 16
experiments 17
field research 18
hypothesis 16
independent variable 16

institution 4
latent functions 9
manifest functions 9
norms 6
objective element of a social problem 2
primary groups 4
roles 5
sample 17
sanctions 7
secondary groups 4
social group 4

social movement 21
social problem 3
sociological imagination 7
status 5
structure 4
subjective element of a social problem 3
survey research 17
symbol 7
theory 8
values 5
variable 16

2

© Gerry Boughan/Shutterstock.com

Physical and Mental Health and Health Care

"America's health care system is neither healthy, caring, nor a system."

—Walter Cronkite

IN A 2013 *TIME MAGAZINE* special report titled "Bitter Pill: Why Medical Bills are Killing Us," journalist Steven Brill (2013) investigated the high cost of U.S. medical care. His report included the story of a patient he called Steven D.

"Soon after he was diagnosed with lung cancer . . . Steven D. and his wife Alice knew that they were only buying time. The crushing question was, How much is time really worth?" Alice, who earned about $40,000 a year running a child care center, explained, "[Steven] kept saying he wanted every last minute he could get, no matter what. But I had to be thinking about the cost and how all this debt would leave me and my daughter." Less than a year after his diagnosis, Steven D. died at his home in northern California, leaving his wife Alice with medical bills totaling more than $900,000. Although the couple had medical insurance, they had exceeded their policy's $50,000 limit. After receiving the first hospital bill for $348,000, Alice hired a billing advocate who negotiated with the hospital to discount Steven's medical bill. By the one-year anniversary of Steven's death, after getting help from Medicaid and negotiating with the hospital and a slew of doctors, clinics, and other providers, Alice had paid about $30,000 of her own money on Steven's medical bills, and she still owed $142,000. "I think about the $142,000 all the time. It just hangs over my head,"

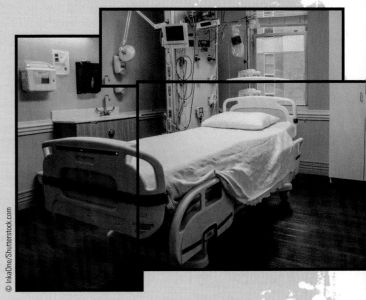

On Steven D's hospital bill, the charge for being in an intensive care room, like the one pictured here, was $13,225 a day.

Alice said, adding that this whole ordeal had taught her a lesson: "I'm never going to remarry. I can't risk the liability" (quoted in Brill 2013, p. 38).

In this chapter, we address problems in health care, focusing on issues related to access, cost, and quality of health care. Using a sociological approach to health issues, we examine why some social groups experience more health problems than others and how social forces affect and are affected by health and illness.

The World Health Organization (1946) defined **health** as "a state of complete physical, mental, and social well-being" (p. 3). One could argue that the study of social problems is, essentially, the study of health problems, as each social problem affects the physical, mental, and social well-being of humans and the social groups of which they are a part.

> One could argue that the study of social problems is, essentially, the study of health problems, as each social problem affects the physical, mental, and social well-being of humans and the social groups of which they are a part.

The Global Context: Health and Illness around the World

Most people are concerned about their own health and the health of their loved ones. Sociologists are concerned not only about their own health, but also about the health of different populations within and between nations. In making international comparisons, social scientists commonly classify countries according to their level of economic development. (1) **Developed countries**, also known as *high-income countries,* have relatively high gross national income per capita; (2) **less developed** or **developing countries**, also known as *middle-income countries,* have relatively low gross national income per capita; and (3) **least developed countries** (known as *low-income countries*) are the poorest countries of the world. As we discuss in the following section, how long people live and what causes their death varies across the globe.

health According to the World Health Organization, "a state of complete physical, mental, and social well-being."

developed countries Countries that have relatively high gross national income per capita, also known as high-income countries.

developing countries Countries that have relatively low gross national income per capita, also known as less-developed or middle-income countries.

least developed countries The poorest countries of the world.

TABLE 2.1 Life Expectancy by Country Income Level, 2011

Country Income Level	Life Expectancy
High	80
Higher middle	74
Lower middle	66
Low	60
WORLD	70

Source: World Health Organization 2013.

Life Expectancy and Mortality in Low-, Middle-, and High-Income Countries

Life expectancy—the average number of years that individuals born in a given year can expect to live—is significantly greater in high-income countries than in low-income countries (see Table 2.1). The World Health Organization (WHO) reports that in some of the poorest countries of the world (primarily in Africa), life expectancy is less than 50 years, compared to an average of 80 in high-income countries.

The leading causes of death, or **mortality**, also vary around the world (see Table 2.2). Deaths caused by parasitic and infectious diseases, such as HIV/AIDS, tuberculosis, diarrheal diseases, and malaria are much more common in less developed countries compared with the more developed countries. Parasitic and infectious diseases spread more easily in poor and overcrowded housing conditions, and in areas with lack of clean water and sanitation (see also Chapter 6).

Worldwide, nearly two-thirds of deaths are due to noncommunicable diseases, primarily heart disease, stroke, cancer, and respiratory diseases. These noninfectious, nontransmissible diseases are also the leading causes of death in wealthy countries such as the United States and are largely caused by behavioral risk factors, including tobacco use, physical inactivity, unhealthy diet, and alcohol abuse (World Health Organization 2013). In recent decades, noncommunicable diseases—particularly heart disease—have also become leading causes of death in low- and middle-income countries, as rising incomes and emerging middle classes in countries such as China and India have led to (1) increased use of tobacco (linked to cancer and respiratory diseases); (2) increased access to automobiles,

What Do You Think? Data on deaths from international terrorism and tobacco-related deaths in 37 developed and eastern European countries revealed that tobacco-related deaths outnumbered terrorist deaths by about a whopping 5,700 times (Thomson & Wilson 2005). The number of tobacco deaths was equivalent to the impact of a September 11, 2001–type terrorist attack every 14 hours! Given that tobacco-related deaths grossly outnumber terrorism-related deaths, why hasn't the U.S. government waged a "war on tobacco" on a scale similar to its "war on terrorism"?

TABLE 2.2 Leading Causes of Death, by Country Income Level

Low Income	Middle Income	High Income
1. Respiratory infections	Heart disease	Heart disease
2. Diarrheal diseases	Stroke and other cerebrovascular disease	Stroke and other cerebrovascular disease
3. HIV/AIDS	Chronic obstructive pulmonary disease	Trachea, bronchus, lung cancers
4. Heart disease	Respiratory infections	Alzheimer and other dementias
5. Malaria	Diarrheal diseases	Respiratory infections
6. Stroke and other cerebrovascular disease	HIV/AIDS	Chronic obstructive pulmonary disease
7. Tuberculosis	Road traffic accidents	Colon and rectum cancer
8. Prematurity/low birth weight	Tuberculosis	Diabetes

Source: World Health Organization 2012b.

life expectancy The average number of years that individuals born in a given year can expect to live.

mortality Death.

televisions, and other technologies that contribute to a sedentary lifestyle, and (3) increased consumption of processed foods high in sugar and fat (linked to obesity).

Mortality among Infants and Children. The rates of **infant mortality** (death of live-born infants under 12 months of age), and **under-5 mortality** (death of children under age 5) provide powerful indicators of the health of a population. The *infant mortality rate,* the number of deaths of live-born infants under 1 year of age per 1,000 live births (in any given year), ranges from an average of 5 in high-income nations to 63 in low-income nations (World Health Organization 2013). The *under-5 mortality rate,* or death rate of children under age 5, similarly is much lower in high-income countries (6 in 2011) than in low-income countries (95 in 2011). Nearly 8 million children died in 2010 before they reached their 5th birthday, mostly from diarrhea, pneumonia, birth complications, and malnutrition (UNICEF 2012a). Diarrhea, which can lead to life-threatening dehydration, often results from contaminated drinking water and lack of sanitation, or the unavailability of toilets or other hygienic means of disposing of human waste. More than one-third of the world's population—2.6 billion people—does not have access to adequate sanitation facilities, and more than one in 10 people on the planet don't have access to safe drinking water (World Health Organization 2013) (see also Chapter 6).

> When Tanzanian mothers are in labor, they often say to their older children, "I'm going to go and fetch the new baby; it is a dangerous journey and I may not return."

Maternal Mortality. Women in the United States and other developed countries generally do not experience pregnancy and childbirth as life threatening. But for women ages 15 to 49 in developing countries, **maternal mortality**—death that results from complications associated with pregnancy and childbirth—is a leading cause of death and disability. When Tanzanian mothers are in labor, they often say to their older children, "I'm going to go and fetch the new baby; it is a dangerous journey and I may not return" (Grossman 2009). The top causes of maternal mortality are hemorrhage (severe loss of blood), infection, high blood pressure during pregnancy, and unsafe abortion. More than a half million women die every year from childbirth or pregnancy-related causes. And for every maternal death, about 20 women suffer from disability or medical problems related to pregnancy or childbirth (Biset 2013).

Rates of maternal mortality show a greater disparity between rich and poor countries than any of the other societal health measures. Nearly all (99 percent) maternal deaths occur in low-income countries (World Health Organization 2013). High maternal mortality rates in less developed countries are related to poor-quality and inaccessible health care; most women give birth without the assistance of trained personnel (see Table 2.3). High maternal mortality rates are also linked to malnutrition and poor sanitation and to pregnancy and childbearing at early ages. Women in many countries also lack access to family planning services and/or do not have the support of their male partners to use contraceptive methods such as condoms. Consequently, many women resort to abortion to limit their childbearing, even in countries where abortion is illegal and unsafe.

TABLE 2.3 Skilled Childbirth Assistance and Lifetime Risk of Maternal Mortality by Development Level

	Percentage of births attended by skilled personnel	Lifetime risk of maternal mortality
Least developed	46 percent	1 in 37
Developing	66 percent	1 in 120
Industrialized	No data available*	1 in 4300

*We can assume nearly 100%, as standard childbirth practices in industrialized countries involve a skilled birth attendant.

Source: UNICEF 2012a.

infant mortality Deaths of live-born infants under 1 year of age.

under-5 mortality Deaths of children under age 5.

maternal mortality Deaths that result from complications associated with pregnancy, childbirth, and unsafe abortion.

Globalization, Health, and Medical Care

Globalization, broadly defined as the growing economic, political, and social interconnectedness among societies throughout the world, has had both positive and negative effects on health and medical care.

Effects of Globalization on Health. Global trade agreements have expanded the range of goods available to consumers, but at a cost to global health. The international trade of tobacco, alcohol, and sugary drinks and high-calorie processed foods, and the expansion of fast-food chains across the globe, are associated with a worldwide rise in cancer, heart disease, stroke, obesity, and diabetes (Hawkes 2006; World Health Organization 2013). Globalization has resulted in rising incomes in the developing world, and although it has improved quality of life for many people, it has also increased access to unhealthy foods and beverages and decreased levels of physical activity. As poor populations move toward the middle class, they can afford to buy televisions, computers, automobiles, and processed foods—products that increase caloric intake and decrease physical activity, leading to increased rates of obesity around the world. Indeed, a new word has emerged to refer to the high prevalence of obesity around the world: **globesity** (see this chapter's Photo Essay titled "Globesity").

Another aspect of globalization that affects health is global travel and transportation, which (1) contributes to harmful pollution caused by the burning of fossil fuels, and (2) can speed the spread of infectious disease. In just the first two months of the swine flu pandemic of 2009, the disease spread to infect nearly 600,000 people in more than 70 countries around the world.

On the positive side, globalized communications technology is helpful in monitoring and reporting outbreaks of disease, disseminating guidelines for controlling and treating disease, and sharing medical knowledge and research findings (Lee 2003).

Medical Tourism. The globalization of medical care involves increased international trade in health products and services. **Medical tourism**—a growing multibillion-dollar global industry—involves traveling, primarily across international borders, for the purpose of obtaining medical care. Health care consumers travel to other countries for medical care for three primary reasons: (1) to obtain medical treatment that is not available in their home country; (2) to avoid waiting periods for treatment; and/or (3) to save money on the cost of medical treatment. Steve Jobs reportedly traveled to Switzerland for special cancer treatment, and National Football League quarterback Peyton Manning

globalization The growing economic, political, and social interconnectedness among societies throughout the world.

globesity The high prevalence of obesity around the world.

medical tourism A global industry that involves traveling, primarily across international borders, for the purpose of obtaining medical care.

flew to Europe for a stem cell procedure to treat his injured neck (Turner & Hodges 2012). Popular medical tourism destinations that lure health care consumers with competitively priced medical care include Mexico, Singapore, Thailand, and India, among other countries. Medical tourism companies offer packages that bundle air travel, ground transportation, hotel accommodations, and guided tours, along with arranging medical treatment, such as organ transplants, dental work, stem cell therapies, cosmetic surgery, reproductive assistance, weight loss surgery, cardiac surgery, and many other medical treatments and services.

Although medical tourism can benefit some patients in providing timely, reduced-cost, quality medical care, there are a number of risks and problems involved. Unlike the highly regulated health care industry in the United States, medical services, products, and facilities in other countries may not be regulated, so quality control is a concern. Medical travel may contribute to the spread of infectious disease, and the medical tourism industry may encourage the illegal market for human organs, as the very poor are vulnerable to being coerced into selling one of their kidneys for transplantation. Although medical tourism can benefit local and national economies, it also takes dollars away from other health care providers. Finally, medical tourism raises ethical concerns about health equity, as health services in popular medical tourism destinations flow not to the local population, but to foreigners.

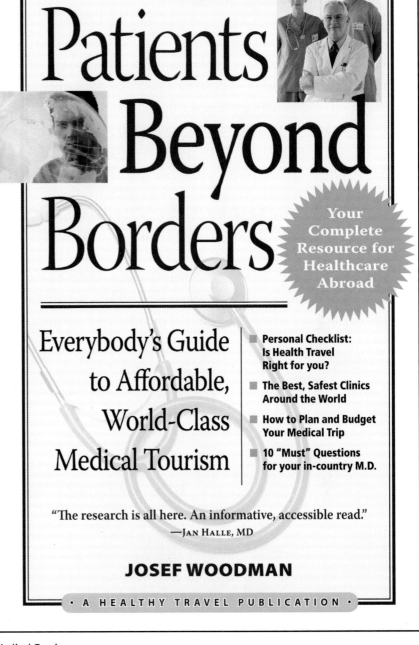

Medical Tourism.

Are Americans the Healthiest Population in the World?

Many Americans view the United States as the best country in the world—the country with the best system of democracy, the most freedoms, the highest standard of living . . . and the best health. As this chapter deals with health issues, we will address here only the latter assumption: that Americans have the best health in the world. Is this true? In a word, the answer is "no." The United States is one of the wealthiest countries in the world, but it is not one of the healthiest.

The United States is one of the wealthiest countries in the world, but it is not one of the healthiest.

In the last few decades, globalization has led to increased consumption of snacks, sugary drinks, supersized portions, and processed and "fast food" (often high in fat and calories). At the same time, people are engaging in more sedentary behavior, such as sitting at their job workstation, watching television, surfing the Internet, and riding in automobiles or on mopeds, instead of walking or cycling. About a third of the world's population does not meet the minimum recommendations for physical activity (Hallal et al., 2012). These factors have contributed to a worldwide increase in overweight and obesity. Between 1980 and 2008, the worldwide prevalence of obesity nearly doubled: In 2008, 10 percent of men and 14 percent of women were obese*, compared with 5 percent of men and 8 percent of women in 1980 (World Health Organization 2013).

▲ Many low- and middle-income countries face the dual burden of obesity and underweight.

▲ Type 2 diabetes, a medical condition linked to overweight and obesity, can cause nerve damage and result in amputation.

Until recently, obesity was a public health problem only in Western industrialized countries. But over the last couple of decades, obesity has become a global problem affecting countries of every income and development level. The World Health Organization has used the term *globesity* to refer to the high prevalence of obesity around the world, and the accompanying health problems associated with obesity, specifically diabetes, heart disease, and certain cancers. Indeed, for the first time in human history, the world has more overweight than underweight people (Harvard School of Public Health 2013).

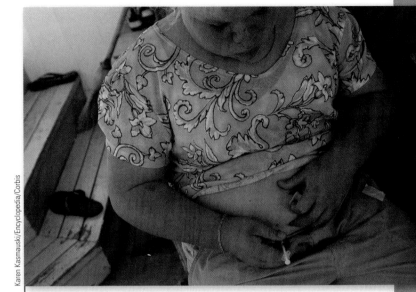

Karen Kasmauski/Encyclopedia/Corbis

▲ This child is giving herself an insulin shot to control her type 2 diabetes. Previously a disease only found in adults, type 2 diabetes is now appearing in some overweight/obese children.

Martyn Evans/Alamy

▲ Globesity is related in part to the increased global access to sugar-sweetened beverages and processed, high-calorie, fast foods.

Andre Jenny/Alamy

▲ Mexico has the highest per capita consumption of sugary drinks in the world. Sugar-sweetened beverages have been linked to obesity.

*Overweight is defined as having a body mass index (BMI) of 25 or more, and obese is defined as having a BMI of 30 or more, with BMI being calculated by taking a person's weight (in kilograms) and dividing it by the square of that person's height (in meters).

A report by the U.S. National Research Council and Institute of Medicine (2013) compared health outcomes in the United States with those of 16 other high-income, industrialized countries: Austria, Australia, Canada, Denmark, Finland, France, Germany, Italy, Japan, Norway, Portugal, Spain, Sweden, Switzerland, the Netherlands, and the United Kingdom. How did the United States compare with its peer countries on measures of health? Let's start with the good news: Compared with people in other industrialized countries, Americans are less likely to smoke and drink heavy alcohol, and they have better control over their cholesterol levels. The United States also has higher rates of cancer screening and survival, and has higher survival after age 75.

But the report's main finding was that despite the fact that the United States spends more on health care per person than any other industrialized country, Americans die sooner and have higher rates of disease or injury. Life expectancy of U.S. men is shorter than for men in any of the other 16 countries, and only one country (Denmark) has a lower life expectancy for women than that of U.S. women. The United States ranked last or near the bottom in nine key areas of health: infant mortality and low birth weight; injuries and homicide; teenage pregnancy and sexually transmitted infections; HIV/AIDS prevalence; drug-related deaths; obesity and diabetes; heart disease; lung disease; and disability. This "U.S. health disadvantage" has been getting worse for three decades, especially among women. And although health outcomes are generally worse among socially disadvantaged members of a population, even advantaged Americans—those who are college educated, upper income, or insured—have poorer health than similar individuals in other industrialized countries.

The following summarizes the multiple explanations for the U.S. health disadvantage:

- **Health systems.** Americans are more likely to find health care inaccessible or unaffordable—problems we will discuss later in this chapter. Unlike other industrialized countries, the United States has a relatively large uninsured population and more limited access to health care.
- **Unhealthy behaviors.** Compared with other industrialized populations, Americans have higher rates of prescription and illegal drug abuse, are more likely to use firearms in acts of violence, are less likely to use seat belts, and are more likely to be involved in traffic accidents that involve alcohol. Americans also consume the most calories per person and have the highest obesity rates.
- **Social and economic conditions.** Although the income of Americans is higher on average than in other countries, the United States has higher rates of poverty (especially child poverty), more income inequality, and less social mobility (see also Chapter 6). The United States also lags behind other countries in the education of youth, which also negatively affects health. And compared with other industrialized populations, Americans benefit less from social safety programs that help buffer the adverse health effects of poverty and low educational attainment.
- **Physical and social environment.** The physical environment in most U.S. communities discourages physical activity, as the environment is designed for automobiles rather than pedestrians. Indeed, U.S. adults take the fewest steps of any industrialized nation, averaging slightly over 5,000 steps a day compared with adults in Australia and Switzerland who average nearly 10,000 steps a day (America's Trust for Health 2012). And in the absence of other transportation options, greater reliance on automobiles in the United States contributes to higher traffic fatalities (U.S. National Research Council and Institute of Medicine 2013). The social environment adversely affects Americans' health in a number of ways: (1) Americans' unhealthy patterns of food consumption are shaped by the agricultural and food industries, grocery store and restaurant offerings, and marketing; (2) the higher rate of firearm-related deaths in the United States is at least in part due to the fact that firearms are more available in the United States than in peer countries; (3) Americans' higher rates of substance abuse, physical illness, and family violence may be related to the higher-stress lifestyle in the United States. For example, Americans tend to work more hours and have less vacation time compared to workers in other industrialized countries (see also Chapter 7).

TABLE 2.4 Mental Disorders Classified by the American Psychiatric Association

Classification	Description
Anxiety disorders	Disorders characterized by anxiety that is manifest in phobias, panic attacks, or obsessive-compulsive disorder
Dissociative disorders	Problems involving a splitting or dissociation of normal consciousness, such as amnesia and multiple personality
Disorders first evident in infancy, childhood, or adolescence	Disorders including mental retardation, attention deficit/hyperactivity disorder, and stuttering
Eating or sleeping disorders	Disorders including anorexia, bulimia, and insomnia
Impulse control disorders	Problems involving the inability to control undesirable impulses, such as kleptomania, pyromania, and pathological gambling
Mood disorders	Emotional disorders such as major depression and bipolar (manic-depressive) disorder
Organic mental disorders	Psychological or behavioral disorders associated with dysfunctions of the brain caused by aging, disease, or brain damage (such as Alzheimer's disease)
Personality disorders	Maladaptive personality traits that are generally resistant to treatment, such as paranoid and antisocial personality types
Schizophrenia and other psychotic disorders	Disorders with symptoms such as delusions or hallucinations
Somatoform disorders	Psychological problems that present themselves as symptoms of physical disease, such as hypochondria
Substance-related disorders	Disorders resulting from abuse of alcohol and/or drugs, such as barbiturates, cocaine, or amphetamines

© Cengage Learning

In sum, the U.S. health disadvantage has multiple causes involving inadequate health care, unhealthy behaviors, adverse social and economic conditions, physical and social environmental factors, and the cultural values and public policies that shape these factors. Unless these conditions change, Americans will continue to have shorter lives and poorer health than people in other industrialized countries.

What Do You Think? Do you think that most Americans would be surprised to learn that their chances for living a long and healthy life are not as good as for people living in other high-income countries? How might greater public awareness of the U.S. health disadvantage affect the national dialogue on health care in the United States?

Mental Illness: The Hidden Epidemic

What it means to be mentally healthy varies across cultures. In the United States, **mental health** is defined as the successful performance of mental function, resulting in productive activities, fulfilling relationships with other people, and the ability to adapt to change and to cope with adversity (U.S. Department of Health and Human Services 2001). **Mental illness** refers collectively to all mental disorders, which are characterized by sustained patterns of abnormal thinking, mood (emotions), or behaviors that are accompanied by significant distress and/or impairment in daily functioning (see Table 2.4).

It is important to recognize that physical and mental health are connected and affect each other. For example, people with type 2 diabetes are twice as likely to experience depression as the general population, and people with diabetes who are depressed have

mental health The successful performance of mental function, resulting in productive activities, fulfilling relationships with other people, and the ability to adapt to change and to cope with adversity.

mental illness Refers collectively to all mental disorders, which are characterized by sustained patterns of abnormal thinking, mood (emotions), or behaviors that are accompanied by significant distress and/or impairment in daily functioning.

more difficulty with self-care. Up to half of cancer patients have a mental illness, especially anxiety and depression, and some evidence suggests that treating depression in cancer patients may improve survival time. People with mental illness are twice as likely to smoke cigarettes as other people. Depression also increases the risk for having a heart attack, but treating the symptoms of depression in people who have had a heart attack improves their survival (Kolappa, Henderson, & Kishore 2013).

Mental illness is a "hidden epidemic" because the shame and embarrassment associated with mental problems discourage people from acknowledging and talking about them. Being labeled as "mentally ill" is associated with a **stigma**—a discrediting label that can negatively affect an individual's self-concept and disqualify that person from full social acceptance. (Originally, the word *stigma* referred to a mark burned into the skin of a criminal or slave.) Negative stereotypes of people with mental illness contribute to its stigma.

> Although untreated mental illness can result in violent behavior, the vast majority of people with mental illness are not violent and they are involved in only about 4 percent of violent crimes.

One of the most common stereotypes of people with mental illness is that they are dangerous and violent. In recent years, there have been a number of incidents involving people with mental illness using guns in mass shootings: In 2007, in Blacksburg, Virginia, a student with a severe anxiety disorder shot and killed 32 people and wounded 25 others on the campus of Virginia Tech. In 2011, a young man who was subsequently diagnosed with schizophrenia shot U.S. Representative Gabrielle Giffords and 18 others in Tucson, Arizona, killing 6. In 2012, a prior graduate student with a history of mental problems killed 12 people and wounded 58 others in a movie theater in Aurora, Colorado, and in the same year, a young man (with suspected mental problems) went on a shooting spree at Sandy Hook Elementary School in Newtown, Connecticut, killing 20 children and 6 adults. Although untreated mental illness can result in violent behavior, the vast majority of people with severe mental illness are not violent and they are involved in only about 4 percent of violent crimes. And people with mental illness are 11 or more times as likely to be victims of violence than members of the general population (Goode & Healy 2013).

Extent and Impact of Mental Illness

Of 17 nations included in the World Mental Health Survey, the United States has the highest rate of mental illness (Shern & Lindstrom 2013). Among the noninstitutionalized civilian population in 2011, one in five U.S. adults had a diagnosable mental, behavioral, or emotional disorder (excluding developmental and substance abuse disorders) either currently or within the past year. One in 20 U.S. adults had a serious mental illness—a mental illness that has resulted in serious impairment in daily functioning (Substance Abuse and Mental Health Services Administration 2012). Among children ages 4 to 17, about 5 percent experience serious emotional or behavioral difficulties and just over 8 percent of adolescents (12 to 17) experience major depression. About half of all Americans will experience some form of mental disorder in their lifetime, with first onset usually occurring in childhood or adolescence (Shally-Jensen 2013).

Untreated mental illness can lead to poor educational achievement, lost productivity, unsuccessful relationships, significant distress, violence and abuse, incarceration, unemployment, homelessness, and poverty. Each year about 35,000 Americans kill themselves, making suicide the tenth leading cause of death in the United States. For every person who commits suicide, 10 more try but fail. Most people who commit suicide are suffering with a mental disorder—most commonly depression or substance abuse—at the time of their death (Shally-Jensen 2013). One population at high risk for suicide is veterans. In recent years, about 18 to 22 veterans have committed suicide each day (Smith-McDowell 2013).

stigma A discrediting label that can negatively affect an individual's self-concept and disqualify that person from full social acceptance.

Causes of Mental Illness

Stigma surrounding mental illness is partly due to misconceptions about their causes, such as the misconception that mental illness is caused by personal weakness, or results from engaging in immoral behavior. In some cultures, people with mental illness are viewed as being possessed by evil spirits or supernatural forces.

Biomedical explanations of mental illness focus on genetic, neurological conditions, and hormonal factors that can cause mental illness. Social and environmental influences that can trigger mental illness include physical, emotional, and sexual abuse; poverty and homelessness; job loss; divorce; the death of a loved one; devastation from a natural disaster such as flood or earthquake; the onset of illness or disabling injury; and the trauma of war. Broadly speaking, "mental health is impacted detrimentally when civil, cultural, economic, political, and social rights are infringed" (World Health Organization 2010, p. xxvi).

Mental Illness among College Students

Mental health problems are not uncommon among college students (see Table 2.5). More than one in four college students has been diagnosed or treated by a professional for a mental health problem within the past year—12 percent of college students were diagnosed or treated for anxiety, 11 percent for depression, 2 percent for bipolar disorder, and 6 percent for panic attacks (American College Health Association 2012). The National Alliance on Mental Illness (2012) (NAMI) surveyed 765 individuals with a mental health condition who were enrolled in college currently or within the past five years, and found that only half had disclosed their diagnosis to their college, even though disclosure is legally required to receive accommodations in college. Table 2.6 lists the five top reasons why students chose either to disclose or not disclose their diagnosis.

> More than one in four college students has been diagnosed or treated by a professional for a mental health problem within the past year.

More than half of respondents in the NAMI survey did not access their college or university's Disabilities Resource Center to request accommodations such as excused absences for treatment and adjustments in test settings and test times. The top reason? Students were unaware that they qualified for and had a right to receive accommodations. Students also cited fear of stigma as a reason for not requesting accommodations. The NAMI survey found that 40 percent of students (both currently and previously enrolled) with mental health conditions did not seek mental health services and supports on campus; the number one reason students did not seek clinical services is concern about the stigma associated with mental illness. Students also cited busy schedules as a barrier to seeking services. Students with mental health problems may also not seek help because they do not realize that they have mental illness (see this chapter's *Self and Society* feature).

TABLE 2.5 **Percentage of College Students Experiencing Selected Mental Health Difficulties Anytime in the Past 12 Months**

Mental Health Difficulty	Percentage
Felt so depressed it was difficult to function	31
Felt overwhelming anxiety	51
Felt very lonely	57
Felt things were hopeless	45
Seriously considered suicide	7

Adapted from: American College Health Association. 2012. *American College Health Association National College Health Assessment II: Reference Groups Executive Summary Spring 2012.* Hanover, MD: American College Health Association.

Note: Percentages are rounded.

TABLE 2.6 Top Reasons for Mental Health Diagnosis Disclosure or Nondisclosure among College Students

Top Five Reasons Why Students Disclose:
To receive accommodations
To receive clinical services and supports on campus
To be a role model and to reduce stigma
To educate students, staff, and faculty about mental health
To avoid disciplinary action by the school and to avoid losing financial aid

Top Five Reasons Why Students Do Not Disclose:
Fear or concern for the impact disclosure would have on how students, faculty, and staff perceive them, especially in mental health degree programs
No opportunity to disclose
Diagnosis does not impact academic performance
Lack of knowledge that disclosing could help secure accommodations
Mistrust that medical information will remain confidential

Source: Based on National Alliance on Mental Illness. 2012. *College Students Speak: A Survey Report on Mental Health*. Available at www.nami.org

Sociological Theories of Illness and Health Care

The three major sociological theories—structural functionalism, conflict theory, and symbolic interactionism—each contribute to our understanding of illness and health care.

Self and Society | Warning Signs for Mental Illness

Do you or someone you know, such as a roommate, friend, or family member have a mental illness and not realize it? Read each of the following warning signs for mental illness, and put a check mark next to each one that applies to yourself or to someone you are concerned about. According to the National Institute of Mental Health, "a person who shows any of these signs should seek help from a qualified health professional" (BSCS 2005, p. 33).

Warning Sign	You	Someone You are Concerned About
1. Marked personality change	___	___
2. Inability to cope with problems and daily activities	___	___
3. Strange or grandiose ideas	___	___
4. Excessive anxieties	___	___
5. Prolonged depression and apathy	___	___
6. Marked changes in eating patterns	___	___
7. Marked changes in sleeping patterns	___	___
8. Thinking or talking about suicide or harming oneself	___	___
9. Extreme mood swings (high or low)	___	___
10. Abuse of alcohol or drugs	___	___
11. Excessive anger, hostility, or violent behavior	___	___

Source: Adapted from *The Science of Mental Illness*. 2005. Colorado Springs: BSCS.

Structural-Functionalist Perspective

According to the structural-functionalist perspective, health care is a social institution that functions to maintain the well-being of societal members and, consequently, of the social system as a whole. Thus, this perspective points to how failures in the health care system affect not only the well-being of individuals, but also the health of other social institutions, such as the economy and the family.

The structural-functionalist perspective examines how changes in society affect health. As societies develop and provide better living conditions, life expectancy increases and birthrates decrease (Weitz 2013). At the same time, the main causes of death and disability shift from infectious disease, and infant, child, and maternal mortality to chronic, noninfectious illness and disease such as cancer, heart disease, Alzheimer's disease, and arthritis.

Just as social change affects health, health concerns may lead to social change. The emergence of HIV and AIDS in the U.S. gay male population helped unite and mobilize gay rights activists. Concern over the effects of exposure to tobacco smoke—the greatest cause of disease and death in the United States and other developed countries—led to legislation banning smoking in public places.

Finally, the structural-functionalist perspective draws attention to latent dysfunctions, or unintended and often unrecognized negative consequences of social patterns or behavior. For example, the use of antibiotics—in prescriptions; soaps, hand wipes, and cleaning agents; and factory farm animal feed—has produced a serious unintended consequence: the emergence of antibiotic-resistant bacteria, or "superbugs" that make treating infections more difficult and costly. In 2011, 29.9 million pounds of antibiotics were sold in the United States for meat and poultry production—nearly four times the amount of antibiotics sold to treat sick people in the United States (Pew Health Initiatives 2013). The Centers for Disease Control and Prevention (2013a) reported a sharp rise in the number of cases of a rare but deadly type of antibiotic-resistant infection known as CRE (carbapenem-resistant *Enterobacteriaceae*). CRE infections lead to death in up to half of all cases, and the CRE infections are highly contagious and can spread like wildfire.

Another example of a latent function is a New York law, passed just weeks after the Sandy Hook shooting in Newtown, Connecticut, requiring mental health practitioners to inform authorities about potentially dangerous patients, enabling law enforcement officials to confiscate any firearm owned by such a patient. New York's law also allows guns to be taken from people who voluntarily commit themselves to hospitalization for mental health treatment. Critics of the law argue that it will have unintended consequences—that it will deter people from seeking treatment and prevent people in treatment from talking about violence (Goode & Healy 2013).

Most antibiotics sold in the United States are used in meat and poultry production. Human consumption of animal products that contain antibiotics has contributed to the rise of "super bugs": infections that are resistant to antibiotic drug treatment.

Visual Mozart/Getty images

Conflict Perspective

The conflict perspective focuses on how wealth, status, power, and the profit motive influence illness and health care. Worldwide, the poor experience more health problems and have less access to quality medical care. The conflict perspective points to ways in which powerful groups and wealthy corporations influence health-related policies and laws through lobbying and financial contributions to politicians and political candidates. Private health insurance companies have much to lose if the United States adopts a national public health insurance program or even a public insurance option, and have spent millions of dollars opposing such proposals (Mayer 2009). The "health care industrial complex," which includes pharmaceutical and health care product industries, and organizations representing doctors, hospitals, nursing homes, and other health services industries, spends more than three times what the military-industrial complex spends on lobbying in Washington, DC (Brill 2013). Corporations also hire public relations (PR) companies to influence public opinion about health care issues. In his book *Deadly Spin* (2010), insurance industry insider Wendell Potter describes how the insurance industry hired a PR firm to manipulate public opinion on health care reform in part by discrediting Michael Moore's 2007 documentary *Sicko.*

The conflict perspective criticizes the pharmaceutical and health care industry for placing profits above people. "Drugmakers, device makers, and insurers decide which products to develop based not on what patients need, but on what their marketers tell them will sell—and produce the highest profit" (Mahar 2006, p. xviii). For example, not enough drugs are being developed to combat the growing public health threat of antibiotic-resistant infections, in part because pharmaceutical companies do not have a financial incentive. "Antibiotics . . . have a poor return on investment because they are taken for a short period of time and cure their target disease. In contrast, drugs that treat chronic illness, such as high blood pressure, are taken daily for the rest of a patient's life" (Braine 2011).

Many industries place profit above health considerations of workers and consumers. Chapter 7, "Work and Unemployment," discusses how employers often cut costs by neglecting to provide adequate safety measures for their employees. Chapter 13, "Environmental Problems," looks at how corporations often ignore environmental laws and policies, exposing the public to harmful pollution. The food industry is more concerned about profits than about public health. For example, most meat and dairy producers routinely feed antibiotics to animals, which humans then consume. Antibiotics in animal feed have contributed to the development of strains of antibiotic-resistant bacteria in humans. Efforts to limit the use of antibiotics in animal feed have been blocked by the pharmaceutical and livestock industry lobbies, whose profits would be threatened if antibiotic use in food animals was limited (Katel 2010).

Symbolic Interactionist Perspective

Symbolic interactionists focus on (1) how meanings, definitions, and labels influence health, illness, and health care; and (2) how such meanings are learned through interaction with others and through media messages and portrayals. According to the symbolic interactionist perspective of illness, "there are no illnesses or diseases in nature. There are only conditions that society, or groups within it, has come to define as illness or disease" (Goldstein 1999, p. 31). Psychiatrist Thomas Szasz (1961/1970) argued that what we call "mental illness" is no more than a label conferred on those individuals who are "different," that is, those who do not conform to society's definitions of appropriate behavior.

Defining or labeling behaviors and conditions as medical problems is part of a trend known as **medicalization**. Behaviors and conditions that have undergone medicalization include post-traumatic stress disorder, premenstrual syndrome, menopause, childbirth, attention deficit/hyperactivity disorder, and even the natural process of dying. Conflict theorists view medicalization as resulting from the medical profession's domination and pursuit of profits. A symbolic interactionist perspective suggests that medicalization results from the efforts of sufferers to "translate their individual experiences of distress into shared experiences of illness" (Barker 2002, p. 295).

According to symbolic interactionism, conceptions of health and illness are socially constructed. It follows, then, that definitions of health and illness vary over time and from society to society. In some countries, being fat is a sign of health and wellness; in others, it is an indication of mental illness or a lack of self-control. Among some cultural groups, perceiving visions or voices of religious figures is considered a normal religious experience, whereas such "hallucinations" would be indicative of mental illness in other cultures. In 18th- and 19th-century America, masturbation was considered an unhealthy act that caused a range of physical and mental health problems (Allen 2000). Today, most health professionals agree that masturbation is a normal, healthy aspect of sexual expression.

Symbolic interactionism draws attention to the effects that meanings and labels have on health and health risk behaviors. For example, among white Americans, having a

medicalization Defining or labeling behaviors and conditions as medical problems.

"tan" is culturally defined as youthful and attractive, and so many white Americans sunbathe and use tanning beds—behaviors that increase one's risk of developing skin cancer. Meanings and labels also affect health policies. After the International Agency for Research on Cancer issued a 2009 report labeling tanning beds as "carcinogenic to humans," many states proposed and/or enacted legislation to restrict the use of tanning beds among minors by, for example, requiring parental permission (National Conference of State Legislators 2010; Reinberg 2009).

What Do You Think? The risk of developing melanoma (skin cancer) increases 75 percent when individuals use tanning beds before age 30 (Reinberg 2009). Some health advocates are calling for a total ban on the use of tanning beds by minors. Would you support such a ban? Why or why not?

Symbolic interactionists also focus on the stigmas associated with certain health conditions. Individuals with mental illness, drug addiction, physical deformities and impairments, missing or decayed teeth, obesity, and HIV infection and AIDS are often stigmatized, and consequently, discriminated against. For example, the HIV/AIDS-related stigma, which stems from societal views that people with HIV/AIDS are immoral and shameful, results in discrimination in employment, housing, social relationships, and medical care. The stigma associated with health problems implies that individuals—rather than society—are responsible for their health. In U.S. culture, "sickness increasingly seems to be construed as a personal failure—a failure of ethical virtue, a failure to take care of oneself 'properly' by eating the 'right' foods or getting 'enough' exercise, a failure to get a Pap smear, a failure to control sexual promiscuity, genetic failure, a failure of will, or a failure of commitment—rather than society's failure to provide basic services to all of its citizens" (Sered & Fernandopulle 2005, p. 16).

Social Factors and Lifestyle Behaviors Associated with Health and Illness

Health problems are linked to lifestyle behaviors such as excessive alcohol consumption and cigarette smoking (see Chapter 3), unprotected sexual intercourse, physical inactivity, and unhealthy diet. For example, only 5 percent of college students in a national survey reported that they ate the recommended five or more servings of fruits and vegetables daily; less than half met physical activity guidelines established by the American College of Sports Medicine and American Heart Association. These lifestyle behaviors help to explain why one-third of college students are either overweight or obese (American College Health Association 2012). Obesity and other health problems also vary by socioeconomic status, race/ethnicity, and gender. Next we discuss how these social factors are associated with health and illness.

Socioeconomic Status and Health

Socioeconomic status refers to a person's position in society based on that person's level of educational attainment, occupation, and household income (see also Chapter 6). These three factors tend to be related: Income tends to go up as the level of education increases, in part because higher levels of education enable individuals to enter higher-paying occupations. (As discussed in later chapters, higher family income also enables children and young adults to achieve higher levels of education.)

One's socioeconomic status—one's income level, education, and occupation—greatly influences one's health. People living in poverty are more likely to suffer from malnutrition; hazardous environmental, housing and working conditions; lack of clean water and sanitation; and lack of access to medical care (see also Chapters 6 and 13). In low-income countries, for example, people with cancer lack access to medical treatment and therefore have lower survival rates compared with cancer patients in high-income countries (Farmer et al., 2010). Most pain medication (about 80 percent) is used by people

socioeconomic status A person's position in society based on the level of educational attainment, occupation, and income of that person or that person's household.

In less developed countries, few people can afford or access pain medication.

living in developed countries; few people in less developed countries can afford pain medication or access it when needed. About 4.8 million people with severe cancer pain go untreated annually (Bafana 2013).

In the United States, low socioeconomic status is associated with higher incidence and prevalence of health problems, and lower life expectancy. A study of mortality rates in 3,140 U.S. counties found that the factors most strongly associated with higher mortality were poverty and lack of college education (Kindig & Cheng 2013). In the United States, rates of overweight and obesity are higher among people living in poverty. This is, in part, because high-calorie processed foods tend to be more affordable than fresh vegetables, fruits, and lean meats or fish. And residents of low-income areas often live in areas where they lack access to large grocery stores that sell a variety of foods, and instead rely on convenience stores and fast-food chains that sell mostly high-calorie processed food.

In addition, members of the lower class are subjected to the most stress and have the fewest resources to cope with it (Cockerham 2007). U.S. adults living below the poverty threshold are nearly eight times more likely to report experiencing serious psychological distress as adults in families with an income at least four times the poverty level (National Center for Health Statistics 2012). Stress has been linked to a variety of physical and mental health problems, including high blood pressure, cancer, chronic fatigue, and substance abuse. Poor U.S. adults ages 45 to 64 are five times more likely to experience depression (24 percent) as adults whose family income is 400 percent or more of the poverty level (National Center for Health Statistics 2012).

U.S. adults who have completed college live an average of ten more years than adults who do not have a high school diploma (Hummer & Hernandez 2013). And although life expectancy has generally been rising in the United States, between 1990 and 2008, life expectancy among U.S. white adults *dropped* five years for women without a high school diploma and three years for men without a high school diploma (Tavernise 2012).

Why is higher educational attainment linked to better health? First, higher education can lead to higher-paying jobs that provide income to afford better health care, a safer living environment, physically active recreation, and a diet with more fresh (and healthy) foods. A second explanation is that higher levels of education lead to greater knowledge about health issues, and encourage the development of cognitive skills that enable individuals to make better health-related choices, such as the choice to exercise, avoid smoking and heavy drinking, use contraceptives and condoms to avoid sexually transmissible infections, seek prenatal care, and follow doctors' recommendations for managing health problems.

Just as socioeconomic status affects health, health affects socioeconomic status. Physical and mental health problems can limit one's ability to pursue education or vocational training and to find or keep employment. The high cost of health care not only deepens the poverty of people who are already barely getting by but also can financially devastate middle-class families. Based on data from 89 countries around the world, an estimated 100 million people each year are pushed into poverty as a result of paying for needed health services (World Health Organization 2012a). Later in this chapter, we look more closely at the high cost of health care and its consequences for individuals and families.

Gender and Health

In many societies, women and girls are viewed and treated as socially inferior, and are denied equal access to nutrition and health care. Traditional family responsibilities also affect women's health. Women do most of the food preparation and, in many areas of the world where solid fuels are used for cooking indoors, women are more likely than men to suffer from respiratory problems due to exposure to indoor air pollution. Gender inequality also exposes women to domestic and sexual exploitation, increasing women's risk of physical injury, and of acquiring HIV and other sexually transmissible infections. Globally, 30 percent of women ages 15 and over have experienced physical and/or sexual violence perpetrated by an intimate partner (Devries et al., 2013). "Although neither health care workers nor the general public typically thinks of battering as a health problem, it is a major cause of injury, disability, and death among American women, as among women worldwide" (Weitz 2013, p. 66).

In the United States before the 20th century, the life expectancy of U.S. women was shorter than that of men because of the high rate of maternal mortality that resulted from complications of pregnancy and childbirth. But today, life expectancy of U.S. women (80.9 years) is greater than that of U.S. men (76 years) (National Center for Health Statistics 2012). Lower life expectancy for U.S. men is due to a number of factors. Men tend to work in more dangerous jobs than women, such as agriculture, construction, and the military. In addition, "beliefs about masculinity and manhood that are deeply rooted in culture . . . play a role in shaping the behavioral patterns of men in ways that have consequences for their health" (Williams 2003, p. 726). Men are socialized to be strong, independent, competitive, and aggressive and to avoid expressions of emotion or vulnerability that could be construed as weakness. These male gender expectations can lead men to take actions that harm themselves or to refrain from engaging in health-protective behaviors. For example, socialization to be aggressive and competitive leads to risky behaviors (such as dangerous sports, fast driving, and violence) that contribute to men's higher risk of injuries and accidents. Men are more likely than women to smoke cigarettes and to abuse alcohol and drugs but are less likely than women to visit a doctor and to adhere to medical regimens (Williams 2003).

Regarding mental health, women in the United States are more likely than men to have a mental disorder (23 percent versus 16 percent in 2011) (Substance Abuse and Mental Health Services Administration 2012). Biological differences can account for some gender differences in mental health. For example, hormonal changes after childbirth can result in some women suffering from postpartum depression—one study found that one in seven women experienced postpartum depression (Wisner, et al. 2013). Gender differences in mental health can also be attributed to gender roles. For example, the unequal status of women and the strain of doing the majority of housework and child care may predispose women to experience greater psychological distress.

Race, Ethnicity, and Health

Many racial and ethnic minorities get sick at younger ages and die sooner than non-Hispanic whites. U.S. black women and men have a lower life expectancy compared with their white counterparts. Black males are significantly more likely to die from homicide (see Chapter 4), and black men and women are much more likely than whites to die of heart disease and stroke than are those from other racial and ethnic groups. Black women have the highest rate of U.S. infant mortality, followed by American Indian/Alaskan Native, with Asian/Pacific Islander women having the lowest infant mortality rate. These are just a few examples of the health disparities among U.S. racial/ethnic groups.

As shown in Table 2.7, Hispanic women and men live longer than either blacks or whites, and Hispanics have higher survival rates for conditions such as cancer, heart disease, HIV/AIDS, kidney disease, and stroke (Ruiz, Steffen, & Smith 2013). The Hispanic health advantage—known as the "Hispanic paradox"—is puzzling because Hispanics share some of the same risk factors that contribute to poorer health among blacks—higher rates of poverty and obesity, and lower educational attainment compared with non-Hispanic whites (Kindig & Cheng 2013). One theory for this "Hispanic paradox" is that Hispanics who immigrate to the United States are among the

TABLE 2.7 **Life Expectancy at Birth by Race/Hispanic Origin and Sex: United States, 2010**

	White	Black	Hispanic
Female	81.3	78	83.8
Male	76.5	71.8	78.5

Source: Centers for Disease Control and Prevention. 2013b. "Deaths: Final Data for 2010." *National Vital Statistics Report* 61(4): Table 7.

healthiest from their countries. Another reason for Hispanic longevity is that Hispanic cultural values promote close and supportive family and community relationships and build strong social support, which is associated with better health outcomes.

Health disparities are largely due to substantial racial/ethnic differences in income, education, housing, and access to health care. Racial and ethnic minorities are less likely than whites to have health insurance and so are less likely to receive preventive services (such as colon cancer screening), medical treatment for chronic conditions, and prenatal care. Minorities are also more likely than whites to live in environments where they are exposed to hazards such as toxic chemicals and other environmental hazards (see also Chapter 13). However, although high-income blacks and whites live longer than their low-income counterparts, whites at every income level live at least three years longer than blacks. And the infant mortality rate of college-educated African American women is more than 2.5 times as high as for college-educated whites and Hispanics. In fact, black female college graduates have a higher infant mortality rate than Hispanic and white women who have not completed high school (Williams 2012). One explanation for the effect of race on health, independent of socioeconomic status, is that health is affected not only by one's current socioeconomic status, but also by social and economic circumstances experienced over the life course. Minorities are more likely than whites to have experienced social and economic adversity in childhood that can affect their health in adulthood (Williams 2012).

Health disparities are sometimes explained by differences in lifestyle behaviors. Compared with white Americans, Native Americans/Alaskan Natives have the highest death rate from motor vehicle crashes, because they have the highest rates of alcohol-impaired driving as well as seatbelt nonuse (Centers for Disease Control and Prevention 2011). Black Americans have the highest rate of obesity in part because of racial differences in eating behavior. But lifestyle behaviors are often influenced by social factors. Blacks, on average, have lower incomes than whites, and so are more likely to choose foods that are more affordable, and junk foods and fast foods tend to be cheaper than fruits, vegetables, and lean sources of protein. However, racial disparities in obesity persist even after controlling for family income (Centers for Disease Control and Prevention 2011).

Another factor that contributes to racial/ethnic health disparities is prejudice and discrimination. Stress associated with prejudice and discrimination can raise blood pressure, which may help to explain why African Americans have higher blood pressure than whites (Fischman 2010). Prejudice and discrimination can adversely affect health by reducing minorities' access to jobs and safe housing (see also Chapter 9) (Williams 2012).

Regarding mental health, research finds no significant difference among races in their overall rates of mental illness (Cockerham 2007). Differences that do exist are often associated more with social class than with race or ethnicity. However, some studies suggest that minorities have a higher risk for mental disorders, such as anxiety and depression, in part because of racism and discrimination, which adversely affect physical and mental health. Minorities also have less access to or are less likely to receive needed mental health services, often receive lower-quality mental health care, and are underrepresented in mental health research (U.S. Department of Health and Human Services 2001).

Problems in U.S. Health Care

In an analysis of the world's health systems, the World Health Organization (2000) found that, although the United States spends a higher portion of its gross domestic product (GDP) on health care than any other country, it ranks 37 out of 191 countries according to its performance. The analysis concluded that France provides the best overall health care among major countries, followed by Italy, Spain, Oman, Austria, and Japan. In another study of six countries—Australia, Canada, Germany, New Zealand, the United Kingdom, and the United States—the United States ranked last on dimensions of access, patient safety, efficiency, and equity (Davis et al. 2007).

A more recent comparison of health care in 13 industrialized countries found that the United States spends far more on health care than any other country, but "despite being more expensive, the quality of health care in the U.S. does not appear to be notably superior to other industrialized countries" (Squires 2012, p. 10). For example, the United States ranks in the bottom quartile in life expectancy among Organization for Economic Cooperation and Development (OECD) countries (Squires 2010). After presenting a brief overview of U.S. health care, we address some of the major health care problems in the United States related to problems of access, cost, and quality of health care.

U.S. Health Care: An Overview

In the United States, there is no one health care system; rather, health care is offered through various private and public means (see Figure 2.1).

In traditional health insurance plans, the insured choose their health care providers, whose fees are reimbursed by the insurance company. Insured individuals typically pay an out-of-pocket "deductible" (usually ranging from a few hundred to a thousand dollars or more per year per person) as well as a percentage of medical expenses (e.g., 20 percent) until a maximum out-of-pocket expense amount is reached (after which insurance will cover 100 percent of medical costs). Most insurance companies control costs through **managed care**, which involves monitoring and controlling the decisions of health care providers. The insurance company may, for example, require doctors to receive approval before they can hospitalize a patient, perform surgery, or order an expensive diagnostic test.

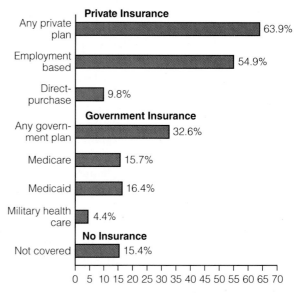

Figure 2.1 Coverage by Type of Health Insurance, 2012
Source: DeNavas-Walt et al. 2013.

Medicare, Medicaid, and Military Health Care. **Medicare** is funded by the federal government and reimburses the elderly and people with certain disabilities for their health care. Individuals contribute payroll taxes to Medicare throughout their working lives and generally become eligible for Medicare when they reach 65, regardless of their income or health status. Medicare consists of four separate programs: Part A is hospital insurance for inpatient care, which is free, but enrollees may pay a deductible and a co-payment. Part B is a supplementary medical insurance program, which helps pay for physician, outpatient, and other services. Part B is voluntary and is not free; enrollees must pay a monthly premium as well as a co-payment for services. Medicare does not cover long-term nursing home care, dental care, eyeglasses, and other types of services, which is why many individuals who receive Medicare also enroll in Part C, which allows beneficiaries to purchase private supplementary insurance that receives payments from Medicare. Part D is an outpatient drug benefit that is also voluntary and requires enrollees to pay a monthly premium, meet an annual deductible, and pay coinsurance for their prescriptions.

Medicaid, which provides health care coverage for the poor, is jointly funded by the federal and state governments. Eligibility rules and benefits vary from state to state, and, in many states, Medicaid provides health care only for those who are well below the federal poverty level. The **State Children's Health Insurance Program (SCHIP)** provides health coverage to children without insurance, many of whom come from families with income too high to qualify for Medicaid but too low to afford private health insurance. Under this initiative, states receive matching federal funds to provide medical insurance to children without insurance.

Military health care includes Civilian Health and Medical Program of the Uniformed Services (CHAMPUS), Civilian Health and Medical Program of the Department of Veterans Affairs (CHAMPVA), and care provided by the Department of Defense and the Department of Veterans Affairs.

managed care Any medical insurance plan that controls costs through monitoring and controlling the decisions of health care providers.

Medicare A federally funded program that provides health insurance benefits to the elderly, disabled, and those with advanced kidney disease.

Medicaid A public health insurance program, jointly funded by the federal and state governments, that provides health insurance coverage for the poor who meet eligibility requirements.

State Children's Health Insurance Program (SCHIP) A public health insurance program, jointly funded by the federal and state governments, that provides health insurance coverage for children whose families meet income eligibility standards.

Workers' Compensation. **Workers' compensation** (also known as *workers' comp*) is an insurance program that provides medical and living expenses for people with work-related injuries or illnesses. Employers pay a certain amount into their state's workers' compensation insurance pool, and workers injured on the job can apply to that pool for medical expenses and for compensation for work days lost. In exchange for that benefit, workers cannot sue their employers for damages. However, not all employers acquire workers' compensation insurance, even in states where it is legally required. Further, many employees with work-related illness or injuries do not apply for workers' compensation benefits because (1) they fear getting fired for making a claim, (2) they are not aware that they are covered by workers' comp, and/or (3) the employer offers incentives (i.e., bonuses) to employees when no workers' comp claims are filed in a given period of time (Sered & Fernandopulle 2005).

Complementary and Alternative Medicine (CAM). In Western nations such as the United States, the conventional or mainstream practice of medicine is known as **allopathic**, or Western, medicine. **Complementary and alternative medicine (CAM)** refers to a broad range of health care approaches, practices, and products that are not considered part of conventional medicine. Types of CAM include herbal and homeopathic remedies, dietary supplements, meditation, Pilates, yoga, tai chi, acupuncture, chiropractic care, massage therapy, Reiki and other energy work, and the use of traditional healers. Some people consider prayer a form of traditional healing.

More than one-third of U.S. adults use CAM to improve their health and well-being (National Center for Complementary and Alternative Medicine 2012). Americans spend $9 billion on CAM each year (M. Davis et al. 2013). The cost of CAM is paid by the consumer primarily out of pocket, although some services, such as chiropractic care, are covered by most health insurance.

A controversial form of CAM is the medical use of cannabis, or marijuana. Marijuana was first used for medicinal purposes in 2737 B.C. In 1851, marijuana was classified as a legitimate medical compound in the United States, but was criminalized in 1937, against the advice of the American Medical Association. In 1996, California became the first state to legalize the medical use of marijuana, and as of September 2013, 20 states and the District of Columbia have legalized medical marijuana (see Table 2.8) and several states have pending legislation to legalize medical marijuana.

Medical marijuana can be smoked in cigarettes and can also be delivered in a variety of nonsmoked preparations, including patches, suppositories, vaporizations, tinctures, nasal sprays, and edible preparations. Medical marijuana is most frequently used to alleviate pain and muscle spasms, but has also been effective for a variety of other conditions, including nausea and vomiting (a common side effect of cancer

workers' compensation Also known as workers' comp, an insurance program that provides medical workers' compensation and living expenses for people with work-related injuries or illnesses.

allopathic medicine The conventional or mainstream practice of medicine; also known as Western medicine.

complementary and alternative medicine Refers to a broad range of health care approaches, practices, and products that are not considered part of conventional medicine.

TABLE 2.8 **States with Legalized Medical Marijuana (as of September 2013)**

State	Year Legalized	State	Year Legalized	State	Year Legalized
Alaska	1998	Hawaii	2000	New Hampshire	2013
Arizona	2010	Illinois	2013	New Jersey	2010
California	1996	Maine	1999	New Mexico	2007
Colorado	2000	Massachusetts	2012	Oregon	1998
Connecticut	2012	Michigan	2008	Rhode Island	2006
DC	2010	Montana	2004	Vermont	2004
Delaware	2011	Nevada	2000	Washington	1998

Source: ProCon.org. 2013.

Bill Delany has Crohn's disease, an inflammatory bowel disease that causes abdominal pain, chronic diarrhea and malnutrition, and can result in death (Delany 2012).

My Crohn's reached a level that I was visiting my toilet about 30 times a day/night, for about a year. . . . I had lost 50 to 60 pounds in a matter of weeks. . . . I had to apply for Social Security Disability in 2008—after dealing with a severe case of Crohn's (four surgeries) since 1999—it was quickly granted. . . . I had lived (and suffered) in Colorado for three and a half years before I heard of the medical marijuana law. . . . I was desperate. I called a doctor and he was willing to sign for me to get a medical marijuana card. At the time the only source I knew of for the medicine was a shadowy guy in Durango, literally in an alleyway apartment. I put on my Depends and drove over there. . . . Medical marijuana was supposed to be legal, but it sure didn't feel like it to me back then before Colorado dispensaries became actual stores instead of alleyways. I went home and medicated. The vapors immediately started to open some of the blockages in my intestinal tract and I knew there was some hope, for the first time in 11 years. The VA recommended removing my colon and rectum, leaving a pouch, just months earlier—I'm certainly glad that I didn't allow it. Within three months I was able to wean myself off of prescription drugs and I knew that I was going to reclaim my life from this disease. My VA doctors were initially skeptical about my new therapy, but they had no better suggestions—now they are pleased that I'm no longer dying a slow death, at taxpayer expense. (Delany 2012)

Sherry Smith suffers from multiple sclerosis, a progressive autoimmune disease that affects the central nervous system and produces symptoms such as numbness, impaired muscular coordination and speech, blurred vision, and severe fatigue.

I've been disabled with MS for so long that I got to the point where I didn't know which was worse, the dozens of medications prescribed by my doctors or the disease itself. . . . My prescription drugs were making me depressed and sick. . . . Now I take hemp oil capsules three times a day, which completely controls my muscle spasticity and pain, and then supplement, when necessary every once in a while, with smokable herb. (Quoted in Mannix 2009)

Vicki Burk, a grandmother and member of Idaho Moms for Marijuana and the Southern Idaho Cannabis Coalition describes how marijuana has provided relief from stomach distress and chronic pain ("Testimonials" 2013).

My story began back in 1997–1998. I was having severe stomachaches after I ate. It didn't matter what it was. I went through a barrage of tests and no cause was ever found. In 2005, my liver enzymes were increased. I was told I had a faulty liver. In 2006 I was diagnosed with PBC (Primary biliary cirrhosis). I was so sick I threw up everything. My boyfriend at the time told me to take a few hits and it would calm my stomach down so that I could eat and drink. That way I would not become dehydrated. I did as he suggested and it was like night & day. What a relief! I now use it to relieve my chronic pain from arthritis and fibromyalgia as well. It works better than the morphine pills like Kadian, which I take. I also take oxycodone. The marijuana works better than both of them together.

I have always been a supporter of marijuana. I thought it was great for cancer patients and glaucoma patients. Now I realize it works for so much more. My friend educated me through my daughter and then I became involved with Moms for Marijuana and the Southern Idaho Cannabis Coalition. I also met with Rep. Tom Trail to discuss his legislative bill for Medical Marijuana in Idaho. I hope that more people become educated on the many uses of Cannabis/Marijuana for medical purposes.

AP Images/SIPA/Kennell Krista

A medical marijuana dispensary

chemotherapy), and anorexia and weight loss in patients with AIDS and other conditions that reduce appetite. Adverse side effects of medical marijuana are typically not serious, with the most common being dizziness, and some users experience anxiety and paranoia (Borgelt et al. 2013). Safety concerns regarding medical marijuana include impairment in memory and cognition, and, among frequent adolescent users of marijuana, there is an increased risk of developing schizophrenia. Another concern is the accidental consumption of food products containing marijuana (e.g., cookies or brownies) among children. Unlike many conventional prescription drugs, no deaths from overdoses of marijuana have been reported. This chapter's *The Human Side* features testimonies from people with chronic health problems who found enormous relief through the use of medical marijuana.

Inadequate Health Insurance Coverage

Many other countries, including 31 European countries and Canada, have national health insurance systems that provide **universal health care**—health care to all citizens. National health insurance is typically administered and paid for by government and funded by taxes or Social Security contributions. Despite differences in how national health insurance works in various countries, typically, the government (1) directly controls the financing and organization of health services, (2) directly pays providers, (3) owns most of the medical facilities (Canada is an exception), (4) guarantees universal access to health care, (5) allows private care for individuals who are willing to pay for their medical expenses, and (6) allows individuals to supplement their national health care with private insurance as an upgrade to a higher class of service and a larger range of services (Cockerham 2007; Quadagno 2004). Any rationing of health care in countries with national health insurance is done on the basis of medical need, not ability to pay.

Before the Affordable Care Act (discussed later in this chapter) took effect in 2014, 15.4 percent of Americans (48 million people) did not have health insurance coverage in 2012 (DeNavas-Walt et al. 2013). Non-Hispanic whites are more likely than racial and ethnic minorities to have health insurance. Hispanics are the least likely to have insurance, with more than one in four Hispanics lacking health insurance in 2012 (DeNavas-Walt et al. 2013). Of all age groups, young adults aged 19 to 34 are the least likely to have health insurance. In 2012, more than one in four adults in these age ranges was uninsured (DeNavas-Walt et al. 2013).

Employed individuals and individuals with higher incomes are more likely to have health insurance. However, employment is no guarantee of health care coverage; in 2012, 15.5 percent of full-time workers were uninsured (DeNavas-Walt et al. 2013). Some businesses do not offer health benefits to their employees; some employees are not eligible for health benefits because of waiting periods or part-time status, and some employees who are eligible may not enroll in employer-provided health insurance because they cannot afford their share of the premiums.

An estimated 45,000 deaths per year in the United States are attributable to lack of health insurance.

universal health care A system of health care, typically financed by the government, that ensures health care coverage for all citizens.

Consequences of Inadequate Health Insurance. An estimated 45,000 deaths per year in the United States are attributable to lack of health insurance (Park 2009). Individuals who lack health insurance are less likely to receive preventive care, are more likely to be hospitalized for avoidable health problems, and are more likely to have disease diagnosed in the late stages. The uninsured are three times more likely than the insured to be unable to pay for basic necessities because of their medical bills (Kaiser Commission on Medicaid and the Uninsured 2010).

Because most health care providers do not accept patients who do not have insurance, many individuals without insurance resort to using the local hospital emergency room (Scal & Town 2007). The federal Emergency Medical Treatment and Active Labor Act requires hospitals to assess all patients who come to their emergency rooms to determine whether an emergency medical condition exists and, if it does, to stabilize patients before transferring them to another facility. Hospital patients without insurance are almost always billed at a much higher cost than the prices negotiated by insurance companies.

Individuals who lack dental insurance commonly have untreated dental problems, which can lead to or exacerbate other health problems:

> Because they affect the ability to chew, untreated dental problems tend to exacerbate conditions such as diabetes or heart disease. . . . Missing and rotten teeth make it painful if not impossible to chew fruits, whole grain foods, salads, or many of the fiber-rich foods recommended by doctors and nutrition experts. (Sered & Fernandopulle 2005, pp. 166–67)

In their book *Uninsured in America*, Sered and Fernandopulle (2005) described one interviewee who "covered her mouth with her hand during our entire interview because she was embarrassed about her rotting teeth" and another interviewee "used his pliers to yank out decayed and aching teeth" (p. 166). The authors note that "almost every time we asked interviewees what their first priority would be if the president established universal health coverage tomorrow, the immediate answer was 'my teeth'" (p. 166).

The High Cost of Health Care

Health care spending in the United States is far greater than in other industrialized countries. Yet nearly every other wealthy nation has better health outcomes, as measured by life expectancy and infant mortality.

It is widely believed that U.S. health costs have gone up due to the aging of the population; people are living longer today than in previous generations and older people have greater health care needs. However, the United States has a relatively young population compared with many other high-income countries that spend much less on health care than does the United States (Squires 2012). One study found that the high rate of obesity in the United States—about a third of the adult population—accounts for nearly 10 percent of medical spending (reported in Squire 2012). Compared to other OECD countries, hospitalizations in the United States are less frequent and shorter. Yet, spending per hospital discharge in the United States (more than $18,000) is nearly three times higher than the OECD median ($6,222) (Squires 2012). The United States has high rates of performing medical tests and procedures involving advanced medical technology, such as diagnostic imaging, coronary procedures (angioplasty, stenting, and cardiac catheterizations), knee replacements, and dialysis, and charges for these medical tests and procedures are much higher in the United States than in other countries. Consider the advancements in the treatment of preterm babies, for which very little could be done in 1950. By 1990, special ventilators and neonatal intensive care became standard treatment for preterm babies in the United States (Kaiser Family Foundation 2007).

Prescription drug prices in the United States are 50 percent more than comparable drugs sold in other developed countries (Brill 2013). Drug prices are not regulated by the U.S. government; they are regulated in other countries. The pharmaceutical industry, which is among the most profitable industries in the United States, argues that U.S. drug prices are high because of the high cost of researching and developing (R & D) new drugs. But a critical analysis reveals that the industry purposely overestimates R & D costs to justify their high drug prices (Light & Warburton 2011). Further, most large drug companies pay substantially more for marketing, advertising, and administration than for research and development (Families USA 2007).

Health insurance in the United States is another health care expense. In 2012, the average annual premiums for employer-sponsored coverage were $5,615 for an individual (worker's share was $951) and $15,745 for a family (worker's share was $4,316). From 2002 to 2012, average premiums for family coverage rose 97 percent, outpacing increases in both workers' wages and inflation (Kaiser Family Foundation/HRET 2012). U.S. health insurance is costly largely because of high administrative expenses, which per capita are six times higher than in western European nations (National Coalition on Health Care 2009). Harrison (2008) explains the reason:

> The United States has the most bureaucratic health care system in the world, including over 1,500 different companies, each offering multiple plans, each with its own marketing program and enrollment procedures, its own paperwork and policies, its CEO salaries, sales commissions, and other nonclinical costs—and, of course, if it is a for-profit company, its profits.

Consequences of the High Cost of Health Care for Individuals and Families. One study found that medical bills, as well as income lost due to illness, contributed to two-thirds of all bankruptcies in 2007 (Himmelstein et al. 2009). Most medical debtors were well-educated homeowners with middle-class jobs, and three-fourths had health insurance. Having insurance does not guarantee that one is protected against financial devastation resulting from illness or injury, because even the insured typically must pay co-payments, deductibles, and exclusions. In 2011, one-third of U.S. households reported problems paying medical bills (Cohen, Gindi, & Kirzinger 2012).

> Having insurance does not guarantee that one is protected against financial devastation resulting from illness or injury, because even the insured typically must pay co-payments, deductibles, and exclusions.

Many individuals forgo needed medicine and/or medical care when they cannot afford to pay for it (see Table 2.9). Forgoing medicine or medical care often exacerbates a medical condition, leading to even higher medical costs, or tragically, leading to death.

Inadequate Mental Health Care

In the 1960s, the U.S. model for psychiatric care shifted from long-term inpatient care in institutions to drug therapy and community-based mental health centers. This transition, known as **deinstitutionalization**, has resulted in a significant decrease in the number of mental health facilities with 24-hour or residential treatment and the number of psychiatric treatment beds available. Deinstitutionalization removed patients from facilities where they were sometimes treated in a neglectful or inhumane manner, and restored freedom of choice to mental health consumers, including the right to refuse treatment. During the deinstitutionalization era, a variety of laws were passed making it

deinstitutionalization The removal of individuals with psychiatric disorders from mental hospitals and large residential institutions to outpatient community mental health centers.

TABLE 2.9 Cutting Back on Medical Care Due to Cost

In the past 12 months, because of cost, have you or another family member living in your household . . . ?

	Percent Saying "Yes"
Relied on home remedies or over-the-counter (OTC) drugs instead of seeing a doctor	37%
Skipped dental care or checkups	35%
Put off or postponed getting health care you needed	31%
Skipped recommended medical test or treatment	27%
Not filled a prescription for medicine	26%
Cut pills in half or skipped doses of medicine	19%
Had problems getting mental health care	8%
Did ANY of the above	55%

Source: Adapted from Kaiser Family Foundation 2009.

illegal to commit psychiatric patients against their will unless they posed an immediate threat to themselves or to others. However, community mental health programs have not adequately met the need for care, and millions of Americans with mental disabilities go without needed care, or rely on hospital emergency room care when their condition deteriorates into a major mental health breakdown.

Among the 11.5 million U.S. adults with serious mental illness in 2011, only 60 percent received mental health services, leaving 40 percent receiving no mental health care (Substance Abuse and Mental Health Services Administration 2012). The most frequently used source of care for mental health problems has become primary care and general doctors and nurses. Other "nonspecialty" care providers include community health centers, schools, nursing homes, correctional institutions, and emergency rooms. This fragmented system of mental health care leaves many people with mental health problems to fall through the cracks.

Mental health services are often inaccessible, especially in rural areas. In most states, services are available from "9 to 5"; the system is "closed" in the evenings and on weekends when many people with mental illness experience the greatest need. Across the nation, people with severe mental illness end up in jails and prisons, homeless shelters, and hospital emergency rooms. Many children with untreated mental disorders drop out of school or end up in foster care or the juvenile justice system. Given the increasing growth of minority populations, another deficit in the mental health system is the inadequate number of mental health clinicians who speak the client's language and who are aware of cultural norms and values of minority populations (U.S. Department of Health and Human Services 2001).

Strategies for Action: Improving Health and Health Care

Two broad approaches to improving the health of populations are selective primary health care and comprehensive primary health care (Sanders & Chopra 2003). **Selective primary health care** focuses on interventions that target specific health problems, such as promoting condom use to prevent HIV infections, providing immunizations against childhood diseases to promote child survival, and encouraging individuals to be more physically active and eat a healthier diet. **Comprehensive primary health care** focuses on the broader social determinants of health, such as poverty and economic inequality, gender inequality, racial/ethnic discrimination, and environmental pollution. Many strategies to alleviate social problems in subsequent chapters of this textbook are also important elements to a comprehensive primary health care approach.

Improving Health in Low- and Middle-Income Countries

Efforts to improve health in low- and middle-income countries include improving access to adequate nutrition, clean water and sanitation, and medical care. More targeted interventions include increasing immunizations for diseases such as measles, distributing mosquito nets to prevent malaria, and promoting the use of condoms to prevent the spread of HIV/AIDS.

Efforts to reduce maternal mortality—a major cause of death among women in the poorest countries of the world—have focused largely on providing access to good-quality reproductive care and family planning services. Worldwide, about one in 10 married or partnered women ages 15 to 49 want to postpone or avoid childbearing, but are not using contraception (World Health Organization 2013). Some women's health advocates are fighting to pass legislation aimed at preventing "child marriage," as pregnant girls ages 15 to 19 are twice as likely to die during pregnancy than women in their 20s, and pregnant girls between the ages of 10 and 14 are five times as likely to die during pregnancy or childbirth than women in their 20s (Biset 2013). Nearly half of women aged 20 to 24 in the least developed countries were married before age 18 (UNICEF 2012b).

selective primary health care An approach to health care that focuses on using specific interventions to target specific health problems.

comprehensive primary health care An approach to health care that focuses on the broader social determinants of health, such as poverty and economic inequality, gender inequality, environment, and community development.

Childbearing at an early age involves higher health risks for women and infants.

Another strategy to improve the health of women and children in low-income countries is to provide women with education and income-producing opportunities. Promoting women's education increases the status and power of women to control their reproductive lives, exposes women to information about health issues, and also delays marriage and childbearing. In many developing countries, women's lack of power and status means that they have little control over health-related decisions. Men make the decisions about whether or when their wives (or partners) will have sexual relations, use contraception, or use health services.

These and other efforts have had some success: Between 1990 and 2010, the mortality rate of children under age 5 declined by 35 percent and maternal deaths were reduced by nearly 50 percent (World Health Organization 2012a). However, a number of countries with the highest rates of maternal mortality have made little or no progress.

Essential health services in low-income countries—those focused on HIV, tuberculosis, malaria, maternal and child health, and prevention of noncommunicable diseases—are estimated to cost only $44 per person per year. But in 2009, 29 countries spent less than this per capita. Less developed countries need financial assistance to improve the health their populations. However, "higher levels of funding might not translate into better service coverage or improved health outcomes if the resources are not used efficiently or equitably" (World Health Organization 2012a, p. 40).

Fighting the Growing Problem of Obesity

Sociological strategies to reducing and preventing obesity involve enacting programs and policies that encourage people (1) to eat a diet with sensible portions, with lots of high-fiber fruits and vegetables, and with minimal sugar and fat, and (2) to engage in regular physical activity.

School Nutrition and Physical Activity Programs. School-based efforts to reduce childhood overweight and obesity have focused on improving the quality of food served and sold in schools, limiting availability of less nutritious food and beverages, and improving physical and health education. In 2012, the federal Healthy Hunger-Free Kids Act established new standards for the National School Lunch and School Breakfast programs that require more fruits and vegetables and whole grain foods; only fat-free or low-fat milk; and less saturated fats, trans fats, and sodium.

Although every state has some physical education requirement for students, these requirements are often limited or not enforced (Trust for America's Health 2012). Some states have enacted laws requiring schools to devote a certain number of minutes to physical activity.

Regulation of Food Marketing to Youth. Food marketing influences children's food preferences, and "children . . . play an important role in which products their parents purchase at the store, and which restaurants they frequent" (Federal Trade Commission 2012, p. ES-9). Recognizing how powerful advertising is in influencing the food and beverage choices of youth, the federal government has proposed guidelines for food marketing to children. But the powerful food industry has fought government regulations on advertising, and instead, self-regulates its food marketing to youth according to guidelines set out by the Children's Food and Beverage Advertising Initiative (CFBAI). But critics of the program claim that self-regulation has led to minimal improvement and that "the overwhelming majority of foods advertised to kids is still of poor nutritional quality. Under the Children's Food and Beverage Advertising Initiative nutritional standards, Cocoa Puffs, Popsicles, SpaghettiOs, and Fruit Roll-Ups are considered nutritious foods" (Wootan 2012).

In 2012, Walt Disney Company announced that all food and beverage products advertised on its child-focused TV channels, radio stations, and websites must comply with a strict set of nutritional standards. Under Disney's new rules, sugared cereals and drinks, and fast food will no longer be advertised in programming aimed at children under 12. Disney also introduced a "Mickey Check" program in which Disney-licensed grocery store products that meet criteria for limited fat, sodium, sugar, and calories can display a logo on their packaging: Mickey Mouse ears and a check mark (Barnes 2012).

Chip Somodevilla/Getty Images

In 2012, the Walt Disney Company introduced the "Mickey Check" logo on the labels of Disney-licensed grocery store products that meet specific nutritional standards.

Local and State Antiobesity Policies.

To encourage physical activity as well as green transportation, many state and local governments are adopting "Complete Street" policies. Complete Streets are roads designed with sidewalks, bike lanes, and other features that encourage walking, bicycling, and use of public transportation.

In New York City, the adult obesity rate increased from 18 percent in 2002 to 24 percent in 2013, costing tax-funded health programs about $2.8 billion a year from obesity-related illnesses alone (Huffington Post 2013). Concerned about the growing obesity epidemic in New York City, Mayor Michael Bloomberg passed a law in 2012 banning restaurants, movie theaters, stadiums, and other food establishments (not including grocery stores) from selling sugar-sweetened soda and other sugary drinks in cups or containers larger than 16 ounces. In 2013, a New York State Supreme Court judge struck down the law one day before it was to take effect.

What Do You Think? Opponents of Bloomberg's soda limit argued that such a plan would infringe on consumers' freedom of choice. Supporters of the proposed ban pointed to the harmful health effects of such drinks: In 2010, 184,000 adult deaths worldwide—due mostly to diabetes, but also other diseases including heart disease and cancer—were linked to sugar-sweetened soda and other sugary beverages (Kaiser 2013). Soda and other sweetened beverages are *the single biggest source of calories* in the American diet, contributing to obesity, diabetes, heart disease, gout, tooth decay, and some cancers. The American Heart Association recommends that women consume no more than 6 teaspoons and men no more than 9 teaspoons of added sugar per day; a 20-ounce regular soda contains about 15 teaspoons of sugar (Center for Science in the Public Interest 2012). Do you think a limit on the serving size of sugary beverages constitutes a valid public health policy, or a violation of civil liberties? Would you be in favor of a policy that requires warning labels on sugar-sweetened soda informing consumers of the health risks of soda consumption? Would you support an added tax, or surcharge, on soft drinks as a strategy to curb sugar consumption and provide revenue to offset the economic costs of treating obesity-related health problems? Why or why not?

Workplace Wellness Programs.

Some workplaces have employee wellness programs that encourage employees to exercise, make healthy food choices, and engage in other health promotion behaviors such as quitting smoking. Some workplaces have onsite gyms, reimburse employees for gym membership, or organize lunchtime walking or jogging activities. Some employers have contests, awarding prizes to employees who

With overweight and obesity being a major public health issue, there is growing concern over the lifestyle choices of teenagers with regard to their eating habits and levels of physical activity. In a national survey, less than 10 percent of U.S. teens reported eating the recommended number of daily servings of fruits and vegetables. Many U.S. teens are consuming foods with excessive sugars and fats, drinking more sweetened beverages, and snacking on high-calorie foods, and at the same time, many adolescents have low levels of physical activity and high levels of sedentary behavior, such as watching television, spending time on Facebook, or surfing the Internet.

Another sedentary behavior teens engage in is texting: 75 percent of youth ages 12 to 17 own a cell phone, with half of those teens sending 50 or more texts each day. Indeed, the majority of U.S. teens use texting, also known as short message service, or SMS, as their primary method of communication. Given the high prevalence of teen texting, Researcher Melanie Hingle at the University of Arizona investigated whether text messages could be a useful tool in influencing the nutrition and physical activity knowledge, attitudes, and behavior of adolescents (Hingle et al. 2013).

Sample and Methods

Research participants included a total of 177 teens ages 12 to 18 enrolled in one of 11 youth programs (such as YMCA teen groups, a middle school student leadership group, and a youth cycling club). These teens included nearly equal numbers of males and females representing diverse socioeconomic and racial/ethnic backgrounds, and included a large proportion of Hispanic youth and youth from lower-income families.

The teens in this study participated either in focus groups (n = 59), classroom discussions (n = 86), or an 8-week pilot study (n = 32). Phase I of this study involved two goals: (1) to identify topics related to physical activity and nutrition that are of interest to teens, and (2) to construct sample messages around these topics of interest. Several strategies were used to identify potential topics, including a literature search conducted by the research team to identify behaviors associated with teen obesity and overweight and a survey of 100 college freshmen who submitted their top three questions about nutrition and physical activity. Over 300 messages were developed around such topics as increased caloric intake, increased consumption of sweetened beverages, low intake of fruits and vegetables, large portions, frequent consumption of fast food and food away from home, physical activity, and infrequent consumption of breakfast. Messages were constructed using a variety of formats and styles to test which teens preferred. Types of messages included short "factoids," polls, scenarios, quizzes, and recipes.

In Phase II, focus groups of 6 to 10 teens were asked to give feedback on the various health-related messages that were developed in Phase I. The teens were asked what they thought about using text messaging as a vehicle for improving health, and which content and message styles they thought were appealing and relevant. Four classroom discussions were held to evaluate whether teens could read and comprehend the message content, and which message style or "voice" teens found most appealing. Students were shown 25 messages and were asked to rate each message as fitting into 1 of 3 categories: 1 = "Cool, I want to know more!;" 2 = "Okay, but . . ." (indicating they liked the message, but it needed an adjustment to make it more youth-friendly); or 3 = "Next!" indicating they did not like the message. In classroom discussion, an interviewer asked students to explain their ratings and any suggestions they had for revising messages.

Based on Phase II findings, the researchers developed a message delivery protocol for testing in Phase III. Teens who participated in Phase III were provided with a mobile phone and received nutrition and physical activity–related health messages via SMS for a period of 8 weeks. Teens in Phase III were interviewed at the end of the 8-week testing period to assess their experiences with receiving health-related texts and to explore the extent to which the teens read, liked/disliked, acted upon, or shared the text messages with others.

lose the most weight or spend the most time exercising. Research suggests that for every dollar a company spends on wellness programs, it saves $3.27 in medical costs and $2.75 in absentee costs (Trust for America's Health 2012).

Public Education and Awareness. In 2010, First Lady Michelle Obama launched the Let's Move! Initiative to emphasize healthy eating and increased physical activity at school, at home, and in the community. Also in 2010, the federal government adopted a menu labeling law (as part of the Affordable Care Act) that requires chain restaurants, other retail eating establishments, and certain vending machine owners/operators to post calorie information on menus, menu boards, or machines and to state that other written nutritional information (such as sodium, cholesterol, sugar, saturated fat, and so on) is available upon request.

Researchers are exploring ways in which social media and communication technology could be used to convey educational messages about nutrition and physical activity. This chapter's *Social Problems Research Up Close* feature looks at the use of texting as a means of delivering healthy lifestyle messages to teens.

Teens in this study expressed enthusiasm about the idea of receiving nutrition and physical activity messages through texting. The two favorite message formats among the teens were the short "factoid" format and the quizzes. An example of a factoid message is "A can of soda has 10 teaspoons of added sugar." An example of a quiz item is "Guarana is an ingredient in a lot of popular energy drinks. What is it?" Participants tended to like texts that relate specifically to their age group (e.g., "American girls aged 12 to 19 years old drink an average of 650 cans of soda a year!" or "2 out of 5 teens don't eat breakfast!"). They also preferred messages that used personal pronouns (e.g., "Eating foods high in protein helps you feel full. Want to see examples of foods that contain protein?"), which teens perceived as "speaking directly to us." Teens also wanted some of the messages to contain information that they referred to as "random" (e.g., "Carrots were originally purple in color"; "Ears of corn have even numbers of rows"). Teens perceived such "random" information as unique and fun to read and share with others. Teens also liked messages that were "translatable" into behaviors related to physical activity, nutrition, or body weight (e.g., "Walking can burn 80 to 100 calories per mile"). Many youth said that simple messages made them curious to know more about certain exercise and nutrition

topics, and suggested that the researchers develop "teaser" messages for future participants (e.g., "Too little sleep can lead to weight gain. Click HERE to learn more").

Teens also commented on message style or "voice." "Teens explained that they did not want to be told what to do and did not like message tones that they perceived as authoritarian" (Hingle et al. 2013, p. 15). For example, they suggested that the message "Teens should get 9 hours of sleep per day" be revised to "Nine hours of sleep each night is recommended for teens." Although participants reported that they enjoyed receiving the nutrition and physical activity text messages, they said they would prefer to receive no more than two health-related messages per day, and they wanted the messages to come from a credible source, such as from a nutritionist. Participants enjoyed being able to text questions back to the researchers and receive responses. For example, after receiving the text, "The typical American teenage girl drinks 650 cans of soda per year!" one participant texted back and asked "Do you know if teens in other countries drink more or less soda than they do in the U.S., like how many cans of soda will a Mexican girl drink a year?"

This study suggests that text messaging can be an effective way to convey health-related messages to teenagers, and that involving youth in developing nutrition and physical activity messages is an effective means of making sure the messages are appealing,

relevant, and meaningful to their age group. This study also reveals that influencing teens to change their nutrition and physical activity behavior is best achieved by avoiding phrases like "you should" and "you need to" because, as the teens in this study expressed, "kids don't like being told what to do" (p. 18). Overall, this study suggests that texting— already a favorite activity among many teens—offers an opportunity to reach teens with information that could help them make healthier lifestyle choices. Hingle noted that promoting good nutrition and physical activity during the teen years is critical, as the risk for developing obesity increases during adolescence. "They're at the age right now that they start making decisions for themselves with regard to food and physical activity. . . . Up until about middle school, parents are a lot more involved in making those decisions, so from a developmental standpoint, it's a good time to intervene" (quoted in "Promoting Health and Physical Activity Via Carefully Worded Text Messages to Teenagers" 2013).

Sources: Hingle, Melanie, Mimi Nichter, Melanie Medeiros, and Samantha Grace. 2013. "Texting for Health: The Use of Participatory Methods to Develop Healthy Lifestyle Messages for Teens." *Journal of Nutrition Education and Behavior* 45(1):12–19.

"Promoting Health and Physical Activity Via Carefully Worded Text Messages to Teenagers." 2013. *Medical News Today,* January 22. Available at www.medicalnewstoday.com

Strategies to Improve Mental Health Care

In the wake of the 2012 tragic shooting at Sandy Hook Elementary School in Newtown, Connecticut, President Obama called for a national dialogue on mental illness. His proposals included (1) training teachers to identify signs of mental illness and provide assistance; (2) improving mental health and substance abuse treatment for teens and young adults; (3) training more mental health professionals to serve students and young adults; and (4) improving understanding of mental illness and the importance of mental health treatment. Next, we discuss other strategies for improving mental health care in the United States, including eliminating the stigma associated with mental illness, improving access to mental health services, and supporting mental health needs of college students.

Eliminating the Stigma of Mental Illness. Eliminating the stigma of mental illness is important because the negative label of "mental illness" and the social rejection and stigmatization associated with mental illness discourages individuals from seeking mental health treatment. The National Alliance on Mental Illness (NAMI) has fought against negative portrayals of mental illness in movies and television, and stigmatizing and inaccurate language in the news media. A major victory in the fight against the stigma of mental illness occurred in 2013, when the Associated Press—a global news network seen or heard by more than half the world's population—adopted new rules on how editors and reporters report about mental illness. The new rules include:

- Mental illness is a general condition. Specific disorders or types of mental illness should be used whenever possible.
- Do not use derogatory terms, such as insane, crazy/crazed, nuts, or deranged, unless they are part of a quotation that is essential to the story.
- Whenever possible, rely on people with mental illness to talk about their own diagnoses.
- Avoid using mental health terms to describe non–health issues. For example, don't say that an awards show was schizophrenic.
- Do not assume that mental illness is a factor in a violent crime, and verify statements to that effect. Research has shown that the vast majority of people with mental illness are not violent, and most people who are violent do not suffer from mental illness. (Carolla 2013)

The Department of Defense has launched an anti-stigma campaign called "Real Warriors, Real Battles, Real Strength" designed to assure military personnel that seeking mental health treatment will not harm their career and to publicize stories of military personnel who have been successfully treated for mental health problems (Dingfelder 2009). Effective anti-stigma campaigns not only focus on eradicating negative stereotypes of people with mental illness, but also emphasize the positive accomplishments and contributions of people with mental illness.

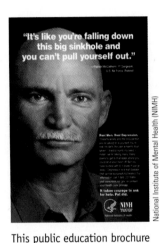

National Institute of Mental Health (NIMH)

This public education brochure on men and depression is available from the National Institute of Mental Health (NIMH), www.nimh.nih.gov.

What Do You Think? The Pentagon has ruled that the Purple Heart medal given to soldiers wounded or killed in combat may not be awarded to war veterans with post-traumatic stress disorder (PTSD) because it is not a physical wound (Alvarez & Eckholm 2009). Do you think veterans with PTSD should be eligible to receive the Purple Heart medal? Why or why not?

Improving Access to Mental Health Care. About a third of U.S. adults with major depression received no mental health treatment in the past year. Of those who did receive treatment, most (61 percent) received it from a family doctor or general practitioner, not from a mental health care specialist (Substance Abuse and Mental Health Services Administration 2012). Primary care physicians who try to obtain outpatient mental health services for their patients are often unsuccessful because of shortages in mental health professionals (e.g., psychiatrists, psychologists, psychiatric nurses, social workers) and lack of adequate health insurance. Thus, improving access to mental health services involves (1) recruiting more mental health professionals, especially those willing to serve in rural and impoverished communities and who have cultural competency to work with clients from diverse cultural backgrounds, and (2) improving health insurance coverage for mental health problems.

The 2010 Affordable Care Act included a new program—the Mental and Behavioral Health Education and Training Grant program—that provides funds to institutions of higher education to recruit and train students pursuing graduate degrees in clinical mental and behavioral health. And in response to an executive order signed by President Obama in 2012, the Department of Veterans Affairs (VA) announced in early 2013 that it had added more than 1,000 mental health professionals and was working to hire and train mental health peer specialists and increase the capacity of the veterans' telephone crisis line.

The Affordable Care Act (ACA) has improved mental health care by greatly expanding mental health and substance use disorder insurance coverage. Under the ACA, all new small-group and individual market insurance plans are required to cover 10 Essential

Health Benefit categories, including mental health and substance use disorder services, and must cover them with the same benefits as medical and surgical benefits—a concept known as **parity**. The Affordable Care Act has expanded mental health and substance use insurance coverage and federal parity protections to 62 million Americans (Beronio et al. 2013).

Another strategy to improve access to mental health care involves making mental health screening a standard practice reimbursed by insurance, just like mammograms and other screening tests are reimbursed. And although most schools screen children and adolescents for hearing and vision problems, some schools also screen for mental health problems such as depression.

What Do You Think? The American Red Cross teaches courses in first aid to millions of people each year. In these courses, participants learn how to respond to first aid emergencies including broken bones, cuts, burns, and other injuries, as well as how to respond to cardiac and breathing emergencies, including how to do CPR (cardiopulmonary resuscitation) and the use of automated external defibrillators (AEDs). The American Red Cross courses in first aid do not include mental health emergencies. To help the public recognize, understand, and respond to signs of mental illness, Mental Health First Aid USA—a 12-hour training that originated in Australia in 2001—teaches participants how to respond to mental health emergencies, such as a stranger having a panic attack, a person who is experiencing visual or auditory hallucinations, or a friend who is expressing hopelessness and despair or who is threatening suicide. Do you think that American Red Cross first aid courses should also cover how to respond to mental health emergencies? Why or why not?

Mental Health Support for College Students. Most colleges and universities offer mental health services to students, and provide accommodations for students with documented mental health conditions (such as adjustments in test setting and times and excused absences for treatment). Colleges can improve these services in a variety of ways, such as making sure that students know these services exist, employing more mental health professionals, and offering extended and flexible hours of service. In a survey of college students diagnosed with a mental health condition, respondents were asked, on average, how long they had to wait for an appointment to access campus mental health services; nearly 4 in 10 waited more than five days for an appointment (National Alliance on Mental Illness 2012).

As discussed earlier in this chapter, many college students with mental health conditions do not disclose their condition and do not seek accommodations or services largely because of the stigma involved in being identified as having a mental illness. To reduce stigma of mental illness on campuses, it is critical to provide information to the campus community on how common mental health conditions are and to emphasize the importance of getting help.

On more than 25 college campuses throughout the United States, students are getting involved in clubs that offer support and advocacy for students with mental illness. NAMI on Campus clubs, which are affiliated with the National Alliance on Mental Illness, are student-led clubs that raise awareness of mental health issues, educate the campus community, and support students. Aaron Chen, a student at the University of Illinois at Urbana-Champaign, explained that he chose to start a NAMI on Campus because he had personally experienced and witnessed the difficulties associated with mental illness and because "it is extremely difficult for those affected by mental illness to speak out about their struggles, especially on a college campus" (quoted in Crudo 2013). Megan Rogers,

parity In health care, a concept requiring equality between mental health care insurance coverage and other health care coverage.

A growing trend in mental health services involves the use of therapy animals, primarily dogs, cats, and horses, to improve the functioning of individuals with a variety of mental health conditions, including anxiety, depression, family problems, autism, eating disorders, post-traumatic stress, attention deficit disorders, and many others (Fine 2010; Peters 2011). Animal-assisted therapy (AAT) is used in a variety of mental health care settings, including private therapy offices, mental health clinics, hospitals, and long-term care and residential facilities. Some school systems use AAT to assist children with behavioral disorders (e.g., attention deficit/hyperactivity disorder, autism spectrum disorders). In the United States, AAT was used as early as 1919, when dogs were used with psychiatric patients at St. Elizabeth's Hospital in Washington, DC. The use of animals in mental health treatment gained increased attention with the 1969 publication of *Pet-Oriented Child Psychotherapy* by Dr. Boris Levinson, who had discovered the therapeutic value of animals quite by accident when a young therapy patient arrived early for his appointment and met and embraced Levinson's dog, Jingles, who was in the office that day. Levinson observed the powerful impact the dog had on the boy and how the dog helped Levinson develop a rapport with his client (Urichuk with Anderson 2003).

Animals can contribute to therapy in a variety of ways, including (1) reducing anxiety, (2) helping the trust- and rapport-building process between the client and the therapist, (3) increasing motivation to attend and participate in therapy because of the desire to spend time with the therapy animal, and (4) stimulating conversation about difficult topics. One clinician reported an experience working with a child who had been traumatized by sexual abuse:

> I told one child that Buster [a dog] had a nightmare. I then asked the child, "What do you think Buster's nightmare was about?" The child said, "The nightmare was about being afraid of getting hurt again by someone mean." (Cited in Kruger & Serpell 2010, p. 39)

AAT is also used with individuals with severe mental disorders. Marsha was a 23-year-old woman who was diagnosed with catatonic schizophrenia. She was treated with medication and electroshock therapy, without improvement. She was withdrawn, frozen, and nearly mute. A therapy dog was introduced into her treatment:

> At first there was no improvement in Marsha's behavior. . . . She remained very withdrawn and the only signs of communication were when she was with the dog. When the dog was taken away, she would get off her chair and go after it. She began to walk the dog a little and . . . was given a written schedule of the hours when the dog would come and visit her; she began to look forward to the visits and to talk about the dog with the other patients. Six days after the introduction of the dog Marsha suddenly showed marked improvement and shortly thereafter she was discharged. (Cited in Urichuk with Anderson 2003, pp. 107–108)

The most common use of AAT in mental health services involves therapists working in partnership with their own pet that has been evaluated and certified as appropriate for therapy work (Chandler 2005). Dogs used as therapy animals must meet rigorous requirements through organizations such as Therapy Dogs Inc., Delta Society®, and Therapy Dogs International. For the dogs, this includes a temperament test, obedience class training, and additional AAT training in which the dogs learn things like not reacting to loud noises, how to ride on elevators, and being comfortable around patients who use wheelchairs or walkers. In addition, the dogs must maintain good health and remain current on vaccinations.

Not all AAT involves using specially trained or certified animals. Some individuals achieve improvements in mental health functioning and well-being as a result of interacting with and taking care of farm animals (Arehart-Treichel 2008). One case study describes how Mark, a young teenager with autism, benefited through his interactions with donkeys:

> At first Mark could not get near the donkeys. Naturally wary of people, the donkeys ran away from Mark as he marched after them. . . . But Mark was motivated, and with guidance and patience has learnt to approach the donkeys slowly and gently. He has become aware of the donkeys' feelings and of how his actions impact them; he has built a relationship with the donkeys based upon mutual trust and respect. He is now rewarded each week by Ceilidh running up to him to have her face rubbed. This is a new experience for Mark

Therapy dogs are used in a variety of settings, including clinics, private medical offices, nursing homes, schools, and hospitals.

> that we are working on transferring to his human relationships. (Cited in Urichuk with Anderson 2003, p. 80)

Small animals such as rabbits, guinea pigs, and birds can also be used as therapy animals:

> Marta was an 8-year-old diagnosed as an emotionally disturbed child of a strict and abusive mother. She was aggressive and hyperactive, sexually precocious, and had temper tantrums. In her first few months at the residential school, no one could get her to talk about her relationship with her mother. In the first session with a small furry rabbit, she held him in her lap and stroked him, telling the therapist that the rabbit's ears had been chewed by the mother rabbit. (The rabbit's ears were normal.) The therapist asked her why this was so. Marta responded, "The mother rabbit chewed the baby rabbit's ears all up. She wanted the baby to leave home." The therapist then asked, "How did the baby rabbit feel?" In answering, Marta said "Sad. The baby rabbit loves the mother rabbit but the mother rabbit no longer loves the baby." This dialogue about the rabbit was an opener for Marta to then talk of her own feelings about the mother who badly beat her. (Cited in Urichuk with Anderson 2003, pp. 64–65)

It is important to note that AAT is not appropriate for individuals who are fearful of or allergic to animals. Even with careful selection and training of therapy animals, there is some risk that a therapy animal could injure a client, and also risk that a client could hurt the animal. But with close supervision and careful management, animal-assisted therapy provides opportunities for improving the well-being and functioning of children and adults with a variety of mental health problems.

president of NAMI on Campus North Carolina State explained that she became involved in a NAMI on Campus club "to connect with those similar to myself in a supportive environment" and to "contribute to ending the stigma and isolation associated with having a mental illness" (quoted in Crudo 2013).

Stress can exacerbate mental health problems, and even trigger their onset. To help students cope with the stress of exam week, some colleges and universities provide students with access to "therapy dogs" as a way to ease stress. The University of Connecticut provides therapy dogs at the library during exam week in a program called "Paws to Relax." At Harvard Medical School and Yale Law School, students can borrow therapy dogs through the library's card catalogue, just like they borrow books. This chapter's *Animals and Society* feature describes how dogs and other animals can be useful in mental health counseling or therapy.

U.S. Federal Health Care Reform: The Affordable Care Act of 2010

Since 1912, when Theodore Roosevelt first proposed a national health insurance plan, the Truman, Nixon, Carter, Clinton, and Obama administrations have advocated the idea of health care for all Americans. Following much heated debate, in 2010, the Patient Protection and Affordable Care Act, commonly referred to as the **Affordable Care Act (ACA)** or "Obamacare," was passed by Congress and signed into law by President Obama, with the overarching goal of increasing health insurance coverage to Americans. Just a few of the many provisions of ACA include:

- Establishing an "individual mandate" that requires U.S. citizens and legal residents to have health insurance or pay a penalty (waivers granted for financial hardship).
- Creating health insurance exchanges: online marketplaces where consumers can shop for, compare, and enroll in insurance plans.
- Providing tax credits to businesses that provide insurance to their employees.
- Requiring health insurance plans to provide dependent coverage for children up to age 26.
- Prohibiting health insurance plans from placing lifetime limits on the dollar value of coverage; restricting annual limits on coverage; prohibiting insurers from canceling coverage except in cases of fraud; and prohibiting denial of insurance due to preexisting conditions.
- Requiring insurance companies to use a certain percentage of the premiums they collect on medical care, as opposed to administrative expenses and profits.
- Expanding Medicaid to cover more low-income individuals/families.
- Providing discounts on brand-name prescription drugs and free preventive services and annual wellness exams for Medicare enrollees, and raising Medicare premiums for some higher-income seniors.

Challenges to the constitutionality of the ACA eventually led to the Supreme Court, which, in a 5–4 vote, ruled in 2012 that the individual mandate is a constitutional exercise of Congress's power to levy taxes (Musumeci 2012). But the court barred the federal government from withholding Medicaid funds from states that refuse to participate in the Medicaid expansion provision of the ACA, thus allowing states to give up the federal Medicaid expansion grants without losing their existing Medicaid funding (Clemmitt 2012).

In 2013, three years after the ACA became law, the public continues to be divided over the ACA, with 37 percent having a favorable opinion, 40 percent having an unfavorable opinion, and 23 percent declining to offer an opinion (Kaiser Family Foundation 2013). When asked about the reasons for their views, the most common response among those who favor the ACA involves expanded access to health care and insurance, while those with unfavorable views of the ACA express concern over the cost, and are opposed to the individual mandate and government involvement in health care.

Some Americans who criticize the ACA for not going far enough to ensure access to health care call for further health care reform to create a **single-payer health care** system

Affordable Care Act (ACA) Health care reform legislation that Obama signed into law in 2010, with the goal of expanding health insurance coverage to more Americans; also known as the Patient Protection and Affordable Care Act, or "Obamacare."

in which a single tax-financed public insurance program replaces private insurance companies. Advocates of single-payer health care financing argue that administration costs of private insurance consume nearly a third of Americans' health dollars. These costs include insurance company overhead, underwriting, billing, sales and marketing departments, exorbitant executive pay, and profits. In addition, hospitals and doctors must pay administrative staff to deal with the various billing policies and procedures of different insurers. Supporters argue that single-payer health care would save more than $400 billion per year—enough to provide health coverage to everyone without any additional expense.

The United States National Health Care Act, also known as the Expanded and Improved Medicare for All Act, introduced by Representative John Conyers (D-Michigan), would expand Medicare to every U.S. resident. In 2011, Vermont became the first state to pass legislation to establish state-level single-payer health care.

Opponents argue that a single-payer national health insurance program would amount to a "government takeover" of health care and would result in higher costs, less choice, rationing, and excessive bureaucracy—the very outcomes that have resulted from corporatized medicine (Nader 2009). The rise of the grassroots Tea Party political movement has fueled opposition to "big government," viewing government "takeover" of health care as an intrusion into individual freedoms.

Dr. Marcia Angell (2003), lecturer at Harvard Medical School and former editor of the *New England Journal of Medicine,* makes the following plea for a national health program:

> We live in a country that tolerates enormous disparities in income, material possessions, and social privilege. That may be an inevitable consequence of a free-market economy. But those disparities should not extend to denying some of our citizens certain essential services because of their income or social status. One of those services is health care. Others are education, clean water and air, equal justice, and protection from crime, all of which we already acknowledge are public responsibilities. We need to acknowledge the same thing for health care.

There is considerable support for single-payer health care among doctors, nurses, and the general public. The insurance industry opposes the adoption of such a system because the private health insurance industry would be virtually eliminated, and so spends a great deal of money on lobbying, political contributions, and public relations to influence the health reform debate.

Understanding Problems of Illness and Health Care

Although human health has probably improved more over the past half-century than over the previous three millennia, the gap in health between rich and poor remains wide. Poor countries need economic and material assistance to alleviate problems such as HIV/AIDS, high maternal and infant mortality rates, and malaria. Cancer, once viewed as a disease that affects primarily wealthy countries, has now become prevalent in low-income countries where treatment is either not available or is not affordable (Farmer et al. 2010). Obesity and its associated health problems has also spread throughout the world, adding to the burden of infectious diseases that already plague low-income countries.

Although poverty may be the most powerful social factor affecting health, other social factors that affect health include globalization, increased longevity, family structure, gender, education, and race or ethnicity. Although individuals make choices that affect their health—choices such as whether to smoke, exercise, eat a healthy diet, engage in risky sexual activity, wear a seat belt, and so on—those choices are also influenced by

single-payer health care A health care system in which a single tax-financed public insurance program replaces private insurance companies.

social, economic, and political forces that must be addressed if the goal is to improve the health not only of individuals but also of entire populations. By focusing on individual behaviors that affect health and illness, we often overlook social causes of health problems (Link & Phelan 2001). A sociological view of illness and health care looks not only at the social causes, but also at the social *consequences* of health problems—consequences that potentially affect us all. In *Uninsured in America* (2005), Sered and Fernandopulle explain:

> If millions of American children do not have reliable, basic health care, all children who attend American schools are at risk through daily exposure to untreated disease. If millions of restaurant and food industry workers do not have health insurance, people preparing food and waiting tables are sharing their health problems with everyone they serve. . . . If tens of millions of Americans go without basic and preventive care, we all pay the bill when their health problems turn into complex medical emergencies necessitating expensive . . . treatment. (p. 20)

A sociological approach to illness and health care also looks at social solutions such as federal, state, and local government policies and laws designed to improve public health. In a survey of U.S. adults, more than 75 percent of adults indicated support for government policies that would (1) require food manufacturers and chain restaurants to significantly reduce sodium in their foods; (2) require public school children to participate in at least 45 minutes of daily physical activity; (3) increase the affordability of fruits and vegetables; (4) require postings of calorie counts; and (5) prevent use of food stamps for soda and other sugary beverages (Morain & Mello 2013).

Improving public health is a complex endeavor. A comprehensive approach to health is informed by the fact that "there is no single silver bullet for population health improvement. Investments in all determinants of health—including health care, public health, health behaviors, and residents' social and physical environments—will be required" (Kindig & Cheng 2013, p. 456). A comprehensive approach to improving the health of a society requires addressing diverse issues such as poverty and economic inequality, gender inequality, population growth, environmental issues, education, housing, energy, water and sanitation, agriculture, and workplace safety. Despite the significance of recent health care reform efforts to improve Americans' health insurance coverage, access to health care is only one piece of the puzzle: "Health and longevity are also profoundly influenced by where and how Americans live, learn, work, and play" (Williams et al. 2010, p. 1,481).

Improving the health of the world also means seeking nonmilitary solutions to international conflicts. In addition to the deaths, injuries, and illnesses that result from combat, war diverts economic resources from health programs, leads to hunger and disease caused by the destruction of infrastructure, causes psychological trauma, and contributes to environmental pollution (Sidel & Levy 2002). Thus, "the prevention of war . . . is surely one of the most critical steps mankind can make to protect public health" (White 2003, p. 228).

The tragic and senseless shooting deaths at Sandy Hook Elementary School in 2012 renewed public concern for affordable and accessible mental health care—another critical but often neglected aspect of public health. The Sandy Hook tragedy also ignited debates over gun control in the United States (see also Chapter 4), raising the question of whether widespread availability of guns is a public health problem. The rate of firearm-related deaths in the United States is nearly 20 times that in other developed nations (Shern & Lindstrom 2013). In other developed and civilized countries that grant their citizens the right to universal health care, owning a gun is a privilege and not a right. In the United States, Americans have a constitutional right to own a gun, but no similar right to health care. Until the United States joins the rest of the developed world as well as the United Nations in declaring access to health care to be a basic human right, it may be easier for many Americans to access a gun than it is to access health care.

- **How does the World Health Organization define health?**
Health, according to the World Health Organization, is "a state of complete physical, mental, and social well-being." Based on this definition, we suggest that the study of social problems is, essentially, the study of health problems, because each social problem affects the physical, mental, and social well-being of humans and the social groups of which they are a part.

- **What are some major differences in the health of populations living in high-income countries compared with the health of populations living in low-income countries?**
Life expectancy is significantly greater in high-income countries compared with low-income countries. Although the majority of deaths worldwide are caused by noncommunicable diseases such as heart disease, stroke, cancer, and respiratory disease, low-income countries have a comparatively higher rate of infectious and parasitic diseases, infant and child deaths, and maternal mortality.

- **How has globalization affected health worldwide?**
Globalization is linked to the rise in obesity worldwide due to increased access to unhealthy foods and beverages, and to televisions, computers, and motor vehicles, which are associated with increased sedentary behavior. These factors have contributed to *globesity*—a worldwide increase in overweight and obesity and the accompanying health problems associated with obesity. Increased global transportation and travel contribute to unhealthy levels of air pollution from burning fossil fuel, and with speeding the spread of infectious disease. On the positive side, globalized communications technology is helpful in monitoring and reporting on outbreaks of disease, disseminating guidelines for controlling and treating disease, and sharing medical knowledge and research findings. Another aspect of globalization and health is the growth of medical tourism—a multibillion-dollar global industry that involves traveling, primarily across international borders, for the purpose of obtaining medical care.

- **Are Americans the healthiest population in the world?**
The United States is one of the wealthiest countries in the world, but it is not one of the healthiest. In a comparison of health outcomes in the United States with those of 16 other high-income, industrialized countries, Americans are less likely to smoke and drink heavy alcohol, and they have better control over their cholesterol levels. The United States also has higher rates of cancer screening and survival, and has higher survival after age 75. But despite the fact that the United States spends more on health care per person than any other industrialized country, Americans die sooner and have higher rates of disease or injury. The United States ranked last or near the bottom in nine key areas of health: infant mortality and low birth weight; injuries and homicide; teenage pregnancy and sexually transmitted infections; HIV/AIDS prevalence; drug-related deaths; obesity and diabetes; heart disease; lung disease; and disability.

- **Why is mental illness referred to as a "hidden epidemic"?**
Mental illness is a "hidden epidemic" because the shame and embarrassment associated with mental problems discourage people from acknowledging and talking about them. The stigma of being labeled as "mentally ill" can negatively affect an individual's self-concept and disqualify that person from full social acceptance. Negative stereotypes of people with mental illness contribute to its stigma. One of the most common stereotypes of people with mental illness is that they are dangerous and violent. Although untreated mental illness can result in violent behavior, the vast majority of people with severe mental illness is not violent and is involved in only about 4 percent of violent crimes. In fact, people with mental illness are much more likely to be victims of violence than members of the general population.

- **How common is mental illness in the United States?**
Among the noninstitutionalized civilian population in 2011, one in five U.S. adults had a diagnosable mental, behavioral, or emotional disorder (excluding developmental and substance abuse disorders), either currently or within the past year. One in 20 U.S. adults had a serious mental illness—a mental illness that has resulted in serious impairment in daily functioning. About half of all Americans will experience some form of mental disorder in their lifetime. More than one in four college students has been diagnosed or treated by a professional for a mental health problem within the past year.

- **Which theoretical perspective criticizes the pharmaceutical and health care industry for placing profits above people?**
The conflict perspective criticizes the pharmaceutical and health care industry for placing profits above people. For example, pharmaceutical companies' research and development budgets are spent not according to public health needs but rather according to calculations for maximizing profits. Because the masses of people in developing countries lack the resources to pay high prices for medication, pharmaceutical companies do not see the development of drugs for diseases of poor countries as a profitable investment.

- **What are three main social factors that are associated with health and illness?**
Three main social factors associated with health and illness are socioeconomic status, race/ethnicity, and gender.

- **How does health care in the United States compare with that of many other high-income nations?**
Many other advanced countries have national health insurance systems—typically administered and paid for by government—that provide universal health care (health care to all citizens). The United States does not have a health care system per se, but rather has a patchwork that includes both private insurance (purchased individually or through employers or other groups), and public insurance plans such as Medicare and Medicaid.

- **What is the difference between selective primary health care and comprehensive primary health care?**
Selective primary health care focuses on using specific interventions to target specific health problems, such as promoting condom use to prevent HIV infections and

providing immunizations against childhood diseases to promote child survival. In contrast, comprehensive primary health care focuses on the broader social determinants of health, such as poverty and economic inequality, gender inequality, environment, and community development.

• **What is the main goal of the Patient Protection and Affordable Care Act, commonly referred to as the Affordable Care Act (ACA) or "Obamacare"?**
The main goal of the Affordable Care Act is to increase health insurance coverage to Americans.

TEST YOURSELF

1. Worldwide, the leading cause of death is
   a. HIV/AIDS
   b. traffic accidents
   c. heart disease
   d. cancer

2. In the United States, _____ of adults are either overweight or obese.
   a. about 10 percent
   b. nearly a quarter
   c. about half
   d. more than two-thirds

3. Americans are the healthiest population in the world.
   a. True
   b. False

4. Which age group has the highest rate of suicidal thoughts?
   a. 12 to 16
   b. 18 to 25
   c. 45 to 55
   d. Over age 65

5. A study of mortality rates in 3,140 U.S. counties found that the factors most strongly associated with higher mortality were poverty and
   a. gender
   b. race
   c. divorce
   d. lack of college education

6. The United States spends far more per person on health care than does any other industrialized nation.
   a. True
   b. False

7. In 2012, the average annual cost of health insurance for a U.S. family was
   a. more than $15,000
   b. nearly $5,000
   c. about $500.00
   d. $150.00

8. In the United States, the most frequently used source of care for mental health problems is
   a. psychologists
   b. psychiatrists
   c. primary care physicians
   d. social workers

9. One study found that medical bills, as well as income lost due to illness, contributed to two-thirds of all bankruptcies in 2007. Most medical debtors
   a. were educated homeowners with middle-class jobs
   b. had health insurance
   c. both a and b
   d. neither a nor b

10. A health care system in which a single tax-financed public insurance program replaces private insurance companies is called which of the following?
    a. Constitutional care
    b. Managed care
    c. Obamacare
    d. Single-payer health care

Answers: 1. C; 2. D; 3. B; 4. B; 5. D; 6. A; 7. A; 8. C; 9. C; 10. D.

KEY TERMS

Affordable Care Act (ACA)   59
allopathic medicine   46
complementary and alternative medicine (CAM)   46
comprehensive primary health care   51
deinstitutionalization   50
developed countries   27
developing countries   27
globalization   30
globesity   30
health   27

infant mortality   29
least developed countries   27
life expectancy   28
managed care   45
maternal mortality   29
Medicaid   45
medicalization   40
medical tourism   30
Medicare   45
mental health   35
mental illness   35

mortality   28
parity   57
selective primary health care   51
single-payer health care   60
socioeconomic status   41
State Children's Health Insurance Program (SCHIP)   45
stigma   36
under-5 mortality   29
universal health care   48
workers' compensation   46

Bob Thomas/Getty Images

3

Alcohol and Other Drugs

"Substance abuse, the nation's number one preventable
health problem, places an enormous burden on American
society, harming health, family life, the economy, and
public safety, and threatening many other aspects of life."

—The Robert Wood Johnson Foundation,
Institute for Health Policy,
Brandeis University

ALL THAT MARKED the death of 20-year-old ballerina Elena Shapiro was a teddy bear, a dimly lit candle, flower petals, and a ballet slipper. This was the intersection where it happened—where Raymond Cook, a prominent plastic surgeon, drove his black Mercedes into Elena Shapiro's Elantra at close to 80 mph. Her car crumpled like a piece of paper, and Elena, pinned inside, was hit with such force that strands of her long blond hair were found in the back seat. The trunk had been pushed into the cabin of her car; a flip-flop remained on the console; the keys dangled from the ignition; and blood stains covered the grey upholstery.

Raymond Cook voluntarily surrendered his medical license and went into an alcohol rehabilitation facility. He was charged with second-degree murder. The jury never heard that this was not the first time he had been charged with drunk driving or the first time he had caused an accident. He also had a history of speeding and, on that awful night, a radar detector was found in his car.

Throughout the trial, he remained emotionless as witness after witness testified that they had tried to stop him from drinking and driving and that he was visibly drunk at the country club that afternoon and, later that evening, at a bar and restaurant.

In the end he was convicted of involuntary manslaughter, felony death by motor vehicle, and driving while impaired. He was sentenced to three to four years, the maximum sentence the jury could impose. Less than three years into his sentence, Raymond Cook was permitted to work at a pharmaceutical company and have weekend home visits.

Andre Jenny/Alamy

Elena Shapiro was just one of many people killed on the road by drunk drivers in 2009, and despite Raymond Cook's relatively light sentence, at least he was caught and prosecuted. The average drunk driver has driven over 80 times before being arrested for the first time.

Drug-related deaths are just one of the many negative consequences that can result from alcohol and drug abuse. The abuse of alcohol and other drugs is a social problem when it interferes with the well-being of individuals and/or the societies in which they live—when it jeopardizes health, safety, work and academic success, family, and friends. But managing the drug problem is a difficult undertaking. In dealing with drugs, a society must balance individual rights and civil liberties against the personal and social harm that drugs promote—fetal alcohol syndrome, suicide, drunk driving, industrial accidents, mental illness, unemployment, and teenage addiction. When to regulate, what to regulate, and who should regulate are complex social issues. Our discussion begins by looking at how drugs are used and regulated in other societies.

The Global Context: Drug Use and Abuse

Pharmacologically, a **drug** is any substance other than food that alters the structure or functioning of a living organism when it enters the bloodstream. Using this definition, everything from vitamins to aspirin is a drug. Sociologically, the term *drug* refers to any chemical substance that (1) has a direct effect on users' physical, psychological, and/or intellectual functioning; (2) has the potential to be abused; and (3) has adverse consequences for individuals and/or society. Societies vary in how they define and respond to drug use. Thus, drug use is influenced by the social context of the particular society in which it occurs.

drug Any substance other than food that alters the structure or functioning of a living organism when it enters the bloodstream.

Globally, 3.6 to 6.9 percent of the world's population between the ages of 15 and 64—between 167 and 315 million people—reported using at least one illicit drug in the previous year (WDR 2013). According to the most recent report, cannabis (i.e., marijuana and hashish) remains by far the most widely used illegal drug, followed by amphetamine-type stimulants (ATS), cocaine, and opiate-based drugs.

Worldwide, 55 percent of adults have consumed alcohol at some time in their life, and 2.3 million people die from alcohol-related use every year (WHO 2011a; WHO 2013). Alcohol consumption is highest in Greenland, the Russian Federation, and western Europe, and lowest in North Africa, the Middle East, and Asia (Pereltsvaig 2013). To understand regional variation in alcohol consumption rates, one must examine a host of social variables. For example, Russian's high alcohol consumption, although descending recently, reflects access and low cost, high unemployment and the resulting boredom, and peer pressure (Jargin 2012). Similarly, rates are low in the Middle East and Asia where religious dictates discourage alcohol use, availability is difficult, and punishment is severe.

> Globally, nearly 20 percent of the adult population smokes cigarettes . . . and 80 percent of the people who smoke cigarettes are from low- and middle-income countries.

Globally, nearly 20 percent of the adult population smokes cigarettes (Eriksen, Mackay, & Ross 2012) and 80 percent of the people who smoke cigarettes are from low- and middle-income countries (WHO 2013). Manufactured cigarettes account for 96 percent of tobacco sales worldwide although other important tobacco products, particularly in the United States, include cigars, smokeless tobacco, and pipes. The top five cigarette-consuming countries are China, the Russian Federation, the United States, Indonesia, and Japan (Eriksen, Mackay, & Ross 2012).

Illicit drug use varies by location as well. For example, 85 million adult Europeans report using an illicit drug in their lifetime—about 25 percent of the population. However, there ". . . is considerable variation in levels of lifetime drug use reported in Europe, ranging from around a third of adults in Denmark, France, and the United Kingdom, to less than one in ten in Bulgaria, Greece, Hungary, Romania, and Turkey" (EMCDDA 2013, p. 29). Further, lifetime use of *any* illicit drug other than marijuana by 15- and 16-year-olds is the highest in the United States (16 percent) when compared to 36 European countries, which averaged a lifetime use of 6 percent (Wadley 2012).

Finally, drug use varies over time. As Figure 3.1 indicates, in general, alcohol use and smoking among the three age groups has decreased over the last two decades. Illicit drug use, although variable over the time period, was at its highest level in 2012. It should also be noted that Figure 3.1 compares 8th, 10th, and 12th grader drug use across drug type. In each case, drug use increased with age.

Some have argued that differences in drug use can be attributed to variations in drug policies. The Netherlands, for example, has had an official government policy of treating the use of "soft" drugs such as marijuana and hashish as a public health issue rather than a criminal justice issue since the mid-1970s. Treatment of the drug user and prevention of future drug use are prioritized over the more punitive response of imprisonment found in many other countries.

In the first decade of the policy, drug use did not appear to increase. However, increases in marijuana use were reported in the early 1990s with the advent of "cannabis cafés." More recently, however, despite coffee shop sales of marijuana for personal use, research indicates that cannabis use in the Netherlands does not significantly differ from that in other European countries (MacCoun 2011). Nonetheless, concerns over the "tolerance" policy and the millions of drug tourists who visit the Netherlands annually has led to a new law that requires that coffee shop customers must be Dutch residents only (Haasnoot 2013).

Historically, Great Britain has also adopted a "medical model," particularly in regard to heroin and cocaine. As recently as the 1960s, English doctors prescribed opiates and cocaine for their drug-addicted patients who were unlikely to quit using drugs on their own and for the treatment of withdrawal symptoms. By the 1970s, however, British laws

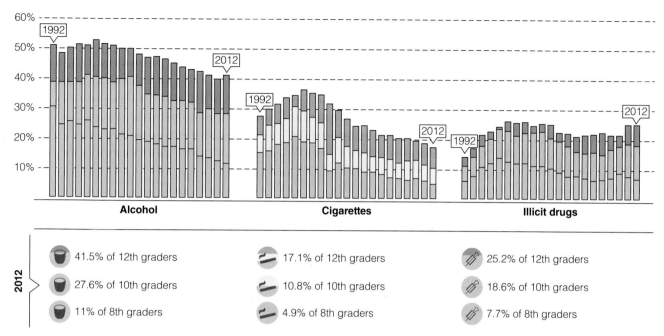

🪣 41.5% of 12th graders	🚬 17.1% of 12th graders	💉 25.2% of 12th graders
🪣 27.6% of 10th graders	🚬 10.8% of 10th graders	💉 18.6% of 10th graders
🪣 11% of 8th graders	🚬 4.9% of 8th graders	💉 7.7% of 8th graders

Figure 3.1 Alcohol, Cigarette, and Illicit Drug Use over the Last Decade for 8th, 10th, and 12th Graders, 2012
Source: NIDA 2012a.

had become more restrictive, making it difficult for either physicians or users to obtain drugs legally. The new emphasis on drug prohibition, criminalization, and incarceration continued for decades and, as in the United States, became known as "war on drugs." Today, British officials are weighing the relative merits of returning to a medical model that embraces drug addiction as a disorder (BMA 2013; Feilding 2013). Similarly, the European Union, an organization of 27 countries, has adopted a 2013–2020 drug policy that for the first time includes the objective of "reduc[ing] . . . the health and social risks and harms caused by drugs . . . " (EMCDDA 2013, p. 2).

What Do You Think? In 2012, initiatives in Colorado and Washington legalized the recreational use of marijuana in small amounts for people over the age of 21. Additionally, medical marijuana is legal in 18 states. Nonetheless, to date, the drug remains illegal under federal law. As a result of the inconsistencies between state and federal law, a Colorado court of appeals found that employees may be fired for after-work use of marijuana even if it's prescribed for a medical condition and the person isn't impaired at work (Ingold 2013). Did the court make the right decision? Why or why not?

In stark contrast to such health-based policies, many other countries execute drug users and/or dealers or subject them to corporal punishment that may include whipping, stoning, beating, and torture. In Iran alone, 450 people were executed in 2011, five times the number executed for drug offenses in 2008, and death sentences in Pakistan have tripled during the same time period (Gallahue et al. 2012). Thirty-three countries or territories have the death penalty for drug violations, including Saudi Arabia, where violators are beheaded.

Drug Use and Abuse in the United States

In the United States, cultural definitions of drug use are contradictory—condemning it on the one hand (e.g., heroin), yet encouraging and tolerating it on the other (e.g., alcohol). At various times in U.S. history, many drugs that are illegal today were legal and readily available. In the 1800s and the early 1900s, opium was routinely used in medicines as a pain reliever, and morphine was taken as a treatment for dysentery and fatigue. Amphetamine-based inhalers were legally available until 1949, and cocaine was an ingredient in Coca-Cola until 1906, when it was replaced with another drug—caffeine (Abadinsky 2013; Witters et al. 1992). Not surprisingly, Americans' concerns with drugs have varied over the years. According to a national poll of U.S. adults, the number of people who say they worry a "great deal" about drugs has decreased from 58 percent in 2001 to 35 percent in 2013, the lowest level in over a decade. Not surprisingly, during the same time period, attitudes favorable toward the legalization of marijuana increased from 34 percent to 48 percent (Gallup Poll 2013).

Use of illicit drugs in the United States is a fairly common phenomenon. According to the most recent National Survey on Drug Use and Health (NSDUH 2013) available, in 2012, over 24 million Americans over the age of 12 had used an illicit drug in the month prior to the survey year, representing 9.2 percent of the 12 and older population. For purposes of the survey, illicit drugs included marijuana/hashish, cocaine (including crack), heroin, hallucinogens, inhalants, or prescription-type psychotherapeutics (pain relievers, tranquilizers, stimulants, and sedatives) used non-medically (NSDUH 2013). Further, 52 percent of the American 12 and older population reported alcohol consumption in the previous month and 26.7 percent reported tobacco use in the previous month (NSDUH 2013).

Sociological Theories of Drug Use and Abuse

Drug abuse occurs when acceptable social standards of drug use are violated, resulting in adverse physiological, psychological, and/or social consequences. When an individual's drug use leads to hospitalization, arrest, or divorce, such use is usually considered abusive. Drug abuse, however, does not always entail drug addiction. Drug addiction, or **chemical dependency**, refers to a condition in which drug use is compulsive—users are unable to stop because of their dependency. The dependency may be psychological (the individual needs the drug to achieve a feeling of well-being) and/or physical (withdrawal symptoms occur when the individual stops taking the drug). For example, withdrawal from marijuana includes a sense of restlessness, mood swings, depression, anger, decreased appetite, and restlessness (Allsop et al. 2012; Zickler 2003).

In 2012, more than 22.2 million Americans, 8.5 percent of the population 12 or older, were defined as being dependent on or abusers of alcohol and/or other drugs. Of that number, 14.9 million were dependent on or abused alcohol only, 4.5 million were dependent on or abused illicit drugs but not alcohol, and 2.8 million were dependent on or abused both illicit drugs and alcohol. The most common illicit drug to be dependent on was marijuana followed by pain relievers, cocaine, tranquilizers, stimulants, heroin, and hallucinogens. Individuals who were dependent on or abuse illicit drugs and/or alcohol were disproportionately male, American Indians or Alaska Natives, unemployed, and between the ages of 18 and 25. Among adults 18 and over, respondents with college degrees had the lowest rates of dependency, and those without a high school degree had the highest (NSDUH 2013).

Various theories provide explanations for why some people use and abuse drugs. Drug use is not simply a matter of individual choice. Theories of drug use explain how structural and cultural forces as well as biological and psychological factors influence drug use and society's responses to it.

Structural-Functionalist Perspective

Structural functionalists argue that drug abuse is a response to weakening societal norms. As society becomes more complex and as rapid social change occurs, norms and values become unclear and ambiguous, resulting in anomie—a state of normlessness. Anomie may exist at the societal level, resulting in social strains and inconsistencies that lead to drug use. For example, research indicates that increased alcohol consumption in the 1830s and the 1960s was a response to rapid social change and the resulting stress (Rorabaugh 1979). Anomie produces inconsistencies in cultural norms regarding drug use. For example, although public

drug abuse The violation of social standards of acceptable drug use, resulting in adverse physiological, psychological, and/or social consequences.

chemical dependency A condition in which drug use is compulsive and users are unable to stop because of physical and/or psychological dependency.

health officials and health care professionals warn of the dangers of alcohol and tobacco use, advertisers glorify the use of alcohol and tobacco, and the U.S. government subsidizes the alcohol and tobacco industries. Furthermore, cultural traditions, such as giving away cigars to celebrate the birth of a child and toasting a bride and groom with champagne, persist.

Anomie may also exist at the individual level, as when a person suffers feelings of estrangement, isolation, and turmoil over appropriate and inappropriate behavior. An adolescent whose parents are experiencing a divorce, who is separated from friends and family as a consequence of moving, or who lacks parental supervision and discipline may be more vulnerable to drug use because of such conditions. Thus, from a structural-functionalist perspective, drug use is a response to the absence of a perceived bond between the individual and society and to the weakening of a consensus regarding what is considered acceptable.

Consistent with this perspective, a recent study on attitudes and drug use concluded that the dramatic increase in teenage use of prescription drugs is, in part, the result of the ". . . lax attitudes and beliefs of parents and caregivers. . . . Parents are not effectively communicating the dangers of Rx medicine misuse and abuse to their kids, nor are they safeguarding their medications at home and disposing of unused medications properly" (PATS 2013, p. 1). The importance of the family in deterring drug use was highlighted in the national youth media campaign—"Parents. The Anti-Drug" (ONDCP 2009).

> Anomie produces inconsistencies in cultural norms regarding drug use. . . . [P]ublic health officials and health care professionals warn of the dangers of alcohol and tobacco use, advertisers glorify the use of alcohol and tobacco, and the U.S. government subsidizes the alcohol and tobacco industries.

Conflict Perspective

The conflict perspective emphasizes the importance of power differentials in influencing drug use behavior and societal values concerning drug use. From a conflict perspective, drug use occurs as a response to the inequality perpetuated by a capitalist system. Societal members, alienated from work, friends, and family as well as from society and its institutions, turn to drugs as a means of escaping the oppression and frustration caused by the inequality they experience. Furthermore, conflict theorists emphasize that the most powerful members of society influence the definitions of which drugs are illegal and the penalties associated with illegal drug production, sales, and use.

For example, alcohol is legal because it is often consumed by those who have the power and influence to define its acceptability—white males (NSDUH 2013). This group also disproportionately profits from the sale and distribution of alcohol and can afford powerful lobbying groups in Washington, DC, to guard the alcohol industry's interests. Because this group also commonly uses tobacco and caffeine, societal definitions of these substances are also relatively accepting. Conversely, minority group members disproportionately use crack cocaine rather than powder cocaine (Mauer 2009). Although the pharmacological properties of the two drugs are the same, possession of 5 grams of crack cocaine carries the same penalty under federal law as possession of 500 grams of powdered cocaine (Taifia 2006). In 2010, Congress voted to change the 1986 law that established the 100 to 1 ratio sentencing disparity but also eliminates the five-year mandatory minimum for first-time possession of crack cocaine. Inequities in state laws sentencing continue but are slowly changing (Crisp 2012).

The use of opium by Chinese immigrants in the 1800s provides a historical example. The Chinese, who had been brought to the United States to work on the railroads, regularly smoked opium as part of their cultural tradition. As unemployment among white workers increased, however, so did resentment of Chinese laborers. Attacking the use of opium became a convenient means of attacking the Chinese, and in 1877, Nevada became the first of many states to prohibit opium use. As Morgan (1978) observed:

> The first opium laws in California were not the result of a moral crusade against the drug itself. Instead, it represented a coercive action directed against a vice that was merely an appendage of the real menace—the Chinese—and not the

Foiled by Moderation!
THE HEARTLESS SHADOW
that threatens the modern figure

"COMING EVENTS CAST
THEIR SHADOWS BEFORE"
(Thomas Campbell, 1777-1844)

**AVOID THAT
FUTURE SHADOW**

by refraining from over-
indulgence, if you would
maintain the modern fig-
ure of fashion

We do not represent that
smoking **Lucky Strike** Cigarettes will bring modern figures
or cause the reduction of flesh.
We do declare that when tempted to do yourself too well, if
you will "Reach for a **Lucky**"
instead, you will thus avoid
over-indulgence in things that
cause excess weight and, by
avoiding over-indulgence, maintain a modern, graceful form.

When Tempted
**Reach
for a
LUCKY**
instead

CIGARETTES

"It's toasted"

Advertising Archives

Advertising for tobacco products has a long history of targeting minorities. Here, a magazine article from the early 1900s appeals to traditional female concerns about femininity, beauty, and body image.

Chinese per se, but the laboring "Chinamen" who threatened the economic security of the white working class. (p. 59)

The criminalization of other drugs, including cocaine, heroin, and marijuana, follows similar patterns of social control of the powerless, political opponents, and/or minorities. In the 1940s, marijuana was used primarily by minority group members, and users faced severe criminal penalties. However, after white, middle-class college students began to use marijuana in the 1970s, the government reduced the penalties associated with its use. Although the nature and pharmacological properties of the drug had not changed, the population of users was now connected to power and influence. Thus, conflict theorists regard the regulation of certain drugs, as well as drug use itself, as a reflection of differences in the political, economic, and social power of various interest groups.

Symbolic Interactionist Perspective

Symbolic interactionism, which emphasizes the importance of definitions and labeling, concentrates on the social meanings associated with drug use. If the initial drug use experience is defined as pleasurable, it is likely to recur, and the individual may earn the label of "drug user" over time. If this definition is internalized so that the individual assumes an identity of a drug user, the behavior will probably continue and may even escalate. Conversely, Copes et al. (2008) observed that respondents who self-identified as "hustlers" rather than "crackheads" were less likely to fall prey to the debilitating effects of the drug, for ". . . [s]lipping into uncontrollable addiction is antithetical to the hustler identity . . ." (p. 256).

What Do You Think? The meaning assigned to alcohol and other drugs is learned not only in small-group interaction with others but through media. One important source of media messages about tobacco use is in the movies. Despite movie studio polices that restrict the portrayal of cigarette use in films, incidents of tobacco imagery increased between 2010 and 2012 (CDC 2013a). What other mediums, often without our awareness, define our attitudes and beliefs about drugs?

Drug use is also learned through symbolic interaction in small groups. In a study of over 7,000 adolescents in 231 different schools in Australia, researchers found that alcohol use is predicted by, regardless of grade or age, the number of peers who consume alcohol (Kelly et al. 2012). Further, in a recent survey of 12- to 17-year-olds, 75 percent reported that "seeing pictures on social networking sites like Facebook and MySpace of kids partying with alcohol and marijuana encourages other teens to want to party like that" (CASA 2012, p. 8). There is also evidence that through viewing peers, first-time users learn not only the motivations for drug use and the techniques but also what to experience. Becker (1966) explained how marijuana users learn to ingest the drug. A novice being coached by a regular user reported the experience:

I was smoking like I did an ordinary cigarette. He said, "No, don't do it like that." He said, "Suck it, you know, draw in and hold it in your lungs . . . for a period of time." I said, "Is there any limit of time to hold it?" He said, "No, just till you feel that you want to let it out, let it out." So I did that three or four times. (p. 47)

Marijuana users not only learn the way to ingest the smoke but also to label the experience positively. When peers define certain drugs, behaviors, and experiences as not only acceptable but also pleasurable, drug use is likely to continue.

Interactionists also emphasize that symbols can be manipulated and used for political and economic agendas. The popular D.A.R.E. (Drug Abuse Resistance Education) program, with its antidrug emphasis fostered by local schools and police, carries a powerful symbolic value with which politicians want the public to identify. "Thus, ameliorative programs which are imbued with these potent symbolic qualities (like D.A.R.E.'s links to schools and police) are virtually assured widespread public acceptance (regardless of actual effectiveness) which in turn advances the interests of political leaders who benefit from being associated with highly visible, popular symbolic programs" (Wysong et al. 1994, p. 461). Ironically, a meta-analysis of the program led West and O'Neal (2004) to conclude that the D.A.R.E. program, as originally conceived, did not significantly prevent drug use among school-aged children. The D.A.R.E. curriculum was modified as a result of its relative lack of effectiveness; however, recent assessments are unclear regarding its effectiveness (Earhart et al. 2011).

Biological and Psychological Theories

Drug use and addiction are probably the result of a complex interplay of social, psychological, and biological forces. For example, some researchers suggest that drug use and addiction are caused by a "biobehavioral disorder," which combines biological and psychological factors (Margolis & Zweben 2011). Biological research has primarily concentrated on the role of genetics in predisposing an individual to drug use. Research indicates that severe, early-onset alcoholism may be genetically predisposed, with some men having 10 times the risk of addiction as those without a genetic predisposition. Interestingly, other problems such as depression, chronic anxiety, and attention deficit disorder are also linked to the likelihood of addiction. Nonetheless, researchers warn, "Nobody is predestined to be an alcoholic" (Firshein 2003).

Psychological explanations focus on the tendency of certain personality types to be more susceptible to drug use. Individuals who are particularly prone to anxiety may be more likely to use drugs as a way to "self-medicate." Research indicates that child abuse or neglect, particularly among females, contributes to alcohol and drug abuse that extends into adulthood (Gilbert et al. 2009). Research also indicates that stressful life events (e.g., death of a loved one) also impact the onset of illicit drug use.

Psychological theories of drug abuse also emphasize that drug use may be maintained by positive or negative reinforcement. Thus, for example, cocaine use may be maintained as a result of the rewarding "high" it produces—a positive reinforcement. Alternatively, heroin use, often associated with severe withdrawal symptoms, may continue as a result of a negative reinforcement, that is, the distress the user feels when faced with withdrawal. Positive and negative reinforcement may come from a variety of sources including family, peers, music, movies, the Internet, and television.

Frequently Used Legal Drugs

Social definitions regarding which drugs are legal or illegal vary over time, circumstance, and societal forces. In the United States, two of the most dangerous and widely abused drugs, alcohol and tobacco, are legal. Compare, for example, the number of people who report past-month use (i.e., current use) of various illicit drugs to the 69.5 million current tobacco users and the 1335.5 million current alcohol drinkers (NSDUH 2013).

Alcohol: The Drug of Choice

Americans' attitudes toward alcohol have a long and varied history. Although alcohol was a common beverage in early America, by 1920, the federal government had prohibited its manufacture, sale, and distribution through the passage of the Eighteenth Amendment to the U.S. Constitution. Many have argued that Prohibition, like the opium

Indicate whether you have or have not experienced any of the following in the last 12 months as a consequence of your own drinking. When finished, compare your responses to those of a national sample of college students.

Consequence	Yes	No
1. Did something you later regretted	___	___
2. Forgot where you were or what you did	___	___
3. Got in trouble with the police	___	___
4. Had sex with someone without giving your consent	___	___
5. Had sex with someone without getting their consent	___	___
6. Had unprotected sex	___	___
7. Physically injured yourself	___	___
8. Physically injured another person	___	___
9. Seriously considered suicide	___	___
10. Reported one or more of the above	___	___

These survey items are from the American College Health Association's (ACHA) National College Health Assessment II (2013). The following data are from 2012. All students are from public or private, two- or four-year colleges and universities. The average age of respondents was 21, 71 percent of the sample was white, 65.6 percent of the sample was female, 40 percent lived in campus residence halls, and 88.2 percent were single at the time of the survey (ACHA 2013).

Consequence	Percentage Reporting Consequence		
	Females	Males	Total
1. Did something you later regretted	33.8	34.9	34.1
2. Forgot where you were or what you did	28.4	32.3	29.6
3. Got in trouble with the police	2.3	4.3	3.0
4. Had sex with someone without giving your consent	2.0	1.1	1.8
5. Had sex with someone without getting their consent	0.5	0.6	0.6
6. Had unprotected sex	17.5	20.6	18.6
7. Physically injured yourself	13.7	15.7	14.4
8. Physically injured another person	1.3	2.9	1.9
9. Seriously considered suicide	2.0	2.1	2.1
10. Reported one or more of the above	49.5	53.3	50.7

Source: American College Health Association. 2013. *American College Health Association National College Health Assessment II*. Reference Group Executive Summary Fall 2012. Hanover, MD: American College Health Association.

regulations of the late 1800s, was in fact a "moral crusade" (Gusfield 1963) against immigrant groups who were more likely to use alcohol. The amendment had little popular support and was repealed in 1933. Today, the U.S. population is experiencing a resurgence of concern about alcohol. What has been called a "new temperance" has manifested itself in federally mandated 21-year-old drinking age laws, warning labels on alcohol bottles, increased concern over fetal alcohol syndrome and underage drinking, stricter enforcement of drinking and driving regulations (e.g., checkpoint traffic stops), and zero-tolerance policies. Such practices may have had an effect on drinking norms, particularly for young people. Between 2002 and 2012, the rate of current alcohol use by 12- to 20-year-olds steadily declined (NSDUH 2013).

Despite such restrictive policies, alcohol remains the most widely used and abused drug in the United States. According to a recent poll, 66 percent of U.S. adults drink alcohol, averaging four drinks a week (Saad 2012). Although most people who drink alcohol do so moderately and experience few negative effects (see this chapter's *Self and Society* feature), people with alcoholism are psychologically and physically addicted to alcohol and suffer various degrees of physical, economic, psychological, and personal harm.

The National Survey on Drug Use and Health, conducted by the U.S. Department of Health and Human Services, reported that in 2011 about half of Americans age 12 and older consumed alcohol at least once in the month preceding the survey; that is, they were *current users* (NSDUH 2013). Of this number, 6.5 percent reported **heavy drinking**, and 22.6 percent—59.7 million people—reported **binge drinking**.

Even more troubling were the nearly 10 million current users of alcohol who were 12 to 20 years old—underage drinkers—many of whom got their alcohol for free from adults—often their parents. Nearly 15.3 percent were binge drinkers and 4.3 percent were heavy drinkers (NSDUH 2013). White males from the Northeast region of the United States were the most likely to be underage drinkers, and 80.1 percent of all 12- to 20-year-old current drinkers reported being with two or more people when consuming alcohol (NSDUH 2013).

heavy drinking As defined by the U.S. Department of Health and Human Services, five or more drinks on the same occasion on each of five or more days in the past 30 days prior to the National Survey on Drug Use and Health.

binge drinking As defined by the U.S. Department of Health and Human Services, drinking five or more drinks on the same occasion on at least one day in the past 30 days prior to the National Survey on Drug Use and Health.

Although teen drinking has decreased in recent years, in part as a result of reduced perceived availability of alcohol (MTF 2013), binge drinking in college continues to attract the public's attention. The likelihood of a college student binge drinking is impacted by environmental variables including place of residence (e.g., on campus versus off campus); cost and availability of alcohol; campus, local, and state alcohol policies; age, gender, and ethnic and racial makeup of the student population; prevention strategies; and the college drinking culture (Wechsler & Nelson 2008). Moreover, research indicates that college students in fraternities and sororities drink more than non-Greek students (Chauvin 2012).

What Do You Think? There are many different drinking games but until recently, just how many was unknown. Using a web-based survey of over 3,400 college students who consumed at least one drink a week, LaBrie et al. (2013) identified 100 drinking games and classified them into five categories: communal games, chance games, targeted and skill games, competition games, and extreme consumption games, the latter being the most dangerous because they encourage binge drinking. If there were no drinking games, do you think college students would drink just as heavily, or do drinking games simply provide the opportunity?

Many binge drinkers began drinking in high school, with almost one-third having their first drink before age 13. Research indicates that the younger the age of onset, the higher the probability that an individual will develop a drinking disorder at some time in his or her life (Behrendt et al. 2009; Hingson et al. 2006; NSDUH 2013). For example, an individual's chance of becoming dependent on alcohol is 40 percent if the person's drinking began before the age of 13. Additional results from the National Survey on Drug Use and Health (2012) include the following:

Research indicates that the younger the age of onset, the higher the probability that an individual will develop a drinking disorder at some time in his or her life.

- The highest levels of current, heavy, and binge drinking are among 21- to 25-year-olds; those between 12 and 17 years of age and older than the age of 65 had the lowest rates of binge drinking.
- Rates of alcohol use are higher among full-time employed adults than among the unemployed; however, the rates of binge drinking are similar between the employed and the unemployed.
- Current, heavy, and binge drinking rates for full-time college students is higher than for those not enrolled in college full-time.
- Asians are the least likely to report binge drinking, and American Indians or Alaska Natives the most likely to report it (see Figure 3.2).
- Underage current drinkers, i.e., those between the ages of 12 and 20, were most likely from the Northeast, followed by the Midwest, West, and South.
- More males than females age 12 to 20 reported binge drinking, heavy drinking, and current alcohol use.

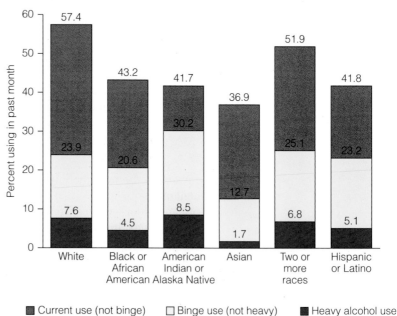

Note: Due to low precision, estimates for Native Hawaiians or Other Pacific Islanders are not shown.

Figure 3.2 Current, Binge, and Heavy Alcohol Use among People Aged 12 or Older, by Race/Ethnicity: 2012
Source: NSDUH 2013.

Researchers have long questioned the relationship between gender and drinking behavior, including binge drinking. After administering two measures of gender identity to a sample of college students, Peralta et al. (2010) concluded that, "[M]ales, who are socialized to be masculine, may rely on heavy alcohol use to coincide with other forms of male-associated behaviors (e.g., sport, risk-taking). Women, who are socialized to be feminine, on the other hand, may not engage in heavy drinking practices because this is not a part of normative femininity expression rituals" (p. 377).

What Do You Think? It's clear that drinking norms vary between males and females but there is some evidence that may be changing. A report by the Centers for Disease Control and Prevention indicates that one in five high school girls report binge drinking, and one in eight adult women report binge drinking, the latter averaging three binges a month and six drinks per binge (CDC 2013b). Why do you think binge drinking has increased among women and girls?

The Tobacco Crisis

Native Americans first cultivated tobacco and introduced it to the European settlers in the 1500s. The Europeans believed that tobacco had medicinal properties, and its use spread throughout Europe, ensuring the economic success of the colonies in the New World. Tobacco was initially used primarily through chewing and snuffing, but smoking became more popular in time, even though scientific evidence that linked tobacco smoking to lung cancer existed as early as 1859 (Feagin & Feagin 1994). However, the U.S. Surgeon General did not conclude that tobacco products are addictive and that nicotine causes dependency until 1989.

Globally, over 80 percent of the nearly one billion smokers in the world live in low- or middle-income countries (WHO 2013). Using what is called **meta-analysis**, researchers examined 125 scientific papers that included 31,146,096 respondents worldwide (WHO 2011a). The results are counterintuitive. Despite the obvious additional cost of using tobacco products, the analysis found strong evidence of an inverse relationship between income and smoking, i.e., as income goes down, the prevalence of smoking goes up—a relationship that is stronger for younger age groups than older. The authors suggest a four-stage explanatory model:

> In earlier stages, smoking disseminates among higher-income groups who are more open to innovation. During the intermediate stages, smoking diffuses to the rest of the population. Later, smoking declines among the high-income level strata, as they are concerned with health, fitness, and the harm of smoking. Only after a long history of cigarette consumption, when all SES [socioeconomic status] groups have been similarly exposed to smoking, does the inverse social status gradient emerge. (p. 27)

Tobacco is one of the most widely used drugs in the United States. According to a U.S. Department of Health and Human Services survey, 69.5 million Americans—26.7 percent of those 12 and older—were current tobacco users in 2012 (NSDUH 2013). Current use of all tobacco products, including smokeless tobacco (9 million users), cigars (13.4 million users), pipe tobacco (1 million users), and cigarettes (57.5 million), is higher for high school graduates than for college graduates, for unemployed rather than employed, and males, Americans Indians and Alaska Natives. In general, with the exception of pipe smoking and smokeless tobacco use, tobacco use has steadily decreased in each tobacco category since 2002 (NSDUH 2013).

In 2012, 8.6 percent of the 12- to 17-year-old population reported use of a tobacco product in the past month (NSDUH 2013). Research evidence suggests that youth develop attitudes and beliefs about tobacco products at an early age (Freeman et al. 2005). Advertising of tobacco products continues to have an influence on youth despite the 2009 Family Smoking Prevention and Tobacco Control Act which, among other provisions, outlawed flavored cigarettes most often marketed to children (see "Government Regulation" in the

meta-analysis Meta-analysis combines the results of several studies addressing a research question; i.e., it is the analysis of analyses.

"Strategies for Action" section). Tobacco company executives, however, have argued that the 2009 act only covers "cigarettes, cigarette tobacco, roll-your-own tobacco, and smokeless tobacco," and that cigars are excluded from the control of the Food and Drug Administration (Myers 2011, p. 1). Several tobacco companies are now selling small, inexpensive, sweet-flavored cigars and, between 2000 and 2012, cigar use more than doubled (Campaign for Tobacco Free Kids 2013). The success of the tobacco industry in marketing "cigarillos" or "blunts" to children and young adults is clear. High school students are now twice as likely as adults to report current cigar use (Campaign for Tobacco Free Kids 2013).

There is also considerable evidence that tobacco advertisers target minorities. Primack et al. (2007) found that tobacco advertisements in African American communities were 2.6 times higher per person than in white communities. Further, the likelihood of tobacco-related billboards was 70 percent higher in African American than in white communities. The Food and Drug Administration's Tobacco Products Scientific Advisory Committee reports that "menthol cigarettes are marketed disproportionately to younger smokers" *and* "disproportionately marketed per capita to African Americans" (FDA 2011, p. 40–41). Finally, the tobacco industry has a long history of targeting women. Researchers at Stanford University collected over 1,500 cigarette advertisements from historical and contemporary magazines and newspapers. They concluded that marketing cigarettes to women and, today, girls has always been tied to body image and the evolving role of women. For example, during the second wave of feminism in the 1960s, Philip Morris's "You've Come a Long Way, Baby" campaign was developed for Virginia Slims. One researcher noted that, even today, ". . . women-targeted cigarette brands are almost universally promoted as slender, thin, slim, lean, or light. Some brands have even gone so far as to recommend 'cigarette diets'" (quoted in Marine-Street 2012, p. 1). Women in developing countries are also being targeted:

> Because most women currently do not use tobacco, the tobacco industry aggressively markets to them to tap this potential new market. Advertising, promotion, and sponsorship, including charitable donations to women's causes, weaken cultural opposition to women using tobacco. Product design and marketing, including the use of attractive models in advertising and brands marketed specifically to women, are explicitly crafted to encourage women to smoke. (WHO 2008, p. 16–17)

AP Images/Hans Pennink

Small cigars are not regulated by the Food and Drug Administration, and marketing often targets youth. In 2013, legislation was introduced into the U.S. House of Representatives that, if passed, would also exclude traditionally large and "premium" cigars from FDA oversight—flavored or not.

Frequently Used Illegal Drugs

More than 23.9 million people in the United States are current illicit drug users (see Figure 3.3), representing 9.2 percent of the population age 12 and older. Users of illegal drugs, although varying by type of drug used, are more likely to be male, to be young, and to be a member of a minority group (NSDUH 2013).

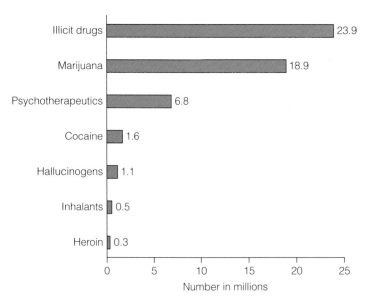

Illicit drugs — 23.9
Marijuana — 18.9
Psychotherapeutics — 6.8
Cocaine — 1.6
Hallucinogens — 1.1
Inhalants — 0.5
Heroin — 0.3

Number in millions

Figure 3.3 Past-Month Illicit Drug Use among People Aged 12 or Older, 2012
Source: NSDUH 2013.

According to the most recent Monitoring the Future survey (MTF 2013), between 2001 and 2007, use of any illicit drug in the past year declined for 8th, 10th, and 12th grade students, but has generally increased since that time.

Marijuana Madness

Marijuana is the most commonly used and most heavily trafficked illicit drug in the world (see Table 3.1 for a list of commonly abused drugs, their commercial and street names, and their intoxication and health effects). Globally, there are over 180 million marijuana users representing nearly 4 percent of the world's 15- to 64-year-old population. Regionally, marijuana is also the most dominant illicit drug, and its consumption is particularly high in west and central Africa, Australia and New Zealand, and North America. The largest producers of marijuana in the world are Morocco, Afghanistan, India, Lebanon, and Pakistan (WDR 2013).

Marijuana's active ingredient is THC (Δ^9-tetrahydrocannabinol), which in varying amounts can act as a sedative or a hallucinogen. When just the top of the marijuana plant is sold, it is called hashish. Hashish is much more potent than marijuana, which comes from the entire plant. Marijuana use dates back to 2737 B.C. in China, and marijuana has a long tradition of use in India, the Middle East, and Europe. In North America, hemp, as it was then called, was used to make rope and as a treatment for various ailments. Nevertheless, in 1937, Congress passed the Marijuana Tax Act which restricted the use of marijuana; the law was passed as a result of a media campaign that portrayed marijuana users as "dope fiends." In the only congressional hearing on the act, the then commissioner of narcotics testified that:

> There are 100,000 total marijuana users in the U.S. and most are Negroes, Hispanics, Filipinos, and entertainers. Their Satanic music, jazz and swing, result from marijuana use. This marijuana causes white women to seek sexual relations with Negroes, entertainers, and any others. The primary reason to outlaw marijuana is its effect on degenerative races. Marijuana is an addictive drug which produces in its users insanity, criminality, and death. You smoke a joint and you're likely to kill your brother. Marijuana is the most violence-causing drug in the history of mankind. (quoted in Rabinowitz & Lurigio 2009)

More than 18.9 million current marijuana users are in the United States, representing 7.3 percent of the U.S. population age 12 and older (NSDUH 2013). According to the most recent Monitoring the Future (MTF) survey, between 2010 and 2012, *daily marijuana use* decreased for 8th and 10th graders and remained the same for high school seniors (MTF 2013). Perceived risk of marijuana use decreased in all three grades. However, when teens are asked about the regular use of marijuana, 71 percent say that it may "mess up their life" (PATS 2013).

In the present debate over the legalization of marijuana, many express fears that it is a **gateway drug**, the use of which causes progression to other drugs. "The gateway hypothesis holds that consumption of abusable drugs progresses in orderly fashion through several discrete stages. The entire sequence, which is exhibited by only a small minority of drug users, begins with beer or wine and moves progressively through hard liquor or tobacco, marijuana, and finally hard drugs" (Tarter et al. 2006, p. 2,134). Most research suggests, however, that people who experiment with one drug are more likely to experiment with another. Indeed, most drug users use several drugs at the same time. As Lee and Abdel-Ghany (2004) note, there is a strong "contemporaneous relationship between smoking cigarettes, drinking alcohol, smoking marijuana, and using cocaine" (p. 454). That in mind, in 2012, use of illicit drugs was predictive of smoking cigarettes and alcohol consumption (MTF 2013).

Cocaine: From Coca-Cola to Crack

Cocaine is classified as a stimulant and, as such, produces feelings of excitation, alertness, and euphoria. Although prescription stimulants such as methamphetamine and dextroamphetamine are commonly abused, societal concern over drug abuse has focused on cocaine over the last 20 years. Its increased use, addictive qualities, physiological

gateway drug A drug (e.g., marijuana) that is believed to lead to the use of other drugs (e.g., cocaine).

TABLE 3.1 Commonly Abused Drugs

Substances: Category and Name	Examples of *Commercial* and Street Names	DEA Schedule*/ How Administered**	*Acute Effects*/Health Risks
Tobacco			
Nicotine	Found in cigarettes, cigars, bidis, and smokeless tobacco (snuff, spit tobacco, chew)	Not scheduled/smoked, snorted, chewed	*Increased blood pressure and heart rate*/chronic lung disease; cardiovascular disease; stroke; cancers of the mouth, pharynx, larynx, esophagus, stomach, pancreas, cervix, kidney, bladder, and acute myeloid leukemia; adverse pregnancy outcomes; addiction
Alcohol			
Alcohol (ethyl alcohol)	Found in liquor, beer, and wine	Not scheduled/swallowed	*In low doses, euphoria, mild stimulation, relaxation, lowered inhibitions; in higher doses, drowsiness, slurred speech, nausea, emotional volatility, loss of coordination, visual distortions, impaired memory, sexual dysfunction, loss of consciousness*/increased risk of injuries, violence, fetal damage (in pregnant women); depression; neurologic deficits; hypertension; liver and heart disease; addiction; fatal overdose
Cannabinoids			
Marijuana	Blunt, dope, ganja, grass, herb, joint, bud, Mary Jane, pot, reefer, green, trees, smoke, sinsemilla, skunk, weed	I/smoked, swallowed	*Euphoria; relaxation; slowed reaction time; distorted sensory perception; impaired balance and coordination; increased heart rate and appetite; impaired learning, memory; anxiety; panic attacks; psychosis*/cough, frequent respiratory infections; possible mental health decline; addiction
Hashish	Boom, gangster, hash, hash oil, hemp	I/smoked, swallowed	
Opioids			
Heroin	*Diacetylmorphine:* smack, horse, brown sugar, dope, H, junk, skag, skunk, white horse, China white; cheese (with OTC cold medicine and antihistamine)	I/injected, smoked, snorted	*Euphoria; drowsiness; impaired coordination; dizziness; confusion; nausea; sedation; feeling of heaviness in the body; slowed or arrested breathing*/constipation; endocarditis; hepatitis; HIV; addiction; fatal overdose
Opium	*Laudanum, paregoric:* big O, black stuff, block, gum, hop	II, III, V/swallowed, smoked	
Stimulants			
Cocaine	*Cocaine hydrochloride:* blow, bump, C, candy, Charlie, coke, crack, flake, rock, snow, toot	II/snorted, smoked, injected	*Increased heart rate, blood pressure, body temperature, metabolism; feelings of exhilaration; increased energy, mental alertness; tremors; reduced appetite; irritability; anxiety; panic; paranoia; violent behavior; psychosis*/weight loss, insomnia; cardiac or cardiovascular complications; stroke; seizures; addiction
Amphetamine	*Biphetamine, Dexedrine:* bennies, black beauties, crosses, hearts, LA turnaround, speed, truck drivers, uppers	II/swallowed, snorted, smoked, injected	**Also, for cocaine**—nasal damage from snorting
Methamphetamine	*Desoxyn:* meth, ice, crank, chalk, crystal, fire, glass, go fast, speed	II/swallowed, snorted, smoked, injected	**Also, for methamphetamine**—severe dental problems
Club Drugs			
MDMA (methylenedioxy-methamphetamine)	Ecstasy, Adam, clarity, Eve, lover's speed, peace, uppers	I/swallowed, snorted, injected	**MDMA**—*mild hallucinogenic effects; increased tactile sensitivity; empathic feelings; lowered inhibition; anxiety; chills; sweating; teeth clenching; muscle cramping*/sleep disturbances; depression; impaired memory; hyperthermia; addiction
Flunitrazepam***	*Rohypnol:* forget-me pill, Mexican Valium, R2, roach, Roche, roofies, roofinol, rope, rophies	IV/swallowed, snorted	**Flunitrazepam**—*sedation; muscle relaxation; confusion; memory loss; dizziness; impaired coordination*/addiction

(Continued)

TABLE 3.1 *(Continued)*

Substances: Category and Name	Examples of *Commercial* and Street Names	DEA Schedule*/ How Administered**	*Acute Effects*/Health Risks
GHB***	*Gamma-hydroxybutyrate:* G, Georgia home boy, grievous bodily harm, liquid Ecstasy, soap, scoop, goop, liquid X	I/swallowed	**GHB**—*drowsiness; nausea; headache; disorientation; loss of coordination; memory loss*/unconsciousness; seizures; coma

Dissociative Drugs

Ketamine	*Ketalar SV:* cat Valium, K, Special K, vitamin K	III/injected, snorted, smoked	*Feelings of being separate from one's body and environment; impaired motor function*/anxiety; tremors; numbness; memory loss; nausea
PCP and analogs	*Phencyclidine:* angel dust, boat, hog, love boat, peace pill	I, II/swallowed, smoked, injected	**Also, for ketamine**—*analgesia; impaired memory; delirium; respiratory depression and arrest; death*
Salvia divinorum	Salvia, Shepherdess's Herb, Maria Pastora, magic mint, Sally-D	Not scheduled/chewed, swallowed, smoked	**Also, for PCP and analogs**—*analgesia; psychosis; aggression; violence; slurred speech; loss of coordination; hallucinations*
Dextromethorphan (DXM)	Found in some cough and cold medications: Robotripping, Robo, Triple C	Not scheduled/swallowed	**Also, for DXM**—*euphoria; slurred speech; confusion; dizziness; distorted visual perceptions*

Hallucinogens

LSD	*Lysergic acid diethylamide:* acid, blotter, cubes, microdot yellow sunshine, blue heaven	I/swallowed, absorbed through mouth tissues	*Altered states of perception and feeling; hallucinations; nausea*
Mescaline	Buttons, cactus, mesc, peyote	I/swallowed, smoked	**Also, LSD and mescaline**—*increased body temperature, heart rate, blood pressure; loss of appetite; sweating; sleeplessness; numbness, dizziness, weakness, tremors; impulsive behavior; rapid shifts in emotion*
Psilocybin	Magic mushrooms, purple passion, shrooms, little smoke	I/swallowed	**Also, for LSD**—*Flashbacks, Hallucinogen Persisting Perception Disorder* **Also for psilocybin**—*nervousness; paranoia; panic*

Other Compounds

Anabolic steroids	*Anadrol, Oxandrin, Durabolin, Depo-Testosterone, Equipoise:* roids, juice, gym candy, pumpers	III/injected, swallowed, applied to skin	**Steroids**—*no intoxication effects*/hypertension; blood clotting and cholesterol changes; liver cysts; hostility and aggression; acne; in adolescents—premature stoppage of growth; in males—prostate cancer, reduced sperm production, shrunken testicles, breast enlargement; in females—menstrual irregularities, development of beard and other masculine characteristics
Inhalants	*Solvents (paint thinners, gasoline, glues); gases (butane, propane, aerosol propellants, nitrous oxide); nitrites (isoamyl, isobutyl, cyclohexyl):* laughing gas, poppers, snappers, whippets	Not scheduled/inhaled through nose or mouth	**Inhalants** *(varies by chemical)*—*stimulation; loss of inhibition; headache; nausea or vomiting; slurred speech; loss of motor coordination; wheezing*/cramps; muscle weakness; depression; memory impairment; damage to cardiovascular and nervous systems; unconsciousness; sudden death

Prescription Medications

CNS Depressants			
Stimulants	For more information on prescription medications, please visit http://www.nida.nih.gov/DrugPages/PrescripDrugsChart.html.		
Opioid Pain Relievers			

* Schedule I and II drugs have a high potential for abuse. They require greater storage security and have a quota on manufacturing, among other restrictions. Schedule I drugs are available for research only and have no approved medical use; Schedule II drugs are available only by prescription (unrefillable) and require a form for ordering. Schedule III and IV drugs are available by prescription, may have five refills in 6 months, and may be ordered orally. Some Schedule V drugs are available over the counter.

** Some of the health risks are directly related to the route of drug administration. For example, injection drug use can increase the risk of infection through needle contamination with staphylococci, HIV, hepatitis, and other organisms.

*** Associated with sexual assaults.

Source: NIDA 2012a.

effects, and worldwide distribution have fueled such concerns. More than any other single substance, cocaine led to the early phases of the war on drugs.

Cocaine, which is made from the coca plant, has been used for thousands of years. Coca leaves were used in the original formula for Coca-Cola but, in the early 1900s, anti-cocaine sentiment emerged as a response to the heavy use of cocaine among urban blacks, poor whites, and criminals (Friedman-Rudovsky 2009; Thio 2007; Witters et al. 1992). Cocaine was outlawed in 1914, by the Harrison Narcotics Tax Act, but its use and effects continued to be misunderstood. For example, a 1982 *Scientific American* article suggested that cocaine was no more habit-forming than potato chips (Van Dyck & Byck 1982). As demand and then supply increased, prices fell from $100 a dose to $10 a dose, and "from 1978 to 1987, the U.S. experienced the largest cocaine epidemic in history" (Witters et al. 1992, p. 256).

According to the National Survey on Drug Use and Health, 1.6 million Americans 12 years and older are current cocaine (including crack) users, representing 0.6 percent of that population (NSDUH 2013). **Crack** is a crystallized product made by boiling a mixture of baking soda, water, and cocaine. Over the last decade, cocaine use decreased for 8th, 10th, and 12th grade students as did use of perceived risk and availability (MTF 2013). The percentage of 12th graders indicating that getting cocaine is "fairly easy" or "very easy" is roughly half what it was in 1988, standing at 30 percent. Over 90 percent of seniors disapprove of using cocaine "once or twice," and 45 percent of seniors believe that such use would put them at "great risk" (MTF 2013).

Methamphetamine: The Meth Epidemic

Methamphetamine is a central nervous system stimulant that is highly addictive. Although the drug has only recently become popular, it is not new:

> During the Second World War, soldiers on both sides used it to reduce fatigue and enhance performance. Hitler was widely believed to be a meth addict. Later, in the 1960s, President John Kennedy also used the drug and soon it caught on among so-called "speed freaks." But, because it was extremely expensive as well as difficult to obtain, meth was never close to being as widely used as cocaine. (Thio 2007, p. 276)

The use of methamphetamine has steadily decreased over the last decade, peaking in 1981. Today, use is fairly low, with about 1.0 percent of 8th, 10th, and 12th graders reporting lifetime use (MTF 2013). Use of crystal meth, the crystalline rather than powdered form of methamphetamine, is also quite low as is its perceived availability. Adolescent beliefs that crystal meth is a "great risk" if used once or twice has increased over the last decade (MTF 2013). The prevalence of monthly methamphetamine use in the United States has remained stable between 2007 and 2012, remaining at about 0.2 percent of the 12-year-old and older population (NSDUH 2013).

Because methamphetamine can be made from cold medications such as Sudafed, the U.S. Congress passed the Comprehensive Methamphetamine Control Act of 1996 that made obtaining the chemicals needed to make methamphetamine more difficult (ONDCP 2006; Thio 2007). In 2006, the Combat Methamphetamine Epidemic Act, which further articulated standards for selling over-the-counter medications used in methamphetamine production, went into effect. Since that time, several states have passed regulations that require a prescription to obtain ephedrine and pseudoephedrine, the key

crack A crystallized illegal drug product produced by boiling a mixture of baking soda, water, and cocaine.

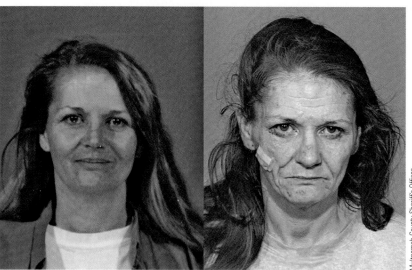

For those who use methamphetamine, the physical transformation is remarkable. The time lapse between the before (left) and after (right) pictures of this methamphetamine user is only three years, five months.

ingredients in methamphetamine, and with some success. The number of methamphetamine labs has declined in these states (GAO 2013).

> . . . [S]everal states have passed regulations that require a prescription to obtain ephedrine and pseudoephedrine, the key ingredients in methamphetamine, and with some success. The number of methamphetamine labs has declined in these states.

Globally, seizures of crystalline methamphetamine rose to their highest levels since 2007, sparking fears of increased use of the drug, particularly in East and Southeast Asia. For example, over one ton of crystal meth was seized in Malaysia, Indonesia, and Thailand in 2012. Further, although use of methamphetamine in western and central Europe was predominantly in the Czech Republic and Slovakia, reports of methamphetamine smoking and increased availability of crystal meth has spread to the Baltic States (e.g., Albania) and to northern Europe (WDR 2013).

Heroin: The White Horse

Heroin is an analgesic—that is, a painkiller—and is the most commonly abused class of drugs called opiates. Highly addictive, heroin can be injected, snorted, or smoked, and when used in conjunction with cocaine, it is called a "speedball." Use of the "white horse"—as it is often called—is highest in Afghanistan, a major opium-producing country, and Iran. Opium is trafficked from Afghanistan through the Balkan States and on to Europe and the coasts of Africa. Heroin use in western Europe is declining in large part due to an aging population, the number of people in treatment, and decreased availability (WDR 2013).

In the United States, heroin is used by approximately 0.1 percent of the population 12 and over and has steadily increased since 2009 (NSDUH 2013). However, since 2005, heroin use among 8th, 10th, and 12th graders has declined, as has the perception of availability. Nonetheless, heroin use in the suburbs has increased dramatically (Murray 2012), and not by accident. Rather, it is the plan of drug lords from Mexico and Colombia, who strategically market the drug to Middle America with new, sophisticated techniques. Packets of heroin are now stamped with popular brand names like Chevrolet or Prada, or marketed using blockbuster movie names aimed at young people, like the *Twilight* series. Dealers even give it away for free in the suburbs at first. Once the kids are hooked, they sell it to them, dirt cheap. In fact, kids can buy a small bag of heroin for as little as $5. It's cheaper than movie tickets or a six-pack of beer (Alfonsi & Siegel 2010, p. 1). Further, recent interviews with teen heroin addicts suggest an emerging pattern of addiction. Teens become addicted to prescription drugs and, once they can no longer afford the price of the pills, they changed to the cheaper drug—heroin (Murray 2012).

Prescription Drugs

Worldwide, the use of psychotherapeutic drugs is a growing health problem (WDR 2013). However, U.S. lifetime use of **psychotherapeutic drugs**—that is, nonmedical use of any prescription pain reliever, stimulant, sedative, or tranquilizer—has remained stable since 2002 (NSDUH 2013). Approximately 6.8 million people, 2.6 percent of the U.S. population 12 and older, reported current use of a psychotherapeutic drug in 2012. Of these users, 4.9 million used pain relievers, 2.1 million used tranquilizers, 1.2 million used stimulants, and 270,000 used sedatives.

A 2013 report by the MetLife Foundation in conjunction with the Partnership at Drug-Free focuses on the nonmedical use of prescription drugs among teens and parental attitudes toward psychotherapeutic drugs (PATS 2013). The sample included 3,884 teenagers in grades 9 through 12 and 817 parents. The results indicated that nonmedical prescription drug use among teens is a real and growing concern. In 2012:

psychotherapeutic drugs The nonmedical use of any prescription pain reliever, stimulant, sedative, or tranquilizer.

- Nearly a quarter of the teens sampled reported lifetime nonmedical use of a prescription drug—a 33 percent increase from 2008.
- Of those who reported nonmedical use of a prescription drug, 20 percent reported using a prescription drug before the age of 14.

- One in eight teens reported nonmedical lifetime use of Ritalin or Adderall.
- Twenty-five percent of teens and 29 percent of parents believe that some prescription drugs can help students in school.
- Both parents and teens believe that nonmedical use of prescription drugs is safer than using "street drugs."
- When teens were asked why they use prescription drugs, most said to relax, to have fun, and because it feels good.
- Ten percent of teens reported nonmedical use of a pain reliever, including Vicodin and OxyContin, in the previous year.

AP Images/The Republic/Andrew Laker

Too many people flush their unused prescriptions down the toilet, where they may enter public sources, or put them in the trash where they can be found. To avoid these and other problems with improper prescription drug disposal, law enforcement in many cities and counties have organized "take-back days."

As Figure 3.4 indicates, over half of the U.S. population 12 and older who reported abuse of pain relievers, the largest single category of abused prescription drugs, reported receiving the drug from friends or relatives for free (NSDUH 2013). Of those who received the drug from a friend or relative for free, 82.2 percent of the friends or relatives reported obtaining the pills from a physician. Other lesser sources include taking them or receiving them for free from a friend or relative, multiple doctors, buying them on the Internet, and buying them from a drug dealer/stranger.

What Do You Think? Inhalants are in everyone's home. They are adhesives (e.g., rubber cement), food products (e.g., vegetable cooking spray), aerosols (e.g., hair spray and air fresheners), anesthetics (ether), gases (e.g., butane), and cleaning agents (e.g., spot remover)—over 1,000 household products in total. "Huffing" is a serious problem and can lead to sudden sniffing death syndrome but, nonetheless, the availability of inhalants cannot be legally controlled. What would you do to attack this potentially deadly practice?

Synthetic Drugs

Concern over **synthetic drugs**, i.e., a category of drugs that are "designed" in laboratories rather than naturally occurring in plant material, is growing, and has led President Obama to sign the Synthetic Drug Abuse Prevention Act of 2012. Although there are many types of synthetic drugs, some of the more popular are synthetic marijuana ("K2" or "Spice"), synthetic stimulants ("bath salts"), and synthetic hallucinogens (LSD, Ecstasy/MDMA).

Synthetic Marijuana. Globally, the use of synthetic marijuana is higher in the United States than in other parts of the world. In a national survey in Spain, of 25,000 14- to 18-year-olds, past-year use of synthetic marijuana was a little under one percent (EMCDDA 2013). The comparable rate for U.S. 9th through 12th grade students was 12 percent, 3 percent of whom had not used organic marijuana during the same time period (PATS 2013). As with the use of marijuana in general, Hispanic teens are more likely to use synthetic marijuana than non-Hispanics teens.

"Bath Salts." Often sold under such innocuous-sounding names as Zoom, Cloud Nine, Blue Silk, and Hurricane Charlie, bath salts are highly addictive synthetic stimulants (Bellum 2013). Past-year use of bath salts by 9th through 12th graders, at 3 percent, was less than past-year use of synthetic marijuana (PATS 2013). The use of bath salts is associated with feelings of euphoria but also results in paranoia, rapid heart rate and chest pains, hallucinations, and suicidal thoughts (McMillen 2011). Calls to poison control centers about bath salts increased dramatically between 2010 and 2011, but have since declined although they remain above

synthetic drugs A category of drugs that are "designed" in laboratories rather than naturally occurring in plant material.

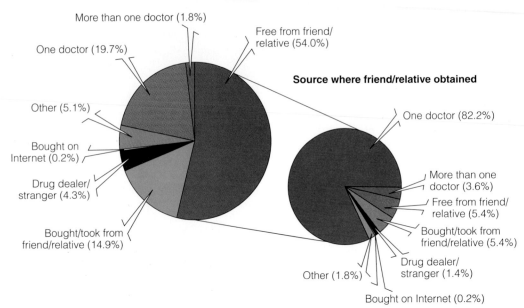

Source where user obtained

More than one doctor (1.8%)

One doctor (19.7%)

Other (5.1%)

Bought on Internet (0.2%)

Drug dealer/ stranger (4.3%)

Bought/took from friend/relative (14.9%)

Free from friend/ relative (54.0%)

Source where friend/relative obtained

One doctor (82.2%)

More than one doctor (3.6%)

Free from friend/ relative (5.4%)

Bought/took from friend/relative (5.4%)

Drug dealer/ stranger (1.4%)

Other (1.8%)

Bought on Internet (0.2%)

Figure 3.4 Where Pain Relievers Were Obtained for Most Recent Nonmedical Use among Past-Year Users Aged 12 or Older, 2012
Source: NSDUH 2013.

Throughout much of the developing world, homeless children "huff" glue to escape hunger pains and the horrors of living on the street. Globally, UNICEF estimates that there are more than 200 million homeless children—the equivalent of two-thirds the population of the United States.

2010 rates. After the Drug Enforcement Administration (DEA) banned some of the chemicals used to produce bath salts, calls to poison control centers declined (McLaughlin 2012). Bath salts sell for between $25 and $50 per 50-milligram packet (Goodnough & Zezima 2011).

Hallucinogens. Hallucinogen use among adults has remained fairly constant over the last decade (NSDUH 2013). LSD is a synthetic hallucinogen, although many other hallucinogens are produced naturally (e.g., salvia). It became popular in the 1960s and 1970s, as part of the counterculture revolution. Use has declined significantly since that time, with lifetime use for 8th, 10th, and 12th graders combined at 2.5 percent, less than half of lifetime use of Ecstasy (MTF 2013). Recent investigations have revealed that between 1955 and 1975, U.S. military officials tested experimental drugs such as LSD on Army volunteers who were told "they would be test[ing] new Army field jackets, clothing, weapons, and things of that nature . . . no mention of drugs or chemicals" (quoted in Martin 2012).

Globally, over 10.5 million people between the ages of 15 and 64 use Ecstasy, the most common name for MDMA, at least once a year. Although use of Ecstasy is declining worldwide, the most recent World Drug Report (2013) available notes that it appears to be increasing in Europe. In 2012, 3.8 percent of U.S. high school seniors reported past-year use of Ecstasy, the most common hallucinogen to be used (MTF 3013). A fairly new drug, "Molly," short for molecule, is thought to be a purer form of MDMA than Ecstasy, although as of this writing, its chemical makeup is unclear and there are no national surveys available to assess use rates. There is some evidence that users are young, most often between 16 to 24 years old (Csomor 2012).

Societal Consequences of Drug Use and Abuse

Drugs are a social problem not only because of their adverse effects on individuals but also because of the negative consequences their use has for society as a whole. Everyone is a victim of drug abuse. Drugs contribute to problems within the family (although families can also lower the likelihood of abusive drug use—see this chapter's *Social Problems Research Up Close*) and to escalating crime rates, are tremendously costly, and place a heavy strain on the environment. Drug abuse also has serious consequences for health at both the individual and the societal level.

The Cost to Children and Family

The cost of drug abuse to families is incalculable. It is estimated that, in the United States, one in 10 children under the age of 18 lives with at least one parent in need of treatment for drug or alcohol dependency (SAMHSA 2009). Children raised in such homes are more likely to (1) live in an environment riddled with conflict, (2) have a higher probability of physical illness including injuries or death from an automobile accident, and (3) be victims of child abuse and neglect (SAMHSA 2007; 2009). Children of alcoholics, the clear majority of children who live in homes with a drug dependent parent(s), are at risk for anxiety disorders, depression, and problems with cognitive and verbal skills (SAMHSA 2012). Children of alcoholics, and particularly female children of alcoholics, suffer from "significant mental health consequences . . . that persist far into adulthood" (Balsa et al. 2009, p. 55).

Parents who report abusing alcohol in the past year are also more likely to report cigarette and illicit drug use than parents who did not report alcohol abuse in the previous year. They were also more likely to report "household turbulence," including yelling, serious arguments, and violence (NSDUH 2004). Moreover, alcohol abuse is the single most common variable associated with wife abuse (Flanzer 2005). For example, a study of 634 couples who were assessed at the time of their marriage and again at the time of their first, second, and fourth anniversaries, indicates that both husbands and wives were impacted by the alcoholism of their opposite-sex parent (Kearns-Bodkin & Leonard 2008):

> For husbands, alcoholism in the mother was associated with lower marital satisfaction. For wives, alcoholism in the father was related to lower marital intimacy. Husbands' physical aggression was influenced by his mother's and father's alcoholism; high levels of physical aggression were present among men with alcoholic mothers and nonalcoholic fathers. Interestingly, wives' experience of her husband's aggression was also highest among women with alcoholic mothers and nonalcoholic fathers.

The family, however, is not the only environment in which drug use is harmful to children. A survey of over 1,000 teens between the ages of 12 and 17 indicates that 32 percent of middle school children and 60 percent of high school students attend schools that are "drug infected," and attending "drug-infected" schools was associated with higher self-reported drug use. High levels of stress, the most common source being academic pressure, were also associated with higher student drug use. Not surprisingly, when asked, teens responded that the most important problem they face today is alcohol, tobacco, and illicit drug use (CASA 2012).

Crime and Drugs

An examination of the drug use among probated and paroled offenders indicates much higher use and dependence rates than in nonoffender populations. In 2011, over a quarter of offenders who were on parole (or otherwise supervised) or on probation were current illicit drug users (NSDUH 2013). Further, juvenile delinquency is associated with drug use. In an interdisciplinary study of 1,354 male and female serious juvenile offenders between the ages of 14 and 18 seven years after conviction, Mulvey (2011) concluded that

There's some debate over the relationship between onset of drinking, parents' alcohol consumption patterns, and the context in which drinking first occurs. For example, some research suggests that the more alcohol consumed in the home, the higher the likelihood that children will grow up and engage in excessive drinking behavior. Another body of research suggests that when children use alcohol in the home, they're less likely to abuse it in the future. The following ethnography looks at the relationship between onset of drinking, parental drinking behavior, type of alcohol consumed, and the social context in which drinking occurs.

Sample and Methods

The researchers used "anthropological ethnographic interviewing" to "understand the world as seen by the respondent within the context of the respondent's everyday life" (Strunin et al. 2010, p. 346). The sample was composed of 160 Italian youth and young adults—half female and half male—divided into two age groups, 16- to 18-year-olds and 25- to 30-year-olds, individuals in each age group having a variety of consumption rates.

To identify prospective subjects, an alcohol screening test was administered in two high school locations for current and former students. Further, to assure the collection of comparable data, an interview guide was developed to make sure that all topics were covered and that the questions were asked in a specific order. A quantity–frequency index was used to measure self-reported drinking behavior. Lastly, students were asked about their family histories, extended family members, as well as questions about their beliefs, norms, and behaviors related to alcohol use. The interviews were audiotaped, transcribed, and translated into English.

Findings and Conclusions

Youth who described drinking as part of their Italian heritage also described it in terms of being "culturally normative." For example, a young female adult described her parents' attitude toward alcohol in the following way: ". . . a simple attitude. Alcohol is a pleasure, and has to be lived as something positive, this is the type of values we have at home. A wine culture, wine at meals for the pleasure of it, no abuse because that can cause health problems" (p. 349).

This research points to the importance of the social context in which drinking takes place. For example, even though parents who allowed their children to drink with meals may be seen as permissive, they were also more likely to talk to their children about the dangers of alcohol, including such health issues as "addiction, liver, and heart diseases . . . and the dangers of drinking and driving" (p. 352). Further, youth who were allowed to drink with meals in their homes were more likely to have their first drink within a family setting and less likely to mix types of alcohol.

Alternatively, the first drink experiences of those not allowed to drink at home included drinking large amounts of more than one type of alcohol and getting drunk.

There are also differences in the likelihood of binge drinking, defined as drinking five or more drinks on one occasion, and age of getting drunk for the first time. Those who consumed alcohol with meals in the home delayed binge drinking and getting drunk when compared to respondents who were not allowed to drink in the home. Further, only those who did not drink alcohol in the home got drunk on their first occasion of drinking.

The authors conclude that, "the introduction of alcohol, typically wine, in a family setting may protect against harmful drinking including binge drinking and drunkenness" (Strunin et al. 2010, p. 354). The researchers also suggest that many of their findings can be generalized to other populations. For example, the United States does not have a tradition of permitting children to consume alcohol with meals and, perhaps consequently, we have one of the highest adolescent consumption rates in the world. Moderate drinking at home within the context of the family setting could lower excessive drinking rates in the United States and teach healthier drinking patterns.

Source: Strunin, Lee, Kirstin Lindeman, Enrico Tempesta, Pierluigi Ascani, Simona Avan, and Luca Parisi. 2010. "Familial Drinking in Italy: Harmful or Protective Factors?" *Addiction Research and Theory* 18(3):344–358.

although substance abuse is a strong predictor of both self-reported delinquent behavior and number of arrests, substance abuse treatment reduces the incidences of drug use and nondrug-related crime.

The relationship between crime and drug use, however, is complex. Sociologists disagree as to whether drugs actually "cause" crime or whether, instead, criminal activity leads to drug involvement. Alternatively, criminal involvement and drug use can occur at the same time; that is, someone can take drugs and commit crimes out of the desire to engage in risk-taking behaviors. Furthermore, because both crime and drug use are associated with low socioeconomic status, poverty may actually be the more powerful explanatory variable.

In addition to the hypothesized crime–drug use link, some criminal offenses are defined by use of drugs: possession, cultivation, production, and sale of controlled substances; public intoxication; drunk and disorderly conduct; and driving while intoxicated. Driving while intoxicated is one of the most common drug-related crimes. In 2012, approximately 11.2 percent of the 12 and older population, over 29.1 million people, drove under the influence of alcohol at least once in the previous year (NSDUH 2013). In 2010, over 10,000 people died in alcohol-related crashes—one-third of all traffic-related deaths.

Alcohol is not the only drug that impairs driving. In 2010, 18 percent of motor vehicle deaths were related to drugs other than alcohol (CDC 2013c), and driving while under the influence of marijuana has surpassed driving while under the use of alcohol among 8th through 12th graders (Liberty Mutual and SADD 2012).

The High Price of Alcohol and Other Drug Use

Smoking-related spending by the government costs the American taxpayers over $70 billion annually—$616 per household (Campaign for Tobacco Free Kids 2013). The cost of alcohol abuse to society exceeded $224 billion in 2006, and nearly 75 percent of that cost was due to binge drinking (CDC 2011a). Further, a comprehensive report by the National Center on Addiction and Substance Abuse at Columbia University (CASA 2009) set the total annual cost of substance abuse and addiction in the United States at $467.7 billion. More importantly, the report contends that, for every dollar spent on drug abuse by federal and state governments

> . . . 95.6 cents went to shoveling up the wreckage and only 1.9 cents on prevention and treatment, 0.4 cents on research, 1.4 cents on taxation or regulation, and 0.7 cents on interdiction. Under any circumstances, spending more than 95 percent of taxpayer dollars on the consequences of tobacco, alcohol, and other drug abuse and addiction and less than 2 percent to relieve individuals and taxpayers of this burden would be considered a reckless misallocation of public funds. In these economic times, such upside-down-cake public policy is unconscionable. (CASA 2009, p. i)

At the federal level, the cost of "shoveling up the wreckage" includes the cost of (1) health care due to substance abuse and addiction (the highest proportion of wreckage spending), (2) adult and juvenile crime (e.g., corrections), (3) child and family assistance programs (e.g., welfare), (4) education (e.g., Safe School Initiatives), (5) public safety (e.g., drug enforcement), (6) mental health and developmental disabilities (e.g., treatment of addiction), and (7) the federal workforce (e.g., loss of productivity). The report concludes that prevention programs must become a priority to reduce the economic costs of drug abuse.

What Do You Think? Although the law was originally designed to protect children from the dangers of methamphetamine labs, in 2013, the Alabama Supreme Court upheld that women who consume drugs during their pregnancy can be arrested for "chemical endangerment" of a child (Weiss 2013). Courts in the majority of other states have avoided similar rulings. Do you think women who use drugs during pregnancy, even if giving birth to a healthy child, should be criminally prosecuted?

Physical Health and Mental Health

Tobacco use is the leading preventable cause of disease and death in the world. Half of all tobacco users—6 million people—will die from its use, and the World Health organization (2013) warns that if something does not change soon, that number could increase to 8 million by 2030 (WHO 2013). Americans who smoke cigarettes are more than twice as likely to develop coronary heart disease and/or to have a stroke, and both men and women have increased risks of developing lung cancer, 23 times more likely and 13 times more likely, respectively (CDC 2012a). It should be noted that the health impact of smoking goes beyond consumption and the effects of secondhand smoke. For example, children who work in tobacco fields are at risk for "green tobacco sickness," a disease caused by the absorption of nicotine through the skin from handling wet tobacco leaves (WHO 2011b).

> Tobacco use is the leading preventable cause of disease and death in the world. Half of all tobacco users—six million people—will die from its use . . . [and] if something does not change soon, that number could increase to 8 million by 2030.

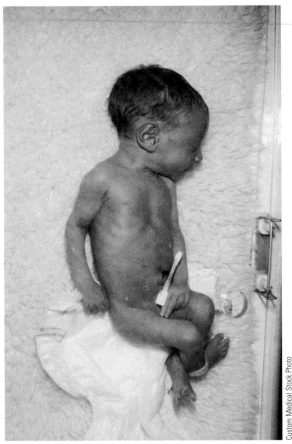

Annually, alcohol abuse is responsible for over 2.5 million deaths, 4 percent of all deaths worldwide (Nebehay 2011). Alcohol kills more people than AIDS, tuberculosis, or violence, and is responsible for 80,000 deaths annually in the United States (WHO 2011c). Excessive alcohol use is associated with a variety of diseases including cirrhosis of the liver, cancer, hypertension, and psychological disorders (CDC 2012b).

Maternal prenatal alcohol use is associated with fetal alcohol spectrum disorders (FASD) in children, a broad term used for a variety of preventable birth defects and developmental disabilities. **Fetal alcohol syndrome** is the most serious of the FASDs and is characterized by serious mental and physical handicaps including facial deformities, growth problems, difficulty communicating, short attention spans, and hearing and vision problems (CDC 2011b).

Use of alcohol during pregnancy does more damage than many illicit drugs. A study of 119 low-income, urban 11-year-olds who had been exposed to tobacco, alcohol, cocaine, and marijuana in the womb indicated that of the four drugs, alcohol alone was associated with lower academic achievement (Rose-Jacobs et al. 2012). That is not to say that ingestion of illegal or prescription drugs during pregnancy isn't harmful. **Neonatal abstinence syndrome** (NAS) is the result of the fetus receiving such drugs during pregnancy through the shared blood stream with the mother, creating drug dependency before the child is even born. At birth, in the absence of the drug, the child goes through withdrawal.

Globally, the use of illicit drugs was associated with an estimated 211,000 deaths in 2011. Of the world's 15- to 64-year-old population, illicit drug use accounts for 1 in 20 deaths in North America and Oceania, 1 in 100 deaths in Asia, 1 in 110 deaths in Europe, 1 in 150 deaths in Africa, and 1 in 200 deaths in South America (WDR 2012). Deaths from illicit drug use are disproportionately of relatively young people; 90 percent of overdose deaths in Europe are of people under the age of 25 (EMCDDA 2013). Illicit drug use is also associated with a variety of diseases. For example, injecting drug users are at a higher risk for hepatitis B, hepatitis C, and HIV (WDR 2013).

Custom Medical Stock Photo

Fetal alcohol syndrome includes a mix of birth defects: mild to severe mental retardation, low birth weight, central nervous system dysfunction, and malformations of the skull and face. This syndrome is a result of women consuming large amounts of alcohol during pregnancy. One in eight women who are pregnant continue to consume alcohol.

The Cost of Drug Use on the Environment

Although not something usually considered, the production of illegal drugs has a tremendous impact on the environment. The production of methamphetamine serves as an example. According to the Drug Enforcement Administration (DEA),

> . . . for every pound of meth produced, 5 to 6 pounds of toxic waste are produced. Common practices by meth lab operators include dumping this waste into bathtubs, sinks, or toilets, or outside on surrounding grounds or along roads and creeks. Some may place the waste in household or commercial trash or store it on the property. In addition to dumped waste, toxic vapors from the chemicals used and the meth-making process can permeate walls and ceilings of a home or building or the interior of a vehicle, potentially exposing unsuspecting occupants. (GAO 2013, p. 19)

The cultivation of marijuana, cocaine, and opium have an additional environmental impact. For example, the Colombian government estimates that during the decades of 1988 to 2008, nearly 5.4 million acres of rainforest (an area the size of New Jersey) were destroyed because of illegal drug production. Further, studies have shown that as much as 25 percent of the deforestation that takes place in Peru is "associated with clear-cutting and burning for planting coca bushes" (DEA 2010, p. 1).

In the United States and Mexico, outdoor cannabis cultivation leads to contaminated water, clear-cutting of natural vegetation, the disposable of nonbiodegradable materials, and

fetal alcohol syndrome A syndrome characterized by serious physical and mental handicaps as a result of maternal drinking during pregnancy.

neonatal abstinence syndrome (NAS) A condition in which a child, at birth, goes through withdrawal as a consequence of maternal drug use.

For most of us, how people begin to use drugs or why they don't stop or the effects of drugs on their lives is difficult to understand. The following "real-life stories" are from those who abuse alcohol and other drugs and the people who care about them. Although not a random sample of users, their stories are not dissimilar to others you might have heard.

"I didn't think I had a 'drug problem'—I was buying the tablets at the chemist [drug store]. It didn't affect my work. I would feel a bit tired in the mornings, but nothing more. The fact that I had a problem came to a head when I took an overdose of about forty tablets and found myself in the hospital. I spent twelve weeks in the clinic conquering my addiction." —Alex

"My friend was on drugs for four years, three of which were on hard drugs such as cocaine, LSD, morphine, and many antidepressants and painkillers. Actually anything he could get his hands on. He complained all the time of terrible pains in his body and he just got worse and worse till he finally went to see a doctor. . . .The doctor told him that there was nothing that could be done for him and that due to the deterioration of his body, he would not live much longer. Within days—he was dead." —Wayne, friend

"My goal in life wasn't living . . . it was getting high. I was falling in a downward spiral towards a point of no return. Over the years, I turned to cocaine, marijuana, and alcohol under a false belief it would allow me to escape my problems. It just made things worse. I had everything, a good job, money, a loving family, yet I felt so empty inside. As if I had nothing. Over twenty years of using, I kept saying to myself, I'm going to stop permanently after using this last time. It never happened. There were even moments I had thought of giving up on life." —John

"I lived with a crack addict for nearly a year. I loved that addict—who was my boyfriend—with all my heart, but I couldn't stick [with] it any more. . . . My 'ex' stole incessantly and couldn't tear himself away from his pipe. I think crack is more evil than heroin—one pipe can be all it takes to turn you into an immoral monster." —Audrey, girlfriend

"I was given my first joint in the playground of my school. I'm a heroin addict now, and I've just finished my eighth treatment for drug addiction." —Christian

"Ecstasy made me crazy. One day I bit glass, just like I would have bitten an apple. I had to have my mouth full of pieces of glass to realize what was happening to me. Another time, I tore rags with my teeth for an hour." —Ann

"When I was thirteen, friends would make fun of me if I didn't have a drink. I just gave in because it was easier to join the crowd. . . . I was really unhappy and just drank to escape my life. I went out less and less, so started losing friends. The more lonely I got, the more I drank. I was violent and out of control. I never knew what I was doing. I was ripping my family apart. . . . Kicked out of my home at age sixteen, I was homeless and started begging for money to buy drinks. After years of abuse, doctors told me there was irreparable harm to my health. . . . I was only sixteen but my liver was badly damaged and I was close to killing myself from everything I was drinking." —Samantha

"I started peeing blood. I felt sick. . . . My body felt weak. . . . I gave up everything because I was obsessed with using. . . . All I cared about was getting high. . . . I thought I could just use Coricidin for fun, that it didn't matter. I never expected to get hooked. . . . I'll never be able to get that time back. If I could erase it and make it go away, I would." —Charlie

"Jason had been at a friend's house, sniffing glue or lighter fluid, maybe both. On the way back to school, Jason kept blacking out. Finally, he fell and never got up. By the time we were able to get him to the hospital, it was too late." —Cathy, parent

Source: Foundation for a Drug Free World 2013.

the diversion of natural waterways often polluted with toxic chemicals, which endanger fish and other wildlife (U.S. Department of Justice 2010). Fish and wildlife aren't the only ones who should worry about water quality. A study of the quality of water at a Spanish national park found traces of eight illegal drugs in the waterways including Ecstasy, cocaine, and methamphetamines (Sohn 2010).

Treatment Alternatives

In 2012, nearly 4 million Americans aged 12 or older were treated for some kind of problem associated with the use of alcohol or illicit drugs (see this chapter's *The Human Side*). In the same year, it is estimated that 20.6 million people 12 years old and older were in need of substance abuse treatment but did not receive it (NSDUH 2013). Treatment

> . . . has been shown to reduce associated health and social costs by far more than the cost of the treatment itself. Treatment is also much less expensive than its alternatives, such as incarcerating addicted persons. For example, the average cost for 1 full year of methadone maintenance treatment is approximately $4,700 per patient, whereas 1 full year of imprisonment costs approximately $24,000 per

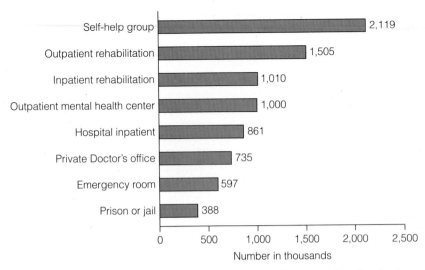

Figure 3.5 **Locations Where Past-Year Substance Use Treatment Was Received among People Aged 12 or Older, 2012**
Source: NSDUH 2013.

person. . . . According to several conservative estimates, every dollar invested in addiction treatment programs yields a return of between $4 and $7 in reduced drug-related crime, criminal justice costs, and theft. (NIDA 2012b)

Individuals who are interested in overcoming problem drug use have a number of treatment alternatives from which to choose (see Figure 3.5). Some options include family therapy, counseling, private and state treatment facilities, community care programs, pharmacotherapy (i.e., use of treatment medications), behavior modification, drug maintenance programs, and employee assistance programs. Three commonly used techniques are inpatient or outpatient treatment, peer support groups, and drug courts.

Inpatient and Outpatient Treatment

Inpatient treatment refers to treatment of drug dependence in a hospital and, most importantly, includes medical supervision of detoxification. Most inpatient programs last between 30 and 90 days and target individuals whose withdrawal symptoms require close monitoring (e.g., alcoholics, cocaine addicts). Some drug-dependent patients, however, can be safely treated as outpatients. Outpatient treatment allows individuals to remain in their home and work environments and is often less expensive. In outpatient treatment, patients are under the care of a physician who evaluates their progress regularly, prescribes needed medication, and watches for signs of a relapse.

The extent to which treatment is successful depends on a number of variables. A study of the effectiveness of inpatient and outpatient treatment of nearly 400 parole violators who opted for a year-long substance abuse program rather than returning to prison found that five variables predicted failure to complete the program: The participants (1) had a history of significant problems with their mothers, (2) had problems with their sexual partners in the 30 days prior to admission to the program, (3) had longer periods of incarceration, (4) had used heroin in the 30 days prior to admission to the program, and (5) were younger in age than those who successfully completed the program. Sadly, less than 33 percent of the participants completed the 12-month program (NIDA 2006).

Further, Saloner and LeCook (2013) report that race is also a significant predictor of treatment completion. Using national data, the researchers found that blacks and Hispanics are less likely to complete alcohol and drug treatment, and that Native Americans are less likely to complete alcohol treatment. Only Asians were more likely to complete both treatments than whites. Because the failure of blacks and Hispanics to complete treatment was largely due to socioeconomic variables, the authors suggest that the Affordable Care Act may mediate differences in completion rates (see Chapter 2).

Peer Support Groups

Twelve-Step Programs. Both Alcoholics Anonymous (AA) and Narcotics Anonymous (NA) are voluntary associations whose only membership requirement is the desire to stop drinking or taking drugs. AA and NA are self-help groups in that nonprofessionals operate them, offer "sponsors" to each new member, and proceed along a continuum of 12 steps to recovery. Members are immediately immersed in a fellowship of caring individuals with whom they meet daily or weekly to affirm their commitment. Some have argued that AA and NA members trade their addiction to drugs for feelings of interpersonal connectedness by bonding with other group members. In a survey of recovering addicts, more than 50 percent reported using a self-help program such as AA in their recovery (Willing 2002). AA boasts over 114,000 groups where over 2.1 million members meet in 170 countries (Alcoholics Anonymous 2013).

Symbolic interactionists emphasize that AA and NA provide social contexts in which people develop new meanings. Others who offer positive labels, encouragement, and social support for sobriety surround abusers. Sponsors tell the new members that they can be successful in controlling alcohol and/or drugs "one day at a time" and provide regular interpersonal reinforcement for doing so. Some research indicates that mutual support programs work. For example, Kelly and Hoeppner (2013) found that AA has some impact on the number of days a respondent was abstinent and the number of drinks per drinking days. The authors conclude, however, that the ". . . recovery benefits derived from AA differ in nature and magnitude between men and women and may reflect differing needs based on recovery challenges related to gender-based social roles and drinking contexts" (p. 1).

Therapeutic Communities. In **therapeutic communities**, which house between 35 and 500 people for up to 15 months, participants abstain from drugs, develop marketable skills, and receive counseling. Synanon, which was established in 1958, was the first therapeutic community for people with alcoholism and was later expanded to include other drug users. More than 400 residential treatment centers are now in existence, including Daytop Village and Phoenix House, the largest therapeutic communities in the country. Phoenix Houses serve more than 6,000 men, women, and teens annually at over 120 locations in 11 states (Phoenix House 2013). The longer a person stays at such a facility, the greater the chance of overcoming dependency. Living with a partner before entering the program and having a strong self-concept are also predictive of success (Dekel et al. 2004). Symbolic interactionists argue that behavioral changes appear to be a consequence of revised self-definition and the positive expectations of others. An assessment of 33 therapeutic community facilities in Peru indicated that therapeutic communities had a significant positive impact on the over 500 former clients interviewed six months after release (Johnson et al. 2008).

Drug Courts

Concern over the punitive treatment of drug offenders and the failure of the criminal justice system to reduce recidivism rates led to the development of **drug courts**. A recent report by the Sentencing Project (King & Pasquarella 2009), entitled *Drug Courts: A Review of the Evidence*, identifies two types of drug courts—deferred prosecution programs and post-adjudication programs.

In a deferred prosecution or diversion setting, defendants who meet certain eligibility requirements are diverted into the drug court system prior to pleading to a charge. Defendants are not required to plead guilty, and those who complete the drug court program are not prosecuted further. Failure to complete the program, however, results in prosecution. Alternatively, in the post-adjudication model, defendants must plead guilty to their charges, but their sentences are deferred or suspended while they participate in the drug court program. Successful completion of the program results in a waived sentence and sometimes an expungement of the offense. However, in cases where individuals fail to meet the requirements of the drug court (such as a habitual recurrence of drug use), they will be returned to the criminal court to face sentencing on the guilty plea.

A National Institute of Justice (2013) study of drug courts found that successful participants are less involved in criminal activity, have fewer rearrest rates, and are less likely to use drugs than comparable

therapeutic communities Organizations in which approximately 35 to 500 individuals reside for up to 15 months to abstain from drugs, develop marketable skills, and receive counseling.

drug courts Special courts that divert drug offenders to treatment programs in lieu of probation or incarceration.

> Drug use is a complex social issue that is exacerbated by the structural and cultural forces of society. . . . [T]he structure perpetuates a system of inequality, creating in some the need to escape, [and] the culture of society, through the media and normative contradictions, sends mixed messages about the acceptability of drug use.

offenders. Although the initial investment of drug courts is higher than incarceration, because drug court participants are less like to reoffend, the average savings per drug court participant was between $5,680 to $6,208. Drug courts, however, serve very few offenders because of their strict eligibility requirements, limited capacity, and potential participants' fears of the legal consequences of failure to complete treatment (Sevigny et al. 2013).

Strategies for Action: America Responds

Drug use is a complex social issue that is exacerbated by the structural and cultural forces of society that contribute to its existence. Although the structure of society perpetuates a system of inequality, creating in some the need to escape, the culture of society, through the media and normative contradictions, sends mixed messages about the acceptability of drug use. Thus, trying to end drug use by developing programs, laws, or initiatives may be unrealistic. Nevertheless, numerous social policies have been implemented or proposed to help control drug use and its negative consequences with various levels of success.

Alcohol and Tobacco

Although there may be some overlap (e.g., education), strategies to deal with alcohol and tobacco abuse are often different from those initiated to deal with illegal drugs. Prohibition, the largest social policy attempt to control a drug in the United States, was a failure, and criminalizing tobacco is likely to be just as successful. However, research has identified several promising strategies in reducing alcohol and tobacco use including economic incentives, government regulations, legal sanctions, and education and treatment.

Economic Incentives. One method of reducing alcohol and tobacco use is to increase the cost of the product. After examining the cost of cigarettes by state, Dinno and Glantz (2009) concluded that increased cigarette prices are associated with fewer people smoking and, if they smoke, fewer cigarettes are consumed.

Other incentives include covering the cost of treatment and rewarding success. For example, in a study of smoking cessation, 828 employees of a large corporation were randomly assigned to one of two groups. Both groups were provided with information on smoking cessation programs but one group also received financial incentives including ". . . $100 for completion of a smoking cessation program, $250 for cessation of smoking within 6 months after study enrollment as confirmed by a biochemical test, and $400 for abstinence for an additional 6 months after the initial cessation as confirmed by a biochemical test" (Volpp et al. 2009, p. 1). The results indicated that financial incentives reduced employees' smoking rates more than information alone. Similarly, Williams et al. (2005) note that "increasing the price of alcohol which can be achieved by eliminating price specials and promotions, or by raising price excise taxes, would lead to a reduction in both moderate and heavy drinking by college students" (p. 88). Other examples of economic incentives include reduced health insurance premiums for nonsmokers, reduced car insurance premiums for nondrinkers, and membership fee discounts at fitness centers.

Government Regulation. Federal, state, and local governmental regulations have each had some success in reducing tobacco and alcohol use and the problems associated with them. In 1984, states raised the legal drinking age to 21, under threat of losing federal highway funds. According to Mothers Against Drunk Driving (MADD), the minimum legal drinking age prevented over 25,000 drunken driving deaths and decreased alcohol-related car accidents by 16 percent (MADD 2011). Further, an analysis of drinking and driving of high school students 16 years old and older between 1991 and 2011 indicates that high school student drunk driving has declined over 50 percent. The researchers suggest that the extension of the learner's driving license period, nighttime driving restrictions, and stricter drunk driving laws may explain the decline (CDC 2012c). Similarly, clean air laws restrict smoking in the workplace, bars, restaurants, and the like, and reduce consumption rates as well as secondhand smoke exposure.

One of the most important pieces of legislation in recent years was the Family Smoking Prevention and Tobacco Control Act of 2009 (see the preceding "The Tobacco Crisis" section). The law gives authority to the Food and Drug Administration to regulate the manufacturing (e.g., tobacco companies must now disclose ingredients in their products), marketing (e.g., tobacco names or logos may no longer be used to sponsor sporting or entertainment events), and sale of tobacco products (e.g., terms like *light, mild,* and *low tar* may no longer be used). As a result of the FDA's regulations, the tobacco industry has shifted its attention to products that are not covered by the 2009 act (e.g., cigars, pipes, and e-cigarettes) and to overseas markets (Sifferlin 2013). It is estimated that tobacco companies spend $24 million *a day* in tobacco product advertising and marketing (Schmidt 2013).

What Do You Think? The popularity and use of e-cigarettes is growing for obvious reasons—lowered nicotine levels, reduced cost, and a variety of flavors to choose from (Alderman 2013). But the safety of "vaping" as it is called, led the U.S. Food and Drug Administration to successfully claim the right to regulate e-cigarettes, and officials of the European Union are considering restrictions on "liquid nicotine" sales. Do you think that e-cigarettes should be regulated by the government? Why or why not?

Legal Action. Federal and state governments, as well as smokers, ex-smokers, and the families of victims of smoking have taken legal action against tobacco companies. The 1990s brought billion-dollar judgments against the tobacco industry on behalf of states suing for reimbursement of smoking-related health care costs and families seeking punitive damages for the deaths of loved ones (Timeline 2001). In 1998, tobacco manufacturers reached a settlement with 46 states, agreeing to pay billions of dollars for reimbursement of state smoking-related health costs. The settlement also restricted the marketing, promotion, and advertising of tobacco products directed toward minors (Wilson 1999).

After years of litigation, the District of Columbia U.S. Court of Appeals held that "the tobacco industry had and continues to engage in a massive, decades-long campaign to defraud the American public . . . including falsely denying that nicotine is addictive, falsely representing that 'light,' and 'low tar' cigarettes present fewer health risks, falsely denying that they market to kids, and falsely denying that secondhand smoke causes disease" (Myers 2009, p. 1). In 2012, tobacco companies were ordered to publically admit their deception. Among other required statements to be placed in advertisements, some tobacco companies must state that "A Federal Court has ruled that the Defendant tobacco companies deliberately deceived the American public about designing cigarettes to enhance the delivery of nicotine, and has ordered those companies to make this statement. Here is the truth: Smoking kills, on average, 1,200 Americans. Every day" (Mears 2012).

Suits against retailers, distributors, and manufacturers of alcohol are more recent and often modeled after tobacco litigation. These suits primarily concern accusations of unlawful marketing, sales to underage drinkers, and failure to adequately warn of the risks of alcohol. In 2012, five inmates filed a $1 billion lawsuit against eight alcohol manufacturers including Anheuser-Busch, Coors, Miller Brewing, American Brands, and the owners of Jim Beam whiskey. The inmates' suit argues that their addiction to alcohol was responsible for their crimes and that alcohol companies should have put warning labels on their products (Jennings 2013).

Prevention. So what can be done about the nation's number one cause of preventable deaths? To a large extent, experts' recommendations mirror those made in a 2010 federal report entitled *Ending the Tobacco Epidemic.* After reviewing successful evidence-based tobacco interventions, the report recommended: (1) the establishment of countermarketing

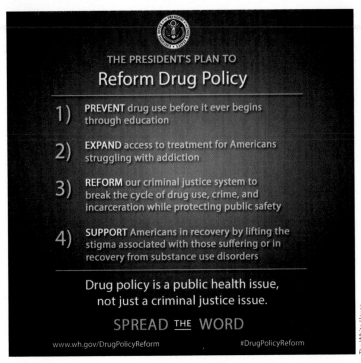

THE PRESIDENT'S PLAN TO
Reform Drug Policy

1) **PREVENT** drug use before it ever begins through education

2) **EXPAND** access to treatment for Americans struggling with addiction

3) **REFORM** our criminal justice system to break the cycle of drug use, crime, and incarceration while protecting public safety

4) **SUPPORT** Americans in recovery by lifting the stigma associated with those suffering or in recovery from substance use disorders

Drug policy is a public health issue, not just a criminal justice issue.

SPREAD THE WORD

www.wh.gov/DrugPolicyReform   #DrugPolicyReform

Since 2009, the government has spent over $30 billion on controlling drugs, including $10.7 billion for treatment and prevention in funding for 2013. In the last three years, the percentage of drug control federal monies spent on "harm reduction" compared to "supply reduction" has increased.

campaigns directed toward youth, (2) adoption of comprehensive smoke-free laws, (3) the development of affordable and accessible cessation programs, (4) an increase in the retail price of tobacco products, and (5) restriction of tobacco advertising and marketing (U.S. Department of Health and Human Services 2010).

There is also a body of research that suggests that school-based interventions may reduce alcohol and/or tobacco use. For example, Champion et al. (2013), in a review of computer- and Internet-based prevention programs housed in schools, reported that of the seven studies reviewed, six ". . . achieved a reduction in alcohol or drug use at post-intervention and/or follow-up, two were associated with decreased intentions to smoke in the future, and two programs significantly increased alcohol- and drug-related knowledge" (p. 120). The authors are quick to note that technology-based interventions avoid a number of problems associated with traditional school delivery systems (e.g., cost, limited teacher time) while assuring a high degree of consistency over time, location, and student population.

Lastly, media campaigns have shown some success in preventing drug use. The Office of National Drug Policy youth anti-drug campaign, "Above the Influence," boasts a 88 percent recognition rate among teens and has 1.7 million "likes" on its Facebook page (ONDCP 2013). Further, using data from Neilsen Media Research and a self-report survey of adolescent drug use (i.e., Monitoring the Future), Terry-McElrath et al. (2010) examined the relationship between anti-drug television messages and drug use. In general, media exposure to anti-drug messages six months prior to the self-report survey were effective and in the desired direction. The effect, of course, varied over time, age, and type of drug used.

Illegal Drugs

Although the greatest costs to individuals and society come from the use and abuse of legal drugs, there is little doubt that the use and abuse of illicit drugs take a tremendous human and financial toll. In this section, we highlight the war on drugs, deregulation and legalization, and other federal and state initiatives in the effort to control illegal substances.

The War on Drugs. In the 1980s, the federal government declared a "war on drugs," which was based on the belief that controlling drug availability would limit drug use and, in turn, drug-related problems. In contrast to a position of **harm reduction**, which focuses on minimizing the costs of drug use for both user and society (e.g., distributing clean syringes to decrease the risk of HIV infection), this "zero-tolerance" approach advocated get-tough law enforcement policies, and is responsible for the dramatic increase in the jail and prison population. In 1980, there were an estimated 41,000 drug offenders in jail or prison; in 2011, there were nearly half a million.

Race, ethnicity, and the war on drugs. The harsher penalties enacted as part of the "war on drugs" required prison sentences for almost all drug offenders—first-time or repeat—and limited judicial discretion in deciding what best served the public's interest. The "Rockefeller drug laws" as they were called resulted in a disproportionate number of Hispanics and African Americans receiving "excessively long and unnecessary prison sentences" for even the most minor drug offenses (Human Rights Watch 2007, p. 1). In response to public outcries and accusations of institutional racism (see Chapter 9), reform of such laws began (Peters 2009). Nonetheless, discriminatory practices continue. For

harm reduction A recent public health position that advocates reducing the harmful consequences of drug use for the user as well as for society as a whole.

example, African American teens are 10 times more likely to be arrested for drug-related crimes than whites (Sanchez-Moreno 2012), despite research that indicates they are less likely to use drugs and less likely to have drug problems than their white counterparts (Wu et al. 2011).

What Do You Think? To recruit members and intimidate enemies, Mexican drug cartel leaders have posted videos on YouTube (Hastings 2013; Jervis 2009). Many are so gruesome that they are preceded by a warning that the content may be inappropriate for viewers under the age of 18. Gang members also post crimes by their rival gangs to publicize their misdeeds (Cattan 2010). Although YouTube officials have alerted law enforcement agencies about these videos, it could be argued that allowing them to be broadcast is "aiding and abetting" the enemy in the war against drugs. Do you think drug cartel leaders should be able to post such videos on the YouTube website? Why or why not?

Gender and the war on drugs. According to the Drug Policy Alliance, in 2010, nearly a quarter of women in state prisons and over half of women in federal prisons were incarcerated for drug violations; over half were mothers (DPA 2013). As with racial and ethnic minorities in general, women sentenced to prison for drug offenses were disproportionately black and Hispanic. As a result, children with drug-related incarcerated mothers are more likely to be racial and ethnic minorities.

Using conflict theory, Merolla (2008) argues that the increasing arrest rate of women is, in part, a function of the war on drugs rather than behavioral changes in women, i.e., their increased use of drugs. Using data from government sources, Merolla concludes that the war on drugs, in and of itself, increased arrest rates of women in two ways. First, the war on drugs redefined through the media and in many other ways who a criminal is in nongendered terms (i.e., men and women). Second, as a consequence of this redefinition, law enforcement practices changed and more aggressively target female drug users.

Cost of the war on drugs. Although the focus here is on the war in drugs in the United States, there remains a global initiative to control drug production, trafficking, and use. One collaborative project, *Count the Costs,* is dedicated to changing the punitive approach to drug control policy and, with it, the reduction of costs associated with such an emphasis (CTC 2013). *Count the Costs* has identified seven categories of costs associated with the global war on drugs:

- Undermining development and creating conflict (e.g., use of military to fight drug cartels)
- Threatening public health (e.g., scarce resources are channeled toward law enforcement rather than treatment)
- Violating human rights (e.g., execution of drug users in some countries)
- Promoting stigma and discrimination (e.g., some drug use is culturally embedded)
- Perpetuating crime and criminals (e.g., prohibiting illicit drugs may reduce supply, but with constant demand, prices increase)
- Deforestation and pollution (e.g., aerial spraying of drug crops)
- Wasting billions of dollars (e.g., spending on law enforcement)

The U.S. government spent between $20 billion and $25 billion a year in fighting the war on drugs over the last decade (Porter 2012).

Losing the War: Drug Policy Reform. There is little debate ". . . that the global prohibition of certain drugs and the war on drugs have largely failed to reach their stated goals" (Chouvy 2013, p. 216). In fact, a CNN article called the war on drugs the "trillion-dollar failure" (Branson 2012), and 82 percent of Americans agree with this assessment (Rasmussen 2012). Consequently, beginning in 2010, the federal government began to reform national drug policy. The U.S. policy on fighting drugs is two-pronged. First is **demand reduction**, which entails reducing the demand for drugs through treatment and prevention

demand reduction One of two strategies in the U.S. war on drugs (the other is supply reduction), demand reduction focuses on reducing the demand for drugs through treatment, prevention, and research.

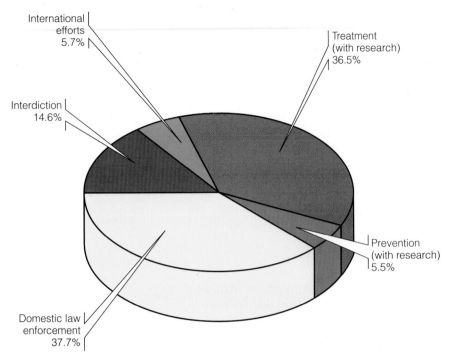

International efforts 5.7%

Interdiction 14.6%

Treatment (with research) 36.5%

Prevention (with research) 5.5%

Domestic law enforcement 37.7%

Figure 3.6 **Federal Drug Control Spending by Function, Fiscal Year 2014**
Source: ONDCP 2013.

(see Figure 3.6). The second strategy is **supply reduction**. A much more punitive strategy, supply reduction relies on international efforts, interdiction, and domestic law enforcement to reduce the supply of illegal drugs.

The 2014 National Drug Control Strategy (see funding, Figure 3.6) describes the present ". . . Administration's vision for a 21st-century drug policy that is based on science and evidence, encompassing prevention, early intervention, treatment, recovery, criminal justice reform, effective law enforcement, and international cooperation" (ONDCP 2013, p. 2). The allocation of funds for each of these strategies has changed over time as a more balanced and less punitive approach has been adopted. For example, the budget allocation for preventive efforts increased by 5 percent between 2012 and 2014. Although the allocation difference appears to be quite small, a change of five percentage points in this case results in a nearly $70 million increase for prevention efforts (ONDCP 2013).

Reform of U.S. drug policy is in part a reflection of the failure of the war on drugs, but is also a response to increasing international pressure. In 2013, a collaborative organization between North and South American countries issued a "game-changing report" (Doward 2013). The report (OAS 2013) describes the negative impact of U.S. drug policies on Latin American countries, including increased crime and violence, economic hardship, and institutional corruption. There is little evidence to the contrary. Of the ". . . world's eight most murderous countries, seven lie on the cocaine-trafficking route from the Andes to the United States and Europe. Only war zones are more violent than Honduras. More than 7,000 of its 8 [million] citizens are murdered each year. In the European Union, with a 500 [million] population, the figure is under 6,000" (*The Economist* 2013).

Deregulation or Legalization. Given the questionable successes of the war on drugs, it is not surprising that many advocate alternative measures to the rather punitive emphasis of the last several decades. **Deregulation** is the reduction of government control over certain drugs. For example, although individuals must be 21 years old to purchase alcohol and 18 to purchase cigarettes, both substances are legal and can be purchased freely. In some states, possession of marijuana in small amounts is a misdemeanor rather than a felony, and marijuana is lawfully used for medical purposes in 18 states (see Chapter 2). Deregulation is popular in other countries as well. For example, personal possession of any drug, even those considered the most dangerous, is legal in Spain, Italy, the Baltic States, and the Czech Republic (*The Economist* 2009).

Proponents for the **legalization** of drugs affirm the right of adults to make informed choices. They also argue that the tremendous revenues realized from drug taxes could be used to benefit all citizens, that purity and safety controls could be implemented, and that legalization would expand the number of distributors, thereby increasing competition and reducing prices. Drugs would thus be safer, drug-related crimes would be reduced, and production and distribution of previously controlled substances would be taken out of the hands of the underworld.

Those in favor of legalization also suggest that the greater availability of drugs would not increase demand, pointing to countries where some drugs have already been

supply reduction One of two strategies in the U.S. war on drugs (the other is demand reduction), supply reduction concentrates on reducing the supply of drugs available on the streets through international efforts, interdiction, and domestic law enforcement.

deregulation The reduction of government control over, for example, certain drugs.

legalization Making prohibited behaviors legal; for example, legalizing drug use or prostitution.

decriminalized. **Decriminalization,** which entails removing state penalties for certain drugs, promotes a medical rather than criminal approach to drug use that encourages users to seek treatment and adopt preventive practices. As mentioned earlier, in 2012, Colorado and Washington decriminalized the recreational use of marijuana. Portugal "became the first European country to officially abolished all criminal penalties for personal possession of drugs, including marijuana, cocaine, heroin, and methamphetamine" (Szalavitz 2009, p. 1). Despite fears of negative consequences, a report by the Cato Institute concluded that decriminalization of personal possession of drugs was responsible for reducing new cases of HIV infections, decreasing drug use among Portuguese teens, and doubling the number of people seeking treatment for drug addiction (Greenwald 2009).

Opponents of legalization argue that it would be construed as government approval of drug use and, as a consequence, drug experimentation and abuse would increase. Furthermore, although the legalization of drugs would result in substantial revenues for the government, drug trafficking and black markets would still flourish because all drugs would not be decriminalized (e.g., methamphetamine). Legalization would also require an extensive and costly bureaucracy to regulate the manufacture, sale, and distribution of drugs. Finally, the position that drug use is an individual's right cannot guarantee that others will not be harmed. It is illogical to assume that a greater availability of drugs will translate into a safer society.

Understanding Alcohol and Other Drug Use

In summary, substance abuse—that is, drugs and their use—is socially defined. As the structure of society changes, the acceptability of one drug or another changes as well. As conflict theorists assert, the status of a drug as legal or illegal is intricately linked to those who have the power to define acceptable and unacceptable drug use. There is also little doubt that rapid social change, anomie, alienation, and inequality further drug use and abuse. Symbolic interactionism also plays a significant role in the process: If people are labeled "drug users" and are expected to behave accordingly, then drug use is likely to continue. If people experience positive reinforcement of such behaviors and/or have a biological predisposition to use drugs, the probability of their drug involvement is even higher. Thus, the theories of drug use complement rather than contradict one another.

There are two issues that need to be addressed in understanding drug use. The first is at the micro level—why does a given individual use alcohol or other drugs? Many individuals at high risk for drug use have been "failed by society"—they are living in poverty, unemployed, victims of abuse, dependents of addicted and neglectful parents, and the like. Despite the social origins of drug use, many treatment alternatives, emanating from a clinical model of drug use, assume that the origin of the problem lies within the individual rather than in the structure and culture of society. Although the problem may admittedly lie within the individual when treatment occurs, policies that address the social causes of drug abuse must be a priority in dealing with the drug problem in the United States.

[A] . . World Health Organization survey of 17 countries concluded that there is no link between the harshness of drug policies and the consumption rates of its citizenry.

The second question, related to the first, asks: Why does drug use vary so dramatically across societies, often independent of a country's drug policies? The United States metes out some of the most severe penalties for drug violations in the world, but has one of the highest rates of marijuana and cocaine use. On the other hand, as mentioned earlier, Portugal decriminalized personal possession of all drugs, and youth drug use is down (Szalavitz 2009). Most compellingly, a 2008 World Health Organization survey of 17 countries concluded that there is no link between the harshness of drug policies and the consumption rates of its citizenry (Degenhardt et al. 2008).

That said, what is needed is a more balanced approach—one that acknowledges that not all drugs have the same impact on society or on the individuals who use them. The present administration appears to be leaning this way. For example, in a break from the previous administration, President Obama supported federally funded needle exchange

decriminalization Entails removing state penalties for certain drugs, and promotes a medical rather than criminal approach to drug use that encourages users to seek treatment and adopt preventive practices.

programs—a staple of harm reduction advocates—and has stated that it is "entirely appropriate" to use marijuana for the same purposes and under the same controls as other drugs prescribed by a physician (Dinan & Conery 2009; Heinrich 2009). Further, the 2014 proposed distribution of funding for harm reduction versus supply reduction is the most balanced in U.S. history. Only time will tell if this new approach to drug control, one that reflects both a public health and criminal justice position, will be more successful than the policies of previous administrations.

CHAPTER REVIEW

- **What is a drug, and what is meant by drug abuse?**
Sociologically, the term *drug* refers to any chemical substance that (1) has a direct effect on the user's physical, psychological, and/or intellectual functioning; (2) has the potential to be abused; and (3) has adverse consequences for the individual and/or society. Drug abuse occurs when acceptable social standards of drug use are violated, resulting in adverse physiological, psychological, and/or social consequences.

- **How do the three sociological theories of society explain drug use?**
Structural functionalists argue that drug abuse is a response to the weakening of norms in society, leading to a condition known as anomie or normlessness. From a conflict perspective, drug use occurs as a response to the inequality perpetuated by a capitalist system as societal members respond to alienation from their work, family, and friends. Symbolic interactionism concentrates on the social meanings associated with drug use. If the initial drug use experience is defined as pleasurable, it is likely to recur, and over time, the individual may earn the label of "drug user."

- **What are the most frequently used legal and illegal drugs?**
Alcohol is the most commonly used and abused legal drug in the United States, with 66 percent of the adult population reporting current alcohol use. Although tobacco use in the United States has been declining, the use of tobacco products is globally very high, with 80 percent of the world's over one billion smokers living in low- or middle-income countries.

- **What are the consequences of drug use?**
The consequences of drug use are fivefold. First is the cost to the family and children, often manifesting itself in higher rates of divorce, spouse abuse, child abuse, and child neglect. Second is the relationship between drugs and crime. Those arrested have disproportionately higher rates of drug use. Although drug users commit more crimes, sociologists disagree as to whether drugs actually "cause" crime or whether, instead, criminal activity leads to drug involvement. Third are the economic costs (e.g., loss of productivity), which are in the billions. Then there are the health costs of abusing drugs, including shortened life expectancy; higher morbidity (e.g., cirrhosis of the liver and lung cancer); exposure to HIV infection, hepatitis, and other diseases through shared needles; a weakened immune system; birth defects such as fetal alcohol syndrome; drug addiction in children; and higher death rates. Finally, illegal drug production takes its toll on the environment, which impacts all Americans.

- **What treatment alternatives are available for drug users?**
Although there are many ways to treat drug abuse, two methods stand out: The inpatient–outpatient model entails medical supervision of detoxification and may or may not include hospitalization. Twelve-step programs such as Alcoholics Anonymous (AA) and Narcotics Anonymous (NA) are particularly popular, as are therapeutic communities. Therapeutic communities are residential facilities where drug users learn to redefine themselves and their behavior as a response to the expectations of others and self-definition. Finally, drug courts are used as an alternative to the traditional punitive methods of probation and incarceration.

- **What can be done about the drug problem?**
First, there are government regulations limiting the use (e.g., the law establishing the 21-year-old drinking age) and distribution (e.g., prohibitions about importing drugs) of legal and illegal drugs. The government also imposes sanctions on those who violate drug regulations and provides treatment facilities for other offenders. Economic incentives (e.g., cost) and prevention programs have also been found to impact consumption rates. Finally, legal action holding companies responsible for the consequences for their product—for example, class-action suits against tobacco producers—have been fairly successful.

1. "Cannabis cafés" are commonplace throughout England.
   a. True
   b. False
2. The most used illicit drug in the world is
   a. heroin
   b. marijuana
   c. cocaine
   d. methamphetamine
3. What theory would argue that the continued legality of alcohol is a consequence of corporate greed?
   a. Structural functionalism
   b. Symbolic interactionism
   c. Reinforcement theory
   d. Conflict theory
4. Cigarette smoking is
   a. the third leading cause of preventable death in the United States
   b. not addictive
   c. the most common use of tobacco products
   d. increasing in the United States
5. In the United States, drinking is highest among young, nonwhite males.
   a. True
   b. False
6. Which European country decriminalized personal possession of all drugs, including marijuana, cocaine, heroin, and methamphetamine?
   a. Spain
   b. France
   c. Portugal
   d. The Netherlands
7. The active ingredient in marijuana, THC, can act as a sedative or a hallucinogen.
   a. True
   b. False
8. According to the proposed 2014 budget, most federal drug control dollars are allocated to
   a. international efforts
   b. domestic law enforcement
   c. prevention and research
   d. treatment and research
9. Decriminalization refers to the removal of penalties for certain drugs.
   a. True
   b. False
10. The two-pronged drug control strategy of the U.S. government entails supply reduction and harm reduction.
    a. True
    b. False

Answers: 1. B; 2. B; 3. D; 4. C; 5. B; 6. C; 7. A; 8. A; 9. B; 10. B.

binge drinking   72
chemical dependency   68
crack   79
decriminalization   95
demand reduction   93
deregulation   94
drug   65

drug abuse   68
drug courts   89
fetal alcohol syndrome   86
gateway drug   76
harm reduction   92
heavy drinking   72
legalization   94

meta-analysis   74
neonatal abstinence syndrome (NAS)   86
psychotherapeutic drugs   80
supply reduction   94
synthetic drugs   81
therapeutic communities   89

Scott Olson/Getty Images

4

Crime and Social Control

"Unjust social arrangements are themselves a kind of extortion, even violence."

—John Rawls, A Theory of Justice

SHANIYA DAVIS WAS just five years old and already had suffered more than most in her short life. Her mother was charged with first-degree murder and "…indecent liberties with a child, felony child abuse, felony sexual servitude, rape of a child, sexual offense of a child by an adult offender, [and] human trafficking…" (CBS 2013, p. 1). She was also charged with misdemeanor making a false police report for telling the police that her little girl was "missing." Davis, 27, had sold her daughter to Mario McNeill to pay off a drug debt. He sexually assaulted her, suffocated her, and then dumped her body in a field. In July 2013, McNeill was convicted of murder and sentenced to death. Annette Davis, Shaniya's mother, awaits trial.

Super Bowl XLVII—football, fun, the halftime show, television ads, and human trafficking? In fact, "[T]he Super Bowl is actually estimated to be one of the largest human trafficking events in the world … because anytime there is increased demand, there will be increased supply," according to one expert (quoted in Murphy 2013, p. 1). To deal with such concerns, Immigration and Customs Enforcement, the Louisiana State Police, the New Orleans Police Department, and the Federal Bureau of Investigation (FBI) formed a task force to increase enforcement during the 2013 Super Bowl. As a result, several women who were being held captive by a sex trafficking ring were rescued (Murphy 2013).

Four law enforcement agencies working together in one state to solve one problem at one event. Such is the bureaucracy of the criminal justice system, which includes not only state and federal law enforcement agencies, but courts and corrections as well. In this chapter, we examine the criminal justice system as well as theories, types, and demographic patterns of criminal behavior. The economic, social, and psychological costs of crime are also examined. The chapter concludes with a discussion of social control, including policies and prevention programs designed to reduce crime in the United States.

The Global Context: International Crime and Social Control

Several facts about crime are true throughout the world. First, crime is ubiquitous—there is no country where crime does not exist. Second, most countries have the same components in their criminal justice systems: police, courts, and prisons. Third, adult males make up the largest category of crime suspects worldwide. Fourth, in all countries, theft is the most common crime committed, whereas violent crime is a relatively rare event.

Despite these similarities, dramatic differences exist in international crime rates although comparisons are made difficult by variations in measurement and crime definitions. Nonetheless, looking at global statistics as a whole, some statements about crime can be made with confidence.

First, violent crime rates vary significantly by region and by country. Examining homicide rates is a case in point. Because homicide rates are correlated with other violent crimes, for example, as robbery rates increase, homicide rates increase, global homicide rates are very telling. In 2010, globally, there were over 450,000 homicides i.e., the intentional killing of another, according to the most recent global surveys available (UNODC 2012). About one-third of the *number* of homicides are estimated to have occurred in Africa, one-third in the Americas, and one-third in Asia, Europe, and Oceania.

However, crime *rates,* usually expressed per 100,000 population, take into consideration differences in population size and allow more accurate comparisons. Of late, homicide rates have been declining in Asia, North America, and Europe, while increasing in Central America and Caribbean countries. Countries with relatively high homicide rates tend to be those with higher levels of income inequality and lower levels of development, for example, Venezuela, Jamaica, and Swaziland (rate >40), while countries that are more egalitarian and have higher levels of development, such as Switzerland, Germany, and the United States (rate <5), have lower homicide rates (UNODC 2012).

Globally, property crimes, most often entailing some kind of theft, are distributed very differently than violent crimes. A comparison of developed countries, the 36 countries that belong to the Organization for Economic Cooperation and Development (OECD),

TABLE 4.1 Types of Transnational Crime

Provision of Illicit Goods	Provision of Illicit Services	Infiltration of Business or Government
Drug trafficking	Human trafficking	Extortion and racketeering
Stolen property	Cybercrime and fraud	Money laundering
Counterfeiting	Commercial vices (e.g., illegal sex and gambling)	Corruption

Source: Albanese 2012.

indicates that where violent crime as measured by homicide rates are high, property crimes tended to be low (Civitas 2012). For example, New Zealand, Sweden, and Denmark have relatively low homicide rates, but each ranks in the top ten countries for burglary and car theft.

Violent crime and property crimes represent just two types of crime that take place worldwide. Although we are concerned about these types of crimes and the possibility of victimization, globalization and technological know-how have fueled the development of new categories of crimes (e.g., human trafficking) and expanded the prevalence of others (e.g., counterfeiting, cybercrime).

As defined by the present administration, **transnational crime** refers to organized criminal activity that crosses one or more national boundaries for the "... purpose of obtaining power, influence, monetary and/or commercial gains...through a pattern of corruption and/or violence" (U.S. Department of Justice 2012). The significance of transnational crime should not be minimized. As Shelley (2007) states:

> Transnational crime will be a defining issue of the 21st century for policy makers—as defining as the Cold War was for the 20th century and colonialism was for the 19th. Terrorists and transnational crime groups will proliferate because these crime groups are major beneficiaries of globalization. They take advantage of increased travel, trade, rapid money movements, telecommunications and computer links, and are well positioned for growth. (p. 1)

Although human trafficking is often thought of as something that exists outside of the United States, it is estimated that 100,000 children annually are sexually exploited, and thousands more are forced into domestic servitude and involuntary labor in the United States.

As Table 4.1 indicates, transnational crime primarily functions to provide illicit goods and services, and to infiltrate businesses or government entities (Albanese 2012). Internet child pornography is an example of transnational crime. Prichard et al. (2013) note that Internet child pornography is increasing and that such code words as *Pthc* (preteen hard core), *Lolita, teen,* and *12yo* are routinely found in the top 1,000 terms used in an international search engine.

Human Trafficking is another example of transnational crime. According to federal law, human trafficking includes "[T]he recruitment, harboring, transportation, provision, or obtaining of a person for

- sex trafficking in which a commercial sex act is induced by force, fraud, or coercion, or in which the person induced to perform such act has not attained 18 years of age; or
- labor or services, through the use of force, fraud, or coercion for the purpose of subjection to involuntary servitude, peonage, debt bondage, or slavery." (U.S. Department of Health and Human Services 2012, p.1)

transnational crime Criminal activity that occurs across one or more national borders.

It is estimated that as many as 27 million men, women, and children at any given time are human trafficking victims (U.S. Department of State 2013). Although human trafficking is often thought of as something that exists outside of the United States, it is estimated that over 100,000 children annually are sexually exploited, and thousands more are

A young boy, a victim of human trafficking, is forced to work in a balloon factory outside of Dhaka, Bangladesh. Human trafficking is estimated to be a multibillion-dollar enterprise, approaching the international value of drug and arms trafficking (Interpol 2011).

forced into domestic servitude and involuntary labor in the United States (Polaris Project 2013). It is also estimated that about 80 percent of the victims of human trafficking are women and girls, and that 50 percent are minors (U.S. Department of State 2013).

What Do You Think? According to the International Maritime Bureau (IMB), 297 ships were attacked at sea in 2012—174 were boarded by pirates, 28 were fired upon, and 28 were hijacked (ICC 2013). In 2013, three Somali pirates were convicted for killing four Americans after hijacking their yacht, and were sentenced to life in prison (Daugherty 2013). In federal law, death resulting from aircraft piracy or attempted hijacking of an aircraft carries the death penalty. Do you think that deaths that result from maritime hijacking should carry the death penalty? What if the death is not of a hostage but of a fellow hijacker?

Sources of Crime Statistics

The U.S. government spends millions of dollars annually to compile and analyze crime statistics. A **crime** is a violation of a federal, state, or local criminal law. For a violation to be a crime, however, the offender must have acted voluntarily and with intent and have no legally acceptable excuse (e.g., insanity) or justification (e.g., self-defense) for their behavior. The three major types of statistics used to measure crime are official statistics, victimization surveys, and self-report offender surveys.

Official Statistics

Local sheriffs' departments and police departments throughout the United States collect information on the number of reported crimes and arrests and voluntarily report them to

crime An act, or the omission of an act, that is a violation of a federal, state, or local criminal law for which the state can apply sanctions.

the FBI. The FBI then compiles these statistics annually and publishes them, in summary form, in the Uniform Crime Reports (UCR). The UCR lists **crime rates** or the number of crimes committed per 100,000 population, the actual number of crimes, the percentage of change over time, and clearance rates. **Clearance rates** measure the percentage of cases in which an arrest and official charge have been made and the case has been turned over to the courts.

These statistics have several shortcomings. For example, many incidents of crime go unreported. Between 2006 and 2010, more than half of the nation's violent crimes—3.4 million in total—were not reported to the police (U.S. Department of Justice 2012).

Even if a crime is reported, the police may not record it. Alternatively, some rates may be exaggerated. Motivation for such distortions may come from the public (e.g., demanding that something be done), from political officials (e.g., election of a sheriff), and/or from organizational pressures (e.g., budget requests). For example, a police department may "crack down" on drug-related crimes in an election year. The result is an increase in the recorded number of these offenses for that year. Such an increase reflects a change in the behavior of law enforcement personnel, not a change in the number of drug violations. Thus, official crime statistics may be a better indicator of what police are doing rather than of what criminals are doing.

In the 1970s, the "law enforcement community called for a thorough evaluation of the UCR Program to recommend an expanded and enhanced data collection system to meet the needs of law enforcement in the 21st century" (FBI 2009, p. 1). The result, the National Incident-Based Reporting System (NIBRS), requires that law enforcement agencies provide extensive information on each criminal incident and arrest for 22 offenses (Group A) and arrestee information on 11 lesser offenses (Group B). Thus far, the FBI has certified 31states for NIBRS participation. The hope is that, once implemented nationally, the NIBRS will provide more reliable and comprehensive crime data.

Victimization Surveys

Acknowledging "the dark figure of crime," that is, the tendency for so many crimes to go unreported and thus undetected by the UCR, the U.S. Department of Justice conducts the National Crime Victimization Survey (NCVS). Begun in 1973 and conducted annually, the NCVS interviews over 135,000 people 12 and older about their experiences as victims of crime. Interviewers collect a variety of information, including the victim's background (e.g., age, race and ethnicity, sex, marital status, education, and area of residence), relationship to the offender (stranger or nonstranger), and the extent to which the victim was harmed (U.S. Census Bureau 2013).

In 2011, the latest year for which victimization data are available, there were 5.8 million violent crimes and 147.1 million property crimes, an increase from the previous year of 17 percent and 11 percent, respectively (Planty & Truman 2012). The majority of the increase in violent crimes was due to greater victimization of whites, Hispanics, young people, and males, and the increase in the number of reported assaults.

Although adding an important dimension to the study of crime, the NCVS is not without problems. In 2008, a panel of experts offered recommendations "…to improve its methodology, assure its sustainability, increase its value to national and local stakeholders, and better meet the challenges of measuring the extent, characteristics, and consequences of criminal victimization" (BJS 2013a, p. 1). The Bureau of Justice Statistics, in response to the recommendations, is presently assessing the various methodologies used to collect self-report data.

> In 2011, the latest year for which victimization data are available, there were 5.8 million violent crimes and 147.1 million property crimes, an increase from the previous year of 17 percent and 11 percent, respectively.

crime rate The number of crimes committed per 100,000 population.

clearance rate The percentage of crimes in which an arrest and official charge have been made and the case has been turned over to the courts.

Read each of the following questions. Since the age of 16, if you have ever engaged in the behavior described, place a "1" in the space provided. If you have not engaged in the behavior, put a "0" in the space provided. After completing the survey, read the section on interpretation to see what your answers mean.

Questions

1. Have you ever been in possession of drug paraphernalia? ___
2. Have you ever lied about your age or about anything else when making application to rent an automobile? ___
3. Have you ever obtained a false ID to gain entry to a bar or event? ___
4. Have you ever tampered with a coin-operated vending machine or parking meter? ___
5. Have you ever shared, given, or shown pornographic material to someone under 18? ___
6. Have you ever begun and/or participated in a basketball, baseball, or football pool? ___
7. Have you ever used "filthy, obscene, annoying, or offensive" language while on the telephone? ___
8. Have you ever given or sold a beer to someone under the age of 21? ___
9. Have you ever been on someone else's property (land, house, boat, structure, and so on) without that person's permission? ___
10. Have you ever forwarded a chain letter with the intent to profit from it? ___
11. Have you ever improperly gained access to someone else's e-mail or other computer account? ___
12. Have you ever written a check for over $150 when you knew it was bad? ___

Interpretation

Each of the activities described in these questions represents criminal behavior that was subject to fines, imprisonment, or both under the laws of Florida in 2008. For each activity, the following table lists the maximum prison sentence and/or fine for a first-time offender. To calculate your "prison time" and/or fines, sum the numbers corresponding to each activity you have engaged in.

Offense	Maximum Prison Sentence	Maximum Fine
1. Possession of drug paraphernalia	1 year	$1,000
2. Fraud	5 years	$5,000
3. Possession of false ID or driver's license	5 years	$5,000
4. Larceny	2 months	$500
5. Protection of minors from obscenity	5 years	$5,000
6. Illegal gambling	2 months	$500
7. Harassing/obscene telecommunications	2 months	$500
8. Illegal distribution of alcohol	2 months	$500
9. Trespassing	1 year	$1,000
10. Illegal gambling	1 year	$1,000
11. Illegal misappropriation of cyber communication	5 years	$5,000
12. Worthless checks	5 years	$5,000

Source: Florida Criminal Code 2009.

Self-Report Offender Surveys

Self-report surveys ask offenders about their criminal behavior. The sample may consist of a population with known police records, such as a prison population, or it may include respondents from the general population, such as college students.

Self-report data compensate for many of the problems associated with official statistics but are still subject to exaggerations and concealment. The Criminal Activities Survey in this chapter's *Self and Society* feature asks you to indicate whether you have engaged in a variety of illegal activities.

Self-report surveys reveal that virtually every adult has engaged in some type of criminal activity. Why then is only a fraction of the population labeled criminal? Like a funnel, which is large at one end and small at the other, only a small proportion of the total population of law violators are ever convicted of a crime. For individuals to be officially labeled criminals, (1) their behavior must become known to have occurred; (2) the behavior must come to the attention of the police, who then file a report, conduct an investigation, and make an arrest; and finally, (3) the arrestee must go through a preliminary hearing, an arraignment, and a trial and may or may not be convicted. At every stage of the process, offenders may be "funneled" out.

Sociological Theories of Crime

Some explanations of crime focus on psychological aspects of the offenders, such as psychopathic personalities, unhealthy relationships with parents, and mental illness. Other crime theories focus on the role of biological variables, such as central nervous system malfunctioning, stress hormones, vitamin or mineral deficiencies, chromosomal abnormalities, and a genetic predisposition toward aggression. Sociological theories of crime and violence emphasize the role of social factors in criminal behavior and societal responses to it.

Structural-Functionalist Perspective

According to Durkheim and other structural functionalists, crime is functional for society. One of the functions of crime and other deviant behavior is that it strengthens group cohesion: "The deviant individual violates rules of conduct that the rest of the community holds in high respect; and when these people come together to express their outrage over the offense…they develop a tighter bond of solidarity than existed earlier" (Erikson 1966, p. 4).

Crime can also lead to social change. For example, an episode of local violence may "achieve broad improvements in city services…be a catalyst for making public agencies more effective and responsive, for strengthening families and social institutions, and for creating public-private partnerships" (National Research Council 1994, pp. 9–10).

Although structural functionalism as a theoretical perspective deals directly with some aspects of crime, it is not a theory of crime per se. Three major theories of crime have developed from structural functionalism, however. The first, called strain theory, was developed by Robert Merton (1957) and uses Durkheim's concept of *anomie*, or normlessness. Merton argued that, when the structure of society limits legitimate means (e.g., a job) of acquiring culturally defined goals (e.g., money), the resulting strain may lead to crime. For example, rapid economic social change in Russia, and specifically high rates of unemployment, has lead to increases in the homicide rates (Pridemore & Kim 2007).

Individuals, then, must adapt to the inconsistency between means and goals in a society that socializes everyone into wanting the same thing but provides opportunities for only some (see Table 4.2). *Conformity* occurs when individuals accept the culturally defined goals and the socially legitimate means of achieving them. Merton suggested that most individuals, even those who do not have easy access to the means and goals, remain conformists. *Innovation* occurs when an individual accepts the goals of society but rejects or lacks the socially legitimate means of achieving them. Innovation, the mode of adaptation most associated with criminal behavior, explains the high rate of crime committed by uneducated and poor individuals who do not have access to legitimate means of achieving the social goals of wealth and power.

Another adaptation is *ritualism*, in which, for example, individuals accept a lifestyle of hard work but reject the cultural goal of monetary rewards. Ritualists go through the motions of getting an education and working hard, yet they are not committed to the goal of accumulating wealth or power. *Retreatism* involves rejecting both the cultural goal of success and the socially legitimate means of achieving it. Retreatists withdraw or retreat from society and may become alcoholics, drug addicts, or vagrants. Finally, *rebellion* occurs when individuals reject both culturally defined goals and means and substitute new goals and means. For example, rebels may use social or political activism to replace the goal of personal wealth with the goal of social justice and equality.

Whereas strain theory explains criminal behavior as a result of blocked opportunities, subcultural theories argue that certain

TABLE 4.2 Merton's Strain Theory

Mode of Adaptation	Culturally Defined Goals	Structurally Defined Means
1. Conformity	+	+
2. Innovation	+	−
3. Ritualism	−	+
4. Retreatism	−	−
5. Rebellion	±	±

+ = acceptance of/access to; − = rejection of/lack of access to; ± = rejection of culturally defined goals and structurally defined means and replacement with new goals and means

Source: From Robert K. Merton's *Social Theory and Social Structure* (1957). Adapted with permission of The Free Press, of Simon & Schuster Adult Publishing Group. Copyright© 1957 by The Free Press; copyright renewed 1985 by Robert K. Merton. All rights reserved.

groups or subcultures in society have values and attitudes that are conducive to crime and violence. Members of these groups and subcultures, as well as other individuals who interact with them, may adopt the crime-promoting attitudes and values of the group. For example, Kubrin and Weitzer (2003) found that retaliatory homicide is a response to subcultural norms of violence that exist in some neighborhoods.

However, if blocked opportunities and subcultural values are responsible for crime, why don't all members of the affected groups become criminals? Control theory may answer that question. Consistent with Durkheim's emphasis on social solidarity, Hirschi (1969) suggests that a strong social bond between individuals and the social order constrains some individuals from violating social norms. Hirschi identified four elements of the social bond: attachment to significant others, commitment to conventional goals, involvement in conventional activities, and belief in the moral standards of society. Several empirical tests of Hirschi's theory support the notion that the higher the attachment, commitment, involvement, and belief, the higher the social bond and the lower the probability of criminal behavior. Ford (2005), using data from the National Youth Survey, concludes that a strong family bond lowers the probability of adolescent substance use and delinquency, and Bell (2009) reports that a weaker attachment to parents is associated with a greater likelihood of gang membership for both males and females. Similarly, Gault-Sherman (2012), after analyzing the responses of a national sample of over 12,500 adolescents, reports a negative association between a youth's attachment to parents and self-reported delinquency—the higher the attachment, the lower the prevalence of delinquency.

Conflict Perspective

Conflict theories of crime suggest that deviance is inevitable whenever two groups have differing degrees of power; in addition, the more inequality there is in a society, the greater the crime rate in that society. Social inequality leads individuals to commit crimes such as larceny and burglary as a means of economic survival. Other individuals, who are angry and frustrated by their low position in the socioeconomic hierarchy, express their rage and frustration through crimes such as drug use, assault, and homicide. In Argentina, for example, the soaring violent crime rate is hypothesized to be "a product of the enormous imbalance in income distribution...between the rich and the poor" (Pertossi 2000). Further, an examination of homicide rates in 165 countries indicates that the greater the income inequality, the higher the homicide rate (Ouimet 2012).

According to the conflict perspective, those in power define what is criminal and what is not, and these definitions reflect the interests of the ruling class. Laws against vagrancy, for example, penalize individuals who do not contribute to the capitalist system of work and consumerism. Furthermore, D'Alessio and Stolzenberg (2002, p. 178) found that "in cities with high unemployment, unemployed defendants have a substantially higher probability of pretrial detention" than employed defendants. Rather than viewing law as a mechanism that protects all members of society, conflict theorists focus on how laws are created by those in power to protect the ruling class. Wealthy corporations contribute money to campaigns to influence politicians to enact tax laws that serve corporate interests (Reiman & Leighton 2012).

In addition, conflict theorists argue that law enforcement is applied differentially, penalizing those without power and benefiting those with power. For example, a 2009 report by the National Council on Crime and Delinquency found that "...in arrests, court processing and sentencing, new admissions and ongoing populations in prison and jails, probation and parole, capital punishment, and recidivism...persons of color, particularly African Americans, are more likely to receive less favorable results than their white counterparts" (Hartney & Vuong 2009, p. 2). Female prostitutes are more likely to be arrested than are the men who seek their services and, unlike street criminals, corporate criminals are most often punished by fines rather than lengthy prison terms.

Societal beliefs also reflect power differentials. For example, "rape myths" are perpetuated by the male-dominated culture to foster the belief that women are to blame for their own victimization, thereby, in the minds of many, exonerating the offenders. Such beliefs have very real consequences. Men who believe that "women secretly want

ANOEK DE GROOT/AFP/Getty Images

Under a plan called *Coalition Project 2012*, prostitutes can no longer advertise by standing or sitting in residential windows or storefronts and, thus, use art objects as symbolic representations. Governmental officials in the Netherlands, which legalized prostitution in 2000, are limiting the number of prostitutes and cannabis cafés in the hopes of reducing crime.

to be raped," "rape is not harmful," and "some women deserve to be raped" are more likely to commit rape than men who do not endorse such beliefs (Mouilso & Calhoun 2013).

Symbolic Interactionist Perspective

Two important theories of crime emanate from the symbolic interactionist perspective. The first, labeling theory, focuses on two questions: How do crime and deviance come to be defined as such, and what are the effects of being labeled criminal or deviant? According to Howard Becker (1963):

> Social groups create deviance by making rules whose infractions constitute deviance, and by applying those rules to particular people and labeling them as outsiders. From this point of view, deviance is not a quality of the act a person commits, but rather a consequence of the application by others of rules and sanctions to an "offender." The deviant is one to whom the label has successfully been applied; deviant behavior is behavior that people so label. (p. 238)

Labeling theorists make a distinction between **primary deviance**, which is deviant behavior committed before a person is caught and labeled an offender, and **secondary deviance**, which is deviance that results from being caught and labeled. After a person violates the law and is apprehended, that person is stigmatized as a criminal. This deviant label often dominates the social identity of the person to whom it is applied and becomes the person's "master status," that is, the primary basis on which the person is defined by others.

Being labeled as deviant often leads to further deviant behavior because (1) the person who is labeled as deviant is often denied opportunities for engaging in nondeviant behavior, and (2) the labeled person internalizes the deviant label, adopts a deviant self-concept, and acts accordingly. For example, a teenager who is caught selling drugs at school may be expelled and thus denied opportunities to participate in nondeviant school activities (e.g., sports and clubs) and to associate with nondeviant peer groups. The labeled and stigmatized teenager may also adopt the self-concept of a "druggie" or "pusher" and continue to pursue drug-related activities and membership in the drug culture. A review of the literature on labeling and juvenile delinquency lends support for the theory. One researcher concludes that, "[R]ather than discourage participation in conventional activities by labeling and isolating offenders, juvenile crime policy should be remedial and foster reintegration . . ." (Ascani 2012, p. 83).

The assignment of meaning and definitions learned from others is also central to the second symbolic interactionist theory of crime, differential association. Edwin Sutherland (1939) proposed that, through interaction with others, individuals learn the values and attitudes associated with crime as well as the techniques and motivations for criminal behavior. Individuals who are exposed to more definitions favorable to law violation (e.g., "crime pays") than to unfavorable ones (e.g., "do the crime, you'll do the time") are more likely to engage in criminal behavior. Thus, children who see their parents benefit from crime or who live in high-crime neighborhoods where success is associated with illegal behavior are more likely to engage in criminal behavior.

Unfavorable definitions come from a variety of sources. Of particular concern in recent years is the role of video games in promoting criminal or violent behavior such as the gratuitously violent video game, *Postal 2*, in which players can set harmless bystanders on fire. In response to this and other violent video games, many states now require a video rating system that differentiates between cartoon violence, fantasy violence, intense

primary deviance Deviant behavior committed before a person is caught and labeled an offender.

secondary deviance Deviant behavior that results from being caught and labeled as an offender.

violence, and sexual violence. In 2010, the U.S. Supreme Court heard arguments concerning the constitutionality of a California law that bans the sale of violent video games to minors. The court concluded that, like books, movies, or other art, "...video games are a creative, intellectual, emotional form of expression and engagement, as fundamentally human as any other" (Schiesel 2011, p. 1).

Individuals who are exposed to more definitions favorable to law violation (e.g., "crime pays") than to unfavorable ones (e.g., "do the crime, you'll do the time") are more likely to engage in criminal behavior.

What Do You Think? Toy guns that look realistic are illegal to sell in New York City (Marsh 2013). In 2011, the owner of a novelty shop that sold gun-shaped lighters that could fit in the palm of your hand and, arguably, looked like a real gun was fined $5,000 for each of the 12 lighters in his store. The owner of the store filed court papers to block the $60,000 fine, stating that he could not afford it. Would you find in favor of the store owner or the Manhattan Supreme Court? Why or why not?

Types of Crime

The FBI identifies eight index offenses as the most serious crimes in the United States. The **index offenses**, or street crimes as they are often called, can be against a person (called violent or personal crimes) or against property (see Table 4.3).

Other types of crime include vice crime (such as drug use, gambling, and prostitution), organized crime, white-collar crime, computer crime, and juvenile delinquency. Hate crimes are discussed in Chapter 9.

Street Crime: Violent Offenses

The most recent data available from the FBI's Uniform Crime Report indicate that the 2012 violent crime rate increased from the previous year (FBI 2013). Remember, however, that crime statistics represent only those crimes *reported* to the police: 1.2 million violent crimes in 2012 (FBI 2013). The reasons why people do not report violent victimization are varied (see Figure 4.1).

Violent crime includes homicide, assault, rape, and robbery. Between 2011 and 2012, of the four violent offenses, only the number of reported robberies declined (FBI 2013). *Murder* refers to the willful or nonnegligent killing of one human being by another individual or group of individuals. Although murder is the most serious of the violent crimes, it is also the least common, accounting for 1.2 percent of all violent crimes in 2012 (FBI 2013). A typical homicide scenario includes a male killing a male with a handgun after a heated argument. The victim and offender are disproportionately young and of minority status. When a woman is murdered and the victim–offender relationship is known, she is most likely to have been killed by her husband or boyfriend (FBI 2013).

Mass murders have more than one victim in a killing event. In 2013, Adam Lanza, after murdering his mother, entered Sandy Hook Elementary School in Newtown, Connecticut, and killed 20 first grade students and six staff members. The Sandy Hook killings, in conjunction with the 2012 murder of 12 theater goers in Aurora, Colorado, and the murder of six people attending a constituent meeting held by then Arizona

TABLE 4.3 Index Crime Rates, Percentage Change, and Clearance Rates, 2012

	Rate Per 100,000 2012	Percentage Change in Rate 2003–2012	Percentage Cleared, 2012
Violent crime			
Murder	4.7	−16.9	62.5
Forcible rape	26.9	−16.7	40.1
Robbery	112.9	−20.7	28.1
Aggravated assault	242.3	−18.0	55.8
Total	386.9	−18.7	46.8
Property crime			
Burglary	670.2	−9.6	12.7
Larceny/theft	1,959.3	−18.9	22.0
Motor vehicle theft	229.7	−47.0	11.9
Arson	N/A*	N/A	20.4
Total†	2,859.2	−20.4	19.0

Source: FBI 2013.

*Arson rates per 100,000 are calculated independently because population coverage for arson is lower than for the other index offenses.

†Property crime totals do not include arson.

index offenses Crimes identified by the FBI as the most serious, including personal or violent crimes (homicide, assault, rape, and robbery) and property crimes (larceny, motor vehicle theft, burglary, and arson).

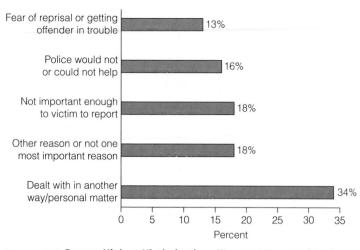

Figure 4.1 Reason Violent Victimizations Were Not Reported to the Police, 2006–2010
Source: Langston et al. 2012.

representative "Gabby" Giffords, led to a national debate about gun control and specifically the availability and use of high-capacity ammunition magazines and assault weapons (see the section titled "Gun Control").

Unlike mass murder, serial murder is the "unlawful killing of two or more victims by the same offender(s), in separate events" (U.S. Department of Justice 2008, p. 1). The most well-known serial killers, who were responsible for some of the most horrific episodes of homicide, are Ted Bundy, Kenneth Bianchi, and Jeffrey Dahmer. One of the most prolific serial killers is Gary Ridgeway, also known as the "Green River Killer," who was convicted of killing 49 women in 2003. As recently as 2012, victims of Ridgeway were being identified by DNA evidence from the remains of unidentified people (Myers 2012).

What Do You Think? The number of serial killers varies over time. According to one criminologist, serial killings were low in the 1960s and before, rose in the 1970s, and then again in the 1980s, but then began to decline again in the 1990s, and has remained relatively low since that time (Fox & Levin 2011). What do you think explains the differences in serial killings over the last 50 years?

In this picture from the video game *Postal 2*, a man holding a shovel has just killed innocent bystanders, predominantly women. *Postal 2* was one of the video games considered as part of the deliberation of the U.S. Supreme Court on whether or not video game content is protected under the First Amendment's right to free speech. In 2011, the Court ruled that video games, as books and plays, are "art" and, therefore, are protected by the First Amendment.

acquaintance rape Rape committed by someone known to the victim.

classic rape Rape committed by a stranger, with the use of a weapon, resulting in serious bodily injury to the victim.

Another form of violent crime, *aggravated assault*, involves attacking a person with the intent to cause serious bodily injury. Like homicide, aggravated assault occurs most often between members of the same race and, as with violent crime in general, is more likely to occur in warm-weather months. In 2012, the assault rate was over 50 times the murder rate, with assaults making up an estimated 62.6 percent of all violent crimes (FBI 2013).

It is estimated that nearly 80 percent of all rapes are **acquaintance rapes**—rapes committed by a family member, intimate partner, friend, or acquaintance (BJS 2013b). Although acquaintance rapes are the most likely to occur, they are the least likely to be reported and the most difficult to prosecute. Unless the rape is what Williams (1984) calls a **classic rape**—that is, the rapist was a stranger who used a weapon and the attack resulted in serious bodily injury—women hesitate to report the crime out of fear of not being believed. The increased use of "rape drugs," such as Rohypnol, may lower reporting levels even further. This chapter's *The Human Side* poignantly describes the impact of rape on one woman's life.

Robbery, unlike simple theft, also involves force or the threat of force or putting a victim in fear and is thus considered a violent crime. Officially, in 2012, more than 350,000 robberies took place in the United States. Robberies are most often committed using "strong-arm" tactics and occur disproportionately in southern states (FBI 2013). Robbers and thus robberies vary dramatically in type, from opportunistic robberies whose

In many jurisdictions, victims are allowed to make statements to the court describing the impact of their victimization on their life and the lives of friends and family. Here, 23-year-old Jessica poignantly describes the often unseen consequences of violent crime, in this case, the impact of a brutal rape.

My name is Jessica. I once knew what that meant. Now all I can tell you is that I am still Jessica; however, this no longer holds any meaning to me. I have lost my identity in the cruelest of ways.

I was a 23-year-old single mother, a sister, a daughter, a girlfriend, partner, friend, and confidant to many people. I was strong, independent, and willing to make an effort. I was attempting to hold down a second job to make a better life for my child and a better person out of me. I was destroyed.

I was raped. Not once, not twice, but so many times and in so many ways that it all becomes a blur. These images haunt my days, my nights, my dreams, and my realities. I am no longer the person I was before. I was once the person that people could rely on. Now I am a shell of my former self, a speck of the brave person that was Jessica. I had my way of life, my self-esteem, my respect, and my dignity stripped from me in the most terrifying of situations.

My trust in people is all but destroyed. I even have trouble enjoying a quiet drink with my partner, family, or friends without feeling anxious and wary. I am constantly looking over my shoulder, fearing there is someone there who wants to hurt me. This is only the beginning. I fluctuate from extreme insomnia to extreme fatigue. My motivation is gone. My joy of motherhood is waning. My ability to love and care for others are disappearing. My trust in the justice system is all but gone. I feel like I tried so hard and was beaten down. I feel weak and vulnerable.

I suffered physical trauma to my shoulder, knees, and feet from being dragged along carpet. I suffered cuts and bruising to my genitals from continuous rapes. I suffered marks to my ankles and wrists from being bound, and I lost chunks of hair from being gagged with tape around my head. I thought I was going to die. I take medication to sleep, to relax, and to stay relatively sane. I live as a corpse with no visible future to aim for. I have cut myself several times to try and take away the pain in my heart. I don't have the strength to end my own life.

What I experienced was like looking into the eyes of Satan himself. I am not a religious person but I know that I have seen the very depths of hell.

I wish that my loved ones didn't know what happened to me. To see the horror in their eyes and feel the pain in their hearts is unbearable. My beautiful daughter doesn't know the beginning of what I suffered but she suffered too. She feels like she was abandoned by her mother and required counseling.

The trauma happened to my partner. I can see it in the way he looks at me. Our communication, love, sex, and friendship is going to take a long time to repair. He has been so deeply affected by my experience but I can't help him and he can't help me. Nobody can.

I was burgled at work on my third day of my second job. I was bound and gagged. I was abducted. I was lied to. I was confused. I was cold and alone. I WAS RAPED. This man, who does not deserve a name, hurt me in the most unimaginable of ways. Yes I survived. Yes I am alive, I just don't live. I am existing. I am empty. . . .

I am a real person who went through torture. I am not a statistic or a nameless face on the street. I am your mother, your sister, your daughter, and your friend. What happened to me was real. At least give me the satisfaction of seeing this man put in jail for the maximum term of 25 years. After all, if I don't qualify as a victim of the most violent, serious, and heinous of crimes, then who the hell does?

I am Jessica.

Source: *The Herald Sun* 2008

---

victims are easily accessible and that yield only a small amount of money, to professional robberies of commercial establishments, such as banks, jewelry stores, and convenience stores. According to the FBI, in 2012, the average dollar value lost per robbery was $1,167, with bank robberies reporting the highest average loss per incident—$3,810 (FBI 2013). In 2013, eight robbers dressed as police officers and carrying machine guns stole over 100 packages of diamonds valued at $467 million from the tarmac of the Brussels international airport (Bartunek 2013).

Street Crime: Property Offenses

Property crimes are those in which someone's property is damaged, destroyed, or stolen; they include larceny, motor vehicle theft, burglary, and arson. The number of property crimes has declined since 1998, with a nearly one percent decrease between 2011 and 2012 (FBI 2013). **Larceny**, or simple theft, accounts for more than two-thirds of all property arrests (FBI 2013), and is the most common index offense. In 2012, the average dollar value lost per larceny incident was $987. Examples of larcenies include purse snatching, theft of a bicycle, pickpocketing, theft from a coin-operated machine, and shoplifting.

larceny Larceny is simple theft; it does not entail force or the use of force, or breaking and entering.

In 2012, an estimated 6.1 million larcenies were reported in the United States (FBI 2013). Larcenies involving automobiles and auto accessories are the largest known category of thefts. However, because of the cost involved, *motor vehicle theft* is considered a separate index offense. Numbering over 700,000 in 2012, the motor vehicle theft rate has decreased 47 percent since 2003 (FBI 2013). Because of insurance requirements, vehicle thefts are one of the most highly reported index crimes and, consequently, estimates between the FBI's Uniform Crime Reports and the National Crime Victimization Survey are fairly compatible. Less than 12 percent of motor vehicle thefts are cleared. Of the top ten car theft cities in 2012, eight were in California and two were in Washington State.

Burglary, which is the second most common index offense after larceny, entails entering a structure, usually a house, with the intent to commit a crime while inside. The burglary rate decreased by 4.4 percent between 2011 and 2012 (FBI 2013). Official statistics indicate that, in 2012, more than 2 million burglaries occurred (FBI 2013). Most burglaries are residential rather than commercial and take place during the day when houses are unoccupied. The most common type of burglary is forcible entry, followed by unlawful entry. Between 2005 and 2010, a six-year period, an estimated 1.4 million guns were stolen during burglaries and other property crimes (Langton 2012).

Arson involves the malicious burning of the property of another. Estimating the frequency and nature of arson is difficult given the legal requirement of "maliciousness." Of the reported cases of arson, almost half involved a structure, most of which were residential, and about a quarter involved movable property (e.g., boat or car), with the remainder being miscellaneous property (e.g., crops or timber). In 2012, the average dollar amount of damage as a result of arson was $12,796 (FBI 2013).

Vice Crime

Vice crimes, often thought of as crimes against morality, are illegal activities that have no complaining participant(s) and are often called **victimless crimes**. Examples of vice crimes include using illegal drugs, engaging in or soliciting prostitution, illegal gambling, and pornography.

Most Americans view illicit drug use, with the exception of marijuana, as socially disruptive (see Chapter 3). There is less consensus, however, nationally or internationally, that gambling and prostitution are problematic. For example, the Netherlands fully legalized prostitution in 2000, hoping to cut the ties between the sex trade and organized crime—a link that remains. Germany legalized prostitution in 2002, and it is estimated that there are now more than 400,000 sex workers, two-thirds of whom are from overseas. That said, one of the arguments commonly voiced against prostitution is that it makes it easier for women and girls to be forced into prostitution by traffickers. In a study of 116 countries, Cho, Dreher, and Neumayer (2013) conclude that countries where prostitution is legal have higher rates of human trafficking than countries in which prostitution is not legal.

In the United States, prostitution is illegal with the exception of several counties in Nevada. Despite its illegal status, it is a multimillion-dollar industry, with over 55,000 people arrested for prostitution and commercial vice in 2012 (FBI 2013). Human trafficking for purposes of prostitution occurs both *between* the United States and other countries, as well as *within* the United States. Children are particularly vulnerable.

> Pimps and traffickers sexually exploit children through street prostitution, in adult strip clubs, brothels, sex parties, motel rooms, hotel rooms, and other locations throughout the United States. Many recovered American victims are runaways or throwaway youth who often suffer from a history of physical abuse, sexual abuse, and family abandonment issues. This population is seen as an easy target by pimps because the children are generally vulnerable, without dependable guardians, and suffer from low self-esteem. (USDOJ 2012)

In 2003, the FBI established the Innocence Lost National Initiative to address domestic sex trafficking of children. Federal, state, and local law enforcement efforts have resulted

victimless crimes Illegal activities that have no complaining participant(s) and are often thought of as crimes against morality, such as prostitution.

in the recovery of 2,300 child victims. It is estimated that there are "tens of thousands" of child sex trafficking victims in the United States (Mertz 2013).

Gambling is legal in many U.S. states including casinos in Nevada, New Jersey, Connecticut, North Carolina, and other states, as well as state lotteries, bingo parlors, horse and dog racing, and jai alai. In 2013, a bill was introduced into Congress that would clarify the federal government's relationship to Internet gambling, even as three states—Delaware, Nevada, and New Jersey—with many more to join, authorized gambling online (NCSL 2013). Some have argued that there is little differences between gambling and other risky ventures such as investing in the stock market, other than societal definitions of acceptable and unacceptable behavior. Conflict theorists are quick to note that the difference is who is making the wager.

Pornography, particularly Internet pornography, is a growing international problem. Regulation is made difficult by fears of government censorship and legal wrangling as to what constitutes "obscenity." For many, the concern with pornography is not its consumption per se but the possible effects of viewing or reading pornography—increased sexual aggression. Although the literature on this topic is mixed, Conklin (2007, p. 221) concludes that there is no "consistent evidence that nonviolent pornography causes sex crimes." Supporting this contention, Diamond, Jozifkova, and Weiss (2011) report that there were no significant differences in the number of rapes and other sex crimes before and after the legalization of pornography in the Czech Republic.

TABLE 4.4 **The Evolution of Organized Crime**	
Original Activity	Modern Version
Local numbers and lottery gambling	Internet gambling at international sites outside national regulation
Heroin and cocaine tracking	Synthetic drugs (less vulnerable to supply problems)
Street prostitution	Internet prostitution and trafficking in human beings
Extortion of local businesses for protection money	Extortion of corporations, kidnappings, and piracy for profit
Loan sharking (exchanging money at interest rates above the rate permitted by law)	Laundering of money, precious stones, and commodities
Theft and fencing stolen property	Theft of intellectual property, Internet scams, and trafficking globally available goods (e.g., weapons, natural resources)

Source: Albanese 2012.

Organized Crime

Traditionally, **organized crime** refers to criminal activity conducted by members of a hierarchically arranged structure devoted primarily to making money through illegal means. For many people, organized crime is synonymous with the Mafia, a national band of interlocked Italian families, or the "Irish mob" made famous by such movies as *The Road to Perdition* and *Gangs of New York*. The Irish mob is one of the oldest organized crime groups in the United States. Fitting the stereotype of the "gangster" is James (Whitey) Bulger who, in 2013, was charged with 19 murders, drug trafficking, extortion, bribery, bookmaking, and loan sharking, just to name a few of the crimes listed in the 32-count indictment. He was convicted of 31 counts and sentenced to two consecutive life terms in prison (Seelye 2013; Murphy and Valencia 2013).

Organized crime also occurs in other countries. With nearly 80,000 members, the Japanese Yakuza is a network of 22 gangs divided into factions, each competing for power, wealth, and influence (McCurry 2012). Although membership in the syndicate remains legal, a 2011 law prohibits payments to the Yakuza, signaling the end of Japan's acceptance of the group. As a consequence of this law and other ordinances, organized crime members are having to rethink their traditional sources of revenue.

Although traditional organized crime groups still exist, Jay Albanese, a noted criminologist, contends that globalization and technology have facilitated a change from the "old" organized crime groups to *transnational* organized crime (TOC) groups (see Table 4.4). These groups continue to supply many of the same products and services (e.g., prostitution, drugs, gambling) and engage in many of the same behaviors (e.g., extortion, money laundering, theft) as their predecessors (Albanese 2012). Transnational crime organizations are a growing threat to the United States and to global security. According to the National Security Council:

Transnational organized crime (TOC) poses a significant and growing threat to national and international security, with dire implications for public safety,

organized crime Criminal activity conducted by members of a hierarchically arranged structure devoted primarily to making money through illegal means.

public health, democratic institutions, and economic stability across the globe. Not only are criminal networks expanding, but they also are diversifying their activities, resulting in the convergence of threats that were once distinct and today have explosive and destabilizing effects. (2011, p. 5)

White-Collar Crime

White-collar crime includes *occupational crime,* in which individuals commit crimes in the course of their employment, and *corporate crime,* in which corporations violate the law in the interest of maximizing profit. Occupational crime is motivated by individual gain. Employee theft of merchandise, or pilferage, is one of the most common types of occupational crime. Other examples include embezzlement, forgery and counterfeiting, and insurance fraud. Price fixing, antitrust violations, and security fraud are all examples of corporate crime, that is, crime that benefits the organization (see Figure 4.5).

In recent years, several officers of major corporations, including Enron, WorldCom, Adelphia, and ImClone, have been charged with securities fraud, tax evasion, and **insider trading**. In 2011, a federal investigation found that Goldman Sachs, an investment banking firm, had defrauded their customers and misled Congress. The results of the investigation led to a New York prosecutor subpoenaing the bank and investment firm's records (Hilzenrath 2011). Federal authorities, however, ended the investigation in 2012, because there was simply "...not a viable basis to bring a criminal prosecution" (Protess & Ahmed 2012, p. 1). In 2013, the U.S. Securities and Exchange Commission and the Department of Justice stepped up their investigation of Microsoft, alleging that corporate executives bribed government officials in China, Italy, and Romania, in the hopes of getting contracts (Blackden 2013).

No doubt triggered by investigations of mortgage fraud on the heels of the subprime mortgage crisis and a host of other financial scandals, the Obama administration has increased its attention on white-collar crime. In President Obama's 2012 State of the Union Address, he "...vowed...to establish a new financial crimes unit dedicated to investigating and prosecuting 'large-scale' financial fraud" (quoted in Carter & Berlin 2012). Nonetheless, since 2008, no executive associated with the top six banks tied to the economic collapse has been criminally prosecuted (Carter & Berlin 2012).

Many white-collar criminals go unpunished for a variety of reasons. First, many companies, not wishing the bad publicity surrounding a scandal, simply dismiss the parties involved rather than press charges. Second, many white-collar crimes, as with traditional crimes, go undetected. In a survey of a representative sample of 2,503 U.S. households, the National White Collar Crime Center (NWCCC) found that nearly one in four households (24 percent) had been the victim of some type of white-collar crime in the previous year. However, only 11.7 percent of the crimes were brought to the attention of a law enforcement agency (NWCCC 2010). Americans' perception that white-collar crime is not "real crime" and, therefore, is less serious (Holtfreter et al. 2008) likely contributed to this lack of reporting.

Third, federal prosecutions of white-collar criminals have generally decreased recently. Few believe the decrease is a result of a lower prevalence of white-collar crime offenses. Two forces appear to be in operation. First, white-collar crimes are becoming increasingly complex, making prosecution a time- and resource-intensive endeavor. Second, the decrease in white-collar crime prosecutions represents a shift in priorities (Marks 2006). After the events of 9/11, nearly one-third of FBI agents were moved from criminal programs to terrorism and intelligence duties, leaving "the bureau seriously exposed in investigating areas such as white-collar crime . . ." (Lichtblau et al. 2008, p. 1).

> Corporate violence is the result of negligence, the pursuit of profit at any cost, and intentional violations of health, safety, and environmental regulations.

white-collar crime Includes both *occupational crime,* in which individuals commit crimes in the course of their employment, and *corporate crime,* in which corporations violate the law in the interest of maximizing profit.

insider trading The use of privileged (i.e., nonpublic) information by an employee of an organization that gives that employee an unfair advantage in buying, selling, and trading stocks or other securities.

Corporate violence, a form of corporate crime, refers to the production of unsafe products and the failure of corporations to provide a safe working environment for their employees. Corporate violence is the result of negligence, the pursuit of profit at any cost, and intentional violations of health, safety, and environmental regulations. Two incidents, both having international repercussions, serve as examples.

The first entails the accusation of sudden acceleration problems in Toyota vehicles and, the second, the Gulf Oil spill. Although recent reports suggest there is little scientific evidence tying car accidents to sudden acceleration issues in Toyota vehicles, documents have surfaced that do detail Toyota Motor Corporation's delay in recalling millions of vehicles thought to be defective (Frank 2011; Maynard 2010). In 2013, Toyota recalled 1.3 million cars because of concerns over the operation of their air bags (Rankin 2013).

Second, the British Petroleum (BP) oil spill off the coast of Louisiana epitomizes, perhaps more than any other case in history, the public's increasing concern with corporate violence and the need for corporate responsibility. Eleven workers were killed and more than 4 million barrels of oil spilled into the Gulf of Mexico. Soon after the spill, a Gallup Poll indicated that the majority of Americans believe that BP should "...pay for all financial losses resulting from the Gulf Coast oil spill, including wages of workers put out of work, even if those payments ultimately drive the company out of business" (Newport 2010). A report by the National Research Council found that the government's efforts to assess the costs of the "Deepwater Horizon" disaster has failed and, consequently, the settlement of environmental claims will be difficult (Goldenberg 2013).

TABLE 4.5 **Types of White-Collar Crime**	
CRIMES AGAINST CONSUMERS	**CRIMES AGAINST EMPLOYEES**
Deceptive advertising	Health and safety violations
Antitrust violations	Wage and hour violations
Dangerous products	Discriminatory hiring practices
Manufacturer kickbacks	Illegal labor practices
Physician insurance fraud	Unlawful surveillance practices
CRIMES AGAINST THE PUBLIC	**CRIMES AGAINST EMPLOYERS**
Toxic waste disposal	Embezzlement
Pollution violations	Pilferage
Tax fraud	Misappropriation of government funds
Security violations	Counterfeit production of goods
Police brutality	Business credit fraud

© Cengage Learning 2013

Computer Crime

Computer crime refers to any violation of the law in which a computer is the target or means of criminal activity. Sometimes called cybercrime, computer crime is one of the fastest-growing types of crime in the United States. Hacking, or unauthorized computer intrusion, is one type of computer crime. In 2013, hackers in more than two dozen countries coordinated with one another and successfully stole $45 million from thousands of ATMs (Santora 2013).

The Consumer Sentinel Network logged over 2 million consumer complaints in 2013, with identity theft being the most common complaint category (FTC 2013). **Identity theft** is the use of someone else's identification (e.g., Social Security number or birth date) to obtain credit or other economic rewards. Although mail theft has traditionally been the means of obtaining the needed information (e.g., dumpster diving), new technologies have contributed to the increased rate of identity theft. According to the National Crime Victimization Survey, in 2010, 7 percent of households reported at least one incident of identity theft and, of that number, nearly two-thirds experienced either the misuse or attempted misuse of a credit card (BJS 2012a).

Identity theft is just one category of computer crime. Another category is Internet fraud. According to the Internet Crime Complaint Center (ICCC 2013), common types of Internet fraud include nondelivery of merchandise or payment, real estate fraud, impersonation of an FBI agent, and auto fraud (e.g., selling a car that isn't yours). Other Internet fraud categories include check fraud, confidence fraud, Nigerian letter fraud (i.e., a letter offering the recipient the "opportunity" to share in millions of dollars being

corporate violence The production of unsafe products and the failure of corporations to provide a safe working environment for their employees.

computer crime Any violation of the law in which a computer is the target or means of criminal activity.

identity theft The use of someone else's identification (e.g., Social Security number, birth date) to obtain credit or other economic rewards.

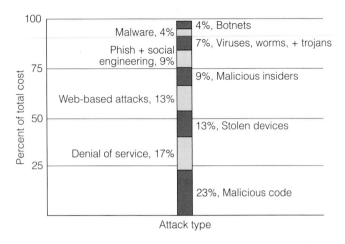

Figure 4.2 Distribution of Annual Cost of Cybercrime by Attack Type, 2011*

Source: NCVC 2013.

*Sample of 50 large organizations.

Edward Snowden, presently seeking permanent asylum, was a computer analyst who, in 2013, leaked classified documents revealing U.S. telephone and Internet surveillance programs to the public. He has been charged with espionage.

illegally transferred to the United States—just send us your bank account numbers!), computer fraud, and credit/debit card fraud.

Another type of computer crime is online child sexual exploitation. As of December 2012, the National Center for Missing and Exploited Children had reported 1.7 million cases of suspected child sexual exploitation to law enforcement agencies and analyzed over 80 million images of child pornography (NCMEC 2013). One such law enforcement agency is the Internet Crimes Against Children (ICAC) Task Force, which includes over 3,000 law enforcement officers and prosecutorial agencies working together in 61 coordinated networks. In 2011, the ICAC investigations led to nearly 6,000 arrests (OJJDP 2012).

Finally, it should be noted that individuals are not the only victims of computer crime. Organizational cyberattacks are also very common, with "malicious code" intrusions being the most frequent (NCVC 2013) (see Figure 4.2). Verizon's (2013) *Data Breach Investigations Report*, which analyzed 621 data breaches and over 47,000 security incidents from 27 countries, details attacks against organizations—governments, corporations, banks, and the like. The results indicate that

- ATMs, followed by desktop computers, file servers, and laptops are the most vulnerable sources for data breaches;
- One-third of data breaches used "social tactics" (e.g., e-mail, phone calls) to gather information;
- Seventy-five percent of breaches took advantage of weak or stolen credentials;
- Most attacks are opportunistic and financially motivated;
- The majority of attacks do not involve employees or "other insiders";
- The majority of attacks require little skill beyond basic methods, and no resources; and
- Nearly one in five attacks were connected to government employees (i.e., were espionage).

What Do You Think? In 2013, Edward Snowden, a former employee of a federal contractor, released classified documents that revealed secret government programs that collected telephone and Internet data on American citizens. A public opinion poll of U.S. registered voters in the same year found that the majority of Americans, 55 percent, viewed Snowden as a whistle-blower rather than a traitor, i.e., a criminal (Salant 2013). What do you think? Is Snowden a traitor as the government contends or a hero for exposing what he claims are civil liberty violations?

Juvenile Delinquency and Gangs

In general, children younger than age 18 are handled by the juvenile courts, either as status offenders or as delinquent offenders. A *status offense* is a violation that can be committed only by a juvenile, such as running away from home, truancy, and underage drinking. A *delinquent offense* is an offense that would be a crime if committed by an adult, such as the eight index offenses. The most common status offenses handled in juvenile court are underage drinking, truancy, and running away. In 2012, 10.8 percent of all arrests (excluding traffic violations) were of offenders younger than age 18 (FBI 2013). As is the case with adults, juveniles commit more property crimes than violent crimes.

The number of juveniles *arrested* for violent crimes decreased by 11.1 percent between 2011 and 2012 (FBI 2013). Nonetheless, the National Gang Intelligence Center (NGIC) estimates that there has been a 40 percent increase in gang membership since 2009 (NGIC 2012). The growth of gangs is, in part, a function of two interrelated social forces: the increased availability of guns in the 1980s, and the lucrative and expanding drug trade. Recently, however, gangs have expanded into nontraditional gang activities such as smuggling, prostitution, and human trafficking (NGIC 2012).

In 2011, there were over 33,000 prison, motorcycle, and violent street gangs with over 1.4 million members in the United States (NGIC 2012). Averaged across jurisdictions, gang members were responsible for nearly half of all violent crime and are becoming more and more violent, "...acquiring high-powered, military-style weapons and equipment which poses a significant threat because of the potential to engage in lethal encounters with law enforcement officers and civilians" (NGIC 2012, p. 1).

The National Gang Center surveys a representative sample of over 2,500 U.S. law enforcement agencies about the characteristics of youth gangs in their jurisdictions (NGC 2012). The results indicate that (1) there has been an increase in the number of adult gang members in recent years, (2) the percentage of female gang members has remained stable over time, and (3) gang members are disproportionately Hispanic/Latino and African American in all jurisdictions represented. Figure 4.3 indicates that when asked about

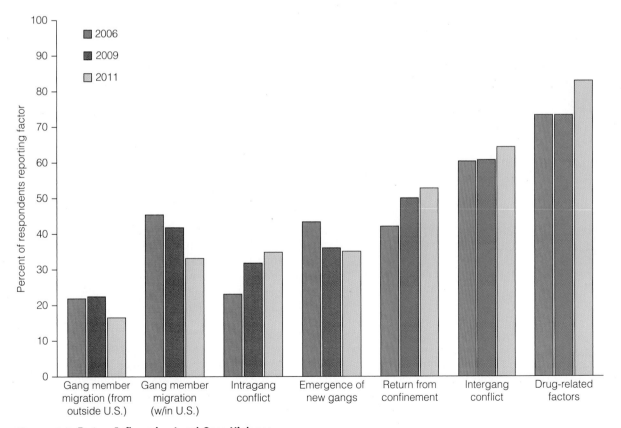

Figure 4.3 Factors Influencing Local Gang Violence
Source: National Gang Center 2012.

factors influencing youth gang violence, "drug-related factors" not only received the highest ranking in each of the three survey years, the proportion of respondents selecting drugs as a cause of gang violence has increased over time.

Demographic Patterns of Crime

Although virtually everyone violates a law at some time, individuals with certain demographic characteristics are disproportionately represented in the crime statistics. Victims, for example, are disproportionately young, lower-class, minority males from urban areas. Similarly, the probability of being an offender varies by gender, age, race, social class, and region. This section ends with a discussion of crime and victimization.

Gender and Crime

It is a universal truth that women everywhere are less likely to commit crime than men. In the United States, both official statistics and self-report data indicate that females commit fewer violent crimes than males. In 2011, males accounted for 73.8 percent of all arrests, 80.1 percent of all arrests for violent crime, and 62.6 percent of all arrests for property crimes (FBI 2013). Not only are females less likely than males to commit serious crimes, but also the monetary value of female involvement in theft, property damage, and illegal drugs is typically far less than that for similar offenses committed by males.

Nonetheless, rates of female criminality have increased dramatically over the last decade. Between 2003 and 2012, arrest rates for women increased for burglary and larceny, contributing to the 24.7 percent increase of women arrested for property crimes (FBI 2013). Over the same time period, female arrests increased dramatically, over 20 percent, for robbery, driving while intoxicated, and public drunkenness. Heimer et al. (2005) hypothesize that such increases are a function of the *economic marginalization* of women relative to men, that is, female criminality goes up when women's economic circumstances in relation to men's decline.

The recent increase in crimes committed by females has led to the growth of feminist criminology. **Feminist criminology** focuses on how the subordinate position of women in society affects their criminal behavior and victimization. For example, Chesney-Lind and Shelden (2004) reported that arrest rates for runaway juvenile females are higher than those for males not only because girls are more likely to run away as a consequence of sexual abuse in the home but also because police with paternalistic attitudes are more likely to arrest female runaways than male runaways. Feminist criminology, concentrating on gender inequality in society, thus adds insights into understanding crime and violence that are often neglected by traditional theories of crime. Feminist criminology has also had an impact on public policy. Mandatory arrest for domestic violence offenders, the development of rape shield laws, public support for battered women's shelters, laws against sexual harassment, and the repeal of the spousal exception in rape cases are all, according to Winslow and Zhang (2008), outcomes of feminist criminology.

> [T]he development of rape shield laws, public support for battered women's shelters, laws against sexual harassment, and the repeal of the spousal exception in rape cases are all . . . outcomes of feminist criminology.

feminist criminology An approach that focuses on how the subordinate position of women in society affects their criminal behavior and victimization.

Age and Crime

In general, criminal activity is more prevalent among younger people than among older people. In 2012, 39.5 percent of all arrests in the United States were of people younger than age 25 (FBI 2013). Although those younger than age 25 made up over half of all arrests in the United States for crimes such as robbery, burglary, vandalism, and arson, those younger than age 25 were significantly less likely to be arrested for white-collar crimes such as embezzlement, fraud, forgery, and counterfeiting. Those 65 and older made up less than 1 percent of total arrests for the same year.

Why is criminal activity more prevalent among individuals in their late teens and early 20s and rapidly declining thereafter? One reason is that juveniles are insulated from many of the legal penalties for criminal behavior. Younger individuals are also more likely to be unemployed or employed in low-wage jobs. Thus, as strain theorists argue, they have less access to legitimate means for acquiring material goods. On the other hand, the decline in criminal offenses associated with aging may be a function of the transition to conventional roles—employee, spouse, and parent.

Other hypothesized reasons for the age–crime relationship are also linked to specific theories of criminal behaviors. For example, conflict theorists would argue that teenagers and young adults have less power in society than their middle-aged and elderly counterparts. One manifestation of this lack of power is that the police, using a mental map of who is a "typical offender," are more likely to have teenagers and young adults in their suspect pool. With increased surveillance of teenagers and young adults comes increased detection of criminal involvement—a self-fulfilling prophecy.

In the hopes of resolving the theoretical debate over the relationship between age and crime, Sweeten, Piquero, and Steinberg (2013) examined over 40 independent variables and their association with self-reported delinquency. Using a sample of 1,300 serious offenders, the researchers interviewed youth at the age of 16 and every six months thereafter for a period of seven years. Although there was some level of support for each of the theories tested, the strongest theoretical explanation of the relationship between age and delinquency was Sutherland's learning theory—the differential association theory—confirming the positive association between criminality and antisocial peers.

What Do You Think? Hundreds of thousands of people watched the trial of George Zimmerman as news channels across the nation broadcast the event live or summarized the happenings with late-night specials and "breaking news" updates. After the verdict determined that George Zimmerman killed Trayvon Martin in self-defense, discussions concerning the issue of race and, specifically, racial profiling, abound in the United States. Do you think the outcome would have been different if George Zimmerman had been black and Trayvon Martin had been white?

Race, Social Class, and Crime

Race is a factor in who gets arrested. Minorities are disproportionately represented in official statistics. For example, African Americans represent about 13 percent of the population but account for 38.5 percent of all arrests for violent index offenses, and 29.3 percent of all arrests for property index offenses (FBI 2013). They have 3.7 times the arrest rate for possession of marijuana, are six times more likely to be admitted to prison and, if admitted to prison for a violent crime, receive longer sentences than their white counterparts (ACLU 2013; Hartney & Vuong 2009).

Nevertheless, it is inaccurate to conclude that race and crime are causally related. First, official statistics reflect the behaviors and policies of criminal justice actors. Thus, the high rate of arrests, conviction, and incarceration of minorities may be a consequence of individual and institutional bias against minorities. For example, **racial profiling**—the practice of targeting suspects on the basis of race—may be responsible for their higher arrest rates. Proponents of the practice argue that because race, like gender, is a significant predictor of who commits crime, the practice should be allowed. Opponents hold that racial profiling is little more than discrimination, often based on stereotypes, and thus should be abolished. A survey of Seattle residents

racial profiling The law enforcement practice of targeting suspects on the basis of race.

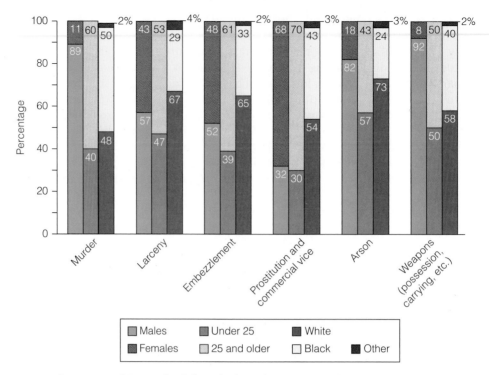

Figure 4.4 **Percentage of Arrests for Selected Crimes by Sex, Age, and Race, 2012**
Source: FBI 2013.

The finding of George Zimmerman as not guilty in the death of Trayvon Martin sparked demonstrations across the United States. Many celebrities, including Jay-Z, Madonna, Stevie Wonder, Usher, Alicia Keys, and Justin Timberlake, responded to the verdict by saying they would not perform in Florida until the "stand your ground law" had been changed.

lends support to such a contention (Drakulich 2013). The results indicate that "...crime stereotypes about racial and ethnic minorities are associated with reduced perceptions of neighborhood safety and increased anxieties about victimization..." among white respondents (p. 322). Presently, more than 25 states have laws that prohibit racial profiling and/or require that state jurisdictions collect data on police stops and searches (Resource Center 2013).

In a 2013 statement, President Obama acknowledged that he, as many other African American males, had been the victim of racial profiling (see also Chapter 9). Referring to the Trayvon Martin shooting, the President went on to say that "...the African-American community is knowledgeable that there is a history of racial disparities in the application of our criminal laws, everything from the death penalty to enforcement of our drug laws" and that history impacts people's perceptions of the case (Obama 2013, p.1). After the verdict, hundreds of protests occurred across the country with people carrying signs reading "Justice for Trayvon" and "Could I be the next?"

Second, race and social class are closely related in that nonwhites are overrepresented in the lower classes. Because lower-class members lack legitimate means to acquire material goods, they may turn to instrumental, or economically motivated, crimes. In addition, although the "haves" typically earn social respect through their socioeconomic status, educational achievement, and occupational role, the "have-nots" more

A large body of research documents that fighting in adolescence is a fairly common event, although frequencies vary significantly by race, age, and sex. For example, black youth are more likely to report being involved in at least one physical fight in the previous year when compared to their white counterparts (CDCP 2008). Thus, the present study is an important one for it examines the relationship between select explanatory variables (both risks and assets) and the likelihood of fighting among a sample of at-risk minority youth (Wright & Fitzpatrick 2006).

Sample and Methods

All respondents were in grades 5 through 12 and were enrolled in a central Alabama school system (Wright & Fitzpatrick 2006). The school district sent home a letter detailing the purpose of the study, and parents were asked to give permission for their child's inclusion in the study and referred to a sample questionnaire that was on file and available for review. The final sample consisted of 1,642 African American youth (51 percent female) with a median age of 14 years. Participation in the survey was voluntary, and the response rate was 65 percent.

The dependent variable is fighting and was measured by asking respondents the frequency of their fighting in the last 30 days. Sociodemographic variables include sex, age, mother's and father's education, and mother's and father's occupational status. Risk factors, that is, factors associated with an increase in the likelihood of fighting, include academic performance (poor grades), family intactness (zero or one parent in the home),

parental violence (physical assault from parent or other adult guardian), and a composite measure of gang affiliation (being a member, being asked to be a member, or having friends who are members).

Asset variables are variables that are predicted to decrease incidents of fighting. The eight asset variables were divided into three categories. The first category is self-esteem measured by a respondent's sense of satisfaction, pride, worth, and respect. Parental involvement was measured by a respondent's (1) parents monitoring of where their child goes with their friends, (2) frequency of talking to parents about problems, and (3) frequency of eating dinner with the family. School involvement includes teacher attention, respondent's involvement in school activities and clubs, and self-reported happiness with school.

Results and Conclusions

Frequency of fighting was higher for elementary and middle school students than for high school students, with the highest fighting frequency occurring in middle schools. Across school type, 15 percent or more of the students reported the highest response category of fighting—six or more times over the last 30 days.

Data analysis also indicated that fighting is negatively associated with family intactness and self-esteem, with two-parent homes and high self-esteem leading to lower probabilities of fighting. Alternatively, parental violence and gang affiliation are associated with increased probabilities of fighting. Talking to parents about problems and having parents who monitor

activities with friends are significantly associated with decreased rates of fighting. Additional asset variables associated with decreased levels of fighting include being happy with school, attention from teachers, and involvement with school clubs. As the authors note, interestingly, higher involvement in sports was associated with higher rates of fighting.

When multiple variables were analyzed at the same time, three of the four risk factors were significantly associated with higher rates of fighting—lower grades, exposure to violence in the family, and gang affiliation. Of the asset variables, a lack of parental monitoring and being unhappy at school were predictive of increased fighting behavior. Note that low self-esteem, when in the presence of other variables, is not associated with youthful fighting.

The authors concluded that the "risk and asset" model they present has practical implications in terms of organizing intervention techniques. Risk factors need to be "suppressed or eliminated," and asset factors need to be "encouraged or facilitated" (Wright & Fitzpatrick 2006, p. 260). For example, results from the present study show that "parental monitoring and being happy at school were associated with lower frequency of fighting, suggesting the importance of continued support for outreach to parents and further efforts to reduce or eliminate the community factors that promote proliferation of gangs" (Wright & Fitzpatrick 2006, p. 251).

Source: Wright & Fitzpatrick 2006.

often live in communities where respect is based on physical strength and violence, as subcultural theorists argue. For example, Kubrin (2005) examined the "street code" of inner-city black neighborhoods by analyzing rap music lyrics. Her results indicate that "lyrics instruct listeners that toughness and the willingness to use violence are central to establishing viable masculine identity, gaining respect, and building a reputation" (p. 375). This chapter's *Social Problems Research Up Close* feature examines violence and "being tough" in a sample of minority youth.

A third hypothesis is that criminal justice system contact, which is higher for nonwhites, may actually act as the independent variable; that is, it may lead to a lower position in the stratification system. Kerley and colleagues (2004) found that "contact with the criminal justice system, especially when it occurs early in life, is a major life event that has a deleterious effect on individuals' subsequent income level" (p. 549).

Some research indicates, however, that even when social class backgrounds of blacks and whites are comparable, blacks have higher rates of criminality. Using data

> [African Americans]…have 3.7 times the arrest rate for possession of marijuana, are six times more likely to be admitted to prison, and, if admitted to prison for a violent crime, receive longer sentences than their white counterparts.

from nearly 3,000 respondents aged 18 to 25, Sampson et al. (2005) found that the likelihood of self-reported violence by blacks was 85 percent higher than for whites. Interestingly, the likelihood of Latino self-reported violence was 10 percent less than that reported by whites.

Region and Crime

In general, crime rates and, in particular, violent crime rates, increase as population size increases. For example, in 2012, the violent crime rate in metropolitan statistical areas (MSA) was 409.4 per 100,000 inhabitants; in cities outside of metropolitan areas, it was 380.4 per 100,000 inhabitants; and, finally, in nonmetropolitan counties, the crime rate was 177 per 100,000 inhabitants (FBI 2013). Further, the National Crime Victimization Survey indicates that people who live "…in urban areas continued to experience the highest rate of total and serious violence" (Truman & Planty 2012).

Higher crime rates in urban areas result from several factors. First, social control is a function of small intimate groups that socialize their members to engage in law-abiding behavior, expressing approval for their doing so and disapproval for their noncompliance. In large urban areas, people are less likely to know one another and thus are not influenced by the approval or disapproval of strangers. Demographic factors also explain why crime rates are higher in urban areas: Large cities have large concentrations of poor, unemployed, and minority individuals. Some cities, including the 10 most dangerous cities in the United States, have crime rates as much as five times the national average.

Crime rates also vary by region of the country. In 2012, both violent and property crimes were highest in southern states with 43.6 percent of all murders, 37.6 percent of all rapes, and 42.9 percent of all aggravated assaults recorded in the South (FBI 2013). The high rate of southern lethal violence has been linked to high rates of poverty and minority populations in the South, a southern "subculture of violence," higher rates of gun ownership, and a warmer climate that facilitates victimization by increasing the frequency of social interaction.

Crime and Victimization

In contrast to the Uniform Crime Report, the National Crime Victimization Survey (NCVS) indicates a 17 percent increase in violent crime and an 11 percent increase in property crime between 2010 and 2011 (Truman & Planty 2012). Males have a higher rate of violent victimization than females, racial and ethnic minorities have a higher rate of violent victimization than white non-Hispanics, and people between the ages of 18 and 24 have the highest rate of violent victimization when compared to other age groups.

Because minorities are disproportionately offenders and violent victimization is often directed toward family members, friends, and acquaintances, it is not surprising that African Americans, Hispanics, and American Indian/Alaska natives are disproportionately victimized. Familiarity also explains the high victimization rates of males and young adults between the ages of 18 and 24.

According to the National Criminal Victimization Survey, violent crimes were more likely to be reported than property crimes. Rape and sexual assault were the least likely violent crimes to be reported (27 percent), followed by aggravated assault (40 percent). Theft was the least likely reported property crime (30 percent). Between 2002 and 2011, the number of rape and sexual assaults reported to law enforcement decreased by 51 percent, the largest decline during the time period.

Although males are more likely than females to be victims of violent crime, females are particularly vulnerable to certain kinds of crimes, most notably those that entail sexual violence. In 2013, Ariel Castro was charged "…with 512 counts of kidnapping, 446 cases of rape, seven counts of gross sexual imposition, six counts of felonious assault, three counts of child endangerment, and one count of possessing criminal tools" (Welsh-Huggins 2013). Castro, who had abducted three young women, ages 14, 16, and 20, after each had accepted a ride, imprisoned them for nearly 10 years, allegedly causing several miscarriages through assault and starvation of the pregnant women.

Sexual violence against women is not unique to the United States. In 2013, a series of brutal rapes in India, including the gang rape of a 23-year-old college student who died from the assault, sparked public outrage around the world. At issue is not only the failure of the Indian criminal justice system to treat such attacks seriously, but the way in which gender inequality fosters violence against women (see Chapter 10).

Children too are especially vulnerable, both directly and indirectly. In 2010, 10 percent of children, i.e., people under the age of 18, were victims of homicide and, of that number, approximately a third were between the ages of 13 and 16, and nearly 15 percent were under the age of one (NCVC 2013). Violent victimization of children was higher in single-headed households, households with a reported annual income of less than $15,000, and households in urban areas (Truman & Smith 2012).

A less obvious form of victimization is the exposure of children to violence in their homes, schools, and communities. In a comprehensive survey of children's exposure to violence, data on more than 4,500 children and adolescents age 17 and younger were collected from a national sample. Respondents between the ages of 10 and 17 were interviewed over the phone, and information about children aged 9 and younger was gathered by interviews with adult caregivers. The results indicate that 61 percent of the sample had been exposed to violence in the previous year, either as victims directly (e.g., abused) or indirectly (e.g., threatened with assault), or as witnesses (e.g., saw bullying) of violence (Finkelhor et al. 2011; see Figure 4.5). Consumption of media violence was not included in the study. The authors conclude that exposure to violence in children is associated with increased depression, anxiety, and behavioral problems including juvenile delinquency.

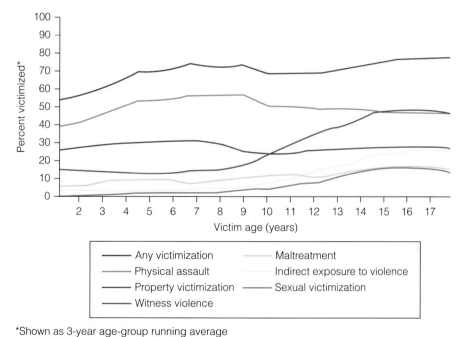

*Shown as 3-year age-group running average

Figure 4.5 Past Year Victimization by Type and Victim Age
Source: Finkelhor et al. 2011.

The Societal Costs of Crime and Social Control

As this chapter has demonstrated, there are a variety of types of crimes and criminals. Nonetheless, often based on media stereotypes, many people think only of street crimes and criminals as predominantly "...young black men living in poor urban neighborhoods committing violent and drug-related crimes" (Leverentz 2012, p. 348). The costs of crime, and many of them are incalculable, go far beyond those perpetrated by what is thought to be the "typical offender." For example, transnational organized crime, perhaps the most insidious of all crime categories, "...threatens peace and human security, leads to human rights being violated, and undermines the economic, social, cultural, political, and civil development of societies around the world (UNODC 2012). Although these costs are impossible to quantify, the following section discusses other costs associated with crime and criminals, including physical injury and loss of life, economic losses, and social and psychological costs.

Physical Injury and the Loss of Life

Crime often results in physical injury and the loss of life. For example, homicide is the second most common cause of death among 15- to 25-year-olds, exceeded only by accidental death (U.S. Census Bureau 2012). In 2012, 14,827 people were the victims of known homicides in the United States (FBI 2013). That number is dwarfed, however, by the deaths that take place as a consequence of white-collar crime. Criminologist Steven Barkan (2006), who collected data from a variety of sources, reports that there are annually (1) 56,425 workplace-related deaths from illness or injury; (2) 9,600 deaths from unsafe products; (3) 35,000 deaths from environmental pollution; and (4) 12,000 deaths from unnecessary surgery. Adding these figures together, 113,025 people a year die from corporate and professional crime and misconduct (p. 388).

"Green criminologists" study one example of the impact of white-collar crime and, specifically, corporate violence. In a review of the literature, Katz (2012) identifies several important findings from the existing research on corporate pollution and health and, specifically, cancer mortality:

> First, a variety of research illustrates multinational corporate culpability in the proliferation of environmental pollution both in the USA and globally. Second, this has resulted in increasing cancer mortality rates across the globe among minority communities as well as non-western developing nation states.... Third, corporate environmental pollution has been facilitated through the proliferation of international free trade agreements and international financial loans from international financial institutions....

Using a conflict theory perspective, with its emphasis on the problems created by capitalism, Katz (2012) examines cancer mortality rates and the location of the richest transnational corporations in the world. She reports that 10 of the 152 countries studied have significantly higher cancer mortality rates than the remainder and, of that number, 6 are the locations of the headquarters of the largest transnational corporations in the world. On closer examination, Katz (2012) argues, the data reveal that most of the industries in the 10 nations where the cancer mortality rate is the highest are chemical, energy, water, and oil. Further, Dow Chemical Company, a U.S.-based corporation that Katz (2012) calls "the primary global polluter," has operations in 9 of the 10 nations.

Finally, the U.S. Public Health Service now defines violence as one of the top health concerns facing Americans. Health initiatives related to crime include reducing drug and alcohol use and the deaths and diseases associated with them, lowering rates of domestic violence, preventing child abuse and neglect, and reducing violence through public health interventions. It must also be noted that crime has mental as well as physical health consequences (see the section titled "Social and Psychological Costs").

The High Price of Crime

Conklin (2007, p. 50) suggests that the financial costs of crime can be classified into at least six categories. First are *direct losses* from crime, such as the destruction of buildings through arson, of private property through vandalism, and of the environment by polluters. In 2011, the average dollar loss of destroyed or damaged property as a result of arson was $12,796 (FBI 2013). White-collar crime also has direct costs for those who are victimized. According to the Internet Crime Complaint Center, the average cost of crimes perpetrated on the Internet (e.g., fraud, nondelivery of goods) was $1,813 (ICCC 2013).

Second are costs associated with the *transferring of property*. Bank robbers, car thieves, and embezzlers have all taken property from its rightful owner at tremendous expense to the victims and society. The total cost of stolen goods in 2012 was estimated to be $15.5 billion. Of the total cost of stolen goods, the largest single category can be attributed to motor vehicle theft, which includes, for example, the theft of "...sport utility vehicles, automobiles, trucks, buses, motorcycles, motor scooters, all-terrain vehicles, and snowmobiles." The average dollar loss per vehicle was $6,019 (FBI 2013).

A third major cost of crime is that associated with *criminal violence*, including the medical cost of treating crime victims. The National Crime Prevention Council (NCPC 2005) estimates that the average cost for *each* criminal incident of rape or sexual assault is $7,700, including expenses related to medical and mental health care, law enforcement, and victim and social services. Recently, Anderson (2012) estimates that crime-related deaths and injuries cost society, directly and indirectly, over $750 billion annually.

Fourth are the costs associated with the production and sale of illegal goods and services, that is, *illegal expenditures*. The expenditure of money on drugs, gambling, and prostitution diverts funds away from the legitimate economy and enterprises, and lowers property values in high-crime neighborhoods. Fifth is the cost of *prevention and protection*—the billions of dollars spent on locks and safes, surveillance cameras, self-defense products, guard dogs, insurance, counseling and rehabilitation programs, and the like. For example, more specifically, it is estimated that Americans spend $16.2 billion annually on security systems alone (U.S. Department of Commerce 2011).

Finally, there is the cost of *controlling crime*—billions of dollars, and the costs are escalating. Figure 4.6 indicates, however, that criminal justice spending varies by function, with spending on law enforcement increasing over the years and spending on criminal justice assistance (i.e., victim services) decreasing (Austin 2013). If passed, a recent bill in Vermont—similar to a Missouri law—would have required that judges are informed of the costs of the sentences they impose (Eisen 2013). The "...cost of incarcerating gang members actually exceeds annual expenses at top private universities, which can total about $60,000 per student for tuition, room, and board" (NCPC 2012).

Although the costs from "street crimes" are staggering, the costs from "crimes in the suites," such as tax evasion, fraud, embezzlement, false advertising, and antitrust violations, are greater than the cost of the FBI index crimes combined (Reiman & Leighton 2012). Further, Barkan (2006), using an FBI estimate, reported that the total cost of property crime and robbery is $17.1 billion annually. This is less than the $44 billion price tag for employee theft alone.

Social and Psychological Costs

Crime entails social and psychological costs as well as economic costs. One such cost—fear—is dependent upon individual perceptions of crime as a problem. For example, surveys since 2001 indicate that Americans' fear of victimization has increased even as violent and property crime rates have decreased (Saad 2010). Such misconceptions are fueled by media presentations that may not accurately reflect the crime picture.

Kohm et al. (2012) researched fear of criminal victimization in relation to media consumption in samples of American and Canadian undergraduate students. Respondents were asked their primary source of crime news, which was then classified into six

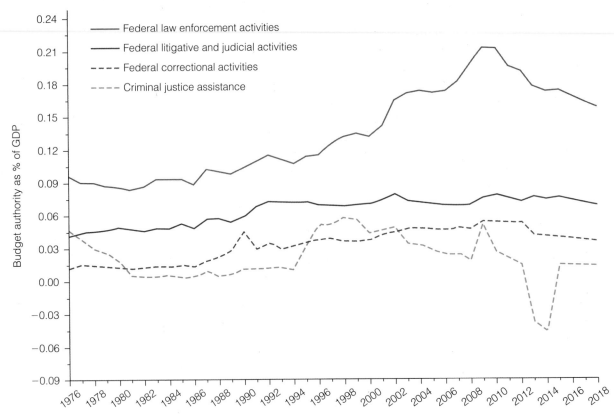

Figure 4.6 Administration of Justice Costs, by Function, 1976–2018 (2014–2018 projected)*

Source: Austin 2013.

*Discretionary Budget Authority as a Percentage of GDP, FY1976–FY2018; FY2014–FY2018 levels projected.

categories: local television, national television, newspapers, magazines, the Internet, and other. Media salience was measured by the number of hours or the number of times a particular media was consumed. The results indicate that women are more fearful than men, that Canadian students are more fearful than American students, and that higher rates of local television news consumption is associated with higher rates of fear.

Not only do Americans worry about crime at the aggregate level, but they also worry about crime at the individual level. The Gallup Organization regularly asks a sample of Americans whether they "feel safe walking alone at night in the city or area where you live?" (Jones 2013a). Gallup then "…compiled the data for residents of the 50 most populous metropolitan statistical areas [MSA]…" (p. 1). Based on respondents' answers, the three safest MSAs are Minneapolis-St. Paul-Bloomington (Minnesota), Denver-Aurora (Colorado), and Raleigh-Cary (North Carolina). The three least safe MSAs of the 50 largest in the United States are Memphis, TN-MS-AR (includes parts of Tennessee, Mississippi, and Arkansas), New Orleans-Metairie-Kenner (Louisiana), and Riverside-San Bernardino-Ontario (California).

What Do You Think? In 2012, Jerry Sandusky, former assistant coach at Pennsylvania State University (PSU), was convicted of sexual abuse of young boys and was sentenced to 30 to 60 years in prison. Male victims of sexual abuse often suffer from anxiety, depression, low self-esteem, a sense of shame, and self-destructive behaviors (Male Survivor 2012). How could you quantify the cost of crime for these victims? Do you think that any adequate monetary settlement can be made?

Shapland and Hall (2007) identify other social and psychological costs such as a sense of shock, a loss of trust, feelings of guilt for being victimized, anger, and a sense of vulnerability. Although these responses vary by type of offense, it should not be concluded that white-collar crimes do not carry a social and psychological toll. In 2009, Bernard Madoff pled guilty to 11 felony counts related to a massive Ponzi scheme run through his investment firm. The scheme to defraud investors of $65 billion took place over a 20-year period and involved thousands of victims. The 71-year-old Madoff was sentenced to 150 years in prison, leaving behind him a trail of misery. Following are excerpts from just seven of the thousands of victims impacted by his crime. They attest to the social and psychological costs of white-collar crime (Victim Statements 2009):

- He robbed us not only of our money, but of our faith in humanity, and in the systems in place that were supposed to protect us.
- I can't tell you how scattered we feel—it goes beyond financially. It reaches the core and affects your general faith in humanity, our government and basic trust in our financial system.
- I am constantly nervous and anxious about my future. I jump at the slightest noise. I can't sleep and all I do is worry about what will happen to us.
- I don't know which emotion is more destructive, the fear and anxiety or the major depression that I experienced daily.
- ...when you sentence Madoff...I trust that you touch on the loss of money, the loss of dignity, the loss of freedom from financial worries and possible financial ruin....
- At this point...we cannot trust anyone.
- How do I live the rest of my life?

Strategies for Action: Crime and Social Control

Clearly, one way to combat crime is to attack the social problems that contribute to its existence. Moreover, when a random sample of Americans were asked which of two views—increasing law enforcement or resolving social problems—came closer to their own in dealing with the crime problem, the majority of respondents (65 percent) selected resolving social problems (Gallup Poll 2007).

In addition to policies that address social problems, numerous social programs have been initiated to alleviate the crime problem. These policies and programs include local initiatives, criminal justice policies, legislative action, and international efforts in the fight against crime.

Local Initiatives

Youth programs such as the Boys and Girls Club, and community programs that involve families and schools are an effective "first line of defense" against crime and juvenile delinquency.

Youth Programs. Early intervention programs acknowledge that preventing crime is better than "curing" it once it has occurred. Fight Crime: Invest in Kids is a nonpartisan, nonprofit anticrime organization made up of more than 5,000 law enforcement leaders, prosecutors, and violence survivors (Fight Crime 2013). The organization holds that "the best way to prevent violence is through investments in quality early childhood education, volunteer in-home parent coaching, afterschool programs, and interventions for troubled youth" (Fight Crime 2013, p. 1). Fight Crime advocates four major areas of activity:

- Strategic recruitment and education of law enforcement leaders and crime survivors
- Analysis of research and policy and production of reports
- Earned (not paid) public education media campaign
- Education of policy makers and opinion leaders

This approach not only benefits society by reducing crime rates, but it also saves the taxpayers money. Cohen and Piquero (2009) report that programs that prevent a child from using drugs, dropping out of school, and becoming a career criminal save society more than $2.5 million per person over the course of their lifetime.

The Perry Preschool Project is a good example of an early childhood intervention program. After a sample of 123 African American children was randomly assigned to either a control group or an experimental group, the experimental group members received academically oriented interventions for one to two years, frequent home visits, and weekly parent–teacher conferences. The control group members received no interventions. The control group members and the experimental group members were then compared at various points in time between the ages of 3 and 40. As adults, experimental group members had higher employment and homeownership rates, and significantly lower violent and property crime rates (Schweinhart 2007).

Finally, many youth programs are designed to engage juveniles in noncriminal activities and integrate them into the community. Because the hours between 3:00 and 6:00 p.m. and the summer months are peak times for delinquency, engaging students in afterschool programs is not only essential but successful (Afterschool Alliance 2013). For example, an evaluation of a California afterschool program called LA's BEST concluded that children who attended the program were 30 percent less likely to engage in criminal behavior than those who did not attend the program (Afterschool Alliance 2013).

Neighborhood watch programs involve local residents in crime prevention strategies. For example, MAD DADS (Men Against Destruction—Defending Against Drugs and Social Disorder) patrol the streets in high-crime areas of the city on weekend nights, providing positive adult role models and fun community activities for troubled children. Members also report crime and drug sales to police, paint over gang graffiti, organize gun buyback programs, and counsel incarcerated fathers. At present, 75,000 men, women, and children are in MAD DADS in 60 chapters in 17 states throughout the country (MAD DADS 2013).

The National Association of Town Watch (NATW) is a nonprofit organization "…dedicated to the development and promotion of various crime prevention programs including neighborhood watch groups, law enforcement agencies, state and regional crime prevention associations, businesses, civic groups, and individuals, devoted to safer communities" (NATW 2013, p. 1). NATW began the "National Night Out" event in 1984, and today, over 35 million people in 15,000 communities in 50 states participate in the event in which citizens, businesses, neighborhood organizations, and local officials join together in outdoor activities to heighten awareness of neighborhood problems, promote anticrime messages, and strengthen community ties (NATW 2013).

Mediation and victim–offender dispute resolution programs are also increasing, with thousands of such programs worldwide. The growth of these programs is a reflection of their success rate: Two-thirds of cases referred result in face-to-face meetings, 95 percent of these cases result in a written restitution agreement, and 90 percent of the written restitution agreements are completed within one year (VORP 2012).

Criminal Justice Policy

The criminal justice system is based on the principle of **deterrence**—the use of harm or the threat of harm to prevent unwanted behaviors. The criminal justice system assumes that people rationally choose to commit crime, weighing the rewards and consequences of their actions. Thus, "get-tough" measures hold that maximizing punishment will increase deterrence and cause crime rates to go down. Yet, most recently, incarceration rates have declined and crime has not increased, calling into question the principle of deterrence. Further, 30 years of "get-tough" policies have not significantly reduced recidivism rates but have merely led to prison overcrowding and a host of criminal justice problems. In 2011, the U.S. Supreme Court ordered the state of California to reduce its prison population, saying that the overcrowded conditions were a violation of the Eighth Amendment's prohibition against cruel and unusual punishment. In 2013, the governor of California requested that a court order to release 10,000 inmates be set aside (Martinez

deterrence The use of harm or the threat of harm to prevent unwanted behaviors.

2013). Given overcrowding and high recidivism, experts are scratching their collective heads and asking, "what works?"

Law Enforcement Agencies. In 2011, Anders Behring Breivik killed 69 people at a youth camp where over 700 teenagers and young adults were meeting before Norwegian police shot him. A 2012 commission investigating the mass murder concluded that "the police and security services could and should have done more to avert the crisis" (Greene 2012).

The United States had nearly 1 million full-time law enforcement officers and full-time civilian employees in 2012 (e.g., clerks, meter attendants, correctional guards), yielding an estimated 3.4 law enforcement employees per 1,000 inhabitants and 2.4 sworn officers per 1,000 inhabitants (FBI 2013). There are over 14,000 law enforcement agencies in the United States, including municipal (e.g., city police), county (e.g., sheriff's department), state (e.g., highway patrol), and federal agencies (e.g., FBI), often with overlapping jurisdictions. Equal numbers of Americans have high confidence in the criminal justice system as those who have low confidence in the criminal justice system, and there are few differences on the basis of gender, political party, race, and education. However, those between the ages of 18 and 34 have significantly more confidence in the criminal justice system than those 35 years and older (Saad 2011).

In 2011, the latest year for which national data are available, 26 percent of the 16 and older U.S. resident population had contact with the police (BJS 2013c). According to the Police-Public Contact Survey, a supplement to the National Crime Victimization Survey (NCVS), contacts were equally initiated by citizens (e.g., request for police assistance) and police (e.g., traffic stop). Results of the survey indicate that blacks and Hispanics were more likely to receive traffic tickets than whites, and to be stopped by the police for a traffic violation. When a traffic stop occurred, black and Hispanic drivers were also more likely to be searched or frisked than white drivers.

For decades, accusations of racial profiling, police brutality, and discriminatory arrest practices have made police–citizen cooperation in the fight against crime difficult. Concerns have been fed by such highly publicized cases as the beating of Rodney King, the arrest of Harvard scholar Henry Louis Gates Jr., and the shooting of Trayvon Martin. There is some empirical support for such contentions. For example, one study concluded that arrests for the possession of marijuana, over 50 percent of all drug arrests, are not only exorbitantly expensive but that "enormous disparities exist in states and counties nationwide between arrest rates of blacks and whites for marijuana possession" (ACLU 2013).

In response to such trends, the Violent Crime Control and Law Enforcement Act of 1994 established the Office of Community Oriented Policing Services (COPS). Community policing

> …emphasizes proactive problem solving in a systematic and routine fashion. Rather than responding to crime only after it occurs, community policing encourages agencies to proactively develop solutions to the immediate underlying conditions contributing to public safety problems. Problem solving must be infused into all police operations and guide decision-making efforts. (COPS 2009, p. 12)

Such an approach often includes "problem-oriented policing," a model that includes analyzing the underlying causes of crime, looking for solutions, and actively seeking out alternatives to standard law enforcement practices. One success story is in Miami, Florida, where, over 20 years ago, the Miami Police Department developed a program for school-age children called "Do the Right Thing." The initiative not only reinforces positive youthful behaviors but it also provides the opportunity for the development of meaningful relationships between children and the police (COPS 2012).

Rehabilitation versus Incapacitation. An important debate concerns the primary purpose of the criminal justice system: Is it to rehabilitate offenders or to incapacitate them through incarceration? Both rehabilitation and incapacitation are concerned with **recidivism** rates, or the extent to which criminals commit another crime.

recidivism A return to criminal behavior by a former inmate, most often measured by rearrest, reconviction, or reincarceration.

Advocates of **rehabilitation** believe that recidivism can be reduced by changing the criminal, whereas proponents of **incapacitation** think that recidivism can best be reduced by placing offenders in prison so that they are unable to commit further crimes against the general public.

Fear of crime has led to a public emphasis on incapacitation and a demand for tougher mandatory sentences, longer time served, a reduction in the use of probation and parole, and support for truth-in-sentencing laws. As a result of these policies, the United States is one of the most punishment-oriented countries in the world. For example, the *punitivity ratio* is calculated "…by contrasting the number of people convicted in a year with the number of prisoners in jail as result of a court sentence" (Civitas 2012). Using this calculation, and comparing the 33 countries studied in the Organisation for Economic Cooperation and Development, the United States had the highest *punitivity ratio*, followed by Mexico, Japan, and Israel.

Of late, however, these get-tough measures and the skyrocketing incarceration rates have come under attack for five reasons. *First,* research indicates that incarceration may not deter crime. Data indicate that,

> …despite the massive increase in corrections spending, in many states there has been little improvement in the performance of corrections systems. If more than four out of 10 adult American offenders still return to prison within three years of their release, the system designed to deter them from continued criminal behavior clearly is falling short. (Pew 2011a, p. 3)

Second is the accusation that get-tough measures, such as California's "three strikes and you're out" policy, are not equally applied. Chen's (2008) analysis of over 170,000 California inmates indicates that African Americans compared to whites and Hispanics are more likely to receive "third-strike sentences," with the greatest racial disparities being for property and drug offenses. Similarly, males are more likely to receive third-strike sentencing than females. In 2013, Californians passed a law that modifies the three strikes policy. As revised, life sentences are only imposed if the third conviction is a serious or violent felony (Donald 2013).

Third, there are concerns that putting people in prison may actually escalate their criminal behaviors once released. Inmates may learn new and better techniques of committing crime and, surrounded by other criminals, further internalize the attitudes toward and motivation for continued criminal behavior. Such a possibility is called the **breeding ground hypothesis**. Further, as labeling theorists suggest, the stigma associated with being in prison makes reentry into society difficult. Unable to find employment, return to school, and meet new friends, former prisoners may feel they have no choice but to return to crime.

Fourth, in an environment of budget deficits and legislative cuts, states simply can no longer afford the policies of decades ago. The average annual cost of incarcerating an offender, more than $29,000 for federal inmates and $30,000 for state inmates, is actually much higher than often stated (BOP 2013; Henrichson & Delaney 2012). Additional costs, often not part of the calculations, include the cost of pensions, health care, and other benefits for correctional employees, construction and renovation costs, administrative and legal expenses, and the cost of rehabilitation programs, hospital care, and private prisons for inmates.

Finally, some are questioning the logic of increased criminal justice spending when crime has declined nearly 40 percent in the last two decades (CBS 2012). At the same time crime has been decreasing, the average prison sentence has been increasing. For example, the average length of time served in prison was nine months longer in 2009 than in 1990 (Pew 2012). For some, the inverse relationship between incarceration rates and crime is an indication that get-tough policies work.

As a response to the economic downturn and concerns over the effectiveness of get-tough policies, many states are rethinking correctional policies—closing prisons, eliminating mandatory sentencing, replacing jail time with community-based programs, shorter terms, commuting sentences, expanding parole, and providing treatment rather than punishment for nonserious drug offenders (Davey 2010; Steinhauer 2009). The result? After 40 years of growth, incarceration rates have begun to decline. States with the greatest declines include California, Hawaii, Michigan, Massachusetts, and New Jersey (Pew 2013).

rehabilitation A criminal justice philosophy that argues that recidivism can be reduced by changing the criminal through such programs as substance abuse counseling, job training, education, and so on.

incapacitation A criminal justice philosophy that argues that recidivism can be reduced by placing offenders in prison so that they are unable to commit further crimes against the general public.

breeding ground hypothesis A hypothesis that argues that incarceration serves to increase criminal behavior through the transmission of criminal skills, techniques, and motivations.

Clearly, sentencing more offenders for longer periods of time to confinement enhances incapacitation. However, faced with budget cuts, states—as well as the Obama administration—are revisiting the ideals of rehabilitation. Rehabilitation assumes that criminal behavior is caused by sociological, psychological, and/or biological forces rather than being solely a product of free will. If such forces can be identified, the necessary change can be instituted. Many rehabilitation programs focus on helping inmates reenter society. In 2008, President Bush signed the Second Chance Act, which supports reentry programs in the hopes of reducing recidivism (Greenblatt 2008). This chapter's *Photo Essay* highlights successful rehabilitation programs, including reentry initiatives.

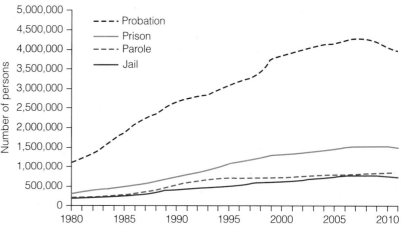

Figure 4.7 Correctional Populations in the United States, 1980–2011
Sources: BJS 2011a, 2011b, 2011c.

What Do You Think? In the wake of rising costs, escalating incarceration rates, and financially strapped communities, private prisons were thought to be a solution—less expensive and better results. However, there is some evidence that privatization is not without problems and, in fact, may not be the most efficient and cost-effective means to house inmates (Kirkham 2013). For example, for-profit prisons want to save money. Recently, cost-saving strategies such as cutting back on correctional personnel, delaying medical treatment of inmates, and failing to repair plumbing led inmates to riot in an Ohio correctional facility run by Corrections Corporation of America, the largest private prison franchise in the United States. What are some of the other problems associated with the privatization of prisons?

Corrections. Prison population rates vary dramatically by regions of the world but share at least one similarity—they are growing (Walmsley 2012). An examination of global rates reveals that the United States has the highest incarceration rate in the world—743 per 100,000 population. The U.S. rate exceeds many times over those of other countries; for example, the rate in Russia is 568; China's rate is 122; the rate for England and Wales is 153; for Canada, 117; and France and Germany, 96 and 85, respectively (Walmsley 2012).

The U.S. prison population grew by over 700 percent between 1970 and 2005, exceeding both population growth and crime (Takei 2013; see Figure 4.7). As a result, the United States, which has only 5 percent of the world's population, now has 25 percent of the world's prisoners.

Probation entails the conditional release of an offender who, for a specific time period and subject to certain conditions, remains under court supervision in the community. **Parole** entails release from prison, for a specific time period and subject to certain conditions, before the inmate's sentence is finished. Although varying by race, age, and gender, nearly 5 million people were on probation or parole in the United States in 2011 (BJS 2012b).

The U.S. prison population grew 700 percent between 1970 and 2005 . . . [and] . . . [A]s a result, the United States, which has only 5 percent of the world's population, now has 25 percent of the world's prisoners.

probation The conditional release of an offender who, for a specific time period and subject to certain conditions, remains under court supervision in the community.

parole Parole entails release from prison, for a specific time period and subject to certain conditions, before the inmate's sentence is finished.

Millions of men and women are behind bars, and recidivism is at an all-time high. Over the next decade, the number of female inmates is projected to increase and there will be more elderly, both groups adding to the already skyrocketing cost of corrections. There are alternatives, however, and many of them have proven to be successful. Each of the described correctional practices has been empirically evaluated and found to be associated with positive changes in the inmate participants. From reducing recidivism rates and enhancing self-esteem, to lowering aggression and increasing the likelihood of post-release employment, these programs not only are cost-effective, but they are also humane.

Shaul Schwarz/Getty Images News/Getty Images

Heather Rowley

◄ Thousands of inmates across the United States participate in postsecondary education programs. Research documents their benefits: reduced recidivism, enhanced problem-solving skills, safer prison conditions, a more marketable post-release inmate, and taxpayer savings (Correctional Association 2009). San Quentin's Prison University Project (PUP) is just such a program. Taught by college professors who volunteer their time, PUP provides twenty courses each semester—classes in the humanities, liberal arts, social sciences, English, math, and science—leading to an associate of arts degree (PUP 2013). Students in the program pay no fees or tuition, using textbooks donated from publishers. Pictured, a San Quentin inmate working on a paper for class.

Twinkle,
Twinkle
little star,
How I wonder
what you are!

◄ Over 200,000 women are incarcerated in the United States and about one-third of them are imprisoned for drug offenses. One in 25 women in state prisons and one in 31 women in federal prisons are pregnant at the time that they are incarcerated. With the exception of 13 states, women can be shackled during labor and delivery. In most instances, the babies born to incarcerated mothers are immediately taken away (The Sentencing Project 2012). Nebraska's Prison Nursery Program provides inmates with prenatal care education, parenting skills, information on child development, "hands-on training" for new and expectant mothers, and community resources upon release. After 10 years of operation, an evaluation of the program revealed lower misconduct and recidivism rates when compared to inmates who were required to give up their infants (Carlson 2009). Pictured, female inmates and their children wait for a parenting class to begin.

Chris Seward/Raleigh News & Observer/MCT via Getty Images

Trevor Kobrin/Prison Entrepreneurship Program

◄ When Catherine Rohr, a Wall Street investor, toured a Texas prison, she had an epiphany—criminals and people in business are a lot alike. They both assess risks, live by their instincts, share profits, network, and compete with one another. It was then she founded the Prison Entrepreneurship Program (PEP). Today, corporate leaders and faculty volunteers teach business skills to PEP participants—former drug dealers, gang leaders, hustlers, and felons—by equipping them with tools for success. An examination of the three-year recidivism rate indicates that less than 5 percent of the PEP graduates returned to prison. Because participants spend less time in prison, the program saves the taxpayers of Texas $10 million for every 200 graduates (PEP 2012). Pictured, PEP students talk with community business leaders as part of their five month series of classes.

▲ Programs such as Puppies Behind Bars, Puppies in Prison, Pen Pals, Project Pooch, and Prison Pet Partnership have been instrumental in changing the lives of inmates, breeder and shelter dogs, and the beneficiaries of the inmate–trainers' months of hard work and discipline. The men and women in the Indiana Canine Assistant and Adolescent Network (ICAAN) program train service and therapy dogs. An empirical evaluation of ICAAN documents the positive impact on the rehabilitation of participating offenders—higher self-esteem and better communication skills, and a marked improvement in patience and trust, leading to better inmate–prison employee relations (Turner 2007). Such programs are also available for some military prisoners. Here a Marine, sitting in the brig, hugs his Dog Eve who he is training as a service dog for returning military vets in need of assistance.

Capital Punishment. With **capital punishment**, the state (the federal government or a state) takes the life of a person as punishment for a crime. In 2012, 21 countries carried out 682 executions excluding the thousands of people who are thought to have been executed in China (Amnesty International 2013). Further, at the end of 2012, 23,000 people were under death sentences. Contrary to in the United States, globally, offenders may be sentenced to death for crimes other than murder including adultery, blasphemy, treason, and drug offense. The trend, however, is away from capital punishment, with ". . . 174 of the hundred and ninety-three members of the United Nations execution free in 2012" (Amnesty International 2013). The United States is the only Western industrialized nation in the world to retain the death penalty.

In 2012, 43 executions were carried out in the United States, 15 in Texas alone (Amnesty International 2013). Of the 33 states that have the death penalty, the majority and the federal government almost exclusively use lethal injection as the method of execution. Three problems, however, have been raised with the use of lethal injection, leading some states to halt executions. First is the question of whether or not death by lethal injection violates the Eighth Amendment's prohibition against cruel and unusual punishment. In 2007, a district court judge held that Tennessee's lethal injection procedures "present a substantial risk of unnecessary pain" that could "result in a terrifying, excruciating death" (Schelzig 2007, p. 1). However, in 2008, the U.S. Supreme Court held that Kentucky's use of lethal injection was not a violation of the Eighth Amendment (*Baze v. Rees* 2008). Since the *Baze* decision, some states have reinstated the death penalty while others have declared a moratorium until the lethal injection debate is resolved.

The second issue concerns the role of physicians in state executions. According to Vu (2007, p. 1), "[t]he American Medical Association is adamant that it is a violation of medical ethics for doctors to participate in, or even be present at, executions." In 2010, the American Board of Anesthesiology (ABA) adopted the position that participation in capital punishment would prohibit a physician from becoming certified by the ABA, arguing that physicians ". . . should not be expected to act in ways that violate the ethics of medical practice, even if these acts are legal" (ABA 2011).

The third issue surrounding the continued use of capital punishment concerns the availability of the drugs used in lethal injection. Pharmaceutical companies throughout Europe, where capital punishment is banned, have refused to supply the drugs under threat of legal sanctions. Beginning in 2011, the two most common sedatives used in the lethal injection process, hospira and pentobarbital, became unavailable as European manufacturers refused to sell to U.S. prisons. Correctional facilities then turned to propofol, the drug used in the connection with the death of Michael Jackson. However, in 2012, the manufacturer released a statement, saying "[W]e understand that one or more departments of corrections in the U.S. are considering amendment of their lethal injection protocols to include propofol. . . . [C]learly, such use is contrary to the FDA-approved indications for propofol and inconsistent with . . . the mission of 'caring for life'" (Kitamura & Narayan 2013, p. 1).

Proponents of capital punishment argue that executions of convicted murderers are necessary to convey public disapproval and intolerance for such heinous crimes. Those against capital punishment believe that no one, including the state, has the right to take another person's life and that putting convicted murderers behind bars for life is a "social death" that conveys the necessary societal disapproval.

capital punishment The state (the federal government or a state) takes the life of a person as punishment for a crime.

Proponents of capital punishment also argue that it deters individuals from committing murder. Critics of capital punishment hold, however, that because most homicides are situational and are not planned, offenders do not consider the consequences of their actions before they commit the offense. Critics also point out that the United States has a higher murder rate than most western European nations that do not practice capital punishment, and that death sentences are racially discriminatory. A study of capital punishment in the United States, between 1973 and 2002, found that ". . . minority death row inmates convicted of killing whites face higher execution probabilities than other capital offenders" (Jacobs et al. 2007, p. 610). Further, there is concern over the representativeness of juries and the integrity of the jury selection process. In 2012, a North Carolina judge ruled that an inmate's death sentence should be commuted to life in prison after it was found that prosecutors ". . . deliberately excluded qualified black jurors from jury service . . . and [that] there was evidence this was happening in courts throughout the state" (Eng 2012, p. 1). This was the first time that the North Carolina Racial Justice Act, which allowed prisoners who were facing execution to present evidence of racial bias to the court, was used. In 2013, the Racial Justice Act was repealed by the North Carolina legislature (Severson 2013).

Capital punishment advocates suggest that executing a convicted murderer relieves taxpayers of the costs involved in housing, feeding, guarding, and providing medical care for inmates. Opponents of capital punishment argue that financial considerations should not determine the principles that decide life and death issues. In addition, taking care of convicted murderers for life may actually be less costly than sentencing them to death because of the lengthy and costly appeals process for capital punishment cases. As of 2011, 17 states have abolished the death penalty including Connecticut, Illinois, New York, Michigan, and Wisconsin. Note, however, no southern state to date has repealed their death penalty statute.

Those in favor of capital punishment argue that it protects society by preventing convicted individuals from committing another crime, including the murder of another inmate or prison official. One study of the deterrent effect of capital punishment concluded that each execution is associated with at least eight fewer homicides (Rubin 2002). Opponents contend that capital punishment may result in innocent people being sentenced to death. According to the Innocence Project, there have been 311 post-conviction exonerations using DNA evidence since 1989. The most common reasons for wrongful convictions, in order, are (1) eyewitness misidentification, (2) faulty forensic science, (3) false confessions or admissions, (4) government misconduct, (5) informants (e.g., paid informants lying), and (6) inadequate legal counsel (Innocence Project 2013).

Legislative Action

Legislative action is one of the most powerful methods of fighting crime. Federal and state legislatures establish criminal justice policy by the laws they pass, the funds they allocate, and the programs they embrace.

What Do You Think? In 2013, 5-year-old Kristian Sparks accidentally shot and killed his 2-year-old sister Caroline with a gun given to him by his parents—a "youth gun" marketed in both pink and blue as "My First Rifle," specifically for children (Gabriel 2013). Although some Americans are horrified by the notion of giving children guns, others value teaching their children the proper way to use them. The local county coroner remarked, "Down in Kentucky where we're from, you know, guns are passed down from generation to generation. . . . You start at a young age with guns for hunting and everything" (quoted in Gabriel 2013). At what age, if any, and under what circumstances should a parent be allowed to give their a child a gun?

Gun Control. According to a report by the Bureau of Justice Statistics, in 2011, firearms were used in 70 percent of the nation's murders, 26 percent of robberies, and 31 percent of aggravated assault (Planty & Truman 2013). Males, African Americans, and people between the ages of 18 and 24 were the most likely to be victims of fatal firearm violence. Firearm homicides are highest in the South and lowest in the Northeast regions of the United States. Globally, the United States has the highest percentage of murders committed with a firearm in the world (See Figure 4.8). The United States also has the highest civilian gun ownership rate in the world with 89 guns for every 100 people. In the United States, of all homicides, 60 percent are committed with a firearm (see Figure 4.8).

There are nearly 130,000 federally licensed gun stores in the United States, over 10 times the number of McDonald's. According to the Bureau of Alcohol, Tobacco, Firearms, and Explosives, 5.5 million new firearms were manufactured in the United States in 2010.

An examination of data collected from interviews with 6,000 U.S. adults between 2007 and 2012 indicates the extent to which gun ownership is predictable using sociological variables. The single best predictor of gun ownership is gender—45 percent of the men surveyed owned a gun, compared to 15 percent of women. Gun ownership was higher in southern compared to northern states and among married rather than nonmarried respondents. A social portrait of the group most likely to own guns included older, married, Republican males from the South, who attend church often but have comparatively lower levels of education. The group least likely to own guns includes unmarried, non-southern women (Jones 2013b). Of those who own guns, 62 percent own more than one gun and 17 percent own a semiautomatic weapon (Infographic 2012).

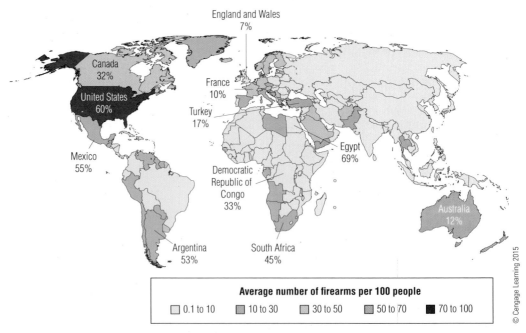

Figure 4.8 **Average Number of Firearms Per 100 People, and Percent of all Murders Committed with a Firearm, Select Countries***

Source: Rogers 2012.

*Percent of Murders by Firearm data not available for all countries; years vary.

Those against gun control argue that not only do citizens have a constitutional right to own guns but also that more guns may actually lead to less crime as would-be offenders retreat in self-defense when confronted (Lott 2003). Advocates of gun control, however, insist that the estimated 300 million privately owned firearms in the United States (Arnold 2013) significantly contribute to the violent crime rate and distinguish the United States from other industrialized nations. Since 9/11, over 300,000 Americans have died from gun violence "... and yet not one single action has been taken in the same time to address gun violence in America" (Arnold 2013, p. 1).

After a seven-year battle with the National Rifle Association (NRA), gun control advocates achieved a small victory in 1993, when Congress passed the Brady Bill. The law initially required a five-day waiting period on handgun purchases so that sellers can screen buyers for criminal records or mental instability. The law was amended in 1998, to include an instant check of buyers and their suitability for gun ownership.

Today, the law requires background checks of not just handgun users but also those who purchase rifles and shotguns. For example, if a person wants to buy a gun, their name and other personal information is entered into the National Instant Criminal Background Check System via the internet or a toll-free number to check whether or not the buyer is eligible. The same information is then run through several FBI-managed databases such as the National Crime Information Center which searches federal and state criminal records for information on the applicant (Jones 2013c).

Fueled by the escalating number of gun deaths in the United States and prompted by the killings at Sandy Hook Elementary school in Newtown, Connecticut, in 2012, President Obama established a task force to examine gun violence in the United States. Although supporters of gun control argued that the task force's recommendations didn't go far enough, recommending only the reinstatement of the assault weapons ban, limiting magazine rounds to 10, and requiring universal background checks, the U.S. Senate nonetheless failed to pass the proposed legislation (Barrett & Cohen 2013).

Fear that the legislation might pass led to record sales of firearms and, with it, record profits for gun manufacturers (O'Toole 2013). The National Rifle Association (NRA) and the firearm industries it represents have spent over $80 million on congressional and presidential elections since 2000. Political candidates, as conflict theorists would argue, are thus hesitant to vote against the interests of such a powerful group, and one that provides financial support for their reelection campaigns (Berlow & Witkin 2013).

The debate over gun control is a divisive one. In fact, when a representative sample of adult Americans were asked how they feel about various public policy areas (e.g., race relations, the environment), the greatest difference between Republicans and Democrats was on the issue of gun control—nearly 60 percent of Republicans and yet only 28 percent of Democrats were satisfied with present U.S. gun policies (Jones 2013c).

Other Crime and Social Control Legislation. There have been several landmark legislative initiatives including the 1994 Violent Crime Control and Law Enforcement Act, which created community policing, and the 2006 Adam Walsh Child Protection and Safety Act, which, when enacted, created a national registry of substantiated cases of child abuse and neglect (Fact Sheet 2012). A sample of significant crime-related legislation presently before Congress includes the following bills:

- *Military Crime Victim's Rights Act of 2013.* If passed, this act would provide rights to victims of offenses under the Uniform Code of Military Justice, including, but not limited to, protection from the accused and notice of public proceedings.
- *Tax Crimes and Identity Theft Prevention Act.* This bill, if passed, would (1) create an identity theft tax fraud prevention program, (2) impose penalties on any person who misuses another person's tax identification number, and (3) amend the Internal Revenue Code to allow tax returns to be shared with law enforcement personnel who are investigating identity theft incidents.
- *Human Trafficking Reporting Act.* Severe forms of human trafficking, if this legislation is passed, would be added to the FBI's list of violent index offenses.

- *Officer Sean Collier Campus Police Recognition Act of 2013.* Named after the MIT campus police officer who was killed during the Boston bombings, this act, if passed, would amend the Omnibus Crime Control and Safe Streets Act of 1968 to include death benefits to campus police officers.
- *Protecting Victims on Campus Act 2013.* This bill, if passed, would require that colleges and universities comply with federal crime reporting guidelines, inform students and employees as to whom crimes should be reported, and maintain a confidential database to be used in institutional reports (U.S. Congress 2013–2014).

International Efforts in the Fight against Crime

Europol is the European law enforcement organization that handles criminal intelligence. Unlike the FBI, Europol officers do not have the power to arrest; they predominantly provide support services for law enforcement agencies of countries that are members of the European Union. For example, Europol coordinates the dissemination of information, provides operational analysis and technical support, and generates strategic reports (Europol 2011). Europol, in conjunction with law enforcement agencies in member states, fights against transnational crimes such as illicit drug trafficking, child pornography, human trafficking, money laundering, illegal immigration, and counterfeiting of the euro.

Interpol, the International Criminal Police Organization, was established in 1923, and is the world's largest international police organization, with 190 member countries (Interpol 2013). Similar to Europol, Interpol provides support services for law enforcement agencies of member nations. It has four core functions: First, Interpol operates a worldwide police communications network that operates 24 hours a day, 7 days a week. Second, Interpol's extensive databases and forensic expertise (e.g., DNA profiles, fingerprints, suspected terrorists) ensure that police get the information they need to investigate existing crime and prevent new crime from occurring. Third, Interpol provides emergency support services and operational activities to law enforcement personnel in the field. Finally, Interpol provides police training and development to help member states better fight the increasingly complex and globalized nature of crime.

Finally, the International Centre for the Prevention of Crime (ICPC 2013) is a consortium of policy makers, academicians, police, governmental officials, and nongovernmental agencies from all over the world (ICPC 2013). Located in Montreal, Canada, ICPC is a "...unique international forum and resource center dedicated to the exchange of ideas and knowledge on crime prevention and community safety" (p. 1). In fulfilling such a task, the ICPC seeks to (1) raise awareness of and access to crime prevention knowledge; (2) enhance community safety; (3) facilitate the sharing of crime prevention information between countries, cities, and justice systems; and (4) respond to calls for technical assistance.

Understanding Crime and Social Control

What can we conclude from the information presented in this chapter? Research on crime and violence supports the contentions of both structural functionalists and conflict theorists. Inequality in society, along with the emphasis on material well-being and corporate profit, produces societal strains and individual frustrations. Poverty, unemployment, urban decay, and substandard schools—the symptoms of social inequality—in turn lead to the development of criminal subcultures and conditions favorable to law violation. Furthermore, criminal behavior is encouraged by the continued weakening of social bonds among members of society and between individuals and society as a whole, the labeling of some acts and actors as "deviant," and the differential treatment of minority groups by the criminal justice system.

Recently, there has been a general decline in crime, making it tempting to conclude that get-tough criminal justice policies are responsible for the reductions. Other valid explanations exist and are likely to have contributed to the falling rates: changing demographics, community policing, stricter gun control, and a reduction in the use of crack cocaine.

Concerns over the cost of "nail 'em and jail 'em" policies, overcrowded prisons, and high recidivism rates have some policy makers looking elsewhere. Many states are

Interpol The largest international police organization in the world.

already expanding the use of community-based initiatives and developing evidence-based reentry programs. Further, the *National Criminal Justice Commission Act of 2013*, if passed, would establish a commission to "...undertake a comprehensive review of the criminal justice system, encompassing current federal, state, local, and tribal criminal justice policies and practices, and make reform recommendations for the President, Congress, State, local, and tribal governments" (HR 446 2013, p. 1).

Rather than getting tough on crime after the fact, some advocate getting serious about prevention. Prevention programs are not only preferable to dealing with the wreckage crime leaves behind, but they are also cost-effective. For example, the Perry Preschool Project, as discussed earlier, cost $15,166 per participant but produced savings of $258,888 per participant. Of that savings, 88 percent was associated with a reduction in costs related to criminal justice (Schweinhart 2007).

Lastly, the movement toward **restorative justice**, a philosophy primarily concerned with repairing the victim–offender–community relation, is in direct response to the concerns of an adversarial criminal justice system that encourages offenders to deny, justify, or otherwise avoid taking responsibility for their actions. Restorative justice holds that the justice system, rather than relying on "punishment, stigma, and disgrace" (Siegel 2006, p. 275), should "repair the harm" (Sherman 2003, p. 10). Key components of restorative justice include restitution to the victim, remedying the harm to the community, and mediation. Restorative justice is increasingly used in schools to resolve conflict and mediate grievances. For example, some schools in Oakland, California, have adopted restorative justice techniques such as "talking circles." Talking circles are facilitated by specially trained teachers who guide students in honest and heartfelt conversations, leading to stronger relationships among students, teachers, and administrators. Talking circles "[E]ncourage young people to come up with meaningful reparations for their wrongdoing, while challenging them to develop empathy for one another..." (Brown 2013, p. 1).

> **restorative justice** A philosophy primarily concerned with reconciling conflict between the victim, the offender, and the community.

CHAPTER REVIEW

- **Are there any similarities between crime in the United States and crime in other countries?**
  All societies have crime and have a process by which they deal with crime and criminals; that is, they have police, courts, and correctional facilities. Worldwide, most offenders are young males, and the most common offense is theft; the least common offense is murder.

- **How can we measure crime?**
  There are three primary sources of crime statistics. First are official statistics, for example, the FBI's Uniform Crime Reports, which are published annually. Second are victimization surveys designed to get at the "dark figure" of crime, crime that official statistics miss. Finally, self-report studies have all the problems of any survey research. Investigators must be cautious about whom they survey and how they ask the questions.

- **What sociological theory of criminal behavior blames the schism between the culture and structure of society for crime?**
  Strain theory was developed by Robert Merton (1957) and uses Durkheim's concept of *anomie*, or normlessness. Merton argued that, when the structure of society limits legitimate means (e.g., a job) of acquiring culturally defined goals (e.g., money), the resulting strain may lead to crime. Individuals, then, must adapt to the inconsistency between means and goals in a society that socializes everyone into

wanting the same thing but provides opportunities for only some.

- **What are index offenses?**
  Index offenses, as defined by the FBI, include two categories of crime: violent crime and property crime. Violent crimes include murder, robbery, assault, and rape; property crimes include larceny, car theft, burglary, and arson. Property crimes, although less serious than violent crimes, are the most numerous.

- **What is meant by white-collar crime?**
  White-collar crime includes two categories: occupational crime, that is, crime committed in the course of one's occupation; and corporate crime, in which corporations violate the law in the interest of maximizing profits. In occupational crime, the motivation is individual gain.

- **How do social class and race affect the likelihood of criminal behavior?**
  Official statistics indicate that minorities are disproportionately represented in the offender population. Nevertheless, it is inaccurate to conclude that race and crime are causally related. First, official statistics reflect the behaviors and policies of criminal justice actors. Thus, the high rate of arrests, conviction, and incarceration of minorities may be a consequence of individual and institutional bias against minorities. Second, race and social class are closely related in

that nonwhites are overrepresented in the lower classes. Because lower-class members lack legitimate means to acquire material goods, they may turn to instrumental, or economically motivated, crimes. Thus, the apparent relationship between race and crime may, in part, be a consequence of the relationship between these variables and social class.

- **What are some of the economic costs of crime?**
First are direct losses from crime, such as the destruction of buildings through arson or of the environment by polluters. Second are costs associated with the transferring of property (e.g., embezzlement). A third major cost of crime is that associated with criminal violence (e.g., the medical cost of treating crime victims). Fourth are the costs associated

with the production and sale of illegal goods and services. Fifth is the cost of prevention and protection. Finally, there is the cost of the criminal justice system, law enforcement, litigation and judicial activities, corrections, and victims' assistance.

- **What is the present legal status of capital punishment in this country?**
Of the 33 states that have the death penalty, the majority and the federal government almost exclusively use lethal injection as the method of execution. Questions concerning the constitutionality of lethal injection and the role of physicians in state executions have led to several court cases.

TEST YOURSELF

1. The United States has the highest violent crime rate in the world.
   a. True
   b. False
2. The Uniform Crime Reports is a compilation of data from
   a. the U.S. Census Bureau
   b. law enforcement agencies
   c. victimization surveys
   d. the Department of Justice
3. According to ___, crime results from the absence of legitimate opportunities as limited by the social structure of society.
   a. Hirschi
   b. Marx
   c. Merton
   d. Becker
4. Which of the following is not an index offense?
   a. Drug possession
   b. Homicide
   c. Rape
   d. Burglary
5. The economic costs of white-collar crime outweigh the costs of traditional street crime.
   a. True
   b. False

6. Women everywhere commit less crime than men.
   a. True
   b. False
7. Probation entails
   a. early release from prison
   b. a suspended sentence
   c. court supervision in the community in lieu of incarceration
   d. incapacitation of the offender
8. Europol is an advisory and support law enforcement agency for European Union members.
   a. True
   b. False
9. Lethal injection
   a. violates the Eighth Amendment of the U.S. Constitution
   b. has been determined to be painless
   c. by physicians is approved by the American Medical Association
   d. is the most common method of execution in the United States
10. The United States has the highest incarceration rate in the world.
    a. True
    b. False

Answers: 1: B; 2: B; 3: C; 4: A; 5: A; 6: A; 7: C; 8: A; 9: D; 10: A.

KEY TERMS

acquaintance rape 108
breeding ground hypothesis 128
capital punishment 132
classic rape 108
clearance rate 102
computer crime 113
corporate violence 113
crime 101
crime rate 102
deterrence 126

feminist criminology 116
identity theft 113
incapacitation 128
index offenses 107
Interpol 136
insider trading 112
larceny 109
organized crime 111
parole 129
primary deviance 106

probation 129
racial profiling 117
recidivism 127
rehabilitation 128
restorative justice 137
secondary deviance 106
transnational crime 100
victimless crimes 110
white-collar crime 112

© Gladskikh Tatiana/Shutterstock.com

5

Family Problems

"We must recognize that there are healthy as well as unhealthy ways to be single or to be divorced, just as there are healthy and unhealthy ways to be married."

—Stephanie Coontz, Family historian

IN 2010, YEARDLEY LOVE, a 22-year-old senior lacrosse player at the University of Virginia, was beaten to death. Her ex-boyfriend, George Huguely, also a 22-year-old senior lacrosse player at UVA, was arrested and charged with her murder. Although mass shootings such as the 2012 shooting at Sandy Hook Elementary School in Newtown, Connecticut, and bombings such as the 2013 bombing at the Boston Marathon are acts of violence that elicit shock, grief, and fear, the violence that occurs between partners and family members—often in the home behind closed doors—gets relatively little public attention, with the exception of occasional high-publicity cases such as that of Yeardley Love.

In this chapter, we turn our attention to family problems, focusing on violence and abuse in intimate and family relationships, and problems associated with divorce and its aftermath. Note that many of the problems families face, such as health problems, poverty, job-related problems, drug and alcohol abuse, discrimination, and military deployment of a spouse are dealt with in other chapters in this text. We begin by sampling the diversity in family life across the globe and identifying patterns and trends in U.S. families.

Yeardley Love, 22-year-old senior at the University of Virginia, was murdered in her off-campus apartment in 2010. Her ex-boyfriend was arrested for the murder.

family A kinship system of all relatives living together or recognized as a social unit, including adopted people.

monogamy Marriage between two partners; the only legal form of marriage in the United States.

serial monogamy A succession of marriages in which a person has more than one spouse over a lifetime but is legally married to only one person at a time.

polygamy A form of marriage in which one person may have two or more spouses.

polygyny A form of marriage in which one husband has more than one wife.

polyandry The concurrent marriage of one woman to two or more men.

bigamy The criminal offense in the United States of marrying one person while still legally married to another.

The Global Context: Family Forms and Norms around the World

The U.S. Census Bureau defines *family* as a group of two or more people related by blood, marriage, or adoption. Sociology offers a broader definition of family: A **family** is a kinship system of all relatives living together, or recognized as a social unit, including adopted people. This broader definition recognizes foster families, unmarried same-sex and opposite-sex couples with or without children, and any relationships that function and feel like a family. As we describe in the following section, family forms and norms vary worldwide.

Monogamy and Polygamy. In many countries, including the United States, the only legal form of marriage is **monogamy**—a marriage between two partners. A common variation of monogamy is **serial monogamy**—a succession of marriages in which a person has more than one spouse over a lifetime but is legally married to only one person at a time.

Polygamy—a form of marriage in which one person may have two or more spouses—is practiced on all continents throughout the world (Zeitzen 2008). The most common form of polygamy, known as **polygyny**, involves one husband having more than one wife. A less common form of polygamy is **polyandry**—the concurrent marriage of one woman to two or more men.

In the United States, Congress outlawed polygamy in 1892; thus, being married to more than one spouse is a crime referred to as **bigamy**. Although the Mormon Church has officially banned polygamy, it is still practiced among members of the Fundamentalist Church of Jesus Christ Latter-Day Saints (FLDS)—a Mormon splinter group with about 10,000 members who believe that a man must marry at least three wives to go to heaven (Zeitzen 2008).

Polygamy in the United States also occurs among some immigrants who come from countries where polygamy is accepted, such as Mali and Ghana and other West African countries. It is estimated, for example, that thousands of New Yorkers are involved in polygamous marriages (Bernstein 2007). Immigrants who practice polygamy generally keep their lifestyle a secret because polygamy is grounds for deportation under U.S. immigration law.

One concern regarding polygamy in the United States is the forcing of underage girls into polygamous marriages. In 2008, Texas state officials raided a FLDS ranch in Eldorado, Texas, and removed more than 400 children, placing them in temporary state custody to protect them from allegedly abusive conditions. Although 31 of these children were girls aged 14 to 17 who had children or were pregnant, a Court of Appeals ruled that the state did not have sufficient evidence of imminent danger to remove the children, and the court ordered the return of the children.

These women and children of the Fundamentalist Church of Jesus Christ of Latter-Day Saints were removed from a compound in Eldorado, Texas, where polygamy is practiced, after allegations of abuse were reported.

What Do You Think? Fundamentalist Mormons, who practice polygamy as an extension of their religious beliefs, have attempted to justify polygamy by referring to the Constitution's First Amendment right to freedom of religion. In 1879, the Supreme Court in *Reynolds v. United States* ruled that, although people are free to believe in their religious principles, religious belief is not justification for acting against the law. Do you think that the criminalization of polygamy violates the right to freedom of religion? Should the government be tolerant of polygamy?

Arranged Marriages versus Self-Choice Marriages. In some parts of the world, young adults do not choose who they want to marry; rather, their parents or other third party arrange their marriage. In arranged marriages, couples are not expected to form relationships or fall in love before they get married, but they are expected to develop emotional attachment following marriage. In many arranged marriages, the bride and groom do not meet until just a few weeks before the wedding, and sometimes they do not meet until the wedding day. Allendorf (2013) reported that, in 2005, less than 5 percent of ever-married women ages 25 to 49 in India had a primary role in choosing their husband, and only 22 percent knew their husband for more than a month before they married. Parents may arrange marriages based on considerations of social class, religion, politics, aristocracy, and/or occupation.

In recent decades, young adults living in parts of the world that have traditionally practiced arranged marriage are increasingly selecting their own marriage partners. Factors associated with this transition from traditional to modern marriage norms include increased education, urbanization, and exposure to Western cultural norms and values (Allendorf 2013).

Division of Power in the Family. In many societies, male dominance in the larger society is reflected in the dominance of husbands over wives in the family (see also Chapter 10, "Gender Inequality"). In many societies, for example, men make decisions about their wife's health care, when their wife may visit relatives, and/or household purchases (Population Reference Bureau 2011). In sub-Saharan Africa, male dominance is linked to widespread wife abuse. According to Nigeria's minister for women's affairs, "It is like it is a normal thing for women to be treated by their husbands as punching bags. . . . The Nigerian man thinks that a woman is his inferior" (quoted by LaFraniere 2005, p. A1).

In some societies, many women and men believe that it is acceptable for a man to beat his wife if she argues with him and/or refuses to have sex with him (Population Reference Bureau 2011).

In developed Western countries, although gender inequality persists (see Chapter 10), marriages tend to be more egalitarian, which means women and men view each other as equal partners who share decision making, housework, and child care. In one study, nearly a third (31 percent) of U.S. couples reported joint decision making for most decisions; 43 percent of couples reported that the woman makes decisions in more areas than the man; and 26 percent of couples reported that men make more of the decisions in the family (Pew Research Center 2008). As couples become more egalitarian, they share more in household and child care responsibilities, with the roles of moms and dads converging: Moms are participating more in earning income for the family and dads are doing more housework and child care. Consider that among parents with children under 18 living in the household, the average number of hours moms spent in the workforce increased from 8 per week in 1965 to 21 per week in 2011. Over the same time, the average number of hours that fathers did housework and child care increased from 6.5 in 1965 to 17 in 2011 (Parker & Wang 2013).

Social Norms Related to Childbearing. In less developed societies, where social expectations for women to have children are strong, women on average have four to five children in their lifetime, and begin having them at an early age, with nearly half of women aged 20 to 24 in the least developed countries getting married before age 18 (UNICEF 2012).

In developed societies, women may view having children as optional—as a personal choice. Among U.S. young adults ages 18 to 29, although 74 percent say they want children, a significant minority aren't sure they want to be parents (19 percent) or say they don't want to have children (7 percent) (Wang & Taylor 2011).

Norms about childbirth out of wedlock also vary across the globe. In many poor countries, nonmarital childbirth is rare because girls often marry at a young age (before they are 18). In India, it is almost unheard of for a Hindu woman to have a child outside marriage; unwed childbearing would bring great shame to a woman and her family (Laungani 2005).

In the United States, 4 in 10 births in 2012 were to unmarried women (Hamilton et al. 2013). Yet, compared with some countries, the United States has a low proportion of births outside of marriage. Two-thirds of births in Iceland and more than half of births in France, Slovenia, Mexico, Estonia, Norway, and Sweden are out of wedlock. At the other end of the spectrum, less than 10 percent of births in Korea, Japan, and Greece are born to parents who are not married or not living in a legally recognized relationship (OECD 2012).

Same-Sex Couples. Norms, policies, and attitudes concerning same-sex intimate relationships also vary around the world. In some countries, homosexuality is punishable by imprisonment or even death. As of November 2013, 16 countries grant same-sex couples the legal right to marry. And two jurisdictions within Mexico allow same-sex marriage. As of this writing, same-sex marriage is legal in 16 U.S. states and the District of Columbia, and in 2013, the Supreme Court struck down part of the Federal Defense of Marriage Act, thus requiring the federal government to recognize same-sex marriages from states where they are legal. Several countries, and some U.S. states, grant same-sex couples legal rights and protections that are more limited than marriage. See Chapter 11, "Sexual Orientation and the Struggle for Equality," for in-depth information about same-sex couples and families.

Looking at families from a global perspective underscores the fact that families are shaped by the social and cultural context in which they exist. As we discuss the family problems related to divorce, violence, and abuse, we refer to social and cultural forces that shape these events and the attitudes surrounding them. Next, we look at changing patterns and trends in U.S. families and households.

Changing Patterns and Trends in U.S. Families

Family forms and norms vary not only across societies; they also change across time. Some of the significant changes in U.S. families and households that have occurred over the past several years or decades include the following:

- *Increased singlehood and older age at first marriage.* U.S. women and men are staying single longer, tying the knot—if they do so at all—at older ages. Between 1980 and 2012, the median age at first marriage for U.S. women rose from 22.0 years to a historic high of 26.6 years; for men, it rose from 24.7 to nearly 29 (Arroyo et al. 2013). Today, 17.5 percent of U.S. adults ages 40 to 44 have never been married—the highest figures in this nation's history (U.S. Census Bureau 2012). One benefit of delayed marriage is that it improves women's financial situation: Women who marry after age 30 earn more per year than their peers who marry before age 20 (Hymowitz et al. 2013). In addition, the divorce rate has been slowly falling, in part, because Americans are getting married at older ages (getting married in one's teens or early 20s is associated with higher risk of divorce than marrying at older ages).

- *Increased heterosexual and same-sex cohabitation.* It is not unusual today for couples to live together without being married. From 2000 to 2010, the number of opposite-sex unmarried couples increased from 4.9 million to 6.8 million, and the number of same-sex unmarried-couple households nearly doubled from 358,000 to 646,000 (Lofquist et al. 2012). The percentage of people who cohabited with their spouses before marriage more than doubled between 1980 and 2000, rising from 16 percent to 41 percent (Amato et al. 2007). Today, more than 60 percent of U.S. marriages are preceded by the couple living together (National Marriage Project and the Institute for American Values 2012).

> **What Do You Think?** Adults with divorced parents are more likely to cohabit before marriage than are adults with continuously married parents (Amato et al. 2007). Why do you think this is so?

Couples live together to assess their relationship, to reduce or share expenses, or to avoid losing pensions or alimony from prior spouses. For some, including many same-sex couples who live in states where they cannot legally marry, cohabitation is an alternative to marriage. Some heterosexual couples do not marry out of solidarity with gays and lesbians who cannot legally wed in most states. Although Brad Pitt announced his engagement to Angelina Jolie in 2012, he had previously said that he and Angelina Jolie would "consider tying the knot when everyone . . . who wants to be married is legally able" (quoted in Davis 2009, p. 58).

Increased cohabitation among adults means that children are increasingly living in families that may function as two-parent families but do not have the social or legal recognition that married-couple families have. More than half of opposite-sex unmarried-partner households have biological children in the home, and about 40 percent of U.S. same-sex households are raising children (U.S. Census Bureau 2012). When children are denied a legal relationship to both parents because of the parents' unmarried status, they may be denied Social Security survivor benefits, health care insurance, or the ability to have either parent authorize medical treatment in an emergency, among other protections. Some states, cities, counties, and employers allow unmarried partners (same-sex and/or heterosexual partners) to apply for a **domestic partnership** designation, which grants them some legal entitlements, such as health insurance benefits and inheritance rights that have traditionally been reserved for married couples.

domestic partnership A status that some states, counties, cities, and workplaces grant to unmarried couples, including gay and lesbian couples, which conveys various rights and responsibilities.

Brad Pitt implied solidarity with gays and lesbians when he said that he and Angelina Jolie will "consider tying the knot when everyone . . . who wants to be married is legally able."

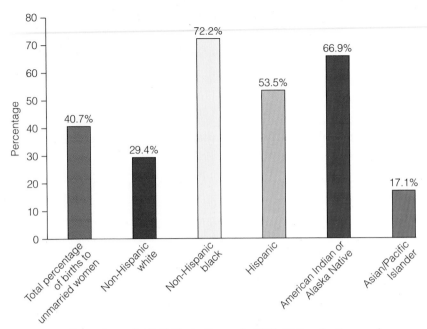

Figure 5.1 Percentage of All Births to Unmarried U.S. Women by Race and Hispanic Origin, 2012

Source: Hamilton et al. 2013.

■ *A new family form: Living apart together.* Some couples live apart in different cities or states because of their employment situation. Known as "commuter marriages," these couples generally would prefer to live together, but their jobs require them to live apart. However, other couples (married or unmarried) live apart in separate residences out of choice. Family scholars have identified this arrangement as an emerging family form known as **living apart together (LAT) relationships**. Couples may choose this family form for a number of reasons, including the desire to maintain a measure of independence, not feeling ready to cohabit, resuming a marital or cohabiting relationship after it broke up, one partner having a job or going to school a distance from the other partner, or prioritizing other responsibilities such as children. This new social phenomenon has been observed in several western European countries as well as in the United States (Lara 2005; Levin 2004). In one study in Britain, nearly 1 in 10 adults was in a living apart together relationship (Duncan et al. 2013.)

> **What Do You Think?** What do you think some of the advantages of living apart together could be? How about the disadvantages? Would you consider living apart together with your partner or spouse?

■ *Increased births to unmarried women.* The percentage of births to unmarried women has risen to historic levels in recent years; about 4 in 10 U.S. births are to unmarried women (Hamilton et al. 2013). The highest rates of nonmarital births are among blacks, Native Americans/Alaskan natives, and Hispanics (see Figure 5.1).

Having a baby outside marriage has become more socially acceptable, with more than half (54 percent) of U.S. adults saying that having a baby outside of marriage is "morally acceptable." However, a significant minority (42 percent) views nonmarital childbearing as "morally wrong" (Newport 2012).

According to family scholar Stephanie Coontz (1997), we must be careful not to overdramatize the increase in births to unmarried women because "much illegitimacy was covered up in the past" (p. 29). In addition, not all unwed mothers are single parents; many unwed mothers are cohabiting with their partners when their children are born. Nevertheless, compared with children living in families with two parents, children in single-parent families tend to have more physical and mental health problems, higher drug use, and earlier onset of sexual behavior (Carr & Springer 2010).

■ *Increased divorce and blended families.* The **refined divorce rate**—the number of divorces per 1,000 married women—is about double today what it was in 1960, but has declined slightly since reaching its peak in the early 1980s. Couples marrying today have a 40 to 50 percent probability of eventually divorcing (The National Marriage Project and the Institute for American Values 2012). Most divorced individuals remarry and create blended families, traditionally referred to as stepfamilies. More than 4 in 10 U.S. adults have at least one step-relative in their family—either a stepparent, a step- or half sibling or a stepchild (see Figure 5.2) (Parker 2011). Seven in 10 adults who have at least

living apart together (LAT) relationships An emerging family form in which couples—married or unmarried—live apart in separate residences.

refined divorce rate The number of divorces per 1,000 married women.

one step-relative say they are very satisfied with their family life. Those who don't have any step-relatives report slightly higher levels of family satisfaction (78 percent very satisfied) (Parker 2011).

Stepparents and stepchildren do not have the same legal rights and responsibilities as biological/adopted children and biological/adoptive parents. Stepchildren do not automatically inherit from their stepparents, and courts have been reluctant to give stepparents legal access to stepchildren in the event of a divorce. In general, U.S. law does not recognize stepparents' roles, rights, and obligations regarding their stepchildren. In case of divorce, stepparents have little or no rights to their stepchildren (Sweeney 2010).

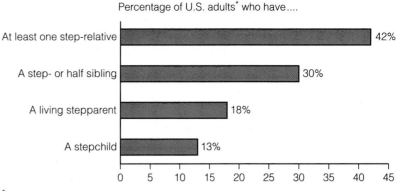

Percentage of U.S. adults* who have....

*based on national sample of 2,691 adults

Figure 5.2 Percentage of U.S. Adults with Step-Relatives
Source: Parker 2011. From "A Portrait of Step Families," January 13, 2011, Social and Demographic Trends, a project of Pew Research Center. Reprinted with permission.

- *Increased employment of mothers.* Labor force participation (either employed or looking for employment) of married women with children under age 18 rose from 47 percent in 1975 to 68 percent in 2012 (Bureau of Labor Statistics 2013a, 2013b). In the majority of married-couple families with children under 18, both parents are employed (59 percent in 2012) (Bureau of Labor Statistics 2013b). Yet work, school, and medical care in the United States tend to be organized around the expectation that every household has a full-time parent at home who is available to transport children to medical appointments, pick up children from school on early dismissal days, and stay home when a child is sick (Coontz 1992). Problems in balancing work and family are discussed in Chapter 7.

- *Increased three-generation family households.* In 2011, 8 percent of U.S. children lived in a three-generation family household—up from 6 percent in 2001 (Pilkauskas 2012). In one study, 9 percent of married mothers, 17 percent of cohabiting mothers, and 45 percent of single mothers lived in a three-generation family household at the time of the child's birth; 60 percent of single mothers and their children lived in a three-generation family household at least once during a nine-year period (Pilkauskas 2012). Reasons for the formation of three-generation family households include economic need, cultural values, and generational needs such as grandparents needing assistance with daily care or, more commonly, parents needing assistance with child care.

Three-generational U.S. families are generally short-lived. However, "given the frequency of their occurrence and their likelihood of reoccurrence, these households likely play an important role in the lives of children, mothers, and grandparents" (Pilkauskas 2012, p. 941).

Public Attitudes toward Changes in Family Life

A national survey found that U.S. adults are sharply divided in their judgments about the changes in U.S. families over the past several decades. A Pew Research Center survey asked a national sample of 2,691 adults whether they considered the following seven trends to be good, bad, or of no consequence to society: (1) more unmarried couples raising children; (2) more gay and lesbian couples raising children; (3) more single women having children without a male partner to help raise them; (4) more people living together without getting married; (5) more mothers of young children working outside the home; (6) more people of different races marrying each other; and (7) more women not ever having children (Morin 2011).

The survey results found that about a third (31 percent) accept the changes, with most saying that changes in family structure either make no difference to society or are good.

A similar share (32 percent) rejects the changing trends in family life, saying that five out of the seven trends are bad for society. The only trends they generally accepted are interracial marriage and fewer women having children. Finally, about a third (37 percent) is skeptical about the changes in American families, as they generally are tolerant of the trends, but also express concern about the impact of these trends on society. The trend they are most concerned about is single mothers raising children.

Marital Decline? Or Marital Resiliency?

Do the recent transformations in American families signify a collapse of marriage and family in the United States? Does the trend toward diversification of family forms mean that marriage and family are disintegrating, falling apart, or even disappearing? Or has family simply undergone transformations in response to changes in socioeconomic conditions, gender roles, and cultural values? The answers to these questions depend on whether we adopt the marital decline perspective or the marital resilience perspective.

According to the **marital decline perspective**, (1) personal happiness has become more important than marital commitment and family obligations, and (2) the decline in lifelong marriage and the increase in single-parent families have contributed to a variety of social problems, such as poverty, delinquency, substance abuse, violence, and the erosion of neighborhoods and communities (Amato 2004). According to the **marital resiliency perspective**, "poverty, unemployment, poorly funded schools, discrimination, and the lack of basic services (such as health insurance and child care) represent more serious threats to the well-being of children and adults than does the decline in married two-parent families" (Amato 2004, p. 960). According to this perspective, many marriages in the past were troubled, but because divorce was not socially acceptable, these problematic marriages remained intact. Rather than the view of divorce as a sign of the decline of marriage, divorce provides adults and children an escape from dysfunctional home environments.

> Whereas once the main purpose of marriage was to have and raise children, today women and men want marriage to provide adult intimacy and companionship.

marital decline perspective A pessimistic view of the current state of marriage that includes the beliefs that (1) personal happiness has become more important than marital commitment and family obligations, and (2) the decline in lifelong marriage and the increase in single-parent families have contributed to a variety of social problems.

marital resiliency perspective A view of the current state of marriage that includes the beliefs that (1) poverty, unemployment, poorly funded schools, discrimination, and the lack of basic services (such as health insurance and child care) represent more serious threats to the well-being of children and adults than does the decline in married two-parent families, and (2) divorce provides adults and children an escape from dysfunctional home environments.

Although the high rate of marital dissolution seems to suggest a weakening of marriage, divorce may also be viewed as resulting from placing a high value on marriage, such that a less than satisfactory marriage is unacceptable. In effect, people who divorce may be viewed not as incapable of commitment but as those who would not settle for a bad marriage. Indeed, the expectations that young women and men have of marriage have changed. Whereas once the main purpose of marriage was to have and raise children, today women and men want marriage to provide adult intimacy and companionship (Coontz 2000).

The high rate of childbirths out of wedlock and single parenting is also not necessarily indicative of a decline in the value of marriage. In interviews with a sample of low-income single women with children, most women said they would like to be married but just have not found "Mr. Right" (Edin 2000). Low-income single mothers in Edin's study were reluctant to marry the father of their children because these men had low economic status, traditional notions of male domination in household and parental decisions, and patterns of untrustworthiness and even violent behavior. Given the low level of trust these mothers have of men and given their view that husbands want more control than the women are willing to give them, women realize that a marriage that is also economically strained is likely to be conflictual and short-lived. "Interestingly, mothers say they reject entering into economically risky marital unions out of respect for the institution of marriage, rather than because of a rejection of the marriage norm" (Edin 2000, p. 130).

Is the well-being of a family measured by the degree to which that family conforms to the idealized married, two-parent, stay-at-home mom model of the 1950s? Or is family well-being measured by function rather than form? As suggested by family scholars Mason et al. (2003), "the important question to ask about American families . . . is not

In *Alone Together: How Marriage in America Is Changing*, Paul Amato and colleagues (2007) examined survey data to see how marital quality changed between 1980 and 2000. One of the key questions the authors attempted to answer is: Has marriage become less satisfying and stable? Or has marriage become stronger and more satisfying for couples today compared to in the past?

Sample and Methods

Data for this study came from two national random samples of married adults 55 years and younger. Response rates for both the 1980 and the 2000 sample were over 60 percent. Data were collected through telephone interviews designed to measure five dimensions of marital quality: (1) marital happiness; (2) marital interaction; (3) marital conflict; (4) marital problems; and (5) divorce proneness. To assess whether marital quality changed between 1980 and 2000, the researchers compared the responses from 1980 with those from 2000. They also looked at gender differences to see whether marital quality for wives and husbands differed.

Selected Findings and Discussion

Analysis of the 1980 and 2000 data suggests that marriages have become stronger and more satisfying in some respects and weaker and less satisfying in other respects. In the following selected findings, note that two dimensions of marital quality improved (conflict and problems), one dimension deteriorated (interaction), and two dimensions were unchanged (happiness and divorce proneness).

1. *Marital happiness:* The percentage of spouses who rated their marriage as better than average was nearly equivalent in 1980 (74 percent) as in 2000 (68 percent).

2. *Marital interaction:* All five measures of marital interaction showed a decline from 1980 to 2000. Compared with spouses in 1980, spouses in 2000 were less likely to report that their husband or wife almost always accompanied them while visiting friends, shopping, eating their main meal of the day, going out for leisure, and working around the home.

3. *Marital conflict:* Of the five items that measure marital conflict, three declined significantly between 1980 and 2000: reports of disagreeing with one's spouse "often" or "very often," reports of violence ever occurring in the marriage, and reports of marital violence within the previous three years. No change was observed in the percentage of spouses who reported arguments over the division of household labor or in the number of serious quarrels in the previous two months.

4. *Marital problems:* Compared with spouses in 1980, spouses in 2000 were less likely to report marital problems in the following four areas: getting angry easily, having feelings that are hurt easily, experiencing jealousy, and being domineering. There was little change in the percentage of spouses who reported that getting in trouble with police, drinking or drugs, or extramarital sex were problems in their marriage.

5. *Divorce proneness:* The average level of divorce proneness was stable between 1980 and 2000. However, the percentage of spouses who reported either a low or a high degree of divorce proneness increased from 1980 to 2000, whereas the percentage reporting moderate divorce proneness declined.

6. *Gender differences:* Responses to items measuring marital interaction, marital problems, and divorce proneness were similar for husbands and wives. Two items measuring marital happiness revealed a gender difference: Between 1980 and 2000, wives (but not husbands) reported greater happiness with the amount of understanding received from their spouses. Wives also reported more happiness with their husbands' work around the house, whereas husbands reported less happiness with their wives' work around the house. Regarding the measure of marital conflict, both husbands and wives reported declines in violence.

In sum, this research suggests that between 1980 and 2000, "marriages became more peaceful, with fewer disagreements, less aggression, and fewer interpersonal sources of tension between spouses" (Amato et al. 2007, p. 68). Although overall levels of marital satisfaction did not seem to change from 1980 to 2000, the results for divorce proneness were mixed, with an increase in the proportion of both stable and unstable marriages. The finding that most concerned the researchers was that, between 1980 and 2000, the lives of husbands and wives became more separate, as spouses shared fewer activities. Citing previous research that has found that spouses who spend less time together are less happy in their marriage and more likely to divorce, Amato et al. suggested that "it is possible that the gradual decline in marital interaction between 1980 and 2000 will erode future marital happiness and increase subsequent levels of marital instability" (p. 69).

how much they conform to a particular image of the family, but rather how well do they function—what kind of love, care, and nurturance do they provide?" (p. 2).

It is also important to have a perspective that takes into account the historical realities of families. Family historian Stephanie Coontz (2004) explained:

> . . . [M]any things that seem new in family life are actually quite traditional. Two-provider families, for example, were the norm through most of history. Stepfamilies were more numerous in much of history than they are today. There have been several times and places when cohabitation, out-of-wedlock births, or nonmarital sex were more widespread than they are today. (p. 974)

In sum, it is clear that the institution of marriage and family has undergone significant changes in the last few generations. What is not as clear is whether these changes are for the better or for the worse. As this chapter's *Social Problems Research Up Close* feature discusses, changes in marriage and family may be viewed as neither all good nor all bad, but are perhaps a more complex mix. Coontz (2005) noted, "Marriage has become more joyful, more loving, and more satisfying for many couples than ever before in history. At the same time it has become optional and more brittle" (p. 306).

Sociological Theories of Family Problems

Three major sociological theories—structural functionalism, conflict theory, and symbolic interactionism—help to explain different aspects of the family institution and the problems in families today.

Structural-Functionalist Perspective

The structural-functionalist perspective views the family as a social institution that performs important functions for society, including producing and socializing new members, regulating sexual activity and procreation, and providing physical and emotional care for family members. This perspective views the high rate of divorce and the rising number of single-parent households as constituting a "breakdown" of the family institution. This supposed breakdown of the family is considered a primary social problem that leads to secondary social problems such as crime, poverty, and substance abuse.

According to the structural-functionalist perspective, traditional gender roles contribute to family functioning: Women perform the "expressive" role of managing household tasks and providing emotional care and nurturing to family members, and men perform the "instrumental" role of earning income and making major family decisions. According to this view, families have been disrupted and weakened by the change in gender roles, particularly women's participation in the workforce—an assertion that has been criticized and refuted, but is still supported by some "family values" scholars and supporters who advocate a return to traditional gender roles in the family.

Structural functionalism also looks at how changes in other social institutions affect families. For example, research has found that changes in the economy—specifically falling wages among unskilled and semiskilled men—contribute to both intimate partner abuse and the rise in female-headed single-parent households (Edin 2000).

Conflict and Feminist Perspectives

Conflict theory focuses on how capitalism, social class, and power influence marriages and families. Feminist theory is concerned with how gender inequalities influence and are influenced by marriages and families. Feminists are critical of the traditional male domination of families—a system known as **patriarchy**—that is reflected in the tradition of wives taking their husband's last name and children taking their father's name. Patriarchy implies that wives and children are the property of husbands and fathers.

The overlap between conflict and feminist perspectives is evident in views on how industrialism and capitalism have contributed to gender inequality. With the onset of factory production during industrialization, workers—mainly men—left the home to earn incomes and women stayed home to do unpaid child care and domestic work. This arrangement resulted in families founded on what Friedrich Engels calls "domestic slavery of the wife" (quoted by Carrington 2002, p. 32). Modern society, according to Engels, rests on gender-based slavery, with women doing household labor for which they receive neither income nor status, whereas men leave the home to earn an income. Times have certainly changed since Engels made his observations, with most wives today leaving the home to earn incomes. However, wives employed full-time still do the bulk of unpaid domestic labor, and women are more likely than men to compromise their occupational achievement to take on child care and other domestic responsibilities. The continuing unequal distribution of wealth that favors men contributes to inequities in power and fosters economic dependence of wives on husbands. When wives do earn

patriarchy A male-dominated family system that is reflected in the tradition of wives taking their husband's last name and children taking their father's name.

more money than their husbands, the divorce rate is higher—the women can afford to leave abusive or inequitable relationships (Jalovaara 2003).

Economic factors have also influenced norms concerning monogamy. In societies in which women and men are expected to be monogamous within marriage, there is a double standard that grants men considerably more tolerance for being nonmonogamous. Engels explained that monogamy arose from the concentration of wealth in the hands of a single individual—a man—and from the need to bequeath this wealth to children of his own, which required that his wife be monogamous. The "sole exclusive aims of monogamous marriage were to make the man supreme in the family and to propagate, as the future heirs to his wealth, children indisputably his own" (quoted by Carrington 2002, p. 32).

Feminist and conflict perspectives on domestic violence suggest that the unequal distribution of power among women and men and the historical view of women as the property of men contribute to wife battering. When wives violate or challenge the male head of household's authority, the male may react by "disciplining" his wife or using anger and violence to reassert his position of power in the family (see Figure 5.3).

Although modern gender relations within families and within society at large are more egalitarian than in the past, male domination persists, even if less obvious. Lloyd and Emery (2000) noted that "one of the primary ways that power disguises itself in courtship and marriage is through the 'myth of equality between the sexes.' . . . The widespread discourse on 'marriage between equals' serves as a cover for the presence of male domination in intimate relationships . . . and allows couples to create an illusion of equality that masks the inequities in their relationships" (pp. 25–26).

Conflict theorists emphasize that powerful and wealthy segments of society largely shape social programs and policies that affect families. The interests of corporations and businesses are often in conflict with the needs of families. Corporations and businesses strenuously fought the passage of the 1993 Family and Medical Leave Act, which gives people employed full-time for at least 12 months in companies with at least 50 employees up to 12 weeks of unpaid time off for parenting leave, illness or death of a family member, and elder care. Government, which corporate interests largely influence through lobbying and political financial contributions, enacts policies and laws that serve the interests of for-profit corporations rather than families.

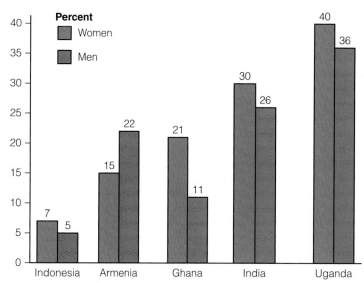

Figure 5.3a **Percent of Women and Men Who Agree that Wife Beating Is Acceptable if a Wife Argues with her Husband**
Source: Population Reference Bureau 2011.

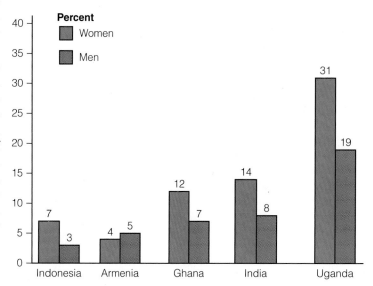

Figure 5.3b **Percent of Women and Men Who Agree That Wife Beating Is Acceptable if a Wife Refuses Sex with Her Husband**
Source: Population Reference Bureau 2011.

Symbolic Interactionist Perspective

The symbolic interactionist perspective emphasizes that interaction with family members, including parents, grandparents, siblings, and spouses, has a powerful effect on our self-concepts. For example, negative self-concepts may result from verbal abuse in the family, whereas positive self-concepts may develop in families in which interactions are supportive and loving. The importance of social interaction in children's developing self-concept suggests a compelling reason for society to accept rather than stigmatize nontraditional family forms. Imagine the effect on children who are called "illegitimate" or who are teased for having two moms or dads.

The symbolic interactionist perspective is useful in understanding the dynamics of domestic violence and abuse. For example, some abusers and their victims learn to define intimate partner violence as an expression of love (Lloyd 2000). Emotional abuse often involves using negative labels (e.g., *stupid, whore*) to define a partner or family member. Such labels negatively affect the self-concept of abuse victims, often convincing them that they deserve the abuse.

The symbolic interactionist insight that labels affect meaning and behavior can be applied to issues related to divorce. For example, when a noncustodial divorced parent (usually a father) is awarded "visitation" rights, he may view himself as a visitor in his children's lives. The meaning attached to the visitor status can be an obstacle to the father's involvement because the label *visitor* minimizes the importance of the noncustodial parent's role (Pasley & Minton 2001). Fathers' rights advocates suggest replacing the term *visitation* with terms such as *parenting plan* or *time-sharing arrangement,* because these terms do not minimize either parent's role.

Problems Associated with Divorce

Divorce is considered problematic because of the negative effects it has on children as well as the difficulties it causes for adults. However, in some societies, legal and social barriers to divorce are considered problematic because such barriers limit the options of spouses in unhappy and abusive marriages. Ireland did not allow divorce under any condition until 1995, and Chile did not allow divorce until 2004.

Even when divorce is a legal option, social barriers often prevent spouses from divorcing. Hindu women, for example, experience great difficulty leaving a marriage, even when the husband is abusive, because divorce leads to loss of status, possible loss of custody of the children, homelessness, poverty, and being labeled a "loose" woman (Laungani 2005, p. 88).

Social Causes of Divorce

When we think of why a particular couple gets divorced, we typically think of a number of individual and relationship factors that might have contributed to the marital breakup: incompatibility in values or goals; poor communication; lack of conflict resolution skills; sexual incompatibility; extramarital relationships; substance abuse; emotional or physical abuse or neglect; boredom, jealousy; and difficulty coping with change or stress related to parenting, employment, finances, in-laws, and illness. However, understanding the high rate of divorce in U.S. society requires awareness of how the following social and cultural factors contribute to marital breakup:

1. *Changing function of marriage.* Before the Industrial Revolution, marriage functioned as a unit of economic production and consumption that was largely organized around producing, socializing, and educating children. However, the institution of marriage has changed over the last few generations:

   > Marriage changed from a formal institution that meets the needs of the larger society to a companionate relationship that meets the needs of the couple and their children and then to a private pact that meets the psychological needs of individual spouses. (Amato et al. 2007, p. 70)

   When spouses do not feel that their psychological needs—for emotional support, intimacy, affection, love, or personal growth—are being met in the marriage, they may consider divorce with the hope of finding a new partner to fulfill these needs.

2. *Increased economic autonomy of women.* Before 1940, most wives were not employed outside the home and depended on their husband's income. Today, the majority of married women are in the labor force. A wife who is unhappy in her marriage is more likely to leave the marriage if she has the economic means to support herself (Jalovaara 2003). An unhappy husband may also be more likely to leave

TABLE 5.1 **Factors that Decrease Women's Risk of Separation or Divorce during the First 10 Years of Marriage**

Factor	Percent Decrease in Risk of Divorce Or Separation
Annual income over $50,000 (versus under $25,000)	−30
Having a baby 7 months or more after marriage (versus before marriage)	−24
Marrying over 25 years of age (versus under 18)	−24
Having an intact family of origin (versus having divorced parents)	−14
Religious affiliation (versus none)	−14
Some college (versus high school dropout)	−25

Source: National Marriage Project and the Institute for American Values 2012.

a marriage if his wife is self-sufficient and can contribute to the support of the children.

3. *Increased work demands and economic stress.* Another factor influencing divorce is increased work demands and the stresses of balancing work and family roles. Some workers are putting in longer hours, often working overtime or taking second jobs, while others face job loss and unemployment. As discussed in Chapters 6 and 7, many families struggle to earn enough money to pay for rising housing, health care, and child care costs. Financial stress can cause marital problems. Couples with an annual income under $25,000 are 30 percent more likely to divorce than couples with incomes over $50,000 (see Table 5.1).

4. *Inequality in marital division of labor.* Many employed parents, particularly mothers, come home to work a **second shift**—the work involved in caring for children and household chores (Hochschild 1989). Wives are more likely than husbands to perceive the marital division of labor as unfair (Nock 1995). This perception of unfairness can lead to marital tension and resentment, as reflected in the following excerpt:

Hill Creek Pictures/UpperCut Images/Getty Images

Wives tend to be less happy in marriage than husbands when they perceive the division of labor to be unfair.

My husband's a great help watching our baby. But as far as doing housework or even taking the baby when I'm at home, no. He figures he works five days a week; he's not going to come home and clean. But he doesn't stop to think that I work seven days a week. Why should I have to come home and do the housework without help from anybody else? My husband and I have been through this over and over again. Even if he would just pick up from the kitchen table and stack the dishes for me, that would make a big difference. He does nothing. . . . He'll help out if I'm not here, but the minute I am, all the work at home is mine. (quoted by Hochschild 1997, pp. 37–38)

Women want to be equal partners in their marriages, not just in earning income but also in sharing the work of household chores, child rearing, marital communication, and in making decisions for the family.

Women want to be equal partners in their marriages, not just in earning income but also in sharing the work of household chores, child rearing, marital communication, and in making decisions for the family. Frustrated by some men's lack of participation in marital work, women who desire relationship egalitarianism may see divorce as the lesser of two evils (Hackstaff 2003).

second shift The household work and child care that employed parents (usually women) do when they return home from their jobs.

5. *Liberalized divorce laws.* Before 1970, the law required a couple who wanted a divorce to prove that one of the spouses was at fault and had committed an act defined by the state as grounds for divorce—adultery, cruelty, or desertion. In 1969, California became the first state to initiate **no-fault divorce**, which permitted a divorce based on the claim that there were "irreconcilable differences" in the marriage. Today, all 50 states recognize some form of no-fault divorce. No-fault divorce law has contributed to the U.S. divorce rate by making divorce easier to obtain.

6. *Increased individualism.* U.S. society is characterized by **individualism**—the tendency to focus on one's individual self-interests and personal happiness rather than on the interests of one's family and community. "Marital commitment lasts only as long as people are happy and feel that their own needs are being met" (Amato 2004, p. 960). Belief in the right to be happy, even if it means getting divorced, is reflected in social attitudes toward divorce: Two-thirds (67 percent) of U.S. adults report that divorce is morally acceptable (Newport 2012).

7. *Weak social ties.* Couples with strong social ties have a network of family and friends who provide support during difficult times, and who express disapproval for behavior that threatens the stability of the marriage. Couples who live in the same community for a long time have the opportunity to develop and maintain strong social ties. But many Americans move from place to place during their adult years, more so than people in other countries, which may help explain why the U.S. divorce rate is higher than in other countries. Cherlin (2009) explains that "moving from one community to another could affect marriages because it disrupts social ties. Migration can separate people from friends and relatives who could help them through family crises" (pp. 148–149).

8. *Increased life expectancy.* Finally, more marriages today end in divorce, in part, because people live longer than they did in previous generations and "till death do us part" involves a longer commitment than it once did. Indeed, one can argue that "marriage once was as unstable as it is today, but it was cut short by death not divorce" (Emery 1999, p. 7).

Consequences of Divorce

When parents have bitter and unresolved conflict and/or if one parent is abusing a child or the other parent, divorce may offer a solution to family problems. However, divorce often has negative effects for ex-spouses and their children and also contributes to problems that affect society as a whole.

> . . . Some men and women experience a decline in well-being after divorce; others experience an improvement.

no-fault divorce A divorce that is granted based on the claim that there are irreconcilable differences within a marriage (as opposed to one spouse being legally at fault for the marital breakup).

individualism The tendency to focus on one's individual self-interests and personal happiness rather than on the interests of one's family and community.

Physical and Mental Health Consequences. Numerous studies show that divorced individuals have more health problems and a higher risk of mortality than married individuals; divorced individuals also experience lower levels of psychological well-being, including more unhappiness, depression, anxiety, and poorer self-concepts (Amato 2003; Kamp Dush 2013). Both divorced and never-married individuals are, on average, more distressed than married people because unmarried people are more likely than married people to have low social attachment, low emotional support, and increased economic hardship (Walker 2001). Some research suggests that divorce leads to higher levels of depressive symptoms for women, but not for men (Kalmijn & Monden 2006), especially when young children are in the family (Williams & Dunne-Bryant 2006). This finding is probably due to the increased financial and parenting strains experienced by divorced mothers who have custody of young children.

However, some studies have found that divorced individuals report higher levels of autonomy and personal growth than married individuals do (Amato 2003). For example, many divorced mothers report improvements in career opportunities, social lives, and happiness after divorce; some divorced women report more self-confidence, and some men report more interpersonal skills and a greater willingness to self-disclose. For people in a poor-quality marriage, divorce has a less negative or even a positive effect on well-being (Amato 2003; Kalmijn & Monden 2006). However, leaving a bad marriage does not always result in increased well-being because "divorce is a trigger for even more problems after the divorce" (Kalmijn & Monden 2006, p. 1210). In sum, some men and women experience a decline in well-being after divorce; others experience an improvement.

Economic Consequences. Following divorce, there tends to be a dramatic drop in women's income and a slight drop in men's income (Gadalla 2009). Compared with married individuals, divorced individuals have a lower standard of living, have less wealth, and experience greater economic hardship, although this difference is considerably greater for women than for men (Amato 2003). The economic costs of divorce are often greater for women and children because women tend to earn less than men (see Chapter 10) and because mothers devote substantially more time to household and child care tasks than fathers do. The time women invest in this unpaid labor restricts their educational and job opportunities as well as their income. Men are less likely than women to be economically disadvantaged after divorce because they continue to profit from earlier investments in education and career.

After divorce, both parents are responsible for providing economic resources to their children. However, some nonresident parents fail to provide child support. In some cases, failure to pay child support is not due to parents being "deadbeats" but rather to the fact that many parents are "dead broke." Parents who are unemployed or who have low-wage jobs may be unable to make child support payments.

Effects on Children and Young Adults. Parental divorce is a stressful event for children and is often accompanied by a variety of stressors, such as continuing conflict between parents, a decline in the standard of living, moving and perhaps changing schools, separation from the noncustodial parent (usually the father), and parental remarriage. These stressors place children of divorce at higher risk for a variety of emotional and behavioral problems. Reviews of research on the consequences of divorce for children have found that children with divorced parents score lower on measures of academic success, psychological adjustment, self-concept, social competence, and long-term health; they also have higher levels of aggressive behavior and depression (Amato 2003; Wallerstein 2003).

Many of the negative effects of divorce on children are related to the economic hardship associated with divorce. Economic hardship is associated with less effective and less supportive parenting, inconsistent and harsh discipline, and emotional distress in children (Demo et al. 2000). Despite the adverse effects of divorce on children, research findings suggest that "most children from divorced families are resilient, that is, they do not suffer from serious psychological problems" (Emery et al. 2005, p. 24). Other researchers conclude that "most offspring with divorced parents develop into well-adjusted adults," despite the pain they feel associated with the divorce (Amato & Cheadle 2005, p. 191).

Divorce can also have positive consequences for children and young adults. In highly conflictual marriages, divorce may actually improve the emotional well-being of children relative to staying in a conflicted home environment (Jekielek 1998). In interviews with 173 grown children whose parents divorced years earlier, Ahrons (2004) found that most of the young adults reported positive outcomes for their parents and for themselves. Although many young adults who have divorced parents fear that they too will have an unhappy marriage (Dennison & Koerner 2008), such a fear can also lead young adults to think carefully about their choices regarding marriage.

Effects on Father–Child Relationships. Children who live with their mothers may have a damaged relationship with their nonresidential father, especially if he becomes disengaged from their lives. Some research has found that young adults whose parents divorced are less likely to report having a close relationship with their father compared with children whose parents are together (DeCuzzi et al. 2004). However, in another study of 173 adult children of divorce, more than half felt that their relationships with their fathers improved after the divorce (Ahrons 2004). Children may benefit from having more quality time with their fathers after parental divorce. Some fathers report that they became more active in the role of father after divorce.

One study has found that the older the child is at the time of parental separation, the greater the amount of time children spend with their fathers (Swiss & Bourdais 2009). This study also found that fathers earning $50,000 or more per year are likely to see their children more often than fathers who earn less than $30,000 per year.

Women who have primary custody of the children serve as gatekeepers for the relationship their children have with their fathers (Trinder 2008). As gatekeepers, custodial mothers can either keep the gate open and encourage contact between the children and their father, or make every effort to keep the gate closed, cutting off the children's contact with their father. Some noncustodial divorced fathers discontinue contact with their children as a coping strategy for managing emotional pain (Pasley & Minton 2001). Many divorced fathers are overwhelmed with feelings of failure, guilt, anger, and sadness over the separation from their children (Knox 1998). Hewlett and West (1998) explained that "visiting their children only serves to remind these men of their painful loss, and they respond to this feeling by withdrawing completely" (p. 69). Divorced fathers commonly experience the legal system as favoring the mother in child-related matters. One divorced father commented:

> I believe that the system [judges, attorneys, etc.] have *[sic]* little or no consideration for the father. At some point the system creates an environment where the father loses any natural desire to see his children because it becomes so difficult, both financially and emotionally. At that point, he convinces himself that the best thing to do is wait until they are older. (quoted by Pasley & Minton 2001, p. 242)

Parental Alienation. **Parental alienation** (PA) occurs when one parent makes intentional efforts to turn a child against the other parent and essentially destroy any positive relationship a child has with the other parent (see this chapter's *The Human Side* feature). Bernet and Baker (2013), two experts on parental alienation, define PA as a "mental condition in which a child, usually one whose parents are engaged in a high-conflict separation or divorce, allies himself strongly with one parent (the preferred parent) and rejects a relationship with the other parent (the alienated parent) without legitimate justification" (p. 99). When a parent influences his or her child to hate or fear the other parent, both the alienated parent and child may be said to be victims of parental alienation. Parents may alienate their child from the other parent by engaging in a variety of behaviors (see Table 5.2) (Baker & Chambers 2011).

Parental alienation is a disputed concept. Critics have argued that there is insufficient research to document its existence, but Bernet and Baker (2013) refute this claim, presenting numerous examples of research documenting parental alienation in both the United States and in other countries. Nevertheless, some lawyers, judges, therapists, and family scholars challenge the validity of parental alienation, claiming that it does not really exist. Baker (2006) explains:

> Perhaps such skeptics hold the belief that a parent must have done *something* to warrant their child's rejection and/or the other parent's animosity. This is the double victimization of [parental alienation]: the shame and frustration of being misunderstood in addition to the grief and anger associated with being powerless to prevent the alienation in the first place. (p. 192)

Other critics fear abusive parents can misuse the concept of "parental alienation" by claiming that children who do not want to visit them (because of past mistreatment) are victims of parental alienation. However, it is unlikely that a judge would ignore evidence of abuse and grant abusive parents unsupervised access to their children.

As we have seen, the effects of divorce on adults and children are mixed and variable. In a review of research on the consequences of divorce for children and adults, Amato (2003) concluded that "divorce benefits some individuals, leads others to experience temporary decrements in well-being that improve over time, and forces others on a downward cycle from which they might never fully recover" (p. 206).

parental alienation The intentional efforts of one parent to turn a child against the other parent and essentially destroy any positive relationship a child has with the other parent.

Amy Baker (2007) conducted interviews with 40 individuals who believed that they had been turned against one parent by the other parent. This *The Human Side* feature presents excerpts from these interviews, providing a glimpse into the experience of parental alienation from the child's perspective.

Larissa (speaking about her mother): *She'd always made both my brother and me feel that our father was somehow to blame for everything. Every day there'd be some attempt by her, some tale she'd tell me, to turn me against my father, so many incidents, it's simply impossible to list them all. . . . I grew to detest him, with a truly visceral hate. I couldn't stand to be in the same room with him, or to even talk to him or have him talk to me.* (pp. 30–31)

Bonnie: My dad tried several different tactics to talk to me but my mom foiled his plans, I guess you could say. At the time I sort of knew what he was doing. . . . It felt like he was trying to contact me but she would stop it. For example, when I was in middle school in 6th grade I think, my dad went up to the school to have lunch with me and she had told the school that if he would ever go up there they were to immediately call the police and I remember it was a big production. They came and got me out of class and they put me in the principal's office and they locked the door. I found out later he was escorted off the grounds by the police. (p. 126)

Maria: Basically I was a chess piece between the two of them, and it was very hard. . . . I started picking up on signals from my mother that showing any kind of affection or love for my father could be a problem for me. If I showed any positiveness toward him, it was, "How could you do this to me? . . . You are betraying me." When I go to visit him and I come back, to this day, my mother still says, "How can you do this to me? You are betraying me. You are slapping me in the face any time you have anything to do with this man. How could you do this to me?" (p. 139) . . . . My dad and I were very close and then after a time and all of this stuff from my mother and then I think I took on a lot of her anger toward him, feeling responsible for her because she was so, "Oh poor me. I am so broken and so wounded." . . . And even if I would have desired anything to do with him, I would not have acted on it because of how my mother would have reacted.* (p. 141)

TABLE 5.2 Parental Alienation Behaviors
Limiting the child's contact with the other parent
Interfering with communication between the child and the other parent
Destroying gifts or memorabilia from the other parent
Failing to display any positive interest in the child's activities or experiences with the other parent
Expressing disapproval or dislike of the child spending time with the other parent
Limiting mention and photographs of the other parent
Withdrawal of love or expressions of anger if the child indicates positive feelings for the other parent
Telling the child that the other parent does not love him or her
Forcing the child to choose between his or her parents
Creating the impression that the other parent is dangerous (when the parent is not)
Forcing the child to reject the other parent
Asking the child to spy on the other parent
Asking the child to keep secrets from the other parent
Referring to the other parent by his or her first name
Referring to a stepparent as "Mom" or "Dad" and encouraging the child to do the same
Withholding medical, social, or academic information from the other parent and keeping the other parent's name off of such records
Changing the child's name to remove association with the other parent
Denigrating the other parent, the other parent's extended family, and any new partner/spouse

© Cengage Learning

Strategies for Action: Strengthening Marriage and Alleviating Problems of Divorce

Two general strategies for responding to the problems of divorce are those that prevent divorce by strengthening marriages and those that strengthen postdivorce families.

Strategies to Strengthen Marriage and Prevent Divorce

A growing "marriage movement" involves efforts to strengthen marriage and prevent divorce through a number of strategies, including premarital and marriage education, covenant marriage and divorce law reform, and provision of workplace and economic supports. Amato et al. (2007) explained:

> Policies to strengthen marital quality and stability are based on consistent evidence that happy and stable marriages promote the health, psychological well-being, and financial security of adults . . . as well as children. . . . Moreover, recent research suggests that a large proportion of marriages that end in divorce are not deeply troubled, and that many of these marriages might be salvaged if spouses sought assistance for relationship problems . . . and stayed the course through difficult times. (pp. 245–46)

Marriage Education. Marriage education, also known as family life education, includes various types of workshops and classes that (1) teach relationship skills, communication, and problem solving; (2) convey the idea that sustaining healthy marriages requires effort; and (3) convey the importance of having realistic expectations of marriage, commitment, and a willingness to make personal sacrifices (Hawkins et al. 2004). An alternative or supplement to face-to-face family life education is web-based education, such as the Forever Families website, a faith-based family education website (Steimle & Duncan 2004).

The Healthy Marriage Initiative provides federal funds to support research and programs to encourage healthy marriages and promote involved and responsible fatherhood. Funds may be used for a variety of activities, including marriage and premarital education, public advertising campaigns that promote healthy marriage, high school programs on the value of marriage, marriage mentoring programs, and parenting skills programs. Some states have passed or considered legislation that requires marriage education in high schools or that provides incentives (such as marriage license fee reductions) to couples who complete a marriage education program.

Covenant Marriage and Divorce Law Reform. In 1996, Louisiana became the first state to offer two types of marriage contracts: (1) the standard marriage contract that allows a no-fault divorce (after a six-month separation), or (2) a **covenant marriage**, which permits divorce only under condition of fault (e.g., abuse, adultery, or felony conviction) or after a two-year separation. Couples who choose a covenant marriage must also get premarital counseling. Variations of the covenant marriage have also been adopted in Arizona and Arkansas.

The covenant marriage option was designed to strengthen marriages and decrease divorce. Research has found that covenant marriages are less likely to end in divorce, but marital satisfaction is only slightly increased, only for husbands, due to the requirement for premarital counseling (DeMaris et al. 2012). Critics argue that covenant marriage may increase family problems by making it more difficult to terminate a problematic marriage and by prolonging the exposure of children to parental conflict (Applewhite 2003). But one study found that "there is no evidence of covenant couples being trapped in unhappier marriages than their standard marriage counterparts by virtue of having made a more binding marital commitment" (DeMaris et al. 2012, p. 1,000).

covenant marriage A type of marriage (offered in a few states) that requires premarital counseling and that permits divorce only under condition of fault or after a marital separation of more than two years.

Workplace and Economic Supports. The most important pro-marriage and divorce-prevention measures may be those that maximize employment and earnings. Given that research finds a link between financial hardship and marital quality, policies to strengthen marriage should include a focus on the economic well-being of poor and near-poor couples and families (Amato et al. 2007). "Policy makers should recognize that any initiative that improves the financial security and well-being of married couples is a pro-marriage policy" (Amato et al. 2007, p. 256). Supports such as job training, employment assistance, flexible workplace policies that decrease work–family conflict, affordable child care, and economic support, such as the earned income tax credit are discussed in Chapters 6 and 7. In addition, policy makers should take a hard look at policies that penalize poor couples for marrying. Poor couples who marry are often penalized by losing Medicaid benefits, food stamps, and other forms of assistance.

> The most important pro-marriage and divorce-prevention measures may be those that maximize employment and earnings.

Strategies to Strengthen Families During and After Divorce

When one or both marriage partners decide to divorce, what can the couple do to minimize the negative consequences of divorce for themselves and their children? According to Ahrons (2004), the postdivorce conflict between parents and not the divorce itself is most traumatic for children. A review of the literature on the effects of parental conflict on children suggests that children who are exposed to high levels of parental conflict are at risk for anxiety, depression, and disruptive behavior; they are more likely to be abusive toward romantic partners in adolescence and adulthood and are likely to have higher rates of divorce and maladjustment in adulthood (Grych 2005). Next, we discuss ways in which divorced (or divorcing) parents can minimize the conflict with their ex-spouse and develop a cooperative parenting relationship.

Forgiveness. Research suggests that divorced parents who forgive each other are more likely to have positive and cooperative co-parenting after divorce (Bonach 2009). Forgiveness does not mean accepting, condoning, or excusing an offender. Rather, forgiveness involves making a choice to think, feel, and behave less negatively toward someone who has hurt or offended you and to act with good will toward the offender (Fincham et al. 2006).

Divorce Education Programs. Divorce education programs are designed to help parents who are divorced or planning to divorce reduce parental conflict and educate them about the factors that affect their children's adjustment. Parents are taught how to respond to their children's reactions to divorce, and how to cooperate in co-parenting. An experimental research study on the effectiveness of a court-ordered education program for divorcing parents found that participation in the program had a significant positive effect on co-parenting skills and the parent relationship for both mothers and fathers (Whitehurst et al. 2008).

Divorce Mediation. In **divorce mediation**, divorcing couples meet with a neutral third party, a mediator, who helps them resolve issues of property division, child support, child custody, and spousal support (i.e., alimony) in a way that minimizes conflict and encourages cooperation. Mediation can also be used with divorcing couples who are in conflict over who gets custody of dogs, cats, and other companion animals (Franklin 2013).

In a longitudinal study, researchers compared two groups of divorcing parents who were petitioning for a court custody hearing: parents who were randomly assigned to try mediation and those who were randomly assigned to continue the adversarial court process (Emery et al. 2005). If mediation did not work, the parents in the mediation group could still go to court to resolve their case. The parents who participated in mediation were much more likely to settle their custody dispute outside court than the parents who did not. The researchers found that mediation can not only speed settlement, save money on attorney and court fees, and increase compliance, it can also result in improved relationships between nonresidential parents and children, as well as between divorced parents 12 years after the dispute settlement.

An increasing number of jurisdictions and states have mandatory child custody mediation programs, whereby parents in a custody or visitation dispute must attempt to resolve their dispute through mediation before a court will hear the case. It should be noted that mediation is not recommended for couples in which one or both partners is unwilling to negotiate in good faith, has a mental illness, or has been violent or abusive in the relationship—the subject we turn to next.

Violence and Abuse in Intimate and Family Relationships

In U.S. society, people are more likely to be physically assaulted, abused and neglected, sexually assaulted and molested, or killed in their own homes rather than anywhere else, and by other family members rather than by anyone else (Gelles 2000). Before reading further, you may want to take the Abusive Behavior Inventory in this chapter's *Self and Society* feature.

Intimate Partner Violence and Abuse

Abuse in relationships can take many forms, including emotional and psychological abuse, physical violence, and sexual abuse. **Intimate partner violence (IPV)** refers to actual or threatened violent crimes committed against individuals by their current or former spouses, cohabiting partners, boyfriends, or girlfriends. Emotional abuse can take many forms including yelling and screaming, withholding physical contact, isolating someone from their family and friends, belittling or insulting a person, restricting a person's activities, controlling a person's behavior (e.g., telling them what to wear, what to eat, where they can go, what time they have to be home, and so on), and exhibiting unreasonable jealousy (Follingstad & Edmundson 2010).

Intimate partner violence is widespread. A study of young U.S. adults ages 17 to 24 found that 4 in 10 experienced physical violence and 5 in 10 experienced verbal abuse in their current or most recent relationship (Halpern-Meekin et al. 2013). Globally, 1 woman in every 3 has been subjected to violence in an intimate relationship (United

divorce mediation A process in which divorcing couples meet with a neutral third party (mediator) who assists the individuals in resolving issues such as property division, child custody, child support, and spousal support in a way that minimizes conflict and encourages cooperation.

intimate partner violence (IPV) Actual or threatened violent crimes committed against individuals by their current or former spouses, cohabiting partners, boyfriends, or girlfriends.

Nations Development Programme 2009). There has been ongoing debate about whether more IPV is perpetrated by women or men. U.S. government data find that about 4 in 5 victims of IPV are female, suggesting men engage in more acts of IPV than do women (Catalano 2012). But other research has found that the rate of female perpetration of IPV is equivalent to or exceeds the rate of male perpetration (Cui et al. 2013; Fincham et al. 2013). Some researchers suggest that women are more likely than men to *report* their acts of IPV because there is less stigma associated with women's violence against men compared to men's violence against women. Anderson (2013) explains that men who abuse women are depicted in U.S. culture as dangerous brutes, whereas women's violence against men is often depicted as funny or as justified. Some research has found that when women assault their male partners, these assaults tend to be acts of retaliation or self-defense (Johnson 2001; Swan et al. 2008). Although the question of which gender engages in more acts of IPV remains controversial, researchers agree that both male and female violence and abuse in intimate relationships are problems that warrant attention.

> In U.S. society, people are more likely to be physically assaulted, abused and neglected, sexually assaulted and molested, or killed in their own homes rather than anywhere else, and by other family members rather than by anyone else.

Four Types of Intimate Partner Violence. One of the reasons why researchers disagree about which gender perpetrates more acts of IPV is that researchers use different methods for defining and measuring IPV, and these methods may not account for the different types of IPV. Johnson and Ferraro (2003) identified the following four types of partner violence:

1. *Common couple violence* refers to occasional acts of violence arising from arguments that get "out of hand." Common couple violence usually does not escalate into serious or life-threatening violence.
2. *Intimate terrorism* is motivated by a wish to control one's partner and involves not only violence but also economic subordination, threats, isolation, verbal and emotional abuse, and other control tactics. Intimate terrorism is almost entirely perpetrated by men and is more likely to escalate over time and to involve serious injury.
3. *Violent resistance* refers to acts of violence that are committed in self-defense, usually by women against a male partner.
4. *Mutual violent control* is a rare pattern of abuse "that could be viewed as two intimate terrorists battling for control" (Johnson & Ferraro 2003, p. 169).

NY Daily News/Getty Images

In 2009, music artist Chris Brown assaulted his then girlfriend, pop singer Rihanna, leaving her with a broken nose and lip and bruises to her face.

Intimate partner abuse also takes the form of sexual aggression, which refers to sexual interaction that occurs against one's will through use of physical force, threat of force, pressure, use of alcohol or drugs, or use of position of authority. In most cases of rape or sexual assault of females—about 8 in 10 cases—the offender is an intimate partner, family member, friend, or acquaintance (Planty et al., 2013).

Three Types of Male Perpetrators of Intimate Partner Violence. Because male perpetrators of IPV tend to inflict more physical harm on victims than do

Circle the number that best represents your closest estimate of how often each of the behaviors happens in your relationship with your partner or happened with a former partner during the previous six months (1, never; 2, rarely; 3, occasionally; 4, frequently; 5, very frequently).

1. Called you a name and/or criticized you   1 2 3 4 5
2. Tried to keep you from doing something you wanted to do (e.g., going out with friends, going to meetings)   1 2 3 4 5
3. Gave you angry stares or looks   1 2 3 4 5
4. Prevented you from having money for your own use   1 2 3 4 5
5. Ended a discussion with you and made the decision himself/herself   1 2 3 4 5
6. Threatened to hit or throw something at you   1 2 3 4 5
7. Pushed, grabbed, or shoved you   1 2 3 4 5
8. Put down your family and friends   1 2 3 4 5
9. Accused you of paying too much attention to someone or something else   1 2 3 4 5
10. Put you on an allowance   1 2 3 4 5
11. Used your children to threaten you (e.g., told you that you would lose custody, said he/she would leave town with the children)   1 2 3 4 5
12. Became very upset with you because dinner, housework, or laundry was not done when he/she wanted it done or done the way he/she thought it should be   1 2 3 4 5
13. Said things to scare you (e.g., told you something "bad" would happen, threatened to commit suicide)   1 2 3 4 5
14. Slapped, hit, or punched you   1 2 3 4 5
15. Made you do something humiliating or degrading (e.g., begging for forgiveness, having to ask permission to use the car or to do something)   1 2 3 4 5
16. Checked up on you (e.g., listened to your phone calls, checked the mileage on your car, called you repeatedly at work)   1 2 3 4 5
17. Drove recklessly when you were in the car   1 2 3 4 5
18. Pressured you to have sex in a way you didn't like or want   1 2 3 4 5
19. Refused to do housework or child care   1 2 3 4 5
20. Threatened you with a knife, gun, or other weapon   1 2 3 4 5
21. Spanked you   1 2 3 4 5
22. Told you that you were a bad parent   1 2 3 4 5
23. Stopped you or tried to stop you from going to work or school   1 2 3 4 5
24. Threw, hit, kicked, or smashed something   1 2 3 4 5
25. Kicked you   1 2 3 4 5
26. Physically forced you to have sex   1 2 3 4 5
27. Threw you around   1 2 3 4 5
28. Physically attacked the sexual parts of your body   1 2 3 4 5
29. Choked or strangled you   1 2 3 4 5
30. Used a knife, gun, or other weapon against you   1 2 3 4 5

Scoring

Add the numbers you circled and divide the total by 30 to find your score. The higher your score, the more abusive your relationship.

Source: Shepard, Melanie F., and James A. Campbell. 1992 (September). "The Abusive Behavior Inventory: A Measure of Psychological and Physical Abuse." *Journal of Interpersonal Violence* 7(3):291–305. Copyright© 1992 by Sage Publications. Used by permission.

female perpetrators, most research on perpetrators has focused on males. Researchers have identified three types of male abusers (Fowler & Westen 2011):

1. The *psychopathic abuser* is generally violent, impulsive, and lacks empathy and remorse. Men in this subgroup often experienced high rates of childhood abuse and neglect.
2. The *hostile/controlling abuser* is suspicious, hypersensitive to perceived criticism, holds grudges, feels misunderstood, and has few friends. He is a "controlling" person who, instead of taking responsibility for his actions, blames and attacks his partner.
3. The *borderline/dependent abuser* is unhappy, depressed, and prone to emotions that spiral out of control. This type of abuser suffers from deep fears of abandonment and lashes out at the person he loves and needs the most. Men in this subgroup appear more fragile than violent, and after abusive episodes, they tend to feel self-loathing and guilt and are likely to beg for forgiveness.

Effects of Intimate Partner Violence and Abuse. IPV results in bruises, broken bones, traumatic brain injury, cuts, burns, other physical injuries, and death. In 2011, 269 husbands or boyfriends and 1,026 wives or girlfriends in the United

States were murdered by their intimate partners (Federal Bureau of Investigation 2011). When women are abused during pregnancy, the result is often miscarriage or birth defects in the child. Victims of intimate partner violence are at higher risk for depression, anxiety, suicidal thoughts and attempts, lowered self-esteem, inability to trust men, fear of intimacy, substance abuse, and harsh parenting practices (Gustafsson & Cox 2012).

Battering also interferes with women's employment. Some abusers prohibit their partners from working. Other abusers "deliberately undermine women's employment by depriving them of transportation, harassing them at work, turning off alarm clocks, beating them before job interviews, and disappearing when they promise to provide child care" (Johnson & Ferraro 2003, p. 508). Battering also undermines employment by causing repeated absences, impairing women's ability to concentrate, and lowering their self-esteem and aspirations.

Abuse is also a factor in many divorces, which often results in a loss of economic resources. Women who flee an abusive home and who have no economic resources may find themselves homeless.

Children who witness domestic violence are at risk for emotional, behavioral, and academic problems as well as future violence in their own adult relationships (Kitzmann et al. 2003; Parker et al. 2000). Children may also commit violent acts against a parent's abusing partner.

Why Do Some Adults Stay in Abusive Relationships? Adult victims of abuse are commonly blamed for tolerating abusive relationships and for not leaving the relationship as soon as the abuse begins. However, from the point of view of the victims, there are compelling reasons to stay, including economic dependency, emotional attachment, commitment to the relationship, guilt, fear, hope that things will get better, and the view that violence is legitimate because they "deserve" it. Having children complicates the decision to leave an abusive relationship. Although some abuse victims leave to protect the children from harm by the abuser, others may stay because they cannot support and/or raise the children alone, or may fear losing custody of the children if they leave (Lacey et al. 2011). Many victims of intimate partner abuse stay because they fear retribution from their abusive partner if they leave. Indeed, the rate of intimate partner violence is highest among separated couples (see Figure 5.4). Some victims also delay leaving a violent home because they fear the abuser will hurt or neglect a family pet (Fogle 2003).

Victims also stay because abuse in relationships is usually not ongoing and constant but rather occurs in cycles. The **cycle of abuse** involves a violent or abusive episode followed by a makeup period when the abuser expresses sorrow and asks for forgiveness and "one more chance." The makeup period may last for days, weeks, or even months before the next violent outburst occurs. Nevertheless, research suggests that most abused women eventually leave their abusers for good, although leaving tends to be a process rather than a one-time event, and the process often involves leaving and returning multiple times (Lacey et al. 2011).

Researchers in one study that investigated the impact of technology on the maintenance of abusive relationships found that undergraduates check their text messages frequently and feel compelled to read their text messages—even those sent by an abusive partner or ex-partner (Halligan et al. 2013). Abusive partners and ex-partners try to maintain or resume the relationship by sending texts that include apologies, recognition of hurt they caused, promises to never be abusive again, and pleas to resume the relationship.

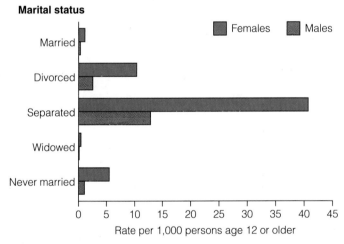

Marital status

Figure 5.4 Average Annual Nonfatal Intimate Partner Victimization Rate, by Gender and Marital Status, 2001–2005
Source: Bureau of Justice Statistics 2011.

cycle of abuse A pattern of abuse in which a violent or abusive episode is followed by a makeup period when the abuser expresses sorrow and asks for forgiveness and "one more chance," before another instance of abuse occurs.

Child Abuse

Child abuse refers to the physical or mental injury, sexual abuse, negligent treatment, or maltreatment of a child under the age of 18 by a person who is responsible for the child's welfare. The most common form of child maltreatment is **neglect**—the caregiver's failure to provide adequate attention and supervision, food and nutrition, hygiene, medical care, and a safe and clean living environment (see Figure 5.5).

The highest rates of victimization are for the youngest children (birth to 1 year), and for children with disabilities. Rates of victimization are also higher for African American, multiracial, and American Indian/Alaska Native children (U.S. Department of Health and Human Services, Administration on Children, Youth, and Families 2012). Children reared in lesbian families are less likely to be abused by a parent or caregiver than are children reared in other family contexts (Gartrell et al. 2010).

Perpetrators of child abuse are most often the parents of the victim. Abuse is more likely to occur in families where there is a lot of stress that can result from a family history of violence, drug or alcohol abuse, poverty, chronic health problems, and social isolation (Centers for Disease Control and Prevention 2012).

Effects of Child Abuse. Physical injuries sustained by child abuse cause pain, disfigurement, scarring, physical disability, and death. In 2011, an estimated 1,570 U.S. children died of abuse or neglect (U.S. Department of Health and Human Services, Administration on Children, Youth, and Families 2012). Most of these children were younger than age 4, and most child deaths were caused by one or both parents. Head injury is the leading cause of death in abused children (Rubin et al. 2003). **Shaken baby syndrome**, whereby a caregiver shakes a baby to the point of causing the child to experience brain

child abuse The physical or mental injury, sexual abuse, negligent treatment, or maltreatment of a child younger than age 18 by a person who is responsible for the child's welfare.

neglect A form of abuse involving the failure to provide adequate attention, supervision, nutrition, hygiene, health care, and a safe and clean living environment for a minor child or a dependent elderly individual.

shaken baby syndrome A form of child abuse whereby a caretaker shakes a baby to the point of causing the child to experience brain or retinal hemorrhage.

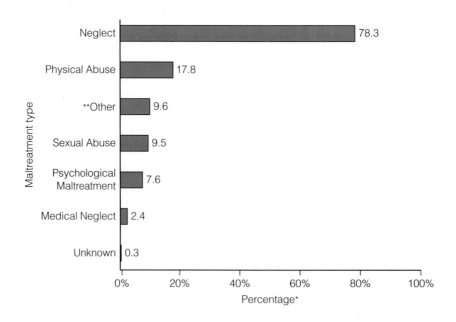

*Percentages sum to more than 100% because children may experience more than one type of maltreatment.

** Includes "abandonment," "threats of harm," and "congenital drug addiction."

Figure 5.5 Types of Child Maltreatment, 2011
Source: U.S. Department of Health and Human Services, Administration on Children, Youth, and Families 2012.

or retinal hemorrhage, most often occurs in response to a baby, typically younger than 6 months, who will not stop crying (Ricci et al. 2003; Smith 2003).

Adults who were abused as children have an increased risk of a number of problems, including depression, smoking, alcohol and drug abuse, eating disorders, obesity, high-risk sexual behavior, and suicide (Centers for Disease Control and Prevention 2012). Sexual abuse of young girls is associated with decreased self-esteem, increased levels of depression, running away from home, alcohol and drug use, and multiple sexual partners (Jasinski et al. 2000; Whiffen et al. 2000). A review of the research suggests that sexual abuse of boys produces many of the same reactions that sexually abused girls experience, including depression, sexual dysfunction, anger, self-blame, suicidal feelings, guilt, and flashbacks (Daniel 2005). Married adults who were physically and sexually abused as children report lower marital satisfaction, higher stress, and lower family cohesion than married adults with no abuse history (Nelson & Wampler 2000). Contrary to prior research findings, a recent study found that physical abuse in childhood is not associated with future violent delinquency, whereas sexual abuse and neglect are strongly related to future violent delinquent behavior (Yun et al. 2011).

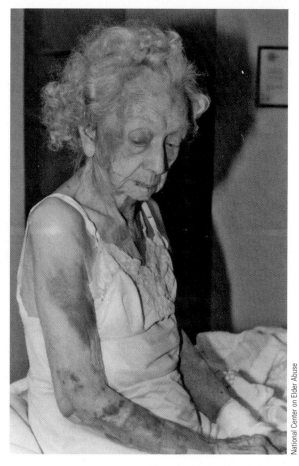

National Center on Elder Abuse

If you suspect elder abuse or are concerned about the well-being of an older person, call your state abuse hotline immediately.

Elder, Parent, Sibling, and Pet Abuse

Domestic violence and abuse may involve adults abusing their elderly parents or grandparents, children abusing their parents, and siblings abusing each other. Pets (or companion animals) are also victimized by domestic violence.

Elder Abuse. **Elder abuse** includes physical abuse, sexual abuse, psychological abuse, financial abuse (such as improper use of the elder person's financial resources), and neglect. The most common form of elder abuse is neglect—failure to provide basic health and hygiene needs, such as clean clothes, doctor visits, medication, and adequate nutrition. Neglect also involves unreasonable confinement, isolation of elderly family members, lack of supervision, and abandonment.

Older women are far more likely than older men to suffer from abuse or neglect. Two out of every three cases of elder abuse reported to state adult protective services involve women. Although elder abuse also occurs in nursing homes, most cases of elder abuse occur in a domestic setting. The most likely perpetrators are adult children, followed by other family members and spouses or intimate partners (Teaster et al. 2006).

Parent Abuse. Some parents are victimized by their children's violence, ranging from hitting, kicking, and biting to pushing a parent down the stairs and using a weapon to inflict serious injury to or even kill a parent. More violence is directed against mothers than against fathers, and sons tend to be more violent toward parents than are daughters (Ulman 2003). In most cases of children being violent toward their parents, the parents had been violent toward the children.

Sibling Abuse. Also referred to as "sibling bullying," sibling abuse is "a widespread and serious problem . . . and, arguably, the most frequent type of aggression in American society" (Skinner & Kowalski, 2013, p. 1,727). A national survey found that nearly 30 percent of children ages 2 to 17 had been physically assaulted by a sibling (Finkelhor et al. 2005). The most common form of sibling abuse is emotional abuse, such as verbal teasing or excluding or ignoring a sibling (Skinner & Kowalski 2013). Sexual abuse also occurs in sibling relationships.

elder abuse The physical or psychological abuse, financial exploitation, or medical abuse or neglect of the elderly.

In 2005, an 81-year-old woman walked nearly a mile from her house to a restaurant, pleading for help to escape the abuse she endured. She said her son repeatedly beat her and her dog. The restaurant owner called the local sheriff's office, and a deputy arrested the son. The woman relocated, but could not take her dog, a 2-year-old 60-pound Catahoula, with her. Kathy Cornwell, one of the victim advocates who helped the woman, adopted the dog. Kathy and the dog were in the bedroom when her husband came home. At the sight of a strange man, the dog jumped onto the bed and laid his body across Kathy. When Kathy later spoke with the abuse victim, she learned the dog had always tried to protect the woman from her abusive son, covering her body and absorbing the blows. As a result of the beatings, the dog had permanent nerve damage in his eyes and must wear goggles when he is outside because his pupils cannot dilate. The dog, known as Little Horatio, goes to universities and other public forums on abuse. Cornwell explains that Horatio gives abuse victims hope: "This dog has gone through so much. And look at how happy he is. He's not stuck in the past. It shows people that it's hard, but you can move on" (quoted in Sullivan 2010, n.p.).

Domestic violence and animal abuse are linked. In a review of the literature, DeGue (2009) noted that between 47 and 71 percent of women seeking services from domestic violence shelters reported that a male abuser had threatened, hurt, or killed their companion animals. One study that compared the reports of women in shelters with a sample of nonabused women found that abused women were 11 times more likely to report that their partner had hurt or killed a pet (54 percent versus 5 percent) (DeGue 2009). Numerous studies have reported that fear of their companion animals being harmed or killed creates a significant barrier that prevents abused women from escaping their violent situations (Ascione 2007). Research has also documented that

Little Horatio was a victim of domestic violence.

children exposed to domestic violence are more likely than nonexposed children to have abused animals (Ascione & Shapiro 2009).

Because animal cruelty and family violence commonly co-occur, cross-reporting and cross-training have been instituted in many communities to teach human service personnel how to recognize and report cases of animal abuse and, conversely, to teach animal welfare personnel to recognize and report child, spousal, and elder abuse. Florida and San Diego County, California, mandate child protective personnel to report suspected animal abuse to animal welfare agencies, and four states require animal care and control personnel to report possible child abuse to child welfare agencies. Several states have laws permitting pets to be included in domestic violence protective orders (Ascione & Shapiro 2009). Finally, some programs provide procedures for housing pets in

domestic violence situations at animal shelters or with animal rescue groups. The American Humane Association PAWS (Pets and Women's Shelters) program provides grants to shelters to create housing where abuse victims and their pets can be together. Carol Wick, CEO of Harbor House, a domestic violence shelter in Orange County, Florida, describes the benefits of providing shelters for abuse victims and their pets:

Every week we receive calls from domestic violence victims who want to leave their situation but are fearful of what will happen to their pets if they are left behind. . . . The new on-site kennel will allow women and children to safely leave an abusive situation and bring their pet with them—so no one gets left behind. (quoted in American Humane Association 2010)

What Do You Think? In recent years, the problem of bullying among peers in schools has received increased attention, and those concerned about bullying have worked to spread the message that bullying is not acceptable behavior. In contrast, sibling bullying is commonly viewed as acceptable and expected (Skinner & Kowalski 2013). Why do you think peer bullying is viewed as a serious social problem, whereas sibling bullying is generally not viewed as a serious social concern?

Pet Abuse. Because pets are often viewed as "members of the family," abused pets (or companion animals) can be considered victims of family violence (see also this chapter's *Animals and Society* feature). Abusers threaten to hurt pets to control their victims, and all too often their threats are carried out. One victim of domestic violence revealed how her abuser killed her cat as a means of inducing fear and intimidation:

> The very last thing he did to my cat hurt my heart so bad. He had me stand here and . . . she was tied to the tree [with] . . . fishing wire or . . . thread or something. And he . . . turned her around, stuffed [fireworks] in her behind and lit it. And I had to stand there and watch my cat explode in my face. And he was like, "That could happen to you." (quoted in Hardesty et al. 2013, p. 9)

Factors Contributing to Intimate Partner and Family Violence and Abuse

In many ways, U.S. culture tolerates and even promotes violence (Walby 2013). Violence in the family stems from the acceptance of violence in our society as a legitimate means of enforcing compliance and solving conflicts at interpersonal, familial, national, and international levels. More specific individual, family, and social factors that contribute to domestic violence and abuse are discussed in the following sections.

Individual, Relationship and Family Factors. Although we all have the potential to "lose our cool" and act abusively toward an intimate partner or family member, some individuals are at higher risk for being abusive to others. Risk factors include having witnessed or been a victim of abuse as a child, past violent or aggressive behavior, lack of employment and other stressful life events or circumstances, and drug and alcohol use (National Center for Injury Prevention and Control 2012). Alcohol use is involved in 50 percent to 75 percent of incidents of physical and sexual aggression in intimate relationships (Lloyd & Emery 2000). Alcohol and other drugs increase aggression in some individuals and enable the offender to avoid responsibility by blaming the violent behavior on drugs or alcohol.

Although abuse in adult relationships occurs among all socioeconomic groups, it is more prevalent among the poor. However, most poor people do not maltreat their children, and poverty, per se, does not cause abuse and neglect; the correlates of poverty, including stress, drug abuse, and inadequate resources for food and medical care, increase the likelihood of maltreatment (Kaufman & Zigler 1992).

Intimate relationships that are characterized by "churning"—a pattern of instability in which the partners break up and then reconcile—are also at higher risk for abuse. A study of young adults ages 17 to 24 found that "relationship churners" were much more likely to report physical and verbal abuse in their most current relationship than were young adults who were either stably together or stably broken up (Halpern-Meekin et al. 2013). This finding is particularly concerning given that nearly half of young adults experience churning in their romantic relationships.

Gender Inequality and Gender Socialization. In the United States before the late 19th century, a married woman was considered the property of her husband. A husband had a legal right and marital obligation to discipline and control his wife through the use of physical force. This traditional view of women as property may contribute to men's doing with their "property" as they wish. In a study of men in battering intervention programs, about half of the men viewed battering as acceptable in certain situations (Jackson et al. 2003).

The view of women and children as property also explains marital rape and father–daughter incest. Historically, the penalties for rape were based on property rights laws designed to protect a man's property—his wife or daughter—from rape by other men (Russell 1990). Although a husband or father "taking" his own property in the past was not considered rape, today, marital rape and incest is considered a crime in all 50 states. Acquaintance and date rape can be explained in part by the fact that, in U.S. culture, "men receive support, even permission, from each other to sexually assault women through their direct encouragement or by ignoring problematic behavior" (Foubert et al. 2010, p. 2,238).

Traditional male gender roles have taught men to be aggressive and to be dominant in male–female relationships. "Because it is so clearly associated with masculinity in American culture, violence is a social practice that enables men to express a masculine identity" (Anderson 1997, p. 667). Traditional male gender socialization also discourages men from verbally expressing their feelings, which increases the potential for violence and abusive behavior (Umberson et al. 2003). Traditional female gender roles have also taught some women to be submissive to their male partner's control.

Acceptance of Corporal Punishment. **Corporal punishment**—the intentional infliction of pain intended to change or control behavior—is widely accepted as a parenting practice. In a review of research on corporal punishment, Straus (2010) concluded that, in the United States, corporal punishment is (1) almost universal—94 percent of toddlers are spanked; (2) chronic—toddlers are often spanked three or more times a week; (3) often severe, with more than one in four parents using an object such as a paddle or belt to punish their children; and (4) of long duration—13 years for a third of U.S. children, and 17 years for 14 percent of U.S. children. Parents of black children are more likely to use corporal punishment than parents of white or Latino children (Grogan-Kaylor & Otis 2007). A study of 35 low- and middle-income countries found that 17 percent of children, on average, experienced severe physical punishment, such as hitting the child on the head, ears, or face, or hitting the child hard and repeatedly (UNICEF 2010).

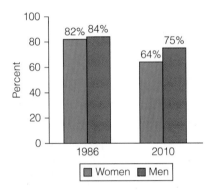

Figure 5.6 **Percent of U.S. Adults Who Agree that Spanking is Sometimes Necessary: 1986 and 2010**
Source: Child Trends 2012.

Although not everyone agrees that all instances of corporal punishment constitute abuse, some episodes of parental "discipline" are undoubtedly abusive. Many mental health and child development specialists advise against physical punishment, arguing that it is ineffective and potentially harmful to the child (Straus 2000). One study of 3,870 families found that children who were spanked at age 1 were more likely to be aggressive at age 3 and depressed, anxious, and/or withdrawn at age 5 (Gromoske & Maguire-Jack 2012). Although the percentage of U.S. women and men who agree that children sometimes need a "good hard spanking" declined from 1986 to 2010 (see Figure 5.6), the majority of U.S. adults still view spanking as sometimes necessary (Child Trends 2012).

Inaccessible or Unaffordable Community Services. Many cases of child and elder abuse and neglect are linked to inaccessible or unaffordable health care, day care, elder care, and respite care facilities. Failure to provide medical care to children and elderly family members (a form of neglect) is sometimes a result of the lack of accessible or affordable health care services in the community. Failure to provide supervision for children and adults may result from inaccessible day care and elder care services. Without elder care and respite care facilities, socially isolated families may not have any help with the stresses of caring for elderly family members and children with special needs.

Strategies for Action: Preventing and Responding to Domestic Violence and Abuse

Next, we look at strategies for preventing and responding to violence and abuse in intimate and family relationships.

Prevention Strategies

Abuse-prevention strategies include public education and media campaigns, which may help to reduce domestic violence by conveying the criminal nature of domestic assault and offering ways to prevent abuse. Other abuse-prevention efforts focus on parent education to teach parents realistic expectations about child behavior and methods of child discipline that do not involve corporal punishment. For example, Mental Health America (2003) distributes a fact sheet on alternatives to spanking (see Table 5.3).

corporal punishment The intentional infliction of pain intended to change or control behavior.

What Do You Think? In 1979, Sweden became the first country in the world to ban corporal punishment in all settings, including the home. By 2013, 33 countries banned corporal punishment in all settings (Global Initiative to End All Corporal Punishment of Children 2012). In the United States, it is legal in all 50 states for a parent to spank, hit, belt, paddle, whip, or otherwise inflict punitive pain on a child, so long as the corporal punishment does not meet the individual state's definition of child abuse.

Thirty-one states and the District of Columbia have passed laws prohibiting corporal punishment in public schools. In Iowa and New Jersey, this ban also covers private schools. Do you think that the United States should ban corporal punishment in the home? In schools? Why or why not?

Another abuse-prevention strategy involves reducing violence-provoking stress by reducing poverty and unemployment and providing adequate housing, child care programs and facilities, nutrition, medical care, and educational opportunities. Strengthening the supports for poor families with children reduces violence-provoking stress and minimizes neglect that results from inaccessible or unaffordable community services.

Responding to Domestic Violence and Abuse

After domestic violence and abuse has occurred, victims may seek safety in a shelter or safe house and take legal action against the abuser. Children may be placed in foster care, and the abuser may participate in mandatory or voluntary treatment.

TABLE 5.3 Effective Discipline Techniques for Parents: Alternatives to Spanking

Punishment is a "penalty" for misbehavior, but discipline is a method of teaching a child right from wrong. Alternatives to physical discipline include the following:

1. Be a positive role model.

Children learn behaviors by observing their parents' actions, so parents must model the ways in which they want their children to behave. If a parent yells or hits, the child is likely to do the same.

2. Set rules and consequences.

Make rules that are fair, realistic, and appropriate to a child's level of development. Explain the rules to children along with the consequences of not following them. If children are old enough, they can be included in establishing the rules and consequences for breaking them.

3. Encourage and reward good behavior.

When children are behaving appropriately, give them verbal praise and occasionally reward them with tangible objects, privileges, or increased responsibility.

4. Create charts.

Charts to monitor and reward behavior can help children learn appropriate behavior. Charts should be simple and should focus on one behavior at a time, for a certain length of time.

5. Give time-outs.

A "time-out" involves removing a child from a situation following a negative behavior. This can help the child calm down, end the inappropriate behavior, and reenter the situation in a positive way. Explain what the inappropriate behavior is, why the time-out is needed, when the time-out will begin, and how long it will last. Set an appropriate length of time for the time-out based on age and level of development, usually just a few minutes.

Source: Based on Mental Health America 2003. *Effective Discipline Techniques for Parents: Alternatives to Spanking.* Strengthening Families Fact Sheet. www.nmha.org. Reprinted with permission.

The National Domestic Violence Hotline (1-800-799-SAFE) is a 24-hour, toll-free service that provides crisis assistance and local domestic violence shelter and safe house referrals for callers across the country. Shelters provide abused women and their children with housing, food, and counseling services. Safe houses are private homes of individuals who volunteer to provide temporary housing to abused people who decide to leave their violent homes. Some communities have abuse shelters for victims of elder abuse. Because one in four victims reports a delay in leaving dangerous domestic situations because of concerns over the safety of a pet, some programs offer a safe shelter for pets of victims of domestic violence.

Arrest and Restraining Order. Domestic violence and abuse are crimes for which individuals can be arrested, jailed, and/or ordered to leave the home or enter a treatment program. About half of the states and Washington, DC, now have mandatory arrest policies that require police to arrest abusers, even if the victim does not want to press charges. Abuse victims may obtain a restraining order that prohibits the perpetrator from going near the abused partner.

What Do You Think? Many instances of intimate partner physical or sexual assault are not reported to the police. Why do you think victims of such IPV do not report to law enforcement authorities?

Foster Care Placement. Children who are abused in the family may be removed from their homes and placed in government-supervised foster care. Foster care placements include other family members, certified foster parents, group homes, and other institutional facilities. Hundreds of thousands of children are in foster care, waiting to be reunited with their families or adopted.

Due to the economic recession, more prospective adoptive parents are considering adopting foster children because they cannot afford private adoptions. However, there continues to be a shortage of people willing to adopt foster children because foster children tend to be older and are more likely to have emotional or physical problems (Koch 2009). Every year, thousands of children who have not been adopted or reunited with their families must leave foster care because they turn 18. In some states, youth who age out of the foster care system can receive government aid such as housing assistance and Medicaid. However, many youth who age out of foster care struggle to fend for themselves, and many become homeless.

Another problem that plagues the foster care system is that, although it is intended to protect children from abuse, foster parents or caregivers sometimes abuse the children. Excluding extreme cases of abuse and neglect, some evidence suggests that children whose families are investigated for abuse or neglect are better off staying with their families than entering foster care, as children in foster care are more likely to drop out of school, commit crimes, abuse drugs, and become teen parents (Doyle 2007).

Treatment for Abusers. Treatment for abusers typically involves group and/or individual counseling, substance abuse counseling, and/or training in communication, conflict resolution, and anger management. Most men in batterer intervention programs are referred by courts, although some may enter treatment voluntarily (Carter 2010). Treatment for men who sexually abuse children typically involves cognitive-behavior therapy (changing the thoughts that lead to sex abuse) and medication to reduce the sex drive (Stone 2004). Men who stop abusing their partners learn to take responsibility for their abusive behavior, develop empathy for their partner's victimization, reduce their dependency on their partners, and improve their communication skills (Scott & Wolfe 2000).

Understanding Family Problems

Family problems can best be understood within the context of the society and culture in which they occur. Although divorce and domestic violence may appear to result from individual decisions, myriad social and cultural forces influence these decisions.

The impact of family problems, including divorce and abuse, is felt not only by family members but also by society at large. Family members experience life difficulties such as poverty, school failure, low self-esteem, and mental and physical health problems. Each of these difficulties contributes to a cycle of family problems in the next generation. The impact on society includes public expenditures to assist single-parent families and victims of domestic violence and neglect, increased rates of juvenile delinquency, and lower worker productivity.

For some, the solution to family problems implies encouraging marriage and discouraging other family forms, such as single parenting, cohabitation, and same-sex unions. However, many family scholars argue that the fundamental issue is making sure that children are well cared for, regardless of their parents' marital status or sexual orientation. Some even suggest that marriage is part of the problem, not part of the solution. Martha Fineman of Cornell Law School said, "This obsession with marriage prevents us from looking at our social problems and addressing them. . . . Marriage is nothing more than a piece of paper, and yet we rely on marriage to do a lot of work in this society: It becomes our family policy, our policy in regard to welfare and children, the cure for poverty" (quoted by Lewin 2000, p. 2).

Strengthening marriage is a worthy goal because strong marriages offer many benefits to individuals and their children. However, "strengthening marriage does not have to mean a return to the patriarchal family of an earlier era. . . . Indeed, greater marital stability will only come about when men are willing to share power, as well as housework and child care, equally with women" (Amato 1999, p. 184). Strengthening marriage does not mean that other family forms should not also be supported. The reality is that the postmodern family comes in many forms, each with its strengths, needs, and challenges. Given the diversity of families today, social historian Stephanie Coontz (2004) suggested that "the appropriate question . . . is not what single family form or marriage arrangement we would prefer in the abstract, but how we can help people in a wide array of different committed relationships minimize their shortcomings and maximize their solidarities" (p. 979). She further argued that

> [i]f we withdrew our social acceptance of alternatives to marriage, marriage itself might suffer. . . . The same personal freedoms that allow people to expect more from their married lives also allow them to get more out of staying single and give them more choice than ever before in history about whether or not to remain together. (Coontz 2005, p. 310)

The family problems emphasized in this chapter—domestic violence and abuse, and problems of divorce—have something in common: Economic hardship and poverty can be a contributing factor and a consequence of each of these problems. In the next chapter, we turn our attention to poverty and economic inequality, problems that are at the heart of many other social ills.

CHAPTER REVIEW

- **What are some examples of diversity in families around the world?**
  Some societies recognize monogamy as the only legal form of marriage, whereas other societies permit polygamy. Societies also vary in their norms and policies regarding arranged marriage versus self-chosen marriage, same-sex couples, childbearing, and the roles of women and men in the family.

- **What are some of the major changes in U.S. families that have occurred in the past several decades?**
  Some of the major changes in U.S. families that have occurred in recent decades include increased singlehood

and older age at first marriage, increased heterosexual and same-sex cohabitation, the emergence of living apart together (LAT) relationships, increased births to unmarried women, increased divorce and blended families, and increased employment of married mothers.

- **What is the marital decline perspective? What is the marital resiliency perspective?**
According to the marital decline perspective, the recent transformations in American families signify a collapse of marriage and family in the United States. According to the marital resiliency perspective, poverty, unemployment, poorly funded schools, discrimination, and the lack of basic services (such as health insurance and child care) are more harmful to the well-being of children and adults than is the decline in married two-parent families.

- **Feminist theories of family are most similar to which of the three main sociological theories: structural functionalism, conflict theory, or symbolic interactionism?**
Feminist theories of family are most aligned with conflict theory. Both feminist and conflict theories are concerned with how gender inequality influences and results from family patterns.

- **What are some of the effects of divorce on children?**
Reviews of recent research on the consequences of divorce for children find that children with divorced parents score lower on measures of academic success, psychological adjustment, self-concept, social competence, and long-term health, and that they have higher levels of aggressive behavior and depression. Such effects are related to the economic hardship associated with divorce, the reduced parental supervision resulting from divorce, and parental conflict during and after divorce. In highly conflictual marriages, divorce may actually improve the emotional well-being of children relative to staying in a conflicted home environment.

- **What is divorce mediation?**
In divorce mediation, divorcing couples meet with a neutral third party, a mediator, who helps them resolve issues of property division, child custody, child support, and spousal support in a way that minimizes conflict and encourages cooperation. In some states, counties, and jurisdictions, divorcing couples who are disputing child custody issues are required to participate in divorce mediation before their case can be heard in court.

- **What are the four types of partner violence that Johnson and Ferraro (2003) identified?**
The four patterns of partner violence are (1) common couple violence (occasional acts of violence arising from arguments that get "out of hand"); (2) intimate terrorism (violence that is motivated by a wish to control one's partner); (3) violent resistance (acts of violence that are committed in self-defense); and (4) mutual violent control (both partners battling for control).

- **Why do some abused adults stay in abusive relationships?**
Adult victims of abuse are commonly blamed for choosing to stay in their abusive relationships. From the point of view of the victim, reasons to stay in the relationship include economic dependency, emotional attachment, commitment to the relationship, guilt, fear, hope that things will get better, and the view that violence is legitimate because they "deserve" it. Some victims with children leave to protect the children, but others stay because they need help raising the children or fear losing custody of them if they leave. Some victims stay because they fear the abuser will abuse or neglect a pet.

- **What is the most common form of child abuse?**
The most common form of child abuse is neglect.

TEST YOURSELF

1. The United States has the highest nonmarital birthrate of any country in the world.
   a. True
   b. False
2. Two perspectives on the state of marriage in the United States are the marital decline perspective and the marital ___ perspective.
   a. health
   b. resiliency
   c. incline
   d. stability
3. A study on how marriage has changed found that, between 1980 to 2000, marriages
   a. have become more prone to divorce
   b. have become more conflictual
   c. involve less interaction between husband and wife
   d. all of the above
4. How many states recognize no-fault divorce?
   a. None
   b. 5
   c. 25 and the District of Columbia
   d. All 50
5. Most U.S. adults believe that divorce is morally acceptable.
   a. True
   b. False
6. In the majority of married-couple families with children under 18, both parents are employed.
   a. True
   b. False
7. The purpose of divorce mediation is to help couples who are considering divorce repair their relationship and stay together.
   a. True
   b. False

8. In the United States, people are more likely to be physically assaulted, sexually assaulted and molested, or killed by ___ than by anyone else.
   a. a family member
   b. an employee
   c. a stranger
   d. a friend

9. Which of the following is the most prevalent form of abuse in families?
   a. Sexual abuse by a father
   b. Sexual abuse by an uncle or cousin
   c. Verbal abuse by a mother
   d. Sibling abuse

10. The majority of U.S. adults view spanking as
   a. appropriate only for children under the age of 12
   b. harmful
   c. sometimes necessary
   d. the only kind of discipline that works

Answers: 1. B; 2. B; 3. C; 4. D; 5. A; 6. A; 7. B; 8. A; 9. D; 10. C.

KEY TERMS

bigamy 140
child abuse 162
corporal punishment 166
covenant marriage 156
cycle of abuse 161
divorce mediation 158
domestic partnership 143
elder abuse 163
family 140
individualism 152

intimate partner violence (IPV) 158
living apart together (LAT) relationships 144
marital decline perspective 146
marital resiliency perspective 146
monogamy 140
neglect 162
no-fault divorce 152
parental alienation 154
patriarchy 148

polyandry 140
polygamy 140
polygyny 140
refined divorce rate 144
second shift 151
serial monogamy 140
shaken baby syndrome 162

6

Economic Inequality, Wealth, and Poverty

"We are the first generation that can look extreme poverty in the eye, and say this and mean it—we have the cash, we have the drugs, we have the science. Do we have the will to make poverty history?"

—Bono, U2 (rock music group)

Marcos Alves/Flickr/Getty Images

IN AN ARTICLE published in *The American Prospect*, Chuck Collins (2013) describes the lives of four young American adults to illustrate how economic inequality creates different life outcomes:

> Two 21-year-old college students sit down in a coffee shop to study for an upcoming test. Behind the counter, a barista whips up their double-shot lattes. In the back kitchen, another young adult washes the dishes and empties the trash. These four young adults have a lot in common. They are the same age and race, each has two parents, and all grew up in the same metropolitan area. They were all strong students in their respective high schools. But as they enter their third decade, their work futures and life trajectories are radically different—and largely determined at this point. (Collins 2013)

The lives of these four young adults unfold so differently, largely as a result of their families' different levels of economic resources. One of the students, Miranda, will graduate with no student loan debt and will have completed three summers of unpaid internships at businesses that will help her advance her career. Miranda's parents will help her buy a car, pay for housing, and will provide her health insurance, and they have a network of family and professional contacts that can help her. Ten years later, Miranda will have a high-paying job, be engaged to another professional, and with the help of a "parental downpayment assistance program," will buy a home that will appreciate in value due to its location.

The other college student, Marcus, will graduate with more than $55,000 in college debt, a maxed-out credit card, and a résumé of part-time food service jobs that he took to pay for school, and which left him less time and energy for studying. When he graduates, Marcus will have no work experience in his field of study, and will work two part-time jobs to pay back his student loans and to afford rent in a shared apartment. Ten years later, Marcus will still be living in a rented apartment, and working in low-paying jobs, feeling frustrated in his attempts to find a good job in the area of his degree. To pay for various health and other expenses, he will take on more debt, and will be unable to buy a home in large part because of a credit history damaged during his early 20s.

Tony, the barista, is fortunate not to have the burden of college student loan debt. Eventually, he will take some classes at a local public university. But without a college degree, his income and employment opportunities will be constrained. After attempting to learn a building trade and start his own business, Tony eventually finds a job with a steady but low income. Although Tony's parents are not college educated or wealthy, they are stable middle class with modest retirement pensions and a debt-free house, and they are able to provide a bedroom for Tony to live in. That home will provide future economic stability to Tony, as he will eventually inherit it.

Cordelia, the kitchen worker, has even less opportunity for mobility and advancement than Tony. Neither of her parents went to college nor have any significant savings, and they rent their housing. Though she was a top student in her urban high school class, she did not consider applying to a selective college, as the costs seemed daunting, and none of her friends were going away to college. There were no adults or guidance counselors to help her explore other options such as financial aid and scholarships. Cordelia did take courses at the local community college. But over time, she will settle into working a steady and low-wage job as she takes on more responsibility for supporting members of her family who are less fortunate (Collins 2013).

Collins (2013) explains that a key determinant of the different life prospects for the four college students is family wealth—"a factor that plays an oversize role in sorting today's coming-of-age generation onto different opportunity trajectories" (n.p.). In this chapter, we examine economic inequality, wealth, and poverty globally and in the United States.

The Global Context: Economic Inequality, Wealth, and Poverty around the World

We live in a world in which some people live in extreme poverty, lacking even the basic necessities of a home to live in; access to food, clean water, and sanitation; and basic health care. In 2013, one in five people on this planet lived in extreme poverty—down from more than half in 1980 (Chandy 2013). This reduction, although significant, represents the extreme poor—those living on less than $1.25 a day. Many more people live on between $1.25 and $2.00 a day, and then many more live on not much more above that.

In stark contrast to the extreme poverty that persists in the world, in 2013, there were 1,426 billionaires in the world (Kroll & Dolan 2013). The wealthiest people around the world live opulent, lavish lifestyles that include luxuries that most of us can only imagine: yachts and private jets, multiple homes around the world, and access to anything money can buy.

Economic inequality—the wide gap that divides the rich and the poor—characterizes nations, communities, families, and individuals, and includes inequality in both income and wealth. **Wealth** refers to the total assets of an individual or household minus liabilities (mortgages, loans, and debts). Wealth includes the value of a home, investment real estate, cars, unincorporated business, life insurance (cash value), stocks, bonds, mutual funds, trusts, checking and savings accounts, individual retirement accounts (IRAs), and valuable collectibles.

The inequality in the world distribution of household wealth is reflected in the following data (Credit Suisse Research Institute 2012):

- The richest one percent of adults (ages 20 and older) in the world own nearly half (46 percent) of global household wealth; the richest 10 percent of adults own 86 percent of total global wealth.
- The poorest half of the world adult population owns barely one percent of global wealth.
- Households with per-adult assets of at least $3,700 (in U.S. dollars) are in the top half of the world wealth distribution; assets of $71,000 per adult place a household in the top 10 percent; and assets of more than $71,000 per adult place a household in the richest one percent worldwide.
- Europe accounts for 31.1 percent of global wealth, North America 30.6 percent, and Asia-Pacific (excluding China and India) accounts for 22.8 percent. The rest of the world, with 60 percent of the global population, owns the remaining 15 percent of global wealth. The United States, with less than 5 percent of the world's population, owns more than a quarter (27.9 percent) of the world's wealth. The next highest share of global wealth—12.6 percent—is owned by the Japanese.

The inequality in wealth is significant, as wealth influences our life options. Researchers at the Urban Institute noted,

> Wealth isn't just money in the bank, it's insurance against tough times, tuition to get a better education and a better job, savings to retire on, and a springboard into the middle class. In short, wealth translates into opportunity. (McKernan et al. 2013, p. 1)

What Do You Think? Researchers asked citizens of various nations about their perceptions of income inequality in their country, and then compared those responses to the actual measurements of income inequality (Forster & d'Ercole 2005). Among 17 advanced countries, U.S. citizens had the largest gap—by a wide margin—between their perception of inequality and its reality. Why do you think Americans perceive the level of inequality in the United States to be much less than what it actually is?

Defining and Measuring Poverty

wealth The total assets of an individual or household minus liabilities.

absolute poverty The lack of resources necessary for material well-being—most importantly, food and water, but also housing, sanitation, education, and health care.

relative poverty The lack of material and economic resources compared with some other population.

extreme poverty Living on less than $1.25 a day.

Absolute poverty refers to the lack of resources necessary for well-being—most importantly, food and water, but also housing, sanitation, education, and health care. In contrast, **relative poverty** refers to the lack of material and economic resources compared with some other population. If you are a struggling college student living on a limited budget, you may feel as though you are "poor" compared with the middle- or upper-middle-class lifestyle to which you may aspire. However, if you have a roof over your head; access to clean water, toilets, and medical care; and enough to eat, you are not absolutely poor; indeed, you have a level of well-being that millions of people living in absolute poverty may never achieve.

Measures of Poverty. The most widely used standard to measure extreme poverty in the developing world is $1.25 per day. Based on this measure, in 2013, about one in five people in the world, or 1.2 billion people, lived in **extreme poverty**—less than $1.25 a day (Chandy 2013).

A measure of relative poverty is based on comparing the income of a household to the median household income in a specific country. According to this relative poverty measure, members of a household are considered poor if their household income is less than 50 percent of the median household income in that country.

Low income is only one indicator of impoverishment. The **Multidimensional Poverty Index** is a measure of serious deprivation in the dimensions of health, education, and living standards that combines the number of deprived and the intensity of their deprivation (see Table 6.1). About 1.7 billion people in 104 countries—a third of their population—live in multidimensional poverty (UNDP 2013). Half of the world's poor, as measured by MPI, live in South Asia, though rates are highest in sub-Saharan Africa.

People in this tented village in New Delhi, India, live in extreme poverty.

What Do You Think? The next section describes how poverty is measured in the United States by comparing the annual pretax income of a household to the official U.S. poverty line—a dollar amount that determines who is considered poor. Before reading further, answer these questions: How much annual income do you think a household with one adult needs to earn to avoid living in poverty? What about a household with two adults? One adult and one child? Two adults and one child? Compare your answers with the official poverty thresholds in Table 6.2.

U.S. Measures of Poverty. In 1964, the Social Security Administration devised a poverty index based on data that indicated that families spent about one-third of their income on food. The official poverty level was set by multiplying food costs by three. Since then, the poverty level has been updated annually for inflation, and differs by the number of adults and children in a household and by the age of the head of household, but is the same across the continental United States (see Table 6.2). Anyone living in a household with pretax income below the official poverty line is considered "poor." Individuals living in households with incomes that are above the poverty line, but not very much above it, are classified as "near-poor," and those living in households with income below 50 percent of the poverty line live in "deep poverty," also referred to as "severe poverty." A common working definition of "low-income" households is households with incomes that are between 100 percent and 200 percent of the federal poverty line or up to twice the poverty level.

The U.S. poverty line has been criticized on several grounds. First, the official poverty line is based on pretax income so tax burdens, as well as tax credits, are disregarded. Family wealth, including savings and property, are also excluded in official poverty calculations, and noncash government benefits that assist low-income families—food stamps, Medicaid, and housing and child care assistance—are not taken into account. In addition, the current poverty measure is a national standard that does not reflect the significant variation in the cost of living from state to state and between urban and rural areas. Finally, the poverty line underestimates the extent of material hardship in the United States because it is based on the assumption that low-income families spend one-third of their household income on food. That was true in the 1950s, but because housing, medical care, child care, and transportation costs have risen more rapidly than food costs, low-income families today spend far less than one-third of their income on food.

TABLE 6.1 Three Dimensions of Multidimensional Poverty

1. Health (nutrition, child mortality)

2. Education (years of schooling, children enrolled)

3. Standard of living (cooking fuel, toilet, water, electricity, floor, assets)

© Cengage Learning

Multidimensional Poverty Index A measure of serious deprivation in the dimensions of health, education, and living standards that combines the number of deprived and the intensity of their deprivation.

TABLE 6.2 Poverty Thresholds, 2012 (Householder Younger than 65 Years)

Household Makeup	Poverty Threshold
One adult	$11,945
Two adults	$15,374
One adult, one child	$15,825
Two adults, one child	$18,480
Two adults, two children	$23,283

Source: U.S. Census Bureau 2013.

When a 2013 Gallup Poll asked the American public, "What is the smallest amount of money a family of four needs to make each year to get by in your community?", the average answer was $58,000—well above the official poverty threshold of $23,283 for a family of four (Saad 2013).

The **Basic Economic Security Tables Index (BEST)**, a measure of the basic needs and income workers require for economic security, finds that a family of four (with two working parents who receive employment-based benefits, and two young children) needs to earn $67,920 a year, or about $16 an hour per worker (Wider Opportunities for Women 2010). Table 6.3 presents the annual pretax income required to meet basic needs for different family types. Notice that workers without employment benefits need more income than those with benefits. Table 6.4 provides a breakdown of specific expenses that are considered in calculating BEST values.

Sociological Theories of Economic Inequality, Wealth, and Poverty

Americans are taught that we live in a **meritocracy**—a social system in which individuals get ahead and earn rewards based on their individual efforts and abilities (McNamee & Miller 2009). In a meritocracy, everyone has an equal chance to succeed; those who are "$ucce$$ful" are smart and talented and have worked hard and deserve their success, while those who fail to "make it" have only themselves to blame. This individualistic perspective views economic inequality as the result of some people developing their potential and working hard and earning their success, while others don't measure up, make bad choices, don't work hard enough, and have only themselves to blame for their predicament. In contrast to the individualistic perspective, structural functionalism, conflict theory, and symbolic interactionism offer sociological insights into the nature, causes, and consequences of poverty and economic inequality.

Basic Economic Security Tables Index (BEST) A measure of the basic needs and income workers require for economic security.

meritocracy A social system in which individuals get ahead and earn rewards based on their individual efforts and abilities.

Structural-Functionalist Perspective

According to the structural-functionalist perspective, poverty results from institutional breakdown: economic institutions that fail to provide sufficient jobs and pay, educational institutions that fail to equip members of society with the skills they need for employment, family institutions that do not provide two parents, and government institutions that do not provide sufficient public support. Sociologist William Julius Wilson explains:

TABLE 6.3 Basic Economic Security Tables, 2010

U.S., by Family Type and Receipt of Employment-Based Benefits

1 Worker		1 Worker, 1 Infant		1 Worker, 1 Preschooler, 1 Schoolchild		2 Workers, 1 Preschooler, 1 Schoolchild	
Workers with Employment-Based Benefits	Workers without Employment-Based Benefits	Workers with Employment-Based Benefits	Workers without Employment-Based Benefits	Workers with Employment-Based Benefits	Workers without Employment-Based Benefits	Workers with Employment-Based Benefits	Workers without Employment-Based Benefits
$30,012	$34,728	$46,438	$53,268	$57,756	$63,012	$67,920	$73,296

Notes: "Benefits" include unemployment insurance and employment-based health insurance and retirement plans.
Source: Wider Opportunities for Women. 2010. *The Basic Economic Security Tables for the United States*. Washington, DC: Wider Opportunities for Women.

Where jobs are scarce . . . and where there is a disruptive or degraded school life purporting to prepare youngsters for eventual participation in the workforce, many people eventually lose their feeling of connectedness to work in the formal economy; they no longer expect work to be a regular, and regulating, force in their lives. . . . (Wilson 1996, pp. 52–53)

More than 60 years ago, Davis and Moore (1945) presented a structural-functionalist explanation for economic inequality, arguing that a system of unequal pay motivates people to achieve higher levels of training and education and to take on jobs that are more important and difficult by offering higher rewards for higher achievements. However, this argument is criticized on the grounds that many important occupational roles, such as child care workers and nurse assistants, have low salaries, whereas many individuals in nonessential roles (e.g., professional sports stars and entertainers) earn outrageous sums of money. The structural-functionalist argument that CEO pay is high to reward high performance is shattered by the fact that CEOs are paid huge salaries and bonuses even when they contribute to the economic failure of their corporation and/or to the problem of unemployment.

In his classic article "The Positive Functions of Poverty" sociologist Herbert Gans (1972) draws on the structural-functionalist perspective to identify ways in which poverty can be viewed as functional for the nonpoor segments of society. For example, having a poor population ensures that society has a pool of low-cost laborers willing to do unpleasant jobs, and provides a labor pool for jobs ranging from the military to prostitution. Poor populations also provide labor for the affluent in the form of domestic work, such as maids and gardeners. The poor help keep others employed in jobs such as policing, prison work, and social work. And the poor provide a pool of consumers for used goods and second-rate service providers. These "functions" of poverty describe how the nonpoor may benefit from poverty, but noting these functions does not imply that poverty is justified on the basis of these functions. As the late Dr. Buford Rhea said to his sociology students, "IS does not equal OUGHT" (author's personal notes).

Conflict Perspective

Karl Marx (1818–1883) proposed that economic inequality results from the domination of the *bourgeoisie* (owners of the factories, or "means of production") over the *proletariat* (workers). The bourgeoisie accumulate wealth as they profit from the labor of the proletariat, who earn wages far below the earnings of the bourgeoisie. Modern conflict theorists recognize that the power to influence economic outcomes arises not only from ownership of the means of production but also from management position, interlocking board memberships, control of media, financial contributions to politicians, and lobbying.

The conflict perspective views money as a tool that can be used to achieve political interests. Wealthy corporations and individuals use financial political contributions to influence political elections and policies in ways that benefit the wealthy. The interests of the wealthy include things like keeping taxes low on capital gains, and the wealthy are

TABLE 6.4 Expenses Involved in Calculating Basic Economic Security Tables*, 2010	
(WORKERS WITH EMPLOYMENT-BASED BENEFITS)	
United States *Monthly Expenses for: 1 Worker, 1 Infant*	
Housing	$821
Utilities	$178
Food	$351
Transportation	$536
Child Care	$610
Personal & Household Items	$364
Health Care	$267
Emergency Savings	$116
Retirement Savings	$73
Taxes	$720
Tax Credits	−$172
Monthly Total	$3,864
Annual Total	$46,368
Hourly Wage	$21.95
Additional Asset-Building Savings	
Children's Higher Education	$43
Homeownership	$130

*For U.S. workers with employment-based benefits

Source: Wider Opportunities for Women. 2010. *The Basic Economic Security Tables for the United States.* Washington, DC: Wider Opportunities for Women.

more likely than the general public to oppose increasing the minimum wage and other policies that would create upward mobility among low-income Americans (Callahan & Cha 2013).

The power of the wealthy to influence political outcomes was reinforced by the 2010 Supreme Court ruling (5 to 4) in *Citizens United v. Federal Election Commission* that corporations have a First Amendment right to spend unlimited amounts of money to support or oppose candidates for elected office. The advent of Super PACs also gave the wealthy increased leverage in the political process. A Super PAC is a political action committee that is allowed to raise and spend unlimited amounts of money for the purpose of supporting or defeating a political candidate, as long as the monies are not given to any political candidate's campaign. During the 2012 election cycle, two wealthy Americans (Sheldon and Miriam Adelson) gave a combined $91.8 million to Super PACs. "The Adelsons gave more to shape the 2012 federal elections than all the combined contributions from residents in 12 states: Alaska, Delaware, Idaho, Maine, Mississippi, Montana, New Hampshire, North Dakota, Rhode Island, South Dakota, Vermont, and West Virginia" (Callahan & Cha 2013, pp. 18–19).

Laws and policies that favor the rich, such as tax breaks that benefit the wealthy, are sometimes referred to as **wealthfare**. For example, the richest fifth of the U.S. population receives housing subsidies through the mortgage interest tax deduction that amounts to nearly four times the housing assistance provided to the poorest fifth (Garfinkel 2013). **Corporate welfare** refers to laws and policies that benefit corporations, such as government subsidies and tax breaks to corporations. A report on corporate "tax dodgers" revealed that although the official federal corporate tax rate is 35 percent, U.S. corporations took advantage of legal tax loopholes that enabled them to pay, on average, only 12.1 percent in federal taxes in 2011 (Anderson et al. 2013). In 1952, under Republican President Eisenhower, corporate income taxes contributed half of federal tax receipts, but declined to less than 10 percent by 2012 (Anderson et al. 2013).

Conflict theorists also note that "free-market" trade and investment economic policies, which some claim to be a solution to poverty, primarily benefit wealthy corporations. Trade and investment agreements enable corporations to (1) expand production and increase economic development in poor countries, and (2) sell their products and services to consumers around the world, thus increasing poor populations' access to goods and services. Yet, such policies also enable corporations to relocate production to countries with abundant supplies of cheap labor, which leads to a lowering of wages, and a resultant decrease in consumer spending, which leads to more industries closing plants, going bankrupt, and/or laying off workers.

Furthermore, the over 3,000 trade and investment agreements in effect include a key provision that gives corporations the right to take legal action against governments with policies that, in the interest of protecting the public, affect corporations' profits. In 2012, in 70 percent of cases where corporations took legal action against governments for violating trade agreements, the World Bank's trade court ruled in favor of the corporation, and governments had to pay tens or even hundreds of millions of dollars—money that could otherwise go toward education, health care, and other public investments to improve the lives of the public, and especially the poor (McDonagh 2013).

When corporations claim that their products or services are essential in the fight against poverty, a conflict perspective might reveal a different story. For example, powerful food and biotech corporations such as Monsanto, Cargill, and Archer Daniels Midland have used their economic and political power to impose a system of agriculture based on intensive chemical use and patented and genetically modified seeds (McDonagh 2013). These corporations assert that their model of agriculture, which requires farmers to purchase their chemicals and seeds, yields more and better food, and thus is important in the global fight against hunger and poverty. Yet, this corporate control of agriculture has resulted in farmers' dependence and debt (and an epidemic of suicides among poor farmers), environmental degradation (through the increased use of chemicals), and health risks associated with chemicals and genetically modified foods.

wealthfare Laws and policies that benefit the rich.

corporate welfare Laws and policies that benefit corporations.

As Americans, we believe that everyone who works hard deserves a chance at opportunity, and that all our citizens deserve some basic measure of security. And so, 50 years ago, President Johnson declared a War on Poverty to help each and every American fulfill his or her basic hopes. We created new avenues of opportunity through jobs and education, expanded access to health care for seniors, the poor, and Americans with disabilities, and helped working families make ends meet. Without Social Security, nearly half of seniors would be living in poverty. Today, fewer than one in seven do. Before Medicare, only half of seniors had some form of health insurance. Today, virtually all do. And because we expanded pro-work and pro-family programs like the Earned Income Tax Credit, a recent study found that the poverty rate has fallen by nearly 40% since the 1960s, and kept millions from falling into poverty during the Great Recession.

These endeavors didn't just make us a better country. They reaffirmed that we are a great country. They lived up to our best hopes as a people who value the dignity and potential of every human being. But as every American knows, our work is far from over. In the richest nation on Earth, far too many children are still born into poverty, far too few have a fair shot to escape it, and Americans of all races and backgrounds experience wages and incomes that aren't rising, making it harder to share in the opportunities a growing economy provides. That does not mean, as some suggest, abandoning the War on Poverty. In fact, if we hadn't declared "unconditional war on poverty in America," millions more Americans would be living in poverty today. Instead, it means we must redouble our efforts to make sure our economy works for every working American. It means helping our businesses create new jobs with stronger wages and benefits, expanding access to education and health care, rebuilding those communities on the outskirts of hope, and constructing new ladders of opportunity for our people to climb.

We are a country that keeps the promises we've made. And in a 21st century economy, we will make sure that as America grows stronger, this recovery leaves no one behind. Because for all that has changed in the 50 years since President Johnson dedicated us to this economic and moral mission, one constant of our character has not: we are one nation and one people, and we rise or fall together.

Source: President Barack Obama, Statement on the 50th Anniversary of the War on Poverty, January 8, 2014.

Symbolic Interactionist Perspective

Symbolic interactionism focuses on how meanings, labels, and definitions affect and are affected by social life. This view calls attention to ways in which wealth and poverty are defined and the consequences of being labeled "poor." Individuals who are poor are often viewed as undeserving of help or sympathy; their poverty is viewed as due to laziness, immorality, irresponsibility, lack of motivation, or personal deficiency (Katz 2013). Wealthy individuals, on the other hand, tend to be viewed as capable, motivated, hardworking, and deserving of their wealth.

The language we use to label things can have a profound influence on how we view things. A Canadian study found that the public expressed higher support for government "spending on the poor" than for spending on "welfare" (Harell et al. 2008). If "welfare" is a "dirty word," as this study suggests, then it makes a difference whether one uses the term "welfare" versus other terms such as "assistance to the poor," "public assistance," or "safety net."

The symbolic interactionist perspective also focuses on how poor individuals label their experience of poverty. A qualitative study of more than 40,000 poor women and men in 50 countries revealed that the experience of poverty involves psychological dimensions such as powerlessness, voicelessness, dependency, shame, and humiliation (Narayan 2000).

Meanings and definitions of wealth and poverty vary across societies and across time. More than 50 years ago, President Johnson started the "War on Poverty" (see this chapter's *The Human Side*). Although many Americans think of poverty in terms of income level, for millions of people, poverty is not primarily a function of income but of their alienation from sustainable patterns of consumption and production. For indigenous women living in the least developed areas of the world, poverty and wealth are determined primarily by access to and control of their natural resources (such as land and water) and traditional knowledge, which are the sources of their livelihoods (Susskind 2005).

Economic Inequality, Wealth, and Poverty in the United States

The United States has the greatest degree of income inequality and the highest rate of poverty of any industrialized nation. In recent years, the U.S. public has become more aware of economic inequality, as the Occupy Wall Street (OWS) movement has brought increased attention to the "99 percent versus the 1 percent." The disparity between the rich and everyone else is striking: In 2012, the wealthiest one percent of U.S. households earned more than 20 percent of the nation's income; the top ten percent of income earners took home more than half of the nation's total income—the highest recorded level in U.S. history (Lowrey 2013). Interestingly, the nation's standard economic indicator—the gross domestic product (GDP)—does not measure inequality, so the GDP can indicate that the economy is "growing" and is thus healthy. At the same time, the adage that "the rich get richer while the poor get poorer" is supported by data: As the U.S. economy recovered during the first two years of economic recovery (2009 to 2011) following the recession of 2007, the average wealth of households in the top 7 percent rose 28 percent, while the average wealth of the bottom 93 percent dropped by 4 percent (Fry & Taylor 2013). The income of the top one percent of wage earners has also grown significantly more than wages in the lower income brackets. From 1979 to 2011, earnings of the top 1 percent grew 134 percent while those of the bottom 90 percent grew 15 (Mishel & Finio 2013).

Another example of economic inequality in the United States is the gap between the compensation (salaries, bonuses, stock options, and so on) of chief executive officers (CEOs) and the average employee. In 2012, CEOs at the top 350 U.S. corporations received, in salaries and other compensation (such as bonuses and stocks), an average of 273 times the average compensation of U.S. workers (Mishel & Sabadish 2013). That means that a typical worker would have to work 273 years to earn what a CEO makes in one year! Table 6.5 shows the dramatic increase in the ratio of CEO pay to average worker pay since 1965.

Disparities in income and wealth show an economic advantage of whites over racial/ethnic minorities. In 2010, the average household income for whites was $89,000—about twice the $46,000 average for black and Hispanic households. The wealth of white households was *six times* that of black and Hispanic households (McKernan et al. 2013) (see Figure 6.1). This is largely related to the fact that whites are more likely than blacks or Hispanics to own homes, which for many Americans, is their most valuable asset.

For most Americans, the super rich are people we will never know personally; our conception of the wealthy is based largely on media representations. After briefly looking at the wealthiest of Americans, we examine patterns of poverty in the United States.

The "One Percent": Wealth in the United States

Who are the wealthiest Americans? In 2011, the top one percent of wage earners were those who earned more than $598,570 (Mishel & Finio 2013).

How do the wealthy become wealthy? The United States is known as the "land of opportunity"—a country where a person can go from "rags to riches." But do wealthy Americans acquire their wealth primarily through hard work, their talents, and abilities? To what extent do the wealthy inherit their wealth or use their family's wealth to achieve educational, occupational, and financial advantages?

In the 2011 Forbes 400 annual list of the wealthiest Americans, more than 70 percent of the 282 billionaires on the list were described as "self-made," suggesting that these individuals achieved financial success on their own, without assistance from family or

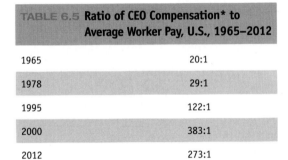

TABLE 6.5	Ratio of CEO Compensation* to Average Worker Pay, U.S., 1965–2012
1965	20:1
1978	29:1
1995	122:1
2000	383:1
2012	273:1

* Rounded; includes pay and stock options; based on the top 350 U.S. firms

Sources: Mishel & Sabadish 2013.

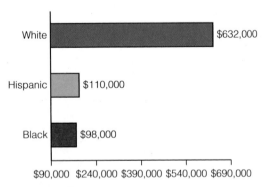

Figure 6.1 Average Household Worth by Race/Hispanic Origin, 2010
Source: McKernan, Signe-Mary, caroline Ratcliffe, c. Eugene Steuerle, and Sisi Zhang, 2013 (April). "Less Than Equal: Racial Disparities in Wealth." Urban Institute. Available at http://www.urban.org

society. But the notion that wealthy individuals have created their own financial success ignores the importance of gender, race, and family background as well as the role that tax policies play in creating wealthy individuals. United for a Fair Economy (2012) examined the 2011 Forbes 400 list of wealthiest Americans and found that:

- 17 percent of the Forbes 400 has family members who are also on the list;
- about 40 percent of the 2011 Forbes 400 list inherited a "sizeable asset" from a spouse or family member;
- more than one in five of the Forbes 400 inherited enough wealth to make the list;
- just one African American is on the list, and of the women on the list (who comprise just 10 percent of the list), 88 percent inherited their fortune; and
- 60 percent of the income owned by those on the Forbes 400 list comes from capital gains (investments) that are taxed at a lower rate than other income.

Photo courtesy of the author

Children are more likely than adults to live in poverty.

There are, indeed, true "rags to riches" success stories in the United States that exemplify the idea that anyone can achieve the American dream. Approximately one-third of the individuals on the 2011 Forbes 400 list came from a lower- or middle-class background. Oprah Winfrey, for example—the only black person and one of 40 women on the 2011 Forbes 400 list—was born to a poor unwed teenage mother, yet she developed a successful career in television, film, and publishing. However, such "rags to riches" stories are the exception rather than the rule.

Patterns of Poverty in the United States

Although poverty is not as widespread or severe in the United States as it is in many other parts of the world, the United States has the highest rate of poverty among wealthy countries belonging to the Organisation for Economic Cooperation and Development. In 2012, 46.5 million Americans—15 percent of the U.S. population—lived below the poverty line (DeNavas-Walt et al. 2013). More than half (58 percent) of Americans between the ages of 20 and 75 will spend at least one year in poverty, and one in three Americans will experience a full year of extreme poverty at some point in adult life (Pugh 2007).

Age and Poverty. If the poverty statistics for adults are troubling, the statistics for children are even worse (see Figure 6.2). While one in six Americans live in poverty and one in three live in "near poverty" (with incomes below twice the poverty level), one in four children under age 5 are poor and nearly half are in the "near-poor" category (Golden 2013). About a third of U.S. children experience poverty for at least part of their childhood, and 10 percent of children are persistently poor, spending at least half their childhood living in poverty (Ratcliffe & McKernan 2010). More than one-third (35 percent) of the U.S. poor population are children (DeNavas-Walt et al. 2013). Compared with other industrialized countries, the United States has the highest child poverty rate. Childhood poverty is particularly problematic because "[g]rowing up in poverty can cast a shadow over the rest of a person's life" (Golden 2013, n.p.).

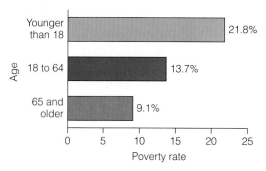

Figure 6.2 U.S. Poverty Rates by Age, 2012
Source: DeNavas-Walt et al. 2013.

Sex and Poverty. Women are more likely than men to live below the poverty line—a phenomenon referred to as the **feminization of poverty**. In 2012, 16.3 percent of women and 13.6 percent of men were living below the poverty line (DeNavas-Walt et al. 2013). As discussed in Chapter 10, women are less likely than men to pursue advanced educational degrees and tend to have low-paying jobs, such as service and clerical jobs. However, even with the same level of education and the same occupational role, women still earn significantly less than men. Women who are racial or ethnic minorities and/or who are single mothers are at increased risk of being poor.

> Education is one of the best insurance policies for protecting an individual against living in poverty.

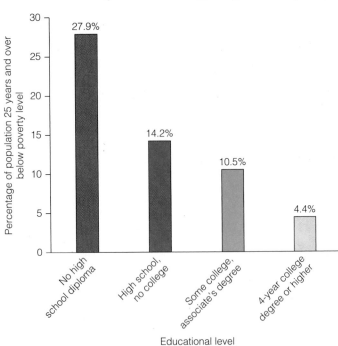

Figure 6.3 Relationship between Education and Poverty, 2011
Source: U.S. Census Bureau, 2012a.

Education and Poverty. Education is one of the best insurance policies for protecting an individual against living in poverty. In general, the higher a person's level of educational attainment, the less likely that person is to be poor (see Figure 6.3). The relationship between educational attainment and poverty points to the importance of fixing our educational system so that students from all socioeconomic backgrounds have access to quality education (see also Chapter 8). But we also need to consider the fact that many jobs do not require advanced education. Indeed, most job growth (68 percent) through 2018 will involve jobs that do not require a four-year college degree (Wider Opportunities for Women 2010). Wright and Rogers (2011) suggest that "poverty in a rich society does not simply reflect a failure of equal opportunity to acquire a good education; it reflects a social failure in the creation of sufficient jobs to provide an adequate standard of living for all people regardless of their education or levels of skills" (p. 224).

Family Structure and Poverty. Poverty is much more prevalent among female-headed single-parent households than among other types of family structures (see Figure 6.4). In other industrialized countries, poverty rates of female-headed families are lower than those in the United States. Unlike the United States, other developed countries offer a variety of supports for single mothers, such as income supplements, tax breaks, universal child care, national health care, and higher wages for female-dominated occupations.

feminization of poverty The disproportionate distribution of poverty among women.

In general, same-sex couples are more likely than heterosexual couples to be poor. Children in same-sex couple families are nearly twice as likely to be poor as children of married different-sex couples (Badgett et al. 2013).

Race or Ethnicity and Poverty. As displayed in Figure 6.5, poverty rates are higher among racial and ethnic minority groups than among non-Hispanic whites. As discussed in Chapter 9, past and present discrimination has contributed to the persistence of poverty among minorities. Other contributing factors include the loss of manufacturing jobs from the inner city, the movement of whites and middle-class blacks out of the inner city, and the resulting concentration of poverty in predominantly minority inner-city neighborhoods (Massey 1991; Wilson 1987, 1996). Finally, blacks and Hispanics are more likely to live in female-headed households with no spouse present—a family structure that is associated with high rates of poverty.

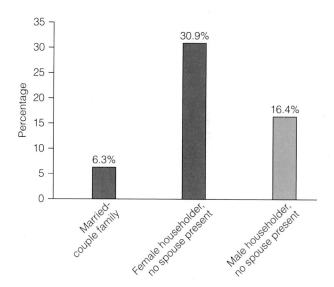

Figure 6.4 **U.S. Poverty Rates by Family Structure, 2012**
Source: DeNavas-Walt et al. 2013.

Labor Force Participation and Poverty. A common image of the poor is that they are jobless and unable or unwilling to work. Although the poor in the United States are primarily children and adults who are not in the labor force, many U.S. poor are classified as **working poor**—individuals who spend at least 27 weeks per year in the labor force (working or looking for work), but whose income falls below the official poverty level.

Region and Poverty. Poverty rates vary considerably by region of the United States, with the highest rates being in the South and West U.S. states, and the lowest rates being in the Northeast. Although the average U.S. poverty rate in 2011 was 15.3, the rates of poverty ranged from a low of 10 in New Hampshire to a high of 24.2 in Mississippi (see Table 6.6).

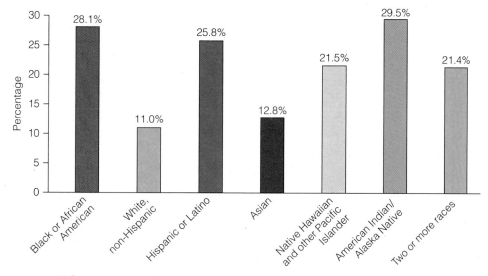

Figure 6.5 **U.S. Poverty Rates by Race and Hispanic Origin, 2011**
Source: U.S. Census Bureau 2012a. *2011 American Community Survey*. Available at www.census.gov

working poor Individuals who spend at least 27 weeks per year in the labor force (working or looking for work) but whose income falls below the official poverty level.

TABLE 6.6 States with the Highest and Lowest Poverty Rates: 2012	
Highest Poverty Rates	Lowest Poverty Rates
Mississippi: 24.2	New Hampshire: 10.0
New Mexico: 20.8	Alaska: 10.1
Louisiana: 19.9	Maryland: 10.3
Arkansas: 19.8	Connecticut: 10.7
Kentucky: 19.4	New Jersey: 10.8
Georgia: 19.2	North Dakota: 11.2
Alabama: 19.0	Minnesota: 11.4
Arizona: 18.7	Hawaii: 11.6
South Carolina: 18.3	Virginia: 11.7
North Carolina: 18.0	Vermont: 11.8
Texas: 17.9	Massachusetts: 11.9
Tennessee: 17.9	Delaware: 12

Source: Alemayehu Bishaw. 2013 (September). "Poverty: 2000 to 2012." *American Community Survey Briefs*. U.S. Census Bureau.

Consequences of Economic Inequality and Poverty

From one point of view, economic inequality and poverty are, in themselves, problematic because they contradict the values of fairness, justice, and equality of opportunity, and constitute a moral violation of basic human rights. Economic inequality and poverty are also viewed as problems because they have economic and social consequences that affect the whole society. For example, when income is concentrated toward the top, less money circulates in the local economy because money earned by low- and middle-income households is more likely to be spent on goods and services that benefit the local economy rather than invested in other regions and spent on luxuries (Talberth et al. 2013). The larger the segment of the population that is in the lowest income brackets, the more our society is affected by problems that plague the poor, but that also affect us all—problems discussed in the following sections.

Health Problems, Hunger, and Poverty

In developing countries, absolute poverty is associated with high rates of maternal and infant deaths, indoor air pollution from heating and cooking fumes, and unsafe water and sanitation (see this chapter's *Photo Essay*) (World Health Organization 2002). Living in poverty is also linked to hunger and malnourishment. In 2012, almost 870 million people globally were chronically undernourished, with most of the world's hungry living in Asia and the Pacific (FAO 2012). Inadequate nutrition hampers the ability to work and generate income, and can produce irreversible health problems such as blindness (from vitamin A deficiency) and physical stunting (from protein deficiency).

Hunger in the United States is measured by the percentage of households that are "food insecure," which means that the household had difficulty providing enough food for all its members due to a lack of resources. In 2011, nearly 15 percent of U.S. households were food insecure at some time during the year (Coleman-Jensen et al. 2012). Assess your own degree of food security in this chapter's *Self and Society* feature.

In the United States, low wage earners have higher rates of obesity, hypertension, diabetes, arthritis, and premature death (Leigh 2013). Low wages and poverty affect health in a variety of ways. Living on limited means can be very stressful, and stress adversely affects health. Poor U.S. children and adults tend to receive inadequate and inferior health care, which exacerbates their health problems. Minimal income means that people may not have funds to purchase medicine to control their cholesterol, high blood pressure, and other health problems. As discussed in Chapter 2, people with limited incomes may not have access to or be able to afford healthier foods such as fresh produce, which tend to be more expensive than processed, convenience foods that are higher in calories, sugar, salt, and fats. Finally, many people partially assess their self-worth based on their income, and long-term feelings of low self-worth also have negative consequences for health (Leigh 2013).

Economic inequality is also linked to health problems. In a comparison of 30 wealthy countries, researchers found an association between greater economic inequality and a higher overall death rate (Kondo et al. 2009). Another study by the U.S. National Research Council and Institute of Medicine (2013) compared health outcomes in the United States with those of 16 other high-income, industrialized countries and found that Americans die sooner and have higher rates of disease or injury. One explanation for this finding is that although the income of Americans is higher on average than in other countries, the United States has higher rates of poverty (especially child poverty), more income inequality, and less social mobility.

The U.S. Department of Agriculture conducts national surveys to assess the degree to which U.S. households experience food security, food insecurity, and food insecurity with hunger. To assess your own level of food security, respond to the following items and use the scoring key to interpret your results:

1. In the last 12 months, the food that (I/we) bought just didn't last, and (I/we) didn't have money to get more.
   a. **Often true**
   b. **Sometimes true**
   c. Never true

2. In the last 12 months, (I/we) couldn't afford to eat balanced meals.
   a. **Often true**
   b. **Sometimes true**
   c. Never true

3. In the last 12 months, did you ever cut the size of your meals or skip meals because there wasn't enough money for food?
   a. **Yes**
   b. No (skip Question 4)

4. If you answered yes to Question 3, how often did this happen in the last 12 months?
   a. **Almost every month**
   b. **Some months but not every month**
   c. Only 1 or 2 months

5. In the last 12 months, did you ever eat less than you felt you should because there wasn't enough money to buy food?
   a. **Yes**
   b. No

6. In the last 12 months, were you ever hungry but didn't eat because you couldn't afford enough food?
   a. **Yes**
   b. No

Scoring and Interpretation

The answer responses in boldface type indicate affirmative responses. Count the number of affirmative responses you gave to the items, and use the following scoring key to interpret your results.

Number of Affirmative Responses and Interpretation

0 or 1 item: *Food secure* (In the last year, you have had access to enough food for an active, healthy life.)

2, 3, or 4 items: *Food insecure* (In the last year, you have had limited or uncertain availability of food and have been worried or unsure you would get enough to eat.)

5 or 6 items: *Food insecure with hunger evident* (In the last year, you have experienced more than isolated occasions of involuntary hunger as a result of not being able to afford enough food.)

If you scored as food insecure (with or without hunger), you might consider exploring whether you are eligible for public food assistance (e.g., food stamps) or whether there is a local food assistance program (e.g., food pantry or soup kitchen) that you could use.

Source: Based on the short form of the 12-month Food Security Scale found in Bickel et al. 2000.

Housing Problems

Housing problems include substandard housing, homelessness, and "the housing crisis."

Substandard Housing. Having a roof over one's head is considered a basic necessity. However, for the poor, that roof may be literally caving in. In addition to having leaky roofs, housing units of the poor often have holes in the floor and open cracks in the walls or ceiling. Low-income housing units often lack central heating and air conditioning, sewer or septic systems, and electric outlets in one or more rooms. Housing for the poor is also often located in areas with high crime rates and high levels of pollution.

Concentrated areas of poverty and poor housing in urban areas are called **slums.** One-third of urban populations in developing regions are living in slums. In sub-Saharan Africa, two-thirds of urban populations are living in slums (UN-Habitat 2010).

AP Photo/Ric Feld

Many Americans would be shocked to see the conditions under which many poor people in this country live.

Homelessness. The National Alliance to End Homelessness (2013) reported that in January 2012, 633,782 people were homeless in the United States: Of those,

slums Concentrated areas of poverty and poor housing in urban areas.

Photo Essay

Lack of Clean Water and Sanitation among the Poor

In wealthy countries, such as the United States, we take for granted the availability of bathrooms and toilets, and access to safe drinking water that is piped into our homes. But in many poor areas of the world, people drink whatever water is available, and they urinate and defecate in the street, along the roadside, in buckets, or in plastic bags that are tied up and thrown in ditches or along the road. Used in this way, plastic bags are known as "flying toilets." More than one-third of the world's population—2.5 billion people—do not use improved sanitation facilities—those that ensure hygienic separation of human excreta from human contact (World Health Organization and UNICEF 2013). Of these, about 1.5 billion use either public or shared sanitation facilities, or they use facilities that don't meet minimum standards of hygiene. The remaining 1 billion—15 percent of the world population—practice open defecation, relieving themselves in open areas such as streets or ditches along the road. In addition, 768 million people worldwide do not have access to safe drinking water.

Lack of access to clean water and sanitation facilities is a major cause of disease and death. For example, poor sanitation causes diarrheal diseases, which are the second leading cause of death among young children in developing countries (Agazzi 2012). Other diseases associated with lack of clean water and sanitation include trachoma (which causes blindness), typhoid fever, intestinal worms, and guinea worm.

Cultural norms in many countries require that women not be seen urinating or defecating, which forces them to limit their food and water intake so they can relieve themselves in the dark of night in fields or roadsides (United Nations Development Programme 2006). Delaying bodily functions can cause liver infection and acute constipation, and going out in darkness to eliminate places women at risk for physical attack.

Neil Cooper/Alamy

▲ The *Musca sorbens* fly, which breeds in human feces, causes 2 million new cases of blindness-causing trachoma each year in the developing world. The flies burrow into human eyes, which causes decades of repeat infections. Victims describe the infections as feeling as if they have thorns in their eyes.

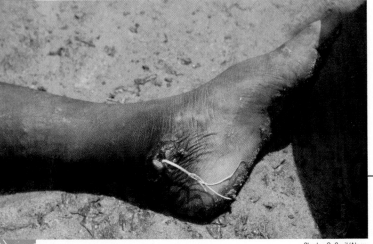

Charles O. Cecil/Alamy

◄ A person gets guinea worm, a parasite, by drinking water contaminated with the larvae. The worm grows up to three feet long in the body and eventually erupts through the skin, causing extremely painful blisters.

◄ Imagine having to share this one toilet with more than 1,000 other people.

Efforts to improve access to clean water and sanitation for poor populations are some of the most important priorities in the fight against poverty. For every $1 invested in water and sanitation, $5 are saved in health spending or through increased productivity (Agazzi 2012). More importantly, access to clean water and sanitation is a basic human right that every woman, man, and child should enjoy. The World Toilet Organization hopes to raise awareness of the crucial issue of sanitation by recognizing World Toilet Day (November 19th), with the slogan, "I give a shit—do you?" ▼

▲ Three-quarters of the world's rural population obtain water from a communal source, which means family members (usually women and girls) must walk to the water source and carry water back to the family. To collect enough water for drinking, food preparation, personal hygiene, house cleaning, and laundry, a household of five needs at least 32 gallons of water per day (a little more than 6 gallons per person) (Satterthwaite & McGranahan 2007). This is equivalent to carrying six heavy suitcases of water every day.

▲ In many poor areas of the world, it is common for residents to defecate in plastic bags that they dump in ditches or throw on the roadside.

38 percent were families with children; nearly 10 percent were veterans; and 16 percent were chronically homeless—living with a disability and homeless repeatedly or for long periods of time. Over the course of a lifetime, an estimated 9 percent to 15 percent of the U.S. population becomes homeless (Hoback & Anderson 2007). Although the majority of the homeless population stays in shelters or transitional housing, more than a third of homeless individuals lives on the street or places not meant for human habitation, such as makeshift dwellings made of a variety of discarded materials such as pieces of wood and boards, cardboard, mattresses, fabric, and plastic tarps (National Alliance to End Homelessness 2013).

Homelessness is essentially caused by the lack of affordable housing. Housing is considered affordable when a household pays no more than 30 percent of its income on housing expenses. The lower the household income, the higher percentage that household spends on housing. Other factors that contribute to homelessness, aside from the inability to afford housing, include unemployment and poverty, eviction, domestic violence, and mental illness and substance abuse and the lack of needed services to treat these problems (U.S. Conference of Mayors 2012).

For people living on the street, every day can be a struggle for survival. In recent years, there has been a surge in unprovoked violent attacks against homeless individuals. The National Coalition for the Homeless (2012) documented 1,289 incidents of violence against the homeless between 1999 and 2011, more than 300 of these attacks resulted in death. Some of the headlines from 2011 include:

Two Young Men Bludgeon a Homeless Man with a Tire Iron "Just for Fun"
Homeless Woman Raped, Strangled, and Set on Fire
Police Leave Homeless Man's Body Mangled
Hearing-Impaired Homeless Teen's Skull Bashed and Fractured as He Tried to Sleep behind a School
Teenager Makes a Game of Knocking a Homeless Man Unconscious (National Coalition for the Homeless, p. 18)

In most cases, the attacks are by teenage and young adult males. Many acts of violence toward the homeless are not reported to the police, so documented cases may be just the tip of the iceberg. During the years he lived homeless on the street, David Pirtle was attacked five times and he did not report the attacks to police. "I was struck on the back, kicked, urinated on, spray-painted. . . . A lot of people who are homeless go through it, and it's just the way it is" (quoted in Dvorak 2009, p. DZ01).

There is also a new fascination with "bum bashing" or "bum fight" videos on YouTube—videos shot by young men and boys who are seen beating the homeless or who pay homeless people a few dollars to fight each other or to do dangerous stunts like banging their heads through glass windows and going down stairs in a shopping cart. More than 15,000 YouTube videos have been tagged with the search phrase "bum fight" with more than 7 million views and 6,000 "likes" by YouTube viewers (National Coalition for the Homeless 2012). Individuals who find the idea of bum bashing entertaining can also purchase bum fight DVDs and play web-based bum-bashing games.

Thousands of people in the United States are homeless on any given day.

Elena Rooraid/PhotoEdit

What Do You Think? Under hate crime laws, violators are subject to harsher legal penalties if their crime is motivated by the victim's race, religion, national origin, or sexual orientation. The number of violent acts against the homeless exceeds the number of hate crime acts toward all other minority groups combined. A handful of states have added homeless status to their hate crime law, and several cities and counties have also taken measures to recognize homeless status in their laws or procedures. Proposed legislation to add homelessness to the federal hate crime law has not, as of this date, passed. Do you think that violent acts toward homeless individuals should be categorized as hate crimes and be subject to harsher penalties? Why or why not?

The "Housing Crisis." In the 1990s and early 2000s, inflated housing values enabled middle-class homeowners with maxed-out credit cards to keep spending by refinancing their mortgages. At the same time, there was an increase in **subprime mortgages**—high-interest or adjustable-rate mortgages that require little money down and are issued to borrowers with poor credit ratings or limited credit history. Subprime mortgage lending enabled low-wage earners to buy a house, and the increased housing demand raised house values further. When the housing bubble burst and house values fell in 2007/08, millions of people were stuck with "upside-down mortgages," in which the amount owed on a mortgage is more than the value of the property. Many homeowners with subprime mortgages could not make their payments, and foreclosures skyrocketed. In this housing crisis, millions of people lost their homes and their credit, and renters in foreclosed dwellings lost their lease with little notice.

Legal Inequality

In 1963, the Supreme Court ruled in *Gideon v. Wainwright* that criminal defendants who cannot afford to hire an attorney have the constitutional right to a public defense. But public defender offices are overworked and underfunded, and often spend only minutes per case due to their unrealistic caseloads (Giovanni & Patel 2013). Without the resources for effective legal representation, poor defendants often accept unfair plea bargains, and "the systemic result is harsher outcomes for defendants and more people tangled in our costly criminal justice system" (Giovanni & Patel 2013, p. 1). Thus, the American ideal of "justice for all" may be more accurately described as "justice for those who can afford to pay for it." The economic inequality embedded in the U.S. legal system is problematic not only for the poor, but for the entire society, as it contributes to the social and economic costs of mass incarceration in the United States (see also Chapter 4).

Political Inequality and Alienation

Economic inequality also contributes to political inequality, as expressed in a version of "the golden rule": "He who has the gold makes the rules." Acemoglu and Robinson (2012, n.p.) explain how the wealthy have a political advantage over the poor:

> The wealthy have greater access to politicians and to media, and can communicate their point of view and interests—often masquerading as "national interest"—much more effectively than the rest of us. How else can we explain that what is on the political agenda for the last several decades has been cutting taxes on the wealthy while almost no attention is paid to problems afflicting the poor, such as our dysfunctional penal system condemning a huge number of Americans to languish in prisons for minor crimes? How else can we explain . . . that U.S. Senators' votes represent the views of their rich constituents but not those of their poor ones?

The 2010 Supreme Court ruling in *Citizens United v. Federal Election Commission* that allows corporations to spend unlimited amounts of money to support or oppose political candidates has contributed to political inequality. Although the United States represents

subprime mortgages High-interest or adjustable-rate mortgages that require little money down and are issued to borrowers with poor credit ratings or limited credit history.

itself as a democracy whose government represents all citizens, the poor and even middle classes often feel that their interests are not represented by their elected politicians. Hence, those in the lower socioeconomic classes are vulnerable to experiencing **political alienation**—a rejection of or estrangement from the political system accompanied by a sense of powerlessness in influencing government. The poor face obstacles in running for political office, as money and connections are needed to run for office. Although low-income Americans are voting at the highest rates since the mid-1960s, they are still less likely than affluent Americans to vote—by as much as 30 percentage points less in the 2008 and 2010 elections (Callahan & Cha 2013). The poor have a lower voting turnout than the wealthier segments of the population, in part due to political alienation, but also because of obstacles such as difficulty taking time off of work and transportation problems getting to the polls.

Economic Inequality, Poverty, War and Social Conflict

Economic inequality and poverty are often root causes of conflict and war within and between nations. Poorer countries are more likely than wealthier countries to be involved in civil war, and countries that experience civil war tend to become and/or remain poor. Armed conflict and civil war are generally more likely to occur in countries with extreme and growing inequalities between ethnic groups (United Nations 2005). Conversely, countries with higher levels of equality are more likely to be peaceful (Institute for Economics & Peace 2013). A United Nations (2005) report suggested that "the most effective conflict prevention strategies . . . are those aimed at achieving reductions in poverty and inequality, full and decent employment for all, and complete social integration" (p. 94).

Not only does poverty breed conflict and war, but war also contributes to poverty. War devastates infrastructures, homes, businesses, and transportation systems. In the wake of war, populations often experience hunger and homelessness.

Natural Disasters, Economic Inequality, and Poverty

Although natural disasters such as hurricanes, tsunamis, floods, and earthquakes strike indiscriminately—rich and poor alike—poverty increases vulnerability to devastation from such disasters. In 2010, both Chile and Haiti experienced major earthquakes, but the damage in Haiti was much more severe, with the death toll magnitudes higher than that in Chile. The reason Haiti suffered more was, in part, due to the fact that Haiti is much poorer than Chile. Chileans had the advantage of having homes and offices with steel skeletons designed to withstand earthquakes—even low-income housing was built to be earthquake resistant. In contrast, there is no building code in Haiti and homes crumbled and collapsed in the earthquake (Bajak 2010). Wealthy countries also have more resources than poor countries for natural disaster relief efforts, such as rebuilding infrastructure, providing medical care for the injured, and providing food and shelter for people who have lost their homes.

But even in wealthy countries, the poor are more vulnerable to natural disasters, while the more affluent have resources that enable them to cope with natural disasters. Columnist David Rohde (2012) wrote about how economic inequality affected people dealing with Hurricane Sandy, which devastated the northeastern United States in 2012:

> Divides between the rich and the poor are nothing new in New York, but the storm brought them vividly to the surface. There were residents like me who could invest all of their time and energy into protecting their families. And there were New Yorkers who could not. Those with a car could flee. Those with wealth could move into a hotel. Those with steady jobs could decline to come into work. But the city's cooks, doormen, maintenance men, taxi drivers, and maids left their loved ones at home. . . . In the Union Square area, New York's privileged—including myself—could have dinner, order a food delivery, and pick up supplies an hour or two before Sandy made landfall. The cooks, cashiers, and hotel workers who stayed at work instead of rushing home made that possible. (n.p.)

political alienation A rejection of or estrangement from the political system accompanied by a sense of powerlessness in influencing government.

Educational Problems and Poverty

In many countries, children from the poorest households have little or no schooling, and enter their adult lives without basic literacy skills (see also Chapter 8). In the United States, children from disadvantaged homes perform less well in school on average than children from more advantaged households (Ladd 2012). Children who grow up in poverty tend to receive lower grades, receive lower scores on standardized tests, are less likely to finish high school, and are less likely to attend or graduate from college than their nonpoor peers.

The poor often attend schools that are characterized by lower-quality facilities, overcrowded classrooms, and a higher teacher turnover rate (see also Chapter 8). Although other rich countries invest more money in education for disadvantaged children, the United States spends more on schools in wealthy districts because public schools are funded largely by local property tax money (Straus 2013).

Children who grow up in poverty suffer more health problems that contribute to their lower academic achievement. Because poor parents have less schooling on average than do nonpoor parents, they may be less able to encourage and help their children succeed in school. Children from poor households have limited access to high-quality preschools, books and computers, and enriching after-school and summer activities including tutoring, travel, lessons (music, dance, sports, and so on), and camps (Ladd 2012; Sobolewski & Amato 2005). With the skyrocketing costs of tuition and other fees, many poor parents cannot afford to send their children to college. Average tuition and room and board at four-year colleges rose from under $9,000 between 1980 and 1981 to nearly $22,000 in 2012 (Stiglitz 2013). Although some students have wealthy parents who write out tuition checks, other students are graduating from college with substantial college debt. Indeed, total student debt in 2012 exceeded total credit card debt in the United States (Stiglitz 2013).

The effects of natural disasters, such as Hurricane Sandy of 2012, are more devastating for the poor.

Family Stress and Parenting Problems Associated with Poverty

The stresses associated with low income contribute to substance abuse, domestic violence, child abuse and neglect, divorce, and questionable parenting practices. Child neglect is more likely to be found with poor parents who are unable to afford child care or medical expenses and leave children at home without adult supervision or fail to provide needed medical care. Poor parents are more likely than other parents to use harsh physical disciplinary techniques, and they are less likely to be nurturing and supportive of their children (Mayer 1997; Seccombe 2001).

Another family problem associated with poverty is teenage pregnancy. Poor adolescent teenagers are at higher risk of having babies than their nonpoor peers. Early childbearing is associated with increased risk of premature babies or babies with low birth weight, dropping out of school, and lower future earning potential as a result of lack of academic achievement. Luker (1996) noted that "the high rate of early childbearing is a measure of how bleak life is for young people who are living in poor communities and who have no obvious arenas for success" (p. 189). For poor teenage women who have been excluded from the American dream and disillusioned with education, "childbearing . . . is one of the few ways . . . such women feel they can make a change in their lives" (p. 182):

> Having a baby is a lottery ticket for many teenagers: It brings with it at least the dream of something better, and if the dream fails, not much is lost. . . . In a few

cases it leads to marriage or a stable relationship; in many others it motivates a woman to push herself for her baby's sake; and in still other cases it enhances the woman's self-esteem, since it enables her to do something productive, something nurturing and socially responsible. . . . (Luker 1996, p. 182)

Intergenerational Poverty

Problems associated with poverty, such as health and educational problems, create a cycle of poverty from one generation to the next. A research study that followed children from birth through age 30 revealed that children who were born into poverty were significantly more likely to be poor as adults (Ratcliffe & McKernan 2010). Poverty that is transmitted from one generation to the next is called **intergenerational poverty**.

Intergenerational poverty creates a persistently poor and socially disadvantaged population, referred to as the underclass. Although the underclass is stereotyped as being composed of minorities living in inner-city areas, the underclass is a heterogeneous population that includes poor whites living in urban and nonurban communities (Alex-Assensoh 1995). Intergenerational poverty and the underclass are linked to a variety of social factors, including the decline in well-paying jobs and their movement out of urban areas, the resultant decline in the availability of marriageable males able to support a family, declining marriage rates and an increase in out-of-wedlock births, the migration of the middle class to the suburbs, and the effect of deteriorating neighborhoods on children and youth (Wilson 1987, 1996).

Strategies for Action: Reducing Poverty and Economic Inequality

Because poverty and economic inequality are primary social problems that cause many other social problems, strategies to reduce or alleviate problems related to poverty and economic inequality include those discussed in other chapters of this text that deal with such issues as health (Chapter 2), work (Chapter 7), education (Chapter 8), racial discrimination (Chapter 9), and gender discrimination (Chapter 10). Here we briefly outline a number of strategies aimed at reducing poverty and economic inequality in the United States, and internationally.

International Responses to Poverty and Economic Inequality

In 2000, leaders from 191 United Nations member countries pledged to achieve eight **Millennium Development Goals**—an international agenda for reducing poverty and improving lives. One of the Millennium Development Goals (MDGs) was to halve, between 1990 and 2015, the proportion of people who live in severe poverty and who suffer from hunger. As can be seen in Table 6.7, several other MDGs involve alleviating problems related to poverty, such as disease, child and maternal mortality, and lack of access to education. The MDG poverty reduction goal was met in 2010—five years ahead of schedule. The World Bank has set a new goal to reduce extreme poverty around the world to less than 3 percent by 2030 (Chandy et al. 2013).

One way to help poor countries provide for the needs of its citizens is for wealthier countries to provide financial aid to cancel debts these countries owe. Another strategy to reduce poverty is to increase public awareness of the magnitude of the problem. For many of our students who have gone on study-abroad programs to poor areas in India, Africa, and South America, witnessing severe poverty firsthand was a transformative experience. Some students came back from the experience with a commitment to raise funds for health care or clean water facilities in a specific village; others changed their career plans to include working with poor populations. Non–student travelers are also interested in learning about areas of the world where extreme poverty persists. Some tourists seek out guides to take them through poor slum areas. Although slum tourism can promote greater awareness of poverty, Kennedy Odede, who was born and raised in the Kibera slum of Nairobi, Kenya, notes that, "Slums will not go away because a

intergenerational poverty Poverty that is transmitted from one generation to the next.

Millennium Development Goals Eight goals that comprise an international agenda for reducing poverty and improving lives.

few dozen Americans or Europeans spent a morning walking around them. There are solutions to our problems—but they won't come about through tours" (Odede 2010, p. A25). Next we discuss other approaches for reducing poverty throughout the world include promoting economic development, investing in human development, and providing microcredit programs that provide loans to poor people.

Economic Development. One approach to alleviating poverty involves increasing the economic output or the gross domestic product of a country. However, economic development does not always reduce poverty; in some cases, it increases it. Policies that involve cutting government spending, privatizing basic services, liberalizing trade, and producing goods primarily for export may increase economic growth at the national level, but the wealth ends up in the hands of the political and corporate elite at the expense of the poor. Economic growth does not help poverty reduction when public spending is diverted away from meeting the needs of the poor and instead is used to pay international debt, finance military operations, and support corporations that do not pay workers fair wages.

Another problem with economic development is that the environment and natural resources are often destroyed and depleted in the process of economic growth. Economic development also threatens the lives and cultures of the 370 million indigenous people who live in 70 countries around the world. Indigenous people who live on land that is rich in natural resources are displaced by corporations that want access to the land and its natural resources, and by government forces that help the corporations expand their activities (Ramos et al. 2009). As remote areas are "developed," many indigenous people are forced to give up their traditional ways of life and become assimilated into the dominant culture.

Human Development. Unlike the economic development approach to poverty alleviation, the human development approach views people—not money—as the real wealth of a nation:

> The central contention of the human development approach . . . is that well-being is about much more than money. . . . Income is critical but so are having access to education and being able to lead a long and healthy life, to influence the decisions of society, and to live in a society that respects and values everyone. (UNDP 2010, p. 114)

In many poor countries, large segments of the population are illiterate and without job skills and/or are malnourished and in poor health. Investments in human development involve programs and policies that provide adequate nutrition, sanitation, housing, health care (including reproductive health care and family planning), and educational and job training.

The human development approach views people—not money—as the real wealth of a nation.

Microcredit Programs. The old saying "It takes money to make money" explains why many poor people are stuck in poverty: They have no access to financial resources and services. **Microcredit programs** refer to the provision of loans to people who are generally excluded from traditional credit services because of their low socioeconomic status. Microcredit programs give poor people the financial resources they need to become self-sufficient and to contribute to their local economies.

The Grameen Bank in Bangladesh, started in 1976, has become a model for the more than 3,000 microcredit programs that have served millions of poor clients (Roseland & Soots 2007). To get a loan from the Grameen Bank, borrowers must form small groups of five people "to provide mutual, morally binding group guarantees in lieu of the collateral required by conventional banks" (Roseland & Soots 2007, p. 160). Initially, only two of the five group members are allowed to apply for a loan. When the initial loans are repaid, the other group members may apply for loans.

microcredit programs The provision of loans to people who are generally excluded from traditional credit services because of their low socio-economic status.

Various public assistance, or "welfare" programs in the United States are aimed at providing a safety net for adults and children who are deemed eligible to receive such assistance. About 32 million U.S. households, or 27 percent of households (in 2011) benefitted from at least one means-tested poverty program (Plumer 2012). A **means-tested program** is one that has eligibility requirements based on income and/or assets. Public assistance programs designed to help the poor include Supplemental Security Income, Temporary Assistance for Needy Families, food programs, housing assistance, medical care, educational assistance, child care, child support enforcement, and the earned income tax credit (EITC).

Supplemental Security Income. Supplemental Security Income (SSI), administered by the Social Security Administration, provides a minimum income to poor people who are age 65 or older, blind, or disabled. SSI is not the same as Social Security: A millionaire can collect Social Security, but a person must be either elderly or disabled AND must have limited income and assets to collect SSI.

Temporary Assistance for Needy Families. In 1996, the U.S. social welfare system was dramatically changed with the passage of the Personal Responsibility and Work Opportunity Reconciliation Act (PRWORA), which replaced the cash assistance program Aid to Families with Dependent Children (AFDC) with a new program, **Temporary Assistance for Needy Families (TANF)**. Although the previous AFCD program provided a more reliable safety net to the poorest of Americans, the current TANF program is a cash assistance program for the poor that offers more limited assistance, with time limits and work requirements. Within two years of receiving benefits, adult TANF recipients must be either employed or involved in work-related activities, such as on-the-job training, job search, and vocational education. A federal lifetime limit of five years is set for families receiving benefits, and able-bodied recipients ages 18 to 50 without dependents have a two-year lifetime limit. Some exceptions to these rules are made for individuals with disabilities, victims of domestic violence, residents in high unemployment areas, and those caring for young children. The success of the TANF program is measured not by how many low-income families move into careers that provide a living wage, but rather by the number of people leaving the TANF program, regardless of their reason for doing so or their well-being thereafter (Green 2013).

Food Assistance. The largest food assistance program in the United States is the **Supplemental Nutrition Assistance Program (SNAP)** (formerly known as the Food Stamp Program), followed by school meals and the Special Supplemental Nutrition Program for Women, Infants, and Children (WIC). SNAP issues monthly benefits through coupons or an electronic debit card. In 2012, the average benefit for an individual receiving SNAP was equal to $133.41 per month (USDA Food and Nutrition Service 2013). To supplement SNAP, school meals, and WIC, many communities have food pantries (which distribute food to poor households), "soup kitchens" (which provide cooked meals on site), and food assistance programs for the elderly population (such as Meals on Wheels).

means-tested programs
Assistance programs that have eligibility requirements based on income.

Temporary Assistance for Needy Families (TANF) A federal cash welfare program that involves work requirements and a five-year lifetime limit.

Supplemental Nutrition Assistance Program (SNAP) The largest U.S. food assistance program.

What Do You Think? The average SNAP benefit in 2012 amounted to about $4.50 per day. If you were dependent on SNAP for food, could you maintain a healthy, adequate diet on $4.50 a day? How would your daily food intake change if you were limited to $4.50 a day for food?

Housing Assistance. Although media and the public tend to focus attention on rising gas and food prices, the biggest expense for most families is housing. In 2011, more than 6.5 million U.S. households spent more than 50 percent of income for housing expenses (National Alliance to End Homelessness 2013). Lack of affordable housing is not just a problem for the poor living in urban areas. "The problem has climbed the income ladder and moved to the suburbs, where service workers cram their families into overcrowded apartments, college graduates have to crash with their parents, and firefighters, police officers, and teachers can't afford to live in the communities they serve" (Grunwald 2006).

Federal housing assistance programs include public housing, Section 8 housing, and other private project-based housing. The **public housing** program, initiated in 1937, provides federally subsidized housing that is owned and operated by local public housing authorities (PHAs). To save costs and avoid public opposition, high-rise public housing units were built in inner-city projects. These have been plagued by poor construction, managerial neglect, inadequate maintenance, and rampant vandalism. Poor-quality public housing has serious costs for its residents and for society:

> Distressed public housing subjects families and children to dangerous and damaging living environments that raise the risks of ill health, school failure, teen parenting, delinquency, and crime—all of which generate long-term costs that taxpayers ultimately bear. . . . These severely distressed developments are not just old, outmoded, or run down. Rather, many have become virtually uninhabitable for all but the most vulnerable and desperate families. (Turner et al. 2005, pp. 1–2)

Section 8 housing involves federal rent subsidies provided either to tenants (in the form of certificates and vouchers) or to private landlords. Unlike public housing that confines low-income families to high-poverty neighborhoods, the aim with Section 8 housing is to disperse low-income families throughout the community. However, because of opposition by residents in middle-class neighborhoods, most Section 8 housing units remain in low-income areas.

A major barrier to building affordable housing is zoning regulations that set minimum lot size requirements, density restrictions, and other controls. Such zoning regulations serve the interests of upper-middle-class suburbanites who want to maintain their property values and keep out the "riffraff"—the lower-income segment of society who would presumably hurt the character of the community. Thus, one answer to the housing problem is to change zoning regulations that exclude affordable housing. Fairfax County, Virginia, is one of more than 100 communities that have adopted "inclusionary zoning," which requires developers to reserve a percentage of units for affordable housing (Grunwald 2006).

Alleviating Homelessness. Programs to temporarily alleviate homelessness include "homeless shelters" that provide emergency shelter beds, and transitional housing programs, which provide time-limited (usually two years) housing and services designed to help individuals gain employment, increase their income, and resolve substance abuse and other health problems. Some areas have clinics that offer free veterinary care to companion animals of homeless individuals. This chapter's *Animals and Society* features research that looks at the important role that companion animals can play in the lives of homeless individuals. Finally, resolving homelessness requires strategies to *prevent* homelessness from occurring in the first place, such as increasing employment and living wages, providing tax benefits to renters (not just to homeowners), providing more affordable housing, and protecting both homeowners and renters against foreclosures.

What Do You Think? The number of spaces in homeless shelters is grossly inadequate to accommodate the numbers of homeless individuals, which means that hundreds of thousands of homeless people have no place to be, except in public. Many cities have passed laws that prohibit homeless people from begging as well as sleeping, sitting, and/or "loitering" in public. Do you think that such laws unfairly punish the homeless? Or are these laws necessary to protect the public?

public housing Federally subsidized housing that is owned and operated by local public housing authorities (PHAs).

Section 8 housing A housing assistance program in which federal rent subsidies are provided either to tenants (in the form of certificates and vouchers) or to private landlords.

Leslie Irvine, professor of sociology at University of Colorado in Boulder, conducted qualitative interviews with homeless individuals who had pets. Irvine found her sample by approaching homeless people in a downtown park and at street clinics that provide veterinarian services for companion animals of homeless individuals. Using an analytic approach called "personal narrative analysis," Irvine uncovered themes in the stories homeless people told about their lives with animals. In the following excerpts, homeless individuals describe how their companion animals were instrumental in improving the quality of their lives in ways that were lifechanging, and in some cases, lifesaving (Irvine 2013).

Donna's Story
. . . Donna's life consisted of homelessness, prostitution, drug addiction, and abusive relationships. . . . Eventually, she "got the virus," referring to HIV, and she still does not know whether it came from "the sex or the needles.". . . "I would never, ever shove a needle in my arm anymore," she told me. I asked her how she quit. She paused while tears sprang up in her eyes. She said softly, "Athena." She paused again, then looked at me and said, "She was the love of my life." Athena, a German shepherd/Labrador retriever mix, was Donna's companion for ten years. . . .

Donna became Athena's guardian through a woman named Sita, long a common denominator between homeless people and homeless animals in San Francisco. About a decade ago, Donna lived with her abusive boyfriend in a garbage-strewn encampment under a freeway. Worn out from addiction and hard living, they began camping in Donna's mother's backyard. Sita and Donna knew each other from the streets, and, as Donna explained, "Sita said, 'You need a dog in your life.'" Sita had rescued three-year-old Athena from death row in a shelter. Although it might not seem that a homeless drug addict in an abusive relationship would make the best guardian for a dog, the match saved two lives. As Donna recalled, "Athena did everything for me. She got me out of an abusive relationship. And it was either the dog or him, and I chose the dog. He used to take my money. My shoes. Everything. The guy used to beat me up, and Sita told me it was either the man or the dog, so I chose Athena. I got the dog. Got rid of the man." With the boyfriend out of the picture, Donna moved into her mother's house, to a space she described as "the upstairs." But Sita had also said, "You have to be clean to have the dog." Her mother agreed, so Donna faced a decision. "I realized Athena meant everything to me," she told me. "I said to myself, 'My dog comes first in my life. Would I rather use drugs, or feed my dog?' And I fell in love with Athena, so I gave up the needle. Gave up the pipe. I gave up liquor. Everything." . . . Donna also credits Athena with improving her HIV status. After becoming clean and sober, she felt better and began taking care of herself.

Trish's Story
I met Trish on a cold December day in Boulder. She stood on the median at the exit of a busy shopping center with her Jack Russell terrier bundled up in a dog bed beside her. She was "flying a sign," or panhandling, with a piece of cardboard neatly lettered in black marker to read, "Sober. Doing the best I can. Please help." . . . Trish's dog Pixel came from a pet store where Trish had worked eight years ago. Then a puppy, Pixel had contracted parvovirus and survived through Trish's diligent care. But the storeowner no longer considered Pixel sellable, and offered him to Trish. Although she could hardly afford to feed herself, she had always loved animals, and the two have remained inseparable ever since. . . . Trish told me that she had been homeless "off and on" for over 10 years. For her, not being homeless meant sleeping in a car or in the back of a store where she briefly held a job. In her younger days, she had followed the Grateful Dead around the country, and eventually landed in Boulder. By then, she had become a heroin addict. . . . She and Pixel slept under a bridge well known in Boulder as a homeless camp. When I met her, she had been clean and sober for two years. She had found an addiction rehabilitation facility that covered the costs of her treatment through a well-timed state program. "It was

Medicaid. Medicaid is a government program that provides medical services and hospital care for the poor through reimbursements to physicians and hospitals (see also Chapter 2). States vary in rules about who is eligible for Medicaid; many low-income individuals and families do not qualify for Medicaid.

Educational Assistance. Educational assistance includes Head Start and Early Head Start programs and college assistance programs (see also Chapter 8). Head Start and Early Head Start programs provide educational services for disadvantaged infants, toddlers, and preschool-age children and their parents, and are designed to improve children's cognitive, language, and social-emotional development and strengthen parenting skills (Administration for Children and Families 2002).

To help low-income individuals wanting to attend college, the federal government offers grants, loans, and work opportunities. The Pell Grant program aids students from low-income families. The federal college work–study program provides jobs for students with "demonstrated need." The guaranteed student loan program enables college students and their families to obtain low-interest loans with deferred interest payments. However, awesome," she said. "I just happened to call when the government was doing this study. They were paying for people's treatment, and then they wanted to follow them for six months. So, I went there, and got sober." She found a friend to care for Pixel, "someone with a house," while she went through what she described as a "severe detox." After rehab, Trish recounted, "I wanted to get off the streets but I had to find a job and try to do all those things I couldn't do before because I was messed up, which was really, really weird. It had been so long since I had to have a schedule. I loved it, though. I was like, 'I can do this.'" She supported herself by working various jobs that paid under the table, mostly cleaning houses. . . . "It took about six months, and I got us off the streets. We lived out of my car."

. . . Trish said that Pixel kept her going, even during her darkest times. She even said that Pixel kept her alive. . . . " I was [on the streets]. I hated it. I was totally at rock bottom. I just wanted to die. . . . But I couldn't give up because I had something else to take care of besides myself. So he kept me alive. . . . I didn't care about myself, but I had to care about him, you know? He got me through a really tough spot. If I would've had to be without him out there before, I don't think I would have made it, at all." In the two years since that "tough spot," Trish credits Pixel with helping her stay sober. "He definitely helps keep me on the straight and narrow,"

she said. She claims that Pixel "hates the smell of alcohol," and he keeps her away from "bad elements, or groups of people, because of the alcohol, the drugs, and all that." She claims that the dog will nip the heels of people who approach smelling of alcohol. "He's an awesome judge of character," she said. "He just knows." She looked down at Pixel and added, "Right, buddy?" The little dog never took his eyes off her.

Denise's Story

Cats, too, figured into the lifesaving role. For example, I met Denise and her cat Ivy at a veterinary clinic in San Francisco. White, middle-aged, slender, and nicely dressed, Denise looked nothing like the stereotypical homeless person. Ivy, a tiny, black-and-white cat, sat sphinx-like in a carrier, her white paws tucked neatly underneath her. . . .

Denise had worked as a self-employed graphic artist, but severe depression caused her to miss deadlines. Major clients lost faith in her, and the accounts gradually dwindled. She fell behind on rent, and her landlord evicted her from her apartment. . . . She put her belongings in storage and moved into her car, where she and Ivy had lived for over eight months when we met. "Half of the car is taken up by her stuff," Denise said of Ivy. "There's a big carrier, which she sleeps in, and then her litter box, and her bowls, so she essentially has the whole back seat, and then I'm in the front seat." . . . The eviction

has made it difficult for Denise to find another apartment. Many landlords simply will not rent to prospective tenants who have an eviction on their records. . . . She never imagined she would live in her car for over eight months.

. . . People often said she had no right to keep a cat in her circumstances. . . . When I asked how she responded to this, she explained: "I have a history with depression up to suicide ideation, and Ivy, I refer to her as my suicide barrier. And I don't say that in any light way. I would say most days, she's the reason why I keep going, because I made a commitment to take care of her when I adopted her. So she needs me, and I need her. She is the only source of daily, steady affection and companionship that I have. The only one. I can't imagine being without her, wanting to go on at all, without her."

. . . Fortunately, Denise would not have to decide between keeping Ivy and finding housing. Her doctor had recently provided the document certifying Ivy as her necessary companion. The emotional support Ivy provides qualifies her as a "reasonable accommodation" for Denise's psychological state. Landlords could not refuse to rent to Denise because of Ivy, even if they usually do not allow pets.

Source: Irvine, Leslie. 2013. "Animals as Lifechangers and Lifesavers: Pets in the Redemption Narratives of Homeless People." *Journal of Contemporary Ethnography* 42(1):3–36.

mounting student debt has reached disturbing levels. Average student loan debt in 2013 was more than $26,000, a 40 percent increase from seven years earlier (Stiglitz 2013).

Child Care Assistance. In the United States, lack of affordable, good child care is a major obstacle to employment for single parents and a tremendous burden on dual-income families and employed single parents. The annual cost of child care for a 4-year-old child ranges from about $4,000 in rural Mississippi to nearly $16,000 in Washington DC (Gould et al. 2013). Child care fees for an infant are even higher.

Some public policies provide limited assistance with child care, such as tax relief related to child care expenses and public funding for child care services for the poor (in conjunction with mandatory work requirements). However, child care assistance is inadequate; many states have waiting lists for child care assistance, and many families earn more than the eligibility limit, but not enough to afford child care expenses.

Child Support Enforcement. To encourage child support from absent parents, the PRWORA requires states to set up child support enforcement programs, and single parents who receive TANF are required to cooperate with child support enforcement efforts. The welfare reform law established a Federal Case Registry and National Directory of New Hires to track delinquent parents across state lines, increased the use of wage withholding to collect child support, and allowed states to seize assets and to revoke driving licenses, professional licenses, and recreational licenses of parents who fall behind in their child support.

Earned Income Tax Credit. The federal **earned income tax credit (EITC)**, created in 1975, is a refundable tax credit based on a working family's income and number of children. The EITC is designed to offset Social Security and Medicare payroll taxes on working poor families and to strengthen work incentives. The federal EITC lifts more children out of poverty than any other program (Llobrera & Zahradnik 2004).

Welfare in the United States: Myths and Realities

During the 2012 presidential election, Mitt Romney famously referred to the 47 percent of Americans who, according to Romney, would vote for President Obama because they rely on public assistance. Romney remarked, "...there are 47 percent...who are dependent upon government, who believe that they are victims, who believe the government has a responsibility to care for them, who believe that they are entitled to health care, to food, to housing, to you-name-it" (quoted in Plumer 2012, n.p.). Negative attitudes toward welfare assistance and welfare recipients, such as those conveyed in Romney's comments, are not uncommon (Epstein 2004). But these negative images are grounded in myths and misconceptions about welfare. For example, Romney was right that about half of Americans live in a household that receives some kind of federal benefit, but a big chunk of these benefits go to support the elderly and disabled. Nearly a third of U.S. households in 2011 received Medicare and Social Security—benefits for the elderly and the disabled (Plumer 2012). Medicare and Social Security are provided to older Americans across the economic spectrum; they are not programs designed to target the poor. Indeed, although government assistance programs are often referred to as "entitlement programs," labeling Social Security an "entitlement program" may be misleading because retirees collect the money that they and their employers have contributed over their work history. As we discuss in Chapter 12, Social Security is a form of retirement insurance administered by the government, which is substantially different from public assistance to the poor that is funded by tax dollars. The following discusses other myths about welfare:

Myth 1. People who receive welfare are lazy, have no work ethic, and prefer to have a "free ride" on welfare rather than work.

Reality. Three-quarters of recipients of TANF are children, and nearly half of TANF cases are child-only cases where no adult is involved in the benefit calculation and only children are aided (Office of Family Assistance 2012). Because we do not expect children to work, we can hardly think of children in need of assistance as "lazy."

Adults receiving public assistance are lazy? Consider that nearly a quarter of adult recipients of TANF worked in 2009, but with average monthly earnings of only $809, they could not survive on the income from their jobs. Other adult welfare recipients are participating in work activities, including job training or education and job searches. Unemployed adult welfare recipients experience a number of barriers that prevent them from working, including disability and poor health, job scarcity, lack of transportation, lack of education, and/or the responsibility of staying home to care for their children (which often stems from the inability to pay for child care or the lack of trust in child care providers) (Zedlewski 2003). And many adult TANF recipients do not work because there are not enough jobs available. Between 2008 and 2013, the ratio of job seekers to job openings has been 3 to 1 or higher, meaning that there are no jobs for more than two out of three unemployed workers (Shierholz 2013) (see also Chapter 7). Finally, most adult welfare recipients would rather be able to support themselves and their families than rely on public assistance. The image of a welfare "freeloader" lounging around enjoying life is far from the reality of the day-to-day struggles and challenges of supporting a household

earned income tax credit (EITC) A refundable tax credit based on a working family's income and number of children.

on a monthly TANF check of $324, which was the average monthly cash assistance to families with one child receiving TANF assistance in 2009 ($408 for families with two children) (Office of Family Assistance 2012). This chapter's *Social Problems Research Up-Close* feature presents research that looks at the challenges that low-income mothers who rely on public assistance face for survival.

Myth 2. Most welfare mothers have large families with many children.

Reality. In 2009, the average number of children in families that receive TANF was only 1.8; half of families receiving TANF had only one child, and less than 8 percent of families had more than three children (Office of Family Assistance 2012).

Myth 3. Welfare benefits are granted to many people who are not really poor or eligible to receive them.

Reality. Although some people obtain welfare benefits through fraudulent means, it is much more common for people who are eligible to receive welfare not to receive benefits. Fewer than one in five individuals living below the poverty line receive cash assistance and fewer than half receive food stamps (see Table 6.8).

TABLE 6.8 Percentage of Individuals Living below Poverty Level in Households that Receive Means-Tested Assistance, 2011

Type of Assistance	Percentage
Any type of assistance	74.3%
Medicaid	61.3%
Food stamps	48.8%
Cash assistance	19.8%
Housing assistance	15.4%

Source: U.S. Census Bureau, 2012b. *Current Population Survey, 2012 Annual Social and Economic Supplement*. Table POV26

One reason for not receiving benefits is lack of information; some people do not know about various public assistance programs, or even if they know about a program, they do not know they are eligible. Another reason that many people who are eligible for public assistance do not apply for it is because they desire personal independence and do not want to be stigmatized as lazy people who just want a "free ride" at the taxpayers' expense. Others have difficulty navigating the complex administrative processes involved in applying for assistance. Assistance programs are administered through separate offices at different locations, have various application procedures and renewal deadlines, and require different sets of documentation.

Green (2013) explains that,

> . . . [a]s low-income mothers struggle to meet the intense demands of balancing work and family, they also have to continue the time-intensive task of piecing together in-kind and cash benefits to pad their low wages. Doing so involves traveling from one office to another; repeatedly disclosing intimate and personal information; and documenting, in a detailed paper trail, the legitimacy of one's story. . . . Although there is little time to spare in this world, each office treats clients as if they have endless time to waste. Furthermore, poor families are often at the mercy of buses that are late, babysitters who do not show up, overworked caseworkers who misplace documents, and other similar barriers to the successful performance of the role of "good client." This situation can lead to extreme levels of personal frustration, which add to the hardship and defeatism experienced while engaging this system. (p. 55)

Navigating through the "system"—getting time off of work to meet with caseworkers and finding child care and transportation—can produce so much frustration that some people who are eligible for assistance just give up on the system.

Finally, some individuals who are eligible for public assistance do not receive it because it is not available. In cities across the United States, thousands of eligible low-income households are on waiting lists for public housing assistance because there are not enough public housing units available, and some cities have even stopped accepting housing applications. Even when people receive benefits, using them may be difficult. For example, individuals with Section 8 vouchers for housing may have a hard time finding a landlord who will accept them, even though it is against the law to refuse a Section 8 renter. Individuals who have Medicaid may have difficulty finding a doctor who will take Medicaid patients. Low-income parents who receive child care assistance

Courtesy of Autumn Green

Autumn Green—sociologist, researcher, and advocate for low-income families.

Autumn Green, a sociologist whose research and advocacy focuses on low-income families, conducted research that explores how poor women who receive public assistance manage to navigate the system—how they piece together, in a patchwork fashion, resources so that they and their children can survive. Patchwork quilting is traditionally women's work, and so the metaphor of "patchwork" reflects the gendered nature of poverty—the highest rates of poverty are among women and single female-headed households with children. The patchwork metaphor also conveys that "the daily work of surviving in poverty can be likened to the creation of a complex tapestry drawing from many resources, performing a multiplicity of tasks and duties, and piecing together one's own subsistence with the work of caregiving and nurturing for others in meaningful and artistic ways" (Green 2013, p. 53).

Green's research and advocacy interests stem from her own life experiences as a former teenage parent, low-income mother, and public assistance recipient. Green describes how she once mapped out the various patches of the "quilt" that represented the responsibilities she faced as a low-income, single mother in graduate school. With the help of a friend, Green composed a 12-page list that she taped together like a quilt spread across the dining room table, on which she wrote out and categorized the different patches that made up the quilt:

One square of paper read "Benefits" at the top. "Assemble packet for Section 8 recertification was written on the first line" with subcategories: "(1) get paperwork from

landlord, (2) collect proof-of-income letter from work, (3) get letter from daughter's therapist for a three-bedroom disability accommodation," each line representing several hours of work to complete. Then the list continued with similarly mapped steps for upcoming deadlines for fuel assistance, SNAP, and MassHealth recertification. And this was only one category at one moment in time listing only those items with immediate deadlines. Work also had a category, then school. Caring for my disabled daughter entailed a long list of tasks, from scheduling school meetings, to therapy appointments, to finding summer camp programs (and, of course, the money to pay for them). And the list went on for 12 pages and was still far from complete. Few low-income women have the training, resources, and connections even to map out their obligations, let alone meet them without feeling frustrated or defeated. (Green 2013, pp. 60–61)

Method and Sample

From 2004 through 2011, Green conducted in-depth interviews with 10 low-income Boston-area single mothers who were current or former college students. These women were recruited through (1) flyers that were distributed to programs serving low-income mothers, and (2) network sampling through Green's personal connections as a local welfare rights advocate. Six participants identified as African American, two as white, one as Latina, and one—an immigrant from Haiti—considered her race to be Haitian. The ages of the participants ranged from 19 to 49 (two were 19, three were in their 20s, three were in their 30s, and two were in their 40s). Five of the participants had only one child; one had two children. The children ranged from age 5 months to 22 years; however, most of the children were younger than age 8. All the participants had at least some college or vocational training beyond high school, but none had completed a four-year degree. Six participants were currently enrolled in college courses, one participant had graduated with a vocational degree, two had previously attended vocational programs and were trying to reenroll in college, and one had recently dropped out of college.

Selected Findings

The interviews Green conducted yielded rich, qualitative data about the lived experiences of poor mothers juggling the various obligations in their roles as parent, worker, and recipient of

various public assistance benefits. After briefly describing just one of Green's participants, we discuss the implications of and policy recommendations that resulted from Green's research.

Research Participant: Jessica's Story

Jessica is a 24-year-old Caucasian mother who recently graduated from a cosmetology program, ended a relationship with her daughter's father, and moved back to Boston from New York. She found a three-bedroom apartment, sight unseen, for $1,000 per month, which is considered very low rent in Boston. Jessica got a job at a salon near her home, but she was paid as an independent contractor, so her income varied depending on the number of clients she served in a given week and the cost of services. Jessica struggled to find child care for her 2-year-old daughter Beth, and first hired a neighbor's daughter to babysit while she was at work, but Jessica's income was low and inconsistent, and it was hard to pay the babysitter. So Jessica began asking friends, family members, and neighbors to watch Beth as a favor on a day-to-day basis. Although Jessica tried to access the state's child care assistance voucher program, she was told that it would be a two-year wait before she could receive a voucher. Green, who combined her advocacy work with her research, informed Jessica that if she received TAFDC benefits (Massachusetts' TANF cash assistance program), she could get a voucher immediately. Jessica applied for TAFDC benefits, and although she worked full time, she was deemed eligible because her wages were low and inconsistent. Through the TAFDC program, she was able to receive cash assistance, SNAP, and MassHealth insurance (i.e., Medicaid) benefits, as well as an immediate child care voucher. Jessica used the voucher to send her daughter to a child care provider in the neighborhood who Jessica said, "isn't perfect, but I just try not to think about it because I don't have another choice" (p. 59).

Jessica had difficulty meeting her monthly expenses, so she found a lower-cost apartment but still struggled to break even, let alone have anything extra. Jessica applied for public housing and Section 8 in the Greater Boston Area, but was told that the waiting list was three to five years for public housing and even longer for Section 8. During the workday, on breaks and between clients, Jessica made phone calls to local charities to inquire about emergency assistance programs to prevent the shutoff of utilities or eviction, but these calls provided nothing but dead ends. With the help of a friend, Jessica figured out how to get an application for fuel assistance. This application required her to produce documentation of her income, so she typed up a letter for her boss to sign, photocopied her paychecks, and took extra time during her lunch break to deliver the paperwork.

Green explains that TAFDC participants and those transitioning from "welfare to work" have the most integrated system of case management, with one caseworker who administers their cash, SNAP, MassHealth, and child care subsidy. But as Jessica's income went up, she transitioned off the TAFDC program and then had to maintain each of her state benefits separately (SNAP, MassHealth, and subsidized child care). For working-poor families not receiving cash assistance, these other social services are administered through separate offices, have various renewal deadlines, and require different sets of documentation. "This situation requires the working poor, whose time is already extremely stretched with the balance of work and family, to engage separate case managers, visit separate locations, individually apply for services, and provide duplicate documentation" (Green 2013, p. 60).

The social service benefits Jessica received, plus her wages, still fell short of supporting her basic needs. So on her days off, Jessica visited family members in the suburbs to do her laundry, traveled to local food pantries, searched for emergency assistance programs, and attended mandatory nutrition classes and recertification appointments for her food benefits. These tasks were in addition to the other generalized tasks of motherhood: cleaning, grocery shopping, cooking, playing with her toddler, scheduling doctor's appointments, and so on. To help with the cost of rent, Jessica decided to bring in a roommate to rent the extra bedroom in her apartment. She had concerns about bringing a stranger into her home, but she had few options available.

Discussion

Although each of Green's participants had a different story, their strategies for survival were similar: They each used a patchwork strategy to piece together resources for survival. Jessica, the research participant profiled here, managed to stay afloat by using a patchwork of strategies, but she did so at great personal cost. She was continuously exhausted; felt torn away from spending time with her daughter; and she felt isolated and lonely with few opportunities to socialize and little time to take care of herself.

Green compared the patchwork strategy she observed her research participants engage in to a literal type of quilt called a *crazy quilt*—a quilt made from random strips of fabric of various sizes and shapes that are pieced together without any cohesive pattern or organization. Green explains, "I know from my first experience trying to make a crazy quilt that sometimes the pieces do not quite fit together, and you get little holes between the patches. . . . [A] patchwork system leaves holes that allow families to fall through the cracks too easily" (p. 60).

Green's interviews with her research participants highlights how the patchwork social service system, with each service having different requirements and guidelines, adds another layer of difficulty for women already struggling against the odds of economic disadvantage.

> If, for example, the utility assistance office is open only from 1 p.m. until 4 p.m. and [a client's] workfare hours are from 1 p.m. to 5 p.m., what are her options for preventing sanction from the Department of Transitional Assistance while still getting the electricity turned back on? If [a client] is reported for bringing [her child] to work with her, she could be sanctioned and lose her cash assistance, but she has been offered no help to find a caregiver who will take her voucher and provide good-quality care for [her child]. Where is he [or she] to go while she completes the work hours required to keep the funds that pay her rent and support her children? (p. 57)

Green concludes that, "[u]ltimately, the patchwork safety net system is ineffective and unreasonably demanding. We eat up women's time with the work of survival and thus make it impossible for women to work on self-improvement, personal empowerment, or education, thus trapping them in perpetual poverty. The system also takes away from caregiving and thus is a detriment for children" (p. 61).

Green's research concludes with a number of broad policy recommendations, a few of which will be mentioned here. First, Green suggests that public assistance should be provided through a more centralized system that would consider benefit levels in relation to one another, and would implement benefit reductions due to wage increases more gradually, allowing a greater opportunity for economic advancement. "Within this system, families would spend less energy obtaining social services and could use their time better toward meeting immediate family needs, nurturing their children, improving job performance (through increased availability and decreased absenteeism), investing in self-care (e.g., receiving counseling or addressing health issues), or meeting long-term goals through education and training" (p. 62).

Another suggestion for welfare reform involves integrating the various forms of public assistance so that clients can simultaneously apply for multiple social services in a single location and/or through a single case manager. Streamlining social services into a one-stop approach would not only be helpful for clients using public assistance, bit it would also cut costs and reduce errors and inconsistencies in program administration.

Given the difficulty Green's participants had in finding landlords who would accept housing vouchers, Green says we should investigate why providers are not willing to accept state assistance vouchers: Low reimbursement rates? Excessive bureaucratic requirements? Negative stereotypes of voucher holders as bad tenants?

Green argues that, given the recent recession and the large number of families struggling for basic survival, we should expand social services, as well as opportunities for living wage work and higher education, and increase our value for women's work in the home. At a minimum, says Green, current funding must be maintained, rather than allow further cuts in programs that are already in crisis.

Finally, Green says that mending the tattered safety net of public assistance "requires embracing a universal perspective on meeting human services needs as a societal duty to all citizens" (p. 62):

> Understanding that these issues address universal human needs, rather than make up for the faults of the "morally deficient," we can change the terms of human services and thus the way in which we provide them. The last of these options ultimately requires a transformative shift in societal values that, given recent political contention, will most likely be an uphill battle. (Green 2013, p. 62)

Source: Green, Autumn R. 2013. "Patchwork: Poor Women's Stories of Resewing the Shredded Safety Net." *Affilia* 28:51–64.

vouchers are often unable to find an available "voucher slot" and may be on a waiting list at a child care center for more than a year.

Myth 4. There is widespread abuse and fraud in SNAP by beneficiaries, who use their food stamp benefits to purchase beer, wine, liquor, cigarettes, and/or tobacco, or who sell their food stamp benefits for cash.

Reality. First, the SNAP program strictly prohibits beneficiaries from purchasing alcoholic beverages and tobacco products, as well as any nonfood items such as pet food, cosmetics, and paper products, with their benefits card. And due to increased government oversight and the introduction of the Electronic Benefit Transfer (EBT) card system, fraud in the SNAP program has decreased considerably. The Government Accountability Office found that "trafficking"—selling SNAP benefits for cash—decreased from 3.8 cents per dollar of benefits in 1993 to about 1 cent per dollar of benefits in 2010 (Blumenthal 2012).

Myth 5. Immigrants place a huge burden on our welfare system.

Reality. Low-income noncitizen immigrants, including adults and children, are less likely to receive public benefits than those who are native born. Moreover, when noncitizen immigrants receive benefits, the value of benefits they receive is lower than the value of benefits received by those born in the United States (Ku & Bruen 2013). Federal rules restrict immigrants' eligibility for public benefit programs, and undocumented immigrants are generally ineligible to receive benefits from Medicaid, SNAP, and TANF, although some benefit programs, such as the National School Lunch Program, the Women, Infants, and Children Nutrition Program (WIC), and Head Start, do not include immigration status as an eligibility factor. And although children born in the United States are considered citizens and are therefore eligible for public assistance, undocumented parents often do not apply for assistance for their children because they either do not know their children can receive benefits, or they fear that applying for benefits for their children will result in their deportation (see also Chapter 9).

Minimum Wage Increase and "Living Wage" Laws

In his 2013 State of the Union address, President Obama, exclaiming that no American who works full time should live in poverty, proposed increasing the minimum wage from $7.25 an hour to $9.00 an hour. In a bolder proposal, Senator Tom Harkin and Congressman George Miller proposed the Fair Minimum Wage Act of 2013, which, if passed, would raise the minimum wage from $7.25 to $10.10 over a period of three years, and would then be raised automatically each year based on inflation. As of January 2013, 19 states and the District of Columbia have mandated a minimum wage that is higher than the federal $7.25.

> **What Do You Think?** Just as there is a federal minimum wage, do you think that there should be a federal maximum wage? Why or why not? If you favor the idea of a maximum wage, what should that wage be?

Many cities and counties throughout the United States have **living wage laws** that require state or municipal contractors, recipients of public subsidies or tax breaks, or, in some cases, all businesses to pay employee wages that are significantly above the federal minimum, enabling families to live above the poverty line. Research findings show that businesses that pay their employees a living wage have lower worker turnover and absenteeism, reduced training costs, higher morale and productivity, and a stronger consumer market (Kraut et al. 2000).

Reducing U.S. Economic Inequality

Reducing economic inequality can be achieved by strategies that increase the economic resources of those at the bottom, such as increasing the minimum wage, providing jobs or support for the unemployed (discussed in Chapter 7), and other strategies already

living wage laws Laws that require state or municipal contractors, recipients of public subsidies or tax breaks, or, in some cases, all businesses to pay employees wages that are significantly above the federal minimum, enabling families to live above the poverty line.

discussed in this chapter. An important area for reducing economic inequality involves improving the quality and *equality* of education to ensure that educational resources and opportunities are not unfairly skewed toward children and young adults from more affluent families (see Chapter 8). In recent years, politicians have made substantial cuts to education while at the same time cutting corporate taxes (Callahan & Cha 2013). Many of the topics addressed in the next chapter, on "Work and Unemployment," are also important in reducing economic inequality in the United States.

A number of tax reforms could also help reduce economic inequality. For example, one way to reduce the gap between the top and the bottom of the economic system is to make the tax system more progressive. **Progressive taxes** are those in which the tax rate increases as income increases, so that those who have higher incomes are taxed at higher rates. A more progressive tax system would increase taxes on the wealthy. Other tax reforms that could help reduce economic inequality include increasing estate taxes (labeled by opponents as "death taxes") and gift taxes, as well as capital gains taxes. McNamee and Miller (2009) explain increasing tax revenue by increasing taxes on the rich does not necessarily result in simple redistribution of income or wealth from the rich to the poor. Rather, revenue from increasing taxes on the rich could be directed to public projects that would provide more equal access to education, health care, public transportation, and other services that would give low-income Americans the resources they need to improve their economic situation. Not surprisingly, the wealthy tend to oppose increasing their own taxes.

The Occupy Wall Street movement has increased public awareness of the problem of economic inequality.

Given the unfair advantage the wealthy have in influencing the political process, another key strategy in reducing economic inequality is to reduce the influence of money in politics. At the time of this writing, 16 states have called for an amendment to the Constitution to overturn the 2010 *Citizens United* ruling that gave corporations unlimited political spending power. Reducing the political inequality that perpetuates economic inequality also necessitates enacting limits on the amount of money that wealthy individuals can spend on politics.

We conclude with a mention of the Occupy Wall Street movement, which has brought public attention to the problem of economic inequality. **Occupy Wall Street (OWS)**, a decentralized protest movement that began in 2011 in Zuccotti Park in the Wall Street financial district of New York City, is concerned with economic inequality, greed, corruption, and the influence of corporations on government. OWS started as a wave of park occupations and spread around the world, but has since branched out into a number of areas, all focused on the concerns of the "99 percent"—i.e., the concerns of regular, hardworking Americans versus the "one percent"—the wealthy. OWS activists have supported striking Wal-Mart workers, and have camped out on front lawns of foreclosed homeowners who refuse to be moved. OWS has been criticized for being nonspecific

progressive taxes Taxes in which the tax rate increases as income increases, so that those who have higher incomes are taxed at higher rates.

Occupy Wall Street A protest movement that began in 2011, and is concerned with economic inequality, greed, corruption, and the influence of corporations on government.

in its mission and demands, but the fact that OWS is not specific in its focus is also a strength. One OWS activist explained, "As soon as there's one issue, then I alienate the two of you who don't have my issue" (Milkman et al. 2013, p. 195). Another activist explained that the lack of specific demands means that, "anyone could come into the movement and see their grievance as equivalent to everyone else. If it's like I don't have a job, I have student debt, I have these medical problems, I'm thrown out of my house, the hydrofracking that's going on, the BP oil spill, it doesn't matter. It's Wall Street. It's the 1%" (p. 197). And although the future of OWS is uncertain, this movement has given rise to a spirit of resistance that has taken hold among a vast number of Americans. As one activist commented:

> Now the genie's out of the bottle. There's this energy! I don't know if they'll be able to put it back in. . . . Whether it's under the Occupy brand or not, people are still going to be organizing. Nobody's going away. There's a lot of work to be done, and we're going to continue tackling it, now that we're all connected, on all these different fronts, in the student movement, in the labor movement, housing, community organizations. (quoted in Milkman et al. 2013, p. 197)

Understanding Economic Inequality, Wealth, and Poverty

As we have seen in this chapter, the quality of our lives is intricately related to the economic resources we have—resources that buy access to goods and services such as housing, food, education, health care, resources that influence virtually every aspect of our lives. On a positive note, significant gains have been made in improving the standard of living for populations living in absolute poverty. But at the same time, economic inequality has reached unprecedented levels in the world, and in the United States, as the "rich get richer."

A common belief among U.S. adults is that the rich are deserving and the poor are failures. Blaming poverty on the individual rather than on structural and cultural factors implies not only that poor individuals are responsible for their plight but also that they are responsible for improving their condition. If we hold individuals accountable for their poverty, we fail to make society accountable for making investments in human development that are necessary to alleviate poverty, such as providing health care, adequate food and housing, education, child care, job training and job opportunities, and living wages. Lastly, blaming the poor for their condition diverts attention away from the recognition that the wealthy—individuals and corporations—receive far more benefits in the form of wealthfare or corporate welfare, without the stigma of welfare.

Efforts to alleviate poverty and reduce economic inequality are often motivated by a sense of moral responsibility. But alleviating poverty and reducing economic inequality also makes sense from an economic standpoint. According to one study, if economic inequality in Maryland was lowered to the level it was in 1968—the year that economic inequality was at its lowest level in modern U.S. history—the economic benefits would be equivalent to adding 22 percent to Maryland's annual gross state product in the form of personal consumption expenditures, decreased social and environmental costs, increased access to higher education, and additional spending by the poor (Talberth et al. 2013). Just as economic inequality has costs that we all bear, so does poverty. The cost of poverty in the United States is more than $500 billion (4 percent of the GDP) per year due to increased health care costs, increased crime-related costs, and lowered productivity (vanden Heuvel 2011). According to one source, the cost of eradicating poverty worldwide would be only about one percent of global income—and no more than 2 percent to 3 percent of national income in all but the poorest countries (UNDP 1997).

Ending or reducing poverty begins with the recognition that doing so is a worthy ideal and an attainable goal. Imagine a world where everyone had comfortable shelter, plentiful food, clean water and sanitation, adequate medical care, and education. If this imaginary world were achieved and if absolute poverty were effectively eliminated, what would be the effects on social problems such as crime, drug abuse, family problems (e.g., domestic violence, child abuse, and divorce), health problems, prejudice and racism, and international conflict? In the current global climate of conflict and terrorism, we might consider that "reducing poverty and the hopelessness that comes with human deprivation is perhaps the most effective way of promoting long-term peace and security" (World Bank 2005). Instead of asking if we can afford to eradicate poverty, we might consider: Can we afford not to?

CHAPTER REVIEW

• **What is the extent of poverty, wealth, and economic inequality in the world?**
In 2013, one in five people on this planet lived in extreme poverty (less than $1.25 a day)—down from more than half in 1980. Many more people live on between $1.25 and $2.00 a day, and then many more live on not much more above that.

In stark contrast to this extreme poverty, in 2013, there were 1,426 billionaires in the world. Although 2 percent of adults own more than half of global wealth, the poorest half of the world adult population owns barely one percent of global wealth.

• **What is the difference between absolute poverty and relative poverty?**
Absolute poverty refers to a lack of basic necessities for life, such as food, clean water, shelter, and medical care. In contrast, relative poverty refers to a deficiency in material and economic resources compared with some other population.

• **How is poverty measured?**
The most widely used standard to measure extreme poverty in the developing world is $1.25 per day. According to measures of relative poverty, members of a household are considered poor if their household income is less than 50 percent of the median household income in that country. Each year, the U.S. federal government establishes "poverty thresholds" that differ by the number of adults and children in a family and by the age of the family head of household. Anyone living in a household with pretax income below the official poverty line is considered "poor." To capture the multidimensional nature of poverty, the Multidimensional Poverty Index measures serious deprivation in the dimensions of health, education, and living standards.

• **Which sociological perspective criticizes wealthy corporations for using financial political contributions to influence politicians to enact policies that benefit corporations and the wealthy?**
The conflict perspective is critical of wealthy corporations that use financial political contributions to influence laws and policies that favor corporations and the rich. Such laws and policies, sometimes referred to as wealthfare or corporate welfare, include low-interest government loans to failing businesses and special subsidies and tax breaks to corporations.

• **In the United States, what age group has the highest rate of poverty?**
U.S. children are more likely than adults to live in poverty. More than one-third of the U.S. poor population is

children. Child poverty rates are much higher in the United States than in any other industrialized country.

• **Who are the wealthiest Americans?**
In 2011, the top one percent of wage earners were those who earned more than $598,570. Of the 282 billionaires on the 2011 Forbes annual list of the wealthiest Americans, about 40 percent of the 2011 Forbes 400 list inherited a "sizeable asset" from a spouse or family member; more than one in five of the Forbes 400 inherited enough wealth to make the list; just one African American is on the list; and 90 percent of those on the list are men.

• **What are some of the consequences of poverty and economic inequality for individuals, families, and societies?**
Poverty is associated with health problems and hunger, increased vulnerability from natural disasters, problems in education, unequal treatment in the legal system, problems in families and parenting, and housing problems. These various problems are interrelated and contribute to the perpetuation of poverty across generations, feeding a cycle of intergenerational poverty. In addition, poverty and economic inequality are associated with political inequality and alienation, social conflict and war.

• **What are some of the U.S. government public assistance programs designed to help the poor?**
Government public assistance programs designed to help the poor include Supplemental Security Income, Temporary Assistance for Needy Families (TANF), food programs (such as school meal programs and SNAP), housing assistance, Medicaid, educational assistance (such as Pell Grants), child care, child support enforcement, and the earned income tax credit (EITC).

• **What are five common myths about welfare and welfare recipients?**
Common myths about welfare and welfare recipients are (1) that welfare recipients are lazy, have no work ethic, and prefer to have a "free ride" on welfare rather than work; (2) that most welfare mothers have large families with many children; (3) that welfare benefits are granted to many people who are not really poor or eligible to receive them; (4) that there is widespread abuse and fraud in SNAP by beneficiaries, who use their food stamp benefits to purchase beer, wine, liquor, cigarettes, and/or tobacco, or who sell their food stamp benefits for cash; and (5) that immigrants place an enormous burden on our welfare system.

- **What are four general approaches for achieving poverty reduction throughout the world?**

  Approaches for achieving poverty reduction throughout the world include promoting economic growth, investing in human development, providing financial aid and debt cancellation to nations, and providing microcredit programs that provide loans to poor people.

- **What are some strategies for reducing economic inequality in the United States?**

  Reducing economic inequality can be achieved by strategies that increase the economic resources of those at the bottom, such as increasing the minimum wage, providing jobs or support for the unemployed, and other general strategies to reduce poverty. In addition, making taxes more progressive could also help reduce economic inequality. Other tax reforms that could help reduce economic inequality include increasing estate taxes and gift taxes, as well as capital gains taxes. Revenue from increasing taxes on the wealthy could be directed to public projects that would provide more equal access to education, health care, public transportation, and other services that would give low-income Americans the resources they need to improve their economic situation. Another strategy to reduce economic inequality involves reducing the influence of money in politics (e.g., overturning *Citizens United* ruling).

TEST YOURSELF

1. The ___ Poverty Index is a measure of serious deprivation in the dimensions of health, education, and living standards that combines the number of deprived and the intensity of their deprivation.
   a. Relative
   b. Human
   c. International
   d. Multidimensional

2. According to the official U.S. poverty threshold guidelines, a single adult earning $12,000 a year is considered "poor."
   a. True
   b. False

3. In 2012, a typical U.S. worker would have to work ___ years to make what a typical CEO of a large U.S. corporation earns in one year.
   a. 15
   b. 99
   c. 211
   d. 273

4. Corporate welfare refers to which of the following?
   a. Taxes corporations pay that provide most of the funding for federal welfare programs for the poor
   b. Tax-deductible contributions that corporations make to charitable organizations
   c. Laws and policies that benefit corporations
   d. Employee assistance programs offered by corporations to help employees who are struggling with debt

5. What age group in the United States has the highest rate of poverty?
   a. Younger than 18
   b. 30 to 44
   c. 45 to 64
   d. Older than 65

6. According to the text, the wealthy are hardest hit by natural disasters because they have more to lose than do the poor.
   a. True
   b. False

7. Subprime mortgages are mortgages that charge very low interest rates and are available only to people with excellent credit and ample collateral.
   a. True
   b. False

8. Which federal program lifts more children out of poverty than any other program?
   a. Public and Section 8 housing
   b. TANF
   c. EITC
   d. SNAP

9. Nearly half of recipients of welfare in the United States are immigrants.
   a. True
   b. False

10. Economic development can result in increased poverty and economic inequality in a country.
    a. True
    b. False

Answers: 1. D; 2. B; 3. D; 4. C; 5. A; 6. B; 7. B; 8. C; 9. B; 10. A.

KEY TERMS

Fan Xia/ZUMA Press/Newscom.

7

Work and Unemployment

"When you go to work, if your name is on the building, you're rich. If your name is on your desk, you're middle class. If your name is on your shirt, you're poor."

—Rich Hall, writer and performer

ON APRIL 5, 2010, 29 coal miners were killed in an explosion at the Massey Energy Upper Big Branch Mine in West Virginia—the worst coal mine disaster in the United States in 40 years. Miner Gary Quarles, who worked at the Massey Mine with his son, Gary Wayne Quarles, said, "I worked the day shift that day, same as he did. . . . I came home. He didn't" (quoted in Dorell 2011). When Quarles attended a meeting with other families of the 29 men killed in the explosion to hear a report about how and why the tragedy happened, he learned that an independent investigation found that the mine explosion was primarily the fault of Massey Energy management, who could have prevented the accident by providing adequate ventilation and by limiting explosive coal dust in the mine (Maher 2011). Massey disputed the report's conclusions and claimed that the explosion was sparked by an uncontrollable inundation of natural gas. According to records from the Mine Safety and Health Administration (MSHA), the mine had more than 500 safety violations issued against it in the last year. The Massey Mine disaster has raised serious questions about the adequacy of mine safety laws, regulations, and oversight, particularly how a mine with so many documented violations could continue to operate.

Jim Young/REUTERS

The West Virginia coal mine tragedy in 2010 was not the only mine accident in the state. In 2006, 12 miners died after a blast at the Sago Mine trapped them in a toxic air shaft. Mine officials claimed that lightning caused the blast, but others blame faulty equipment. Sago Mine had a long list of safety violations and fines, with more than 270 safety citations in the two years prior to the fatal explosion.

Health and safety hazards in the workplace, often due to employers' willful violations of health and safety regulations, are among the work-related problems discussed in this chapter. Other problems we examine include unemployment, forced labor, child labor, sweatshop labor, alienation, work-life conflict, and declining labor strength and representation. We set the stage with a brief look at the global economy.

The Global Context: The New Global Economy

In recent decades, innovations in communication and information technology have spawned the emergence of a **global economy**—an interconnected network of economic activity that transcends national borders and spans the world. The globalization of economic activity means that our jobs, the products and services we buy, and our nation's economic policies and agendas are influenced by economic activities occurring around the world.

In 2007, the economic situation around the world took a downward turn, as banks faltered, credit froze, businesses closed, unemployment rates soared, and investments plummeted. This global financial crisis, which originated in the United States and spread throughout the world, illustrates the globalization of the **economic institution**. Some say that the cause of the global economic crisis was the lack of U.S. financial regulatory oversight that enabled financial institutions to engage in predatory and subprime lending during the housing boom in the early 2000s. As the housing boom turned to bust, and adjustable rate mortgages were reset to higher rates, millions of homeowners were unable to keep up with mortgage payments, and foreclosures skyrocketed. Homeowners lost their homes, renters lost their leases, and banks suffered because the foreclosed homes they now owned were often worth less than the mortgages owed on them. As banks lost revenue, they had less money to lend so credit froze, consumer spending plummeted,

global economy An interconnected network of economic activity that transcends national borders.

economic institution The structure and means by which a society produces, distributes, and consumes goods and services.

businesses went bust, and stockholders watched their investments and retirement accounts take a nosedive. The whole banking system was faltering; some got bailed out, others (e.g., Bear Stearns) went bust. All this happened in the United States, but in this new global economy, what happens in Vegas does not stay in Vegas; the crisis spread around the world. This is because those risky subprime and adjustable rate mortgages were packaged and resold as "mortgage-backed securities" to financial institutions around the world.

> In this new global economy, what happens in Vegas does not stay in Vegas.

The U.S. economic crisis also triggered a huge drop in world trade. Between 2000 and 2007, U.S. consumption accounted for more than a third of the growth in global consumption, and much of that consumption was based on borrowed money (Baily & Elliott 2009). When the United States went into recession and consumer spending declined, all the countries that depended on U.S. consumers to buy their goods and services lost a major source of revenue.

The global economic crisis reignited debate between those who view U.S. capitalism as the cause of economic problems in the world, and those who hail capitalism as "the greatest engine of economic progress and prosperity known to mankind" (Ebeling 2009). After summarizing capitalism and socialism—the two main economic systems in the world—we describe how industrialization and postindustrialization have changed the nature of work, and look at the emergence of free trade agreements and transnational corporations.

Capitalism and Socialism

Under **capitalism**, private individuals or groups invest capital (money, technology, machines) to produce goods and services to sell for a profit in a competitive market. Capitalism is characterized by economic motivation through profit, the determination of prices and wages primarily through supply and demand, and the absence of government intervention in the economy. **Socialism** is an economic system in which the state owns the means of production (factories, machinery, land, stores, offices, etc.) and oversees the distribution of goods and services. In a socialist economy, the government controls income-producing property. Theoretically, goods and services are equitably distributed according to the needs of the citizens. Whereas capitalism emphasizes individualistic pursuit of profit and individual freedom, socialism emphasizes collective well-being and social equality.

In reality, there are no pure socialist or capitalistic economies. Rather, most countries have mixed economies, incorporating elements of both capitalism and socialism. Most developed countries, for example, have both private-owned and state-owned enterprises, as well as a social welfare system. The U.S. economy is dominated by capitalism, but there are elements of socialism in our welfare system and in government subsidies and low-interest loans to industry, fiscal stimulus money, and bailout money to banks that are "too big too fail."

Critics of socialism argue that socialism creates excessive government control, reduces work incentives and technological development, and lowers the standard of living. A national survey of U.S. adults found that 39 percent say they have a positive image of socialism, compared with 54 percent who have a negative image of socialism (Newport 2012).

Capitalistic values are deeply ingrained values in U.S. society; a national poll found that 72 percent of Americans agree that "the strength of this country today is based on the success of American business" (Pew Research Center for the People & the Press 2012). But in the same poll, the majority of Americans criticized business for being too big, too profitable, and failing to serve the public interest. Another poll found that 67 percent of Americans agree that people are better off in a free-market (i.e., capitalistic) economy, but one in four Americans *disagreed* (Pew Research Center Global Attitudes Project 2012). Although a majority (61 percent) of Americans say they have a positive image of capitalism, nearly a third (31 percent) have a negative image of capitalism (Newport 2012). Critics of capitalism point to a number of social ills linked to this economic system, including high levels of inequality; economic instability; job insecurity; pollution

capitalism An economic system characterized by private ownership of the means of production and distribution of goods and services for profit in a competitive market.

socialism An economic system characterized by state ownership of the means of production and distribution of goods and services.

and depletion of natural resources; and corporate dominance of media, culture, and politics. Capitalism is also criticized as violating the principles of democracy by allowing (1) private wealth to affect access to political power (see also Chapter 6); (2) private owners of property to make decisions that affect the public (such as when the owner of a factory decides to move the factory to another country); and (3) workplace dictatorships where workers have little say in their working conditions, thus violating the democratic principle that people should participate in collective decisions that significantly affect their lives (Wright 2013). Wolff (2013a, 2013b) explains that major shareholders and boards of directors within corporations make key decisions about what products the corporation will produce, what technologies will be used, where production will occur, and how the revenues will be distributed—decisions that profoundly affect workers who have no say in these decisions:

> The most important activity of an adult's life in this country is work. It's what we do five days out of every seven. If democracy belongs anywhere, it belongs in the workplace. Yet we accept, as if it were a given, that once we cross the threshold of our store, factory, or office, we give up all democratic rights. (Wolff, quoted in Barsamian 2012, p. 12)

What Do You Think? Economist Richard Wolff points out that capitalism is an institution, like education and health care, but although it is considered appropriate to debate whether our schools and health care system are working properly and meeting our needs, it is taboo to ask whether the way we organize the production and distribution of goods and services is meeting our needs (Barsamian 2012). Do you view Americans who critically examine capitalism as being "un-American" or disloyal to the United States? Why or why not?

Industrialization, Postindustrialization, and the Changing Nature of Work

The nature of work has been shaped by the Industrial Revolution, the period between the mid-18th century and the early 19th century when the factory system was introduced in England. **Industrialization** dramatically altered the nature of work: Machines replaced hand tools, and steam, gasoline, and electric power replaced human or animal power. Industrialization also led to the development of the assembly line and an increased division of labor as goods began to be mass-produced. The development of factories contributed to the emergence of large cities. Instead of the family-centered economy characteristic of an agricultural society, people began to work outside the home for wages.

Postindustrialization refers to the shift from an industrial economy dominated by manufacturing jobs to an economy dominated by service and information technology jobs. In the global economy, jobs in the service sector outnumber jobs in both agriculture and industry. In developed countries and the European Union, the majority of jobs (73 percent) are in services, followed by jobs in industry (25 percent) and agriculture (4 percent) (ILO 2011).

What Do You Think? Virtually all of the products and services produced in today's global economy depend on the use of petroleum. Indeed, most jobs in the world today would not exist without oil. Thus, professors Charles Hall and John Day (2009) note, "We do not live in an information age, or a postindustrial age . . . but a petroleum age" (p. 237). Do you think the world will ever achieve a "post-petroleum" economy? Why or why not?

industrialization The replacement of hand tools, human labor, and animal labor with machines run by steam, gasoline, and electric power.

postindustrialization The shift from an industrial economy dominated by manufacturing jobs to an economy dominated by service-oriented, information-intensive occupations.

McFlexible

We have shift patterns to suit the lifestyles of all our Crew.

Not bad for a McJob

We know it is important to have a life outside of work, so many of our Crew work shift patterns that allow them to pursue their hobbies and interests.
They also enjoy some great discounts as well as thorough training, in addition to access to our free private healthcare scheme after they've been with us for 3 years.
www.mcdonalds.co.uk

© 2006 McDonald's. The Golden Arches and the logo design are trademarks of McDonald's Corporation and affiliates.

McDonalds/PA Photos/Landov

This poster is part of McDonald's advertising campaign to discredit the negative image implied by the term *McJob*.

McDonaldization of the Workplace

Sociologist George Ritzer (1995) coined the term **McDonaldization** to refer to the process by which the principles of the fast-food industry are being applied to more and more sectors of society, particularly the workplace. McDonaldization involves four principles:

Efficiency. Tasks are completed in the most efficient way possible by following prescribed steps in a process overseen by managers.

Calculability. Quantitative aspects of products and services (such as portion size, cost, and the time it takes to serve the product) are emphasized over quality.

Predictability. Products and services are uniform and standardized. A Big Mac in Albany is the same as a Big Mac in Tucson. Workers behave in predictable ways. For example, servers at McDonald's learn to follow a script when interacting with customers.

Control through technology. Automation and mechanization are used in the workplace to replace human labor.

What are the effects of McDonaldization on workers? In a McDonaldized workplace, employees are not permitted to use their full capabilities, be creative, or engage in genuine human interaction. Workers are not paid to think, just to follow a predetermined set of procedures. Because human interactions are unpredictable and inefficient (they waste time), "we're left with either no interaction at all, such as at ATMs, or a 'false fraternization.' Rule number 17 for Burger King workers is to smile at all times" (Ritzer, quoted by Jensen 2002, p. 41). Workers may also feel that they are merely extensions of the machines they operate. The alienation that workers feel—the powerlessness and meaninglessness that characterize a "McJob"—may lead to dissatisfaction with one's job and, more generally, with one's life.

What Do You Think? The slang term *McJob* is found in several dictionaries. For example, the *Oxford English Dictionary* defines McJob as "an unstimulating, low-paid job with few prospects, especially one created by the expansion of the service sector." *Merriam-Webster* defines McJob as "a low-paying job that requires little skill and provides little opportunity for advancement." Not surprisingly, McDonald's is opposed to such characterizations of their employment conditions and has fought back with an advertising campaign in which posters at more than 1,200 restaurants play up "the positive aspects of working for McDonald's," and include the phrase, "Not bad for a McJob." Do you think that the *Oxford* and *Merriam-Webster* dictionary definitions of McJob are accurate and fair?

McDonaldization The process by which principles of the fast-food industry (efficiency, calculability, predictability, and control through technology) are being applied to more sectors of society, particularly the workplace.

The Globalization of Trade and Free Trade Agreements

Just as industrialization and postindustrialization changed the nature of economic life, so has the globalization of trade—the expansion of trade of raw materials, manufactured goods, and agricultural products across national and hemispheric borders. The first set of global trade rules were adopted through the General Agreement on Tariffs and Trade (GATT) in 1947. In 1995, the World Trade Organization (WTO) replaced GATT as the organization overseeing the multilateral trading system.

In the 1980s and early 1990s, U.S. officials began negotiating regional free trade agreements that would open doors to U.S. goods in neighboring countries and reduce the growing U.S. trade deficit. A **free trade agreement (FTA)** is a pact between two countries or among a group of countries that make it easier to trade goods across national boundaries. Free trade agreements reduce or eliminate foreign restrictions on exports, reduce or eliminate tariffs (or taxes) on imported goods, and prevent technology from being copied and used by competitors through protection of "intellectual property rights." Treaties such as the Canada–U.S. Free Trade Agreement, the North American Free Trade Agreement (NAFTA), the Free Trade Area of the Americas (FTAA), and the Central America Free Trade Agreement (CAFTA) were designed to accomplish these trade goals.

Although free trade agreements have expanded trading opportunities, benefiting large export manufacturing and service industries in the global north, they have also undermined the ability of national, state, and local governments to implement environmental and food or product safety policies (see also Chapter 6) (Faux 2008; Schaeffer 2003; Scott & Ratner 2005). Free trade agreements have also hurt both U.S. and foreign workers. Before the U.S.–Korea free trade agreement took effect in 2012, the United States had a trade deficit with Korea, meaning that the United States imported more goods from Korea than it exported to Korea. The U.S.–Korea FTA was implemented with the promise that it would reduce the trade deficit, and that it would lead to increased U.S. exports to Korea, which means more U.S. jobs. But in the year after the U.S.–Korea FTA took effect, the U.S. trade deficit with Korea actually *increased,* which cost U.S. jobs (Public Citizen 2013). The North American Free Trade Agreement (NAFTA) allowed U.S. corn growers to sell their corn in Mexico, but Mexican corn farmers could not complete with the cheap price of U.S. corn, which put many Mexican corn growers out of business. Free trade agreements have also made it easier for U.S. companies to move jobs "offshore," usually to countries where wages are low and there are few environmental, health, or safety regulations with which to comply. Although offshoring jobs increases profits to corporations, it also takes jobs away from U.S. workers.

Transnational Corporations

Although free trade agreements have increased business competition around the world, resulting in lower prices for consumers for some goods, they have also opened markets to monopolies (and higher prices) because they have facilitated the development of large-scale transnational corporations. **Transnational corporations**, also known as *multinational corporations*, are corporations that have their home base in one country and branches, or affiliates, in other countries. Among the world's 50 largest economies in 2011, 8 were multinational corporations (White 2012).

Transnational corporations provide jobs for U.S. managers, secure profits for U.S. investors, and help the United States compete in the global economy. Transnational corporations benefit from increased access to raw materials, cheap foreign labor, and the avoidance of government regulations. They can also avoid or reduce tax liabilities by moving their headquarters to a "tax haven." But the savings that transnational companies reap from cheap labor and reduced taxes are not passed on to consumers. "Corporations do not outsource to far-off regions so that U.S. consumers can save money. They outsource in order to increase their margin of profit" (Parenti 2007). For example, shoes made by Indonesian children working 12-hour days for 13 cents an hour cost only $2.60 but are still sold for $100 or more in the United States.

Transnational corporations contribute to the trade deficit in that more goods are produced and exported from outside the United States than from within. Transnational corporations also contribute to the budget deficit, because the United States does not get tax income from U.S. corporations abroad, yet transnational corporations pressure the government to protect their foreign interests; as a result, military spending increases. Transnational corporations contribute to U.S. unemployment by letting workers in other countries perform labor that U.S. employees could perform. Finally, transnational corporations are implicated in an array of other social problems, such as poverty resulting from

free trade agreement (FTA)
A pact between two or more countries that makes it easier to trade goods across national boundaries by reducing or eliminating restrictions on exports and tariffs (or taxes) on imported goods and protecting intellectual property rights.

transnational corporations
Also known as multinational corporations, corporations that have their home base in one country and branches, or affiliates, in other countries.

Cities across the United States have cancelled the traditional fireworks show on Independence Day to save money.

fewer jobs, urban decline resulting from factories moving away, and racial and ethnic tensions resulting from competition for jobs.

Sociological Theories of Work and the Economy

In sociology, structural functionalism, conflict theory, and symbolic interactionism serve as theoretical lenses through which we may better understand work and economic issues and activities.

Structural-Functionalist Perspective

According to the structural-functionalist perspective, the economic institution is one of the most important of all social institutions. By providing the basic necessities common to all human societies, including food, clothing, and shelter, the economic institution contributes to social stability. After the basic survival needs of a society are met, surplus materials and wealth may be allocated to other social uses, such as maintaining military protection from enemies, supporting political and religious leaders, providing formal education, supporting an expanding population, and providing entertainment and recreational activities. Societal development is dependent on an economic surplus in a society (Lenski & Lenski 1987).

The economic institution can also be dysfunctional when it fails to provide members with the goods and services they need, when the distribution of goods and services is grossly unequal, and when the production, distribution, and consumption of goods and services depletes and pollutes the environment.

The structural-functionalist perspective is also concerned with how changes in one aspect of society affect other aspects. For example, when unemployment rates rise, college enrollments go up, crime increases, and tax revenues decrease (unemployed people pay less in income tax and sales tax), which hurts the government's ability to pay for services such as education, garbage pickup, police and fire services, and road repairs.

What Do You Think? In recent years, many cities across the country have cancelled their annual Fourth of July fireworks to save money. Mayor Bill Cervenik, of Euclid, Ohio, said, "It came down to this: Did we want to spend $150,000 on something that would be over in a few hours? Or did we want to use that money to keep city workers employed?" In the Los Angeles suburb of Montebello, the city council voted to use its $39,000 fireworks budget on donations to local food banks. Mayor Rosemarie Vasquez explained, "We figured that, instead of burning the money in the air, why not give it to people who need it?" (Huffstutter 2009). Protestors against the cancellation of fireworks shows argued that they are an important American tradition. If you were on the city council, and the issue of whether or not to cancel the Fourth of July fireworks was being considered, what would your vote be, and why?

Conflict Perspective

corporatocracy A system of government that serves the interests of corporations and that involves ties between government and business.

According to the conflict perspective, the ruling class controls the economic system for its own benefit and exploits and oppresses the working masses. The conflict perspective is critical of ways that the government caters to the interests of big business at the expense of workers, consumers, and the public interest. This system of government that serves the interests of corporations—known as **corporatocracy**—involves ties between government and business. For example, in Chapter 2, we discussed how the pharmaceutical and health insurance industries influence politicians on matters related to health care.

Corporations influence government through lobbying, and donations to politicians, campaigns, and Super PACs. As we discussed in Chapter 6, in 2010, corporations gained power to influence political outcomes when the Supreme Court ruled (5 to 4) in *Citizens United v. Federal Election Commission* that corporations, as well as unions, have a First Amendment right to spend unlimited amounts of money to support or oppose candidates for elected office. That decision also freed corporations and unions to "educate" workers about elections through voter guides, get-out-the-vote activities, and other tools. So a company can legally inform its workers that if they vote for candidate A, the company will suffer and jobs will be lost. Is this education? Or is it the company intimidating or coercing the workers to vote a certain way? The line is murky (Carney 2012).

The pervasive influence of corporate power in government exists worldwide. The policies of the International Monetary Fund (IMF) and the World Bank pressure developing countries to open their economies to foreign corporations, promoting export production at the expense of local consumption, encouraging the exploitation of labor as a means of attracting foreign investment, and hastening the degradation of natural resources as countries sell their forests and minerals to earn money to pay back loans. In his book *Confessions of an Economic Hit Man,* John Perkins (2004) described his prior job as an "economic hit man"—a highly paid professional who would convince leaders of poor countries to accept huge loans (primarily from the World Bank) that were much bigger than the country could possibly repay. The loans would be used to help develop the country by paying for needed infrastructure, such as roads, electrical plants, airports, shipping ports, and industrial plants. One of the conditions of the loan was that the borrowing country had to give 90 percent of the loan back to U.S. companies (such as Halliburton or Bechtel) to build the infrastructure. The result: The wealthiest families in the country benefit from additional infrastructure and the poor masses are stuck with a debt they cannot repay. The United States uses the debt as leverage to ask for "favors," such as land for a military base or access to natural resources such as oil. According to Perkins, large corporations want "control over the entire world and its resources, along with a military that enforces that control" (quoted by MacEnulty 2005, p. 10).

Symbolic Interactionist Perspective

According to symbolic interactionism, the work role is a central part of a person's self-concept and social identity. When making a new social acquaintance, one of the first questions we usually ask is, "What do you do?" The answer largely defines for us who that person is. An individual's occupation is one of the person's most important statuses; for many, it represents a "master status," that is, the most significant status in a person's social identity. This chapter's *Social Problems Research Up Close* feature describes a study that looks at how job loss of white-collar professionals in midlife affects their self-concepts and attitudes about work and unemployment.

Symbolic interactionism emphasizes the fact that attitudes and behavior are influenced by interaction with others. The applications of symbolic interactionism in the workplace are numerous: Employers and managers use interpersonal interaction techniques to elicit the attitudes and behaviors they want from their employees; union organizers use interpersonal interaction techniques to persuade workers to unionize. And, as noted in the *Social Problems Research Up Close* feature, parents teach their young adult children important lessons about work and unemployment through interaction with them.

Symbolic interaction also focuses on the importance of labels. Although the concepts of capitalism and free enterprise are theoretically similar, Americans tend to view the term "capitalism" more negatively than "free enterprise." Thus, "politicians seeking the most positive overall reaction from voters should choose to use the term 'free enterprise' rather than 'capitalism' in describing America's prevailing economic system" (Newport 2012, n.p.).

In recent years, job losses have increasingly involved middle-class professionals. Research presented here investigates how job loss affected the self-concepts and views about employment among a sample of unemployed professionals in midlife (Mendenhall et al. 2008).

Sample and Methods

The sample consisted of 77 men and women who were recruited through two Chicago-area networking groups that help unemployed managers and executives find employment, and also through announcements at local churches and posted flyers at cafés and public libraries. Participants had to meet four eligibility criteria: They had to (1) have been unemployed for at least three months during the past year; (2) have been married at the time of their job loss; (3) have children between the ages of 12 and 18 living at home; and (4) live in the Chicago area. Most participants were male (83 percent) and white (80 percent). Respondents had been unemployed for an average of 15 months at the time of their first interview. More than half had earned annual salaries of $100,000 or more; no one earned less than $50,000, and most had jobs with generous benefits.

The researchers interviewed participants at libraries, cafés, or on the University of Chicago campus. Interview topics included the job loss event, how the job loss affected family relationships, and educational plans for their children. Participants also completed a survey about their job loss experience, family economic circumstances, family relationships, and their own health and well-being. About a year later, participants were interviewed again, and completed another questionnaire.

Selected Findings

This study revealed some interesting patterns in how job loss among professionals in midlife (1) affected their self-concept, (2) influenced their job-seeking strategies, and (3) shaped the messages about employment they conveyed to their children. The participants viewed their job termination as evidence of a lack of employer loyalty and a change from a lifetime employment contract to one in which even high-level employees can be terminated without warning—a shift in which participants came to view themselves as free agents who effectively "rent" their services to employers. As free agents, there is no expectation of permanent employment or loyalty in the employee–employer relationship.

More than half of the participants 50 years and older said they perceived age discrimination in the job market, in that employers signaled that they were looking for entry-level employees, which is code for "young." In response, participants often de-emphasized their age and experience by omitting graduation dates and some of their work history on their resumes—a phenomenon known as "deprofessionalization." A former CEO explained that, in the first several months of job searching, he listed the date of his college graduation on his resume and got no responses. Then he deleted the date, and got a half dozen responses.

Many participants viewed their job loss experience as providing an opportunity to teach real-life lessons to their children about the world of work. A 50-year-old project manager said:

> I think [that my son getting a good glimpse into the reality of life is] a positive because . . . if he goes out there with rose-tinted glasses, he's going to get

smacked upside the head real hard some day. (p. 203)

One 47-year-old information technology consultant used his job loss experience to teach his children to prepare for a job market in which neither employee nor employer expects a lifetime employment contract. He told his children:

> For most jobs now, you need to view [it] like the movie industry where . . . you're rented[,] . . . you're contracted to do that movie, whether you're the lights or the cameraman or whatever, and then you're out of work again. And that's more the way most jobs are now where you should view a job as just your job until you're out of work again. (p. 204)

Respondents advised their children to take specific steps to prevent being overly dependent on employers. Some advised their children to own their own businesses; others urged their children to develop skills that could be "transferable" to a new employer. Others encouraged their children to choose careers where their potential client base was diversified:

> I've started talking to them about how . . . they should start thinking about [whether] they want to work for big companies or . . . go into a field where their client base is very much diversified, which is lawyers, doctors, CPAs, therapists. . . . Because then if someone fires you, you don't care because you have another 100 [clients] whereas when you work for a company, and . . . your boss . . . fires you . . . you're out of a job. (p. 204)

Source: Mendenhall et al. 2008.

Problems of Work and Unemployment

In this section, we examine unemployment and other problems associated with work. Poverty, minimum wage and living wage issues, workplace discrimination, and retirement and employment concerns of older adults are discussed in other chapters. Here, we discuss problems concerning unemployment and underemployment, child labor, forced labor, sweatshop labor, health and safety hazards in the workplace, alienation, work-life conflict, and labor unions and the struggle for workers' rights.

Unemployment and Underemployment

Measures of **unemployment** consider individuals to be unemployed if they are currently without employment, are actively seeking employment, and are available for employment. In 2012, 197 million people worldwide—about 6 percent of the global labor force—were unemployed. The unemployment rate for youths (ages 15 to 24) was double the overall rate—12.6 percent (ILO 2013).

In the United States, the unemployment rate dipped to a 31-year low of 4 percent in 2000, but the economic recession that began in 2007 pushed the unemployment rate to 10 percent in the last quarter of 2009, as companies went out of business and plants closed. A **recession** refers to a significant decline in economic activity spread across the economy and lasting for at least six months. In communities hardest hit by the recession, unemployment rates rose to over 20 percent. Rates of unemployment are higher among racial and ethnic minorities and among those with lower levels of education (see Chapters 8 and 9).

The **long-term unemployment rate** refers to the share of the unemployed who have been out of work for 27 weeks or more. In June 2013, more than a third (37 percent) of unemployed Americans had been jobless for 27 weeks or more (Bureau of Labor Statistics 2013a).

Unemployment figures do not include (1) "discouraged" workers who have given up on finding a job and are no longer looking for employment; (2) people marginally attached to the labor force who currently are neither working nor looking for work but indicate that they want and are available for a job and have looked for work sometime in the past 12 months; and (3) those who want to work full time but who settle for a part-time job ("involuntary" part-timers). These individuals are counted in the **underemployment** rate, which is always higher than the unemployment rate. This means that official unemployment rates—frequently reported in the media—undercount those whose employment needs are not being met. For example, although the official unemployment rate in June 2013 was 7.6 percent, the underemployment rate was 14.3 percent (Bureau of Labor Statistics 2013a).

Causes of Unemployment. A primary cause of unemployment is lack of available jobs. In December 2000, jobs were plentiful: For every job opening, the ratio of job seekers to job openings was 1 to 1, meaning there was one available job for every person seeking employment. But between 2008 and 2013, the ratio of job seekers to job openings was 3 to 1 or higher. In 2009, for every job opening, there were *more than six unemployed U.S. workers* (Shierholz 2011, 2013).

Another cause of U.S. unemployment is **job exportation**, also referred to as **offshoring**—the relocation of jobs to other countries. Jobs most commonly offshored have been in manufacturing, but offshoring of service jobs, including information technology, human resources, finance, purchasing, and legal services is expected to increase in the next few years (Davidson 2012). Exporting jobs enables corporations to maximize their profits by reducing their costs of raw materials and labor. By offshoring service jobs, employers cut labor costs by about 75 percent (Davidson 2012). **Outsourcing**, which involves a business subcontracting with a third party to provide business services, saves companies money as they pay lower salaries and no benefits to those who provide outsourced services. Many commonly outsourced jobs, including accounting, web development, information technology, telemarketing, and customer support, are outsourced to non-U.S. workers.

Automation, or the replacement of human labor with machinery and equipment, also contributes to unemployment, although automation creates new jobs too. ATMs have reduced the need for bank tellers, but at the same time created jobs for workers who produce and service ATMs. A technology called **3-D printing**—a $2.2 billion global market in 2012—is likely to have a profound effect on manufacturing jobs. The revolutionary 3-D printing involves downloading a digital file containing a design for a product. The printer reads the file and then shoots out the product (made of specialized plastic or other raw materials) through a heated nozzle. For example, General Electric uses 3-D printing

unemployment To be currently without employment, actively seeking employment, and available for employment, according to U.S. measures of unemployment.

recession A significant decline in economic activity spread across the economy and lasting for at least six months.

long-term unemployment rate The share of the unemployed who have been out of work for 27 weeks or more.

underemployment Unemployed workers as well as (1) those working part-time but who wish to work full-time, (2) those who want to work but have been discouraged from searching by their lack of success, and (3) others who are neither working nor seeking work but who want and are available to work and have looked for employment in the last year. Also refers to the employment of workers with high skills and/or educational attainment working in low-skill or low-wage jobs.

job exportation The relocation of jobs to other countries where products can be produced more cheaply.

offshoring The relocation of jobs to other countries.

outsourcing A practice in which a business subcontracts with a third party to provide business services.

automation The replacement of human labor with machinery and equipment.

3-D printing A revolutionary manufacturing technology that involves downloading a digital file containing a design for a product. A printer reads the file and then shoots out the product (made of specialized plastic or other raw materials) through a heated nozzle.

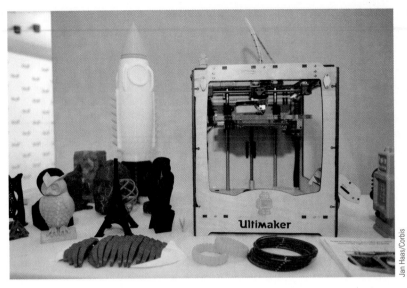

3-D printers, such as the one depicted here, are expected to have a major impact on manufacturing jobs.

to manufacture parts for jet engines; Nike uses it to make soccer cleats. A *Bloomberg View* report concluded that,

> 3-D printing seems likely to throw a lot of people out of work in the medium term, especially in industries that depend on assembly line labor. Eventually, as with most technological breakthroughs, it will probably create new jobs in new industries. But that transition period will be hazardous, and displaced workers will need help to navigate it. (The Editors 2013, n.p.)

Another cause of unemployment is increased global and domestic competition. Mass layoffs in the U.S. automobile industry have occurred, in part, due to competition from makers of foreign cars who, unlike U.S. automakers, do not have the burden of providing health insurance for their employees. Finally, unemployment results from mass layoffs that occur when plants close and companies downsize or go out of business. In 2009, Circuit City went out of business after revenues steadily declined due, in part, to competition from Best Buy, Wal-Mart, and Amazon. Circuit City was also hurt by the recession that started in late 2007, as consumers cut back on spending.

Effects of Unemployment on Individuals, Families, and Societies. One study found that poor-quality jobs—those with high demands, low control over decision making, high job insecurity, and low—pay had more negative effects on mental health than having no job at all (Butterworth et al. 2011). Nevertheless, unemployment has been linked to depression, low-self esteem, and increased mortality rates (Turner & Irons 2009). One study found that, among workers with no preexisting health conditions, losing a job due to a business closure increased the risk for a new health problem by 85 percent, with the most common health problems including hypertension, heart disease, and arthritis (Strully 2009). Unemployment is also a risk factor for homelessness, substance abuse, and crime, as some unemployed individuals turn to illegitimate, criminal sources of income, such as theft, drug dealing, and prostitution.

One study found that poor-quality jobs—those with high demands, low control over decision making, high job insecurity, and low pay—had more negative effects on mental health than having no job at all.

Long-term unemployment can have lasting effects, such as increased debt, diminished retirement and savings accounts (which are depleted to meet living expenses), home foreclosure, and/or relocation from secure housing and communities to unfamiliar places to find a job. But, even when individuals who are laid off find another job in one or two weeks, they still suffer damage to their self-esteem from having been told they are no longer wanted or needed at their workplace. And being fired affects worker trust and loyalty in future jobs. Employees who are not fired during a mass layoff are also affected, as they worry that "their job could be next" (Uchitelle 2006).

In families, unemployment is also a risk factor for child and spousal abuse and marital instability. When an adult is unemployed, other family members are often compelled to work more hours to keep the family afloat. And unemployed noncustodial parents, usually fathers, fall behind on their child support payments.

Plant closings and large-scale layoffs affect communities by lowering property values and depressing community living standards. High numbers of unemployed adults create a drain on societies that provide support to those without jobs. Job

Difficult economic times make people reconsider what is important for them to have in their daily lives, and what they can live without or change to save money. How do your spending habits change in hard economic times? For each of the following spending categories, place a check mark next to those items that describe changes in your spending habits in response to tough economic times.

1. Buy more generic brands. _____
2. Take lunch from home instead of buying lunch out. _____
3. Go to hairdresser/barber less often. _____
4. Stopped purchasing bottled water; switched to refillable water bottle. _____
5. Cancelled one or more magazine subscriptions. _____
6. Cancelled or cut back on cable service. _____
7. Stopped buying coffee in the morning. _____
8. Cut down on dry cleaning. _____
9. Changed or canceled cell phone service. _____
10. Began carpooling or using mass transit. _____

Comparison data: Table 1 presents the percentages of responses to each of the 10 previous items from a 2012 online Harris Poll of 2,383 U.S. adults. The most common changes in spending behavior include purchasing more generic brands and "brown bagging" lunch (taking lunch from home instead of buying lunch out).

TABLE 1	Cutting Back on Spending over Past Six Months, in Percentages		
1.	57	6.	20
2.	41	7.	18
3.	38	8.	18
4.	33	9.	14
5.	21	10.	14

Source: Harris Poll 2012.

displacement of the nonelderly is linked to lower rates of social participation in church groups, charitable organizations, youth and community groups, and civic and neighborhood groups (Brand & Burgard 2008). Globally, the high numbers of young adults without jobs create a risk for crime, violence, and political conflict (United Nations 2005).

Unemployment and underemployment create a vicious cycle: The unemployed and underemployed (as well as those who fear job loss) cut back on spending, which hurts businesses that then must cut jobs to stay afloat. During times of economic stress or uncertainty, have you cut back on everyday spending? See this chapter's *Self and Society* feature.

Employment Concerns of Recent College Grads

Most college students seek a college degree to improve their job opportunities. But having a college degree is no guarantee of employment in a struggling economy. The unemployment rate for young college graduates (without an advanced degree and not currently enrolled in further education) was 5.7 percent in 2007, rose to 10.4 percent in 2010, and was 8.8 percent in March 2013 (Shierholz et al. 2013). Some new college graduates are glad that jobs are scarce. After graduation, some new grads are taking time off to relax and to travel instead of seeking employment (Weiss 2009). Other college grads question whether going to college was worth the expense and the effort. In 2012, just over half of employed college graduates under age 25 were working in jobs that did not require a college degree (Shierholz et al. 2013).

> Most college students seek a college degree to improve their job opportunities. But having a college degree is no guarantee of employment in a struggling economy.

College grads who find employment are not receiving the same wages or benefits as they used to. Between 2000 and 2012, the inflation-adjusted wages of young college grads declined 8.5 percent. And from 2000 to 2011, the share of young college grads who received pension benefits from their employer dropped from 41.5 percent to 27.2 percent (Shierholz et al. 2013). For young college graduates entering the labor force during a sluggish economy, the effects of reduced earnings and more spells of unemployment can last over the next 10 to 15 years (Shierholz et al. 2013). Due to the high unemployment rate, low wages of many jobs, high cost of living, and student loans and other

Forced prison labor is a type of forced labor that is controlled by the state. Forced prison labor is particularly widespread in China.

college debts, many recent college graduates end up living with their parents or other family, instead of on their own. In 2011, 45 percent of recent college grads (ages 24 or younger) were living with their families (parents or aunt/uncle)—up from 31 percent in 2001 (Weissmann 2013).

Forced Labor

Forced labor, also known as *slavery*, refers to any work that is performed under the threat of punishment and is undertaken involuntarily. Forced labor exists all over the world but is most prevalent in South Asia. Most forced laborers work in agriculture, mining, prostitution, and factories. There are more slaves in the world today—27 million—than at any other time in history (Hardy 2013). The median cost of a slave is $140; trafficked sex slaves cost $1,910 (Hardy 2013).

The form of slavery most people are familiar with is **chattel slavery**, in which slaves are considered property that can be bought and sold. Although chattel slavery still exists in some areas, most forced laborers today are not "owned" but are rather controlled by violence, the threat of violence, and/or debt. The most common form of forced labor today is called *bonded labor*. Bonded laborers are poor individuals who take out a loan simply to survive or to pay for a wedding, funeral, medicines, fertilizer, or other necessities. Debtors must work for the creditor to pay back the loan, but they are often unable to repay it. Creditors can keep debtors in bondage indefinitely by charging the debtors illegal fines (for workplace "violations" or for poorly performed work) or charge laborers for food, tools, and transportation to the work site while keeping wages too low for the debt to ever be repaid (Miers 2003).

Another form of forced labor is sex slavery. In South Asia, where sex slavery is most common, girls are either forced into prostitution by their own husbands, fathers, and brothers to earn money to pay family debts, or they are lured by offers of good jobs and then are forced to work in brothels under the threat of violence.

In the United States, each year, 14,000 to 17,000 people are trafficked into the country and forced into slavery, most commonly in domestic work, farm labor, and the sex industry (Skinner 2008). Migrant workers are tricked into working for little or no pay as a means of repaying debts from their transport across the U.S. border. Migrant workers are particularly vulnerable because, if they try to escape and report their abuse, they risk deportation. Traffickers posing as employment agents lure women into the United States with the promise of good jobs and education but then place them in "jobs" where they are forced to do domestic or sex work.

forced labor Also known as slavery, any work that is performed under the threat of punishment and is undertaken involuntarily.

chattel slavery A form of slavery in which slaves are considered property that can be bought and sold.

What Do You Think? In both state and federal prisons in the United States, inmates work for little to no pay, often under conditions that expose the inmates to toxic chemicals without proper safety equipment (Flounders 2013). In some states, prisoners who refuse to work are sent to disciplinary housing and lose canteen privileges as well as "good time" credit that could lower their prison sentence. Some prisoners claim that they were beaten for refusing to work. In U.S. federal prisons, inmates earn 23 cents an hour producing night-vision goggles, camouflage uniforms, body armor, and high-tech components for military aircraft and helicopters, guns, and land mine sweepers. Large corporations purchase these fruits of prison labor for a low price, and then make a huge profit from contracts with the U.S. military. Should large corporations be allowed to profit from prison labor? Should prisoners have the same workplace protections as other workers? Why or why not?

Sweatshop Labor

Millions of people worldwide work in **sweatshops**—work environments that are characterized by less than minimum wage pay, excessively long hours of work (often without overtime pay), unsafe or inhumane working conditions, abusive treatment of workers by employers, and/or the lack of worker organizations aimed at negotiating better working conditions. Sweatshop labor conditions occur in a wide variety of industries, including garment production, manufacturing, mining, and agriculture.

More than 97 percent of all clothing purchased in the United States is imported, often made under sweatshop conditions (Institute for Global Labour & Human Rights 2011). Garment workers at a factory in Ha-Meem, Bangladesh, work 12- to 14-hour shifts, 7 days a week, with one day off a month. Young women making garments for Gap, J.C. Penney, Phillips-Van Heusen (PVH), Target, and Abercrombie & Fitch make 20 to 26 cents an hour (Institute for Global Labour & Human Rights 2011).

Many products in the U.S. consumer market are made under sweatshop conditions. An investigative report on working conditions in five Chinese factories that produce products for Disney, Wal-Mart, Kmart, Mattel, and McDonald's revealed sweatshop conditions that violate Chinese labor laws (Students and Scholars Against Corporate Misbehavior 2005). Workers are forced to work grueling 12- to 15-hour days, earning just 33 cents to 41 cents an hour. Workers are housed in overcrowded dorm rooms and are fed horrible food at the factory canteen. They are charged for the housing and food provided at the factory (even if they live and eat elsewhere), which often costs them one-fifth to one-third of their monthly wages. Some factories have no fans and become oppressively hot. Workers often faint from exhaustion and the unbearably stifling heat. Some workers are exposed to strong-smelling gases from working with glue, with no protective masks or ventilation system. Crushed fingers and other injuries are common in some factory departments. Workers have no health insurance, no pension, and no right to freedom of association or to organize.

Sweatshop labor commonly occurs in the garment industry.

Sweatshop Labor in the United States. Sweatshop conditions in overseas industries have been widely publicized. However, many Americans do not realize the extent to which sweatshops exist in the United States. For example, in 2012, the Department of Labor found widespread "sweatshop-like" labor violations in the Los Angeles garment industry (Miles 2012). Ten garment contractors, producing clothing for more than 20 retailers, were found to be paying less than minimum wage, not paying overtime, and/or falsifying time cards or failing to maintain records of employees' hours. Most garment workers in the United States are immigrant women who typically work 60 to 80 hours a week, often earning less than minimum wage with no overtime, and many face verbal and physical abuse.

Immigrant farmworkers, who process 85 percent of the fruits and vegetables grown in the United States, also work under sweatshop conditions. Many live in substandard and crowded housing provided by their employer and lack access to safe drinking water as well as bathing and sanitary toilet facilities. Farmworkers commonly suffer from heat exhaustion, back and muscle strains, injuries resulting from the use of sharp and heavy farm equipment, and illness resulting from pesticide exposure (Austin 2002). Working 12-hour days under hazardous conditions, farmworkers have the lowest annual family incomes of any U.S. wage and salary workers, and more than 60 percent of them live in poverty (Thompson 2002). Problems associated with immigrant labor are discussed further in Chapter 9.

sweatshops Work environments that are characterized by less than minimum wage pay, excessively long hours of work (often without overtime pay), unsafe or inhumane working conditions, abusive treatment of workers by employers, and/or the lack of worker organizations aimed to negotiate better working conditions.

In most jobs, U.S. law requires children to be at least 16 years old to work, with a few exceptions, such as that 14- and 15-year-olds can work as cashiers, grocery baggers, and car washers. But in agricultural jobs, U.S. law allows children of any age to work on small farms as long as they have a parent's permission. Children ages 12 and older may be hired on any farm with their parent's consent, or if they work on the same farm with their parents. U.S. law allows children at age 14 to work on any farm, even without parental permission.

▼ Hundreds of thousands of U.S. child workers labor on commercial farms (Human Rights Watch 2010). Child farmworkers in the United States are typically of Hispanic origin; many are U.S. citizens. Girls who work on U.S. farms are sometimes victims of sexual harassment.

Melissa Farlow/National Geographic Creative

◄ The Children's Act for Responsible Employment (CARE Act) is a bill that proposes to raise the minimum age of farmworkers from 12 to 14 (except for children working on their parents' farm). The bill also contains provisions to protect young farmworkers from hazardous work and exposure to pesticides. As of this writing, the CARE Act has not passed.

Martin Rogers/Getty Images

▼ Child laborers working on a tobacco farm are especially vulnerable to "green tobacco sickness," a poisoning that occurs when workers absorb tobacco through the skin when they come in contact with the leaves. Symptoms include nausea, vomiting, weakness, dizziness, abdominal pain, diarrhea, and shortness of breath.

LOWELL GEORGIA/National Geographic Creative

◄ Farmwork is considered the most dangerous form of child labor in the United States. Child farmworkers are exposed to chemicals, sun, and temperature extremes; they work with sharp tools and heavy machinery, climb tall ladders, and carry heavy buckets and sacks. They frequently work 12-hour days without adequate access to drinking water, toilets, or hand-washing facilities. In nonagricultural jobs, U.S. law requires workers engaged in hazardous work to be at least 18 years old. But in agricultural jobs, children are permitted to do work that the U.S. Department of Labor deems "particularly hazardous" at age 16, and at any age on farms owned or operated by their parents.

© Stacy Barnett/Shutterstock.com

▼ Children are more susceptible to heat stroke than adults. Working long hours in the hot sun places child farmworkers at the risk of heat stroke and dehydration.

Jim Sugar/CORBIS

Child Labor

Child labor involves a child performing work that is hazardous, that interferes with a child's education, or that harms a child's health or physical, mental, social, or moral development. Even though virtually every country in the world has laws that prohibit or limit the extent to which children can be employed, child labor persists throughout the world. More than half of the 215 million school-age children who are engaged in child labor are exposed to the worst forms of child labor: hazardous environments, forced labor, drug trafficking and prostitution, and armed conflict (ILO 2013).

Child labor is involved in many of the products we buy, wear, use, and eat. Child laborers work in factories, workshops, construction sites, mines, quarries, fields, and on fishing boats. Child laborers make bricks, shoes, soccer balls, fireworks and matches, furniture, toys, rugs, and clothing. They work in the manufacturing of brass, leather goods, and glass. They tend livestock and pick crops. Tens of thousands of children in at least 24 countries and territories are recruited by armed forces where children are used in combat or for sexual exploitation (UNICEF 2009).

Child laborers work long hours with few (or no) breaks or days off, often in unsafe conditions where they are exposed to toxic chemicals and/or excessive heat, and they endure beatings and other forms of mistreatment from their employers, all for as little as a dollar a day. A former textile worker in Bangladesh described conditions at a company called Harvest Rich, where clothing is sewn for U.S. firms including Wal-Mart and J.C. Penney. She testified that hundreds of children, some as young as 11 years old, were illegally working at Harvest Rich, sometimes for up to 20 hours a day: "Before clothing shipments had to leave for the United States, there are often mandatory 19- to 20-hour shifts from 8:00 a.m. to 3:00 or 4:00 a.m. . . . The workers would sleep on the factory floor for a few hours before getting up for their next shift in the morning. If they did anything wrong, they were beaten every day." Workers had two days off a month and were paid $3.20 a week (Tate 2007). At another garment factory that made fleece jackets for Wal-Mart, 14- or 15-year-old kids worked 18- or 20-hour shifts, from 8:00 a.m. to midnight or 4:00 a.m., seven days a week. When they passed out, rulers struck them to wake them up. Some of the girls were raped by management (Tate 2007).

Child labor also exists in the United States in restaurants, grocery stores, meatpacking plants, garment factories, and agriculture. Despite federal prohibitions, U.S. youth employed in service and retail jobs are exposed to harmful conditions and dangerous equipment (e.g., paper balers, box crushers, and dough mixers) (Runyan et al. 2007). One of the most dangerous forms of child labor in the United States is agricultural work. This chapter's *Photo Essay* looks at child labor in U.S. agriculture.

Health and Safety in the U.S. Workplace

Although many workplaces are safer today than in generations past, fatal and disabling occupational injuries and illnesses still occur in troubling numbers. In 2011, 4,693 U.S. workers—most of whom were men—died of fatal work-related injuries (Bureau of Labor Statistics 2013b). The most common type of job-related fatality involves transportation accidents (see Figure 7.1). Although the highest number of fatal injuries occurred in construction, the highest rates (per 100,000 workers) occurred in agriculture/forestry/fishing/hunting, transportation and warehousing, and mining.

Nonfatal occupational injuries and illnesses in U.S. workplaces are not uncommon—nearly 4 million cases were reported in 2011—but due to underreporting, the actual number could be more than 11 million (AFL-CIO 2013). Common workplace injuries include sprains, strains, and cuts. Workers who do repeated motions such as typing or assembly line work are prone to repetitive strain injuries (see this chapter's *The Human Side* feature).

The incidence of illnesses resulting from hazardous working conditions is probably much higher than the reported statistics show, because long-term latent illnesses caused by, for example, exposure to carcinogens often are difficult to relate to the workplace and are not adequately recognized and reported. In addition, employees don't always report workplace injuries or illnesses for fear of losing their jobs, or in the case of undocumented immigrants, fear of being deported. And companies don't always maintain accurate

child labor Involves a child performing work that is hazardous, that interferes with a child's education, or that harms a child's health or physical, mental, social, or moral development.

records of workplace injuries and illnesses, to avoid scrutiny and fines for possible violations of health and safety regulations.

In many cases, workplace injuries, illnesses, and deaths are due to employers' willful violations of health and safety regulations. The Occupational Safety and Health Administration (OSHA), created in 1970, develops, monitors, and enforces health and safety regulations in the workplace. But inadequate funding leaves OSHA unable to hire enough workplace inspectors to do the job effectively. The International Labour Organization (ILO) recommends one inspector per 10,000 workers. To meet the ILO's benchmark, we would need 12,941 inspectors, but instead, OSHA had just 1,938 inspectors (in 2012) (AFL-CIO 2013). At current staffing levels, it would take OSHA inspectors 113 years to inspect the nation's 8 million workplaces. Nevertheless, in 2012, federal OSHA inspectors documented more than 78,000 violations of health and safety regulations in the workplace, most of which were classified as "serious" (AFL-CIO 2013).

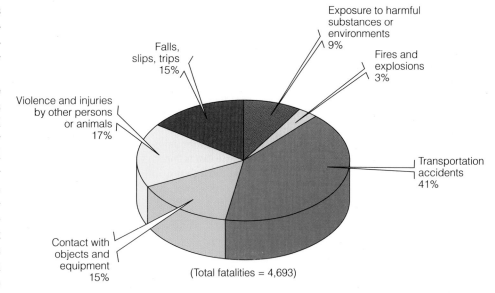

Falls, slips, trips 15%

Exposure to harmful substances or environments 9%

Fires and explosions 3%

Violence and injuries by other persons or animals 17%

Transportation accidents 41%

Contact with objects and equipment 15%

(Total fatalities = 4,693)

Figure 7.1 Causes of Workplace Fatalities
Source: Bureau of Labor Statistics 2013.

What Do You Think? Suppose that a corporation is guilty of a serious violation of health and safety laws, in which "serious violation" is defined as one that poses a substantial probability of death or serious physical harm to workers. What penalty do you think that corporations should pay for such a violation? Serious violations of workplace health and safety laws in 2012 carried an average federal penalty of $1,052 or a state penalty of $974 (AFL-CIO 2013). Even when violations result in worker fatalities, the average federal penalty in 2012 was only $6,625, and average state penalty was only $4,900. Under federal law, willful violation that results in a worker's death is considered a misdemeanor and carries a maximum prison sentence of only six months. Contrast this with the crime of harassing a wild burro on federal lands, which is punishable by one year in prison (Barstow & Bergman 2003). Why do you think penalties for violating workplace health and safety laws are so weak?

Job Stress. Another work-related health problem is job stress. A Gallup Poll found that one-third of U.S. workers is "totally dissatisfied" with the amount of stress in their jobs, making job stress workers' most common complaint (dissatisfaction with the amount of money they make is workers' second most common complaint) (Saad 2012). Workers can be stressed by the number of hours they work, as the 40-hour workweek has become, for many workers, a 60-hour (and more) workweek. Workers may also be stressed from working in a **toxic workplace**, which is a work environment in which employees are subjected to co-workers and/or bosses who engage in a variety of negative, stress-inducing behaviors such as intimidation and workplace bullying, gossiping, and "backstabbing."

Prolonged job stress, also known as **job burnout**, can cause or contribute to physical and mental health problems, such as high blood pressure, ulcers, headaches, anxiety, and depression. Taking time off to heal and "recharge one's batteries" is not an option for many workers who do not have paid sick leave or paid vacation. The United States is the only advanced nation that does not mandate a minimum number of vacation days; in

toxic workplace A work environment in which employees are subjected to co-workers and/or bosses who engage in a variety of negative, stress-inducing behaviors such as intimidation and workplace bullying, gossiping, and "backstabbing."

job burnout Prolonged job stress that can cause or contribute to high blood pressure, ulcers, headaches, anxiety, depression, and other health problems.

Poultry production is big business in the United States. Americans consume more chicken than anyone else in the world—nearly 84 pounds per year per capita, and more than any other type of meat. According to the National Chicken Council (2012), the United States has the largest broiler chicken industry in the world (broiler chickens are chickens raised for their meat).

To investigate the working conditions of the poultry industry in Alabama—one of the top three states in production of broilers—the Southern Poverty Law Center and Alabama Appleseed conducted interviews with 302 workers currently or previously employed in Alabama's poultry industry. Participants were asked about safety procedures and equipment, safety rights and enforcement, their experience with injuries and employer response to injuries, workplace discrimination and harassment, and about working conditions in general. The following excerpts describe the experiences of six poultry industry workers.*

Oscar's Story: When Oscar heard that a poultry processing plant in Alabama was looking for workers, he thought he could apply the skills he learned from studying mechanical engineering in Cuba. "I thought maybe . . . that I could work with the machinery, given my abilities and my hands," he said. But after the 47-year-old arrived in Alabama from Miami, his hopes were dashed. He was denied two positions where he could apply his mechanical skills and instead was asked to fold chicken wings on the production line. As bird carcasses sped by him on the line, Oscar had to grab the wings and twist them into the position the company wanted, folding fast enough to meet a quota of approximately 40 chicken wings per minute—or roughly 18,000 wings per day. "I did my job well,"

he said. "But little did I know I was harming myself in the process. They don't warn you that this can happen." As he repeated this motion thousands of times, it put pressure on his hands and wrists. After about a month, he developed serious hand and wrist pain, which he had never experienced before starting work at the plant. Oscar was diagnosed with tendinitis and carpal tunnel syndrome. When his injuries made him no longer useful to the company, he was fired.

Natashia's Story: Natashia Ford had been a healthy person all her life. But after spending six years deboning chickens at a poultry processing plant in North Alabama, she's a different person. She's been diagnosed with histoplasmosis, a lung disease similar to tuberculosis that's caused by breathing airborne spores at the plant. Eight nodules are growing within her lungs, and they cannot be removed. The company she worked for resisted paying for any of her medical expenses, such as her inhaler and medications. When she worked, Natashia was always coughing inside the frigid plant. She said the company didn't provide her or her co-workers with face masks as they worked on the processing line. Chicken juices would get into Natashia's ears, nose, and mouth. "You couldn't pay me to go back there," she said. Workers would process 30,000 to 60,000 birds per shift as they raced to keep pace with the mechanized line. If a chicken became lodged in the machinery, the line would stop so it could be dislodged. Hurt workers couldn't count on the same mercy. The processing line never slowed or stopped for them, she said. It didn't matter if they were cut, hurt, or

sick. It didn't matter if a worker's muscles stiffened and locked from standing and repeating the same motions for hours. The machinery kept churning—even when Natashia was so sick that she had to be picked up and carried off. Natashia eventually sued her employer, a rare occurrence in this industry. The company ultimately paid for some of her medical bills, but not all of them. Today, Natashia can't stand for more than 15 minutes. She wears knee and back braces and walks with a cane. "No line shut down for a human, but it'd shut down for a bird," she said.

Diane's Story: When Diane was diagnosed with severe carpal tunnel syndrome, her doctor was clear about what caused the condition—her work at the poultry processing plant. It was news her employer didn't want her to hear. In fact, Diane knew she could be fired if her employer learned she had sought treatment from her own doctor instead of the company doctor or nurse. Her boss had made it clear how she should deal with her pain. "My supervisor let me know that if my hands hurt and I go see the nurse, I should tell her that the pain comes from something that happened at home," the 38-year-old African American woman said. "I shouldn't say it's work-related. If I say my pain comes from something I did at work, then I will be laid off without pay and three days later get fired. So, when I go to the nurse I tell her that I hurt my hands at home." But Diane knew the treatment recommended by the nurse—taking a Tylenol and soaking her hands in water—wouldn't be enough to address a serious injury. So, she secretly saw a doctor and was diagnosed with carpal tunnel syndrome.

The United States is the only advanced nation that does not mandate a minimum number of vacation days.

the European Union countries, employers are required to give workers at least four weeks of vacation each year (some countries mandate five or six weeks of vacation) (Greenhouse 2008). And, even when the workday is officially over, many workers are connected to their jobs by their smartphones and laptops.

Work-Life Conflict

A major source of stress for U.S. workers is the day-to-day struggle to simultaneously meet the demands of work and other life responsibilities, including family and education. Work-life conflicts are common among U.S. workers: Seventy percent of U.S. women and men report some interference between work and nonwork responsibilities (Schieman et al. 2009).

Horacio's Story: Horacio was only 18 when he began working as a chicken catcher in Alabama. . . . Chicken catchers—the workers who catch birds in chicken houses and load them onto trucks bound for processing plants—encounter many of the same problems as plant workers. These problems include repetitive motion injuries, respiratory ailments and supervisors who have little concern for their safety. Horacio and his crew worked at night because the chickens are calmer then. It's also not as hot—though the heat inside the houses is still intense. His crew typically filled 14 or 15 trailers with chickens during each shift. Each trailer held about 4,400 chickens. Horacio would carry about seven chickens at a time—roughly 63 pounds total. It's a feat he would perform more than 100 times for each trailer. For Horacio to carry seven chickens at a time, he had to pick the birds up by their feet and place the feet between the fingers of his hand until he held four live, squawking, scratching, pecking chickens. He then had to grab three more birds and secure their feet between the fingers of his other hand without dropping the first four chickens. Given these conditions, it's no surprise that chicken catchers often develop the same types of back, arm, wrist, and hand injuries other poultry workers suffer, though the damage is often more severe. Chicken catchers have reported their hands have swollen to twice their normal size. Horacio's hands, fingers, and wrists would swell to the point where he couldn't completely close his hands. They also would often go numb at night, a common symptom of carpal tunnel syndrome. Chicken catchers may also develop respiratory ailments, due to the poor air quality in chicken houses. . . . Horacio, like other chicken catchers, said the dust and fecal matter in the air made his eyes burn and skin itch. He frequently had a rash from his work. He had a protective mask to wear, but it was so heavy he didn't use it. None of the workers wore the masks because they inhibited their breathing—preventing them from working fast enough to meet their boss's demands. Today, after 19 years as a chicken catcher, 37-year-old Horacio exhibits a telltale sign of the profession. Both of his arms are in constant pain. He also walks with a limp—a painful reminder of the time a truck ran over his foot as it backed into a chicken house. His boss insisted that he keep working through the pain.

Juan's Story: Juan, a Latino father of three living in Alabama since 1999, worked for six years in a poultry plant. He worked primarily in stacking jobs that required him to lift, carry, and stack two 80-pound boxes of chicken a minute. While lifting a box of chicken, he became dizzy, slipped, and fell to the floor. He was told to go right back to working despite being in great pain. Juan's back pain worsened and the swelling became constant. He was unable to sleep. When he was finally able to get X-rays, they revealed that he had two lumbar vertebrae fractures from the fall. He was eventually fired. Juan has yet to recover. His employer never paid for any medical treatments.

Marta's Story: Marta couldn't take it anymore. She picked up the phone and called her company's human resources hotline. She had endured several years of sexual harassment from her supervisor at the processing plant in southeast Alabama where she was a sanitation worker. He had repeatedly pressured the 48-year-old Latina to have sex with him, telling her that she could have any job she wanted—if she gave in to his advances. She was finally reporting him. But Marta's phone call didn't end her ordeal. In fact, it made matters worse. She was accused of inventing the story and was transferred to a lower-paying job. Her two sons, who also worked at the plant, received job transfers that cut their pay as well. A year later, Marta was fired. She was told she was fired over her immigration status—after seven years at the company. Her harasser, who kept his job, made it clear that immigration wasn't the real issue: He told her that if she had agreed to sleep with him, she'd still have her job.

Conclusion: The problems experienced by the workers described here are not uncommon in the poultry industry. One in five workers in the survey cited here said they or someone they knew was subjected to unwelcome touching of a sexual nature, and about a third of the workers said they or someone they knew had been subjected to unwelcome sexual comments. Nearly three-quarters of the poultry workers interviewed for this report described suffering from some type of significant work-related injury or illness, such as debilitating pain in their hands and gnarled fingers from carpal tunnel syndrome, chemical burns, and respiratory problems.

*Some names have been changed to protect the identity of the participants.

Source: Based on Southern Poverty Law Center and Alabama Appleseed. 2013. *Unsafe at These Speeds: Alabama's Poultry Industry and Its Disposable Workers*. Available at www.splc.org. Used by permission.

Employed spouses strategize ways to coordinate their work schedules to have vacation time together, or simply to have meals together. Among dual-income spouses with children, 56 percent of moms and 50 percent of dads report that it is very or somewhat difficult to juggle work and family life (Parker & Wang 2013). Employed parents with young children must find and manage arrangements for child care and negotiate with their employers about taking time off to care for a sick child, or to attend a child's school event or extracurricular activity. Some employers post their workers' weekly schedules only a few days in advance, making it difficult to arrange child care. In some two-parent households, spouses or partners work different shifts so that one adult can be home with the children. However, working different shifts strains marriage relationships, because the partners rarely have time off together. Some employed parents who cannot find or afford child care leave their children with no adult supervision.

Employees with elderly and/or ill parents worry about how they will provide care for their parents, or arrange for and monitor their care, while putting in a 40-hour (or more) workweek. About 2 in 10 employees are caregivers providing care to a person over age 50 (Council of Economic Advisers 2010).

Women tend to experience their jobs as interfering with their nonwork lives more so than do men; however, the level of work-life conflict women experience has remained stable over the past three decades, whereas men's reported level of work-life conflict has increased, probably due to the increased involvement of men in child care and household responsibilities (Galinsky et al. 2009).

Balancing the responsibilities of a job with the demands of school is also a challenge for many adults. Millions of college students, both traditional age (18 to 24) and over age 25 work either part-time or full-time jobs while pursuing their college degree.

Alienation

Work in industrialized societies is characterized by a high degree of division of labor and specialization of work roles. As a result, workers' tasks are repetitive and monotonous and often involve little or no creativity. Limited to specific tasks by their work roles, workers are unable to express and utilize their full potential—intellectual, emotional, and physical. According to Marx, when workers are merely cogs in a machine, they become estranged from their work, the product they create, other human beings, and themselves. Marx called this estrangement "alienation." As we discussed earlier, the McDonaldization of the workplace also contributes to alienation.

Alienation has four components: (1) *Powerlessness* results from working in an environment in which one has little or no control over the decisions that affect one's work; (2) *meaninglessness* results when workers do not find fulfillment in their work; (3) workers may experience *normlessness* if workplace norms are unclear or conflicting, such as when companies have nondiscrimination policies yet they practice discrimination; and (4) *self-estrangement* may stem from the workers' inability to realize their full human potential in their work roles.

Labor Unions and the Struggle for Workers' Rights

In times of high unemployment, many people with jobs are thankful that they are employed. But having a job is no guarantee of having favorable working conditions and receiving decent pay and benefits. **Labor unions** are worker advocacy organizations that developed to protect workers and represent them at negotiations between management and labor.

Benefits and Disadvantages of Labor Unions to Workers. Labor unions have played an important role in fighting for fair wages and benefits, healthy and safe work environments, and other forms of worker advocacy. Compared with nonunion workers, unionized workers tend to have higher earnings, better insurance and pension benefits, and more paid time off. Jobs that have higher rates of unionization have a lower gender pay gap (International Trade Union Confederation 2013).

Labor unions are also influential in achieving better working conditions. For example, the United Food and Commercial Workers (UFCW), the country's largest union representing poultry-processing workers, was instrumental in the formation of an OSHA rule that established a federal workplace "potty" policy governing when employees can use the bathroom while on the job. Prior to this rule, workers in food-processing industries were denied the right to go to the bathroom when needed, and often had no other choice but to relieve themselves while standing on the assembly line because their boss would not let them leave their work station ("New OSHA Policy Relieves Employees" 1998). Now, OSHA mandates that employers must make toilet facilities available so that employees can use them when they need to.

One of the disadvantages of unions is that members must pay dues and other fees, and these dues have been rising in recent years. Union members resent the high salaries

labor unions Worker advocacy organizations that developed to protect workers and represent them at negotiations between management and labor.

that many union leaders make. Another disadvantage for unionized workers is the loss of individuality. Unionized workers are members of an overall bargaining unit in which the majority rules. Decisions made by the majority may conflict with individual employees' specific employment needs.

Declining Union Density.

The strength and membership of unions in the United States have declined over the last several decades. **Union density**—the percentage of workers who belong to unions—grew in the 1930s and peaked in the 1940s and 1950s, when 35 percent of U.S. workers were unionized. In 2012, the percentage of U.S. workers belonging to unions had fallen to just under 11.3 percent, down from 20.1 percent in 1983 (Bureau of Labor Statistics 2013c).

One reason for the decline in union representation is the loss of manufacturing jobs, which tend to have higher rates of unionization than other industries. Job growth has occurred in high technology and financial services, where unions have little presence. In addition, globalization has led to layoffs and plant closings at many unionized work sites as a result of companies moving to other countries to find cheaper labor. A major reason why union representation has declined is that corporations take active measures to keep workers from unionizing, and weak U.S. labor laws fail to support and protect unionization.

Corporate Antiunion Activities.

At least 23,000 workers each year are fired or discriminated against at their workplace because of involvement in union-related activity (Bonior 2006). Some employers faced with organizing campaigns have engaged in the following antiunion strategies: (1) firing pro-union workers, (2) threatening to close a work site when workers try to form a union, (3) coercing workers into opposing unions with bribery or favoritism, (4) hiring high-priced union-busting consultants to fight union-organizing drives, and (5) forcing employees to attend one-on-one antiunion meetings with their supervisors (Bonior 2006; Mehta & Theodore 2005).

Weak U.S. Labor Laws.

The 1935 National Labor Relations Act (NLRA) is the primary federal labor law in the United States. The NLRA guarantees the right to unionize, bargain collectively, and to strike against private-sector employees. However, in addition to excluding public-sector workers, the law excludes agricultural and domestic workers, supervisors, railroad and airline employees, and independent contractors. As a result, millions of workers do not have the right under U.S. law to negotiate their wages, hours, or employment terms.

In addition, changes in U.S. labor law over the years have eroded workers' rights to freedom of association. Originally, labor law required employers to grant a demand for union recognition if a majority of workers signed a card indicating they wanted a union. But since 1947, employers can reject workers' demand for unionization and force a National Labor Relations Board (NLRB) election, which requires about one-third of workers to petition for the board to hold the election. "The company then uses the time leading up to the election to focus its campaign against union formation, while disallowing opportunities for opposing views" (Human Rights Watch 2007, p. 18).

> Penalties for violating U.S. labor law are so weak that many employers consider them as a cost of doing business and a small price to pay for defeating workers' attempts to organize.

In the United States, the NLRB and the courts play an important role in upholding workers' rights to unionize and sanctioning employers who violate these rights. The NLRB has the authority to issue job reinstatement and "back pay" orders or other remedial orders to workers wrongfully fired or demoted for participating in union-related activities. Although it is illegal to fire workers for engaging in union activities, there are few consequences for employers that do so. Penalties for violating U.S. labor law are so weak that many employers consider them as a cost of doing business and a small price to pay for defeating workers' attempts to organize (Human Rights Watch 2009).

union density The percentage of workers who belong to unions.

If you get fired for trying to organize, for example, you can apply to the NLRB. If the NLRB finds that you were illegally fired, the employer has to give you back pay for the time you were fired—minus any money that you may have earned at another job. As you can imagine, most people who are fired for trying to organize will, in fact, get another job somewhere, so there's no compensation for them at all. Then all the employer has to do is post on a bulletin board at the work site that they won't do it again. So there are effectively no sanctions, and it is in the employers' interest to fire people. They really don't suffer many consequences for doing so, and firing a leading union supporter sends a very powerful message to the rest of the employees. The message is: If you too try to lead an organizing campaign, you are going to lose your job; and if you vote for a union, you could lose your job (Bonior 2006).

In addition, there is a backlog of thousands of cases of unfair labor practices by employers, and workers often wait years from the filing of a charge until the NLRB resolves a case, discouraging many workers from filing charges (Greenhouse 2008).

Antiunion Legislation. In 2011, Governor Scott Walker (Wisconsin) signed legislation to weaken unions representing state and local government employees. The legislation prohibited collective bargaining by most public workers for issues beyond wages, required unions to hold annual votes on whether they should remain in existence, and increased workers' contributions for pensions and health care. Although Judge Marilyn Sumi struck down the legislation on procedural grounds, the Wisconsin Supreme Court later upheld the new law, and a federal judge ruled that the law was constitutional and did not violate First Amendment rights of union members.

Labor Union Struggles around the World. In 1949, the International Labour Office established the Convention on the Right to Organise and Collective Bargaining. About half of the world's workforce lives in countries that have not ratified this convention, including China, India, Mexico, Canada, and the United States. According to the International Trade Union Confederation (2013), union members face violence in at least 24 countries, and between 75 and several hundred trade unionists are killed each year. Several thousands more are imprisoned, beaten in demonstrations, tortured by security forces or others, and often sentenced to long prison terms. And each year, hundreds of thousands of workers lose their jobs merely for attempting to organize a trade union. Guatemala has become the most dangerous country in the world; 53 trade unionists were killed over the past 6 years.

Strategies for Action: Responses to Problems of Work and Unemployment

Government, private business, human rights organizations, labor organizations, college student activists, and consumers play important roles in responding to problems of work and unemployment.

Reducing Unemployment

Efforts to prepare high school students for work include the establishment of technical and vocational high schools and high school programs and school-to-work programs. School-to-work programs involve partnerships between business, labor, government, education, and community organizations that allow high school students to explore different careers, and provide job skill training and work-based learning experiences (Bassi & Ludwig 2000; Leonard 1996).

Although educational attainment is often touted as the path to employment and economic security, more than a third of college-educated workers under age 25 are working at jobs that don't require a college degree (Luhby 2013). More than half of all U.S. jobs require only a high school diploma or less (Lockard & Wolf 2012); thus, it is important to ensure that all jobs pay a minimum "living wage" (see also Chapter 6). As long as our economy allows people who work full-time to earn poverty-level wages, having a job is not necessarily the answer to economic self-sufficiency. As David Shipler (2005) explained in his book *The Working Poor*:

> A job alone is not enough. Medical insurance alone is not enough. Good housing alone is not enough. Reliable transportation, careful family budgeting, effective parenting, effective schooling are not enough when each is achieved in isolation from the rest. There is no single variable that can be altered to help working people move away from the edge of poverty. Only where the full array of factors is attacked can America fulfill its promise. (p. 11)

Workforce Development. The 1998 **Workforce Investment Act (WIA)** provides a wide array of programs and services designed to assist individuals to prepare for and find employment, such as skill assessment, job search, and placement assistance; individual career planning and counseling; resume preparation; English as a second language instruction; computer literacy; wage subsidies for on-the-job training; and support services such as transportation and child care to enable individuals to participate in WIA programs. Some workforce development programs focus on strategies to improve the employability of hard-to-employ individuals through providing targeted interventions such as substance abuse treatment, domestic violence services, prison release reintegration assistance, mental health services, and homelessness services, in combination with employment services (Martinson & Holcomb 2007).

Job Creation and Preservation. In a national poll, one in four Americans said that the best way to create more U.S. jobs is to keep manufacturing jobs in the United States and stop sending work overseas (Newport 2011). Although both Democrat and Republican respondents cited keeping manufacturing jobs in the United States as the best way of creating jobs, Republicans also favored job-creation strategies that involve reducing taxes and limiting government involvement in regulating business, whereas Democrats were more likely to favor using government to create jobs through infrastructure projects.

In 2010, Obama signed into law the Small Business Jobs Act, which provides tax breaks and better access to credit for small businesses so that small businesses can create new jobs. The 2010 Hiring Incentives to Restore Employment (HIRE) Act provided tax incentives to companies that hired employees who had been looking for work for 60 days or more. Obama also launched the National Export Initiative with a goal of supporting new jobs through doubling exports.

The Global Jobs Pact. Improving workers' lives and the economy as a whole requires a coordinated and comprehensive effort that combines a number of goals and strategies. In 2009, the International Labour Organization adopted a Global Jobs Pact designed to guide national and international policies aimed at stimulating the economy, creating jobs, and providing protection to workers and their families.

The Global Jobs Pact calls for a wide range of measures to retain workers, sustain businesses, create jobs, and provide social protections to workers and the unemployed. The pact also calls for more stringent supervision and regulation of the financial industry so that it better serves the economy and protects individuals' savings and pensions. The pact urges a number of measures, including (1) a shift to a low-carbon, environmentally friendly economy that will help create new jobs; (2) investments in public infrastructure; and (3) increases in social protection and minimum wages (to reduce poverty, increase demand for goods, and stimulate the economy). The Global Jobs Pact provides a vision for a healthier economy that meets the needs of workers and consumers. The hard work of translating the Global Jobs Pact into reality falls to employers, trade unions, and especially governments.

Workforce Investment Act (WIA) Legislation passed in 1998 that provides a wide array of programs and services designed to assist individuals to prepare for and find employment.

Worker Cooperatives: An Alternative to Capitalism

The global financial crisis that began in 2007 stimulated public discussions about the problems that stem from capitalism and renewed efforts to consider alternative business models. Economist Richard Wolff explains:

> We believe the capitalist organization of production has now finished its period of usefulness in human history. It is now no longer able to deliver the goods. It's bringing profits and prosperity to a tiny portion of the population, and delivering not the goods but the "bads" to most people. Jobs are steadily more insecure, unemployment is high[,] . . . benefits are increasingly being reduced, the prospects for our children are even worse, as more of them go deeper and deeper in debt to get the degrees that do not provide them with the jobs and incomes to get out of that debt. . . . It's longer overdue that we face honestly that the crisis we endure is the product of an economic system whose organization is something we should question, debate, and change. (quoted in Rampell 2013, n.p.)

One alternative to the capitalistic economic model is the cooperative workplace model that uses democratic methods in its business organization. **Worker cooperatives** or **workers' self-directed enterprises**—also known as "co-ops"—are businesses that are owned and democratically governed by their employees; workers participate in deciding what, how, and where to produce, and they would decide how to distribute the surpluses (or profits) generated in and by their enterprise. One of the most successful cooperative businesses is Mondragon, a corporation in Spain that started out as a small 7-member cooperative and, over 60 years, has grown into a corporation with over 10,000 workers. In 2013, the Mondragon Corporation won the Drivers of Change category at the annual Boldness in Business awards, sponsored by the London newspaper *The Financial Times.* At Mondragon, the workers hire the managers, the exact opposite of a capitalistic corporation. The workers decided to adopt a rule specifying that the highest paid worker cannot be paid more than six and a half times what the lowest paid worker earns, unlike many U.S. CEOs who make up to a few hundred times more than the *average* worker (not even the lowest paid). Richard Wolff is not surprised by this rule:

> If all the workers in any office, store, or factory, got together and had the power—which in a co-op they would—to decide what the wages and salaries of everybody are, do you think they'd give a handful of people at the top tens of millions of dollars, while everybody else is scrambling and unable to pay for their kids' college education, etc.? That's not going to happen. Even if you decide to pay some people more you're not going to live in the world of extreme inequality the way that's normal and typical for capitalism. . . . (quoted in Rampell 2013, n.p.)

Efforts to End Slavery and Child Labor

More than 50 years ago, the United Nations stated in Article 4 of its Universal Declaration of Human Rights, that "no one shall be held in slavery or servitude; slavery and the slave trade shall be prohibited in all their forms." Yet slavery persists throughout the world. The international community has drafted treaties on slavery, but many countries have yet to ratify and implement the different treaties.

In at least 25 countries, slave trafficking is actively prosecuted and treated as a serious crime. However, slave traffickers often avoid punishment because, as a former official of the U.S. Agency for International Development explained, "government officials in dozens of countries assist, overlook, or actively collude with traffickers" (quoted by Cockburn 2003, p. 16). In many countries, the justice system is more likely to jail or expel sex slaves than to punish traffickers ("Sex Trade Enslaves Millions of Women, Youth" 2003). In the United States, the Victims of Trafficking and Violence Protection Act, passed by Congress in 2000, protects slaves against deportation if they testify against their former owners. Convicted slave traffickers in the United States are subject to prison sentences.

worker cooperatives Democratic business organizations controlled by their members, who actively participate in setting their policies and making decisions; also known as *workers' self-directed enterprises.*

workers' self-directed enterprises See *worker cooperatives.*

In 1989, the General Assembly of the United Nations adopted the Convention on the Rights of the Child, which asserts the right that children should not be engaged in work deemed to be "hazardous or to interfere with the child's education, or to be harmful to the child's health." The International Labour Office has taken a leading role in enforcing these rights, leading efforts to prevent and eliminate child labor. Although almost every country has laws prohibiting the employment of children below a certain age, some countries exempt certain sectors—often the very sectors where the highest numbers of child laborers are found. Efforts to prevent child labor also focus on increasing penalties for violating child labor laws, which are often weak and poorly enforced.

Education is a primary means to combat child labor. Children with no access to education have little choice but to enter the labor market. In most countries, primary education is not free and parents must pay for costs such as uniforms and books. Parents who cannot afford these fees are also likely to view their children's labor as a necessary source of income to the household. Promoting Education for All—a global movement advocating that every child has access to free, basic education (primary school plus two or three years of secondary school)—is critical in the effort to reduce child labor.

Responses to Sweatshop Labor

The Fair Labor Association (FLA), established in 1996, is a coalition of companies, universities, and nongovernmental organizations (NGOs) that works to promote adherence to international labor standards and improve working conditions worldwide. More than 30 leading companies that sell brand-name apparel, footwear, and other goods voluntarily participate in FLA's monitoring system, which inspects their overseas factories and requires them to meet minimum labor standards, such as not requiring workers to work more than 60 hours a week. In addition, more than 200 colleges and universities require their collegiate licensees (companies that manufacture logo-carrying goods for colleges and universities) to participate in FLA's monitoring system.

The FLA is criticized, however, for having low standards in allowing below-poverty wages and excessive overtime and for requiring that only a small percentage of a manufacturer's supplier factories be inspected each year. Critics also suggest that companies use their participation in the FLA as a marketing tool. Once "certified" by the FLA, companies can sew a label into their products saying that the products were made under fair working conditions (Benjamin 1998). In 2006, United Students Against Sweatshops created "FLA Watch" to "expose the truth about the Fair Labor Association . . . and . . . the FLA's ongoing failure to defend the rights of workers" (FLA Watch 2007a). FLA Watch accuses the Fair Labor Association of being "nothing more than a public relations mouthpiece for the apparel industry. Created, funded, and controlled by Nike, Adidas, and other leading sweatshop abusers, the FLA is a classic case of the 'fox guarding the hen house'" (FLA Watch 2007b).

Student Activism. United Students Against Sweatshops (USAS), formed in 1997, is a grassroots organization of youth and students who fight against labor abuses and for the rights of workers around the world, particularly campus workers and garment workers who make collegiate licensed apparel. USAS student activists have influenced more than 180 colleges and universities to affiliate with the Worker Rights Consortium (WRC), which investigates factories that produce clothing and other goods with school logos to make sure that the factory meets the code of conduct developed by each school. A typical code of conduct includes fair wages, a safe working environment, a ban on child labor, and the right to be represented by a union or other form of employee representation. If the WRC investigation finds that a factory fails to meet the code of conduct, the companies—often well-known international brands—who purchase

Kirschner/United Students Against Sweatshops

College student groups across the country have participated in boycotts against Coca-Cola in protest of the violence against union leaders at Colombian Coca-Cola plants.

items from that factory are warned that their contract with the school will be terminated if working conditions at the factory do not improve.

> **What Do You Think?** Do you know where the clothing with your college or university logo is made? Do you think most students care if the college or university logo clothing or products they buy are made under sweat-shop conditions?

Sodexo—a profitable multinational food service company—serves food to more U.S. college students than any other company. In 2009, United Students Against Sweatshops began a Kick Out Sodexo campaign after discovering that Sodexo paid its campus food service and janitorial workers poverty-level wages and interfered with employees' attempts to form a union. Students who are active in the Kick Out Sodexo campaign have written letters to Sodexo, expressing their concerns about the company's labor practices, and have campaigned to convince college and university administrators to cancel their food service contracts with Sodexo and to do business instead with a food service company that meets certain labor standards.

Legislation. Perhaps the most effective strategy against sweatshop work conditions is legislation. Some U.S. states, cities, counties, and school districts have passed "sweat-free" procurement laws that prohibit public entities (such as schools, police, and fire departments) from purchasing uniforms and apparel made under sweatshop conditions (SweatFree Communities n.d.).

Establishing and enforcing labor laws to protect workers from sweatshop labor conditions is difficult in a political climate that offers more protections to corporations than it does to workers. Companies have demanded and won all sorts of intellectual property and copyright laws to defend their corporate trademarks, labels, and products. Yet, the corporations have long said that extending similar laws to protect the human rights of a 16-year-old girl in Bangladesh who sews the garment would be "an impediment to free trade." Under this distorted sense of values, the label is protected, but not the human being, the worker who makes the product (National Labor Committee 2007).

Responses to Workplace Health and Safety Concerns

In 2009, the Protecting America's Workers Act (PAWA) was introduced in Congress—a bill that would strengthen OSHA by extending coverage to uncovered workers, including state and local public employees; enhance whistle-blower protections so that workers who report workplace safety violations would have job protection; and increase penalties for serious and willful violations and in cases of worker death. PAWA was reintroduced in Congress in 2013, but as of this writing, has not passed. Other proposed legislation would strengthen OSHA's authority to shut down operations that pose an imminent danger to workers; require large corporate employers to provide regular reports to OSHA on work-related injuries, illnesses, and fatalities; mandate OSHA to issue a standard on safe patient handling to protect health care workers from injuries; and issue a standard to protect workers from explosions and fires (AFL-CIO 2013). Efforts have also been made to strengthen the Mine Safety and Health Administration (MSHA) to give it more authority to close down mines that are in violation of safety standards and to prevent "black lung disease" in miners by establishing stricter rules on the amount of coal dust miners inhale. These and other initiatives to strengthen workplace health and safety are opposed by industry groups, and Republicans in Congress who are against government regulations argue that excessive regulation is burdensome to business, hampers investment, and hurts job creation.

In developing countries, governments fear that strict enforcement of workplace regulations will discourage foreign investment. Investment in workplace safety in developing countries, whether by domestic firms or foreign multinationals, is far below that in the rich countries. Unless global standards of worker safety are implemented and enforced in all countries, millions of workers throughout the world will continue to suffer under hazardous work conditions. Low unionization rates and workers' fears of losing their jobs—or their lives—if they demand health and safety protections leave most workers powerless to improve their working conditions.

Behavior-Based Safety Programs. A controversial health and safety strategy used by business management is behavior-based safety programs. Instead of examining how work processes and conditions compromise health and safety on the job, **behavior-based safety programs** direct attention to workers themselves as the problem. Behavior-based safety programs claim that most job injuries and illnesses are caused by workers' own carelessness and unsafe acts (Frederick & Lessin 2000). These programs focus on teaching employees and managers to identify, "discipline," and change unsafe worker behaviors that cause accidents and encourage a work culture that recognizes and rewards safe behaviors.

Critics contend that behavior-based safety programs divert attention away from the employers' failures to provide safe working conditions. They also say that the real goal of behavior-based safety programs is to discourage workers from reporting illness and injuries. Workers whose employers have implemented behavior-based safety programs describe an atmosphere of fear in the workplace, such that workers are reluctant to report injuries and illnesses for fear of being labeled "unsafe workers."

Work-Life Policies and Programs

Policies that help women and men balance their work and family responsibilities are referred to by a number of terms, including *work-family, work-life*, and *family-friendly* policies. As shown in Table 7.1, the United States lags far behind many other countries in national work-family provisions.

Federal and State Family and Medical Leave Initiatives. In 1993, President Clinton signed into law the first national policy designed to help workers meet the dual demands of work and family. The **Family and Medical Leave Act (FMLA)** requires all public agencies and private-sector employers (with 50 or more employees who worked at least 1,250 hours in the preceding year) to provide up to 12 weeks of job-protected, *unpaid* leave so that they can care for a seriously ill child, spouse, or parent; stay home to care for their newborn, newly adopted, or newly placed foster child; or take time off when they are seriously ill. A 2008 amendment to the FMLA requires employers to provide up to 26 weeks of unpaid leave to employees to care for a seriously ill or injured family member who is in the armed forces, including the National Guard or Reserves. However, 41 percent of employees are not eligible for the FMLA benefit because they work for companies with fewer than 50 employees, they work part-time, or they have not met the requirement of having worked at least 1,250 hours in the past year (Klerman et al. 2013). Some employers do not comply with the FMLA either because they are unaware of their responsibilities under FMLA, or because they are deliberately violating the law. Some eligible

behavior-based safety programs A strategy used by business management that attributes health and safety problems in the workplace to workers' behavior, rather than to work processes and conditions.

Family and Medical Leave Act (FMLA) A federal law that requires public agencies and companies with 50 or more employees to provide eligible workers with up to 12 weeks of job-protected, unpaid leave so that they can care for an ill child, spouse, or parent; stay home to care for their newborn, newly adopted, or newly placed child; or take time off when they are seriously ill, and up to 26 weeks of unpaid leave to care for a seriously ill or injured family member who is in the armed forces, including the National Guard or Reserves.

TABLE 7.1 How Do U.S. Work Policies Compare with Other Countries?

Policy	United States	Other Countries
Paid childbirth leave	No federal policy	168 countries offer paid leave to women; 98 countries offer 14 or more weeks of paid leave
Right to breast-feed at work	No federal policy	107 countries protect working women's right to breast-feed
Paid sick leave	No federal policy	145 countries provide paid sick leave
Paid annual leave	No federal policy	137 countries require employers to provide paid annual leave; 121 countries guarantee 2 weeks or more
Guaranteed leave for major family events (e.g., weddings, funerals)	No federal policy	49 countries guarantee leave for major family events (leave is paid in 40 countries)

Source: Based on Heymann et al. 2007.

TABLE 7.2 Employer-Based Work-Life Benefits and Policies, U.S., 2012

Benefit Or Policy	Percentage of Employers* That Provide Benefit/Policy
Full pay for maternity leave	9%
Partial pay for maternity leave	63%
Private space for breast-feeding	79%
Some paid time off for spouses/partners of women who give birth	14%
Health insurance for unmarried partners	13%
Allow some employees to periodically change their arrival and departure time	77%
Provide information about elder care services	41%
Child care at or near worksite	7%
Dependent Care Assistance Programs that help employees pay for child care with pretax dollars	62%

Source: Matos & Galinsky 2012.
*Employers with 50 or more employees

workers do not use their FMLA benefit because they cannot afford to take leave without pay, and/or they fear they will lose their job if they take time off.

As of this writing, California and New Jersey are the only states that have enacted paid family leave insurance programs. These programs are financed through small payroll deductions (no employer contribution) and offer eligible employees a portion of their salary for up to six weeks.

As shown in Table 7.1, there is no U.S. federal policy requiring employers to provide workers with any paid sick leave; however, Connecticut and a handful of cities (New York City, Portland, Seattle, San Francisco, and Washington, DC) passed laws requiring some employers (e.g., with 50 employees or more) to provide paid sick leave to their workers (McGregor 2013). The Healthy Families Act is a bill proposing to establish a federally mandated paid sick day policy. Employees could take the benefit if they or a family member is sick. Supporters of the Healthy Families Act point out that, when sick employees go to work because they cannot afford to take unpaid sick days, or fear losing their job by taking a sick day, they risk spreading infectious diseases at the workplace. And sending sick children to school because their parents cannot afford to take unpaid sick days to stay home with them risks spreading infectious diseases at school.

Employer-Based Work-Life Policies. Aside from government-mandated work-family policies, some corporations and employers have "family-friendly" work policies and programs, including unpaid or paid family and medical leave, child care assistance, assistance with elderly parent care, and flexible work options such as **flextime**, **compressed workweek**, and **telecommuting** (see Table 7.2). Studies show that flexible work arrangements reduce work-life conflict and increase job satisfaction. For example, Best Buy Co., Inc. developed a Results-Only Work Environment (ROWE), which allows employees and managers to control when and where they work as long as they get the job done. A controlled study found that employees participating in ROWE benefitted from reduced work-family conflict (Kelly et al. 2011).

Flexible work arrangements also benefit employers in reducing absenteeism, lowering turnover, improving the health of workers, and increasing productivity. Yet, less than one-third of full-time workers report having flexible work hours and only 15 percent report working from home at least once a week (Council of Economic Advisers 2010).

Efforts to Strengthen Labor

About two-thirds of U.S. adults agree that labor unions are necessary to protect working people (Pew Research Center for the People & the Press 2012). Although efforts to strengthen labor are viewed as problematic to corporations, employers, and some governments, such efforts have the potential to remedy many of the problems facing workers.

In an effort to strengthen their power, some labor unions have merged with one another. Labor union mergers result in higher membership numbers, thereby increasing the unions' financial resources, which are needed to recruit new members and to withstand long strikes. Because workers must fight for labor protections within a globalized economic system, their unions must cross national boundaries to build international cooperation and solidarity. Otherwise, employers can play working and poor people in different countries against each other.

flextime A work arrangement that allows employees to begin and end the workday at different times so long as 40 hours per week are maintained.

compressed workweek A work arrangement that allows employees to condense their work into fewer days (e.g., four 10-hour days each week).

telecommuting A work arrangement involving the use of information technology that allows employees to work part- or full-time at home or at a satellite office.

Strengthening labor unions requires combating the threats and violence against workers who attempt to organize or who join unions. One way to do this is to pressure governments to apprehend and punish the perpetrators of such violence. Another tactic is to stop doing business with countries where government-sponsored violations of free trade union rights occur.

Proposed legislation called the Employee Free Choice Act would allow workers to sign a card stating that they want to be represented by a union. If a majority of the employees in any workplace sign such a card, the company would then have to recognize the union and bargain over terms and conditions of employment. If the company does not negotiate a first contract in a timely manner after workers unionize, the Employee Free Choice Act requires binding arbitration. This legislation would also strengthen U.S. labor law enforcement by increasing penalties for violations. In 2007, the U.S. House of Representatives passed the Employee Free Choice Act and, as of this writing, the bill is pending in the U.S. Senate.

Understanding Work and Unemployment

On December 10, 1948, the General Assembly of the United Nations adopted and proclaimed the Universal Declaration of Human Rights. Among the articles of that declaration are the following:

> Article 23. Everyone has the right to work, to free choice of employment, to just and favourable conditions of work and to protection against unemployment.
>
> Everyone, without any discrimination, has the right to equal pay for equal work.
>
> Everyone who works has the right to just and favourable remuneration ensuring for himself and his family an existence worthy of human dignity, and supplemented, if necessary, by other means of social protection.
>
> Everyone has the right to form and to join trade unions for the protection of his interests.
>
> Article 24. Everyone has the right to rest and leisure, including reasonable limitation of working hours and periodic holidays with pay.

More than a half century later, workers around the world are still fighting for these basic rights as proclaimed in the Universal Declaration of Human Rights.

To understand the social problems associated with work and unemployment, we must first recognize that corporatocracy—the ties between government and corporations—serves the interests of corporations over the needs of workers. We must also be aware of the roles that technological developments and postindustrialization play on what we produce, how we produce it, where we produce it, and who does the producing. With regard to what we produce, the United States has moved away from producing manufactured goods to producing services. With regard to production methods, the labor-intensive blue-collar assembly line has declined in importance, and information-intensive white-collar occupations have increased. Although some people argue that the growth of multinational corporations brings economic growth, jobs, lower prices, and quality products to consumers throughout the world, others view global corporations as exploiting workers, harming the environment, dominating public policy, and degrading cultural values.

Decisions made by U.S. corporations about what and where to invest influence the quantity and quality of jobs available in the United States. As conflict theorists argue, such investment decisions are motivated by profit, which is part of a capitalist system. Profit is also a driving factor in deciding how and when technological devices will be used to replace workers and increase productivity. If goods and services are produced too efficiently, however, workers are laid off and high unemployment results. When people have no money to buy products, sales slump, recession ensues, and social welfare programs are needed to support the unemployed. When the government increases spending to pay for its social programs, it expands the deficit and increases the national debt. Deficit spending and a large national debt make it difficult to recover from the recession, and the cycle continues.

What can be done to break the cycle? Those adhering to the classic view of capitalism argue for limited government intervention on the premise that business will regulate itself by means of an "invisible hand" or "market forces." But Americans are growing increasingly skeptical of the notion that big business, which has caused many of the economic problems in our country and throughout the world, can solve the very problems it creates. New models of business, such as the workplace cooperative or workers' self-directed enterprises, offer visions of what the workplace could look like. The Occupy Wall Street social movement (see also Chapter 6) is part of a growing consciousness that job insecurity, worker exploitation, and extreme levels of wage inequality are not consistent with the American dream; there must be a better way. Participants in the Occupy Wall Street protests include students, the unemployed, union members, professionals, and others who are fed up with corporate greed and the widening gap between the rich and the poor. They are frustrated by high unemployment and the erosion of workers' salaries, benefits, and rights. Although this growing movement has not formulated any specific demands, what is certain is that Americans want change in the economy and in the workplace. Many of us spend much of our time at our jobs, working. And how we spend our time is how we spend our lives.

CHAPTER REVIEW

- **The economy has become globalized. What does that mean?**
In recent decades, innovations in communication and information technology have led to the globalization of the economy. The global economy refers to an interconnected network of economic activity that transcends national borders and spans the world. The globalized economy means that our jobs, the products and services we buy, and our nation's economic policies and agendas are influenced by economic activities occurring around the world.

- **What are some of the criticisms of capitalism?**
Capitalism is criticized for creating high levels of inequality; economic instability; job insecurity; pollution and depletion of natural resources; and corporate dominance of media, culture, and politics. Capitalism is also criticized as violating the principles of democracy by allowing (1) private wealth to affect access to political power, (2) private owners of property to make decisions that affect the public (such as when the owner of a factory decides to move the factory to another country), and (3) workplace dictatorships where workers have little say in their working conditions, thus violating the democratic principle that people should participate in collective decisions that significantly affect their lives.

- **The United States is described as a "postindustrialized" society. What does that mean?**
Postindustrialization refers to the shift from an industrial economy dominated by manufacturing jobs to an economy dominated by service-oriented, information-intensive occupations.

- **What are transnational corporations?**
Transnational corporations are corporations that have their home base in one country and branches, or affiliates, in other countries.

- **What are the four principles of McDonaldization?**
The four principles of McDonaldization are (1) efficiency, (2) predictability, (3) calculability, and (4) control through technology.

- **What are some of the causes of unemployment?**
Causes of unemployment include not enough jobs, job exportation or "offshoring" (the relocation of jobs to other countries), automation (the replacement of human labor with machinery and equipment), increased global competition, and mass layoffs as plants close and companies downsize or go out of business.

- **Does slavery still exist today? If so, where?**
Forced labor, commonly known as slavery, exists today all over the world, including the United States. Most forced laborers work in agriculture, mining, prostitution, and factories.

- **What is the most common cause of job-related fatality and nonfatal job-related illness or injury?**
The most common type of job-related fatality involves transportation accidents. Sprains and strains are the most common nonfatal occupational injury or illness involving days away from work.

- **According to a Gallup Poll, what is the most common complaint among U.S. workers?**
A Gallup Poll found that one-third of respondents is "totally dissatisfied" with the amount of stress in their jobs, making job stress the most common complaint of workers (followed by the amount of money they earn).

- **What are some of the challenges that workers face in balancing work and family?**
Workers often struggle to find the time and energy to care for elderly parents and children and to have time with their families.

- **Why has union membership in the United States declined over the last several decades?**
Union membership has declined for several reasons: the loss of manufacturing jobs, which tend to have higher rates of unionization than other industries; layoffs and plant closings at many unionized work sites as a result of companies moving to other countries to find cheaper labor; active measures by corporations to discourage unionization; weak

U.S. labor laws that fail to support and protect unionization; and anti-union legislation.

- **What is the federal Family and Medical Leave Act?**
In 1993, President Clinton signed into law the Family and Medical Leave Act (FMLA), which requires all companies with 50 or more employees to provide eligible workers with up to 12 weeks of job-protected, unpaid leave so that they can care for a seriously ill child, spouse, or parent; stay home to care for their newborn, newly adopted, or newly placed child; or take time off when they are seriously ill. The United States is the only advanced nation that does not mandate paid leave for family and medical reasons.

TEST YOURSELF

1. The global financial crisis that began in 2007 started in what country?
   a. China
   b. India
   c. United States
   d. Mexico
2. According to this text, corporations outsource labor to other countries so that U.S. consumers can save money on cheaper products.
   a. True
   b. False
3. Under federal law, willful violation that results in a worker's death is considered which of the following?
   a. A misdemeanor with a maximum prison sentence of 6 months
   b. A misdemeanor with a minimum prison sentence of 6 years
   c. A felony with a maximum prison sentence of 26 years
   d. A felony with a minimum prison sentence of 6 years
4. The most common type of job-related fatality involves which of the following?
   a. Mining accidents
   b. Construction accidents
   c. Transportation accidents
   d. Homicide
5. The United States is the only advanced nation that does not mandate a minimum number of vacation days.
   a. True
   b. False

6. The 1935 National Labor Relations Act (NLRA) gives all U.S. workers the right to unionize, bargain collectively, and to strike.
   a. True
   b. False
7. Which is the most dangerous country in the world to be involved in trade union activities?
   a. China
   b. Ireland
   c. Egypt
   d. Guatemala
8. The Decent Working Conditions and Fair Competition Act is the first proposed federal legislation in the United States aimed at which of the following?
   a. Combating sweatshop labor
   b. Preventing outsourcing of jobs
   c. Strengthening unions
   d. Prohibiting age discrimination in the workplace
9. The United States is the only country in the world that protects working women's right to breast-feed at work.
   a. True
   b. False
10. The majority of U.S. adults have an unfavorable opinion of labor unions.
   a. True
   b. False

ANSWERS: 1. C; 2. B; 3. A; 4. C; 5. A; 6. B; 7. D; 8. A; 9. A; 10. B.

KEY TERMS

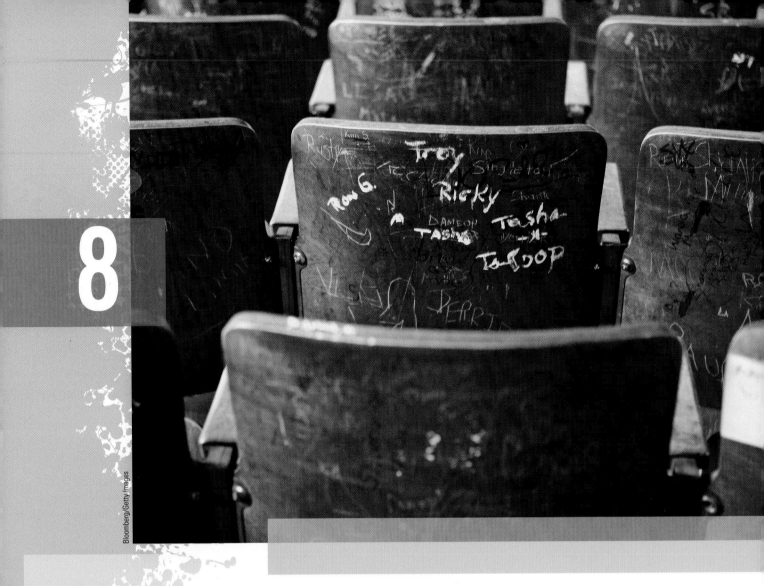

Bloomberg/Getty Images

Problems in Education

8

IN 2012, MALALA YOUSAFZAI was on her way home with friends when several masked men stopped the van they were riding in and asked "Which one is Malala?" (Almond 2013, p. 1). After identifying the 15-year-old Pakistani student, the men opened fire, injuring her and two of her girlfriends. Despite being shot in the head, Malala survived the attack with no permanent brain injuries and, after several surgeries, now advocates on behalf of young girls around the world. Why was Malala brutally attacked? She defied the Taliban's prohibition against girls attending school.

The attempted murder of Malala Yousafzai, although extreme, represents just one of the many problems in education today—inequality and discrimination based on the ascribed statuses of gender, race, and ethnicity. Too often, graduates are ill prepared for the 21-century workplace. Frustrated, teachers leave the profession for better salaries and the fulfillment they once thought teaching would provide. The public questions whether their children are safe at school, and dropout rates for minority and poor students remain high. Solutions to these problems seem out of reach as budgets are slashed, political wrangling over the best way to "fix" schools heats up, and an ever-growing student population puts additional strain on school resources.

Yet, education is often claimed as a panacea—the cure-all for poverty and prejudice, drugs and violence, war and hatred, and the like. Can one institution, riddled with problems, be a solution for other social problems? In this chapter, we focus on this question and on what is being called the educational crisis. We begin with a look at education around the world.

The Global Context: Cross-Cultural Variations in Education

Looking only at the American educational system might lead one to conclude that most societies have developed some method of formal instruction for their members.

After all, the United States has more than 142,000 schools, 5.2 million primary and secondary schoolteachers and college faculty, 5.6 million administrators and support staff, and 76.3 million students (NCES 2013a). In reality, many societies have no formal mechanism for educating the masses. One in five adults cannot read or write—796 million people worldwide—two-thirds of them women (UNESCO 2012a; World Literacy Foundation 2012).

> One in five adults cannot read or write—796 million people worldwide—two-thirds of them women.

Illiteracy rates vary dramatically by region of the world. Adult illiteracy is highest in Africa and in some areas of Asia. Between 60 percent and 70 percent of adults in Chad, Mali, and Niger are illiterate. However, fewer than 10 percent of adults are illiterate in Mexico, Brazil, South Africa, China, and the Russian Federation (UNESCO 2012a).

Despite the unacceptably high number of illiterate adults in the world, there is some reason for optimism. Three countries—China, Indonesia, and Iran—are expected to reduce their illiteracy rate in half by 2015—an international goal set by the United Nations Educational, Scientific, and Cultural Organization (UNESCO). To reach that target, ". . . 6 percent of the world's population, or more than 360 million people, will have to become literate. That's like teaching the entire population of the United States and Canada to read and write, in just three years" (Hammer 2012, p. 1).

Education at a Glance, a publication of the Organization for Economic Cooperation and Development (OECD), reports education statistics on over 34 countries (OECD 2013a). Some interesting findings are revealed. First, in general, educational levels are rising. For example, many countries saw an increase in the proportion of young people attending colleges and universities. However, large numbers of people do not graduate from tertiary institutions. On the average, only 35 percent of all adults attain a college degree (see Figure 8.1). Those most likely to do so are women, and people between 25 and 34 years of age.

Second, there is a clear link between education and income, and between education and employment. In general, across the OECD-participating countries, the more education people have, the higher their income and the greater the likelihood they will be employed.

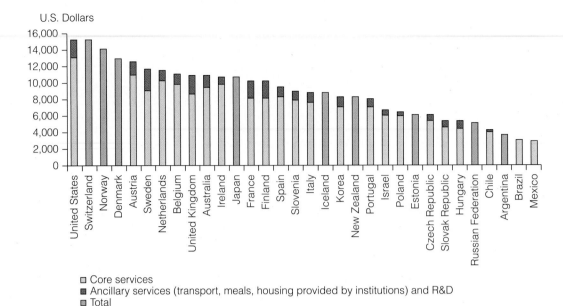

U.S. Dollars

Chart y-axis: 16,000 / 14,000 / 12,000 / 10,000 / 8,000 / 6,000 / 4,000 / 2,000 / 0

Countries (x-axis): United States, Switzerland, Norway, Denmark, Austria, Sweden, Netherlands, Belgium, United Kingdom, Australia, Ireland, Japan, France, Finland, Spain, Slovenia, Italy, Iceland, Korea, New Zealand, Portugal, Israel, Poland, Estonia, Czech Republic, Slovak Republic, Hungary, Russian Federation, Chile, Argentina, Brazil, Mexico

☐ Core services
■ Ancillary services (transport, meals, housing provided by institutions) and R&D
☐ Total

Figure 8.1 Global Spending Per Student in U.S. Dollars, Elementary School through College
Source: OECD 2013a.

Of college-educated people, 80 percent are employed compared to 60 percent of people with less than a high school degree. Further, investment in a college degree has an "**earnings premium**"; that is, the benefits of having a college degree far outweigh the cost of getting one (OECD 2012). For example, men and women with high school degrees earn, on the average, 77 percent and 74 percent, respectively, of what a male and female college graduate earns (OECD 2012).

Third, across all OECD-participating countries, an average of $7,637 is spent per student each year they are in school from elementary school to college. Spending, although increasing in general, varies widely by country (see Figure 8.1). In Mexico, Argentina, and Brazil, less than $4,000 is spent per student per year. However, Australia, France, Japan, Sweden, the United Kingdom, and the United States spend over $10,000, on the average, per student per year (OECD 2013a).

Fourth, in reference to teachers, the average student–teacher ratio in elementary schools in OECD-participating countries is 22:1. Average elementary school class sizes range from a high of 38 students in China to a low of 14 students in Luxembourg (OECD 2013a).

Teachers' salaries represent the single largest category in educational spending. Between 2000 and 2011, teachers' salaries rose in most countries with the notable exceptions of France and Japan (OECD 2013a). In general, the higher the level of class taught, the higher the teacher's salary. The average salary per teaching-hour after 15 years of experience is $49.00 for elementary school teachers, $58.00 for middle school teachers, and $66.00 for high school teachers (OECD 2013a).

Despite the fact that public school teachers' salaries in the United States averaged $56,069 in 2011 (NCES 2013a), near the OECD average, teachers in the United States work more hours than in any of the other OECD-participating countries. On the average, teachers report working 53 hours a week (Primary Sources 2012).

Fifth, educational attainment in OECD-participating countries has increased over the last decade, i.e., the percentage of people without a high school education has declined and the percentage of people with a college education has grown. However, generational and regional differences exist. For example, in Japan, Poland, and Korea, the percent of

Globally, few countries have the quality of schools or accessibility to education as enjoyed in the United States. Here, students in Lahtora, India, have outdoor lessons because their classroom is too crowded and too cold to have classes inside.

The New York Times/Redux Pictures

earnings premium The benefits of having a college degree far outweigh the cost of getting one.

25- to 34-year-olds who have a college degree is over 25 percent higher than the percent of 55- to 64-year-olds who have completed a college degree. In the United States, the difference is slightly more than one percent (OECD 2013a).

Finally, the Program for International Student Assessment (PISA) is designed to "... assess to what extent students at the end of compulsory education, can apply their knowledge to real-life situations and be equipped for full participation in society" (OECD 2012, p. 1). A random sample of over 400,000 15-year-old students in 65 countries participates in the PISA. The results of the recent study indicate that of the 31 OECD-participating countries, four—Finland, Japan, the Republic of Korea, and Canada—outperformed all others in science and mathematics literacy. More recently, in the Trends in International Mathematics and Science Study (TIMSS), which uses data from over 50 countries, the United States scored higher than the international TIMSS average for both 4th and 8th graders in mathematics and in science (Kastberg et al. 2013).

Sociological Theories of Education

The three major sociological perspectives—structural functionalism, conflict theory, and symbolic interactionism—are important in explaining different aspects of American education.

Structural-Functionalist Perspective

According to structural functionalism, the educational institution serves important tasks for society, including instruction, socialization, the sorting of individuals into various statuses, and the provision of custodial care (Sadovnik 2004). Many social problems, such as unemployment, crime and delinquency, and poverty, can be linked to the failure of the educational institution to fulfill these basic functions (see Chapters 4, 6, and 7). Structural functionalists also examine the reciprocal influences of the educational institution and other social institutions, including the family, political institution, and economic institution.

Instruction. A major function of education is to teach students the knowledge and skills that are necessary for future occupational roles, self-development, and social functioning. Although some parents teach their children basic knowledge and skills at home, most parents rely on schools to teach their children to read, spell, write, tell time, count money, and use computers. As discussed later, many U.S. students display low levels of academic achievement. The failure of many schools to instruct students in basic knowledge and skills both causes and results from many other social problems.

Socialization. The socialization function of education involves teaching students to respect authority—behavior that is essential for social organization (Merton 1968). Students learn to respond to authority by asking permission to leave the classroom, sitting quietly at their desks, and raising their hands before asking a question. Students who do not learn to respect and obey teachers may later disrespect and disobey employers, police officers, and judges.

The educational institution also socializes youth into the dominant culture. Schools attempt to instill and maintain the norms, values, traditions, and symbols of the culture in a variety of ways, such as celebrating holidays (e.g., Martin Luther King Jr. Day and Thanksgiving); requiring students to speak and write in standard English; displaying the American flag; and discouraging violence, drug use, and cheating.

As the number and size of racial and ethnic minority groups have increased, American schools are faced with a dilemma: Should public schools promote only one common culture, or should they emphasize the cultural diversity reflected in the U.S. population? Some evidence suggests that most Americans believe that schools should do both—they should promote one common culture and emphasize diverse cultural traditions. **Multicultural education**—that is, education that includes all racial and ethnic groups in the school curriculum—promotes awareness and appreciation for cultural diversity (also see Chapter 9).

multicultural education
Education that includes all racial and ethnic groups in the school curriculum, thereby promoting awareness and appreciation for cultural diversity.

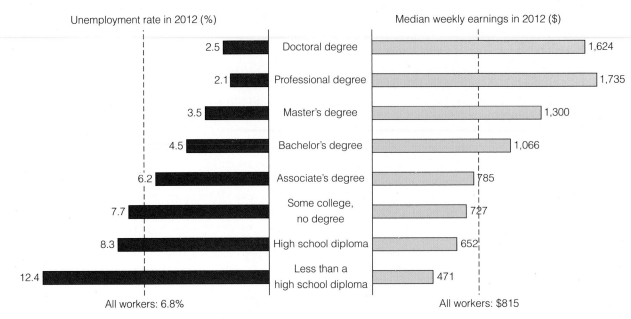

Unemployment rate in 2012 (%)		Median weekly earnings in 2012 ($)
2.5	Doctoral degree	1,624
2.1	Professional degree	1,735
3.5	Master's degree	1,300
4.5	Bachelor's degree	1,066
6.2	Associate's degree	785
7.7	Some college, no degree	727
8.3	High school diploma	652
12.4	Less than a high school diploma	471

All workers: 6.8%   All workers: $815

Figure 8.2 Unemployment Rate and Median Weekly Earnings for Individuals 25 and Over by Highest Level of Education, 2012
Source: BLS 2013.

Sorting Individuals into Statuses. Schools sort individuals into statuses by providing credentials for individuals who achieve various levels of education at various schools within the system. These credentials sort people into different statuses—for example, "high school graduate," "Harvard alumna," and "English major." In addition, schools sort individuals into professional statuses by awarding degrees in fields such as medicine, engineering, and law. The significance of such statuses lies in their association with occupational prestige and income—the higher one's education, the higher one's income. Furthermore, unemployment rates and earnings are tied to educational status as seen in Figure 8.2.

Custodial Care. The educational system also serves the function of providing custodial care by providing supervision and care for children and adolescents until they are 18 years old (Merton 1968). Despite 12 years and almost 13,000 hours of instruction, some school districts are increasing the number of school hours and/or days beyond the "traditional" schedule. In 2013, Arne Duncan, U.S. Secretary of Education and advocate for a longer school year, argued that, "whether educators have more time to enrich instruction or students have more time to learn how to play an instrument and write computer code, adding meaningful in-school hours is a critical investment that better prepares children to be successful in the 21st century" (quoted in Smyth 2013). To date, over 1,000 U.S. schools have extended their academic year.

Conflict Perspective

Conflict theorists emphasize that the educational institution solidifies the class positions of groups and allows the elite to control the masses. Although the official goal of education in society is to provide a universal mechanism for achievement, in reality, educational opportunities and the quality of education are not equally distributed.

Conflict theorists point out that the socialization function of education is really indoctrination into a capitalist ideology (Sadovnik 2004). In essence, students are socialized to value the interests of the state and to function to sustain it. Such indoctrination begins in early childhood education. Rosabeth Moss Kanter (1972) coined the term the *organization child* to refer to the child in nursery school who is most comfortable with supervision, guidance, and adult control. Teachers cultivate the organization child by providing daily routines and rewarding those who conform. In essence, teachers train future bureaucrats to be obedient to authority.

In addition, to conflict theorists, education serves as a mechanism for **cultural imperialism**, or the indoctrination into the dominant culture of a society. When cultural imperialism exists, the norms, values, traditions, and languages of minorities have been historically ignored. A Mexican American student recalls his feelings about being required to speak English:

> When I became a student, I was literally "remade"; neither I nor my teachers considered anything I had known before as relevant. I had to forget most of what my culture had provided, because to remember it was a disadvantage. The past and its cultural values became detachable, like a piece of clothing grown heavy on a warm day and finally put away. (Rodriguez 1990, p. 203)

Tim Boyle/Getty Images News/Getty Images

What Do You Think? Some schools, in need of additional funding sources, have entered into corporate partnerships. However, as conflict theorists argue, ". . . many school business partnerships are little more than marketing arrangements that have few benefits for schools while carrying with them the potential to harm children in a variety of ways" (Molnar et al. 2013). One health threat is in-school advertising that encourages children to consume food high in fat, sugar, and/or salt. Do you think corporations, should be able to advertise in schools and, if so, should there be some restrictions on what and how products can be promoted?

To cover financial costs associated with the increasing needs of educational programs, school systems often find it necessary to contract with major corporations such as Coca-Cola. However, a recent analysis of the impact of commercialization in schools concluded that ". . . the potential threat to children posed by marketing in schools is great enough that . . . the default assumption must be that marketing in schools is harmful unless explicitly proven otherwise."

The conflict perspective also focuses on what Kozol (1991) called the "savage inequalities" in education that perpetuate racial disparities. Kozol documented gross inequities in the quality of education in poorer districts, largely composed of minorities, compared with districts that serve predominantly white middle-class and upper-middle-class families. Kozol revealed that schools in poor districts tend to receive less funding and to have inadequate facilities, books, materials, equipment, and personnel. For example, disadvantaged children are more likely to attend schools with fewer certified or experienced teachers and to have fewer advanced placement courses than schools attended by more advantaged children (Viadero 2006).

Finally, it should be noted that such inequities exist in other countries as well. Using data from *Education at a Glance,* researchers assessed the differences between reading scores of advantaged and disadvantaged youth (OECD 2013b). Knowledge of effective strategies for summarizing information such as writing a book report was correlated with higher test scores. Advantaged students demonstrated greater knowledge of summarizing techniques when compared to disadvantaged students explaining, in part, the "reading gap."

Symbolic Interactionist Perspective

Whereas structural functionalism and conflict theory focus on macro-level issues, such as institutional influences and power relations, symbolic interactionism examines education from a micro-level perspective. This perspective is concerned with individual and small-group issues, such as teacher–student interactions and the self-fulfilling prophecy.

Teacher–Student Interactions. Symbolic interactionists have examined the ways that students and teachers view and relate to one another. For example, children from economically advantaged homes may be more likely to bring to the classroom social and verbal skills that elicit approval from teachers. From the teachers' point of view, middle-class children are easy and fun to teach. They grasp the material quickly, do their homework, and are more likely to "value" the educational process. Children from economically

cultural imperialism The indoctrination into the dominant culture of a society.

disadvantaged homes often bring fewer social and verbal skills to those same middle-class teachers, who may, inadvertently, hold up social mirrors of disapproval. Teacher disapproval contributes to lower self-esteem among disadvantaged youth.

Self-Fulfilling Prophecy. The **self-fulfilling prophecy** occurs when people act in a manner consistent with the expectations of others. For example, a teacher who defines a student as a slow learner may be less likely to call on that student or to encourage the student to pursue difficult subjects. The teacher may also be more likely to assign the student to lower ability groups or curriculum tracks (Riehl 2004). As a consequence of the teacher's behavior, the student is more likely to perform at a lower level.

A classic study by Rosenthal and Jacobson (1968) provided empirical evidence of the self-fulfilling prophecy in the public school system. Five elementary school students in a San Francisco school were selected at random and identified for their teachers as "spurters." Such a label implied that they had superior intelligence and academic ability. In reality, they were no different from the other students in their classes. At the end of the school year, however, these five students scored higher on their intelligence quotient (IQ) tests and made higher grades than their classmates who were not labeled as spurters. In addition, the teachers rated the spurters as more curious, interesting, and happy, and more likely to succeed than the non-spurters. Because the teachers expected the spurters to do well, they treated the students in a way that encouraged better school performance.

For years, ability grouping—which takes place within classes in contrast to tracking which takes place between classes—has been criticized for its potential negative effect on students and, consequently, went out of favor several decades ago (Loveless 2013). The fear, still held by some today, was that ability grouping not only reflects differences in race and class but also contributes to those differences, that is, perpetuates inequality. Nonetheless, a recent study by the Brown Center on Education Policy reports a resurgence of grouping students based on ability. For example, using data from the United States, 61 percent of fourth grade teachers and 76 percent of eighth grade teachers report creating math groups based on student achievement (Loveless 2013).

> **What Do You Think?** In the wake of federal immigration reform (see Chapter 9), New Jersey legislators have proposed the Tuition Equality Act, which would allow undocumented graduates of New Jersey high schools to pay in-state tuition when attending New Jersey public colleges and universities (Kalet 2013). If passed, the legislation would apply to new as well as returning students. Do you think the Tuition Equality Act should be passed? If so, do you think other states should pass a similar act?

Who Succeeds? The Inequality of Educational Attainment

Figure 8.3 shows the extent of the variation in highest level of education attained by individuals 25 years of age and older in the United States. As noted earlier, conflict theory focuses on such variations in discussions of educational inequalities. Educational inequality is based on social class and family background, race and ethnicity, and gender. Each of these factors influences who succeeds in school.

Social Class and Family Background

One of the best predictors of educational success and attainment is socioeconomic status. Children whose families are in middle and upper socioeconomic brackets are more likely to perform better in school and to complete more years of education than children from families of lower socioeconomic classes. Using parents' education as a proxy for socioeconomic status, global statistics indicate that students whose parents attended college are significantly more likely to also attend college (OECD 2013b).

self-fulfilling prophecy A concept referring to the tendency for people to act in a manner consistent with the expectations of others.

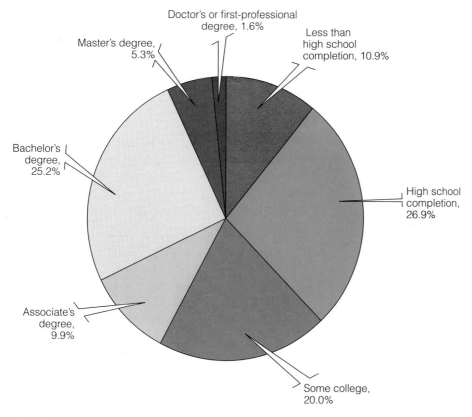

Note: Percentages do not sum to 100 percent because of rounding.

Figure 8.3 **Highest Level of Education Attained by Individuals Age 25 Years Old and Older, 2011**
Source: NCES 2012.

Socioeconomic status also predicts academic achievement. On standardized tests such as the SAT and the ACT, "children from the lowest-income families have the lowest average test scores, with an incremental rise in family income associated with a rise in test scores" (Corbett et al. 2008, p. 3). Muller and Schiller (2000) reported that students from higher socioeconomic backgrounds are more likely to enroll in advanced courses for mathematics credit and to graduate from high school—two indicators of future educational and occupational success. In addition, a report to Congress on educational inequality acknowledged that students from high-income families compared to low- and middle-income families are not only more likely to attend college, but they are also more likely to complete it (ACSFA 2010).

One of the best predictors of educational success and attainment is socioeconomic status. Children whose families are in middle and upper socioeconomic brackets are more likely to perform better in school and to complete more years of education than children from families of lower socioeconomic classes.

What Do You Think? Reardon (2013) argues that the growing achievement gap between the rich on the one hand, and middle- and lower-class students on the other, is not a consequence of disparities in test scores, failing schools, or a widening racial gap in school performance. The gap, he argues, is a consequence of ". . . rich students . . . increasingly entering kindergarten much better prepared to succeed in school . . ." (p. 1), thus creating a "rug rat race" (Ramey & Ramey 2010). Do you think the federal government should establish a nationalized program of preschools for all 3- and 4-year-olds to level the playing field?

Families with low incomes have fewer resources to commit to educational purposes—less money to buy books or computers or to pay for tutors. Disadvantaged parents are less involved in learning activities. As parental education and income levels increase, the likelihood of a parent taking a child to a library, play, concert or other live show, art gallery, museum, or historical site also increases. Further, parents who have less education and lower income levels are less likely to be involved in school-related activities such as attending a school event or volunteering at a school (NCES 2011).

Many poor children have to depend on donations for much-needed school supplies. Here, a first grader smiles as he opens his box of school supplies, including a backpack from "Kits for Kidz" of Chicago.

Head Start and Early Head Start. In 1965, Project **Head Start** began to help preschool children from the most disadvantaged homes. Head Start provides an integrated program of health care, parental involvement, education, and social services. In 2012, over one million preschoolers in 50 states participated in the program, 82 percent between the ages of 3 and 4 years old. The majority of families served were white non-Hispanic (Office of Head Start 2013).

Assessments of Head Start and Early Head Start, a program for infants and toddlers from low-income families, generally conclude that children and families benefit from inclusion in the programs. There are, however, some concerns as to whether or not those benefits are sustainable. Results from a ". . . large-scale, randomized, controlled study of nearly 5,000 children from low-income families . . . found that the positive effects on literacy and language development demonstrated by children who entered Head Start at age 4 had dissipated by the end of the third grade" (Maxwell 2013, p. 1).

School District Funding. Local dollars make up 43 percent of school funding (Dixon 2013), and vary by socioeconomic status of the district (NCES 2011). For example, local expenditures on schools come from taxes, usually property taxes, and as housing prices decline, property taxes decline. The U.S. system of decentralized funding for schools has several additional consequences:

- School districts with low socioeconomic status are more likely to be in urban areas and in inner cities where the value of older and dilapidated houses has depreciated; less desirable neighborhoods are hurt by "white flight," with the result that the tax base for local schools is lower in deprived areas.
- School districts with low socioeconomic status are less likely to have businesses or retail outlets where revenues are generated; such businesses have closed or moved away.
- Because of their proximity to the downtown area, school districts with low socioeconomic status are more likely to include hospitals, museums, and art galleries, all of which are tax-free facilities. These properties do not generate revenues.
- Neighborhoods with low socioeconomic status are often in need of the greatest share of city services; fire and police protection, sanitation, and public housing consume the bulk of the available revenues, often leaving little for education.
- In school districts with low socioeconomic status, a disproportionate amount of the money has to be spent on maintaining the school facilities, which are old and in need of repair.

Although states provide additional funding to supplement local taxes, this funding is not always enough to lift schools in poorer districts to a level that even approximates the funding available to schools in wealthier districts. For example, in *Leandro v. State* (1997), the North Carolina Supreme Court held that the state constitution required that all schools must provide adequate resources to fully educate disadvantaged students (i.e., those who are poor, in special education, and have limited English proficiency). Yet, over nearly two decades, problems continue to exist. One county in North Carolina was doing such a poor job of educating its disadvantaged students

Head Start Begun in 1965 to help preschool children from the most disadvantaged homes, Head Start provides an integrated program of health care, parental involvement, education, and social services for qualifying children.

that a judge has called it "academic genocide" (Waggoner 2009). In 2011, attorneys for several poor counties argued that legislative budget cuts in education would prevent North Carolina children from getting a "basic education" in violation of the *Leandro* decision (Stancill 2011). In 2012, the North Carolina Court of Appeals concurred (North Carolina Justice Center 2012).

Race and Ethnicity

It is projected that by 2021, racial and ethnic minorities will comprise 52 percent of the pre-kindergarten through 12th grade student population (NCES 2013b). To a large extent, this demographic shift in the racial and ethnic makeup of public school students is a result of the growth in the number of Hispanic students. The white student population is expected to decrease from 61 percent in 2000 to 48 percent in 2021. Alternatively, the number of Hispanic students in public schools is projected to increase from 16 percent in 2000 to 27 percent in 2021 (NCES 2013b).

In comparison to whites, Hispanics and blacks are less likely to succeed at every level. As early as the start of kindergarten, Hispanics and blacks have lower mean achievement scores than other racial or ethnic groups in reading and mathematics (NCES 2013b). By fourth grade, over 80 percent of blacks and Hispanics are reading below grade level compared to 58 percent of whites. Similarly, by eighth grade, over 83 percent of black and Hispanic students compared to 57 percent of white students are below grade level in mathematics (CDF 2012a). Further, as Table 8.1 indicates, although educational attainment has increased over time for all groups, racial and ethnic disparities remain.

It is important to note that socioeconomic status interacts with race and ethnicity. Because race and ethnicity are so closely tied to socioeconomic status, i.e., a disproportionate number of racial and ethnic minorities are poor, it *appears* that race or ethnicity determines school success. For example, cities have the highest proportion of minority students as well as the highest proportion of students receiving free or reduced-price lunch, a proxy for socioeconomic status (NCES 2013b). Although race and ethnicity may have an independent effect on educational achievement, their relationship is largely a result of the association between race and ethnicity and socioeconomic status.

In addition to the socioeconomic variables, there are several reasons why minority students have academic difficulty. First, minority children may be English language learners (ELL). The U.S. Department of Education estimates that the number of English language learners in the United States is over 4.5 million, an increase of over 50 percent over the last decade (Ferlazzo & Sypnieski 2012). ELL students come from over 400 different language backgrounds, the overwhelming majority being native Spanish speakers. Now imagine, as Goldenberg (2008) suggests,

> . . . you are in second grade and don't speak English very well and are expected to learn in one year . . . irregular spelling patterns, diphthongs, syllabication rules, regular and irregular plurals, common prefixes and suffixes, antonyms and synonyms; how to follow written instructions, interpret words with multiple meanings, locate information in expository texts . . . read fluently

> Although race and ethnicity may have an independent effect on educational achievement, their relationship is largely a result of the association between race and ethnicity and socioeconomic status.

TABLE 8.1 **Educational Attainments by Race, Ethnicity, and Sex, 1970 and 2012**

	1970		2012	
	Males	Females	Males	Females
High school completion or higher				
White	54.0	55.0	92.2	92.7
Black	30.1	32.5	85.1	86.1
Hispanic	37.9	34.2	64.6	66.0
Asian and Pacific Islander	61.3	63.1	90.6	87.9
Total	51.9	52.8	87.3	88.0
Bachelors or higher degree				
White	14.4	8.4	35.5	33.5
Black	4.2	4.6	19.5	22.9
Hispanic	7.8	4.3	13.3	15.8
Asian and Pacific Islander	23.5	17.3	53.1	48.6
Total	13.5	8.1	31.4	30.6

Source: NCES 2013a.

and correctly at least 80 words per minute, add approximately 3,000 words to your vocabulary . . . and write narratives and friendly letters using appropriate forms, organization, critical elements, capitalization, and punctuation, revising as needed. (p. 8).

Not surprisingly, ELL students score significantly below non-ELL students on standardized tests in both reading and mathematics (NCES 2013b). To help ELL students, some educators advocate **bilingual education**, teaching children in both English and their non-English native language.

Advocates claim that bilingual education results in better academic performance of minority students, enriches all students by exposing them to different languages and cultures, and enhances the self-esteem of minority students. Critics argue that bilingual education limits minority students and places them at a disadvantage when they compete outside the classroom, reduces the English skills of minorities, costs money, and leads to hostility with other minorities who are also competing for scarce resources.

What Do You Think? Independent of language skills, students from backgrounds other than the white-Anglo majority often have subtle challenges when interacting with teachers from the dominant culture. For some students, talking all at once, i.e., "overlapping conversations," indicates enthusiasm for the topic at hand but may be interpreted as disrespectful by the teacher. A student looking at the ground may lead a teacher to exclaim, "Look at me when I'm talking to you!" but is a sign of deference to authority in some cultures (Kugler 2013). Do you think teachers should be required to take cultural sensitivity courses?

The second reason why racial and ethnic minorities don't perform well in school, and compounding the difficulty ELL students have, is that many of the tests used to assess academic achievement and ability are biased against minorities. Questions on standardized tests often require students to have knowledge that is specific to the white middle-class culture, knowledge that racial and ethnic minorities may not have. The disadvantages that minority students face on standardized tests have not gone unnoticed. In 2012, a complaint was filed with the U.S. Department of Education alleging civil rights violations in New York City's elite public high schools that base admissions decisions solely on standardized tests scores (Rooks 2012a).

A third factor that hinders minority students' academic achievement is overt racism and discrimination. Discrimination against minority students may take the form of unequal funding, as discussed earlier, as well as racial profiling, school segregation, and teacher and peer bias. A study of 668 Latino students and their perceptions of discrimination found that boys, language minority speakers, and students in more ethnically diverse populations but with a less diverse teaching staff were more likely to perceive discrimination. Most importantly, perceptions of discrimination were linked to negative academic outcomes via perceptions of a limiting school climate (Brenner & Graham 2011).

Studies indicate that minority students, and specifically black students, may be the victims of what is being called "learning while black" (Morse 2002). The allegation is often supported by an examination of school discipline statistics. African American children represented 18 percent of public school children in 2010, but were 40 percent of all students who experienced corporal punishment, 46 percent of all children who received multiple out-of-school suspensions, and 39 percent of all students expelled from school (CDF 2012b). The debate, of course, is whether these differences can be attributed to overt discrimination or a reflection of differences in behavior.

Racial Integration. In 1954, the U.S. Supreme Court ruled in *Brown v. Board of Education* that segregated education was unconstitutional because it was inherently unequal. In 1966, a landmark study titled *Equality of Educational Opportunity* (Coleman et al. 1966) revealed that almost 80 percent of all U.S. schools attended by whites contained 10 percent

bilingual education In the United States, teaching children in both English and their non-English native language.

or fewer blacks and that, with the exception of Asian Americans, whites outperformed minorities on standardized tests. Coleman and colleagues emphasized that the only way to achieve quality education for all racial groups was to desegregate the schools. This recommendation, known as the **integration hypothesis**, advocated busing to achieve racial balance.

Despite the Coleman report, court-ordered busing, and a societal emphasis on the equality of education, U.S. public schools remain largely segregated. The majority of black and Hispanic students attend schools that are predominantly minority in enrollment. This is particularly true in large urban areas where minorities comprise over 85 percent of the student population in such cities as New York City, Chicago, Philadelphia, and Los Angeles (Rooks 2012b).

A study by the Civil Rights Project at UCLA indicates that not only does school segregation exist, it's increasing (Orfield et al. 2012). Nationwide, 15 and 14 percent of black and Latino students, respectively, attend schools in which white students makeup between zero and one percent of the school population (Orfield et al. 2012). Attending a desegregated school, although having no negative achievement effect on white students, has been found to have a have positive effect on black and Hispanic students in terms of learning and graduation rates (Orfield & Lee 2006).

Increasingly, school classrooms will be characterized by racial and ethnic diversity. By 2040, less than half of all school-age children will be non-Hispanic whites.

Socioeconomic Integration. In 2007, the U.S. Supreme Court held that public school systems "cannot seek to achieve or maintain integration through measures that take explicit account of a student's race" (Greenhouse 2007, p. 1). At the time, the court's decision reflected a general trend toward using socioeconomic or income-based integration rather than race-based integration variables.

Kahlenberg (2006, 2013) has long advocated this approach for several reasons. *First,* "socioeconomic integration more directly and effectively achieves the first aim of racial integration: raising the achievement of students" (Kahlenberg 2006, p. 10). *Second,* socioeconomic integration, because of the relationship between race and income, achieves racial integration, and racial integration, in turn, fosters racial tolerance and social cohesion. *Third,* unlike race-based integration that is subject to "strict scrutiny" by the government, school assignments based on socioeconomic status are perfectly legal. *Fourth,* the problem of low-income students in schools, regardless of race and ethnicity, is growing as poverty spreads beyond urban areas and into suburban neighborhoods. *Finally,* there is evidence that suggests socioeconomic integration is a more cost-effective means of raising student achievement than spending additional dollars in high-poverty schools (Kahlenberg 2013).

Gender

Not only do women comprise two-thirds of the world's illiterate, girls comprise more than 70 percent of the 125 million children who don't attend school (Save the Children 2011). Although progress in reducing the education gender gap has been made, gender parity in primary and secondary schools has not been achieved. Globally, of the 167 countries with data available, 68 had not achieved gender parity in elementary school education; girls were disadvantaged in 60 of the 68 countries. However, in the 97 countries where secondary school gender disparity existed, boys were less likely to be in school in over half of the cases (UNESCO 2012b).

Historically, U.S. schools have discriminated against women. Before the 1830s, U.S. colleges accepted only male students. In 1833, Oberlin College in Ohio became the first college to admit women. Even so, in 1833, female students at Oberlin were required to wash male students' clothes, clean their rooms, and serve their meals and were forbidden to speak at public assemblies (Fletcher 1943; Flexner 1972).

In the 1960s, the women's movement sought to end sexism in education. Title IX of the Education Amendments of 1972 states that no person shall be discriminated against on

integration hypothesis A theory that the only way to achieve quality education for all racial and ethnic groups is to desegregate the schools.

the basis of sex in any educational program receiving federal funds. These guidelines were designed to end sexism in the hiring and promoting of teachers and administrators. Title IX also sought to end sex discrimination in granting admission to college and awarding financial aid. Finally the guidelines called for an increase in opportunities for female athletes by making more funds available for their programs.

Although gender inequality in education continues to be a problem worldwide, the push toward equality has had considerable effect in the United States. For example, in 1970, nearly twice as many men than women had four or more years of college; by 2012, those differences significantly declined (see Table 8.1). Further, scores on the National Assessment of Educational Progress (NAEP) exam indicate that the gender gap in both mathematics and reading scores has decreased over the last few decades. Where differences exist, for example, in a specific grade level, they generally follow educational stereotypes—boys outscore girls in mathematics and girls outscore boys in reading (NAEP 2013).

Traditionally, the gender gap in achievement has been explained by differences in gender role socialization (see Chapter 10). Recently, however, popular portrayals of male and female differences in achievement have been attributed to "hardwiring," as though there are "pink and blue brains." As neuroscientist Lise Eliot (2010) notes

> . . . A closer look reveals that the gaps vary considerably by age, ethnicity, and nationality, for example, among the countries participating in PISA, the reading gap is more than twice as large in some countries (Iceland, Norway, and Austria) as in others (Japan, Mexico, and Korea); for math, the gap ranges from a large male advantage in certain countries (Korea and Greece) to essentially no gap in other countries—or even reversed in girls' favor (Iceland and Thailand). What's more, a recent analysis of PISA data found that higher female performance in math correlates with higher levels of gender equity in individual nations. (p. 32)

> Boys are more likely to lag behind girls in the classroom, be diagnosed with attention-deficit/hyperactivity disorder (ADHD), have learning disabilities, feel alienated from the learning process, and drop out or be expelled from school.

Much of the research on gender inequality in the schools focuses on how female student are disadvantaged in the educational system. But what about male students? Boys are more likely to lag behind girls in the classroom, be diagnosed with attention-deficit/hyperactivity disorder (ADHD), have learning disabilities, feel alienated from the learning process, and drop out or be expelled from school (Dobbs 2005; Mead 2006; Tyre 2008). Further, black and Latino males compared to white males score lower on the NAEP, are less likely to be in gifted and talented programs or to be in advanced placement (AP) classes, and are less likely to graduate from high school or college (Schott Report 2012). Thus, the problems boys have in school may indeed require schools to devote more resources and attention to them (e.g., recruit male teachers).

Problems in The American Educational System

For years, the PDK/Gallup annual survey on education began by asking a random sample of U.S. adults to "grade" our nation's schools (Bushaw & Lopez 2013). In 2013, 25 percent of respondents assigned a letter grade of D or F to the nation's schools, and another 53 percent assigned a C. Given budget cuts, low academic achievement, high dropout rates, questionable teacher training, school violence, and the challenges of higher education, such concern may be warranted.

Lack of Financial Support

When a national sample of public school parents was asked the biggest problems facing the public schools in their communities, the most common response was a "lack of financial support" (Bushaw & Lopez 2013). Despite the importance the American public places on education, efforts to increase state funding are often rejected. For example,

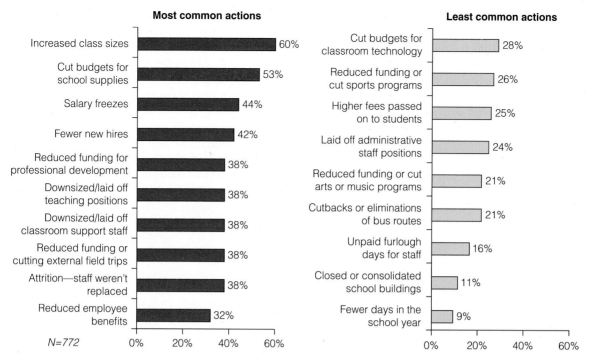

Most common actions

Increased class sizes	60%
Cut budgets for school supplies	53%
Salary freezes	44%
Fewer new hires	42%
Reduced funding for professional development	38%
Downsized/laid off teaching positions	38%
Downsized/laid off classroom support staff	38%
Reduced funding or cutting external field trips	38%
Attrition—staff weren't replaced	38%
Reduced employee benefits	32%

N=772

Least common actions

Cut budgets for classroom technology	28%
Reduced funding or cut sports programs	26%
Higher fees passed on to students	25%
Laid off administrative staff positions	24%
Reduced funding or cut arts or music programs	21%
Cutbacks or eliminations of bus routes	21%
Unpaid furlough days for staff	16%
Closed or consolidated school buildings	11%
Fewer days in the school year	9%

Figure 8.4 **Actions Most and Least Likely to Occur as a Result of Budget Cuts, Teachers and School Employees: June, 2013**
Source: Horace Mann Educator Survey 2013.

in Arizona, where there was a $183 billion budget cut in public school funding, voters defeated a proposed permanent increase in the sales tax to support K–12 education (Molnar et al. 2013).

Figure 8.4 represents the responses to an online survey completed by 1,700 educators, most of whom were public elementary and secondary school teachers. Respondents were asked what actions were taken at their schools as a result of budget cuts in the 2012–2013 school year. Results indicate that among the most common actions were increased class size, a reduction in school supplies, and salary freezes for and reduction of school personnel (Horace Mann Educator Survey 2013).

A survey of those who make budgetary decisions, that is, school district superintendents, further details the impact of state and federal budget cuts (AASA 2012). More than three-quarters of the superintendents surveyed described their districts as "inadequately funded," resulting in the elimination of teaching positions, an increase in class size, the elimination or delay of instructional initiatives, the elimination of summer school, and a four-day school week.

Low Levels of Academic Achievement

The Educational Research Center uses three indicators to measure achievement in public elementary and secondary schools: current levels of performance, improvement over time, and achievement gap between poor and nonpoor learners called the poverty gap (Hightower 2013a). Based on a 100-point scale, the achievement average for the nation was 69.7 (C-), ranging from a high of 85.9 (B) for Massachusetts and a low of 56.6 (F) for Mississippi (Hightower 2013a).

One way to measure performance is to look at the results of the National Assessment of Educational Progress (NAEP) of public and private school students, over time. National trends indicate that reading and mathematics scores between 1971 and 2012 have increased for 9-year-olds and 13-year-olds but have remained stagnant for 17-year-olds. In general, black and Hispanic students made larger gains during the same time period than white students, narrowing the reading and math gap between whites and minorities.

Nonetheless, blacks and Hispanics remain significantly behind their white counterparts in both reading and math. The gender gap has also narrowed over time, although in 2012, females still outperformed males in reading and males still outperformed females in mathematics (NAEP 2013).

It is important to note that these gains, even if statistically significant, may mask poor performances. Nine-year-olds are typically in the fourth grade, 13-year-olds in the eighth grade, and 17-year-olds in the 11th grade. Because long-term assessments are based on age not grade level, NAEP is able to assess the percentage of students at each age who are performing below their typical grade level. In 2012, for example, 73 percent of 11th graders performed below grade level in reading, a trend that has increased over time.

U.S. students are also outperformed by many of their foreign counterparts—something particularly troubling in a knowledge-based global economy that emphasizes STEM (i.e., science, technology, engineering, and math) proficiency. The most recent results from the *Trends in International Mathematics and Science Study* (TIMSS) documents that U.S. students, although scoring above the global average, are consistently outperformed by students in several other countries (Kastberg et al. 2013). For example, nearly half of eighth graders in South Korea, Singapore, and Taiwan, scored in the "advanced" level in math in 2011, compared to only 7 percent of American students (Mullis et al. 2012).

Lastly, the results of *Education at a Glance* indicate that the United States ranks 14th among OECD-participating countries in the percentage of 25- to 34-year-olds with a college education. Countries that surpass the United States in 25- to 34-year-olds' college education rates include Korea, Japan, Canada, the Russian Federation, and Ireland. The report also concludes that although academic achievement is quite high in the United States, other countries' educational attainment is growing at a faster rate than that of the United States (OECD 2012).

What Do You Think? California, as roughly half of all states, requires an exit exam for high school graduation. Researchers found that after the exam was put into place, (1) lower-achieving students had significantly higher dropout rates, and (2) the negative effect of the exit exam was stronger for minority and female students than for nonminority and male students, even when levels of academic achievement were held constant (Reardon et al. 2009). These findings support the "**stereotype threat**" hypothesis, i.e., the tendency for minorities and women to perform poorly on high-stakes tests because of the anxiety created by fear that a poor performance will validate negative societal stereotypes. Do you think high schools should or should not have exit exams as a requirement for graduation? Do you think exit exams unfairly target minorities and females?

. . . [E]very 26 seconds, a U.S. student drops out of high school— over 1 million students annually.

School Dropouts

The *status dropout rate* is the percentage of 16- to 24-year-olds that is not in school and has not earned a high school degree or its equivalent. In the last several decades, the status dropout rate has significantly declined, dropping from 12 percent in 1990 to 7 percent in 2011 (NCES 2012). Nonetheless, every 26 seconds, a U.S. student drops out of high school—over 1 million students annually (AFE 2011; Segal 2013). Dropout rates vary by race and ethnicity as indicated by Figure 8.5.

Dropout rates are associated with increased costs of public assistance, crime, and health care. Because 45 percent of graduates with a GED (general educational development) smoke compared to 10 percent of college graduates, high school dropouts cost Medicare $20 billion a year (Segel 2013). The dropout rate is also associated with lost income and lower tax revenues (AFE 2011; Tyler & Lofstrom 2009). For example, if all of the students who dropped out of high school in the class of 2011 had graduated, it is estimated that the economy would have benefited from an additional $154 billion over their lifetimes (AFE 2011).

stereotype threat The tendency of minorities and women to perform poorly on high-stakes tests because of the anxiety created by the fear that a negative performance will validate societal stereotypes about one's member group.

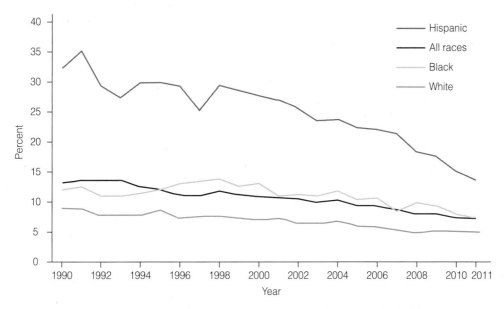

Figure 8.5 **Status Dropout Rates of 16- through 24-Year-Olds, by Race and Ethnicity, 1990 to 2011**
Source: NCES 2012.

Second-chance initiatives such as GED certification allow students to complete their high school requirements. In an article by Heckman et al. (2012), titled "Taking the Easy Way Out," researchers investigated the impact of three different state GED policies on the high school graduation rates: The results indicate that (1) increasing GED testing requirements is associated with lower dropout rates, (2) when the GED is integrated into high school programs, dropout rates increase, and (3) after a GED program is introduced as an option, the dropout rate increases. As the authors conclude, "[T]aken together, these studies suggest that the GED program induces students to drop out of school" (p. 517).

Other dropout interventions include early or middle school college programs that allow dropouts to enroll in community colleges or, in some cases, four-year degree programs. There, they receive a secondary school education, earn a high school diploma, and often accrue college credits (Manzo 2005). There are also efforts, at the state and federal level, to increase the age of compulsory school attendance in the hopes that it will deter students from dropping out of school. In nearly half the states, students ranging from 14 to 18 years old can be exempted from mandatory attendance under certain circumstances such as employment, completion of the eighth grade, parental permission, alternative education or training, or meeting the requirements of an exit interview (Mikulecky 2013).

What Do You Think? After the horrific events at Sandy Hook Elementary School, lawmakers around the country introduced bills designed to increase protection of students and school personnel. Examples of proposed legislation include increasing the police presence at school, arming teachers and principals, and loosening the laws that prohibit guns from being carried on school grounds. Do you agree with such proposals?

Crime, Violence, and School Discipline

Despite the horrors of high-publicity school killings such as those at Columbine, Virginia Tech, and Sandy Hook Elementary School, the chance of a student dying at school is quite rare. Less than 2 percent of the total youth homicide rate, and proportionately even fewer suicides, take place on school property (NCES 2013c). The unlikelihood of such an event is reflected in students' perceptions of safety. In 2008, 93 percent of students said they felt very or somewhat safe while at school, a number that has changed little over the last 15 years (MetLife 2008).

For many, the killings at Sandy Hook Elementary School on December 14, 2012, were a "wake-up" call about gun violence in the United States and the need for better mental health care. But for the parents, friends, and relatives of those killed, it was so much more. Here, the parents of Dylan Hockley, 6 years old, express their gratitude to others.

Statement from the Hockley Family following the Sandy Hook School Shooting

We want to give sincere thanks and appreciation to the emergency services and first responders who helped everyone on Friday, December 14. It was an impossible day for us, but even in our grief we cannot comprehend what other people may have experienced.

The support of our beautiful community and from family, friends, and people around the world has been overwhelming and we are humbled. We feel the love and comfort that people are sending and this gives our family strength. We thank everyone for their support, which we will continue to need as we begin this long journey of healing.

Our thoughts and prayers are with the other families who have also been affected by this tragedy. We are forever bound together and hope we can support and find solace with each other. Sandy Hook and Newtown have warmly welcomed us since we moved here two years ago from England. We specifically chose Sandy Hook for the community and the elementary school. We do not and shall never regret this choice. Our boys have flourished here and our family's happiness has been limitless.

We cannot speak highly enough of Dawn Hochsprung and Mary Sherlach, exceptional women who knew both our children and who specifically helped us navigate Dylan's special education needs. Dylan's teacher, Vicki Soto, was warm and funny and Dylan loved her dearly. We take great comfort in knowing that Dylan was not alone when he died, but was wrapped in the arms of his amazing aide, Anne Marie Murphy. Dylan loved Mrs. Murphy so much and pointed at her picture on our refrigerator every day. Though our hearts break for Dylan, they are also filled with love for these and the other beautiful women who all selflessly died trying to save our children.

Everyone who met Dylan fell in love with him. His beaming smile would light up any room and his laugh was the sweetest music. He loved to cuddle, play tag every morning at the bus stop with our neighbors, bounce on the trampoline, play computer games, watch movies, the color purple, seeing the moon, and eating his favorite foods, especially chocolate. He was learning to read and was so proud when he read us a new book every day. He adored his big brother Jake, his best friend and role model.

There are no words that can express our feeling of loss. We will always be a family of four, as though Dylan is no longer physically with us, he is forever in our hearts and minds. We love you Mister D, our special gorgeous angel.

Source: Dylan Hockley Memorial web page.

In 2011, 4 percent of the student population between 12 and 18 years of age reported being the victim of a nonviolent crime, resulting in 1.2 million incidents (NCES 2013c). The most common offense reported was theft. Public schools had higher rates of victimization than private schools, and there were no significant differences based on student characteristics. There were, however, significant differences in the likelihood of being threatened or injured by a weapon on school property, with nonwhites, males, and students in public schools having higher rates than their student counterparts (NCES 2013c).

Teachers may also be victimized in schools (NCES 2013c). In 2010, the most recent year for which data are available, teachers were more likely to be threatened or injured in urban schools than in suburban or rural schools. Male teachers were more likely to be victimized than female teachers, and student–teacher victimizations were higher in public schools than in private schools. Surprisingly, a greater percentage of elementary school teachers reported being physically attacked than high school teachers (NCES 2013c).

As a consequence of student misbehaviors, teachers, principles, and administrators institute disciplinary policies. The likelihood of out-of-school suspension and expulsion varies dramatically by race, gender, and ethnicity, leading to concerns over differential treatment. In an analysis of U.S. Department of Education data, Shah and McNeil (2013) determined that African Americans and Hispanics were disproportionally represented in out-of-school suspensions and expulsions. For example, blacks represented 18 percent of the sample but comprised 41.5 percent of the expulsions. Similarly, Losen and Skiba's (2010) analysis of middle school data indicates that males compared to females, and racial and ethnic minorities compared to whites were disproportionately suspended.

Disciplinary practices differ between schools and school districts, with some expelling or suspending 90 percent of the student body at least once. Such rates are the result of *zero-tolerance disciplinary policies*. Although the origin of the phrase is from the 1980s and fears over drug use and students bringing guns to school, today's zero-tolerance policies may include suspensions or expulsions for ". . . bringing a cell phone to school, public displays of affection, truancy, or repeated tardiness" (NPR 2013, p. 1). In 2012, these offenses accounted for more than half of all suspensions in California.

Increasingly, educators, counselors, teachers, and psychologists ". . . denounce such practices [suspensions and expulsions] as harmful to students academically and socially, useless as prevention tools, and unevenly applied" (Shah 2013, p. 1). Further motivation for school reform comes from the **school-to-prison pipeline**—the established relationship between severe disciplinary practices, increased rates of dropping out of school, lowered academic achievement, and court or juvenile detention involvement. The school-to-prison pipeline disproportionately hurts minority students who are more likely to be suspended or expelled and, when ". . . forced out of school become stigmatized and fall behind in their studies[,] . . . drop out of school altogether, and . . . may commit crimes in the community" (Amurao 2013, p. 1)

Bullying. When a random sample of American adults was asked about bullying in school, 45 percent said they had been bullied by another student and 16 percent said that they had bullied one of their classmates (Bushaw & Lopez 2012). **Bullying** is characterized by an "imbalance of power that exists over a long period of time between two individuals, two groups, or a group and an individual in which the more powerful intimidate or belittle others" (Hurst 2005, p. 1; see this chapter's *Social Problems Research Up Close* feature). Bullying may be direct (e.g., hitting someone) or indirect (e.g., spreading rumors) and may be considered a type of aggression (Wong 2009). Sometimes called cyberbullying, bullying can also take place remotely, i.e., through electronic communication devices (e.g., cell phone) (see Chapter 14).

Research by Wong (2009), AASA (2009), and NCES (2013c), indicates that:

- In 2010, 23 percent of public schools reported that bullying occurred on a daily or weekly basis.
- Students who bully often perform poorly academically, have high dropout rates, and are more likely to get into fights, drink alcohol, vandalize property, and be truant.
- Victims of bullying report, in order of frequency, being the subject of rumors, being pushed, shoved, tripped, spit on, being threatened with harm, and being intentionally excluded from activities.
- Most bullying takes place inside the school building and is more often directed at females than males. Being bullied is associated with lower self-esteem, anxiety, depression, alcohol and drug use, running away, and suicide.

What Do You Think? In 2011, Robert Champion, a drum major for Florida A&M's renowned "100" marching band, died of injuries sustained from a hazing incident. Twelve former members of the band were charged with manslaughter (Hightower 2013b). Hazing is a situation or circumstance that is intentionally created to embarrass, ridicule, harass, or otherwise demean the recipient, often as part of an initiation process. What do you think is the difference between bullying and hazing?

Inadequate School Facilities

Almost half of all schools in the United States were built in the 1940s and 1950s, and many are in need of costly repairs and renovations (ASCE 2013). At a time when school enrollment is projected to increase, state and local building funds remain in

school-to-prison pipeline The established relationship between severe disciplinary practices, increased rates of dropping out of school, lowered academic achievement, and court or juvenile detention involvement.

bullying Bullying "entails an imbalance of power that exists over a long period of time in which the more powerful intimidate or belittle others" (Hurst 2005, p. 1).

Researchers often study the variables associated with being a student bully or victim. The hope is that, if we can identify the characteristics of a student bully or victim early on, we can develop successful interventions that will reduce the prevalence of the behavior. To that end, Peskin et al. (2006) investigate variables associated with being a bully, a victim, or both, in a sample of middle and high school students.

Sample and Methods

Students from eight predominantly black and Hispanic secondary schools located in a large urban school district in Texas were selected for study. Classes were sampled by grade, resulting in a sample size of 1,413 respondents and a response rate of 52 percent. Nearly 60 percent of the sample was female. Middle school students (6th, 7th, and 8th graders) comprised 56 percent of the sample, 9th graders comprised 11 percent of the sample, and 10th through 12th graders comprised 32 percent of the sample. Sixty-four percent of the sample described themselves as Hispanic, with the remainder self-identifying as African American.

Students were asked about their participation in bullying and their rates of victimization in the last 30 days. Response options included two categories: 0 to 2 times, and 3 or more times. Students were asked the frequency of their bullying behavior: (1) upsetting other students for the fun of it, (2) group teasing, (3) harassing, (4) teasing, (5) rumor spreading, (6) starting arguments, (7) getting others to fight, and (8) excluding others. There were four student victimization variables: (1) called names, (2) picked on by others, (3) made fun of, and (4) got hit and pushed.

In addition to measuring the frequency of bullying and victimization events, students were classified into one of four categories:

. . . [A] student was classified as a bully if he/she participated in at least two of the "bullying" behaviors at least three times in the last 30 days. Victims were classified as those students who reported that at least one of the "victim" behaviors happened to them at least three times in the last 30 days. Four mutually exclusive categories were constructed: (1) bullies; (2) victims; (3) those who reported both bullying and being a victim (bully-victim); and (4) students reporting neither behavior. (p. 471)

Findings and Conclusions

Seven percent of the sample was defined as bullies, 12 percent as victims, and 5 percent as bully-victims. Grade level was found to be significantly related to bully or victim status. The prevalence of bullying is highest in the 9th grade (11.5 percent) and lowest in the 6th grade (4.9 percent) and 10th grade (6.2 percent). The highest rate of victimization is in the 6th grade, with 20.8 percent of 6th graders meeting the criteria for being a victim. The lowest rate of being a victim was among 11th and 12th graders (7.5 percent). Bully-victims also varied by grade level, with the highest levels occurring in the 11th and 12th grades (7.9 percent) and lowest levels in the 6th grade (3.8 percent) and 9th grade (3.8 percent).

There were no significant relationships between the dependent variable and gender. Males and females were equally likely to report being bullies, victims, and bully-victims. However, race and ethnicity were significantly associated with the dependent variable. Blacks, compared with Hispanics, were more likely to be bullies, victims, and bully-victims. For example, although only 3.7 percent of Hispanics reported being bully-victims, 8.6 percent of blacks reported being bully-victims.

When specific bullying behaviors are examined, "upsetting students for the fun of it" was the most common kind of bullying activity, followed by teasing, group teasing, and starting arguments. The lowest prevalence of bullying behaviors was getting others to fight. Males were significantly more likely to participate in harassing and teasing behaviors than females and, with the exception of rumor spreading, African Americans were more likely to be involved in bullying behaviors than Hispanics.

Victims report that the most common form of bullying is name calling, followed by being made fun of. Males were significantly more likely to be hit and pushed than females, and blacks were significantly more likely to be made fun of, called names, or be hit and pushed than their Hispanic counterparts. Further, 11th and 12th graders were statistically more likely to report being picked on or made fun of than students in other class levels.

The authors conclude that the results of the study suggest the need for early intervention and the direction it should take:

While steps to decrease physical types of bullying may be targeted largely at males, steps to reduce verbal and relational types should be targeted at all students. . . . Interventions should be developed in middle school as the prevalence of these behaviors seems to peak as students begin high school. . . . In our study, teasing and name calling were most prevalent; thus, targeted actions for the reduction of these behaviors may be a focus for intervention activities. (p. 479)

Finally, the authors note that future research should concentrate on the role of race and ethnicity in the prevalence of bullying and victimization. Rather than simply noting racial and ethnic differences, however, research must begin to articulate the way in which racial and ethnic differences impact "the content" of bullying behavior. Only then can we fully appreciate the role of social factors "in the development of bullying and victimization problems" (p. 480).

Source: Peskin et al. 2006.

decline, further exacerbating the problem. According to the American Society of Civil Engineers:

National spending on school construction has diminished to approximately $10 billion in 2012, about half the level spent prior to the recession, while the condition of school facilities continues to be a significant concern for communities. Experts now estimate the investment needed to modernize and maintain our nation's school facilities is at least $270 billion or more. However, due to the absence of national data on school facilities for more than a decade, a complete picture of the condition of our nation's schools remains mostly unknown. (ASCE 2013, p. 104)

Older schools have greater needs than newer ones, and are more often located in disadvantaged neighborhoods. Because of development patterns, many of the most disadvantaged schools are in large urban areas. A study of the condition of New York City public schools found that it is directly associated with poverty—the more impoverished the neighborhood, the worse the school (SEIU 2013). In 2013, the city of Philadelphia closed 23 public schools, some because of underuse and others in need of millions of dollars in repair (Hurdle 2013). Because courts have consistently held that the quality of building facilities is part and parcel of equal educational opportunities, state governments monitor spending on infrastructure needs to ensure equitable distributions of funds (AFT 2009).

More school buildings and facilities are in need of repair. Mold, defective ventilation systems, faulty plumbing, and the like are not uncommon. Still, quality education is expected to continue in the classrooms despite such deplorable conditions.

There is a considerable body of evidence that documents the relationship between the school physical environment and academic achievement. Milkie and Warner (2011), after interviewing over 10,000 parents and teachers of first grade students, conclude that students' stress levels are negatively impacted by deteriorated school facilities. Tanner (2008) found a significant relationship between school environment (e.g., space, movement patterns, light, etc.) and academic achievement of third graders even when controlling for socioeconomic status of the school. Air quality, noise, overcrowding, inadequate space, environmental contaminants, and lighting all affect a child's ability to learn, a teacher's ability to teach, and a staff member's ability to be effective.

Recruitment and Retention of Quality Teachers

School districts with inadequate funding and facilities, low salaries, lack of community support, and minimal professional development have difficulty attracting and retaining qualified school personnel. According to the U.S. Department of Education, the number of schoolteachers leaving the profession has increased over the last 20 years, reaching 8 percent in 2011 (NCES 2013b). Teachers leaving the profession tend to be at either end of the experience continuum, i.e., they have less than three years experience or more than 20 years experience. High teacher turnover is a problem in a number of ways (Boyd et al. 2009). First, newer teachers are less experienced and often less effective. Second, teacher turnover contributes to a lack of continuity in programs and educational reforms. Finally, recruiting and training expenses, in addition to the time and effort devoted to replacing teachers who have left the profession, is considerable.

> School districts with inadequate funding and facilities, low salaries, lack of community support, and minimal professional development have difficulty attracting and retaining qualified school personnel.

Using global data from several large-scale learning assessments, Dalton and Marcenaro-Gutierrez (2011) conclude that there is a direct association between teachers' salaries and student performances—the higher one, the higher the other. Unfortunately, in the United States, because teacher salaries are the largest component of a school district's costs, poor school districts have less money to offer qualified teachers (McKinsey & Company 2009). Thus, they are more likely to employ beginning teachers with less than three years of experience, and are more likely to assign these teachers to areas outside of their specialty.

Knowledge of subject taught is one of the key characteristics of an effective teacher (Stotsky 2009). Students in poorer school districts are twice as likely to be taught by substitute teachers, and to have less effective teachers, when compared to students in more affluent districts (Barton 2004; Sawchuk 2011). Further, when a high-quality teacher leaves a middle-class school, one in six new recruits will be of the same high quality; however, when a high-quality teacher leaves a disadvantaged school, only one in eleven possible replacements will be of the same high quality (TNTP 2013).

Recruiting and retaining quality teachers in poverty-level schools is critical to the success of its students. Hanushek et al. (2005) report that if a child from a poor family has a good teacher for five consecutive years, the achievement gap between that child and a child from a higher-income family would be closed. Teacher job satisfaction, however, is at its lowest point in 25 years, making the retention of good teachers difficult, particularly in a challenging school environment (MetLife 2013). Further, because minority students disproportionately populate poor school districts, it is also important to recruit and retain teachers who meet the needs of children from diverse backgrounds and of varying abilities. The number of minority teachers who can serve as role models, have similar life experiences, and have similar language and cultural backgrounds is far too few for the number of minority students.

Teacher Effectiveness. Recruiting and retaining quality teachers may be made more difficult with the recent emphasis on accountability and implementation of **value-added measurement (VAM)**. VAM is the use of student achievement data to assess a teacher's effectiveness. The American Recovery and Reinvestment Act of 2009, passed to stimulate the economy, provided competitive funds for school districts willing to "race to the top." Use of VAM was a requirement for receiving those funds. Although some would argue that quantitative measures of accountability are a necessary evil in a time of budget shortfalls, critics are quick to note that assessing teachers based on student performance assumes all else constant, and ignores the reality of student differences in such nonschool factors as family life, poverty, emotional and physical obstacles, and the like (Mitchell 2010). Further, there are concerns that teachers, fearing for their jobs and concerned about merit-based pay, may begin "teaching to the test" and worse. In 2013, 35 Georgia teachers, principles, and administrators were indicted in a case where ". . . test answers were altered, fabricated, and falsely certified" (Carter 2013).

What Do You Think? In 2012, teachers in Chicago, the third largest school district in the country, went on strike. Among other things, teachers objected to a proposal that would tie their evaluations to their students' academic performance. What do you think? Should teachers' salaries be determined by how well their students perform?

Even seasoned teachers may not be effective. There is evidence that those who choose teaching as a career, on average, have lower college entrance exam scores than the average college student (Tucker 2013). Further, there is a documented relationship between college entrance scores and the likelihood of continued teaching—the *lower* the college entrance scores, the *higher* the probability of still teaching 10 years postbaccalaureate (NCES 2007). When asked to think about training teachers, 57 percent of a sample of American adults responded that entrance requirements into college teacher preparation programs should be more rigorous. In fact, the majority of respondents held that the entrance requirements into college teacher preparation programs should be as selective as those used for business school, prelaw, engineering, and premed (Bushaw & Lopez 2012).

To place quality teachers in the classroom, many states have implemented mandatory competency testing (e.g., the Praxis Series). The need for teachers who are officially classified as "highly qualified" is tied to federal mandates that place an emphasis on the importance of having licensed teachers in the classroom. Additionally, teachers who have a bachelor's degree and have been in the classroom for three or more years are also eligible for national board certification. Some studies indicate that students of "highly

value-added measurement (VAM) VAM is the use of student achievement data to assess teacher effectiveness.

qualified" teachers and/or board-certified teachers perform better on standardized tests and have shown greater testing gains than students of teachers who are not "highly qualified" and/or board certified (NBPTS 2012; Viadero 2005).

In an effort to meet the demands of placing teachers in classrooms while facing teacher shortages due to baby boomer retirements, states are now allowing skilled professionals who have an interest in teaching but did not receive a teaching degree to enter the teaching profession. Called lateral entry by some states, the program allows the person to obtain a lateral entry teaching license while actually teaching in the classroom. In addition, more than half of the states have adopted **alternative certification programs**, whereby college graduates with degrees in fields other than education can become certified if they have "life experience" in industry, the military, or other relevant jobs. Teach for America (TFA), a program originally conceived by a Princeton University student in an undergraduate honors thesis, is an alternative teacher education program with the aim of recruiting liberal arts graduates into teaching positions in economically deprived and socially disadvantaged schools.

Critics argue that the program may place unprepared personnel in schools. However, an analysis of TFA teachers versus traditional teachers concludes that TFA teachers are more effective in the classroom than traditional teachers as measured by student achievement (Xu et al. 2007). Since 1990, Teach for America has trained over 20,000 teachers (Martin 2012).

> . . . [I]f a child from a poor family has a good teacher for five consecutive years, the achievement gap between that child and a child from a higher-income family would be closed.

The Challenges of Higher Education in America

Although there are many types of postsecondary education, higher education usually refers to two- or four-year, public or private, degree-granting institutions. In 2013, there were nearly 5,373 colleges and universities in the United States (NCES 2013d). Of the 29 million undergraduate and graduate students enrolled in U.S. colleges and universities, full-time students, women, younger students, minorities, and students enrolled at four-year schools have disproportionately contributed to enrollment growth over the years (NCES 2011; NCES 2013d). However, in the 2012–2013 academic year, as the economy began to recover, college student enrollment fell by 2 percent, the first significant drop since the 1990s, as students returned to the workforce (Perez-Pena 2013).

Higher education employs an estimated 2.9 million professional staff (e.g., administrators, faculty, nonteaching professional staff, etc.) and 0.9 million nonprofessional staff (e.g., clerical, service, maintenance, etc.) (NCES 2013a). Over the last decade, there has been a significant decrease in full-time tenured or tenure-track faculty, once the "core" of academia, and significant increases in noninstructional staff (e.g., administrators) and nontenure-track, part- or full-time instructors. In general, as academic rank increases, that is, from instructor to assistant professor, to associate professor, to full professor, salaries increase and the proportion of women and minorities decreases (NCES 2013a).

Cost of Higher Education. The average tuition and required fees for attending a four-year degree-granting institution as an undergraduate increased by 7 percent for in-state students and 4 percent for out-of-state students between 2011 and 2012 (NCED 2013c). The average cost of tuition and fees for a four-year U.S. college in 2013 was $8,655 for in-state students and $21,706 for out-of-state students (NCES 2013d). With rising costs, it is not surprising that parent and student loans for tuition and other fees have increased by 24 percent over the last five years (Adams 2012).

In 2013, President Obama signed the Student Loan Certainty Act, a law that will lower interest rates on student loans (Associated Press 2013). As student debt surpasses the $1 trillion mark, a total larger than the sum of all mortgage debt or all credit card debt, several other legislative actions have been introduced into Congress. Each of these measures are designed to give relief to those burdened by student debt, but may not pass

alternative certification programs Programs whereby college graduates with degrees in fields other than education can become certified if they have "life experience" in industry, the military, or other relevant jobs.

as Sallie Mae, ". . . the largest private student loans lender and one of the chief profiteers of student debt . . ." lobbies legislators in the hopes of squashing the proposed legislation (Gupta 2013, p. 1).

Racial and Ethnic Minorities. Access to higher education among minority and/or low-income students is particularly problematic. Only 14.6 percent of students at four-year degree-granting institutions are black, and only 11.3 percent are Hispanic. Minority representation at two-year institutions, 44.8 percent in total, is significantly higher than at four-year institutions (NCES 2013a). It should be noted, however, that a fairly high proportion of black students, approximately 9 percent, attend "historically black universities and colleges" (HBCU) (NCES 2013b).

College enrollments of racial and ethnic minorities have increased over the last several decades but, as Carnevale and Strohl (2013) note, the increases have been accompanied by "separate and unequal" access to ". . . selective and well-funded four-year colleges" (p. 6). For example, between 1995 and 2009, 82 percent of incoming white college students were enrolled at the top 468 colleges and universities in the United States, compared to 9 percent of African American college students and 13 percent of Hispanic college students.

Further, African American and Hispanic students with "A" averages in high school are more likely to be enrolled in community colleges compared to their similarly qualified white counterparts (Carnevale & Strohl 2013). The 468 most selective four-year colleges where white students disproportionately attend have (1) more financial resources, (2) higher graduation rates, (3) higher enrollment in and completion of graduate degrees, and (4) graduates with higher future earnings (Carnevale & Strohl 2013). (See Chapter 9 for a discussion of race, ethnicity, and affirmative action.)

Community Colleges. Despite being called the nation's "unsung heroes of the American education system" (Obama quoted in Adams 2011, p. 1), just over half of the respondents in a survey of U.S. adults agreed with the statement that "community colleges offer high-quality education" (Lumina 2013). In contrast to such perceptions, enrollment in community colleges continues to increase, with 7.8 million students in 2011. The number of associate degrees awarded in the same year was nearly twice that awarded a decade ago (NCES 2013a).

Community colleges play a vital role in U.S. educational policy. They are starting points for many American who hope to eventually transfer to baccalaureate institutions. Minority and low-income students and women are more likely to enroll in community colleges than their white, male middle-class counterparts for a variety of reasons. Community colleges are often closer to where students live, and they offer a more flexible schedule, allowing students to work full-or part-time while attending school and avoiding the cost of room and board by living at home. According to the American Association of Community Colleges, 15 percent of community college students are 40 years old or older, 80 percent of full-time students work full-or part-time outside of the classroom, 40 percent are first-generation college students, and nearly half are receiving some kind of financial aid (AACC 2013). Thus, community colleges provide a pathway to education that might not otherwise be available for a segment of the population.

Strategies for Action: Trends and Innovations in American Education

Americans consistently rank improving education as one of their top priorities. Recent attempts to improve schools include raising graduation requirements, barring students from participating in extracurricular activities if they are failing academic subjects, providing three-year bachelor's degree programs, lengthening the school day, prohibiting dropouts from obtaining driver's licenses, implementing year-round schooling, and extending the number of years permitted to complete a high school degree.

However, educational reformers on both sides of the political aisle continue to call for changes that go beyond these get-tough policies. On the one hand, many Republicans believe that increasing competition and accountability lead to better schools, teachers, and students. Democrats, on the other hand, argue that increased funding and a commitment to ending educational inequality is what is needed. Differences aside, all agree, something needs to be done.

Educational Policy across the States

The challenges facing national educational policies are considerable. In 2011, President Obama, speaking at Kenmore Middle School in Arlington, Virginia, stated that ". . . unfortunately too many students aren't getting a world-class education today. As many as a quarter of American students aren't finishing high school. The quality of our math and science education lags behind many other nations. And America has fallen to 9th in the proportion of young people with a college degree. Understand, we used to be first, and we now rank 9th. That's not acceptable" (quoted in Adams 2011, p. 1). Although it may be difficult to predict the totality of this administration's educational policy and its impact, there is evidence of both significant changes and "things as they were" politics.

Reforming No Child Left Behind. The No Child Left Behind (NCLB) Act of 2001 was signed into law in January 2002. However, soon after the act was signed, it became clear that its implementation was problematic and empirical support was mixed. In 2010, President Obama issued his "blueprint" for educational reform, which addressed many of the issues NCLB created. It also provided ". . . flexibility regarding specific requirements of [NCLB] . . . in exchange for rigorous and comprehensive state-developed plans designed to improve educational outcomes for all students, close achievement gaps, increase equity, and improve the quality of instruction" (U.S. Department of Education 2012, p. 1).

NCLB is the most current version of the Elementary and Secondary Education Act (ESEA) originally passed in 1965. As of 2013, the act has not been reauthorized predominantly because of political wrangling. Points of contention include the adoption of common core state standards, standardized testing and accountability, and privatization of schools.

Common Core State Standards. In 2008, the National Governors Association and the Council of Chief State School Officers convened to adopt a common set of academic standards to be used across the states as a means of standardizing educational requirements. Known as the Common Core State Standards (CCSS), the initiative was motivated by concerns that students in different states were not being prepared equally for postsecondary education and/or jobs in the global workforce.

Intended to be fully operational in the 2014–2015 school year, CCSS was initially adopted by 47 states (PARCC 2013; SBAC 2013). To help assure its success, the federal government provided funds to two consortia with which states could collaborate in the development of assessment measures. But as expenses associated with the implementation of the assessments for CCSS increased, several states dropped out of the consortia with others likely to follow (Moxley 2013; Ujifusa 2013).

Testing and Accountability. Most Americans believe that the CCSS will improve the quality of education, allow students to compete globally, and provide more consistency in the quality of education (Bushaw & Lopez 2013). There is less consensus, however, over the development and implementation of assessment measures. The recent emphasis on teacher and school accountability requires that students take multiple standardized tests to make comparisons within and between school districts, and between states. Teachers, students, and even schools are then "graded" based on the results.

Criticisms of "high-stakes tests" and their uses are numerous. *First,* and perhaps most importantly, standardized tests encourage rote learning, superficial thinking, and memorization rather than critical thinking skills (Harris et al. 2011). That said, *second,* there is concern that using standardized tests scores, either in whole or in part, to determine student placements, teachers' salaries, or school closures is highly unreliable (Rothstein et al. 2010). For example, a teacher might be highly rated one year, dropped to the very bottom the next, and then reclaim the top ranking in the third year.

Third, there are concerns about the content of standardized tests. The National Center for Fair and Open Testing argues that ". . . standardized tests are not objective. . . . [D]ecisions on what to include, how questions are worded, [and] which answers are 'correct' . . . are made by subjective human beings" (NCFOT 2012). Often, questions reflect the cultural biases of the test maker.

Finally, standardized testing is not cost-effective, a particularly salient concern in the time of budget cuts and sequestering. The cost of tests, testing services, and test preparatory materials, according to one estimate, is more than $2.3 billion a year and is rapidly increasing (Gardner 2013).

As a result of these and other criticisms, a 30-member group called the Gordon Commission (2013) issued a report on the future of testing and assessment. After an exhaustive review of the literature, the Commission concluded that "radically different tests" were needed and recommended that standardized tests

- provide meaningful information to policy makers, administrators, teachers, and students
- make a distinction between testing for evaluative purposes and testing geared toward improving instructional practices
- allow for the interpretation of results ". . . informed by the understanding of the context in which the student lives, learns, was taught, and was assessed" (p. 130)
- factor ". . . fairness into the test design, delivery, scoring, analysis, and use" of educational assessments (p. 133)
- be just one of multiple sources of information when used for student placement and teacher salaries

The commission also recommends that federal and state governments, in conjunction with agencies, nonprofit organizations, businesses, and educational leaders, commit to the development of valid and reliable assessment instruments.

Advocacy and Grassroots Movements

Concerns about failing schools, standardized testing, and value-added measurement (VAM), i.e., the practice of teachers' salaries being tied to students' performances, has led to a growing movement of parents and educators "opting out" of high-stakes testing. There have been writing protests, new campaigns by teacher unions, boycotts of standardized testing by teacher and student groups, and even Facebook pages with names like "Parents and Kids Against Standardized Testing" and "Scrap the Map"—the Measurement of Academic Progress (Toppo 2013).

Several states have passed **parent trigger laws**. The principle behind parent trigger laws is that, with a sufficient number of signatures, parents can intervene on behalf of their children by taking one or more actions as statutorily permitted. In general, these actions include replacing administrators, principals, and/or teachers, converting the school to a charter school (see the section titled "The Debate over School Choice" in this chapter), or closing the school altogether (NCSL 2013a).

To date, at least 25 states have considered such legislation and seven have enacted parent trigger law policies (NCSL 2013a). For example, Ohio schools that rank ". . . in the lowest 5 percent in performance statewide for three or more consecutive years . . . " are eligible for intervention based on the state's trigger laws (NCSL 2013a, p. 1). Options for action include converting the school to a charter school, replacing ". . . at least 70 percent of the school's personnel related to its poor performance," turning control of the school over to the state, and privatization (NCSL 2013a, p. 1).

As a structural functionalist would argue, trigger laws empower parents who heretofore had little to say about the quality of their children's education. Compatible with conflict theory, opponents hold that trigger laws are part of a political agenda funded by wealthy donors to privatize schools (Lu 2013, p. 1). Ironically, because privatization ". . . outsources school governance to educational management organizations who have no obligation to (and often no physical presence in) the community, the parent trigger ultimately thwarts continued, sustained community and parental involvement" (Lubienski et al. 2012, p. 2).

parent trigger laws State legislation that allows parents to intervene in their children's education and schooling.

Below, indicate whether you have never engaged in the following behaviors, engaged in the behaviors just once, or engaged in the behaviors more than once while in the 9th, 10th, 11th, or 12th grade. When finished, compare your responses to those from a national survey of more than 23,000 American high school students.

Behaviors	Never	Once	More Than Once
1. Lied to a parent about something significant	____	____	____
2. Lied to a teacher about something significant	____	____	____
3. Copied an Internet document for a classroom assignment	____	____	____
4. Cheated during a test at school	____	____	____
5. Copied another's homework	____	____	____
6. Stole something from parents or relatives	____	____	____
7. Stole something from a friend	____	____	____
8. Stole something from a store	____	____	____

Source: Josephson Institute of Ethics 2012

Results from the survey are below. Sample statistics indicate that the majority of students, 67 percent, attended public school, 52 percent were female, students were equally divided between the four grade levels, and nearly 90 percent expected to attend college.

Behaviors	Percentage in Each Category		
	Never	Once	More Than Once
1. Lied to a parent about something significant	24	28	48
2. Lied to a teacher about something significant	45	26	29
3. Copied an Internet document for a classroom assignment	68	16	16
4. Cheated during a test at school	49	24	28
5. Copied another's homework	25	23	51
6. Stole something from parents or relatives	82	10	8
7. Stole something from a friend	86	9	5
8. Stole something from a store	80	11	9

Character Education

For the first time in a decade, incidences of students lying, cheating, and stealing declined, as self-reported on the Biennial Report Card on American Youth (Josephson Institute of Ethics 2012). Nonetheless, the survey of more than 23,000 high school students indicates that 14 percent of respondents admitted to stealing from a friend, 76 percent admitted lying to their parents about something important, 52 percent admitted to cheating on a test in the previous year, and 36 percent admitted to plagiarizing an assignment from the Internet (Josephson Institute of Ethics 2012). (See this chapter's *Self and Society* feature and compare your answers to the national sample.) To many educators, results like these signify the need for **character education**.

Character education entails teaching students to act morally and ethically, including the ability to "develop just and caring relationships, contribute to community, and assume the responsibilities of democratic citizenship" (Lickona & Davidson 2005). Despite most schools' emphasis on academic achievement, knowledge without character is potentially devastating. Sanford McDonnell (2009), former CEO of McDonnell Aircraft Corporation and chairman emeritus of the Character Education Partnership, recounts a letter written by a principal and former concentration camp survivor to his teachers at the start of a new school year:

> Despite most schools' emphasis on academic achievement, knowledge without character is potentially devastating.

My eyes saw what no person should witness: gas chambers built by learned engineers, children poisoned by educated physicians, infants killed by trained nurses, women and babies shot and burned by high school and college graduates. Your efforts must never produce learned monsters and skilled psychopaths. Reading, writing, and arithmetic are important only if they serve to make our children more humane. (p. 1)

character education Education that emphasizes the moral and ethical aspects of an individual.

Given the school violence, bullying, discipline problems, and lack of academic integrity plaguing schools today, "character" is an essential ingredient of a positive school environment. Character education in schools is associated with ". . . higher academic performance, improved attendance, reduced violence, fewer disciplinary issues, reduction in substance abuse, and less vandalism" (Lahey 2013, p. 1).

Virtual Education

Computers in the classroom allow students to access large amounts of information. The proliferation of computers both in school and at home means that teachers have become facilitators and coaches rather than just the sole providers of information. Professional development for educators now includes classes on integrating technology into the curriculum. Not only do computers enable students to access enormous amounts of information, but they also allow students to progress at their own pace. However, computer technology is not equally accessible to all students and varies dramatically by parents' education, parents' income, and race and ethnicity. Although there are lingering concerns over the "digital divide" (see Chapter 14) and its impact on students' achievement, some research indicates that having a computer in the home decreases student math and reading scores (Vigdor & Ladd 2010).

E-learning separates students and teachers by time and/or place. They are, however, connected by some communication technology (e.g., videoconferencing, e-mail, real-time chat room, or closed-circuit television). Some degree programs and classes have blended learning, a mix of online and traditional face-to-face learning. E-learning encompasses the full spectrum of computer-based education from online tests to massive open online courses (MOOC). Figure 8.6, based on a national survey of public school districts, indicates the number of school districts with various learning technologies, the percentage of school districts in the process of implementing those technologies, and changes over time (CDE 2012).

Research suggests that, for many children in poor, often urban schools, ". . . learning [is about following] the rules and following directions. Not critical thinking. Not creativity. It's about how to correctly eliminate three out of four bubbles" (Hopkinson 2011, p. 1).

Despite concerns about the rigors of online education, the number of students enrolled in online classes and degree programs continues to rise. In 2011, 6 million U.S. college students took at least one online course, and 65 percent of colleges indicated that online learning is an important part of their long-term strategic planning (Allen & Seaman 2011). Unlike MOOCS that traditionally offer courses at no cost to or credit for the user, universities are increasingly joining with online for-profit providers to offer classes and degree programs (Lytle 2012). For example, Semester Online provides face-to-face online classes for college credit from top-ranked universities such as Emory University, Duke University, Boston College, the University of Notre Dame, and Vanderbilt University (Semester Online 2013).

Online education often serves a segment of the population that would not otherwise be able to attend school—older, married, full-time employees, and people from remote areas. Thus, for example, the number of public school online course offerings is higher in rural areas than in towns, cities, or urban areas (Bausell & Klemick 2007). Further, some research suggests that online learning benefits those who have historically been disadvantaged in the classroom. A study by DeNeui and Dodge (2006) indicates that females in a blended course were more likely to use a learning management system (e.g., Blackboard) than males, and to significantly outperform them as measured by final grades in an introductory psychology class.

The significance of online education lies in its potential to ameliorate problems in American education. A survey of high school administrators indicates that offering online or blended courses is important because they allow schools to offer courses that would otherwise not be available, permit flexibility and schedules, extend the school day and year, are financially beneficial, build linkages with colleges, and prepare students for 21st-century jobs (Picciano & Seaman 2010).

e-learning Learning in which, by time or place, the learner is separated from the teacher.

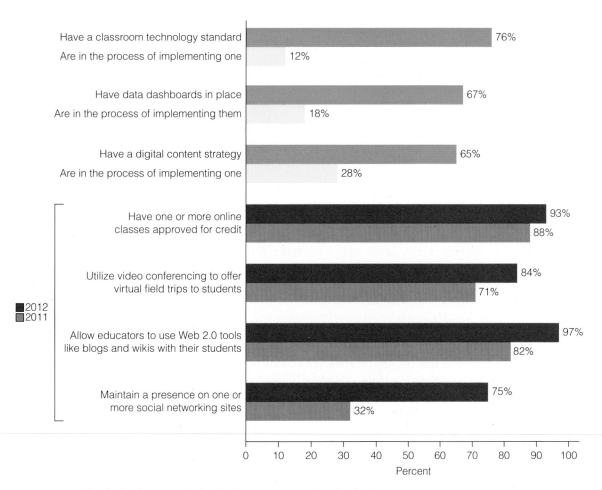

Figure 8.6 School Districts New Technologies, 2011–2012 Academic Year
Source: CDE 2012.

The Debate over School Choice

Traditionally, children have gone to school in the district where they live. School vouchers, charter schools, and private schools provide parents with alternative school choices for their children.

School Vouchers. **School vouchers** are state-funded "scholarships" that allow public school students to attend private schools. Eligibility for vouchers is usually confined to special populations, for example, low-income students, special education students, and students in chronically low-performing schools (NCSL 2013b). According to the National Conference of State Legislatures, 13 states and the District of Columbia presently have school voucher programs.

Proponents of the voucher system argue that it increases the quality of schools by creating competition for students. Those who oppose the voucher system argue that it will drain needed funds and the best students away from public schools. Opponents also argue that vouchers increase segregation because white parents use the vouchers to send their children to private schools with few minorities.

The Milwaukee Parental Choice Program is the ". . . oldest and largest publicly funded group of the voucher program in the United States," beginning in 1991 and continuing to the present (Cowen et al. 2013). Wisconsin state law requires an evaluation of the voucher program and its impact on academic achievement. To assess attainment, students attending voucher schools were matched (e.g., percent black, percent female, etc.) with those attending public schools so that an accurate comparison could be made. The results indicated that students who were exposed to a voucher environment in the

school vouchers Tax credits that are transferred to the public or private school that parents select for their child.

eighth and ninth grades were more likely to graduate from high school, to enroll at a four-year institution, and stay in school beyond their first year of college (Cowen et al. 2013).

Charter Schools. In some states, vouchers can be used for charter schools. **Charter schools** originate in contracts, or charters, which articulate a plan of instruction that local or state authorities must approve. Although foundations, universities, private benefactors, and entrepreneurs can fund charter schools, many are supported by tax dollars. In the 2012–2013 school year, approximately 4 percent of U.S. public school students, 2.3 million in total, attended more than 6,000 charter schools in 41 states. This represents an 80 percent increase in the number of charter school students since 2009. Growth among black and Hispanic students and students living in poverty is particularly high (CREDO 2013).

Charter schools, like school vouchers, were designed to expand schooling options and to increase the quality of education through competition. Like vouchers, charter schools have come under heavy criticism for increasing school segregation, reducing public school resources, and "stealing away" top students. Proponents argue that charter schools encourage innovation and reform, and increase student learning outcomes.

A 27-state analysis of charter schools by the Center for Research on Education Outcomes (CREDO 2013) indicates that, overall, charter schools' performances have improved over time. For example, the average charter school student showed growth in reading and was on par in math with traditional public school students. Additionally, African American students, students living in poverty, and English language learners had achievement performances that were higher than their counterparts in traditional public schools. The authors conclude that their ". . . findings lend support to the education and social policies that focus on education as a mechanism to improve life chances for historically underserved students" (CREDO 2013, p. 18).

Privatization of Schools. Another school choice parents can make is to send their children to a private school. The number of students enrolled in private schools in 2010, the most recent year for which data are available, was 4.7 million, a decrease from previous years. The decline in private school enrollment is associated with an increase in students attending charter schools (Ewert 2013).

Parents send their children to private schools for a variety of reasons including the availability of academic programs and extracurricular activities, smaller classes and a lower student–teacher ratio, religious instruction, and dissatisfaction with public schools (Ewert 2013). Many people believe that private schools are superior to public schools in terms of academic achievement. Contrary to expectations, however, there is evidence that public school students fair as well or better academically as private school students. For example, Lubienski and Lubienski (2006), using National Assessment of Educational Progress (NAEP) data, report that mathematics scores for public school students were higher than mathematics scores for private school students.

Some experts are ". . . alarmed at what they see as increasingly aggressive moves by companies to make money from the K–12 system; others say the expanding role of for-profit ventures is just a natural evolution of the interplay between the private and public sectors in efforts to improve schools" (Davis 2013, p. 52). Regardless of which position you embrace, there's little doubt that corporate entities, through lobbyists and campaign donations, are impacting public policy.

What Do You Think? Another school choice is to educate your children at home. In 2013, an Ohio state legislator introduced a bill that would decrease property taxes for parents who homeschool their children (Blackwell 2013). Do you think parents who homeschool their children should receive financial compensation for doing so? If so, should couples who are childless, or the elderly, have their property taxes reduced also?

charter schools Schools that originate in contracts, or charters, which articulate a plan of instruction that local or state authorities must approve.

Understanding Problems in Education

Educational reform continues to be the focus of legislators and governments across the country. Although there is disagreement as to what needs to be done and how, all can agree that significant reform is needed to meet the needs of a global economy in the 21st century and, perhaps more importantly, to fulfill Horace Mann's dream of education as the "balanced wheel of social machinery," equalizing social differences among members of an immigrant nation.

First, we must invest in teacher education and in teaching practices that have been empirically documented to work in raising student outcomes. Teachers' salaries also need to better reflect the priority Americans place on children, education, and the education of children, and should not be tied to student performance. After all, there are lingering doubts that standardized tests accurately measure student performance, let alone a teacher's performance (Ravitch 2010).

Second, the "savage inequalities" in education, primarily based on race, ethnicity, and socioeconomic status, must be addressed. Segregation, rather than decreasing, is increasing, a reflection of housing patterns, local school districts' heavy reliance on property taxes, and immigration patterns. Public schools should provide all U.S. children with the academic and social foundations necessary to participate in society in a productive and meaningful way; however, for many children, schools perpetuate an endless downward cycle of failure, alienation, and hopelessness.

Third, the general public needs to become involved, not just in their children's education but also in the *institution* of education. An uneducated and unthinking populace hurts all of society, particularly in terms of global competitiveness. As Kohn (2011) observes, children from low-income families continue to be taught "the pedagogy of poverty" (Haberman 1991) or what has been called the "McEducation of the Negro" (Hopkinson 2011). Like Big Macs, children are being packaged in one-size-fits-all wrappers—with learning how to think, explore, question, and debate being replaced by worksheets and standardized tests. Sadly, as education historian Diane Ravitch notes, the present reform movement that "once was an effort to improve the quality of education [has] turned into an accounting strategy" (Ravitch 2010, p. 16).

Fourth, as conflict theorists would note, we must be wary of market principles in schools and of those who advocate them. Educational policy in the United States is increasingly influenced by K–12 corporate providers, many of which have political ties and vested interests. The number of schools operated by for-profit education management companies has increased from 6 in 1996 to 758 in 2012 (Davis 2013).

Finally, as a society, we must attend to and be cognizant of the importance of early childhood development. Poliakoff (2006) notes:

Children's physical, emotional, and cognitive development are profoundly shaped by the circumstances of their preschool years. Before some children are

Students work on their laptops at the Philadelphia School of the Future. Sponsored as part of a public/private partnership between the city of Philadelphia and Microsoft, this model for future schools opened on September 6, 2006. In June 2010, students graduated their first class, and all 117 seniors were accepted at postsecondary institutions.

Tim Shaffer/Microsoft via Getty Images

Like Big Macs, children are being packaged in one-size-fits-all wrappers—with learning to think, explore, question, and debate being replaced by worksheets and standardized tests.

even born, birth weight, lead poisoning, and nutrition have taken a toll on their capacity for academic achievement. Other factors—excessive television watching, little exposure to conversation or books, parents who are absent or distracted, inadequate nutrition—further compromise their early development. (p. 10)

We must provide support to families so that children grow up in healthy, safe, and nurturing environments. Children are the future of our nation and of the world. Whatever resources we provide to improve the lives and education of children are sure to be wise investments in our collective future.

CHAPTER REVIEW

- **Do all countries educate their citizens?**
No. Many societies have no formal mechanism for educating the masses. As a result, millions of adults around the world are illiterate. The problem of illiteracy is greater in developing countries than in developed nations and, worldwide, disproportionately affects women more than men.

- **According to the structural-functionalist perspective, what are the functions of education?**
Education has four major functions. The first is instruction—that is, teaching students knowledge and skills. The second is socialization that, for example, teaches students to respect authority. The third is sorting individuals into statuses by providing them with credentials. The fourth function is custodial care—a babysitting agency of sorts.

- **What is a self-fulfilling prophecy?**
A self-fulfilling prophecy occurs when people act in a manner consistent with the expectations of others.

- **What variables predict school success?**
Three variables tend to predict school success. Socioeconomic status predicts school success: The higher the socioeconomic status, the higher the likelihood of school success. Race and ethnicity predict school success, with nonwhites and Hispanics having more academic difficulty than whites and non-Hispanics. Gender also predicts success, although it varies by grade level.

- **What are the four reasons given as to why black and Hispanic Americans, in general, do not perform as well in school as their white and Asian counterparts?**
First, because race and ethnicity are so closely tied to socioeconomic status, it appears that race or ethnicity alone can determine school success when, in fact, it may be socioeconomic status. Second, many minorities are not native English speakers, making academic achievement significantly more difficult. Third, standardized tests have been demonstrated to be culturally biased favoring those in the upper and middle classes and, finally, racial and ethnic minorities may be the victims of racism and discrimination.

- **What are some of the conclusions of the study summarized in the *Social Problems Research Up Close* feature?**
The results of the study indicate that there are more victims of bullying than bullies, and that bullying behavior is highest in the 9th grade and lowest in the 6th and 10th grades. The highest rate of victimization is in the 6th grade. Males and females were equally likely to report being bullies, victims, and bully-victims. Blacks, compared to Hispanics, were more likely to be bullies, victims, and bully-victims. "Upsetting students for the fun of it" was the most common kind of bullying activity, followed by teasing, group teasing, and starting arguments.

- **What are some of the problems associated with the American school system?**
One of the main problems is the lack of funding and the resulting reduction in programs and personnel. Low levels of academic achievement in our schools are also of some concern—particularly when U.S. data are compared with data from other industrialized countries. Minority dropout rates are high, and school violence, crime, and discipline problems continue to be a threat. School facilities are in need of repair and renovations, and personnel, including teachers, have been found to be deficient. Higher education must also address several challenges.

- **What is meant by value-added measurement (VAM), and why are there concerns about its use?**
VAM is the use of student achievement data to assess teachers' effectiveness. Critics of VAM argue that assessing teachers based on student performance assumes all else constant, and ignores the reality of student differences in such nonschool factors as family life, poverty, emotional and physical obstacles, and the like.

- **What are the arguments for and against school choice?**
Proponents of school choice programs argue that they reduce segregation and that schools that have to compete with one another will be of a higher quality. Opponents argue that school choice programs increase segregation and treat disadvantaged students unfairly. Low-income students cannot afford to go to private schools, even with vouchers. Furthermore, those opposed to school choice are quick to note that using government vouchers to help pay for religious schools is unconstitutional.

1. All societies have some formal mechanism to educate their citizenry.
   a. True
   b. False

2. According to structural functionalists, which of the following is not a major function of education?
   a. Teach students knowledge and skills
   b. Socialize students into the dominant culture
   c. Indoctrinate students into the capitalist ideology
   d. Provide custodial care for children

3. Common Core State Standards have been embraced by all states and will be implemented in the 2015–2016 school year.
   a. True
   b. False

4. Which of the following statements is true about dropouts in the United States?
   a. Dropout rates in the United States are increasing.
   b. Students who drop out of school, on the average, make more money because they've been working longer than their high school graduate counterparts.
   c. Dropout rates in the United States are similar between racial and ethnic subgroups.
   d. There is evidence that GED programs increase dropout rates.

5. Which of the following statements about bullying is not true?
   a. Student bullies tend to be marginal students.
   b. More females are bullied than males.

c. There are serious consequences for the student victims of bullying.
   d. Most bullying takes place outside so it is undetected.

6. Character education is associated with increased student achievement.
   a. True
   b. False

7. Over the next decade, the need for teachers is likely to
   a. decrease because of the baby boom retirements
   b. increase because of immigration patterns
   c. remain stable
   d. decrease because of the aging population

8. Online students compared to face-to-face students are more likely to be
   a. single
   b. part-time employees
   c. inner-city dwellers
   d. older

9. Computer software in the classroom has been shown to significantly increase academic achievement.
   a. True
   b. False

10. School vouchers are state-funded "scholarships" that are transferred to the private school that parents select for their child.
    a. True
    b. False

Answers: 1. B; 2. C; 3. B; 4. D; 5. D; 6. A; 7. B; 8. D; 9. B; 10. A.

alternative certificationrograms 261
bilingual education 250
bullying 257
character education 265
charter schools 268
cultural imperialism 245

earningsremium 242
e-learning 266
Head Start 248
integration hypothesis 251
multicultural education 243
parent trigger laws 264

school-to-prisonipeline 257
school vouchers 267
self-fulfillingrophecy 246
stereotype threat 254
value-added measurement (VAM) 260

AP Images/Kiichiro Sato

9

Race, Ethnicity, and Immigration

"No one is born hating another person because of the color of his skin, or his background, or his religion. People must learn to hate, and if they can learn to hate, they can learn to love..."

—Nelson Mandela

IN 2012, 17-YEAR-OLD TRAYVON MARTIN was shot and killed in Sanford, Florida, by George Zimmerman, a neighborhood watch volunteer who claimed he shot Martin in self-defense. Martin's parents claim that Zimmerman's pursuit and shooting of Martin was based on racial profiling—that Zimmerman assumed Martin, who was unarmed, was dangerous because of the color of Martin's skin. Whether or not race was a crucial factor in Zimmerman's shooting of Martin remains a debated question. What is certain is that racism continues to be a problem in the United States. After George Zimmerman was acquitted of second-degree murder and manslaughter charges in the killing of Trayvon Martin, President Obama delivered remarks to address the questions and issues raised by the Zimmerman/Martin case. Obama emphasized that regardless of whether racism played a role in Trayvon Martin's death, racism in the United States is undeniable. Obama said:

> There are very few African American men in this country who haven't had the experience of being followed when they were shopping in a department store. That includes me. There are probably very few African American men who haven't had the experience of

walking across the street and hearing the locks click on the doors of cars. That happens to me—at least before I was a senator. There are very few African Americans who haven't had the experience of getting on an elevator and a woman clutching her purse nervously and holding her breath until she had a chance to get off. That happens often. (quoted in *Washington Post* staff 2013)

Shortly after the shooting of Trayvon Martin, an unidentified entrepreneur offered Trayvon Martin shooting targets for sale online. These targets, which had a bull's-eye over the heart, did not depict Martin's face, but they clearly resembled Martin before he was shot: wearing a hoodie, with a package of Skittles in the pocket and a can of ice tea in his right hand. The seller of these targets reportedly informed a news reporter that the response to the online ad for the shooting targets was "overwhelming" and that the targets sold out in two days (Follman 2012). A year after Martin's death, Sergeant Ron King of Florida's Port Canaveral Police Department was fired after he brought "Trayvon Martin" shooting targets to a firearms training session (Martinez 2013).

In this chapter, we focus on racism, prejudice, and discrimination, their consequences for racial and ethnic minorities, and the strategies designed to reduce these problems. A **minority group** is a category of people who have unequal access to positions of power, prestige, and wealth in a society and who tend to be targets of prejudice and discrimination. Minority status is not based on numerical representation in society but rather on social status. For example, although Hispanic individuals outnumber non-Hispanic whites in California, Texas, and New Mexico, they are considered a "minority" because they are underrepresented in positions of power, prestige, and wealth, and because they are targets of prejudice and discrimination.

In this chapter, we also examine issues related to U.S. immigration, because immigrants often bear the double burden of being minorities *and* foreigners who are not welcomed by many native-born Americans. We begin by examining racial and ethnic diversity worldwide and in the United States, emphasizing first that the concept of race is based on social rather than biological definitions.

The Global Context: Diversity Worldwide

A first grade teacher asked the class, "What is the color of apples?" Most of the children answered red. A few said green. One boy raised his hand and said, "white." The teacher tried to explain that apples could be red, green, or sometimes golden, but never white. The boy insisted his answer was right and finally said, referring to the apple, "Look inside" (Goldstein 1999). Like apples, human beings may be similar on the "inside," but they are often classified into categories according to external appearance. After examining the social construction of race and ethnicity, we review patterns of interaction among racial and ethnic groups and examine racial and ethnic diversity in the United States.

The Social Construction of Race and Ethnicity

Martin Marger (2012), a researcher and writer on race and ethnic relations, says that "race is one of the most misunderstood, misused, and often dangerous concepts of the modern world" (p. 12). Marger explains that the term *race* has been used to describe people of a particular nationality (the Mexican "race"), religion (the Jewish "race"), skin color (the white

minority group A category of people who have unequal access to positions of power, prestige, and wealth in a society and who tend to be targets of prejudice and discrimination.

"race"), and even the entire human species (the human "race"). Confusion around the term *race* stems from the fact that it has both biological and social meanings.

Race as a Biological Concept. As a biological concept, race refers to a classification of people based on hereditary physical characteristics such as skin color, hair texture, and the size and shape of the eyes, lips, and nose. But there is no scientific basis—no blood test or genetic test—that reveals a person's race. There are also no clear guidelines for distinguishing racial categories on the basis of visible traits. Skin color is not black or white but rather ranges from dark to light with many gradations of shades. Noses are not either broad or narrow but come in a range of shapes. Physical traits come in an infinite number of combinations. For example, a person with dark skin can have a broad nose (a common combination in West Africa), a narrow nose (a common combination in East Africa), or even blond hair (a combination found in Australia and New Guinea). Further, skin color, hair texture, and facial features are only a few of the many traits that vary among human beings.

> What if we classified people into racial categories based on eye color instead of skin color?

Another problem with race as a biological concept is that the physical traits used to mark a person's race are arbitrary. What if we classified people into racial categories based on eye color instead of skin color? Or hair color? Or blood type? What if all dark-haired individuals were considered to belong to one race, and all light-haired people to another race? Is there any scientific reason for selecting certain traits over others in determining racial categories? The answer is "no." As a biological concept, "races are not scientifically valid because there are no objective, reliable, meaningful criteria scientists can use to construct or identify racial groupings" (Mukhopadhyay et al. 2007, p. 5).

The science of genetics also challenges the biological notion of race. Geneticists have discovered that the genes of any two unrelated people, chosen at random from around the globe, are 99.9 percent alike (Ossorio & Duster 2005). Furthermore, "most human genetic variation—approximately 85 percent—can be found between any two individuals from the same group (racial, ethnic, religious, etc.). Thus, the vast majority of variation is within-group variation" (Ossorio & Duster 2005, p. 117). Finally, classifying people into different races fails to recognize that, over the course of human history, migration and intermarriage have resulted in the blending of genetically transmitted traits:

> To summarize, races are unstable, unreliable, arbitrary, culturally created divisions of humanity. This is why scientists . . . have concluded that race, as scientifically valid biological divisions of the human species, is fiction not fact. (Mukhopadhyay et al. 2007, p. 14)

Race as a Social Concept. The idea that race is socially created is one of the most important lessons in understanding race from a sociological perspective. The social construction of race means that "the actual meaning of race lies not in people's physical characteristics, but in the historical treatment of different groups and the significance that society gives to what is believed to differentiate so-called racial groups" (Higginbotham & Andersen 2012, p. 3). The concept of race grew out of social institutions and practices in which groups defined as "races" have been enslaved or otherwise exploited.

People learn to perceive others according to whatever racial classification system exists in their culture. Systems of racial classification vary across societies and change over time. For example, as late as the 1920s, U.S. Italians, Greeks, Jews, Irish, and other "white" ethnic groups were not considered to be white. Over time, the category of "white" changed so that it included these groups. As an example of cross-cultural variation in racial categories, Brazilians use dozens of terms to racially categorize people based on various combinations of physical characteristics, although officially, the major racial categories in Brazil are *branco* (white), *pardos* (brown or mulatto), *pretos* (black), and *amarelos* (yellow).

Incorporating both biological and social meanings of race, we define **race** as a category of people who are perceived to share distinct physical characteristics that are deemed socially significant. The significance of race is not biological but social and political, because race is used to separate "us" from "them" and becomes a basis for unequal treatment of one group

race A category of people who are perceived to share distinct physical characteristics that are deemed socially significant.

by another. Despite the increasing acceptance that "there is no biological justification for the concept of 'race'" (Brace 2005, p. 4), its social significance continues to be evident throughout the world.

What Do You Think? Do you think the time will ever come when a racial classification system will no longer be used? Why or why not? What arguments can be made for discontinuing racial classification? What arguments can be made for continuing it?

Ethnicity as a Social Construction. **Ethnicity**, which refers to a shared cultural heritage, nationality, or lineage, is also socially constructed in part. Ethnicity can be distinguished on the basis of language, forms of family structures and roles of family members, religious beliefs and practices, dietary customs, forms of artistic expression such as music and dance, and national origin or origin of one's parents.

Although the Census Bureau defines Hispanic or Latino as "a person of Cuban, Mexican, Puerto Rican, South or Central American, or other Spanish culture or origin regardless of race" (Ennis et al. 2011, p. 2), when it comes down to collecting census data on the U.S. population, a person is Hispanic if they say they are Hispanic (see Table 9.1). Hence, ethnicity is socially constructed.

Patterns of Racial and Ethnic Group Interaction

When two or more racial or ethnic groups come into contact, one of several patterns of interaction occurs; these include genocide, expulsion, segregation, acculturation, pluralism, and assimilation. Although not all patterns of interaction between racial and ethnic groups are destructive, author and Mayan shaman Martin Prechtel reminded us, "Every human on this earth, whether from Africa, Asia, Europe, or the Americas, has ancestors whose stories, rituals, ingenuity, language, and life ways were taken away, enslaved, banned, exploited, twisted, or destroyed" (quoted by Jensen 2001, p. 13).

Genocide refers to the deliberate, systematic annihilation of an entire nation or people. The European invasion of the Americas, beginning in the 16th century, resulted in the decimation of most of the original inhabitants of North and South America. Some native groups were intentionally killed; others fell victim to diseases brought by the Europeans. In the 20th century, Hitler led the Nazi extermination of 12 million people, including 6 million Jews, in what is known as the Holocaust. In the early 1990s, ethnic Serbs attempted to eliminate Muslims from parts of Bosnia—a process they called "ethnic cleansing." In 1994, genocide took place in Rwanda when Hutus slaughtered hundreds of thousands of Tutsis—an event depicted in the 2004 film *Hotel Rwanda*. Genocide is continuing in the Darfur region of Sudan, where the Sudanese government, using Arab *Janjaweed* militias, its air force, and organized starvation, is systematically killing the African Muslim communities because some among them have challenged the authoritarian rule of the Sudanese government (see also Chapter 15).

Expulsion occurs when a dominant group forces a subordinate group to leave the country or to live only in designated areas of the country. The 1830 Indian Removal Act called for the relocation of eastern tribes to land west of the Mississippi River. The movement, lasting more than a decade, has been called the Trail of Tears because tribes were forced to leave their ancestral lands and endure harsh conditions of inadequate supplies and epidemics that caused illness and death. After Japan's attack on Pearl Harbor in 1941, 120,000 Japanese Americans, who became viewed as threats to national security, were forced from their homes and into evacuation camps surrounded by barbed wire.

TABLE 9.1 Who's Hispanic?

THE U.S. CENSUS BUREAU APPROACH TO DEFINING WHO IS HISPANIC:

Q. *I immigrated to Phoenix from Mexico. Am I Hispanic?*

A. You are if you say so.

Q. *My parents moved to New York from Puerto Rico. Am I Hispanic?*

A. You are if you say so.

Q. *My grandparents were born in Spain but I grew up in California. Am I Hispanic?*

A. You are if you say so.

Q. *I was born in Maryland and married an immigrant from El Salvador. Am I Hispanic?*

A. You are if you say so.

Q. *My mom is from Chile and my dad is from Iowa. I was born in Des Moines. Am I Hispanic?*

A. You are if you say so.

Q. *I was born in Argentina but grew up in Texas. I don't consider myself Hispanic. Does the Census count me as Hispanic?*

A. Not if you say you aren't.

Source: Passel & Taylor 2009.

ethnicity A shared cultural heritage or nationality.

genocide The deliberate, systematic annihilation of an entire nation or people.

expulsion Occurs when a dominant group forces a subordinate group to leave the country or to live only in designated areas of the country.

Segregation refers to the physical separation of two groups in residence, workplace, and social functions. Segregation can be *de jure* (Latin meaning "by law") or *de facto* ("in fact"). Between 1890 and 1910, a series of U.S. laws, which came to be known as *Jim Crow laws,* were enacted to separate blacks from whites by prohibiting blacks from using "white" buses, hotels, restaurants, and drinking fountains. In 1896, the U.S. Supreme Court (in *Plessy v. Ferguson*) supported de jure segregation of blacks and whites by declaring that "separate but equal" facilities were constitutional. Blacks were forced to live in separate neighborhoods and attend separate schools. Beginning in the 1950s, various rulings overturned these Jim Crow laws, making it illegal to enforce racial segregation. Although de jure segregation is illegal in the United States, de facto segregation still exists in the tendency for racial and ethnic groups to live and go to school in segregated neighborhoods.

Acculturation refers to adopting the culture of a group different from the one in which a person was originally raised. Acculturation may involve learning the dominant language, adopting new values and behaviors, and changing the spelling of the family name. In some instances, acculturation may be forced. For decades, the Australian government removed aboriginal children from their families and placed them in missions or foster families, forcing them to abandon their language and traditional aboriginal culture. Authorities targeted children of mixed descent—what they referred to as "half-caste"—because they thought these children could be more easily acculturated into white society. Today, these individuals are known as the "stolen generation" because they had been stolen from their families and their culture.

This is a scene from the 2002 Australian film *Rabbit-Proof Fence*, which tells the story of three aboriginal girls who were taken from their family by the Australian government as part of a program to force Aborigines to adopt the white culture.

Pluralism refers to a state in which racial and ethnic groups maintain their distinctness but respect each other and have equal access to social resources. In Switzerland, for example, four ethnic groups—French, Italians, Swiss Germans, and Romansch—maintain their distinct cultural heritage and group identity in an atmosphere of mutual respect and social equality.

Assimilation is the process by which formerly distinct and separate groups merge and become integrated as one. *Primary assimilation* occurs when members of different racial or ethnic groups are integrated in personal, intimate associations, as with friends, family, and spouses. *Secondary assimilation* occurs when different groups become integrated in public areas and in social institutions, such as neighborhoods, schools, the workplace, and in government.

Assimilation is sometimes referred to as the "melting pot," whereby different groups come together and contribute equally to a new, common culture. Although the United States has been referred to as a melting pot, in reality, many minorities have been excluded or limited in their cultural contributions to the predominant white Anglo-Saxon Protestant culture.

segregation The physical separation of two groups in residence, workplace, and social functions.

acculturation The process of adopting the culture of a group different from the one in which a person was originally raised.

pluralism A state in which racial and ethnic groups maintain their distinctness but respect each other and have equal access to social resources.

assimilation The process by which formerly distinct and separate groups merge and become integrated as one.

Racial and Ethnic Group Diversity in the United States

The first census in 1790 divided the U.S. population into four groups: free white males, free white females, slaves, and other people (including free blacks and Indians). To increase the size of the slave population, the *one-drop rule* specified that even one drop of "negroid" blood defined a person as black and therefore eligible for slavery. The "one-drop rule" is still operative today: Biracial individuals are typically seen as a member of whichever group has the lowest status (Wise 2009).

In 1960, the census recognized only two categories: white and nonwhite. In 1970, the census categories consisted of white, black, and "other" (Hodgkinson 1995). In 1990, the U.S. Census Bureau recognized four racial classifications: (1) white, (2) black, (3) American Indian, Aleut, or Eskimo, and (4) Asian or Pacific Islander. The 1990 census also included the category of "other." Beginning with the 2000 census, racial categories expanded to include Native Hawaiian or other Pacific Islander, and also allowed individuals the option of identifying themselves as being more than one race rather than checking only one racial category (see Figure 9.1).

Figure 9.1 **2010 Census Questions**
Source: Humes et al. 2011.

→ **NOTE: Please answer BOTH Question 5 about Hispanic origin and Question 6 about race. For this census, Hispanic origins are not races.**

5. **Is this person of Hispanic, Latino, or Spanish origin?**

☐ **No,** not of Hispanic, Latino, or Spanish origin
☐ Yes, Mexican, Mexican Am., Chicano
☐ Yes, Puerto Rican
☐ Yes, Cuban
☐ Yes, another Hispanic, Latino, or Spanish origin — *Print origin, for example, Argentinean, Colombian, Dominican, Nicaraguan, Salvadoran, Spaniard, and so on.*

6. **What is this person's race?** Mark ☒ one or more boxes.

☐ White
☐ Black, African Am., or Negro
☐ American Indian or Alaska Native — *Print name of enrolled or principal tribe.*

☐ Asian Indian ☐ Japanese ☐ Native Hawaiian
☐ Chinese ☐ Korean ☐ Guamanian or Chamorro
☐ Filipino ☐ Vietnamese ☐ Samoan
☐ Other Asian — *Print race, for example, Hmong, Laotian, Thai, Pakistani, Cambodian, and so on.* ☐ Other Pacific Islander — *Print race, for example, Fijian, Tongan, and so on.*

☐ Some other race — *Print race.*

Although U.S. citizens come from a variety of ethnic backgrounds, the largest ethnic population in the United States is of Hispanic origin. The Census Bureau began collecting data on the U.S. Hispanic population in 1970.

As you read the following section on U.S. census data on race and Hispanic origin, keep in mind that the use of racial and ethnic labels is often misleading and imprecise. The ethnic classification of "Hispanic/Latino," for example, lumps together disparate groups such as Puerto Ricans, Mexicans, Cubans, Venezuelans, Colombians, and others from Latin American countries. The racial term *American Indian* includes more than 300 separate tribal groups that differ enormously in language, tradition, and social structure. The racial label *Asian* includes individuals from China, Japan, Korea, India, the Philippines, or one of the countries of Southeast Asia. And what about people who are Asian but who live in the United States? The term *Asian American* is used to describe people with Asian racial features who are born in the United States, as well as those who immigrate to the United States. Columbia University professor Derald Wing Sue, an Asian American who was born in the United States, said, "When I get out of a cab after having a conversation with a white cab driver, they'll say something like, 'Boy, you speak excellent English.' . . . From their perspective, that's meant as a compliment, but another hidden meaning is being communicated, and that is that I am a perpetual foreigner in my own land" (quoted in Tanneeru 2007).

U.S. Census Data on Race and Hispanic Origin

Census data show that the U.S. population is becoming increasingly diverse: From 2000 to 2010, the percentage of the population that is non-Hispanic white declined from 69 percent to 64 percent of the total population (Humes et al. 2011). Census estimates predict that by 2042, non-Hispanic whites will no longer outnumber racial/ethnic

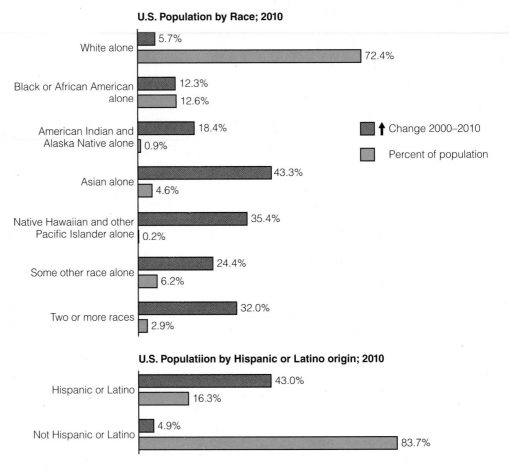

U.S. Population by Race; 2010

White alone — 5.7% / 72.4%

Black or African American alone — 12.3% / 12.6%

American Indian and Alaska Native alone — 18.4% / 0.9%

Asian alone — 43.3% / 4.6%

Native Hawaiian and other Pacific Islander alone — 35.4% / 0.2%

Some other race alone — 24.4% / 6.2%

Two or more races — 32.0% / 2.9%

↑ Change 2000–2010
Percent of population

U.S. Populatiion by Hispanic or Latino origin; 2010

Hispanic or Latino — 43.0% / 16.3%

Not Hispanic or Latino — 4.9% / 83.7%

minorities. Figure 9.2 shows that from 2000 to 2010, the percent increase in minority populations has increased significantly more than the increase in the white population.

In 2010, 16 percent of the U.S. population was Hispanic, up from 13 percent in 2000, with the majority being Mexican (see Figure 9.3). More than half of the growth in the total U.S. population between 2000 and 2010 was due to the increase in the Hispanic population (Humes et al. 2011). More than half of the U.S. Hispanic population lives in just three states: California, Texas, and Florida.

The current Census Bureau classification system does not allow people of mixed Hispanic or Latino ethnicity to identify themselves as such. Individuals with one Hispanic and one non-Hispanic parent still must say that they are either Hispanic or not Hispanic.

Among the general public, one of the most common confusions about Hispanic origin is the question of whether "Hispanic" is a race or an ethnicity. According to

Figure 9.3 Percent Distribution of U.S. Hispanics by Type, 2010
Source: Ennis et al. 2011.

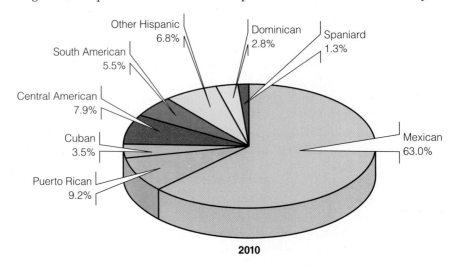

Other Hispanic 6.8%
Dominican 2.8%
Spaniard 1.3%
South American 5.5%
Central American 7.9%
Cuban 3.5%
Puerto Rican 9.2%
Mexican 63.0%

2010

the federal government, Hispanic origin is considered an ethnicity, not a race (Ennis et al. 2011). But many people—Hispanic and non-Hispanic—think otherwise. Although the 2010 census included instructions that stated, "for this census, Hispanic origins are not races," many Hispanics identified their race as "Latino," "Mexican," "Puerto Rican," "Salvadoran," or other ethnicity or national origin, which the census classified in the category of "Some Other Race" (see Table 9.2). Because of the widespread view of "Hispanic" as a race, the Census Bureau is considering whether to make "Hispanic" a racial instead of an ethnic category for the 2020 census (Ayala & Huet 2013).

Currently, racial and ethnic minority populations outnumber non-Hispanic whites in four states—California, New Mexico, Hawaii, and Texas. In these states, Hispanics are the largest minority group, except for Hawaii, where the largest minority group is Asian American (Humes et al. 2011). And in 10 states, racial and ethnic minority youth outnumber non-Hispanic white youth (Tavernise 2011).

> One of the most common confusions about Hispanic origin is the question of whether "Hispanic" is a race or an ethnicity.

TABLE 9.2 Racial Identification of Hispanics/Latinos in the United States, 2010	
White	53%
Black	2.5%
American Indian/Alaskan Native	1.4%
Asian	0.4%
Native Hawaiian/Pacific Islander	0.1%
Some Other Race	36.7%
Two or More Races	6.0%

Source: Ennis et al. 2011.

Mixed-Race Identity

As shown in Figure 9.2, 2010 census data found that only a small percentage (2.9 percent) of the U.S. population identify themselves as being of more than one race. But between 2000 and 2010, the mixed-race U.S. population grew 32 percent. And among U.S. children, the multiracial population has grown almost 50 percent, making it the fastest-growing youth group in the country (Saulny 2011).

The multiracial population has grown as mixed-race marriages have increased over recent years. Until 1967, 16 states had **antimiscegenation laws** banning interracial marriage of whites and nonwhites—primarily blacks, but in some cases, also Native Americans and Asians. In 1967, the Supreme Court (in *Loving v. Virginia*) declared these laws unconstitutional. In 2010, 15 percent of new marriages in the United States were between spouses with different racial or ethnic identities—more than double the percentage in 1980 (6.7 percent) (Wang 2012; see Figure 9.4).

Attitudes toward interracial marriages have changed dramatically over the last few generations. The percentage of U.S. adults who disapproved of marriage between blacks and whites decreased from 94 percent in 1958, to 29 percent in 2002, to 11 percent in 2013 (Gallup Organization 2013). More than 4 in 10 Americans view the increase in intermarriages as a change for the better in society; 1 in 10 say it has been a change for the worse (the remaining share say it doesn't make a difference). When asked, "How would you react if a member of your family were going to marry someone of a different race or ethnicity?", 63 percent of Americans say they would be fine with it. Individuals most likely to have positive attitudes about intermarriage are minorities, younger, more educated, liberal, and living in the eastern or western states (Wang 2012).

AP Images/Alan Diaz

Although Barack Obama's background is biracial—he is the son of a black Kenyan immigrant father and a white Kansas native mother—he identifies as a black African American.

> **What Do You Think?** Do you think of Barack Obama as a black person or as a person of mixed race? A Pew Research (2010) poll found that only 24 percent of whites and 23 percent of Hispanics view Obama as black, whereas more than half of blacks (55 percent) view Obama as black. Why do you think blacks are more likely than whites or Hispanics to view Obama as black?

antimiscegenation laws Laws banning interracial marriage until 1967, when the Supreme Court (in *Loving v. Virginia*) declared these laws unconstitutional.

Race and Ethnic Group Relations in the United States

Despite significant improvements over the last two centuries, race and ethnic group relations continue to be problematic. In response to a question asking whether relations between blacks and whites will always be a problem for the United States or whether a solution will eventually be worked out, 4 in 10 U.S. adults said that race relations will always be a problem (Gallup Organization 2013). A Gallup Poll asked a national sample of U.S. adults to rate relations between various groups in the United States. Table 9.3 presents results of this survey. Relations between various racial and ethnic groups are influenced by prejudice and discrimination (discussed later in this chapter). Race and ethnic relations are also complicated by issues concerning immigration—the topic we turn to next.

What Do You Think? A Gallup Organization poll asked U.S. adults, "If blacks and whites honestly expressed their true feelings about race relations, do you think this would do more to bring races together or cause greater racial division?" More than half (56 percent) replied "bring races together," and about a third (37 percent) said "cause greater division" (7 percent had no opinion) (Gallup Organization 2013). What would your answer be? Why?

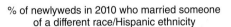
% of newlyweds in 2010 who married someone of a different race/Hispanic ethnicity

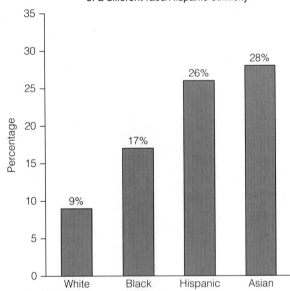

Figure 9.4 U.S. Intermarriage Rates, by Race and Hispanic Origin
Source: Wang 2012.

TABLE 9.3 Perceptions of Race and Ethnic Relations in the United States, 2013		

A national sample of U.S. adults was asked to rate relations between various groups in the United States. The results are depicted in this table.

	Very or Somewhat Good	Very or Somewhat Bad
Whites and blacks	70%	30%
Whites and Hispanics	70%	29%
Blacks and Hispanics	60%	32%
Whites and Asians	87%	10%

Note: Percentages do not add up to 100 because of "no opinion" responses.
Source: Gallup Organization 2013.

Immigrants in the United States

The growing racial and ethnic diversity of the United States is largely the result of immigration as well as the higher average birthrates among many minority groups. Immigration generally results from a combination of "push" and "pull" factors. Adverse social, economic, and/or political conditions in a given country "push" some individuals to leave that country, whereas favorable social, economic, and/or political conditions in other countries "pull" some individuals to those countries.

U.S. Immigration: A Historical Perspective

For the first 100 years of U.S. history, all immigrants were allowed to enter and become permanent residents. The continuing influx of immigrants, especially those coming from nonwhite, non-European countries, created fear and resentment among native-born Americans, who competed with immigrants for jobs and who held racist views toward some racial and ethnic immigrant populations. America's open-door policy on immigration ended in 1882 with the Chinese Exclusion Act, which suspended the entrance of the Chinese to the United States for 10 years and declared Chinese ineligible for U.S. citizenship (this act was repealed in 1943). The Immigration Act of 1917 required all immigrants to pass a literacy test before entering the United States. Immigration legislation passed in the 1920s established a quota system, limiting the numbers of immigrants from specific countries. Under the quota system, 70 percent of immigrant slots were allotted to people from just three countries: United Kingdom, Ireland, and Germany. In 1965, the passage of the Hart-Celler Act abolished the national origins quota system that had been in place since the 1920s, and instituted a system that gave preference to immigrants who had family in the United States or who had needed job skills. The Hart-Celler Act was an extension of the civil rights movement, as it was designed to end discrimination based on race and ethnicity, and it continues to be the basic structure of immigration law today.

In 2011, immigrants represented 13 percent of the U.S. population. More than three out of four immigrants in the United States are either naturalized citizens or are in the United States legally (Pew Hispanic Center 2013). In the 1960s, most immigrants were from Europe, but most immigrants today are from Latin America (predominantly Mexico) or Asia (see Figure 9.5).

Guest Worker Program

The United States has two guest worker programs that allow employers to import unskilled labor for temporary or seasonal work: the H-2A program for agricultural work and the H-2B program for nonagricultural work. H-2 visas generally do not permit guest workers to bring their families to the United States.

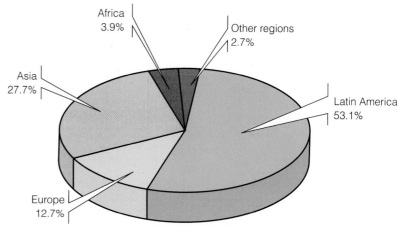

Total foreign-born U.S. population in 2010: 39.5 million

Africa 3.9%
Other regions 2.7%
Asia 27.7%
Latin America 53.1%
Europe 12.7%

Figure 9.5 **U.S. Foreign-Born Residents by Region of Birth, 2010**
Source: Grieco et al. 2012.

Immigrant "guest workers" are hardly treated like "guests"; these workers are systematically exploited and abused. Guest workers are bound to the employers who "import" them so they are not able to change jobs if they are mistreated. Guest workers are often cheated out of pay, forced to live in squalid conditions, and although they perform some of the most difficult and dangerous jobs in the United States, many who are injured on the job are unable to obtain medical treatment and workers' compensation benefits. Immigrant women working at low-wage jobs are often targets of sexual violence. If guest workers complain about mistreatment, they are threatened with deportation or being "blacklisted" and unable to find another job. Rampant with labor and human rights abuses, the guest worker program has been described as a modern-day system of indentured servitude (SPLC 2013b).

As this book goes to press, Congress is considering immigration reform legislation that would create a new class of guest worker visas for lower-skilled workers. The proposed W-visa would allow foreigners to enter the country to work for three years, with one option to renew for another three years. Employers could not hire W-visa holders if U.S. workers are willing to take the job. W-visa holders would be entitled to the same wages and labor rights as U.S. employees in a similar job would have.

Southern Poverty Law Center

This photo depicts the kind of substandard housing in which many immigrant guest workers are forced to live.

Illegal Immigration

Illegal immigration occurs when immigrants enter the United States without going through legal channels such as the H-2 visa program, and when immigrants who were admitted legally stay past the date they were required to leave. In 2011, an estimated 11.5 million immigrants were living illegally in the United States (Hoefer et al. 2012). More than half of the total unauthorized immigrant population lives in just four states: California, Texas, Florida, and New York. The majority of unauthorized immigrants (59 percent) are from Mexico, followed by El Salvador, Guatemala, Honduras, and China.

Border Crossing. U.S. Customs and Border Protection, an agency within the Department of Homeland Security, has more than 21,000 border patrol agents who are responsible for patrolling 6,000 miles of Canadian and Mexican borders and more than 2,000 miles

of coastal waters around Florida and Puerto Rico (U.S. Customs and Border Protection 2013). Despite the border patrol, along with a "fence" at the U.S.–Mexico border, people continue to find ways to illegally cross the U.S. border.

Some people cross (or attempt to cross) the U.S.–Mexican border with the help of a *coyote*—a hired guide who typically charges $3,000 to $5,000 to lead people across the border (Maril 2011). Crossing the border illegally involves a number of risks, including death from drowning (e.g., while trying to cross the Rio Grande) or dehydration. Border crossers also risk en-

Chris Simoux is an example of an American who patrols the U.S.–Mexico border looking for unauthorized immigrants. Such self-appointed border patrol guards sometimes use violence when they encounter suspected undocumented immigrants.

counters with members of **nativist extremist groups**—organizations that not only advocate restrictive immigration policy, but also encourage their members to use vigilante tactics to confront or harass suspected unauthorized immigrants. The number of anti-immigrant nativist extremist groups fell from its peak of 319 groups in 2010 to 38 in 2012 (Beirich 2013).

Unauthorized Immigrants in the Workforce. In 2010, there were 8 million undocumented immigrants in the U.S. labor force, comprising 5.2 percent of the U.S. workforce (Passel & Cohn 2011). Sociologist Robert Maril (2004) noted that "[t]he vast majority of illegal immigrants leave their home countries to work hard, save their money, then return to their homeland. . . . These individuals do not travel their difficult and dangerous journeys searching for a welfare handout; they immigrate to work" (pp. 11–12). Virtually all undocumented men are in the labor force. Their labor force participation exceeds that of men who are legal immigrants or who are U.S. citizens because undocumented men are less likely to be disabled, retired, or in school. Undocumented women are less likely to be in the labor force than undocumented men because they are more likely to be stay-at-home mothers.

> Undocumented workers often do work that U.S. workers are unwilling to do—they routinely work 60 or more hours per week and earn less than the minimum wage, with no paid overtime and no benefits.

Undocumented workers often do work that U.S. workers are unwilling to do—they routinely work 60 or more hours per week and earn less than the minimum wage, with no paid overtime and no benefits.

Policies Regarding Illegal Immigration. The 1986 Immigration Reform and Control Act made hiring unauthorized immigrants an illegal act punishable by fines and even prison sentences. Enforcement of this act occurs primarily through workplace raids, and the E-Verify system, which is a free online service created in 1997 by the Department of Homeland Security (DHS) that allows employers to check the legal statuses of their workers. Employers enter basic information—name, Social Security number, date of birth, alien registration number, etc.—which is then cross-checked against Social Security Administration and DHS databases. If an employee is not in any database as a legally authorized worker, the employee has eight days to prove legal status, or the employer is mandated to fire them.

The Secure Fence Act of 2006 authorized the construction of a 700-mile "fence" along the U.S.–Mexico border to prevent unauthorized immigrant workers, as well

nativist extremist groups
Organizations that not only advocate restrictive immigration policy, but also encourage their members to use vigilante tactics to confront or harass suspected undocumented immigrants.

as drug dealers and terrorists, from entering the United States. Critics of the fence argue that this barrier—miles of double chain-link and barbed wire fences with light and infrared camera poles—is expensive and ineffective in stopping illegal immigration; disrupts the environment and harms wildlife along the border; increases the risk and danger to immigrants trying to cross the border; and damages diplomatic relations with Mexico.

The Development, Relief, and Education for Alien Minors Act (DREAM Act), introduced in Congress in 2009, would provide a path to legal status for undocumented immigrants who were brought to the United States as children. If passed, the DREAM Act would permit certain immigrant students who have grown up in the United States to apply for temporary legal status and to eventually obtain permanent status and become eligible for U.S. citizenship if they go to college or serve in the U.S. military.

On May 1, 2010, tens of thousands of people protested in over 70 cities against Arizona's immigration law.

In 2010, Arizona passed the Support Our Law Enforcement and Safe Neighborhoods Act, known as SB 1070—one of the toughest illegal immigration bills in the country. Federal law requires that legal immigrants carry registration papers with them at all times. Arizona's SB 1070 made failure to carry registration documents a state crime, and required police to verify the legal status of a person during traffic stops, detentions, or arrests if the police suspect that person is in the country illegally. SB 1070 also criminalized anyone who transported unauthorized immigrants, which, according to the law, constituted "smuggling of human beings." Employers who transported unauthorized day laborers to the worksite, as well as church volunteers who transported unauthorized immigrants to church services, were in violation of the law. Following the passage of Arizona's SB 1070, five other states—Alabama, Georgia, South Carolina, Utah, and Indiana—enacted legislation modeled after Arizona's law, and many more states considered similar legislation. But after a series of federal and state court rulings that struck down key provisions of restrictive state immigration policies, states shifted from pursuing restrictive immigration policies to enacting pro-immigrant legislation, such as laws granting unauthorized immigrant students who graduate from state high schools in-state tuition rates at public universities and laws allowing unauthorized immigrants to apply for a driver's license. Another reason for the shift to pro-immigrant state policies is that after Latino voters helped re-elect President Obama in 2012, Republicans reassessed the political viability of anti-immigration policy (Chisti & Hipsman 2013).

As this book goes to press, Congress is considering immigration reform legislation—the Border Security, Economic Opportunity, and Immigration Modernization Act (S. 744). This proposed legislation addresses border security and enforcement issues as well as legal immigration reforms. The bill would also create a "path to citizenship," allowing eligible unauthorized immigrants currently in the United States to legalize their immigration status and eventually become U.S. citizens.

Becoming a U.S. Citizen

Almost half of the nearly 40 million foreign-born U.S. residents in 2010 were **naturalized citizens** (immigrants who applied and met the requirements for U.S. citizenship) (Grieco et al. 2012). Requirements to become a U.S. citizen for most immigrants include (1) having resided continuously as a lawful permanent U.S. resident for at least five years (three years for a spouse of a U.S. resident); (2) being able to read, write,

naturalized citizens
Immigrants who apply for and meet the requirements for U.S. citizenship.

speak, and understand basic English (certain exemptions apply); (3) being "a person of good moral character" (cannot have a record of criminal offenses such as prostitution, illegal gambling, failure to pay child support, drug violations, and violent crime); (4) demonstrating willingness to support and defend the U.S. Constitution by taking the Oath of Allegiance; and (5) passing an examination on English (speaking, reading, and writing) and U.S. government and history (U.S. Citizenship and Immigration Services 2011).

Myths about Immigration and Immigrants

Many foreign-born U.S. residents work hard to succeed educationally and occupationally. The percentage of foreign-born adults (age 25 and older) with at least a bachelor's degree (27 percent) matches that of native-born U.S. adults (28 percent) (Grieco et al. 2012). Despite the achievements and contributions of immigrants, many myths about immigration and immigrants persist, largely perpetuated by anti-immigrant groups and campaigns:

Myth 1. Immigrants increase unemployment and lower wages among native workers.

Reality: Most academic economists agree that immigration has a small but positive impact on the wages of native-born workers because although new immigrant workers add to the labor supply, they also consume goods and services, which creates more jobs (Shierholz 2010). Immigrants also start their own businesses at a higher rate than native U.S. residents, which increases demand for business-related supplies (such as computers and office furniture) and service providers (such as accountants and lawyers) (Pollin 2011).

Myth 2. Immigrants drain the public welfare system and our public schools.

Reality: Unauthorized and temporary immigrants are ineligible for major federal benefit programs, and even legal immigrants may face eligibility restrictions. Two benefit programs that do not have restrictions against unauthorized immigrants are the Special Supplemental Nutrition Program for Women, Infants, and Children (WIC) and the National School Lunch Program.

Regarding public education, a 1982 Supreme Court case (*Plyler v. Doe*) held that states cannot deny students access to public education, even if they are not legal U.S. residents. The court ruled that denying public education could impose a lifetime of hardship "on a discrete class of children not accountable for their disabling status" (Armario 2011). Children of unauthorized immigrants, 73 percent of whom are U.S. citizens, comprise only 6.8 percent of students in elementary and secondary schools (Passel & Cohn 2009), although the percentage is much higher in communities with large immigrant populations.

Although the states bear the cost of education, social services, and medical services for the immigrant population, research suggests that the economic benefits that immigrants provide for the states outweigh the costs associated with supporting them. For example, a study of immigrants in North Carolina found that, over the prior 10 years, Latino immigrants had cost the state $61 million in a variety of benefits—but were responsible for more than $9 billion in state economic growth (Beirich 2007). Half to three-fourths of undocumented immigrants pay federal, state, and local taxes, including Social Security taxes for benefits they will never receive (*Teaching Tolerance* 2011).

Myth 3. Immigrants do not want to learn English.

Reality: The demand for English classes for immigrants exceeds their availability; out of 176 providers of English as a Second Language (ESL) classes, more than half reported waiting lists ranging from a few weeks to more than three years (Bauer & Reynolds 2009). Even so, the majority of the foreign-born U.S. population either speak only English at home (15 percent) or speak English "well" or "very well" (55 percent) (Grieco et al. 2012). Only 1 in 10 foreign-born people does not speak English at all.

Myth 4. Undocumented immigrants have children in the United States as a means of gaining legal status.

Reality: Under the Fourteenth Amendment of the U.S. Constitution, any child born in the United States is automatically granted U.S. citizenship. But having children who are U.S. citizens does not provide immigrants with a means of gaining legal status in the United States. Children under 21 are not allowed to petition for their parents' U.S. citizenship. Nevertheless, some legislators have called for ending birthright citizenship by amending the Constitution or enacting state law to limit citizenship to children who have at least one authorized parent (Dwyer 2011).

Myth 5. Immigrants have high rates of criminal behavior.

Reality: Immigrants are less likely than natives to commit crimes. Because they risk deportation, undocumented immigrants have a strong motivation to avoid involvement with the law. In 2000, the U.S. incarceration rate for native-born men aged between 18 and 39 was 3.5 percent—five times greater than that of their foreign-born counterparts (Bauer & Reynolds 2009). El Paso, Texas, a city with a high immigrant population, is among the safest big cities in the United States. Criminologist Jack Levin said, "If you want to find a safe city, first determine the size of the immigrant population. If the immigrant community represents a large proportion of the population, you're likely in one of the country's safer cities" (quoted in Balko 2009).

> Immigrants are less likely than natives to commit crimes. Because they risk deportation, undocumented immigrants have a strong motivation to avoid involvement with the law.

Sociological Theories of Race and Ethnic Relations

Some theories of race and ethnic relations suggest that individuals with certain personality types are more likely to be prejudiced or to direct hostility toward minority group members. Sociologists, however, concentrate on the impact of the structure and culture of society on race and ethnic relations. Three major sociological theories lend insight into the continued subordination of minorities.

Structural-Functionalist Perspective

The structural-functionalist perspective focuses on how parts of the whole are interconnected. This perspective reminds us that we cannot fully understand the history of U.S. civil rights in a vacuum; we need to consider how forces outside the United States affected U.S. policies and culture regarding race relations. In *Cold War Civil Rights: Race and the Image of American Democracy*, Mary Dudziak (2000) links the 1965 passage of U.S. civil rights legislation to the United States' efforts to win the Cold War. Following World War II, international public opinion was critical of the extreme racial inequality in the United States. Racial discrimination was damaging to the United States' credibility as a democracy and to U.S. foreign relations. Although some legislators who supported the passage of civil rights legislation wanted to end the injustice of discrimination, they were also responding to international pressure and seeking to bolster an image of the United States as a democracy and world leader.

The structural-functionalist perspective considers how aspects of social life are functional or dysfunctional—that is, how they contribute to or interfere with social stability. Racial and ethnic inequality are functional in that keeping minority groups in a disadvantaged position ensures that there are workers who will do menial jobs for low pay. Most sociologists emphasize the ways in which racial and ethnic inequality are dysfunctional—a society that practices discrimination fails to develop and utilize the resources of minority members (Williams & Morris 1993). Prejudice and discrimination aggravate social problems, such as crime and violence, war, unemployment and poverty, health problems, family problems, urban decay, and drug use—problems that cause human suffering as well as impose financial burdens on individuals and society. Picca and Feagin (2007) explained:

> [T]he system of racial oppression in the United States affects not only Americans of color but white Americans and society as a whole. . . . Whites lose when they have to pay huge taxes to keep people of color in prisons because they are not willing to remedy patterns of unjust enrichment and . . . to pay to expand education, jobs, or drug-treatment programs that would be less costly. They lose by driving long commutes so they do not have to live next to people of color in cities. . . . They lose when white politicians use racist ideas and arguments to keep from passing legislation that would improve the social welfare of all Americans. Most of all, whites lose . . . by not having in practice the democracy that they often celebrate to the world in their personal and public rhetoric. (p. 271)

The structural-functionalist analysis of manifest and latent functions also sheds light on issues of race and ethnic relations. For example, the manifest function of the civil rights legislation in the 1960s was, in part, to improve conditions for racial minorities. However, civil rights legislation produced an unexpected negative consequence, or latent dysfunction. Because civil rights legislation supposedly ended racial discrimination, whites were more likely to blame blacks for their social disadvantages and thus perpetuate negative stereotypes such as "blacks lack motivation" and "blacks have less ability" (Schuman & Krysan 1999).

Conflict Perspective

The conflict perspective examines how competition over wealth, power, and prestige contributes to racial and ethnic group tensions. Consistent with this perspective, the "racial threat" hypothesis views white racism as a response to perceived or actual threats to whites' economic well-being or cultural dominance by minorities.

For example, between 1840 and 1870, large numbers of Chinese immigrants came to the United States to work in mining (the California Gold Rush of 1848), railroads (the transcontinental railroad, completed in 1869), and construction. As Chinese workers displaced whites, anti-Chinese sentiment rose, resulting in increased prejudice and discrimination and the eventual passage of the Chinese Exclusion Act of 1882, which restricted Chinese immigration until 1924. More recently, white support for Proposition 209—a 1996 resolution passed in California that ended state affirmative action programs—was higher in areas with larger Latino, African American, or Asian American populations, even after controlling for other factors (Tolbert & Grummel 2003). In other words, opposition to affirmative action programs that help minorities was higher in areas with greater racial and ethnic diversity, suggesting that whites living in diverse areas felt more threatened by the minorities.

In another study, researchers interviewed individuals in white racist Internet chat rooms to examine the extent to which people would advocate interracial violence in response to alleged economic and cultural threats (Glaser et al. 2002). The researchers posed three scenarios that might be perceived as threatening: interracial marriage, minority in-migration (i.e., blacks moving into one's neighborhood), and job competition (i.e., competing with a black person for a job). Respondents' reactions to interracial marriage were the most volatile, followed by in-migration. The researchers concluded that violent ideation among white racists stems from perceived threats to white cultural dominance and separateness rather than from perceived economic threats.

Furthermore, conflict theorists suggest that capitalists profit by maintaining a surplus labor force, that is, by having more workers than are needed. A surplus labor force ensures that wages will remain low because someone is always available to take a disgruntled worker's place. Minorities who are disproportionately unemployed serve the interests of the business owners by providing surplus labor, keeping wages low, and, consequently, enabling them to maximize profits.

Conflict theorists also argue that the wealthy and powerful elite foster negative attitudes toward minorities to maintain racial and ethnic tensions among workers. So long as workers are divided along racial and ethnic lines, they are less likely to join forces to advance their own interests at the expense of the capitalists. In addition, the "haves" perpetuate racial and ethnic tensions among the "have-nots" to deflect attention away from their own greed and exploitation of workers.

Symbolic Interactionist Perspective

The symbolic interactionist perspective focuses on the social construction of race and ethnicity—how we learn conceptions and meanings of racial and ethnic distinctions through interaction with others—and how meanings, labels, and definitions affect racial and ethnic groups and intergroup interaction. We have already explained that contemporary race scholars agree that there is no scientific, biological basis for racial categorizations. However, people have learned to think of racial categories as real, and, as the *Thomas theorem* suggests, if things are defined as real, they are real in their consequences. Ossorio and Duster (2005) explain:

> People often interact with each other on the basis of their beliefs that race reflects physical, intellectual, moral, or spiritual superiority or inferiority. . . . By acting on their beliefs about race, people create a society in which individuals of one group have greater access to the goods of society—such as high-status jobs, good schooling, good housing, and good medical care—than do individuals of another group. (p. 119)

The labeling perspective directs us to consider the role that negative stereotypes play in race and ethnicity. **Stereotypes** are exaggerations or generalizations about the characteristics and behavior of a particular group. Negative stereotyping of minorities leads to a self-fulfilling prophecy—a process in which a false definition of a situation leads to behavior that, in turn, makes the originally falsely defined situation come true (see also Chapter 8). Marger (2012) provides the following example of how the negative stereotype of blacks as less intelligent than whites can lead to a self-fulfilling prophecy:

> If blacks are considered inherently less intelligent, fewer community resources will be used to support schools attended primarily by blacks on the assumption that such support would only be wasted. Poorer-quality schools, then, will inevitably turn out less capable students, who will score lower on intelligence tests. The poorer performance on these tests will "confirm" the original belief about black inferiority. Hence, the self-fulfilling prophecy. (p. 15)

Even stereotypes that appear to be positive can have negative effects. The view of Asian Americans as a "model minority" involves the stereotypes of Asian Americans as excelling in academics and occupational success. These stereotypes mask the struggles and discrimination that many Asian Americans experience and also put enormous pressure on Asian American youth to live up to the social expectation of being a high academic achiever (Tanneeru 2007).

The symbolic interactionist perspective is concerned with how individuals learn negative stereotypes and prejudicial attitudes through language. Different connotations of the colors white and black, for example, may contribute to negative attitudes toward people of color. The white knight is good, and the black knight is evil; angel food cake is white, and devil's food cake is black. Other negative terms associated with black include *black sheep, black plague, black magic, black mass, blackballed,* and *blacklisted.* The continued use of derogatory terms for racial and ethnic groups confirms the power of language in perpetuating negative attitudes toward minority group members. One of the most insulting, demeaning words used to refer to African Americans—the word *nigger*—became a topic of public discussion in 2013, after celebrity chef Paula Deen admitted in a legal deposition that she had used the "N-word." When the story

stereotypes Exaggerations or generalizations about the characteristics and behavior of a particular group.

hit the media, the Food Network canceled its contract, and Wal-Mart and Smithfield Hams severed their business relationships with Deen. The public was divided; some believed that Paula Deen had made a understandable mistake for which she apologized and that she should be forgiven, and others believed her behavior was blatant racism and should not be excused.

What Do You Think? When whites use the "N-word," it is demeaning and insulting to blacks. Yet black comedians and rappers commonly use the N-word. And some blacks, as well as people of other races, use the word *nigger* or *nigga* to refer to a buddy or friend, conveying a neutral or affectionate tone. The late rapper Tupac Shakur defined "nigger" as a black man with a slavery chain around his neck, whereas a "nigga" is a black man with a gold chain on his neck (urbandictionary .com). Tupac is also known for coming up with the acronym "N.I.G.G.A." meaning "Never Ignorant Getting Goals Accomplished." Oprah Winfrey is one of many blacks who oppose the use of the N-word (and any variation of it) by anyone—including by blacks. She explained, "when I hear the N-word, I still think about every black man who was lynched—and the N-word was the last thing he heard" (Winfrey 2009, n.p.). In an interview, Oprah asked rapper Jay-Z if using the N-word is necessary. Jay-Z explained:

Nothing is necessary. It's just become part of the way we communicate. My generation hasn't had the same experience with that word that generations of people before us had. We weren't so close to the pain. So in our way, we disarmed the word. We took the fire pin out of the grenade.

Do you think that the attempt to redefine the N-word by Tupac and Jay-Z will effectively disarm it? Why or why not?

Rapper Jay-Z is among those blacks who use the N-word in a way that conveys a neutral or even friendly meaning.

Advocates for immigrant rights suggest that the terms *illegal aliens* and *illegal immigrants* are derogatory and stigmatize and criminalize people rather than their actions. In 2013, the Associated Press (AP) news agency announced it would not use the term *illegal immigrant*. AP journalists are now instructed to "use *illegal* only to refer to an action, not a person: *illegal immigration*, but not *illegal immigrant* (unless the term is used in direct quotations)" (Colford 2013). Other news agencies, including *USA Today,* have also implemented guidelines that prohibit the use of the terms *illegal alien* and *illegal immigrant.*

Bonilla-Silva (2012) gives examples of what he calls the "racial grammar" that shapes how we see or don't see race, and how we frame matters as racial or not race-related. Racial grammar involves conveying and perpetuating racial meanings not only through what we say, but also by what we *don't* say. For example, "in the USA one can talk about HBCUs (historically black colleges and universities), but not about HWCUs (historically white colleges and universities) or one can refer to black movies and black TV shows but not label movies and TV shows white when in fact most are" (Bonilla-Silva 2012, p. 173).

In the next section, we explore the concepts of racism and prejudice in more depth and discuss ways in which socialization and the media perpetuate negative stereotypes.

Racism and Prejudice

Racism is the belief that race accounts for differences in human character and ability and that a particular race is superior to others. This chapter's *Social Problems Research Up Close* feature suggests that racism in the United States is more common than many people realize. **Institutional racism** refers to the systematic distribution of power, resources, and opportunity in ways that benefit whites and disadvantage minorities. This concept is similar to "institutional discrimination" which is discussed later in this chapter. **Prejudice** refers to negative attitudes and feelings toward or about an entire category of people. Prejudice can be directed toward individuals of a particular religion, sexual orientation, political affiliation, age, social class, sex, race, or ethnicity.

Forms of Racism

Compared with traditional, "old-fashioned" prejudice, which is blatant, direct, and conscious, contemporary forms of prejudice are often subtle, indirect, and unconscious. Three variants of these more subtle forms of prejudice are aversive racism, modern racism, and color-blind racism.

Aversive Racism. **Aversive racism** represents a subtle, often unintentional, form of prejudice exhibited by many well-intentioned white Americans who possess strong egalitarian values and who view themselves as unprejudiced. The negative feelings that aversive racists have toward blacks and other minority groups are not feelings of hostility or hate but rather feelings of discomfort, uneasiness, disgust, and sometimes fear (Gaertner & Dovidio 2000). Aversive racists may not be fully aware that they harbor these negative racial feelings; indeed, they disapprove of individuals who are prejudiced and would feel falsely accused if they were labeled prejudiced. "Aversive racists find blacks 'aversive,' while at the same time find any suggestion that they might be prejudiced 'aversive' as well" (Gaertner & Dovidio 2000, p. 14).

Another aspect of aversive racism is the presence of pro-white attitudes, as opposed to anti-black attitudes. In several studies, respondents did not indicate that blacks were worse than whites, only that whites were better than blacks (Gaertner & Dovidio 2000). For example, blacks were not rated as being lazier than whites, but whites were rated as being more ambitious than blacks. Gaertner and Dovidio (2000) explain that "aversive racists would not characterize blacks more negatively than whites because that response could readily be interpreted by others or oneself to reflect racial prejudice" (p. 27). Compared with anti-black attitudes, pro-white attitudes reflect a more subtle prejudice that, although less overtly negative, is still racial bias.

Modern Racism. Like aversive racism, **modern racism** involves the rejection of traditional racist beliefs, but a modern racist displaces negative racial feelings onto more abstract social and political issues. The modern racist believes that serious discrimination in the United States no longer exists, that any continuing racial inequality is the fault of minority group members, and that demands for affirmative action for minorities are unfair and unjustified. "Modern racism tends to 'blame the victim' and places the responsibility for change and improvements on the minority groups, not on the larger society" (Healey 1997, p. 55). Like aversive racists, modern racists tend to be unaware of their negative racial feelings and do not view themselves as prejudiced.

Color-Blind Racism. Comedian Stephen Colbert, of Comedy Central's *The Colbert Report*, has an ongoing joke about being racially color-blind, not only to others' race, but also to his own. Colbert's color-blind remarks include: "People tell me I'm white and I believe them, because police officers call me 'sir,'" or "because I belong to an all-white country club," or "because I own a lot of Jimmy Buffet albums" (cited in Knowles & Marshburn 2010, p. 134).

racism The belief that race accounts for differences in human character and ability and that a particular race is superior to others.

institutional racism The systematic distribution of power, resources, and opportunity in ways that benefit whites and disadvantage minorities.

prejudice Negative attitudes and feelings toward or about an entire category of people.

aversive racism A subtle form of prejudice that involves feelings of discomfort, uneasiness, disgust, fear, and pro-white attitudes.

modern racism A subtle form of racism that involves the belief that serious discrimination in the United States no longer exists, that any continuing racial inequality is the fault of minority group members, and that the demands for affirmative action for minorities are unfair and unjustified.

In a book titled *Two-Faced Racism*, sociologists Leslie Picca and Joe Feagin present research on white college students' encounters with racism in their everyday lives.

Sample and Methods

Picca and Feagin (2007) recruited 934 college students to participate in research that required students to keep a journal of "everyday events and conversations that deal with racial issues, images, and understandings" (p. 31). Students did not write their name on the journals, but did indicate their gender, race, and age. The majority of students (63 percent) were from colleges and universities in the South, 19 percent were from the Midwest, 14 percent were from the West, and 4 percent were from the Northeast. Most of the participating students were between 18 and 25 years of age, although many were in their late 20s, and a few were older. The results presented here are based on roughly 9,000 journal entries written by 626 white students (68 percent women and 32 percent men).

Selected Findings

About three-quarters of the students' journal entries described racist events. In a few of these accounts, students described their own racist actions or thoughts:

Kristi: When I went to pick up the laundry, I saw a young black man sitting in the driver's side of a minivan with the engine running. My first thought was that he was waiting for a friend to rob the store and he was the getaway driver. . . . I am so embarrassed and saddened by my thinking. . . . (p. 1)

More frequently, students described the racist comments or actions of friends, family members, acquaintances, and strangers. Students wrote about racist accounts targeted at African Americans, Latinos, Asian Americans, Native Americans, Jewish Americans, and Middle Easterners. The most frequently targeted minority group was black Americans. A few journal accounts described antiracist actions by whites.

Picca and Feagin used sociologist Erving Goffman's theatrical metaphors to describe and understand the racist behavior described in the journals. Goffman suggested that when people are "backstage" in private situations, they act differently than they do when they are in "frontstage" public situations. The

expression of racist attitudes has gone backstage to private settings where whites are among other whites, especially family and friends because "many . . . whites realize it is no longer socially acceptable to be blatantly racist in frontstage areas" (p. xi). Consider the following diary entry:

Hannah: Three of my friends (a white girl and two white boys) and I went back to my house to drink a little more before we ended the night. My one friend, Dylan, started telling jokes. . . . Dylan said: "What's the most confusing day of the year in Harlem?" "Father's Day . . . who's your daddy?" Dylan also referred to black people as "porch monkeys." Everyone laughed a little, but it was obvious that we all felt a little less comfortable when he was telling jokes like that. My friend Dylan is not a racist person. He has more black friends than I do, that's why I was surprised he so freely said something like that. Dylan would never have said something like that around anyone who was a minority. . . . It is this sort of "joking" that helps to keep racism alive today. People know the places they have to be politically correct and most people will be. However, until this sort of "behind-the-scenes" racism comes to an end, people will always harbor those stereotypical views that are so prevalent in our country. This kind of joking really does bother me, but I don't know what to do about it. I know that I should probably stand up and say I feel uncomfortable when my friends tell jokes like that, but I know my friends would just get annoyed with me and say that they obviously don't mean anything by it. (pp. 17–18)

Hannah's journal entry illustrates the social norm that defines expressions of racism as acceptable in private "backstage" settings, and inappropriate in public "frontstage" settings. Hannah's journal entry also exemplifies a theme of "white innocence," whereby whites often view racist comments not as "real racism," but as harmless remarks that are not to be taken seriously.

Although white women also display racist behavior, the journal entries in this research suggest that "white men disproportionately make racist jokes and similar racially barbed comments" (p. 133):

Carissa: The white men got on the subject of so-called nicknames for black people. Some mentioned were porch monkeys, jigaboos, tree swingers, etc. The one thing I took notice of was that not one girl made a comment. (p. 133)

Some students made comments that reflected a general lack of awareness of racism. One student, for example, wrote, "This assignment made me realize how many racial remarks are said every single day and I usually never catch any of them or pay close attention" (p. 39).

Many students were offended by the racist remarks they recorded in their journals, but did not express their disapproval and, so, participated in racism as a "bystander." Some students indicated that, after participating in the journal writing assignment, they are more aware of the need to intervene in social situations where racism is being expressed:

Kyle: As my last entry in this journal, I would like to express what I have gained out of this assignment. I watched my friends and companions with open eyes. I was seeing things that I didn't realize were actually there. By having a reason to pick out the racial comments and actions, I was made aware of what is really out there. Although I noticed that I wasn't partaking in any of the racist actions or comments, I did notice that I wasn't stopping them either. I am now in a position to where I can take a stand and try to intervene in many of the situations. (p. 275)

Conclusion

Based on the research presented in *Two-Faced Racism*, Picca and Feagin suggest that "the majority of whites still participate in openly racist performances in the backstage arena . . . and do not define such performances as problematic and deserving of action aimed at eradication" (p. 22). The researchers further note that "most of our college student diarists did not, or could not, see the connection between the everyday racial performances they recorded and the unjust discrimination and widespread suffering endured by people of color in society generally" (p. 28).

Source: Picca & Feagin 2007.

Color-blind racism is based on the belief that paying attention to race is, itself, racism, and therefore, people should ignore race. A college student at Washington University explains how she learned the color-blind mindset (Frieden 2013):

> Growing up as a white girl in New Hampshire, I was raised with this kind of a "color-blind" mindset: a mindset that says that race should be ignored, and that racism exists because some people simply refuse to ignore race like they should. (p. 1)

Color-blindness assumes we are living in a postracial world where race no longer matters, when in reality, race continues to be a significant issue. Color-blindness is a form of racism because it prevents acknowledgment of privilege and disadvantage associated with race, and therefore allows the continuation of cultural and structural forms of racial bias. Interestingly, research has found that, "people who claim to be 'color-blind' and go to great pains to avoid talking about race during social interactions, are in fact perceived as more prejudiced by black observers than people who openly acknowledge race" (Apfelbaum 2011).

Learning to Be Prejudiced: The Role of Socialization and the Media

Psychological theories of prejudice focus on forces within individuals that give rise to prejudice. For example, the frustration-aggression theory of prejudice (also known as the scapegoating theory) suggests that prejudice is a form of hostility that results from frustration. According to this theory, minority groups serve as convenient targets of displaced aggression. The authoritarian-personality theory of prejudice suggests that prejudice arises in people with a certain personality type. According to this theory, people with an authoritarian personality—who are highly conformist, intolerant, cynical, and preoccupied with power—are prone to being prejudiced.

Rather than focus on individuals, sociologists focus on social forces that contribute to prejudice. Earlier, we explained how intergroup conflict over wealth, power, and prestige gives rise to negative feelings and attitudes that serve to protect and enhance dominant group interests. Prejudice is also learned through socialization and the media.

Learning Prejudice through Socialization. In the socialization process, individuals adopt the values, beliefs, and perceptions of their family, peers, culture, and social groups. Prejudice is taught and learned through socialization, although it need not be taught directly and intentionally. White parents who teach their children to not be prejudiced yet live in an all-white neighborhood, attend an all-white church, and have only white friends may be indirectly teaching negative racial attitudes to their children. Socialization can also be direct, as in the case of parents who use racial slurs in the presence of their children or who forbid their children from playing with children from a certain racial or ethnic background. Children can also learn prejudicial attitudes from their peers. Telling racial and ethnic jokes among friends, for example, perpetuates stereotypes that foster negative racial and ethnic attitudes.

Prejudice and the Media. The media—including television, radio, and the Internet—play a role in perpetuating prejudice and hate. In television and movies, racial minorities are underrepresented and appear mostly in stereotypical and secondary roles, such as thugs, buffoons, and angry people. In news media, stories of missing white women and children receive more coverage than missing minority women and children (Bonilla-Silva 2003).

Another form of media that is used to promote hatred toward minorities and to recruit young people to the white power movement is "white power music," which contains anti-Semitic, racist, and homophobic lyrics. A CD that was distributed to thousands of middle and high school students by the neo-Nazi record label Panzerfaust Records contains the following lyrics from a group called the Bully Boys: "Whiskey bottles/baseball bats/pickup trucks/and rebel flags/we're going on the town tonight/hit and run/let's have some fun/we've got jigaboos on the run." And in a song called "Wrecking Ball," the band

color-blind racism A form of racism that is based on the idea that overcoming racism means ignoring race, but color-blindness is, in itself, a form or racism because it prevents acknowledgment of privilege and disadvantage associated with race, and therefore allows the continuation of institutional forms of racial bias.

Byron Calvert operates Panzerfaust Records, one of the nation's largest "white power" music labels.

H8Machine advises kids to "destroy all your enemies," promising that "the best things come to those who hate" (SPLC 2004). Hate groups also use social networking sites such as Facebook, MySpace, and YouTube to spread their message of hate and to recruit new members (Turn It Down 2009). Hate is also spread via Twitter. On November 1, 2012, a Twitter user tweeted "So an 11 year old nigger girl killed herself over my tweets? . . . thats [sic] another nigger off the streets!!" (Kwok & Wang 2013, p. 1,621).

Discrimination against Racial and Ethnic Minorities

Whereas prejudice refers to attitudes, **discrimination** refers to actions or practices that result in differential treatment of categories of individuals. Although prejudicial attitudes often accompany discriminatory behavior or practices, one can be evident without the other.

Individual versus Institutional Discrimination

Individual discrimination occurs when individuals treat other individuals unfairly or unequally because of their group membership. Individual discrimination can be overt or adaptive. In **overt discrimination**, individuals discriminate because of their own prejudicial attitudes. For example, a white landlord may refuse to rent to a Mexican American family because of a prejudice against Mexican Americans. Or a Taiwanese American college student who shares a dorm room with an African American student may request a roommate reassignment from the student housing office because of prejudice against blacks.

Suppose that a Cuban American family wants to rent an apartment in a predominantly non-Hispanic neighborhood. If the landlord is prejudiced against Cubans and does not allow the family to rent the apartment, that landlord has engaged in overt discrimination. But what if the landlord is not prejudiced against Cubans but still refuses to rent to a Cuban family? Perhaps that landlord is engaging in **adaptive discrimination**, or discrimination that is based on the prejudice of others. In this example, the landlord may fear that other renters who are prejudiced against Cubans may move out of the building or neighborhood and leave the landlord with unrented apartments. Overt and adaptive individual discrimination can coexist.

Institutional discrimination refers to institutional policies and procedures that result in unequal treatment of and opportunities for minorities. Institutional discrimination is covert and insidious and maintains the subordinate position of minorities in society. For example, when schools use standard intelligence tests to decide which children will be placed in college preparatory tracks, they are limiting the educational advancement of minorities whose intelligence is not fairly measured by culturally biased tests developed from white middle-class experiences. And the funding of public schools through local tax dollars results in less funding for schools in poor and largely minority school districts (see also Chapter 8).

Institutional discrimination is also found in the criminal justice system, which disproportionately incarcerates people of color (see also Chapter 4). Because the "War on Drugs" has been waged predominately in poor communities of color, most people arrested for drug offenses are black or Latino, even though people of color are no more likely to use or sell drugs than whites (Alexander 2010). Millions of people of

discrimination Actions or practices that result in differential treatment of categories of individuals.

individual discrimination The unfair or unequal treatment of individuals because of their group membership.

overt discrimination Discrimination that occurs because of an individual's own prejudicial attitudes.

adaptive discrimination Discrimination that is based on the prejudice of others.

institutional discrimination Discrimination in which institutional policies and procedures result in unequal treatment of and opportunities for minorities.

color are labeled as felons for relatively minor, nonviolent drug offenses. Due to felon disenfranchisement laws, those labeled felons are denied the right to vote and are automatically excluded from juries. Felons may also be legally discriminated against in employment, housing, access to education, and public benefits, much like their grandparents or great grandparents may have been discriminated against during the Jim Crow era (Alexander 2010).

Another example of institutional discrimination concerns voting rights. In June 2013, the U.S Supreme Court in *Shelby County v. Holder* ruled that Section 4 of the Voting Rights Act—which determines which states and jurisdictions are covered by Section 5— is invalid. Section 5 requires states with a historical record of discriminatory practices to receive preclearance from the federal government before these states could make changes to voting laws or procedures. Within months following the ruling, six southern states passed legislation imposing new voting restrictions that will disproportionately affect minorities. North Carolina—one of the states affected by Section 5—was the first state to approve legislation that will effectively suppress voting among minorities. The North Carolina bill enacted strict voter ID requirements, ended election-day registration, cut early voting, and made it harder to register to vote.

Employment Discrimination

Despite laws against it, discrimination against minorities occurs today in all phases of the employment process, from recruitment to interview, job offer, salary, promotion, and firing decisions. A sociologist at Northwestern University studied employers' treatment of job applicants in Milwaukee, Wisconsin, by dividing job applicant "testers" into four groups: blacks with a criminal record, blacks without a criminal record, whites with a criminal record, and whites without a criminal record (Pager 2003). Applicant testers, none of whom actually had a criminal record, were trained to behave similarly in the application process and were sent with comparable résumés to the same set of employers. The study found that white applicants with no criminal record were the most likely to be called back for an interview (34 percent) and that black applicants with a criminal record were the least likely to be called back (5 percent). But surprisingly, white applicants with a criminal record (17 percent) were more likely to be called back for a job interview than were black applicants *without* a criminal record (14 percent). The researcher concluded that "the powerful effects of race thus continue to direct employment decisions in ways that contribute to persisting racial inequality" (Pager 2003, p. 960).

Discrimination in hiring may be unintended. For example, many businesses rely on their existing employees to refer new recruits when a position opens up. Word-of-mouth recruitment is inexpensive and efficient; some companies offer bonuses to employees who bring in new recruits. But this traditional recruitment practice tends to exclude minority workers, because they often do not have a network of friends and family members in higher positions of employment who can recruit them (Schiller 2004).

Employment discrimination contributes to the higher rates of unemployment and lower incomes of blacks and Hispanics compared with those of whites (see Chapters 6 and 7). Lower levels of educational attainment among minority groups account for some, but not all, of the disadvantages they experience in employment and income. As shown in Figure 9.6, mean earnings of whites are higher than those for blacks, Hispanics, and Asians at most levels of educational attainment. And among young college graduates (who do not have an advanced degree and are not currently in college), unemployment is notably higher among racial and ethnic minorities compared with whites. In a report on the job prospects of young college graduates, the authors explained that "having an equivalent amount of higher education and a virtual blank slate of prior work experience still does not generate parity in unemployment across races and ethnicities. This suggests other factors may be at play, such as minorities not having equal access to the informal professional networks that often lead to job opportunities, and/or discrimination against racial and ethnic minorities" (Shierholz et al. 2013, p. 11).

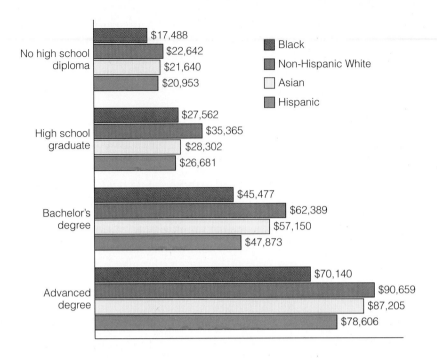

Workplace discrimination also includes unfair treatment and harassment. African American employees working for A.C. Widenhouse—a North Carolina–based freight trucking company—were repeatedly subjected to derogatory racial comments and slurs by employees and managers. These insulting comments and slurs included "n----r," "monkey," and "boy" (EEOC 2011). One employee was approached by a coworker with a noose who said, "This is for you. Do you want to hang from the family tree?" A company manager allegedly told an employee, "We are going coon hunting, are you going to be the coon?"

Housing Discrimination and Segregation

Before the 1968 Fair Housing Act and the 1974 Equal Credit Opportunity Act, discrimination against minorities in housing and mortgage lending was common. Banks and mortgage companies commonly engaged in "redlining"—the practice of denying mortgage loans in minority neighborhoods on the premise that the financial risk was too great, and the ethical standards of the National Association of Real Estate Boards prohibited its members from introducing minorities into white neighborhoods.

Although housing discrimination is illegal today, it is not uncommon. To assess discrimination in housing, researchers use a method called "paired testing." In a paired test, two individuals—one minority and the other nonminority—are trained to pose as home seekers, and they interact with real estate agents, landlords, rental agents, and mortgage lenders to see how they are treated. The testers are assigned comparable or identical income, assets, and debt as well as comparable or identical housing preferences, family circumstances, education, and job characteristics. A paired testing study of housing discrimination in 23 metropolitan areas found that whites in the rental market were more likely to receive information about available housing units and had more opportunities to inspect available units than did blacks and Hispanics (Turner et al. 2002). The incidence of discrimination was greater for Hispanic renters than for black renters. The same study found that, in the home sales market, white home buyers were more likely to be able to inspect available homes and to be shown homes in more predominantly non-Hispanic white neighborhoods than were comparable black and Hispanic buyers. Whites were also more likely to receive information and assistance with financing.

In a study of housing discrimination in the Philadelphia area, Massey and Lundy (2001) found that, compared with whites, African Americans were less likely to have a rental agent return their calls, less likely to be told that a unit was available, more likely to pay application fees, and more likely to have credit mentioned as a potential problem in qualifying for a lease. Sex and class exacerbated these racial effects. Lower-class blacks experienced less access to rental housing than middle-class blacks, and black females experienced less access than black males. Lower-class black females were the most disadvantaged group. They experienced the lowest probability of contacting and speaking to a rental agent and, even if they did make contact, they faced the lowest probability of being told of a housing unit's availability. Lower-class black females also faced the highest chance of paying an application fee. On average, lower-class black females were assessed $32 more per application than white middle-class males.

Despite continued housing discrimination, homeownership rates among minorities and low-income groups increased substantially in the 1990s, reaching record rates in many central cities. However, minority and low-income homeowner rates still lag behind the overall homeownership rate. Also, many of the gains in minority and low-income homeownership rates are due to increases in *subprime lending*—loans with higher fees and higher interest rates that are offered to borrowers who have poor (or nonexistent) credit records (Williams et al. 2005).

Residential segregation of racial and ethnic groups also persists. Almost a quarter of all census tracts within the largest U.S. metropolitan areas are more than 90 percent white, and 12 percent are more than 90 percent minority (Turner & Fortuny 2009).

Educational Discrimination and Segregation

Both institutional discrimination and individual discrimination in education negatively affect racial and ethnic minorities and help to explain why minorities (with the exception of Asian Americans) tend to achieve lower levels of academic attainment and success (see also Chapter 8). Institutional discrimination is evidenced by inequalities in school funding—a practice that disproportionately hurts minority students (Kozol 1991). Because minorities are more likely than whites to live in economically disadvantaged areas, they are more likely to go to schools that receive inadequate funding. Inner-city schools, which serve primarily minority students, receive less funding per student than do schools in more affluent, primarily white areas.

Another institutional education policy that is advantageous to whites is the policy that gives preference to college applicants whose parents or grandparents are alumni. The overwhelming majority of alumni at the highest-ranked universities and colleges are white. Thus, white college applicants are the primary beneficiaries of these so-called legacy admissions policies. About 10 percent to 15 percent of students in most Ivy League colleges and universities are children of alumni. Harvard University accepts about 11 percent of its overall applicant pool, but the admission rate is 40 percent for legacy applicants (Schmidt 2004). As a result of pressure from state lawmakers and minority rights activists, in 2004, Texas A&M University became the first public college to abandon its legacy admittance policy.

Minorities also experience individual discrimination in the schools as a result of continuing prejudice among teachers. In a survey conducted by the Southern Poverty Law Center, 1,100 educators were asked whether they had heard racist comments from their colleagues in the past year. More than a quarter of survey respondents answered "yes" ("Hear and Now" 2000). It is likely that teachers who are prejudiced against minorities discriminate against them, giving them less teaching attention and less encouragement.

Racial and ethnic minorities are also treated unfairly in educational materials, such as textbooks, which often distort the history and heritages of people of color (King 2000). For example, Zinn (1993) observed, "To emphasize the heroism of Columbus and his successors as navigators and discoverers, and to de-emphasize their genocide, is not a technical necessity but an ideological choice. It serves, unwittingly—to justify what was done" (p. 355).

Finally, racial and ethnic minorities are largely isolated from whites in a largely segregated school system. U.S. schools in the 2000–2001 school year were more segregated than they were in 1970 (Orfield 2001). School segregation is largely due to the persistence of housing segregation and the termination of court-ordered desegregation plans. Court-mandated busing became a means to achieve equality of education and school integration in the early 1970s, after the Supreme Court (in *Swann v. Charlotte-Mecklenburg*) endorsed busing to desegregate schools. But in the 1990s, lower courts lifted desegregation orders in dozens of school districts (Winter 2003). And in 2007, the United States Supreme Court issued a landmark ruling, in a bitterly divided 5-to-4 vote, that race cannot be a factor in the assignment of children to public schools. The decision jeopardizes similar plans in hundreds of districts nationwide, and it further restricts how public school systems may achieve racial diversity. Racial and ethnic segregation in U.S. suburban schools has declined slightly, as the percentage of minority students in suburban school districts has increased from 28 percent in the 1993–94 school year to 41 percent in the 2006–07 school year (Fry 2009).

Hate Crimes

In June 2011, 49-year-old James Craig Anderson, an African American auto plant worker, was robbed, beaten, and run over by a truck near a hotel in Jackson, Mississippi. Deryl Dedmon, the 18-year-old who drove the truck that ran over Anderson, and a group of other white teenagers were reportedly out looking for a black person to assault when they found Anderson. During the brutal attack, which was captured on hotel security video, racial slurs were used, and later Dedmon bragged that he "just ran that n----- over" (Associated Press 2011).

The murder of James Craig Anderson is an example of a **hate crime**—an unlawful act of violence motivated by prejudice or bias. Examples of hate crimes, also known as "bias-motivated crimes," include intimidation (e.g., threats), destruction of or damage to property, physical assault, and murder.

According to the Federal Bureau of Investigation (FBI), which has published hate crime statistics annually since 1995, the number of hate crimes has ranged from about 6,000 to 10,000 each year. However, FBI hate crime data undercount the actual number of hate crimes because (1) not all U.S. jurisdictions report hate crimes to the FBI (reporting is voluntary); (2) it is difficult to prove that crimes are motivated by hate or prejudice; (3) law enforcement agencies shy away from classifying crimes as hate crimes because it makes their community "look bad"; and (4) victims are often reluctant to report hate crimes to the authorities. The National Crime Victimization Survey reveals that the incidence of hate crimes is 25 to 40 times higher than the FBI numbers. According to recent data from the National Crime Victimization Survey, more than 250,000 people in the United States ages 12 and over are victims of hate crimes each year, and the majority of these hate crimes—about two out of three—are not reported to the police ("DOJ Study: Hate Crimes More Prevalent than Previously Known" 2013).

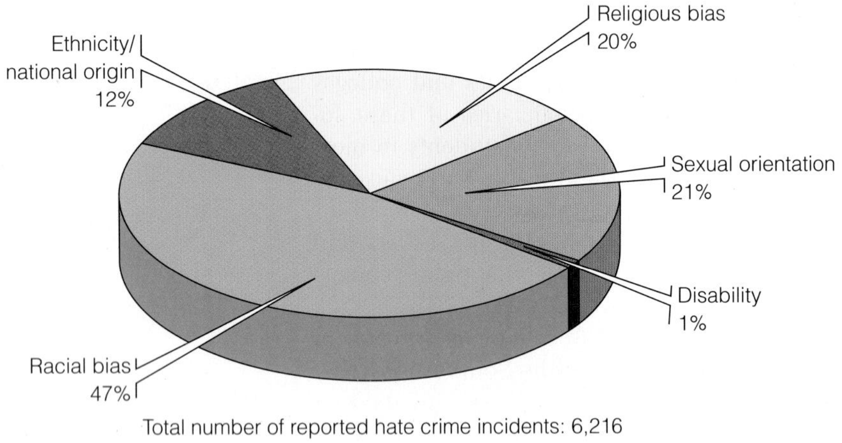

FBI hate crime data reveal that most hate crimes are based on racial bias, primarily against blacks, followed by whites (see Figure 9.7). Most hate crimes motivated by religious bias target the Jewish religion, and the majority of hate crimes based on ethnic bias are anti-Hispanic, and are largely motivated by hate toward immigrants (see this chapter's *The Human Side* feature). In 2013, the U.S. Justice Department announced it would begin collecting data on hate crimes against Sikhs, and six other groups (Hindus, Arabs, Buddhists, Mormons, Jehovah's Witnesses, and Orthodox Christians). This decision followed the 2012 hate crime shooting of six people in a

Religious bias 20%

Ethnicity/ national origin 12%

Sexual orientation 21%

Disability 1%

Racial bias 47%

Total number of reported hate crime incidents: 6,216

Figure 9.7 Hate Crime Incidence by Category of Bias, 2011 (rounded to nearest percent)
Source: FBI 2012.

hate crime An unlawful act of violence motivated by prejudice or bias.

Milwaukee-area Sikh temple by a man with ties to white supremacist groups (Associated Press 2013). After the terrorist attacks of September 11, 2001, hate crimes against individuals perceived to be Muslim or Middle Eastern increased significantly. More than half of the United States' 7 million Muslims (individuals who adhere to the religion of Islam) said they experienced bias or discrimination after September 11 (Morrison 2002). Anti-Muslim and anti-Islam bias—commonly known as **Islamophobia**—was rekindled after the proposal to build an Islamic cultural center at 9/11 Ground Zero, and again after the 2013 Boston Marathon bombing, which was executed by two brothers who reportedly were motivated by an anti-American radical version of Islam. Islamophobia is largely based on ignorance and misunderstanding of Islam. Georgetown professor John L. Esposito (2011) explains:

Islamophobia is largely based on ignorance and misunderstanding of Islam.

> Mainstream American Muslims have too often been equated inaccurately with terrorists and people who reject democracy. . . [M]ajorities of Muslims globally desire democracy and freedom and fear and reject religious extremism and terrorism.

Motivations for Hate Crimes. Levin and McDevitt (1995) found that the motivations for hate crimes were of three distinct types: thrill, defensive, and mission. Thrill hate crimes are committed by offenders who are looking for excitement and attack victims for the "fun of it." Defensive hate crimes involve offenders who view their attacks as necessary to protect their community, workplace, or college campus from "outsiders" or to protect their racial and cultural purity from being "contaminated" by interracial marriage and childbearing. Mission hate crimes are perpetrated by white supremacist group members or other offenders who have dedicated their lives to bigotry. Hate groups known to engage in violent crimes include the Ku Klux Klan, the Identity Church Movement, the neo-Nazis, and the skinheads. The Southern Poverty Law Center identified 1,007 active hate group chapters in 2012—a significant increase since 2000, when there were 602 hate group chapters in the United States (SPLC 2013a). Although some members of hate groups have visible features, such as tattoos and armbands, that identify their hate group membership,

This public demonstration in New York was staged to show support for construction of an Islamic cultural center near Ground Zero.

Islamophobia Anti-Muslim and anti-Islam bias.

Domingo Lopez Vargas left his dirt-poor Guatemalan farm village to come to the United States, where he hoped to earn decent money for his wife and nine children. After picking oranges in Florida, he moved to Georgia, where the booming construction business lured immigrant workers. Unlike many of his compadres, Vargas had legal status, which helped him find steady work hanging doors and windows. When work dried up, Vargas joined the more than 100,000 jornaleros—day laborers—who wait for landscaping and construction jobs on street corners and in front of convenience stores all across Georgia. Usually, plenty of pickup trucks swing by, offering $8 to $12 an hour for digging, planting, painting, or hammering. But, this day, nada. By late afternoon, Vargas had tired of waiting in the cold, so he walked up the street to pick up a few things at a grocery store.

"I got milk, shampoo, and toothpaste," Vargas recalled. "When I was leaving the store, this truck stopped right in front of me and said, 'Do you want to work?'. . . I said, yes, how much? They said nine dollars an hour. I didn't ask what kind of job. I just wanted to work, so I said yes."

Until that afternoon, Vargas said, "Americans had always been very nice to me"—which might explain why he wasn't concerned that the four guys in the pickup truck looked awfully young to be contractors. Or why he didn't think twice about being picked up so close to sunset. "I

Domingo Lopez Vargas, an immigrant day laborer, was brutally beaten in a hate crime in Canton, Georgia.

took the offer because I know sometimes people don't stop working until 9 at night," he says.

The four young men, all high school students, drove Vargas to a remote spot strewn with trash. "They told me to pick up some plastic bags that were on the ground. I thought that was my job, to clean up the trash. But when I bent over to pick it up, I felt somebody

hit me from behind with a piece of wood, on my back." It was just the start of a 30-minute pummeling that left Vargas bruised and bloody from his thighs to his neck. "I thought I was dying," he said. "I tried to stand up but I couldn't." Finally, after he handed over all the cash in his wallet, $260, along with his Virgin Mary pendant, the teenagers sped away.

As a result of injuries incurred by the beating, Vargas could not work for four months, and he was left with $4,500 in medical bills. Sometimes he still puzzles over his attackers' motives. "They were young," he speculates, "and maybe they didn't have enough education. Or maybe their families . . . taught them to kill people, and that is what they have learned."

Having sworn off day labor, Vargas works night shifts now, cutting up chickens at the nearby Tyson plant, wincing through the pain that shoots up his right arm when he lowers the boom on a bird. But it's only temporary, he says. "I called my wife and told her what happened. She told me to move back to Guatemala. I wanted to, but I didn't have enough money to go back and the police officers told me not to move out of the country because they will still need me to work on the case. After the case is finished, I want to go back to my family."

Source: Moser, Bob, "The Battle of 'Georgiafornia,'" *Southern Poverty Law Center's Intelligence Report* 116:40–50. Copyright © 2004. Reprinted by permission of Southern Poverty Law Center.

many hate group members are not so easily identifiable. Criminologist Jack Levin, who studies hate crimes, noted that "many white supremacist groups are going more mainstream . . . eliminating the sheets and armbands. . . . The groups realize if they want to be attractive to middle-class types, they need to look middle class" (quoted in Leadership Council on Civil Rights Education Fund 2009, p. 19). Some anti-immigrant hate groups, such as Federation for American Immigration Reform (FAIR), the Center for Immigration Studies (CIS), and NumbersUSA, portray themselves as legitimate mainstream advocates against illegal immigration, but "some of these organizations have disturbing links to or relationships with extremists in the anti-immigrant movement" (p. 14).

Hate on Campus. In 2013, Oberlin College in Ohio canceled classes to address a string of reported racial incidents on campus, including several posters and fliers placed around campus containing hateful language and a reported sighting of someone dressed as a member of the Ku Klux Klan on campus (Ly 2013). In the same year, a black teddy bear was hung by a noose in the office of a black professor at Northwestern University. According to the FBI, nearly 1 in 10 hate crimes occur at schools or colleges (FBI 2012). In focus groups with students of color at a large, predominantly white university, students reported commonly hearing racial jokes and slurs and encountering racial slurs written on residence hall room doors, study rooms, and elevators (Harwood et al. 2012). Students also reported that residence halls with higher numbers of minority students are inferior (e.g., are older and lack air conditioning) and are labeled with names like "minority central" and "the projects."

Strategies for Action: Responding to Prejudice, Racism, and Discrimination

Because racial and ethnic tensions exist worldwide, strategies for combating prejudice, racism, and discrimination globally require international cooperation and commitment. The World Conference against Racism, Racial Discrimination, Xenophobia, and Related Intolerance, held in Durban, South Africa, in 2001, exemplifies international efforts to reduce racial and ethnic tensions and inequalities and to increase harmony among the various racial and ethnic populations of the world. Unfortunately, the U.S. delegation to this conference withdrew because of the expectation that hateful language would be used against Israel (because of the Israeli-Palestinian conflict).

In the following sections, we discuss the Equal Employment Opportunity Commission's role in responding to employment discrimination and examine the issue of affirmative action in the United States. We also discuss educational strategies to promote diversity and multicultural awareness and appreciation in schools. Finally, we look at apologies and reparations as a means of achieving racial reconciliation.

The Equal Employment Opportunity Commission

The **Equal Employment Opportunity Commission (EEOC)**, a U.S. federal agency charged with ending employment discrimination in the United States, is responsible for enforcing laws against discrimination, including Title VII of the 1964 Civil Rights Act that prohibits employment discrimination on the basis of race, color, religion, sex, or national origin. The EEOC investigates, mediates, and may file lawsuits against private employers on behalf of alleged victims of discrimination. The most frequently filed claims with the EEOC are allegations of race discrimination, racial harassment, or retaliation from opposition to racial discrimination.

In 2007, the EEOC launched a national initiative to combat racial discrimination in the workplace. The goals of this initiative, called E-RACE (Eradicating Racism And Colorism from Employment), are to (1) identify factors that contribute to race and color discrimination, (2) explore strategies to improve the administrative processing and litigation of race and color discrimination cases, and (3) increase public awareness of race and color discrimination in employment.

Affirmative Action

Affirmative action refers to a broad range of policies and practices in the workplace and educational institutions to promote equal opportunity and diversity. Affirmative action is an attempt to compensate for the effects of past discrimination and prevent current discrimination against women and racial and ethnic minorities. Veterans and people with disabilities may also qualify under affirmative action policies. Federal affirmative action policies developed in the 1960s required any employer (universities as well as businesses) who received contracts from the federal government to make "good faith" efforts to increase the pool of qualified minorities and women by expanding recruitment and training programs (U.S. Department of Labor 2002).

In higher education, affirmative action policies seek to improve access to education for those groups that have been historically excluded or underrepresented, such as women and racial/ethnic minorities. Supporters of affirmative action in higher education view the underrepresentation of minorities in higher education as a problem. Consider that, in 1965, only 5 percent of undergraduate students, one percent of law students, and 2 percent of medical students in the country were African American (NCSL 2013). Over the last several decades, colleges and universities have implemented affirmative action policies designed to recruit and admit more minority students, and these policies have been successful in increasing the enrollment of minority students, although racial/ethnic gaps in college enrollment remain (see also Chapter 8).

In 1978, the Supreme Court's ruling in *Regents of the University of California v. Bakke* marked the beginning of a series of several court challenges to affirmative action. Although the court ruled that affirmative action programs could not use fixed quotas

Equal Employment Opportunity Commission (EEOC) A U.S. federal agency charged with ending employment discrimination in the United States and responsible for enforcing laws against discrimination, including Title VII of the 1964 Civil Rights Act that prohibits employment discrimination on the basis of race, color, religion, sex, or national origin.

affirmative action A broad range of policies and practices in the workplace and educational institutions to promote equal opportunity as well as diversity.

in admission, hiring, or promotion policies, the court affirmed the right for universities and employers to consider race as a factor in admission, hiring, and promotion to achieve diversity. Since the *Bakke* case, numerous legal battles have challenged affirmative action. In 2013, the Supreme Court ruled in *Fisher v. University of Texas* that universities may seek racial diversity but they must demonstrate that race-neutral alternatives to achieving diversity are not sufficient (Kahlenberg 2013). A number of colleges and universities have already developed creative alternatives to race-based admissions policies to increase their minority student populations. Some universities are using a "holistic admissions" approach that considers the unique circumstances of each student, prioritizing academics and treating all other factors, including race, equally. Other universities use the "10 percent plan" as a way to maintain minority enrollment. Graduates in the top 10 percent of their high school classes are admitted automatically to the public college or university of their choice; standardized test scores and other factors are not considered. Yet another approach is to increase financial aid packages to low-income students, which benefits all students from economically disadvantaged backgrounds, regardless of race/ethnicity.

Legal battles over affirmative action are likely to continue as affirmative action remains a controversial and divisive issue among Americans. A Gallup Poll found that 51 percent of whites, 76 percent of blacks, and 69 percent of Hispanics generally favor affirmative action programs for racial minorities, with Democrats being twice as likely as Republicans to favor affirmative action (Jones 2013). That same poll, however, found that when questioned about college admission policies, only 28 percent of U.S. adults agree that "an applicant's racial and ethnic background should be considered to help promote diversity on college campuses." Blacks were more likely to agree (48%) than Hispanics (31%) or whites (22%).

Educational Strategies

Schools and universities play an important role in whether minorities succeed in school and in the job market. One way to improve minorities' chances of academic success is to reduce or eliminate disparities in school funding. As noted earlier, schools in poor districts—which predominantly serve minority students—have traditionally received less funding per pupil than do schools in middle- and upper-class districts (which predominantly serve white students). Other educational strategies focus on reducing prejudice, racism, and discrimination, and fostering awareness and appreciation of racial and ethnic diversity. These strategies include multicultural education, "whiteness studies," and efforts to increase diversity among student populations.

Multicultural Education in Schools and Communities. In schools across the nation, multicultural education, which encompasses a broad range of programs and strategies, works to dispel myths, stereotypes, and ignorance about minorities; to promote tolerance and appreciation of diversity; and to include minority groups in the school curriculum (see also Chapter 8). With multicultural education, the school curriculum reflects the diversity of U.S. society and fosters an awareness and appreciation of the contributions of different racial and ethnic groups to U.S. culture. The Southern Poverty Law Center's program Teaching Tolerance publishes and distributes materials and videos designed to promote better human relations among diverse groups. These materials are sent to schools, colleges, religious organizations, and a variety of community groups across the nation.

Many colleges and universities have made efforts to promote awareness and appreciation of diversity by offering courses and degree programs in racial and ethnic studies and by sponsoring multicultural events and student organizations. Many colleges mandate that students take a certain number of required diversity courses to graduate. Positive outcomes for students who take college diversity courses include increased racial understanding and cultural awareness, increased social interaction with students who have backgrounds different from their own, improved cognitive development, increased support for efforts to achieve educational equity, and higher satisfaction with their college experience (Humphreys 1999).

 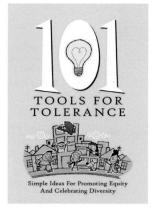  

Teaching Tolerance magazine is a free resource available to teachers for tolerance education in the classroom. Other free materials available from the Southern Poverty Law Center (www.splcenter.org) include *101 Tools for Tolerance, Responding to Hate at School,* and *Ten Ways to Fight Hate.*

Whiteness Studies. Traditionally, studies of and courses on race have focused on the social disadvantages that racial minorities experience, while ignoring or minimizing the other side of the racial inequality equation—the social advantages conferred upon whites. Courses in Whiteness Studies, which are being offered in many colleges and universities, focus on understanding "whiteness" as a social construction and fostering critical awareness of white privilege—an awareness that is limited among white students (Yeung et al. 2013). "White privilege is so fundamental as to be largely invisible, expected, and normalized" (Picca & Feagin 2007, p. 243). This chapter's *Self and Society* feature asks you to think about how you personally explain the white racial advantage.

In an often-cited paper called "White Privilege: Unpacking the Invisible Knapsack," Peggy McIntosh (1990) likened white privilege to an "invisible weightless knapsack" that whites carry with them without awareness of the many benefits inside the knapsack. Just a few of the many benefits McIntosh associates with being white include being able to

- Go shopping without being followed or harassed.
- Be assured that my skin color will not convey that I am financially unreliable when I use checks or credit cards.
- Swear, or dress in secondhand clothes, or not answer letters without having people attribute these choices to the bad morals, poverty, or the illiteracy of my race.
- Avoid ever being asked to speak for all the people of my racial group.
- Be sure that, if a traffic cop pulls me over or if the IRS audits my tax return, it is not because I have been singled out because of my race.
- Take a job with an affirmative action employer without having coworkers on the job suspect that I got it because of race.

Diversification of College Student Populations. Recruiting and admitting racial and ethnic minorities in institutions of higher education can foster positive relationships among diverse groups and enrich the educational experience of all students—minority and nonminority alike (American Council on Education and American Association of University Professors 2000). Psychologist Gordon Allport's (1954) "contact hypothesis" suggested that contact between groups is necessary for the reduction of prejudice between group members. Ensuring a diverse student population can provide students with opportunities for contact with different groups and can thereby reduce prejudice. In a study of 2,000 UCLA students, researchers found that students who were randomly assigned to roommates of a different race or ethnicity developed more favorable attitudes toward students of different backgrounds (Sidanius et al. 2010). Another study found that students with the most exposure to diverse populations during college had the most cross-racial interactions five years after leaving college (Gurin 1999).

Directions: On average, white Americans have better jobs, income, and housing than others. How important are each of the following factors in explaining this racial advantage associated with being white? Indicate your answer using the following key, putting your answer in the space provided.

Very important = 3
Somewhat important = 2
Not very important = 1
Not important at all = 0

ANSWER

1. Prejudice and discrimination in favor of whites   _____

2. Laws and institutions in favor of whites more than other groups   _____

3. Access to better schools and social connections   _____

4. Effort and hard work   _____

Interpretation and Comparison Data: Results from a nationally representative sample of U.S. adults is provided as follows, with the results broken down by race and Hispanic origin (Croll 2013). The first three explanations for white racial advantage are structural explanations that measure a belief in white privilege, and indicate an awareness of how the same factors that disadvantage minorities, benefit whites. The last explanation ("Effort and hard work") is an individualistic explanation, which explains the white racial advantage as being the result of whites working harder than other races.

	Whites	Blacks	Hispanics	Other
1. Prejudice and discrimination in favor of whites				
Very important	18%	41%	25%	29%
Somewhat important	45%	39%	47%	40%
Not very important	21%	6%	17%	16%
Not important at all	16%	14%	11%	15%
2. Laws and institutions in favor of whites more than other groups				
Very important	12%	39%	31%	29%
Somewhat important	35%	43%	42%	41%
Not very important	27%	9%	16%	15%
Not important at all	26%	9%	11%	15%
3. Access to better schools and social connections				
Very important	50%	69%	60%	58%
Somewhat important	33%	23%	26%	31%
Not very important	11%	5%	13%	4%
Not important at all	6%	3%	1%	7%
4. Effort and hard work				
Very important	57%	45%	71%	55%
Somewhat important	32%	30%	15%	26%
Not very important	7%	16%	11%	4%
Not important at all	11%	9%	3%	15%

Source: Adapted from Paul R. Croll. 2013. "Explanations for Racial Disadvantage and Racial Advantage: Beliefs about Both Sides of Inequality in America." *Ethnic and Racial Studies* 36(1):47–74.

Retrospective Justice Initiatives: Apologies and Reparations

In 2003, Brown University President Ruth J. Simmons appointed a Steering Committee on Slavery and Justice to investigate and issue a public report on the university's historical relationship to slavery and the trans-Atlantic slave trade. The steering committee's research concluded that "there is no question that many of the assets that underwrote the University's creation and growth derived, directly and indirectly, from slavery and the slave trade" (Brown University Steering Committee on Slavery and Justice 2007, p. 13).

The committee's final report recommended that Brown University should acknowledge and make amends for its past ties to the slave trade.

Various governments around the world have issued official apologies for racial and ethnic oppression. After World War II, West Germany signed a reparations agreement with Israel in which West Germany agreed to pay Israel for the enslavement and persecution of Jews during the Holocaust and to compensate for Jewish property that the Nazis stole. In 2008, the Australian government issued a formal apology for the treatment of the country's aboriginal people, specifically, for the decades during which the government removed aboriginal children from their families in a forced acculturation program.

In the United States, President Gerald Ford and Congress apologized to Japanese Americans for their internment during World War II, and reparations of $20,000 were granted to each surviving internee who was a U.S. citizen or legal resident alien at time of internment. In 1993, President Bill Clinton apologized to native Hawaiians for overthrowing the government of their nation. In 1994, the state of Florida offered monetary compensation to the survivors and descendants of the 1923 "Rosewood massacre," in which a white mob attacked and murdered black residents in Rosewood, Florida, and set fire to the town. And in 1997, the U.S. government offered monetary reparations to surviving victims of the Tuskegee syphilis study, in which blacks suffering from syphilis were denied medical treatment.

Although various forms of reparations were offered to Native American tribes to compensate for land that had been taken by force or deception, it wasn't until 2008 that the Senate Committee on Indian Affairs passed a resolution apologizing to all Indian tribes for the mistreatment and violence committed against them. In 2008, the U.S. House of Representatives issued a formal apology to African Americans for slavery, and in 2009, the U.S. Senate passed a resolution apologizing for slavery, adding the stipulation that the official apology cannot be used to support claims for restitution. The resolution "acknowledges the fundamental injustice, cruelty, brutality, and inhumanity of slavery and Jim Crow laws," and "apologizes to African Americans . . . for the wrongs committed against them and their ancestors who suffered under slavery and Jim Crow laws" (CNN 2009).

What Do You Think? Some scholars argue that achieving black/white racial reconciliation in the United States requires that the U.S. government not only issue an official apology but also provide substantial monetary and other reparations to African Americans (Brooks 2004). Do you think African Americans should receive monetary or other reparations? Why or why not?

The growing movement to redress past large-scale violations of human rights is based on the moral principles of taking responsibility for and attempting to rectify past wrongdoings. Supporters of the reparative justice movement believe that the granting of apologies and reparations to groups that have been mistreated promotes dialogue and healing, increases awareness of present inequalities, and stimulates political action to remedy current injustices. Some who are opposed to this movement claim that "preoccupation with past injustice is a distraction from the challenge of present injustice" (Brown University Steering Committee on Slavery and Justice 2007, p. 39).

The Brown University Steering Committee on Slavery and Justice (2007) examined "retrospective justice initiatives" from around the world and concluded that the most successful generally combined three elements: (1) formal acknowledgment of an offense; (2) a commitment to truth telling, to ensure that the relevant facts are uncovered, discussed, and properly memorialized; and (3) the making of some form of amends in the present to give material substance to expressions of regret and responsibility. In the committee's view, "reparative justice is not an invitation to 'wallow in the past' but a way for societies to come to terms with painful histories and move forward" (Brown University Steering Committee on Slavery and Justice 2007, p. 39).

Understanding Race, Ethnicity, and Immigration

After considering the material presented in this chapter, what understanding about race and ethnic relations are we left with? First, we have seen that racial and ethnic categories are socially constructed; they are largely arbitrary, imprecise, and misleading. Although some scholars suggest that we abandon racial and ethnic labels, others advocate adding new categories—multiethnic and multiracial—to reflect the identities of a growing segment of the U.S. and world population.

Conflict theorists and structural functionalists agree that prejudice, discrimination, and racism have benefited certain groups in society. But racial and ethnic disharmony has created tensions that disrupt social equilibrium. Symbolic interactionists note that negative labeling of minority group members, which is learned through interaction with others, contributes to the subordinate position of minorities.

Prejudice, racism, and discrimination are debilitating forces in the lives of minorities and immigrants. Despite these negative forces, many minority group members succeed in living productive, meaningful, and prosperous lives. But many others cannot overcome the social disadvantages associated with their minority status and become victims of a cycle of poverty (see Chapter 6). Minorities are disproportionately poor, receive inferior education and health care, and, with continued discrimination in the workplace, have difficulty improving their standard of living.

Intentional and hateful racism still exist, and efforts to socialize people to appreciate diversity, counter negative racial and ethnic stereotypes, and create a culture of social disapproval for individual displays of prejudice and acts of discrimination are well founded. Attempts to be "color-blind" may be well intentioned, but if we don't "see" race, then we also do not see racial injustices or the need to remedy such injustices.

Achieving racial and ethnic equality requires first *seeing* how discrimination is institutionalized in the structure of society, and then making changes in society to eliminate institutionalized discrimination and provide opportunities for minorities—in education, employment and income, and political participation. In addition, policy makers concerned with racial and ethnic equality must find ways to reduce the racial and ethnic wealth gap and to foster wealth accumulation among minorities (Conley 1999). Social class is a central issue in race and ethnic relations. Professor and activist bell hooks (2000) (who spells her name in all lowercase) warned that focusing on issues of race and gender can deflect attention away from the larger issue of class division that increasingly separates the haves from the have-nots. Addressing class inequality must, suggests hooks, be part of any meaningful strategy to reduce inequalities that minority groups suffer.

Making change requires members of society to recognize that change is necessary, that there is a problem that needs rectifying:

> One has to perceive the problem to embrace the solutions. If you think racism
> isn't harmful unless it wears sheets or burns crosses or bars blacks from motels
> and restaurants, you will support only the crudest antidiscrimination laws
> and not the more refined methods of affirmative action and diversity training.
> (Shipler 1998, p. 2)

The historic election and re-election of the first nonwhite U.S. president and the appointment of the first Hispanic woman to the U.S. Supreme Court have raised hopes for a new era in race and ethnic relations in the United States. We end this chapter with an excerpt from the benediction that Dr. Joseph Lowery delivered at President Obama's first inauguration:

> Lord . . . we ask you to help us work for that day when black will not be asked
> to get back, when brown can stick around—when yellow will be mellow—when
> the red man can get ahead, man—and when white will embrace what is right.
> (quoted in Kovac 2009)

- **What is a minority group?**

  A minority group is a category of people who have unequal access to positions of power, prestige, and wealth in a society and who tend to be targets of prejudice and discrimination. Minority status is not based on numerical representation in society but rather on social status.

- **What is meant by the idea that race is socially constructed?**

  The concept of race refers to a category of people who are perceived to share distinct physical characteristics that are deemed socially significant. The significance of race is not biological but social and political, because race is used to separate "us" from "them" and becomes a basis for unequal treatment of one group by another. Races are cultural and social inventions; they are not scientifically valid because there are no objective, reliable, meaningful criteria scientists can use to identify racial groupings. Different societies construct different systems of racial classification, and these systems change over time.

- **What are the various patterns of interaction that may occur when two or more racial or ethnic groups come into contact?**

  When two or more racial or ethnic groups come into contact, one of several patterns of interaction occurs, including genocide, expulsion, segregation, acculturation, pluralism, and assimilation.

- **Beginning with the 2000 census, what are the five race categories used to identify the race composition of the United States?**

  Beginning with the 2000 census, the five race categories are (1) white, (2) black or African American, (3) American Indian or Alaska Native, (4) Asian, and (5) Native Hawaiian or other Pacific Islander. In addition, respondents to federal surveys and the census have the option of officially identifying themselves as being of more than one race, rather than checking only one racial category.

- **What is an ethnic group?**

  An ethnic group is a population that has a shared cultural heritage, nationality, or lineage. Ethnic groups can be distinguished on the basis of language, forms of family structures and roles of family members, religious beliefs and practices, dietary customs, forms of artistic expression such as music and dance, and national origin. The largest ethnic population in the United States is Hispanics or Latinos.

- **What percentage of the U.S. population (in 2011) was born outside the United States?**

  More than 1 in 10 U.S. residents (13 percent) were born in a foreign country.

- **What were the manifest function and latent dysfunction of the civil rights movement?**

  The manifest function of the civil rights legislation in the 1960s was to improve conditions for racial minorities. However, civil rights legislation produced an unexpected consequence, or latent dysfunction. Because civil rights legislation supposedly ended racial discrimination, whites were more likely to blame blacks for their social disadvantages and thus perpetuate negative stereotypes such as "blacks lack motivation" and "blacks have less ability."

- **How does contemporary prejudice differ from more traditional, "old-fashioned" prejudice?**

  Traditional, old-fashioned prejudice is easy to recognize, because it is blatant, direct, and conscious. More contemporary forms of prejudice are often subtle, indirect, and unconscious. In addition, racist expressions have gone "backstage" to private social settings.

- **Is it possible for an individual to discriminate without being prejudiced?**

  Yes. In overt discrimination, individuals discriminate because of their own prejudicial attitudes. But sometimes individuals who are not prejudiced discriminate because of someone else's prejudice. For example, a store clerk may watch black customers more closely because the store manager is prejudiced against blacks and has instructed the employee to follow black customers in the store closely. Discrimination based on someone else's prejudice is called adaptive discrimination.

- **How is color-blindness a form of racism?**

  Color-blindness is a form of racism because it prevents acknowledgment of privilege and disadvantage associated with race, and therefore allows the continuation of institutional racism.

- **Are U.S. schools segregated?**

  Racial and ethnic minorities are largely isolated from whites in an increasingly segregated school system. One study found that U.S. schools in the 2000–2001 school year were more segregated than they were in 1970. The upward trend in school segregation is due to large increases in minority student enrollment, continuing white flight from urban areas, the persistence of housing segregation, and the termination of court-ordered desegregation plans.

- **According to FBI data, the majority of hate crimes are motivated by what kind of bias?**

  Since the FBI began publishing hate crime data in 1992, the majority of hate crimes have been based on racial bias.

- **What is the role of the Equal Employment Opportunity Commission (EEOC) in combating employment discrimination?**

  The Equal Employment Opportunity Commission (EEOC) is responsible for enforcing laws against discrimination, including Title VII of the 1964 Civil Rights Act that prohibits employment discrimination on the basis of race, color, religion, sex, or national origin. The EEOC investigates, mediates, and may file lawsuits against private employers on behalf of alleged victims of discrimination.

- **What group constitutes the largest beneficiary of affirmative action policies?**

  Affirmative action policies are designed to benefit racial and ethnic minorities, women, and, in some cases, Vietnam veterans and people with disabilities. The largest category of affirmative action beneficiaries is women.

- **What are Whiteness Studies?**

  Courses in Whiteness Studies, which are being offered in many colleges and universities, focus on increasing awareness of white privilege—an awareness that is limited among white students.

- **According to the Brown University Steering Committee on Slavery and Justice, successful retrospective justice initiatives contain what three elements?**

  The Brown University Steering Committee on Slavery and Justice examined retrospective justice initiatives from around the world and concluded that the most successful generally combined three elements: (1) formal acknowledgment of an offense; (2) a commitment to truth telling, to ensure that the relevant facts are uncovered, discussed, and properly memorialized; and (3) the making of some form of amends in the present to give material substance to expressions of regret and responsibility.

TEST YOURSELF

1. When it comes down to collecting census data on the U.S. population, a person is Hispanic if they say they are Hispanic.
   a. True
   b. False
2. Which of the following occurs when a person adopts the culture of a group different from the one in which that person was originally raised?
   a. Secondary assimilation
   b. Acculturation
   c. Genocide
   d. Pluralism
3. More than half of the growth in the total U.S. population between 2000 and 2010 was due to the increase in which of the following populations?
   a. Non-Hispanic whites
   b. Blacks
   c. Hispanics
   d. Mixed-race
4. A Southern Poverty Law Center report describes the guest worker program as a modern-day system of
   a. amnesty
   b. colonialism
   c. paid apprenticeship
   d. indentured servitude
5. Which minority group in the United States is considered a "model minority"?
   a. Women
   b. Black women
   c. Scandinavian immigrants
   d. Asian Americans
6. Which of the following has/have moved "backstage"?
   a. Racist behavior
   b. Undocumented immigrants
   c. Multicultural education
   d. Affirmative action
7. The number of hate groups in the United States has been steadily declining over recent years.
   a. True
   b. False
8. Which of the following is responsible for enforcing laws against discrimination?
   a. Local police departments
   b. Equal Employment Opportunity Commission
   c. U.S. Department of Labor
   d. The Supreme Court
9. Colleges and universities are prohibited from requiring students to take diversity courses as a graduation requirement.
   a. True
   b. False
10. The U.S. Congress has issued an official apology to African Americans for slavery and Jim Crow laws.
    a. True
    b. False

Answers: 1. A; 2. B; 3. C; 4. D; 5. D; 6. A; 7. B; 8. B; 9. B; 10. A.

KEY TERMS

Scott Barbour/Getty Images

10

Gender Inequality

"Gender equality is more than a goal in itself. It is a precondition for meeting the challenge of reducing poverty, promoting sustainable development, and building good governance."

—Kofi A. Annan, Nobel Prize winner, 7th Secretary-General of the United Nations

BORN PREMATURE AND LIVING IN FOSTER CARE, M.C. had a tough start in life but, as luck would have it, was adopted by a loving couple who already had two children of their own (SPLC 2013). Three months before Pam and Mark Crawford were able to take their baby home, 16-month-old M.C., who had "ambiguous genitals" and both male and female reproductive organs, underwent sex assignment surgery. Today, at age 8, M.C. continues to live with the sexual anatomy of a girl and the gender identity of a boy. In 2013, his parents filed a lawsuit claiming that the child welfare agencies and doctors involved violated M.C.'s constitutional rights by subjecting him to unnecessary surgery without due process. Further, the lawsuit claims that the ". . . doctors knew that sex assignment surgeries on infants with conditions like M.C.'s pose a significant risk of imposing a gender that is ultimately rejected by the patient" (Jenkins 2013, p. 1).

This groundbreaking lawsuit, *M.C. v. Medical University of South Carolina et al.,* brings to the forefront of Americans' consciousness the significance of and difference between sex and gender. **Sex** refers to one's biological classification, whereas **gender** refers to the social definitions and expectations associated with being female or male. Today, however, researchers, educators, and parents alike are challenging the binary concepts of sex and gender and, just as our evolving notion of sexual orientation (see Chapter 11), have led to feelings of uneasiness for some. For example, words like *transgender* and *transsexual, gender variant, androgyny, intersexed* (the condition M.C. was determined to have)*, boi and birl, metrosexual, two-spirited, gender neutrality, a third sex,* and *gender bender* were not part of the American lexicon just decades ago.

> ... researchers, educators, and parents alike are challenging the binary concepts of sex and gender and, just as our evolving notion of sexual orientation, have led to feelings of uneasiness for some.

In most Western cultures, we take for granted that there are two categories of gender. However, in many other societies, three and four genders have been recognized:

On nearly every continent, and for all of recorded history, thriving cultures have recognized, revered, and integrated more than two genders. Terms such as transgender and gay are strictly new constructs that assume three things: that there are only two sexes (male/female), as many as two sexualities (gay/straight), and only two genders (man/woman). (NPR 2011, p. 1)

A **transgender individual** (sometimes called "trans" or "gender non-conforming") is a person whose sense of gender identity—male or female—is inconsistent with their birth (sometimes called chromosomal) sex. Transgender is not a sexual orientation, and transgender individuals may have any sexual orientation—heterosexual, homosexual, or bisexual. Transsexuals are transgender individuals who are in ". . . the process of changing his or her physical and/or legal sex to conform to his or her internal sense of gender identity" (HRC 2012, p. 1).

Although there is documented movement away from a binary emphasis on sex and gender (see this chapter's *Photo Essay,* "The Gender Continuum"), most Americans still think in terms of females and males, and women and men, much the same way they continue to use such false dichotomies as black and white, gay and straight, and young and old. Although empirically inaccurate, this *social shorthand* makes conversation easier and fulfills our need to "know" and respond "appropriately." Thus, upon meeting someone, we quickly attach the label of gay, white male, or elderly African American female, though knowing little about their social biographies. Similarly, this chapter emphasizes **sexism** and gender inequality as traditionally defined, i.e., as the inequality between women and men but, wherever possible, information on transgender individuals will be included in the discussion. Note that civil rights issues for transgender individuals are discussed in Chapter 11.

sex A person's biological classification as male or female.

gender The social definitions and expectations associated with being female or male.

transgender individual A transgender individual is a person whose sense of gender identity—masculine or feminine—is inconsistent with their birth (sometimes called chromosomal) sex (male or female).

sexism The belief that innate psychological, behavioral, and/or intellectual differences exist between women and men and that these differences connote the superiority of one group and the inferiority of the other.

What Do You Think? Colleges are paying less attention to labels these days. The growing trend is toward more options as society acknowledges that the binaries of sex, gender, and sexual orientation don't represent reality. To that end, a new state law in California allows transgender students to compete in sports, shower in facilities, and use restrooms consistent with their gender identity rather than their biological sex (McGreevy 2013). Do you think that students who are biologically male (or female) should be able to play on women's (or men's) sports teams?

The Global Context: The Status of Women and Men

There is no country in the world in which women and men have equal status. Although much progress has been made in closing the gender gap in areas such as education, health care, employment, and government, gender inequality is still prevalent throughout the world.

The World Economic Forum assessed the gender gap in 132 countries by measuring the extent to which women have achieved equality with men in four areas: economic participation and opportunity, educational attainment, health and survival, and political empowerment (Hausmann et al. 2012). Table 10.1 presents the overall scores and rankings of (1) the 10 countries with the smallest gender gap (i.e., the least gender inequality); (2) the 10 countries with the largest gender gap (i.e., the most gender inequality); and (3) the United States, which did not rank in the top 10 or the bottom 10 but ranked number 22 (a slip from number 19 in the previous year) of the 132 countries studied. Ties were possible. For example, several countries, including the United States, tied for first place on the composite measure of educational attainment. Further, note that the overall score approximates the proportion of the gender gap a country has *closed*—in the United States, 0.7373 or 73.73 percent.

Gender inequality varies across cultures, not only in its extent or degree but also in its forms. For example, in the United States, gender inequality in family roles commonly takes the form of an unequal division of household labor and child care, with women bearing the heavier responsibility for these tasks. In other countries, forms of gender inequality in the family include the expectation that wives ask their husbands for permission to use birth control (see Chapter 12), unequal penalties for spouses who commit adultery, with wives receiving harsher punishment, and the practice of aborting female fetuses in cultures that value male children over female children. In a book entitled, *Unnatural Selection*, author Mara Hvistendahl (2011) documented how medical technology, and specifically the increased availability of ultrasounds, has made sex-selection abortions commonplace around the world. Today, there are over 163 million "missing" females, more than the entire female population of the United States.

A global perspective on gender inequality must also take into account the different ways in which such inequality is viewed. For example, many non-Muslims view the practice of Muslim women wearing a headscarf in public as a symbol of female subordination and oppression. To Muslims who embrace this practice (and not all Muslims do), wearing a headscarf reflects the high status of women and represents the view that women should be respected and not treated as sexual objects.

Similarly, cultures differ in how they view the practice of female genital mutilation, also known as female genital cutting or female circumcision (FGM/C). There are several forms of FGM/C, ranging from a symbolic nicking of the clitoris to removal of

Race and ethnicity, sex and gender, and age are considered core demographic variables because of their effectiveness in predicting life chances—education, occupation, income, and the like. Yet, over the last several decades, the meanings we associate with each of these variables have evolved. Race is no longer black, white, and other in the U.S. Census, and being middle aged extends into the 60s as life expectancy increases. Such is the case with sex and gender. As a society, we are beginning to embrace those who don't fit neatly into the female–male, woman–man binary. For example, in 2013, the state of California instituted a school policy whereby transgender students may use the facilities that match their gender identity.

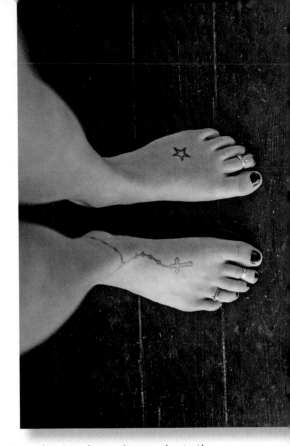

Pep Roig/Alamy

▲ In India's 2011 national census, the census form had three options under the heading of "sex"—male, female, and other (Haub 2011). Similarly, Nepal, after a ruling by their highest court which held that lesbian, gay, bisexual, and transgender (LGBT) people needed greater protection, also used the three gender option in their census (Cohn 2011). Pictured here is a *hijra*, the Indian word for "third sex," institutionalized in India for thousands of years.

▼ In Native American culture, prior to the invasion of the Europeans, there existed three genders—male, female, and male-female. *Berdache*, the Native American word used to describe the male–female gender, was translated into English as "the two-spirited person" (McGill 2011). Rather than being ostracized, the two-spirited person was embraced, revered, and envied as a person who had the "privilege to house both male and female spirits in their body" (p. 1). Here, a We-Wa Zuni man is dressed as a woman, weaving a belt on a waist-high loom with a reed heddle (National Archives 2010).

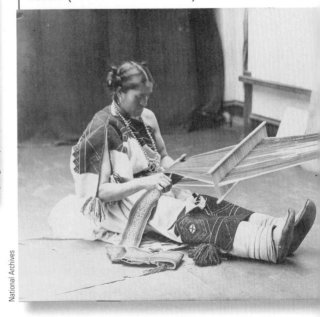

National Archives

◄ In 2011, an advertisement for J. Crew showing the store's creative director painting her son's toenails sparked both accolades and indignation. *The Daily Mail*, a British newspaper, noting that the photo created "gender identity debates across the U.S.," published photographs of young boys with nail polish, including the sons of Gwen Stefani, Jennifer Lopez, and Christina Aguilera (Abraham & Madison 2011). The paper also noted that singer Justin Bieber launched his own brand of nail polish in 2010, which sold over 2 million bottles in the first few months.

▲ Research indicates that in classes where teachers call attention to gender differences, even in seemingly harmless ways such as saying "Good morning boys and girls" or by dividing the class by sex for a spelling bee, children are less likely to interact with opposite-sex children and more likely to use gender stereotypes (Moskowitz 2010). Schools are beginning to acknowledge that sex and gender are not categorical and that some children may identify as female, male, both, or neither. Pictured here is Livvy, a 10-year-old male-to-female transgender. When Livvy returned to school for the first time as a female, there was a special assembly to explain Livvy's situation to the other children. The mood at school was one of acceptance until some concerned parent went to the newspaper and the little girl's story went global.

▼ Whenever social change takes place, requests for interviews with the agents, victims, or beneficiaries of change increase as media outlets search for the latest "news." Transgendered children have made appearances on Lisa Ling's *Our America, 20/20, Dr. Oz, National Public Radio (NPR),* and *Katie.* There are also emerging support groups, conferences, professional workshops, and the like. Here, Chaz Bono (third from the right), child star of Cher and the late Sonny Bono, poses with his cast mates from the popular television show *Dancing with the Stars.*

▲ Andrej Pejic, a transgender model, states that "Sometimes I feel like more of a woman, other times I feel male" (quoted in *Daily Mail* 2011, p. 1). The 21-year-old Serbian who relocated to Australia is one of the top-ranked models in the world. He's been described as the "poster boy for fashion androgyny" (Models 2011, p. 1).

Country	Overall Score*	Overall Ranking	Economic Participation and Opportunity	Educational Attainment	Health and Survival	Political Empowerment
TOP TEN COUNTRIES						
Iceland	0.8640	1	27	1	98	1
Finland	0.8451	2	14	1	1	2
Norway	0.8403	3	4	1	94	3
Sweden	0.8159	4	10	39	73	4
Ireland	0.7839	5	29	30	69	6
New Zealand	0.7805	6	15	1	94	9
Denmark	0.7777	7	16	1	67	11
Philippines	0.7757	8	17	1	1	14
Nicaragua	0.7697	9	88	23	58	5
Switzerland	0.7672	10	28	71	68	13
United States	0.7373	22	8	1	33	55
BOTTOM TEN COUNTRIES						
Oman	0.5986	125	127	96	62	129
Egypt	0.5975	126	124	110	54	125
Iran, Islamic Rep.	0.5927	127	130	101	87	126
Mali	0.5842	128	103	132	57	101
Morocco	0.5833	129	128	115	88	108
Cote d'Ivoire	0.5785	130	111	131	1	104
Saudi Arabia	0.5731	131	133	91	55	133
Syria	0.5626	132	135	107	61	111
Chad	0.5594	133	56	135	111	102
Pakistan	0.5478	134	134	129	123	52
Yemen	0.5054	135	132	133	82	128

*Overall scores are reported on a scale of 0 to 1, with 1 representing maximum gender equality.

Source: Hausmann et al. 2012.

the clitoris and labia and partial closure of the vaginal opening by stitching the two sides of the vulva together, leaving only a small opening for the passage of urine and menstrual blood. After marriage, the sealed opening is reopened to permit intercourse and childbearing.

Nonmedical personnel perform most FGM/C procedures using unsterilized blades or string. Health risks associated with FGM/C include pain, hemorrhage, infection, shock, scarring, and infertility. Worldwide, over 140 million girls and women are estimated to have experienced FGM/C, and an additional 3 million are at risk each year (UNICEF 2011; WHO 2013).

People from countries in which FGM/C is not the norm generally view this practice as a barbaric form of violence against women. An Ethiopian immigrant in the United States was convicted of aggravated battery and cruelty to children for using a pair of scissors to remove his daughter's clitoris. He was sentenced to 10 years in prison (Haines 2006). In countries where it commonly occurs, FGM/C is viewed as an important and useful practice. In some countries, it is considered a rite of passage that enhances a woman's status. In other countries, it is aesthetically pleasing. For others, FGM/C is a moral imperative based on religious beliefs (WHO 2008; Yoder et al. 2004). However, in all cases, it ". . . reflects deep-rooted inequality between the sexes, and constitutes an extreme form of discrimination against women" (WHO 2013, p. 1).

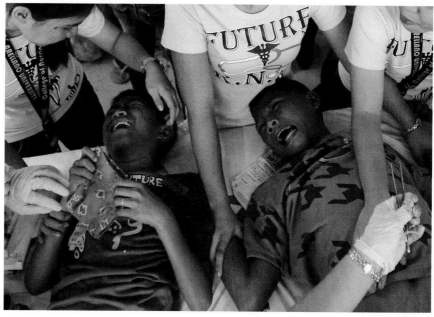

REUTERS/Cheryl Ravelo

In 2011, in Manila, 1,500 boys were forced to take place in a mass circumcision as the organizers of the event tried to get into the *Guinness World Records* for the most circumcisions performed on boys 9 and older. Of late, the medical procedure has come under fire as questions arise about its safety, necessity, and, similar to FGM/C, abuse of a child's anatomical integrity.

Inequality in the United States

Although attitudes toward gender equality are becoming increasingly liberal, the United States has a long history of gender inequality. Women have had to fight for equality: the right to vote, equal pay for comparable work, quality education, entrance into male-dominated occupations, and legal equality. As shown in Table 10.1, the World Economic Forum (Hausmann et al. 2013)—based on its assessment of women's economic participation and opportunities, political empowerment, educational attainment, and health and survival—ranks the United States 22 in the world in terms of gender equality. Most U.S. citizens agree that American society does not treat women and men equally: Women have lower incomes, hold fewer prestigious jobs, earn fewer graduate degrees, and are more likely than men to live in poverty.

Men are also victims of gender inequality. In 1963, sociologist Erving Goffman wrote that in the United States there is only

> one complete unblushing male . . . a young, married, white, urban, northern heterosexual, Protestant father of college education, fully employed, of good complexion, weight and height, and a recent record in sports. . . . Any male who fails to qualify in one of these ways is likely to view himself . . . as unworthy, incomplete, and inferior. (p. 128)

Although standards of masculinity have relaxed, Williams (2000) argued that masculinity is still based on "success"—at work, on the athletic field, on the streets, and at home. Similarly, Vandello et al. (2008) concluded "[t]he view that manhood is tenuous, and therefore requires public proof, is consistent with research across multiple areas" (p. 1, 326). (See this chapter's *Social Problems Research Up Close.*)

When U.S. college students were asked to list the best and worst things about being the opposite sex, the same qualities, although in opposite categories, emerged (Cohen 2001). For example, what males listed as the best thing about being female (e.g., free to be emotional), females list as the worst thing about being male (e.g., not free to be emotional). Similarly, what females listed as the best thing about being male (e.g., higher pay), males listed as the worst thing about being female (e.g., lower pay). As Cohen (2001) noted, although "some differences are exaggerated or oversimplified. . . . [W]e identif[ied] a host of ways in which we 'win' or 'lose' simply because we are male or female" (p. 3).

Willer et al. (2013) test what is called the **masculine overcompensation thesis**—the assertion ". . . that men react to masculinity threats with extreme demonstrations of masculinity. . ." (p. 980). The researchers evaluate this thesis using four research designs, three of which are discussed here. Two are laboratory experiments, and the third looks at the relationship between masculinity threat and attitudes associated with dominance using a national sample of American adults.

Sample and Methods

Experiments are unique methods of research in that the researchers manipulate the independent variable and measure its effect on the dependent variable. In studies 1 and 2, female and male undergraduates participated in a laboratory experiment in which, on the basis of a "gender identity survey" administered, they were randomly assigned to one of two conditions of the independent variable. The independent variable is *gender identity threat*. Males and females were randomly assigned to one of two conditions—feedback that indicated they were in the average female range, or feedback that indicated they were in the average male range. In this way, masculinity or femininity threat was accomplished. In the first study, dependent variables were assessed on a "political views survey." In the second study, which was similarly designed to the first laboratory experiment, social dominance variables were measured.

In the third study, to assess the reliability of the results of the first two studies with a larger and more diverse sample, Willer et al. (2013) used data from the American Values Survey. Masculinity threat was measured by respondents' feelings that social change is threatening the status of men. The dependent variables were a variety of attitudinal questions related to masculinity and dominance.

Findings and Conclusions

Study 1 examined the relationship between masculinity threat and femininity threat and support for the war in Iraq, attitudes toward homosexuality, desirability of owning a SUV, and how much a subject was willing to pay for a SUV. Males who had received feedback that they had scored in the average femininity range, i.e., had their masculinity threatened, were significantly more likely when compared to male subjects who did not have their masculinity threatened to voice support for the war in Iraq and express negative attitudes toward homosexuality.

Further, men whose masculinity was threatened reported that a SUV was more desirable than men who did not have their masculinity threatened, and were willing to pay, on the average, $7,320 more for the SUV than nonthreatened men. Femininity threat was unrelated to any of the dependent variable indicators—support for the war in Iraq, negative views of homosexuality, SUV desirability, or SUV purchase price.

The second study looked at the relationship between gender threat and dominance attitudes, including such items as, "Superior groups should dominate inferior groups" and "In getting what your group wants, it is sometimes necessary to use force against other groups." Participants were also administered scales designed to measure political conservatism (e.g., support for the U.S. military, affirmative action, etc.), system justification (e.g., "Everyone has a fair shot at wealth and happiness"), and traditionalism (e.g., "It's better to stick with what you have than to keep trying new uncertain things").

The results of study 2 indicate that men—although not women—whose gender identity was threatened scored higher on the dominance scale than men whose gender identity was not threatened, but did not score significantly higher on political conservatism, system justification, or traditionalism. Study 3 yielded similar results. Using a national sample of 2,210 American adults, the findings indicate that ". . . [t]he more men felt that the status of their gender was threatened by social changes, the more they tended to support the Iraq War, hold negative views of homosexuality, believe in male superiority, and hold strong dominance attitudes" (p. 1,001).

The results of these three studies lend consistent support for the masculine overcompensation thesis. Alternatively, they do not support a feminine overcompensation thesis. The authors note, however, that this does not mean that feminine overcompensation does not exist. It may simply mean that the dependent variables used in these studies, for example, support for the war on Iraq, were not sufficient to detect feminine overcompensation. Finally, Willer et al. (2013) conclude

. . . that extreme masculine behaviors may in fact serve as telltale signs of threats and insecurity. Perhaps those men who appear most assuredly masculine, who in their actions communicate strength, power, and dominance at great levels, may actually be acting to conceal underlying concerns that they lack exactly those qualities they strive to project. (p. 1016)

Source: Willer et al. 2013.

Sociological Theories of Gender Inequality

Both structural functionalism and conflict theory concentrate on how the structure of society and, specifically, its institutions contribute to gender inequality. However, these two theoretical perspectives offer opposing views of the development and maintenance of gender inequality. Symbolic interactionism, on the other hand, focuses on the culture of society and how gender roles are learned through the socialization process.

Structural-Functionalist Perspective

masculine overcompensation thesis The thesis that men have a tendency to act out in an exaggerated male role when believing their masculinity to be threatened.

Structural functionalists argue that preindustrial society required a division of labor based on gender. Women, out of biological necessity, remained in the home performing functions such as bearing, nursing, and caring for children. Men, who were physically stronger and could be away from home for long periods of time, were responsible for providing food, clothing, and shelter for their families. This division of labor was functional for society and became defined as both normal and natural over time.

Industrialization rendered the traditional division of labor less functional, although remnants of the supporting belief system still persist. With increased control over reproduction (e.g., contraception), declining birthrate, and fewer jobs dependent upon physical size and strength, women's opportunities for education and workforce participation increased (Wood & Eagly 2002). Thus, modern conceptions of the family have, to some extent, replaced traditional ones—families have evolved from extended to nuclear, authority is more egalitarian, more women work outside the home, and greater role variation exists in the division of labor. Structural functionalists argue, therefore, that, as the needs of society change, the associated institutional arrangements also change.

Conflict Perspective

Many conflict theorists hold that male dominance and female subordination are shaped by the relationships men and women have to the production process. During the hunting-and-gathering stage of development, males and females were economic equals, both controlling their own labor and producing needed subsistence. As society evolved to agricultural and industrial modes of production, private property developed and men gained control of the modes of production, whereas women remained in the home to bear and care for children. Inheritance laws that ensured that ownership would remain in their hands furthered male domination. Laws that regarded women as property ensured that women would remain confined to the home.

Thus, unlike structural functionalists, conflict theorists hold that the subordinate position of women in society is a consequence of social inducement rather than biological differences that led to the traditional division of labor.

As industrialization continued and the production of goods and services moved away from the home, the gaps between females and males continued to grow—women had less education, lower incomes, fewer occupational skills, and were rarely owners. World War II necessitated the entry of a large number of women into the labor force, but in contrast with previous periods, many of them did not return to the home at the end of the war. They had established their own place in the workforce and, facilitated by the changing nature of work and technological advances, now competed directly with men for jobs and wages.

Conflict theorists also argue that continued domination by males requires a belief system that supports gender inequality. Two such beliefs are (1) that women are inferior outside the home (e.g., they are less intelligent, less reliable, and less rational); and (2) that women are more valuable in the home (e.g., they have maternal instincts and are naturally nurturing). Thus, unlike structural functionalists, conflict theorists hold that the subordinate position of women in society is a consequence of social inducement rather than biological differences that led to the traditional division of labor.

Symbolic Interactionist Perspective

Although some scientists argue that gender differences are innate, symbolic interactionists emphasize that, through the socialization process, both females and males are taught the meanings associated with being feminine and masculine. Gender assignment begins at birth as a child is classified as either female or male. However, the learning of gender roles is a lifelong process whereby individuals acquire society's definitions of appropriate and inappropriate gender behavior.

Gender roles are taught by the family, in the school, in peer groups, and by media presentations of girls and boys and women and men (see the discussion on the social construction of gender roles later in this chapter). Most importantly, however, gender roles are learned through symbolic interaction as the messages that others send us reaffirm or challenge our gender performances. In an examination of parent–child

In traditional Muslim societies, women are forbidden to show their faces or other parts of their bodies when in public. Muslim women wear a veil to cover their faces and a chador, a floor-length loose-fitting garment, to cover themselves from head to toe. Although some women adhere to this norm out of fear of repercussions, many others believe veiling was first imposed on Muhammad's wives out of respect for women and the desire to protect them from unwanted advances. More than half a million Muslims are living in the United States.

interactions, Tenenbaum (2009) found that discussions regarding course selections for high school followed gender-stereotyped patterns. Here, a father talks to his fifth grade daughter (how the conversation was coded by the researchers is in brackets):

> But you know, spelling is English, right? That's what English is. For the most part generally speaking, girls do better with those kind of skills, they have a harder time with math, generally, you know? [Code: Lack of ability.] Now, of course, you know it's nice to do the things you're really good at too, and you like to do. But, sometimes when you're trying to be, when you grow up to be someone in life, you also gotta take classes that are, not really, how do I say? You're not really good at, you have to put more practice in, right? [Code: Lack of ability.] So, I picked, I selected algebra, cause that's kinda, high school, mathematics. (pp. 458–459)

Although the father encouraged his daughter to take mathematics, he twice conveyed to his daughter that she is (and girls in general are) not very good in math.

Feminist theory, although also consistent with a conflict perspective, incorporates many aspects of symbolic interactionism. Feminists argue that conceptions of gender are socially constructed as societal expectations dictate what it means to be female or what it means to be male. Thus, women are generally socialized into **expressive roles** (i.e., nurturing and emotionally supportive roles), and males are more often socialized into **instrumental roles** (i.e., task-oriented roles). These roles are then acted out in countless daily interactions as boss and secretary, doctor and nurse, football player and cheerleader "do gender."

Ridgeway (2011), in her well-received book *Framed by Gender*, begins by asking the research question, "How, in the modern world, does gender manage to persist as a basis or principal for inequality?" (p. 3). Her answer lies in the socially constructed meanings associated with gender. Despite the tremendous advances that have been made, whenever people encounter gender in a social relationship, circumstance, or situation that they are unsure of, they rely on traditional definitions as an organizing principle whereby ". . . [g]ender inequality is rewritten into new economic and social arrangements as they emerge, preserving that inequality in modified form . . ." (Ridgeway 2011, p. 7).

Noting that the impact of the structure and culture of society is not the same for all women and all men, feminists encourage research on gender that takes into consideration the intersection of class, race, ethnicity, and sexual orientation. In other words, to fully understand the experiences of others, we must acknowledge that it cannot be accomplished by concentrating on just one subordinate status.

What Do You Think? In Afghanistan, little girls in families with no male children often assume the persona of little boys—short hair, traditional male clothing, and an appropriate male name (Nordberg 2010). In addition to the social pressure to have a boy child, the decision to raise a female child as a "bacha posh" is pragmatic—such a child can help "his" father at work, get a better education, and does not have to be chaperoned. The problem is that when it's time to return to being female, usually at puberty, some have difficulty transitioning. Said one young woman, "Nothing in me feels like a girl . . . for always I want to be a boy . . ." (Nordberg 2010, p. 6). Do you think gender is a consequence of nature or nurture?

expressive roles Roles into which women are traditionally socialized (i.e., nurturing and emotionally supportive roles).

instrumental roles Roles into which men are traditionally socialized (i.e., task-oriented roles).

structural sexism The ways in which the organization of society, and specifically its institutions, subordinate individuals and groups based on their sex classification.

Gender Stratification: Structural Sexism

As structural functionalists and conflict theorists agree, the social structure underlies and perpetuates much of the sexism in society. **Structural sexism**, also known as institutional sexism, refers to the ways the organization of society and specifically its institutions subordinate individuals and groups based on their sex classification. Structural sexism has resulted in significant differences in the education and income levels, occupational and political involvement, and civil rights of women and men.

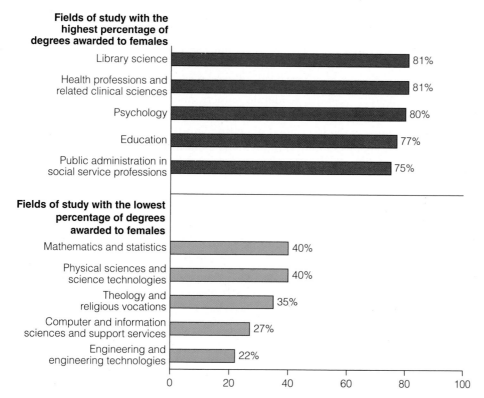

Fields of study with the highest percentage of degrees awarded to females

- Library science — 81%
- Health professions and related clinical sciences — 81%
- Psychology — 80%
- Education — 77%
- Public administration in social service professions — 75%

Fields of study with the lowest percentage of degrees awarded to females

- Mathematics and statistics — 40%
- Physical sciences and science technologies — 40%
- Theology and religious vocations — 35%
- Computer and information sciences and support services — 27%
- Engineering and engineering technologies — 22%

Figure 10.1 Percentage of Master's Degrees Awarded to Females by Degree-Granting Institutions in Selected Fields of Study, Academic Year 2009–2010
Source: NCES 2013.

Education and Structural Sexism

Over 775 million adults over the age of 15 cannot read or write, and nearly two-thirds of them are women. Of the approximately 123 million illiterate youth, 76 million are female (UNESCO 2012) (see Chapter 8).

Because children born to educated mothers are less likely to die at a young age, there is an **education dividend** associated with educating women. For example, if universal primary education was available for all girls in sub-Saharan Africa, 200,000 fewer children would die each year, and universal secondary education for all girls would save as many as 1.8 million lives annually (UNESCO 2011).

In 2012, few differences existed between men and women in their completion rates of high school and college degrees (NCES 2013). In fact, in recent years, most U.S. colleges and universities have had a higher percentage of women than men enrolling directly from high school (see Chapter 8). Similarly, in two-parent families, 23 percent of mothers are better educated than their husbands while only 16 percent of fathers are better educated than their wives, the remainder of spouses having similar educational background (Wang et al. 2013). This trend is causing some concern that many young American men may not have the education they need to compete in today's global economy.

Concern over the continued lack of women in STEM (science, technology, engineering, and mathematics) is also justified. As Figure 10.1 indicates, women earn 81 percent of master's degrees in library science but only 22 percent of master's degrees in engineering and engineering technologies. Reasons for the STEM gender disparity include reliance on gender stereotyping ("Boys are better in math and science than girls!"), a lack of female STEM role models, little encouragement to follow STEM pursuits, and a lack of awareness about women in STEM fields (AAUW 2011).

Women earned 51 percent of doctorate degrees (e.g., JDs, MDs, PhDs) in 2012. Nonetheless, unlike many of their male counterparts, many women "opt out" of labor force participation. Hersch (2013) reports that female graduates of elite institutions, those most likely to find satisfying employment in their field of study, are more likely to

education dividend The additional benefit of universal education for women is that it reduces the death rate of children under 5 years of age.

"opt out" of the workforce than peers from less selective institutions. Further, female graduates from elite institutions who have children under the age of 18, have significantly lower levels of labor force involvement than their academic equals without children under the age of 18.

In higher education, there are also structural limitations that discourage women from advancing. Women seeking academic careers may find that securing a tenure-track position is more difficult for them than it is for men, and that having children pre-tenure negatively impacts the likelihood of getting tenure, that is, there is a "pregnancy penalty" (Ceci et al. 2009). Similarly, in an analysis of data from the *National Study of Postsecondary Faculty*, Leslie (2007) reports that, as the number of children increases, the number of hours a female faculty member works decreases, and the number of hours a male faculty member works increases (see the discussion on household division of labor later in this chapter). Finally, research also indicates gender stereotyping in letters of recommendation. Women are more often described as socio-emotive (e.g., helpful) rather than active/assertive (e.g., ambitious) and, thus, are evaluated less positively (Madera et al. 2009). Similarly, female faculty are disproportionately assigned service duties (e.g., advising) when compared to their male counterparts, which places them at a disadvantage when being considered for tenure and/or promotion (Misra et al. 2011).

Work and Structural Sexism

According to an International Labour Organization (ILO) report, women made up 40 percent of the world's total labor force in 2012 (ILO 2012a). In the same year, the gender gap in unemployment rates increased. Globally, the female unemployment rate was 6.4 percent compared to 5.7 percent unemployment for males. The report concludes that women have higher unemployment rates because of differences in educational attainment between men and women, occupational segregation, and higher rates of exiting and reentering the labor force as result of family obligations.

> No matter what the job, if a woman does it, it is likely to be valued less than if a man does it.

Women are also disproportionately employed in what is called vulnerable employment. Vulnerable employment is characterized by informal working arrangements, little job security, few benefits, and little recourse in the face of an unreasonable demand. Such is the case in developing countries where women are often "contributing family workers," and in developed countries where women are disproportionately working in low-wage service occupations (ILO 2012a).

Women are also more likely to hold positions of little or no authority within the work environment and to receive lower wages than men (ILO 2012b). No matter what the job, if a woman does it, it is likely to be valued less than if a man does it. For example, in the early 1800s, 90 percent of all clerks were men and being a clerk was a prestigious profession. As the job became more routine, in part because of the advent of the typewriter, the pay and prestige of the job declined and the number of female clerks increased. Today, 92.2 percent of clerks are female (U.S. Census Bureau 2013), and the position is one of relatively low pay and prestige.

The concentration of women in certain occupations and men in other occupations is referred to as **occupational sex segregation**. Although occupational sex segregation remains high, as indicated by the first column in Table 10.2, it has decreased in recent years for some occupations. For example, between 1983 and 2010, the percentage of female physicians and surgeons more than doubled from 16 percent to 36 percent, female dentists increased from 7 percent to 30 percent, and female clergy increased from 6 percent to 18 percent (U.S. Census Bureau 2009, 2013).

Although the pace is slower, men are increasingly applying for jobs that women traditionally held. Spurred by the loss of jobs in the manufacturing sector and the recent economic crisis, many jobs traditionally defined as male (e.g., auto worker, construction worker) have been lost. Thus, over the last 20 years, there has been a significant increase in the number of men in traditionally held female jobs; for example, a 50 percent

occupational sex segregation The concentration of women in certain occupations and men in other occupations.

The following remarks by President Obama were on the occasion of signing the Lilly Led-Better Fair Pay Restoration Act and was the first bill he signed after becoming President. Here he recounts the strength, courage, and persistence of Lilly Ledbetter who, against great odds, fought for gender equality. It should be noted that although these remarks were made in 2009, in 2012, a fulltime working woman earned just 81 percent of a fulltime working man (Hegewisch et al. 2012).

THE WHITE HOUSE

January 29, 2009

It is fitting that with the very first bill I sign—the Lilly Ledbetter Fair Pay Restoration Act—we are upholding one of this nation's first principles: that we are all created equal and each deserve a chance to pursue our own version of happiness.

It is also fitting that we are joined today by the woman after whom this bill is named—someone Michelle and I have had the privilege of getting to know for ourselves. Lilly Ledbetter didn't set out to be a trailblazer or a household name. She was just a good hard worker who did her job—and did it well—for nearly two decades before discovering that for years, she was paid less than her male colleagues for the very same work. Over the course of her career, she lost more than $200,000 in salary, and even more in pension and Social Security benefits—losses she still feels today.

Now, Lilly could have accepted her lot and moved on. She could have decided that it wasn't worth the hassle and harassment that would inevitably come with speaking up for what she deserved. But instead, she decided that there was a principle at stake, something worth fighting for. So she set out on a journey that would take more than ten years, take her all the way to the Supreme Court, and lead to this bill which will help others get the justice she was denied.

Because while this bill bears her name, Lilly knows this story isn't just about her. It's the story of women across this country still earning just 78 cents for every dollar men earn—women of color even less—which means that today, in the year 2009, countless women are still losing thousands of dollars in salary, income and retirement savings over the course of a lifetime.

But equal pay is by no means just a women's issue—it's a family issue. It's about parents who find themselves with less money for tuition or child care; couples who wind up with less to retire on; households where, when one bread-winner is paid less than she deserves, that's the difference between affording the mortgage—or not; between keep-ing the heat on, or paying the doctor's bills—or not. And in this economy, when so many folks are already working harder for less and struggling to get by, the last thing they can afford is losing part of each month's paycheck to simple discrimination.

So in signing this bill today, I intend to send a clear message: That making our economy work means making sure it works for everyone. That there are no second class citizens in our workplaces, and that it's not just unfair and illegal—but bad for business—to pay someone less because of their gender, age, race, ethnic-ity, religion or disability. And that justice isn't about some abstract legal theory, or footnote in a casebook—it's about how our laws affect the daily realities of people's lives: their ability to make a liv-ing and care for their families and achieve their goals.

Ultimately, though, equal pay isn't just an economic issue for millions of Americans and their families, it's a question of who we are—and whether we're truly living up to our fundamental ideals. Whether we'll do our part, as generations before us, to ensure those words put to paper more than 200 years ago really mean

something—to breathe new life into them with the more enlightened understandings of our time.

That is what Lilly Ledbetter challenged us to do. And today, I sign this bill not just in her honor, but in honor of those who came before her. Women like my grand-mother who worked in a bank all her life, and even after she hit that glass ceiling, kept getting up and giving her best every day, without complaint, because she wanted something better for me and my sister.

And I sign this bill for my daughters, and all those who will come after us, because I want them to grow up in a nation that values their contributions, where there are no limits to their dreams and they have opportunities their mothers and grand-mothers never could have imagined.

In the end, that's why Lilly stayed the course. She knew it was too late for her—that this bill wouldn't undo the years of injustice she faced or restore the earnings she was denied. But this grandmother from Alabama kept on fighting, because she was thinking about the next generation. It's what we've always done in America—set our sights high for ourselves, but even higher for our children and grandchildren.

Now it's up to us to continue this work. This bill is an important step—a simple fix to ensure fundamental fairness to American workers—and I want to thank this remarkable and bi-partisan group of legislators who worked so hard to get it passed. And this is only the beginning. I know that if we stay focused, as Lilly did—and keep standing for what's right, as Lilly did—we will close that pay gap and ensure that our daughters have the same rights, the same chances, and the same freedom to pursue their dreams as our sons.

Thank you.

Source: Obama 2009.

	PERCENT OF WORKERS IN OCCUPATION THAT ARE FEMALE	MEDIAN WEEKLY EARNINGS FOR WOMEN	MEDIAN WEEKLY EARNINGS FOR MEN	WOMEN'S EARNINGS AS PERCENT OF MEN'S
ALL FULL-TIME WORKERS	44.2%	$691	$854	80.9%
10 MOST COMMON OCCUPATIONS FOR WOMEN				
Secretaries and administrative assistants	95.3%	$665	$803	82.8%
Elementary and middle school teachers	80.9%	$921	$1,128	81.6%
Registered nurses	89.4%	$1,086	$1,189	91.3%
Nursing, psychiatric, and home health aides	88.1%	$445	$508	88.8%
Customer service representatives	67.3%	$585	$684	87.6%
First-line supervisors of retail sales workers	42.6%	$598	$792	75.5%
Cashiers	69.8%	$368	$400	92.0%
Managers, all other	38.1%	$1,078	$1,409	76.5%
Accountants and auditors	60.2%	$996	$1,350	73.8%
First-line supervisors of office and administrative support workers	67.3%	$760	$895	84.9%
10 MOST COMMON OCCUPATIONS FOR MEN				
Driver/sales workers and truck drivers	4.0%	$537	$736	73.0%
Managers, all other	38.1%	$1,078	$1,409	76.5%
First-line supervisors of retail sales workers	42.6%	$598	$792	75.5%
Retail salespersons	38.4%	$436	$678	64.3%
Janitors and building cleaners	25.1%	$408	$511	79.8%
Laborers and freight, stock, and material movers, hand	16.5%	$476	$519	91.7%
Construction laborers	2.6%	–	$609	–
Cooks	32.9%	$361	$403	89.6%
Software developers, applications and systems software	19.6%	$1,362	$1,674	81.4%
Sales representatives, wholesale and manufacturing	26.1%	$822	$1,161	89.6%

Source: Hegewisch & Matite 2013.

glass escalator effect The tendency for men seeking or working in traditionally female occupations to benefit from their minority status.

pink-collar jobs Jobs that offer few benefits, often have low prestige, and are disproportionately held by women.

increase in male telephone operators, 45 percent increase in male tellers, and 40 percent increase in male preschool and kindergarten teachers (Bourin & Blakemore 2008). Some evidence suggests that men in traditionally held female jobs have an advantage in hiring, promotion, and salaries called the **glass escalator effect** (Williams 2007). Table 10.2, for example, indicates that secretaries and administrative assistants are overwhelmingly female—95.3 percent. Yet the average median weekly salary for male secretaries is $803 compared to $665 for female secretaries. Despite the increase of men into traditionally held female occupations, women are still heavily represented in low-prestige, low-wage, **pink-collar jobs** that offer few benefits.

Persistence of the Occupational Sex Segregation. Sex segregation in occupations continues for several reasons. First, cultural beliefs about what is an "appropriate" job for a man or a woman still exist. Snyder and Green's (2008) analysis of nurses in the United States is a case in point. Using survey data and in-depth interviews, the researchers identified patterns of sex segregation. Over 88 percent of all patient-care nurses were in sex-specific specialties (e.g., intensive care and psychiatry for male nurses, and labor or delivery and outpatient services for female nurses). Interestingly, although women rarely mentioned gender as a reason for their choice of specialty, male nurses frequently did so, acknowledging the "process of gender affirmation that led them to seek out 'masculine' positions within what was otherwise construed to be a women's profession" (p. 291).

Second, opportunity structures for men and women differ. For example, women and men, upon career entry, are often channeled by employers into gender-specific jobs that carry different wages and promotion opportunities. However, even women in higher-paying jobs may be victimized by a **glass ceiling**—an often invisible barrier that prevents women and other minorities from moving into top corporate positions. For example, women and minorities have different social networks than do white men, which contributes to this barrier. White men in high-paying jobs are more likely to have interpersonal connections with individuals in positions of authority (Padavic & Reskin 2002). In addition, women often find that their opportunities for career advancement are adversely affected after returning from family leave. Female lawyers returning from maternity leave found their career mobility stalled after being reassigned to less prestigious cases (Williams 2000).

There is also evidence that working mothers pay a price for motherhood. Using an experimental design, Correll et al. (2007) report that, even when qualifications, background, and work experience were held constant, "evaluators rated mothers as less competent and committed to paid work than non-mothers" (p. 1,332). Other examples of the "**motherhood penalty**" include women who feel pressured to choose professions that permit flexible hours and career paths, sometimes known as mommy tracks (Moen & Yu 2000). Thus, women dominate the field of elementary education, which permits them to be home when their children are not in school. Nursing, also dominated by women, often offers flexible hours. Although the type of career pursued may be the woman's choice, it is a **structured choice**—a choice among limited options as a result of the structure of society.

Finally, Blau and Kahn (2013) argue that the comparatively low rates of female labor force participation, and female labor force participation growth, is the result of a lack of U.S. worker-friendly, and perhaps more importantly female-friendly, employment policies. An analysis of policy data indicates that ". . . most other countries have enacted parental leave, part-time work, and child care policies that are more extensive than in the United States, and the gap has grown over time" (p. 4). The authors conclude that such policies make part-time work more attractive, and make it easier for women to "have it all," i.e., combine work and family life.

Income and Structural Sexism

In 2012, full-time working women in the United States earned, on the average, 81 percent of the weekly median earnings of full-time working men (Hegewisch et al. 2012). (See this chapter's *The Human Side.*) Further, cashiers, waitresses, maids and household cleaners, and retail sales workers, four of the most common occupations for women, have 40-hour-a-week median incomes below the federally established poverty level for a family of four (Hegewisch & Matite 2013).

The gender pay gap varies over time. By decade, in 1980, 1990, 2000, and 2010, women's annual earnings as a percentage of men's increased from 60 percent to 72 percent, 74 percent, and 77 percent, respectively. As indicated, closing the gender gap has slowed down since the 1980s and early 1990s (Hegewisch et al. 2012). At the present rate of progress, it is predicted that the gender gap will not be closed until 2057 (IWPR 2013).

glass ceiling An invisible barrier that prevents women and other minorities from moving into top corporate positions.

motherhood penalty The tendency for women with children, particularly young children, to be disadvantaged in hiring, wages, and the like compared to women without children.

structured choice Choices that are limited by the structure of society.

Racial differences also exist. Although women, in general, earn 81 percent as much as men, Black American women earn just 68 percent of white men's salaries, and Hispanic American women earn just 59 percent of white men's salaries (Hegewisch et al. 2012). Even among celebrities, a significant income gap exists. According to *Forbes* magazine, the maximum salary for a Women's National Basketball Association (WNBA) player is $107,000, while Kobe Bryant signed a contract for $30.5 million for the 2013–2014 season with the Los Angeles Lakers (Badenhausen 2013).

Why Does the Gender Pay Gap Exist? There are several arguments as to why the gender pay gap exists. One, the **human capital hypothesis** holds that pay differences between females and males are a function of differences in women's and men's levels of education, skills, training, work experience, and the like. For example, Rose and Hartman (2008), using a longitudinal data set, found that over a 15-year period, women worked fewer years than men and, when they worked, they worked fewer hours per year. Rose and Hartman concluded that over ". . . the 15 years, the more likely a woman is to have dependent children and be married, the more likely she is to be a low earner and have fewer hours in the labor market" (Rose & Hartman 2008, p. 1). Bertrand et al. (2009) reported similar findings, concluding that the "presence of children is associated with less accumulation of job experience, more career interruptions, and shorter work hours for female MBAs but not for male MBAs" (p. 24). Based on their analysis, the authors concluded that a decade after graduation, female MBAs earn an average annual salary of $243,481 and male MBAs earn an average annual salary of $442,353. Lower incomes over time create a significant deficit later in life. This is particularly true given a woman's higher life expectancy and the exhaustion of household savings when her husband becomes ill (see Figure 10.3 later in this chapter).

One variation of the human capital hypothesis is called the *life-cycle human capital hypothesis*. Here it is argued that women have less incentive to invest in education and marketable skills because they know that, as wives and mothers, they will be working less than their male counterparts and that their careers will be interrupted by family responsibilities. Alternatively, men's incentives to acquire marketable skills increase with greater family responsibilities and it is this human capital difference, or so it is argued, that is responsible for the female–male pay gap (Polachek 2006).

Human capital theorists also argue that women make educational choices (e.g., school attended, major, etc.) that limit their occupational opportunities and future earnings. Women, for example, are more likely to major in the humanities, education, or the social sciences rather than science and engineering, which results in reduced incomes (Corbett & Hill 2012). Research also indicates, however, that after controlling for "college major, occupation, economic sector, hours worked, months unemployed since graduation, GPA, type of undergraduate institution, institution selectivity, age, geographical region, and marital status . . . a 7 percent difference in the earnings of male and female college graduates one year after graduation was still unexplained" (Corbett & Hill 2012, p. 8).

Female–male human capital differences are a result of structural constraints (e.g., no national system of child care) as well as expectations that women should remain in the home. The results of a survey indicate that few working mothers (11 percent) or stay-at-home mothers (10 percent) believe that a full-time working mother is the "ideal situation for a child" (Pew Research Center 2007). Further, 72 percent of men with children under 18 years of age responded that working full-time was the ideal situation for them compared to only 20 percent of women with children under the age of 18.

The second explanation for the gender gap is called the **devaluation hypothesis**. It argues that women are paid less because the work they perform is socially defined as less valuable than the work men perform. Guy and Newman (2004) argue that these jobs are undervalued in part because they include a significant amount of **emotion work**—that is, work that involves caring, negotiating, and empathizing with people, which is rarely specified in job descriptions or performance evaluations.

Finally, there is evidence that, even when women and men have equal education and experience (and, therefore, not a matter of human capital differences) and are in the same occupations (and, therefore, not a matter of women's work being devalued), pay

human capital hypothesis The hypothesis that pay differences between females and males are a function of differences in women's and men's levels of education, skills, training, and work experience.

devaluation hypothesis The hypothesis that women are paid less because the work they perform is socially defined as less valuable than the work men perform.

emotion work Work that involves caring for, negotiating, and empathizing with people.

differences remain. Table 10.2 indicates that men make more than women in each of the 10 most sex-segregated occupations for females and males. Even among elementary and middle school teachers, a profession that is 81 percent female, women earn a weekly median salary of $921 and men earn a weekly median salary of $1,128—a difference of $10,350 annually (Hegewisch & Matite 2013).

> **What Do You Think?** Closing the gender pay gap may not only be the result of women's gains but also the result of men's losses. Between 1979 and 2012, men's hourly wages dropped from $19.53 an hour to $18.03 an hour. The decrease in men's salaries was brought about by economic policies that encouraged consumerism (e.g., lower prices, deregulation, and cheaper imports) at the expense of workers (Shierholz 2013). What sociological theory best explains such policies, and who would be motivated to institute them?

Politics and Structural Sexism

Women received the right to vote in the United States in 1920, with the passage of the Nineteenth Amendment. Even though this amendment went into effect more than 90 years ago, women still play a rather minor role in the political arena. In general, the more important the political office is, the lower the probability that a woman will hold it. Although women constitute over half of the population, the United States has never had a woman president or vice president and, until 2010, when Justice Elena Kagan was appointed, had only three female Supreme Court Justices in the history of the Court. The highest-ranking women ever to serve in the U.S. government have been former Secretaries of State Madeleine Albright, Condoleezza Rice, and Hillary Clinton. In 2013, women represented only 10 percent of all governors and held only 18.3 percent of all U.S. congressional seats. Further, women held an average of 24.2 percent of all state legislative seats, with Colorado having the highest female representation at 42 percent and Louisiana the lowest at 12 percent (CAWP 2013a).

Worldwide, women represent just 20 percent of representatives to legislative bodies (Quota Project 2013). The percent, however, varies significantly by region of the world. Note that "developed regions" include the United States and Canada, Europe, the Russian Federation, and Australia. As Figure 10.2 indicates, with the exception of Oceania and Eastern Asia, representation of women throughout the globe has increased between 2000 and 2013.

In response to the underrepresentation of women in the political arena, some countries have instituted electoral quotas. Of the 59 countries that held elections in 2011, 17 of them had legislative quotas. In countries where legislative quotas were in place, women won 27 percent of the seats compared to 16 percent of the seats in countries without legislative quotas (UN 2013).

Quotas are particularly useful in countries where women are underrepresented and their candidacy is threatened by long-standing patriarchal traditions. In Afghanistan, where quotas were established by constitutional provision in 2004, 28 percent of the national parliamentary seats are held by women (Quote Project 2013). Despite our often preconceived notions about the status of women in Middle Eastern countries, note that this number exceeds the 2013 representation of women holding U.S. congressional seats by 10 percent.

The Underrepresentation of Women in Politics. The relative absence of women in politics, as in higher education and in high-paying, high-prestige jobs in general, is a consequence of structural limitations. Running for office requires large sums of money, the political backing of powerful individuals and interest groups, and a willingness of the voting public to elect women. In fact, female state legislators

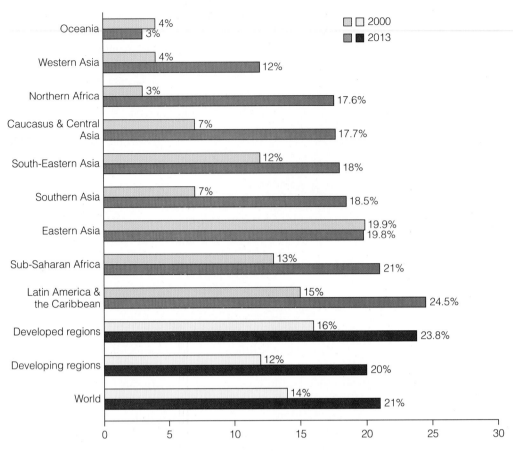

Figure 10.2 Proportion of Seats Held by Women in National Legislatures, 2000 and 2013
Source: United Nations 2013. From *The Millennium Development Goals Report 2013*, p. 22.

identified money as the single greatest barrier to running for higher office (Political Parity 2013). Disproportionately lacking these resources, minority women have even greater structural barriers to election and, not surprisingly, represent an even smaller percentage of elected officials. Of the 535 U.S. congressional representatives, women of color represent just 6 percent, and no minority woman is presently a member of the U.S. Senate (CAWP 2013b).

There is also evidence of gender discrimination against female candidates. In an experiment where two congressional candidates' credentials were presented to a sample of respondents—in one case as Ann Clark and in the other as Andrew Clark—Republican respondents were significantly more likely to say they would vote for a father with young children rather than a mother with young children. They were also more likely to vote for women without small children than women with small children—additional evidence of the "motherhood penalty." The opposite pattern was detected for Democrat respondents (Morin & Taylor 2008).

Not only do voters discriminate on the basis of gender, political parties do as well. When a sample of female state legislators was asked whether or not they believed that their political party encouraged women more, encouraged women less, or equally encouraged women and men, 44 percent of the sample responded that their party was more encouraging to men than women. Only 3 percent responded that their party encouraged women more than men (Political Parity 2013).

> Running for office requires large sums of money, the political backing of powerful individuals and interest groups, and a willingness of the voting public to elect women. Disproportionately lacking these resources, minority women have even greater structural barriers to election and, not surprisingly, represent an even smaller percentage of elected officials.

Finally, arguing that part of the American ethic of politics is whether or not you are "man enough to serve," Faludi (2008) contends that John McCain, who survived years in a prisoner of war camp, typifies the American ideal of what it means to be a man. On the other hand, Don Imus, as Faludi notes, "never to be outdone in the sexual slur department, dubbed Mr. Obama as a 'sissy boy'" (pp. 1–2). Questioning a candidate nomination to the U.S. Supreme Court, conservative talk show host Rush Limbaugh commented that in Kagan's nomination, "Obama has chosen himself a different gender" (quoted in Media Matters 2010, p. 1). Thus, to some, a candidate's gender performance, or at least the perception of it, should play a role in an individual's voting behavior.

What Do You Think? In a report called *Men Rule,* Lawless and Fox (2012) find evidence that traditional gender role socialization significantly contributes to the underrepresentation of women in elected positions. Women are less likely to (1) say they want to seek political candidacy, (2) feel confident about their candidacy, (3) believe they are qualified, (4) say they have the "thick skin" necessary to run for office, and (5) have partners/spouses who are responsible for the majority of household and child care duties when compared to their male counterparts. If gender role socialization explains the lack of women in politics, does it also explain the overrepresentation of men?

Civil Rights, the Law, and Structural Sexism

In many countries, victims of gender discrimination cannot bring their cases to court. This is not true in the United States. The 1963 Equal Pay Act and Title VII of the 1964 Civil Rights Act make it illegal for employers to discriminate in wages or employment on the basis of sex. Nevertheless, such discrimination still occurs as evidenced by the thousands of grievances filed each year with the Equal Employment Opportunity Commission (EEOC)—30,356 grievances in 2012, 31 percent of the total number of individual charges levied (EEOC 2013). Known as gender identity discrimination, differential treatment of individuals because of their transgender status is also a violation of federal law.

One technique employers use to justify differences in pay is the use of different job titles for the same type of work. Repeatedly, the courts have ruled, however, that jobs that are "substantially equal," regardless of title, must result in equal pay. For example, in one of the largest employment discrimination suits ever filed, Wal-Mart Stores, Inc. has been named in a class-action suit in which 2 million former and current female employees allege that management discriminated against them by (1) paying male employees more and (2) denying promotions to women. In court, Wal-Mart attorneys argued that because the stores were independently owned, there could be no company-wide policy or practice of discrimination in place (Kravets 2007). In 2011, the U.S. Supreme Court held for Wal-Mart Stores, Inc. (Stohr 2011). In 2012, 1,975 women in 48 states filed lawsuits against the discount giant. Attorneys for the plaintiffs hope that by pursuing grievances at the regional rather than national level, the systematic pattern of discrimination required by the U.S. Supreme Court decision will be clear (Hines 2012).

Discrimination, although illegal, takes place at both the institutional and the individual levels (see Chapter 9). Institutional discrimination includes screening devices designed for men, hiring preferences for veterans, the practice of promoting from within an organization based on seniority, and male-dominated recruiting networks (Reskin & McBrier 2000). For decades, the Augusta National Golf Club, home of the Masters Golf Tournament and a virtual "who's who of the corporate world," has refused to change its policy, forbidding women from joining, despite years of negative publicity and political pressure from women's groups (Rudnanski 2011). In 2012, former Secretary of State Condoleezza Rice and corporate executive Darla Moore became the first two women to be admitted to the prestigious private club (Ludka 2012). One of the most blatant forms of individual discrimination is sexual harassment, discussed later in this chapter.

Some child brides are as young as 5 years old and rarely know the ramifications of what's happening to them. Here, a 16-year-old screams in protest as she is carried in a cart to her new husband's village.

In the United States, women often have difficulty obtaining home mortgages or rental property because they have lower incomes, shorter work histories, and less collateral. Until fairly recently, husbands who raped their wives were exempt from prosecution. Even today, some states require a legal separation agreement and/or separate residences for a wife who has been raped to receive full protection under the law. Women in the military have traditionally been restricted in the duties they can perform (see Chapter 15) and, finally, since the U.S. Supreme Court's 1973 *Roe v. Wade* decision, which made abortion legal, the right of a woman to obtain an abortion has steadily been limited and narrowed by subsequent legislative acts and judicial decisions (see Chapter 14).

What Do You Think? Throughout the world, as many as 10 to 12 million young girls (Gorney 2011)—some as young as age 5—have been forced into marriage by their parents or as a result of human trafficking. Consequently, they have little or no education, bear children long before they are physically mature and, in many cases, die in childbirth. Do you think that countries where this tradition exists should have anti–child marriage laws? What if such laws contradict religious beliefs?

The Social Construction of Gender Roles: Cultural Sexism

As social constructionists note, structural sexism is supported by a system of cultural sexism that perpetuates beliefs about the differences between women and men. **Cultural sexism** refers to the ways the culture of society—its norms, values, beliefs, and symbols—perpetuate the subordination of an individual or group because of the sex classification of that individual or group.

Cultural sexism takes place in a variety of settings, including the family, the school, and the media, as well as in everyday interactions.

Family Relations and Cultural Sexism

From birth, males and females are treated differently. **Gender roles** are patterns of socially defined behaviors and expectations, associated with being female or male.

Toys are one way that what is considered appropriate gender role behavior is conveyed to children. Coloring books, for example, provide children with the experience of creating

cultural sexism The ways in which the culture of society perpetuates the subordination of an individual or group based on the sex classification of that individual or group.

gender roles Patterns of socially defined behaviors and expectations associated with being female or male.

and experimenting with color and the rewards of conformity ("Good for you! You stayed within the lines!") or the sting of non-conformity ("Oh, you went outside of the lines. Try and do better next time.").

Some toys, however, are clearly gender specific as evidenced by the girls' toy aisle and boys' toy aisle in a discount store. A recently manufactured doll, actually

> . . . eight in all with a variety of skin tones and facial features, looks like many others, until children don the little top with petal appliqués at the nipples. That's where the sensors are located, setting off the suckling noise when the doll's mouth makes contact. It also burps and cries, but those sounds don't require contact at the breast. (Italie 2012)

Even when toys are "educational," there may be subtle biases. A study by Professor Becky Francis of Roehampton University in England, examined the impact of educational toys on 3- to 5-year-olds' learning. When parents were asked their child's favorite toys, boys' toys "involved action, construction, and machinery," whereas girls' toys were more often "dolls and perceived 'feminine' interests, such as hairdressing" (Lepkowska 2008, p. 1). After purchasing and analyzing the toys parents selected, Francis concluded that girls' toys had limited "learning potential" whereas boys' toys "were far more diverse" and propelled boys "into a world of action as well as technology" designed to "be exciting and stimulating" (quoted in Lepkowska 2008, p. 1).

Societal definitions of the appropriateness of gender roles have traditionally restricted women and men in terms of educational, occupational, and leisure-time pursuits. Fifty years ago, little boys playing with dolls and displaying nurturing behaviors would have been unheard of.

Household Division of Labor

Globally, women and girls continue to be responsible for household maintenance including cooking, gathering firewood and fetching water, and taking care of younger siblings (IPC 2008). In a study of household labor in 10 Western countries, Bittman and Wajcman (2000) report that "women continue to be responsible for the majority of hours of unpaid labor," ranging from a low of 70 percent in Sweden to a high of 88 percent in Italy (p. 173). A study in Mexico found that women in the paid labor force performed household chores for an additional 33 hours of work each week; men contributed 6 hours a week to domestic chores (UNICEF 2007).

In the United States, girls and boys work within the home in approximately equal amounts until the age of 18 when girls' household labor begins to increase. Although men's share of the household labor has more than doubled in the last 25 years, the fact that women, even when working full-time, contribute significantly more hours to home care than men is known as the "second shift" (Hochschild 1989).

Explanations for the Traditional Division of Household Labor. Although some changes have occurred, the traditional division of labor remains to a large extent, even in terms of what one "expects" after marriage (Askari et al. 2010). Three explanations emerge from the literature. The first explanation is the *time-availability approach.* Consistent with the structural-functionalist perspective, this position claims that role performance is a function of who has the time to accomplish certain tasks. Because women are more likely to be at home, they are more likely to perform domestic chores.

A second explanation is the *relative resources approach.* This explanation, consistent with a conflict perspective, suggests that the spouse with the least power is relegated the most unrewarding tasks. Because men, on average, often have more education, higher incomes, and more prestigious occupations, they are less responsible for domestic labor. Thus, for example, because women earn less money than men do, on average, they turn down overtime and other work opportunities to take care of children and household responsibilities, which subsequently reduces their earnings potential even further (Williams 2000).

Gender role ideology, the final explanation, is consistent with a symbolic interactionist perspective. It argues that the division of labor is a consequence of traditional socialization and the accompanying attitudes and beliefs. Women and men have been socialized to perform various roles and to expect their partners to perform other complementary roles. Women typically take care of the house, and men take care of the yard. These patterns begin with household chores assigned to girls and boys and are learned through the media, schools, books, and toys.

Simister's (2013) study of the division of household labor in seven countries-Cameroon, Chad, Egypt, India, Kenya, Nigeria, and the United Kingdom—is illustrative. When traditional gender roles are reversed, i.e., wives earn more money than their husbands, men are more resistant to contributing to household labor than when a husband's income is greater. Similarly, Schneider (2012) reports that men in traditional female occupations spend more time on "men's work" at home; women who are in traditionally male occupations spend more time on "women's work" at home.

The findings in both of these studies are consistent with the **gender deviance hypothesis**. When there is "gender deviance" (income or occupation inconsistent with traditional gender roles), techniques to neutralize the deviance (engage in traditional household division of labor), to bring it back into alignment, are employed, thereby reclaiming what is perceived to be what it means to be male and what it means to be female in American society. Note that the findings of both of these studies contradict the *time availability* approach and the *relative resources approach*.

The School Experience and Cultural Sexism

Generations ago, high school girls were required to take "home economics," which included sewing and cooking classes, and boys were required to take "shop," where they learned woodworking and auto mechanics. Although less obvious, sexism exists in schools today, often in the instructional materials used, student–teacher interactions, academic gender stereotypes, and in school programs and activities.

Instructional Material. The bulk of research on gender images in books and other instructional materials documents the way males and females are portrayed stereotypically. In a study of 200 "top-selling" children's picture books, women and girls were significantly underrepresented with twice as many male title and main characters. Males were also more likely to be in the illustrations, to be pictured in the outdoors and, if an adult, to be visibly portrayed as employed outside of the home. Both men and women were over nine times more likely to be pictured in traditional rather than in nontraditional occupations, and "female main characters . . . were more than three times more likely than were male main characters . . . to perform nurturing or caring behaviors" (Hamilton et al. 2006, p. 761). Further, children's coloring books are stereotypical. In an analysis of 889 characters in 54 contemporary American coloring books, Fitzpatrick and McPherson (2010) report that males were significantly more likely to be pictured and, when pictured, significantly more likely to be active, older, and more powerful.

Student–Teacher Interactions. Sexism is also reflected in the way that teachers treat their students. Millions of young girls are subjected to sexual harassment by male teachers, who then fail them when they refuse the teachers' sexual advances (Quist-Areton 2003). There is also convincing evidence that elementary and secondary school teachers pay more attention to boys than to girls—talking to them more, asking them more questions, listening to them more, counseling them more, giving them more extended directions, and criticizing and rewarding them more frequently. In *Still Failing at Fairness,* Sadker and Zittleman (2009) recount a fifth grade teacher's instructions to her students. "There are too many of us here to all shout out at once. I want you to raise your hands, and then I'll call on you. If you shout out, I'll pick somebody else." The discussion on presidents continues, with Stephen calling out:

Stephen: I think Lincoln was the best president. He held the country together during a war.

Teacher: A lot of historians would agree with you.

gender deviance hypothesis
The tendency to overconform to gender norms after an act(s) of gender deviance; a method of neutralization.

Kelvin (seeing that nothing happened to Stephen, calls out): I don't. Lincoln was OK but my Dad liked Reagan. He always said Reagan was a great president.

David (calling out): Reagan? Are you kidding?

Teacher: Who do you think our best president was, David?

David: FDR. He saved us from the Depression.

Max (calling out): I don't think it's right to pick one best president. There were a lot of good ones.

Teacher: That's interesting.

Rebecca (calling out): I don't think the presidents today are as good as the ones we used to have.

Teacher: Ok, Rebecca. But you forgot the rule. You're supposed to raise your hand. (pp. 65–66)

Self-Fulfilling Prophecy. Symbolic interactionists remind us that the expectation of an outcome increases the likelihood of that outcome occurring, i.e., is a self-fulfilling prophecy. The differing expectations and/or encouragement that females and males receive contribute to their varying abilities, as measured by standardized tests, in disciplines such as reading, math, and science. Are such differences a matter of aptitude? Social science research would indicate otherwise. For example, in an experiment at the University of Waterloo, male and female college students, all of whom were good in math, were shown either gender-stereotyped or gender-neutral advertisements. When, subsequently, female students who had seen the female-stereotyped advertisements took a math test, they performed significantly lower than women who had seen the gender-neutral advertisements (Begley 2000).

Further, the results of a study at the University of Virginia document the pervasiveness of gender stereotypes and their impact on science achievement (Nosek et al. 2009). Over 500,000 respondents from 34 countries participated in the research designed to measure *implicit bias*—bias that we are unaware of. About 70 percent of the respondents were found to have stereotypical views of the relationship between gender and science in the predictable direction. In countries where gender stereotypes were the strongest, that is, males were expected to be good in science and females were not, males performed better and females performed worse on standardized math and science tests. The authors suggest that "implicit stereotypes and sex differences in science participation and performance are mutually reinforcing, contributing to the persistent gender gap in science engagement" (Nosek et al. 2009, p. 10, 593).

School Programs and Policies. As discussed in Chapter 8, Title IX of the 1972 Educational Amendments Act prohibits sex discrimination in educational programs and activities that receive federal financial assistance (Office of Civil Rights 2012). An evaluation of Title IX suggests that sex discrimination continues at many levels. Women's and girls' athletic programs lag behind boys' and men's programs in terms of participation, resources, and coaching, and despite significant gains, career and technical education remain sex-segregated, in part because of stereotyped career counseling in high school (NCWGE 2009).

What Do You Think? Caroline Pla had been playing football since she was in kindergarten. But in 2013, at the age of 11, the all-star guard and defensive end was told that a boy-only rule would be enforced (Hoye 2013). Despite a panel's recommendation that the boy-only rule remain in place, the decision was made that the 2013–2014 football season would be coed. The decision, however, is provisional. What would you recommend the final outcome be?

Additionally, in 2006, the U.S. Department of Education granted school districts permission to expand single-sex education if it can be shown to accomplish a specific educational objective. To that end, Park et al. (2012) examined educational outcomes of students randomly assigned to single-sex or coed high schools in South Korea. The results indicated that students in single-sex schools have higher test scores and a higher percentage of graduates who attend four- rather than two-year colleges.

Despite what appears to be unequivocal support for single-sex education, Halpern et al. (2013) argue that ". . . sex-segregated education . . . is deeply misguided, and often justified by weak, cherry-picked, or misconstrued scientific claims rather than by valid scientific evidence" (p. 1,706). The assumption behind single-sex education is that boys and girls learn differently—have "pink and blue brains," if you will—a belief for which there is very little scientific support (see Chapter 8). Single-sex education also furthers sex and gender binaries and ignores that, "[i]n a world of ever-increasing visibility of gender diversity . . . single-sex schooling is an anachronism—one that has the potential to take us back to a time when females and males who behaved outside gender norms were perceived as 'problems' instead of as people" (Jackson 2010, p. 237).

In 2013, a settlement was reached between Wood County Board of Education in West Virginia, and a mother and her daughter, a middle school student (ACLU 2013). According to the lawsuit, boys were allowed to move about in their classrooms freely while girls were required to sit quietly. Further, a video posted online by the principal described seating arrangements in boys' classrooms as side-by-side ". . . because when they look at each other in the eye it becomes more of a confrontational type thing. . . . Girls sit around tables, where they can make eye contact, where they can make relationships, and that sort of thing" (p. 1).

Media, Language, and Cultural Sexism

Another concern social scientists voice is the extent to which the media portray females and males in a limited and stereotypical fashion and the impact of such portrayals. For example, Levin and Kilbourne (2009), in *So Sexy So Soon,* document the sexualizing of young girls and boys. Advertising, books, cartoons, songs, toys, and television shows create:

> [a] narrow definition of femininity and sexuality [that] encourages girls to focus heavily on appearance and sex appeal. They learn at a very young age that their value is determined by how beautiful, thin, "hot," and sexy they are. And boys, who get a very narrow definition of masculinity that promotes insensitivity and macho behavior, are taught to judge girls based on how close they come to an artificial, impossible, and shallow ideal. (p. 2)

Men are also victimized by media images. A study of 1,000 adults found that two-thirds of the respondents thought that women in television advertisements were pictured as "intelligent, assertive, and caring," whereas men were portrayed as "pathetic and silly" (Abernathy 2003). In a study of beer and liquor ads, Messner and Montez de Oca (2005) concluded that, although beer ads of the 1950s and 1960s focused on men in their work roles and only depicted women as a reflection of men's home lives, present-day ads portray young men as "bumblers" and "losers" and women as "hotties" and "bitches."

Realty television shows such as *Toddlers and Tiaras* and the beauty pageants they are based on hypersexualize the contestants, some as young as 6 months old. These distorted images of beauty, and the Barbie dolls and airbrushed models that follow, set an impossibly high standard for female beauty.

Alan Polzner/Barcroft USA/Getty Images

TABLE 10.3 Percent of Men's and Women's Images by Occupations in Family Films and Prime-Time Television, 2012

| | Employed Characters within Work Sector by Highest Clout Position | | | |
| | Family Films | | Prime-Time Programs | |
Industry	Males	Females	Males	Females
Corporate Suite (Officers)	96.0%	3.4%	86.0%	14.0%
Investors, Developers	100.0%	0.0%	57.1%	42.9%
High-Level Politicians	95.5%	4.5%	72.2%	27.8%
Chief Justices, District Attorneys	100.0%	0.0%	100.0%	0.0%
Doctors, Health Care Managers	78.1%	21.9%	70.4%	29.6%
Editors in Chief	100.0%	0.0%	100.0%	0.0%
Academic Administrators	61.5%	38.5%	61.5%	38.5%
Media Content Creators	65.8%	34.2%	72.7%	27.3%

| | Employed Characters Portrayed in STEM-Related Jobs | | | |
| | Family Films | | Prime Time Programs | |
Industry Focused on STEM	Males	Females	Males	Females
STEM Careers*	83.8%	16.3%	78.9%	21.1%
Life/Physical Science	49.3%	65.4%	46.4%	66.7%
Computer Science	23.1%	7.7%	32.1%	33.3%
Engineering	19.4%	7.7%	16.1%	0.0%
Other STEM* Jobs	8.2%	19.2%	5.4%	0.0%

*STEM = Science, Technology, Engineering, and Mathematics
Source: Smith et al. 2012.

Gender differences exist in other media as well. Smith et al. (2013) analyzed the content of nearly 12,000 speaking characters in 129 top grossing family films, 275 prime-time television programs, and 36 children's TV shows (see Table 10.3). The results indicate that females were less likely to be present, narrators, or speaking characters when compared to males. In primetime television, males were most present in comedies followed by dramas, and females most present in reality shows and appeared least often in children's programming. The Geena Davis Institute on Gender in Media, concerned about the relative absence of girls in children's programming, has sponsored public service announcements (PSAs) about "Jane"—a fictional character with a little girl's voice used as part of an awareness campaign. In one PSA, a little girl says:

> Meet Jane. See Jane. See her? She makes up half the world's population. But you wouldn't know it by watching kids' media. On screen, Jane is outnumbered by a ratio of 3 to 1. When she is there, a lot of time it's purely as eye candy. And girls everywhere are watching. On average, over seven hours a day. If they see Jane, it's with little to say, few career options, and even fewer aspirations. But we can change. Meet Jane. See Jane? She is half the world's population. She has important things to say. And she can be anything she wants to be. But to empower girls, we need to see Jane. See Jane.org. If she can see it, she can be it. (See Jane 2013)

As with media images, both the words we use and the way we use them can reflect gender inequality. Radio talk show host Rush Limbaugh, referring to female reporters as "infobabes" and stating that they have "chickified the news," is demeaning to women and men (Media Matters 2009, 2012). The term *nurse* carries the meaning of "a woman who . . ." and the term *engineer* suggests "a man who. . . ." Terms such as *broad, old maid,* and

Please use the following definition in completing this exercise: Sexist language includes words, phrases, and expressions that unnecessarily differentiate between females and males *or* exclude, trivialize, or diminish either gender.

For each of the following statements, choose the descriptor (1 = strongly disagree; 2 = tend to disagree; 3 = undecided; 4 = tend to agree; 5 = strongly agree) that most closely corresponds with your beliefs about language, and enter the number in the blank following each statement.

1. Women who think that being called a "chairman" is sexist are misinterpreting the word *chairman*. ___
2. We should not change the way the English language has traditionally been written and spoken. ___
3. Worrying about sexist language is a trivial activity. ___
4. If the original meaning of the word *he* was "person," we should continue to use *he* to refer to both males and females today. ___

5. When people use the term *man and wife*, the expression is not sexist if the users don't mean it to be. ___
6. The English language will never be changed because it is too deeply ingrained in the culture. ___
7. The elimination of sexist language is an important goal. ___
8. Most publication guidelines require newspaper writers to avoid using ethnic and racial slurs. So, these guidelines should also require writers to avoid sexist language. ___
9. Sexist language is related to sexist treatment of people in society. ___
10. When teachers talk about the history of the United States, they should change expressions, such as "our forefathers," to expressions that include women. ___
11. Teachers who require students to use nonsexist language are unfairly forcing their political views upon their students. ___

12. Although change is difficult, we still should try to eliminate sexist language. ___

Scoring
All items are scored on a 5-point Likert-type scale. Items 1, 2, 3, 4, 5, 6, and 11 are reverse-scored. High scores (4 to 5) indicate a positive attitude toward inclusive language; low scores (1 to 2) indicate a negative attitude toward inclusive language. A score of 3 indicates neutrality or uncertainty. Each respondent's score on the instrument is the total of all the items.

Interpretation
The range of possible total scores on the 12-item IASNL is 12 to 60. Across the 12 items, total scores between 42.1 and 60 reflect a supportive attitude toward nonsexist language; total scores between 12 and 30 reflect a negative attitude toward nonsexist language; and total scores between 30.1 and 42 reflect a neutral attitude.

Sources: Parks & Roberton 2000; Parks & Roberton 2001.

Language is so gender-stereotyped that the placement of female or male before titles is sometimes necessary, as in the case of "female police officer" or "male prostitute."

spinster have no male counterparts. Sexually active teenage females are described by terms carrying negative connotations, whereas terms for equally sexually active male teenagers are considered complimentary.

Language is so gender-stereotyped that the placement of male or female before titles is sometimes necessary, as in the case of "female police officer" or "male prostitute." Furthermore, as symbolic interactionists note, the embedded meanings of words carry expectations of behavior. This chapter's *Self and Society* assesses your attitudes toward language and gender.

Religion and Cultural Sexism

Most Americans claim membership in a church, synagogue, or mosque. Research indicates that women attend religious services more often, rate religion as more important to their lives, and are more likely to believe in an afterlife than are men (Adams 2007; Davis et al. 2002; Smith 2006). In general, religious teachings have tended to promote traditional conceptions of gender. For example, in 2012, the Vatican appointed an American bishop to "rein in" a group of Catholic nuns who, in the church's view, have ". . . radical feminist themes [which are] incompatible with the Catholic faith" (Goodstein 2012, p. 1). Members of the group were cited for advocating ordination of female priests, failure to wear the traditional habit, living outside of the convent, joining advocacy groups, working in academia, and focusing too much on poverty and economic injustice while remaining silent on such issues as abortion and marriage equality (Goodstein 2009, 2012). In 2013, Pope Francis reiterated the church's ban on female priests, saying that ". . . the church has spoken and says no . . . that door is closed" (Strauss 2013). In the same year, however, the lawmaking body of the Anglican Church of England, after a previous year's blocked attempt, voted to ordain women bishops. The decision, which came in 2013, will likely not take effect until 2015 (Jourdan 2013).

Gallagher (2004) found that most evangelical Christians believe that marriage should be considered an equal partnership, but also believe that the male is the head of the household. Female Orthodox Jews are not counted as part of the *minyan* (i.e., a quorum required at prayer services), are not allowed to read from the Torah, and are required to sit separately from men at religious services. Roman Catholic Church doctrine forbids the use of artificial forms of contraception, and Muslim women are required to be veiled in public at all times. In addition to the Catholic Church, women cannot serve as ordained religious leaders in Islamic temples or in Orthodox Jewish synagogues.

What Do You Think? "Women of the Wall," a group of women wearing men-only prayer shawls in Israel, won a court battle allowing them to worship at Jerusalem's Western Wall without threat of being arrested (Bryant 2013). After the decision, however, would-be worshipers were unable to even reach the holy landmark, for groups of ultra-Orthodox schoolgirls filled the women's section. Should the Orthodox schoolgirls, who believe that the activists are trampling on tradition, be arrested for blocking access to the prayer site?

Despite this "stained glass ceiling" (Adams 2007), women are increasingly active in leadership positions in churches and temples across the nation. In 2009, the Lutheran Church in Great Britain ordained a woman bishop—the first female bishop in Great Britain (Lutheran World Federation 2009), and Alysa Stanton became the first U.S. Black American female rabbi (Kaufman 2009). Nonetheless, today, 82 percent of clergy in the United States are male (U.S. Census Bureau 2013) and, even in denominations that allow for the ordination of women, female clergy often do not hold the same status as their male counterparts and are often limited in their duties (Renzetti & Curran 2003).

However, religious teachings are not all traditional in their beliefs about women and men. Quaker women have been referred to as the "mothers of feminism" because of their active role in the early feminist movement. Reform Judaism has allowed ordination of women as rabbis for more than 40 years, and gays and lesbians for more than 15 years. In addition, the women-church movement—a coalition of feminist faith-sharing groups composed primarily of Roman Catholic women—offers feminist interpretations of Christian teachings. Within many other religious denominations, individual congregations choose to interpret their religious teachings from an inclusive perspective by replacing masculine pronouns in hymns, the Bible, and other religious readings (Renzetti & Curran 2003; Religious Tolerance 2011).

Social Problems and Traditional Gender Role Socialization

One of the fundamental questions in the sociology of gender is the extent to which observed differences are a consequence of nature (i.e., innate) or nurture (i.e., environmental). It is likely, as author and neuroscientist Lise Eliot states, "Sex differences are real and some are probably present at birth, but then social factors magnify them." So, she continues, "if we, as a society, feel that gender divisions do more harm than good, it would be valuable to break them down" (quoted in Weeks 2011, p. 1).

The recent acknowledgment that gender is not binary but is best conceived of as a continuum suggests that traditional gender roles are changing. However, to a large extent, the social definitions of what it means to be a woman and what it means to be a man have varied little over the decades. These definitions, in turn, are associated with several social problems, including the feminization of poverty, social-psychological costs, death and illness, conflict in relationships, and gendered violence.

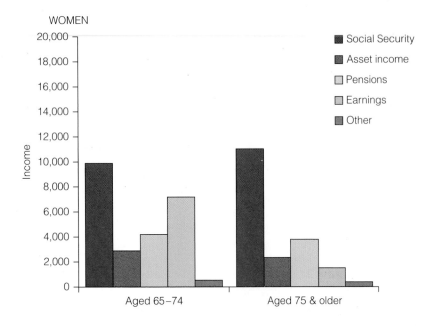

WOMEN

Legend:
- Social Security
- Asset income
- Pensions
- Earnings
- Other

Aged 65–74   Aged 75 & older

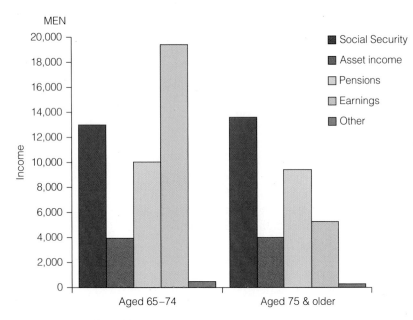

MEN

Legend:
- Social Security
- Asset income
- Pensions
- Earnings
- Other

Aged 65–74   Aged 75 & older

Figure 10.3 Older Persons Average Annual Income From Each Source By Gender and Age
Source: Fischer & Hayes 2013.

The Feminization of Poverty

A large majority of the global poor are women. Women comprise 40 percent of the global workforce yet earn between 10 percent and 30 percent less than their male counterparts, have higher unemployment rates, particularly in Middle Eastern countries, and often find it more difficult to reenter the labor force. Women are more likely than men to work in the agricultural sector in unpaid capacities, and women and girls increasingly belong to such vulnerable groups as migrant and domestic workers (ILO 2012b).

Women make up the majority of minimum wage workers in the United States and are significantly more likely to live in poverty than are men (BLS 2013). Household incomes of transgender individuals are also low and unemployment high. Research indicates that nearly four times as many transgender households live on less than $10,000 a year when compared to the general population (Grant et al. 2011), and unemployment rates are as high as 40 percent (*The Huffington Post* 2013).

Two groups of women are the most likely to be poor in the United States—women over the age of 65 and women with dependent children. As Figure 10.3 indicates, in both of the age brackets, women have fewer assets than men with the exception of the "other" category. Women have less than half the earnings and half the pension funds than men at early and late old age. These differences are reflected in the poverty rate. Over 10 percent of women over the age of 65 are living in poverty compared to 6 percent of men (U.S. Census Bureau 2013).

Women with children are far more likely to work part-time than their male counterparts or women who don't have children (ITUC 2011). Further, 16.9 percent of male-headed households are below the poverty level compared to 29.9 percent of female-headed households (U.S. Census Bureau 2013). Hispanic (38.8 percent) and black female-headed (36.7 percent) households are the poorest of all families, headed by a single woman (U.S. Census Bureau 2013).

It is often assumed that antipoverty programs designed to address overall economic inequality will reduce the feminization of poverty. However, Brady and Kall's (2008) analysis of poverty in 18 affluent countries suggests that not only is the feminization of poverty universal, it is unique in its origins, tied to the percent of children in single-mother families and the male-to-female ratio of the elderly in a country.

The Social Psychological Costs of Gender Socialization

How we feel about ourselves begins in early childhood. Significant others, through the socialization process, expect certain behaviors and prohibit others based upon our birth sex. Both girls and boys, often as a consequence of these expectations, feel varying degrees of self-esteem, autonomy, depression, and life dissatisfaction. For example, Jose and Brown (2009) studied depression, stress, and rumination (i.e., worrying) in a coed sample of 10- to 17-year-olds. Each of the three dependent variables was significantly higher in females when compared to males.

Transgender individuals also suffer from self-esteem issues and depression. In a survey of 6,450 transgender adults, 41 percent reported attempting suicide. The comparable number for the general population is 1.6 percent (Grant et al. 2011). For transgender individuals, family acceptance is an important buffer against societal prejudice and discrimination. Not surprisingly, transgender family members who are accepted and supported by their families have lower rates of risky behavior and negative experiences.

Pressure to conform to traditional gender roles exists not only at the individual level but at the societal level as well. After administering a questionnaire to a sample of over 6,500 undergraduates in 13 countries, Arrindell et al. (2013) conclude that in societies where there is a strong emphasis on masculinity, i.e., in "tough countries," masculine gender role stress is significantly higher than in countries where a masculine identity is less emphasized, i.e., in "soft countries."

The traditional male gender role also places enormous cultural pressure on men to be successful in their work and to earn high incomes. Sanchez and Crocker (2005) found that, among college-age women *and* men, the more participants were invested in traditional ideals of gender, the *lower* their self-concept and psychological well-being. Traditional male socialization also discourages males from expressing emotion and asking for help—part of what William Pollack (2000) calls the **boy code**.

The "Cult of Thinness." Adolescent girls are more likely to be dissatisfied with their looks, including physical attractiveness, appearance, and body weight than adolescent boys. Canadian nonprofit group MediaSmarts summarizes the present state of research on the "cult of thinness" (MediaSmarts 2012): Research indicates that even before being exposed to advertisements for fashion and beauty products, girls as young as 3 years old prefer game pieces that portray people who are thin rather than heavy; by age 7, they are already concerned about their appearance and can identify something they would like to change; and by 16 to 21 years old, half of young women say they would like to have cosmetic surgery to improve their bodies.

According to MediaSmarts, messages that perpetuate insecurities in women are simply a matter of economics. If you liked the way you looked, you wouldn't buy anything. The diet industry alone makes billions of dollars a year selling products that don't work. The social-psychological costs are high. Research links ". . . exposure to images of thin, young, airbrushed female bodies to depression, loss of self-esteem, and on healthy eating habits in girls and young women . . ." (MediaSmarts 2012).

Boys too, from an early age, are concerned about body image, and as adults, their self-esteem is also linked to body shape and weight (Grogan 2008). Using focus groups and interview techniques, Norman (2011) describes how young men between the ages of 11 and 15, on the one hand, have concerns about their body images but, on the other, are anxious about discussing their concerns. As with females, the media often portrays an unrealistic ideal body image for males.

Dallesasse and Kluck (2013) examined muscle mass and percent of body fat of primary male cast members in three types of reality television shows—reality drama, endurance contest, and dating/romance—on MTV, VH1, Spike TV, and the Discovery Channel. After capturing images and taking measurements, the researchers concluded that between 70 to 88 percent of the primary male characters were "somewhat muscular to very muscular," and over 90 percent had "medium to low body fat." This is in stark contrast to the 35.6 percent of 18- to 44-year-old males who strength train sufficiently enough to be considered muscular. The researchers conclude that, although ". . . promotion of the masculine ideal *may* encourage male viewers to adopt more health-related behaviors (e.g., exercise, healthy eating habits), the ideal may also contribute to body dissatisfaction and engagement in a number of unhealthy body investment strategies among some men" (p. 314).

boy code A set of societal expectations that discourages males from expressing emotion, weakness, or vulnerability, or asking for help.

Gender Role Socialization and Health Outcomes

Men are less likely to go to a doctor than women for a variety of structural and cultural reasons (WHO 2010). Men, for example, work longer hours than women and are more likely to be working full-time, making it difficult to see a doctor or attend preventive medicine programs that are often only available during the day. Men are also less likely to have a regular physician or to go to the hospital, even if time permits (White & Witty 2009).

> Men . . . work longer hours than women and are more likely to be working full-time, making it difficult to see a doctor or attend preventive medicine programs. . . . Men are also less likely to have a regular physician or to go to the hospital, even if time permits.

At every stage of life, "American males have poorer health and a higher risk of mortality than females" (Gupta 2003, p. 84). On average, men in the United States die five years earlier than women, although gender differences in life expectancy have shrunk over the years (U.S. Census Bureau 2013). Traditionally defined gender roles for men are linked to high rates of cirrhosis of the liver (e.g., alcohol consumption), many cancers (e.g., tobacco use), and cardiovascular diseases (e.g., stress). Men also engage in self-destructive behaviors—poor diets, lack of exercise, higher drug use, refusal to ask for help or wear a seat belt, and stress-related activities—more often than women. Being married improves men's health more than it does women's (Williams & Umberson 2004), in large part because wives encourage their husbands to take better care of themselves.

Women's health is also gendered. Although men have slightly higher rates of HIV/AIDS worldwide, the disease disproportionately affects women in many areas of the world (see Chapter 2). Of those who are infected in sub-Saharan Africa, approximately 60 percent are female (USAID 2013). Women's inequality contributes to the spread of the disease in several ways. First, in many of these societies, "women lack the power in relationships to refuse sex or negotiate protected sex" (Heyzer 2003, p. 1). Second, gender norms often dictate that men have more sexual partners than women, putting women at greater risk. Third, women are often the victims of rape and sexual assault, with little social or legal recourse. During World War II, hundreds of thousands of women and girls were forced into sexual slavery at what were euphemistically called "comfort stations." A memorial to the ". . . more than 200,000 women and girls who were abducted by the Armed Forces of the Government of Imperial Japan" . . . also reads that they ". . . endured human rights violations that no peoples should leave unrecognized. Let us never forget the horrors of crimes against humanity" (Alvarado 2013).

The World Health Organization (2009, 2011) has identified additional ways traditional definitions of gender impact the health and well-being of women and girls. Every day, over 1,600 women die from preventable complications during pregnancy and childbirth. Primarily responsible for household duties, women are exposed to hundreds of pollutants as they cook that contribute to their disproportionately high rates of death from chronic obstructive pulmonary disease (COPD). Deaths from lung cancer and other tobacco-related illnesses are expected to rise as the tobacco industry targets women in developing countries. Finally, many women and girls throughout the world have a higher probability of suffering or dying from a variety of diseases because of gender: They are more likely to be poor, less likely to be seen as worthy of care when resources are short, and in many countries, they are forbidden to travel unaccompanied by a male, making access to a hospital difficult.

Ninety-nine percent of maternal deaths take place in developing countries. According to the World Health Organization, approximately 800 women die every day while giving birth, and many others ". . . suffer serious complications from pregnancy, labor, and delivery, which can result in long-term disabilities" (p. 1). The United Nations calls for a 75 percent reduction of maternal deaths by 2015 (U.S. Department of State 2012).

Gendered Violence

Men are more likely than women to be involved in violence—to kill and be killed; to wage war and die both as combatants and noncombatants; to take their own lives, usually with the use of a firearm; to engage in violent crimes of all types; to bully, harass, and abuse. As sociologist Michael Kimmel (2012) notes in his discussion of the mass killings at Sandy Hook Elementary School, unlike girls

> …boys learn that violence is not only an acceptable form of conflict resolution, but the one that is admired…They learn it from their fathers….from a media that glorifies it, from sports heroes who commit felonies and get big contracts, from a culture saturated in images of heroic and redemptive violence. They learn it from each other…

Kimmel (2011) also argues that violence against transgender and gay youth is rooted in notions of masculinity and the threat that gender nonconformity poses to "real men." Not surprisingly, a high proportion of transgender adults recount being harassed (78 percent), and physically (35 percent) and sexually (12 percent) assaulted while in school. In some cases, the victimization was so severe that 15 percent of the respondents report having to leave school (Grant et al. 2011).

Women and girls are also the victims of male violence. Worldwide, the United Nations Development Fund estimates that ". . . at least one in every three women globally has been beaten, coerced into sex, or otherwise abused in her lifetime, with rates reaching 70 percent in some countries" (Amnesty International 2013, p. 1; see Chapter 5). Attacks on women's and girls' bodies are thus fairly routine, often taking place in the name of religion, war, or honor. Over 5,000 women and girls are killed each year in **honor killings**—murders, often public, as a result of a female dishonoring, or being perceived to have dishonored, her family or community (Foerstel 2008).

Honor killings, although occurring throughout history, are most likely to take place today in the Middle East (e.g., Syria, Jordan) and Western and Central Asia (e.g., Iraq, India). To assess contemporary attitudes toward honor killing, Eisner and Ghuneim (2013) administered a survey to 856 ninth graders in Amman, a city of 2.5 million and the capital of Jordan. The results indicate that nearly half of the boys and 20 percent of the girls surveyed condone honor killings, that is, believe that killing a woman who has dishonored her family is justifiable. The authors conclude that ". . . three . . . factors are important for understanding attitudes toward honor killings, namely traditionalism, the belief in female chastity as a precious good, and a general tendency to morally neutralize aggressive behavior" (p. 413).

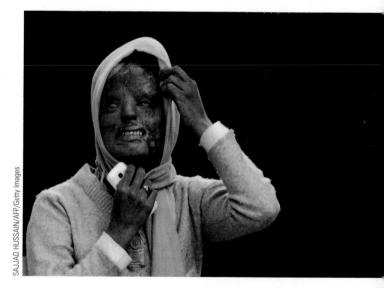

SAJJAD HUSSAIN/AFP/Getty Images

Sonali Mukherjee, an acid attack victim, has had 27 surgeries. Three college classmates threw acid on her after she ignored their advances. The men were released from jail after serving two years.

honor killings Murders, often public, as a result of a female dishonoring, or being perceived to have dishonored, her family or community.

Violence against women and girls, a problem of pandemic proportions, is rooted in gender inequality and the lingering notion that women and girls are property—a belief tied to ancient law and many of the world's religions. Yet, in the United States, being female is not part of the federal hate crime statutes despite the fact that, as one advocate testified before Congress, "women and girls . . . are exposed to terror, brutality, serious injury, and even death because of their sex" (LCCREF 2009, p. 33).

What Do You Think? Worldwide, much of the violence against women is steeped in "harmful traditional practices," including honor killings, female genital mutilation, forced marriage, and dowry killings. Dowry killings involve "a woman being killed by her husband or in-laws because her family is unable to meet their demands for her dowry—a payment made to a woman's in-laws upon her engagement or marriage as a gift to her new family" (UNIFEM 2007, p. 1). Although illegal in India since 1961, the number of women killed in India because of inadequate dowries has recently increased (Bedi 2012). How would you stop dowry killings?

Strategies for Action: Toward Gender Equality

In recent decades, there has been a growing awareness of the need to increase gender equality throughout the world. Strategies to achieve this end have focused on empowering women in social, educational, economic, and political spheres and improving women's access to education, nutrition, health care, and basic human rights. But as we will see in the following section on grassroots movements, there is also a men's movement that is concerned with gender inequities and the issues facing men.

Grassroots Movements

Efforts to achieve gender equality in the United States have been largely fueled by the feminist movement. Despite a conservative backlash, and to a lesser extent men's activist groups, feminists have made some gains in reducing structural and cultural sexism in the workplace and in the political arena.

Feminism and the Women's Movement. **Feminism** is the belief that women and men should have equal rights and responsibilities. The U.S. feminist movement began in Seneca Falls, New York, in 1848, when a group of women wrote and adopted a women's rights manifesto modeled after the Declaration of Independence. Although many of the early feminists were primarily concerned with suffrage, feminism has its "political origins . . . in the abolitionist movement of the 1830s," when women learned to question the assumption of "natural superiority" (Anderson 1997, p. 305). Early feminists were also involved in the temperance movement, which advocated restricting the sale and consumption of alcohol, although their greatest success was the passing of the Nineteenth Amendment in 1920, which recognized women's right to vote.

The rebirth of feminism almost 50 years later was facilitated by a number of interacting forces: an increase in the number of women in the labor force, the publication of Betty Friedan's book *The Feminine Mystique,* an escalating divorce rate, the socially and politically liberal climate of the 1960s, student activism, and the establishment of the Commission on the Status of Women by John F. Kennedy. The National Organization for Women (NOW) was established in 1966, and remains one of the largest feminist organizations in the United States, with more than 500,000 members in 550 chapters across the country.

One of NOW's hardest-fought battles is the struggle to win ratification of the **equal rights amendment (ERA)**, which states that "equality of rights under the law shall not be denied or abridged by the United States, or by any state, on account of sex." The

feminism The belief that men and women should have equal rights and responsibilities.

equal rights amendment (ERA) The proposed 28th amendment to the Constitution, which states that "equality of rights under the law shall not be denied or abridged by the United States, or by any state, on account of sex."

proposed 28th amendment to the Constitution passed both the House of Representatives and the Senate in 1972, but failed to be ratified by the required 38 states by the 1979 deadline, which was later extended to 1982. With the exception of when presidential politics took precedence in 2008, the bill has been reintroduced into Congress every year since 1982 (ERA 2013). To date, 15 states have not ratified the equal rights amendment, some of which have partial or inclusive guarantees of equal rights on the basis of sex within their state constitutions (ERA 2013).

Proponents of the ERA argue that its opponents used scare tactics—saying that the ERA would lead to unisex bathrooms, mothers losing custody of their children, and mandatory military service for women—to create a conservative backlash. However, Susan Faludi, in *Backlash: The Undeclared War against American Women* (1991), contends that contemporary arguments against feminism are the same as those levied against the movement 100 years ago and that the negative consequences predicted by opponents of feminism (e.g., women unfulfilled and children suffering) have no empirical support. Proponents also argue that "without the explicit wording and intention of women's rights documented in the principles of our government, women remain second-class citizens" (Cook 2009, p. 1).

Today, a new wave of feminism is being led by young women and men who grew up with the benefits their mothers or grandmothers won, but who are shocked by the stoning to death of a woman accused of adultery, the selling of an 8-year-old child bride, the continued practice of female genital mutilation, the lack of access to modern contraceptive techniques, and the alarming rates of violence against women. These young feminists are more inclusive than their predecessors, welcoming all who champion the cause of global equality.

The Men's Movement. As a consequence of the women's rights movement, men began to reevaluate their own gender status. As with any grassroots movement, the men's movement has a variety of factions. One of the early branches of the men's movement is known as the mythopoetic men's movement, which began after the publication of Robert Bly's (1990) *Iron John*—a fairy tale about men's wounded masculinity that was on the *New York Times* best-seller list for more than 60 weeks (Zakrzewski 2005). Participants in the men's mythopoetic movement met in men-only workshops and retreats to explore their internal masculine nature, male identity, and emotional experiences through the use of stories, drumming, dance, music, and discussion.

The ManKind Project (MKP), founded in 1985, grew out of the men's mythopoetic movement. It is an international organization with 32 training centers throughout the United States. Their motto, "changing the world one man at a time," reflects the stated core values of the organization: accountability, authenticity, integrity, community, service, and inclusivity (MKP 2013). MKP conducts the New Warrior Training Adventure, "the most recognized modern initiation experience for men in the world. . . ." According to their homepage, over 50,000 men have completed the "new warrior" training.

The men's movement also includes men's organizations that advocate for gender equality and work to make men more accountable for sexism, violence, and homophobia. The National Organization for Men Against Sexism (NOMAS) was founded in 1975, and "advocates a perspective that is pro-feminist, gay affirmative, antiracist, dedicated to enhancing men's lives, and committed to justice on a broad range of social issues including class, age, religion, and physical abilities" (NOMAS 2013, p. 1).

Some men's groups focus on issues concerning children's and fathers' rights. Groups such as the American Coalition of Fathers and Children, Dads Against Discrimination, and Fathers4Justice are attempting to change the social and legal bias against men in divorce and child custody decisions, which tend to favor women. Members of the National Coalition for Men (NCFM) are also concerned with issues surrounding fatherhood, and the way in which ". . . many men perform the

Other concerns on the agenda of men's rights groups include the domestic violence committed against men by women, false allegations of child sexual abuse, wrongful paternity suits, and the oppressive nature of restrictive masculine gender norms.

parent role in only a perfunctory manner, depriving them and their children of the essence of that experience" (NCFM 2013, p. 2).

Other concerns on the agenda of men's rights groups include the domestic violence committed against men by women, false allegations of child sexual abuse, wrongful paternity suits, and the oppressive nature of restrictive masculine gender norms. Just as women have fought against being oppressed by expectations to conform to traditional gender stereotypes, men are beginning to want the same freedom from traditional gender expectations. Unfortunately, men who enter nontraditional work roles, such as nurse and primary school teacher, are often stigmatized for participating in "feminine" work. A study of men in nontraditional work roles found that these men commonly experience embarrassment, discomfort, shame, and disapproval from friends and peers (Sayman 2007; Simpson 2005). Similarly, Dunn et al. (2013), after interviewing working women with stay-at-home husbands, report that the wives' co-workers and relatives often made negative comments about their husbands.

U.S. National Policy

A number of important federal statutes have been passed to help reduce gender inequality. They include the Equal Pay Act of 1963, Title VII of the Civil Rights Act of 1964, Title IX of the Education Amendments of 1972, and the Victims of Trafficking and Violence Protection Act of 2000. In 2009, President Obama signed the Ledbetter Fair Pay Act, which reversed the 2007 U.S. Supreme Court decision that held that victims of pay discrimination had 180 days to file a grievance after the act of discrimination. The act now defines each paycheck as a separate act of discrimination (Mehmood 2009).

One of the most important pieces of legislation to date is the 2013 Violence Against Women Reauthorization Act (VAWA 2013). The act:

- closes jurisdictional gaps that have previously excluded American Indian and Alaska native women;
- provides protection for lesbian, gay, bisexual, and transgender survivors of domestic or sexual violence;
- establishes a stalker database;
- expands victim services;
- protects women and children against human trafficking; and
- adds protection for college student victims of domestic or sexual violence.

As a complement to the Violence Against Women Reauthorization Act, President Obama also signed the Campus Sexual Violence Elimination Act of 2013. The act requires that every college and university that is receiving Title IV funds must compile statistics on domestic violence, sexual assault, dating violence, and stalking on campus. Further, in their annual report, they must include campus prevention and educational programs, as well as the procedures followed once an incident has been reported (Clery Center 2013).

What Do You Think? One of the concerns of all women is sexual assault, and women on college campuses are no exception. Two college students, Annie Clark and Andrea Pino, filed a federal lawsuit alleging that UNC-CH not only failed to handle their sexual assault cases properly, but that the university systematically underreported sexual assault cases in violation of their own campus crime reporting law (Stancill 2013). Do you think there's a sexual assault problem on your campus and, if so, what is being done about it?

In addition to sexual violence, two additional gender-related areas of concern are sexual harassment and affirmative action.

Sexual Harassment. Sexual harassment is a form of sex discrimination that violates Title VII of the 1964 Civil Rights Act. The U.S. Equal Employment Opportunity Commission (EEOC) (2012) defines **sexual harassment** in the workplace as "unwelcome sexual advances, requests for sexual favors, and other verbal or physical conduct of a sexual nature . . . when this conduct explicitly or implicitly affects an individual's employment, unreasonably interferes with an individual's work performance, or creates an intimidating, hostile, or offensive work environment" (p. 1). In 2013, over 15 women accused the mayor of San Diego of inappropriate behavior, leading to a recall effort and to signs in various establishments, including Hooters, that the mayor was not welcome (Smith & Lah 2013).

Any individual—female, male, or transgender—can be a victim of sexual harassment; a victim might also be someone who was not harassed but was affected by offensive conduct. There are two types of sexual harassment: (1) *quid pro quo,* in which an employer requires sexual favors in exchange for a promotion, salary increase, or any other employee benefit, and (2) the existence of a hostile environment that unreasonably interferes with job performance, as in the case of sexually explicit comments or insults being made to an employee.

Common examples of sexual harassment include unwanted touching, the invasion of personal space, making sexual comments about a person's body or attire, and telling sexual jokes (Uggen & Blackstone 2004). Sexual harassment occurs in a variety of settings, including the workplace, public schools, military academies, and college campuses. Women who work in male-dominated occupations and blue-collar jobs, who are young, financially dependent, and single or divorced are the most likely to experience sexual harassment (ILO 2011; Jackson & Newman 2004).

Affirmative Action. As discussed in Chapter 9, **affirmative action** refers to a broad range of policies and practices to promote equal opportunity as well as diversity in the workplace and on campuses. Affirmative action policies, developed from federal legislation in the 1960s, require that any employer (universities as well as businesses) that receives contracts from the federal government must make "good faith efforts" to increase the number of female and other minority applicants. Such efforts can be made through expanding recruitment and training programs and by making hiring decisions on a non-discriminatory basis.

However, a 1996 California ballot initiative, the first of its kind, abolished race and sex preferences in government programs, which included state colleges and universities. Over the next three years, several other states followed suit. In 2003, the U.S. Supreme Court held that universities have a "compelling interest" in a diverse student population and therefore may take minority status into consideration when making admissions decisions. Since that time, several states have addressed the issue of affirmative action in government programs with varying results. For example, in 2006, Michigan voters passed a referendum banning the use of race or gender in state college or university admissions decisions. However, in 2012, an appellate court held that the ban was unconstitutional and, in 2013, the U.S. Supreme Court agreed to hear the case (Richey 2013).

International Efforts

International efforts to address problems of gender inequality date back to the 1979 Convention to Eliminate All Forms of Discrimination Against Women (CEDAW), often referred to as the international women's bill of rights, adopted by the United Nations in 1979. To date, 187 out of 194 countries have ratified the women's bill of rights, including every country in Europe, and South and Central America. Only seven countries have not ratified CEDAW—Iran, Sudan, South Sudan, Somalia, Palau, Tonga, and the United States (Deen 2013).

Another significant international effort occurred in 1995, when representatives from 189 countries adopted the Beijing Declaration and Platform for Action at the Fourth World Conference on Women sponsored by the United Nations. The platform reflects an international commitment to the goals of equality, development, and peace for women everywhere. The platform identifies strategies to address critical areas of concern related to women and girls, including poverty, education, health, violence, armed conflict, and human rights.

sexual harassment In reference to workplace harassment, when an employer requires sexual favors in exchange for a promotion, salary increase, or any other employee benefit and/or the existence of a hostile environment that unreasonably interferes with job performance.

affirmative action A broad range of policies and practices in the workplace and educational institutions to promote equal opportunity as well as diversity.

In addition to the CEDAW and the Beijing Platform, in 2000, all of the members of the United Nations adopted the Millennium Declaration. One of the eight Millennium Development Goals, as stated in the Millennium Declaration, is the promotion of gender equality and women's empowerment by 2015. Progress has been slow. In an evaluation of this goal, a United Nations' Report (2013) concludes that, "whether in the public or private sphere, from the highest levels of government decision making to households, women continue to be denied equal opportunity with men to participate in decisions that affect their lives" (p. 5).

Finally, directed toward preventing the millions of acts of violence against women and girls annually is the International Violence Against Women Act (I-VAWA 2013). The act, which is largely based on the *U.S. Strategy to Prevent and Respond to Gender-Based Violence Globally* that was released in 2012, was written by Amnesty International and a consortium of other concerns. It was most recently, in 2013, introduced into Congress, where it awaits action. If passed, the new law would increase the efficiency and effectiveness of the United States' global response to violence against women.

Understanding Gender Inequality

Gender roles and the social inequality they create are ingrained in our social and cultural ideologies and institutions and are therefore difficult to alter as indicated by the persistence of discriminatory attitudes. Nevertheless, as we have seen in this chapter, growing attention to gender issues in social life has spurred change. The traditional gender roles of men and women, and the binary concepts of sex and gender, are weakening. Women who have traditionally been expected to give domestic life first priority are now finding it more acceptable to seek a career outside the home; the gender pay gap is narrowing, albeit slowly; and significant improvements in educational disparities between men and women, boys and girls, have been made. Most of these improvements, however, are in developed countries. Globally, millions of women and girls continue to be victimized by poverty, gendered violence, illiteracy, and limited legal rights and political representation.

Eliminating gender stereotypes and redefining gender in terms of equality does not simply mean liberating women but liberating men as well in society. Men are also victimized by discrimination and gender stereotypes that define what they "should" do rather than what they are capable, interested, and willing to do. The National Coalition for Men (NCFM 2011) has incorporated this view into their position statement:

> We have heard in some detail from the women's movement how much sex stereotyping has limited the potential of women. More recently, men have become increasingly aware that they too are assigned limiting roles which they are expected to fulfill regardless of their individual abilities, interests, physical/emotional constitutions, or needs. Men have few or no effective choices in many critical areas of life. They face injustices under the law. And typically they have been handicapped by socially defined "shoulds" in expressing themselves in other than stereotypical ways.

Increasingly, people are embracing **androgyny**—the blending of traditionally defined masculine and feminine characteristics. The concept of androgyny implies that both masculine and feminine characteristics and roles are *equally valued.* However, "achieving gender equality . . . is a grindingly slow process, since it challenges one of the most deeply entrenched of all human attitudes" (Lopez-Claros & Zahidi 2005, p. 1).

An international strategy for achieving gender equality, *gender mainstreaming* is "the process of assessing the implications for women and men of any planned action, including legislation, policies or programmes, in all areas and at all levels" (IASC 2009, p. 7). Difficult? Yes. But, regardless of whether traditional gender roles emerged out of biological necessity as the structural functionalists argue, or out of economic oppression as the conflict theorists hold, or both, it is clear that, today, gender inequality carries a high price: poverty, loss of human capital, feelings of worthlessness, violence, physical and mental illness, and death. Perhaps we have reached a time when we need to ask ourselves, are the costs of traditional gender roles too high to continue to pay?

androgyny Having both traditionally defined feminine and masculine characteristics.

- **Does gender inequality exist worldwide?**
  There is no country in the world in which men and women are treated equally. Although women suffer in terms of income, education, and occupational prestige, men are more likely to suffer in terms of mental and physical health, mortality, and the quality of their relationships.

- **How do the three major sociological theories view gender inequality?**
  Structural functionalists argue that the traditional division of labor was functional for preindustrial society and has become defined as both normal and natural over time. Today, however, modern conceptions of the family have replaced traditional ones to some extent. Conflict theorists hold that male dominance and female subordination evolved in relation to the means of production—from hunting-and-gathering societies in which females and males were economic equals to industrial societies in which females were subordinate to males. Symbolic interactionists emphasize that, through the socialization process, both females and males are taught the meanings associated with being feminine and masculine.

- **What is meant by the terms *structural sexism* and *cultural sexism*?**
  Structural sexism refers to the ways in which the organization of society, and specifically its institutions, subordinate individuals and groups based on their sex classification. Structural sexism has resulted in significant differences between education and income levels, occupational and political involvement, and civil rights of women and men. Structural sexism is supported by a system of cultural sexism that perpetuates beliefs about the differences between women and men. Cultural sexism refers to the ways the culture of society—its norms, values, beliefs, and symbols—perpetuate the subordination of an individual or group because of the sex classification of that individual or group.

- **What is the difference between the glass ceiling and the glass escalator?**
  The glass ceiling is an often invisible barrier that prevents women and other minorities from moving into top corporate positions. The glass escalator, on the other hand, refers to the tendency for men seeking traditionally female jobs to have an edge in hiring and promotion practices.

- **What are some of the problems caused by traditional gender roles?**
  First is the feminization of poverty. Women are socialized to put family ahead of education and careers, a belief that is reflected in their less prestigious occupations and lower incomes. Second are social-psychological costs. Women and transgender individuals tend to have lower self-esteem and higher rates of depression than men. Men and boys, on the other hand, are often subject to the emotional restrictions of the "boy code." Third, traditional gender roles carry health costs in terms of death and illness. For example, many traditionally defined male behaviors shorten life expectancy, and women in many regions of the world disproportionately suffer from HIV/AIDS. Finally, gendered violence is responsible for the deaths of men, women, and transgender individuals.

- **What strategies can be used to end gender inequality?**
  Grassroots movements, such as feminism and the women's rights movement and the men's rights movement, have made significant inroads in the fight against gender inequality. Their accomplishments, in part, have been the result of successful lobbying for passage of laws concerning sex discrimination, sexual harassment, and affirmative action. Besides these national efforts, international efforts continue as well. One of the most important is the Convention to Eliminate All Forms of Discrimination Against Women (CEDAW), also known as the international women's bill of rights, which the United Nations adopted in 1979.

TEST YOURSELF

1. The United States is the most gender-equal nation in the world.
   a. True
   b. False

2. Symbolic interactionists argue that
   a. male domination is a consequence of men's relationship to the production process
   b. gender inequality is functional for society
   c. gender roles are learned in the family, in the school, in peer groups, and in the media
   d. women are more valuable in the home than in the workplace

3. Recent trends indicate that more males enter college from high school than females.
   a. True
   b. False

4. The devaluation hypothesis argues that female and male pay differences are a function of women's and men's different levels of education, skills, training, and work experience.
   a. True
   b. False

5. The glass ceiling is an often invisible barrier that prevents women and other minorities from moving into top corporate positions.
   a. True
   b. False

6. Which of the following statements is true about women and U.S. politics?
   a. There are more female than male governors.
   b. In the history of the U.S. Supreme Court, there have only been four female justices.
   c. The highest-ranking woman in the U.S. government is vice president Hillary Clinton.
   d. Women received the right to vote with the passage of the 21st Amendment.

7. Which of the following statements is *not* true?
   a. Females are underrepresented in STEM occupations on prime-time television.
   b. Based on a study of realty TV shows, primary male characters are most often portrayed as overweight and "paunchy."

c. Little boys when compared to little girls are overrepresented in children's media.

d. An analysis of prime-time television indicates that males are most often pictured in comedies.

8. The feminization of poverty refers to
   a. the tendency for women to be caretakers of the poor
   b. feminists' criticism of public policy on poverty
   c. the disproportionate number of women who are poor
   d. gender role socialization of poor women

9. *Quid pro quo* sexual harassment refers to the existence of a hostile working environment.
   a. True
   b. False

10. Feminists would argue which of the following?
    a. The ERA should be part of the Constitution.
    b. The passage of the Nineteenth Amendment was a mistake.
    c. Occupational sex segregation benefits everyone.
    d. NOMAS is a radical, sexist organization.

Answers: 1. B; 2. C; 3. B; 4. B; 5. A; 6. B; 7. B; 8. C; 9. B; 10. A.

KEY TERMS

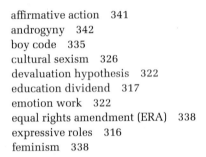

affirmative action   341
androgyny   342
boy code   335
cultural sexism   326
devaluation hypothesis   322
education dividend   317
emotion work   322
equal rights amendment (ERA)   338
expressive roles   316
feminism   338

gender   308
gender deviance hypothesis   328
gender roles   326
glass ceiling   321
glass escalator effect   320
honor killings   337
human capital hypothesis   322
instrumental roles   316
masculine overcompensation thesis   314
motherhood penalty   321

occupational sex segregation   318
pink-collar jobs   320
sex   308
sexism   308
sexual harassment   341
structural sexism   316
structured choice   321
transgender individual   308

Michael S. Williamson/The Washington Post/Getty Images

Sexual Orientation and the Struggle for Equality

"Homophobia alienates mothers and fathers from sons and daughters, friend from friend, neighbor from neighbor, Americans from one another. So long as it is legitimated by society, religion, and politics, homophobia will spawn hatred, contempt, and violence, and it will remain our last acceptable prejudice."

—Byrne Fone, Author, Emeritus Professor

EIGHTY-THREE-YEAR OLD Edith Windsor sued the United States government on the basis of discriminatory treatment under the Defense of Marriage Act (DOMA). Ms. Windsor and her partner had lived together for over 40 years and were married in Canada in 2007. After the death of her partner, Ms. Windsor was assessed over $300,000 in estate taxes, money that she would not have owed if she had been in an opposite-sex marriage (Condon 2013). Ms. Windsor challenged the government's decision on constitutional grounds, and in a 5–4 ruling, the U.S. Supreme Court held that the denial of federal benefits to same-sex married couples that are granted to opposite-sex married couples is a violation of the due process clause of the U.S. constitution. After hearing of the Court's ruling, "Edie" Windsor said, "I feel honored, humbled, and overjoyed" (Condon 2013).

In addition to Edith "Edie" Windsor, who challenged the constitutionality of the Defense of Marriage Act (DOMA), Kris Perry, Sandy Stier, Paul Katami, and Jeff Zarrillo filed suit against California's Proposition 8, which was also held to be unconstitutional. Seen here entering the U.S. Supreme Court, both couples married soon after the decision was announced.

sexual orientation A person's emotional and sexual attractions, relationships, self-identity, and behavior.

heterosexuality The predominance of emotional, cognitive, and sexual attraction to individuals of the other sex.

homosexuality The predominance of emotional, cognitive, and sexual attraction to individuals of the same sex.

bisexuality The emotional, cognitive, and sexual attraction to members of both sexes.

lesbian A woman who is attracted to same-sex partners.

gay A term that can refer to either women or men who are attracted to same-sex partners.

gender non-conforming Often used synonymously with transgender, gender non-conforming (sometimes called gender variant) refers to displays of gender that are inconsistent with society's expectations.

LGBT, LGBTQ, and **LGBTQI** Terms used to refer collectively to lesbian, gay, bisexual, transgender, questioning or "queer," and/or intersexed individuals.

Despite the progress that has been made with respect to equal rights for same-sex–attracted individuals in recent years, the fight for equality on the basis of sexual orientation and gender identity continues to be met with opposition. The term **sexual orientation** refers to a person's emotional and sexual attractions, relationships, self-identity, and behavior. **Heterosexuality** refers to the predominance of emotional, cognitive, and sexual attraction to individuals of the opposite sex. **Homosexuality** refers to the predominance of emotional, cognitive, and sexual attraction to individuals of the same sex, and **bisexuality** is the emotional, cognitive, and sexual attraction to members of both sexes. The term **lesbian** refers to women who are attracted to same-sex partners, and the term **gay** can refer to either women or men who are attracted to same-sex partners.

Much of the current literature on the treatment, and political and social agendas of individuals who are gay, lesbian, and bisexual includes transgender individuals. As discussed in Chapter 10, *transgender individuals* (sometimes called "*trans*") are people whose sense of gender identity—as male or female—is inconsistent with their assigned birth (sometimes called chromosomal) sex. Gender non-conforming is often used synonymously with transgender. **Gender non-conforming** refers to displays of gender that are inconsistent with society's expectations.

The acronyms **LGBT**, **LGBTQ**, and **LGBTQI**, as well as an assortment of other variations, are used to refer collectively to individuals who are lesbian, gay, bisexual, transgender, questioning or "queer," and *intersexed*. Sometimes, one will encounter the acronym accompanied by the letter "A," which stands for *allies*.

This chapter focuses on sexual orientation. However, it is impossible to discuss the struggle for same-sex equality without acknowledging transgender individuals who have been central to the larger LGBT civil rights movement. Thus, some of the research reported in this chapter is on lesbians, gays, and bisexuals, whereas other studies include transgender people as participants in addition to lesbians, gays, and bisexuals. Further, many laws and policies discussed in this chapter do not specifically refer to bisexuals but would, by default, apply if they were in a same-sex relationship.

In this chapter, we focus primarily on Western conceptions of diversity in sexual orientation. It is beyond the scope of this chapter to explore in depth how sexual orientation and its cultural meanings vary throughout the world. The global legal status of lesbians and gay men, however, will be summarized. We also discuss the prevalence of homosexuality, heterosexuality, and bisexuality in the United States, the beliefs about the origins of sexual orientation, and then apply sociological theories to better understand societal reactions to non-heterosexuals. After detailing the ways in which non-heterosexuals are victimized by prejudice and discrimination, we then end the chapter with a discussion of strategies to reduce antigay prejudice and discrimination.

The Global Context: A Worldview of the Status of Homosexuality

Homosexuality has existed throughout human history and in most, perhaps all, human societies (Joannides 2011; Kirkpatrick 2000). A global perspective on laws and social attitudes regarding homosexuality reveals that countries vary tremendously in their treatment of same-sex sexual behavior—from intolerance and criminalization to acceptance and legal protection. Seventy-eight out of 193 countries worldwide continue to criminalize same-sex relations. Legal penalties vary for violating laws that prohibit homosexual acts. In some countries, homosexuality is punishable by prison sentences (e.g., two months of prison in Algeria or a life sentence in Bangladesh) and/or corporal punishment, such as whipping or lashing (e.g., a number of lashes in Iran), and in five nations—Iran, Mauritania, Saudi Arabia, Yemen, and Sudan—people found guilty of engaging in same-sex sexual behavior may receive the death penalty (ILGBTIA 2012).

In general, countries throughout the world are moving toward increased legal protection of non-heterosexuals, as discrimination on the basis of sexual orientation has become part of a broad international human rights agenda. In fact, "global trends towards more approval of same-sex sexual behavior have largely paralleled changes in the United States" (Smith 2011). In 1996, South Africa became the first country in the world to include in its constitution a clause banning discrimination based on sexual orientation (Karimi 2011).

Following South Africa's lead, other countries began amending their constitutions. At least 20 countries—from Ecuador to Greece—also have national constitutions banning discrimination based on sexual orientation. Some countries have laws prohibiting discrimination in some areas of their country, but not in others. Finally, many countries have laws prohibiting discrimination based on sexual orientation specific to the workplace, or regarding access to social services (Bruce-Jones & Itaborahy 2011).

> A global perspective on laws and social attitudes regarding homosexuality reveals that countries vary tremendously in their treatment of same-sex sexual behavior—from intolerance and criminalization to acceptance and legal protection.

In 2009, the United States joined the majority of the United Nations member states when President Barack Obama supported the decriminalization of homosexuality and the international expansion of human rights protections for those who are non-heterosexual or gender non-conforming. The resolution includes language prohibiting harassment, discrimination, exclusion, stigmatization, and prejudice against members of the LGBT population. Eighty-five out of the 192 United Nations members supported the resolution, which passed with 23 votes in favor and 19 opposed by the United Nations Human Rights Council (Dougherty 2011).

Despite these general international trends, there have been recent setbacks in some countries. In 2013, a decision by Singapore's High Court upheld the criminalization of consensual gay sex after a gay couple sought to have the law declared unconstitutional (Oi 2013). Further, although Russia decriminalized homosexuality in 1993, it recently passed legislation that prohibits the "propaganda of nontraditional sexual relations" and bans gay couples from adopting children born in Russia. What constitutes gay propaganda, however, is open to interpretation and could be so broadly defined as to include public displays of affection between same-sex couples. There are also fears that future legislation may require the removal of children from same-sex parent households (Guillory 2013).

The growing legal recognition of same-sex relationships provides evidence of the changing status of homosexuality throughout the world, yet it remains a complicated issue that has divided jurisdictions in many countries. Some countries recognize same-sex **registered partnerships** or "civil unions," which are federally recognized relationships that convey most but not all the rights of marriage. Some countries also offer registered partnerships or civil unions to opposite-sex couples. Registered partnerships are also referred to as "domestic partnerships." Legally recognized registered partnerships for same-sex couples are available in 25 countries in Europe, the Americas, and in specific jurisdictions of Mexico, Australia, and Venezuela (Bruce-Jones & Itaborahy 2011).

In 2000, the Netherlands became the first country in the world to offer full legal marriage to same-sex couples. In 2003, Belgium became the second country to legalize same-sex marriages. Since then, 14 more countries have legalized same-sex marriage: Canada (2005), Spain (2005), South Africa (2006), Norway (2009), Sweden (2009), Portugal (2010), Iceland (2010), Argentina (2010), Denmark (2012), Uruguay (2013), England/Wales (2013), New Zealand (2013), France (2013), and Brazil (2013). In some countries, such as the United States (2003) and Mexico (2009), for example, same-sex marriage is legal only in certain jurisdictions (The Pew Forum 2013a).

Finally, in 2013, the 27 foreign ministers of the European Union ". . . instructed their diplomats around the world to defend human rights of lesbian, gay, bisexual, and transgender and intersex . . . people . . ." around the world (Zweynert 2013, p. 1). This "groundbreaking" move is guided by four priorities:

- elimination of laws and policies that discriminate on the basis of LGBTI status, including the elimination of the death penalty;
- promotion of "equality and nondiscrimination at work, in health care, and in education";
- termination of government and individual violence against LGBTI individuals; and
- protection of and support for those who defend human rights (Zweynert 2013, p. 1).

Homosexuality and Bisexuality in the United States: A Demographic Overview

Before looking at demographic data concerning homosexuality and bisexuality in the United States, it is important to understand the ways in which identifying or classifying individuals as homo-, hetero-, and bisexual is problematic.

Sexual Orientation: Problems Associated with Identification and Classification

The classification of individuals into sexual orientation categories (e.g., gay, bisexual, lesbian, or heterosexual) is problematic for a number of reasons. First, distinctions among sexual orientation categories are simply not as clear-cut as many people would believe. Consider the early research on sexual behavior by Alfred Kinsey and his colleagues (1948, 1953). Although 37 percent of men and 13 percent of women reported at least one homosexual encounter since adolescence, few of the individuals reported exclusive same-sex sexual behavior. These data led Kinsey to conclude that heterosexuality and homosexuality represent two ends of the continuum and that most individuals fall somewhere along this theoretical line.

registered partnerships Federally recognized relationships that convey most but not all the rights of marriage.

More recent research has also indicated that many individuals are not exclusively heterosexual or homosexual. In a study of 243 undergraduates, Vrangalova and Savin-Williams (2010) found that 84 percent of self-identified heterosexual women and 51 percent of self-identified heterosexual men reported at least one homosexual quality—same-sex sexual attraction, fantasy, and/or behavior. Newer research (Savin-Williams & Vrangalova 2013; Vrangalova & Savin-Williams 2012) also supports a sexual orientation continuum and suggests a five-category classification of sexual identity: heterosexual, mostly heterosexual, bisexual, mostly gay/lesbian, and gay/lesbian.

Research using, as traditionally is the case, a three-category classification scheme—heterosexual, bisexual, or gay/lesbian—rather than a nuanced identity system is theoretically and methodologically problematic (Morgan & Thompson 2011; Morgan et al. 2010). For example, a woman who selects "lesbian" as her sexual orientation label when only presented with three options is presumed to identify as lesbian, and therefore only feels attraction to and fantasizes about women, and only has or desires sexual and romantic relationships with women throughout her lifetime. Research has also shown that many people with same-sex attractions or who have engaged in same-sex sexual behavior do not identify or think of themselves as gay, lesbian, or bisexual (Herek & Garnets 2007; Rothblum 2000; Savin-Williams 2006).

The second factor that makes classification difficult is that research with non-heterosexual populations has tended to define sexual orientation based on one of the three components: sexual/romantic attraction or arousal, sexual behavior, and sexual identity (Savin-Williams 2006). However, sexual orientation involves *multiple dimensions*—sexual attraction, sexual behavior, sexual fantasies, emotional attraction, self-identification, and the interrelations between these dimensions (Herek & Garnets 2007).

"*You know, statistically speaking, at least one of these gingerbread men is gay.*"

Cartoonbank.com

Research . . . supports a sexual orientation continuum and suggests a five-category classification of sexual identity: heterosexual, mostly heterosexual, bisexual, mostly gay/lesbian, and gay/lesbian.

Lastly, a third factor complicating the identification and classification of same-sex populations is the social stigma associated with non-heterosexual identities. As a result of the stigma associated with being gay, lesbian, or bisexual, many individuals conceal or lie about their sexual orientation to protect themselves from prejudice and discrimination (Meyer 2003; Röndahl & Innala 2008; Varjas et al. 2008). This complicates the recruitment of participants for research studies on sexual orientation because some people are uncomfortable "outing" themselves as gay, lesbian, or bisexual.

LGBT Individuals and Same-Sex–Couple Households in the United States

Reliable estimates of the percentage of the U.S. population that is gay, lesbian, or bisexual (LGB) are scarce. One study estimates that there are more than 8 million LGB adults in the United States, comprising 3.5 percent of the adult population. In addition, there are approximately 700,000 transgender individuals in the United States. This means that an estimated 9 million Americans identify as LGBT, a number nearly comparable to the total population of New Jersey (Gates 2011).

Self-reported LGBT identity has been found to vary across racial lines, with nonwhites more likely to identify as LGBT—for example, 4.6 percent of African Americans, 4.0 percent of Hispanics, and 4.3 percent of Asian Americans identify as LGBT compared to 3.2 percent of white Americans (Gates 2012). Further, women when compared to men are significantly more likely to identify as bisexual, and estimates of individuals who identify as LGB are significantly lower than estimates of those who have experienced same-sex attraction or have engaged in homosexual behavior (Gates 2011).

A study of 22 public middle schools found that 3.8 percent of students identified as gay, lesbian, or bisexual, 1.3 percent as transgender, and 12.1 percent reported being "not sure" of their sexual orientation (Shields et al. 2012). Table 11.1 displays the sexual orientation distribution among a sample of college students. An estimated 1 million to 3 million American adults over the age of 65 are gay, lesbian, bisexual, or transgender—a number that is expected to double over the next quarter century as baby boomers age (Cahill 2007).

Data on the prevalence of Americans who identify as LGBT are important because they can influence laws and policies that affect lesbian, gay, bisexual, and transgender individuals and their families. The 2010 Census did not ask about sexual orientation or gender identity. However, the 2010 Census became the first census to report the number of same-sex partners and the number of same-sex spouses. LGBT people living with a spouse or partner could identify their relationship by checking either the "husband or wife" or "unmarried partner" box. Based on this data, there are an estimated 581,300 same-sex–couple households in the United States, 28 percent of which are in legally recognized relationships, including marriages or the state-level equivalent (U.S. Census Bureau 2010).

TABLE 11.1	Self-Described Sexual and Gender Identity of U.S. College Students*
Heterosexual	91.6 percent
Gay/lesbian	2.6 percent
Bisexual	3.8 percent
Unsure	2.0 percent
Transgender	0.2 percent

Source: American College Health Association 2012.
*Based on a sample of 28,237 students at 51 campuses.

The Origins of Sexual Orientation

One of the most common questions regarding sexual orientation centers on its origin or "cause." Questions about the causes of sexual orientation are typically concerned with the origins of homosexuality and bisexuality because heterosexuality is considered normative and "natural." A wealth of biomedical and social science research has investigated the possible genetic, hormonal, developmental, social, and cultural influences on sexual orientation, yet no conclusive findings have suggested that sexual orientation is determined by any particular factor or interplay of factors. In the absence of compelling findings, many practitioners and professionals believe that sexual orientation might be determined by the interplay of environmental and biological factors.

Aside from what "causes" homosexuality, sociologists are interested in what people *believe* about the "causes" of homosexuality. Research shows that people rely heavily on ideology, religion, and life experiences to form their beliefs on the causes of homosexuality (Haider-Markel & Joslyn 2008). Recent research (see Figure 11.1) indicates that nearly half (47 percent) of Americans believe that one's sexual orientation is a consequence of birth (i.e., nature), whereas 33 percent believe it is a matter of external variables such as socialization and the environment (i.e., nurture) (Jones 2013).

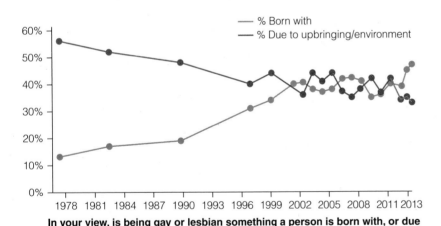

— % Born with
— % Due to upbringing/environment

In your view, is being gay or lesbian something a person is born with, or due to factors such as upbringing and environment?

Figure 11.1 **Public Opinion on the Causes of Homosexuality, 1978–2011**
Source: Jones 2013.

Americans' views on what causes non-heterosexual orientations are associated with their support for same-sex rights, that is, believing that gays and lesbians are born that way is associated with support for gay rights while believing non-heterosexuality is a product of the environment is associated with reduced support for gay rights (Jones 2011). Importantly, the data also indicate that

> Americans' beliefs about the origins of same-sex orientation are much more strongly related to their views of the legality and morality of gay or lesbian relations than to party identification, ideology, religious commitment, age, and other demographic characteristics, taking all those factors into account simultaneously. (Jones 2011)

Can Gays, Lesbians, and Bisexuals Change Their Sexual Orientation?

People who believe that gays, lesbians, and bisexuals choose to be attracted to the same sex tend to believe that LGB people can and should change their sexual orientation (Haider-Markel & Joslyn 2008). Various forms of *reparative therapy*, *conversion therapy*, and *reorientation therapy* are dedicated to changing the sexual orientation of individuals who are non-heterosexual. The American Psychological Association collectively refers to these "treatments" as **sexual orientation change efforts (SOCE)** (APA 2009).

A wealth of research has investigated the possible genetic, hormonal, developmental, social, and cultural influences on sexual orientation, yet no conclusive findings have suggested that sexual orientation is determined by any particular factor or interplay of factors.

Many conversion and reparative therapy programs view homosexuality as inherently immoral and/or pathological and allegedly achieve "conversion" to heterosexuality through embracing evangelical Christianity and being "born again" (Cianciotto & Cahill 2007). Some "treatments" have gone to such unethical extremes as shaming and physical punishment. In other cases, adolescents are "kidnapped" by their parents and sent to "conversion camps" or group homes.

When active, Exodus International, a leader in the religious ex-gay movement, claimed over 500 ministries around the world (Exodus International 2011). Exodus International came under scrutiny after some of its founding members came forth stating that, in fact, they were not "cured" (Besen 2010). In 2013, Exodus president Alan Chambers issued a formal apology in which he stated: "I am sorry for the pain and hurt that many of you have experienced. I am sorry some of you spent years working through the shame and guilt when your attractions didn't change. I am sorry we promoted sexual orientation change efforts and reparative theories about sexual orientation that stigmatized parents" (Exodus International 2013, p. 2).

The National Association for Research and Therapy of Homosexuality (NARTH), a secular organization, is comprised of mental health practitioners who use their training in psychology and related fields to assist clients in ridding themselves of "unwanted" same-sex attractions, behaviors, and identifications (NARTH 2011). Members of NARTH deny that they are "antigay" and assert that they provide non-religious psychosocial services (called *reorientation therapy*) for people who are distressed by their sexual orientation. NARTH claims to have the same success rate for reorientation therapies as for any other problem presented in therapy (Hamilton 2011).

. . . [A]ll agree that homosexuality is *not* a mental disorder, that sexual orientation *cannot* be changed, and that efforts to change sexual orientation may be harmful.

Critics of SOCE include the American Medical Association, American Academy of Pediatrics, American Counseling Association, American Psychological Association, American School Health Association, American Psychiatric Association, National Association of Social Workers, National Education Association, and the American Association of School Administrators, who all agree that homosexuality is *not* a mental disorder, that sexual orientation *cannot* be changed, and that efforts to change sexual orientation may be harmful (Lambda Legal 2011a). In 2013, the United Nations sponsored a panel of experts that directly addressed the issues surrounding SOCE, the first of what organizers hope is a continuing effort to deal with LBGT issues (Shapiro 2013).

sexual orientation change efforts (SOCE) Collectively refers to reparative, conversion, and reorientation therapies, according to the APA.

Sociological Theories of Sexual Orientation Inequality

Sociological theories do not explain the origin or "cause" of sexual orientation diversity; rather, they help to explain societal reactions to homosexuality and bisexuality and ways in which sexual identities are socially constructed.

Structural-Functionalist Perspective

Structural functionalists, consistent with their emphasis on institutions and the functions they fulfill, emphasize the importance of monogamous heterosexual relationships for the reproduction, nurturance, and socialization of children. From a structural-functionalist perspective, homosexual relations, as well as heterosexual nonmarital relations, are "deviant" because they do not fulfill the main function of the family institution—producing and rearing children. Clearly, however, this argument is less salient in a society in which (1) other institutions, most notably schools, have supplemented the traditional functions of the family, (2) reducing (rather than increasing) population is a societal goal, and (3) same-sex couples can and do raise children.

To protest antigay initiatives in several states, activists declared a "Day without a Gay." Gay men and women were encouraged to "call in gay" and stay home from work to emphasize the economic impact of the gay community.

Some structural functionalists argue that antagonisms between individuals who are heterosexual and homosexual disrupt the natural state, or equilibrium, of society. Durkheim (1993 [1938]), however, recognized that deviation from society's norms can also be functional. Specifically, the gay rights movement has motivated many people to reexamine their treatment of sexual orientation minorities and has produced a sense of cohesion and solidarity in the gay population. Gay activism has also been instrumental in advocating HIV/AIDS prevention strategies and health services that benefit the society as a whole.

The structural-functionalist perspective is concerned with how changes in one part of society affect other aspects. For example, research has shown that the worldwide increase in legal and social support of same-sex equality has been influenced by three cultural changes: the rise of individualism, increasing gender equality, and the emergence of a global society in which nations are influenced by international pressures (Frank & McEneaney 1999).

According to Frank and McEneaney (1999), individualism "appears to loosen the tie between sex and procreation, allowing more personal modes of sexual expression" (p. 930). They add:

> Whereas once sex was approved strictly for the purpose of family reproduction, sex increasingly serves to pleasure individualized men and women in society. This shift has involved the casting off of many traditional regulations on sexual behavior, including prohibitions of male-male and female-female sex. (p. 936)

Gender equality involves the breakdown of sharply differentiated sex roles, thereby supporting the varied expressions of male and female sexuality (see Chapter 10). Lastly, globalization permits the international community to influence individual nations. For example, international leaders have expressed serious concerns over Uganda's Anti-Homosexuality Bill that was first proposed in 2009. Uganda lawmakers recently removed the death penalty clause from the bill, but the sections that criminalize homosexuality and people who fail to report a known homosexual remain (NPR 2013).

The structural-functionalist perspective is also concerned with latent functions, or unintended consequences. An unintended consequence of the gay rights movement is increased opposition to gay rights. For example, after California granted same-sex couples the right to be legally married, a fierce campaign was launched to pass Proposition 8—a voter referendum that, in 2008, amended the state constitution to prohibit same-sex marriage. Following the elections, demonstrations and protests over the passage of Proposition 8 spread across California and throughout the United States. Numerous lawsuits were filed with the California Supreme Court challenging the proposition's validity and effect on previously administered same-sex marriages.

In *Strauss v. Horton,* the court upheld Proposition 8, but allowed existing same-sex marriages to stand. On August 4, 2010, in the case of *Perry et al. v. Schwarzenegger,* Judge Vaughn R. Walker overturned Proposition 8, and a federal judge later upheld Vaughn's ruling after questions were raised about his impartiality (Mears 2011). The ruling, however, was stayed pending appeal by proponents of Proposition 8. On June 26, 2013, the Supreme Court heard *Hollingsworth V. Perry* and ruled that supporters of Proposition 8 did not have the legal standing to defend the law in federal court. As a result, the Supreme Court dismissed the appeal and directed the Ninth Circuit Court of Appeals to lift the stay, enabling same-sex marriages to resume in California (Savage 2013).

Conflict Perspective

The conflict perspective frames the gay rights movement and the opposition to it as a struggle over power, prestige, and economic resources. Sexual minorities want to be recognized as full citizens deserving of all the legal rights and protections entitled to individuals who are heterosexual. A major achievement in gaining acceptance of gay and lesbian individuals occurred in 1973, when the American Psychiatric Association (APA) removed homosexuality from its official list of mental disorders (Bayer 1987).

More recently, gay and lesbian individuals have been waging a political battle to win civil rights protections against employment discrimination and to be allowed to marry a same-sex partner. Conflict theory helps to explain why many business owners and corporate leaders support nondiscrimination policies. Gay-friendly work policies help employers maintain a competitive edge in recruiting and maintaining a talented and productive workforce. Companies are also competing for gay and lesbian dollars. Many LGBT as well as heterosexual consumers prefer to purchase products and services from businesses that provide workplace protections for LGBT employees. Recent trends toward increased social acceptance of homosexuality may, in part, reflect the corporate world's competition for the gay and lesbian consumer dollar.

In support of marriage equality, Levi Strauss and Co., with the help of a LGBT activist group, testified before the Supreme Court of California that, "Ending marriage discrimination will improve businesses' ability to attract the best and the brightest to California and enhance California's reputation as a diverse, inclusive, and innovative community, both of which are key factors to continued economic growth and prosperity in this state . . ." (California Supreme Court 2011). After the U.S. Supreme Court DOMA decision in 2013, companies like Apple, Google, GAP, Johnson and Johnson, Nike, Inc., Ben and Jerry's, AT&T, and Marriott Hotels sent tweets or posted statements on their homepages expressing support for LGBT rights and marriage equality (Huffington Post 2013).

Symbolic Interactionist Perspective

Symbolic interactionism focuses on the meanings of heterosexuality, homosexuality, and bisexuality; how these meanings are socially constructed; and how they influence the social status, self-concepts, and well-being of non-heterosexual individuals. The meanings we associate with same-sex relations are learned from society—from family, peers, religion, and the media. The negative meanings associated with homosexuality are reflected in the current slang use of the phrase "That's so gay" or "You're so gay," which is meant to convey that something or someone is stupid or worthless (Burn et al. 2005; GLSEN 2011a; Kosciw & Diaz 2005; Sue 2010). Sociological research has shown that the

Sociological research has shown that the use of such words as *fag* and *queer* by heterosexuals to insult one another facilitates social acceptance by peers, particularly among heterosexual males.

use of such words as *fag* and *queer* by heterosexuals to insult one another facilitates social acceptance by peers, particularly among heterosexual males (Burn 2000; Kimmel 2011). However, this behavior comes at an expense to the well-being of non-heterosexuals (Burn et al. 2005; GLSEN 2011a; Meyer 2003; Sue 2010).

The symbolic interactionist perspective also points to the effects of labeling on individuals. Language is power. Once individuals become identified or labeled as lesbian, gay, or bisexual, that label tends to become their **master status**. In other words, the dominant heterosexual community tends to view "gay," "lesbian," and "bisexual" as the most socially significant statuses of individuals who are identified as such. Esterberg (1997) notes that, "unlike heterosexuals, who are defined by their family structures, communities, occupations, or other aspects of their lives, lesbians, gay men, and bisexuals are often defined primarily by what they do in bed" (p. 377).

Symbolic interactionism draws attention to how social interaction affects our self-concept, behavior, and well-being. When gay and lesbian individuals interact with people who express antigay attitudes, they may develop what is known as **internalized homophobia** (or **internalized heterosexism**)—the internalization of negative messages about homosexuality by lesbian, gay, and bisexual individuals as a result of direct or indirect social rejection and stigmatization. Internalized homophobia has been linked to increased risk for depression, substance abuse, anxiety, and suicidal thoughts (Bobbe 2002; Gilman et al. 2001; Meyer 2003; Newcomb & Mustanski 2010; Szymanski et al. 2008).

Family members also have a powerful effect on the self-concepts, behavior, and well-being of lesbian, gay, and bisexual youth. For example, a study of young lesbian, gay, and bisexual adults found that higher rates of family rejection during adolescence were significantly associated with poorer health outcomes. Further, young LGBT adults who reported higher levels of family rejection were more likely to have attempted suicide, have high levels of depression, use illegal drugs, and engage in unprotected sex compared to LGBT peers from families that reported no or low levels of family rejection (Ryan et al. 2009). Research also indicates that a positive reaction from parents to a youth's coming out is a protective factor that buffers against stigma-related stress and reduces risk for illegal drug use (Padilla et al. 2010). Although newer research is pointing to the resilience of LGBT individuals in the face of sexual and gender identity prejudice (Stanley 2009), research also indicates that the effects of environmental stress combined with internal stress processes, such as internalized homophobia, make a person more vulnerable to developing a mental health disorder (Meyer 2003).

Prejudice Against Lesbians, Gays, and Bisexuals

Oppression refers to the use of power to create inequality and limit access to resources, which impedes the physical and/or emotional well-being of individuals or groups of people. A person or group is **privileged** when they have a special advantage or benefits as a result of cultural, economic, societal, legal, and political factors (Guadalupe & Lum 2005). **Heterosexism** is a form of oppression and refers to a belief system that gives power and privilege to heterosexuals, while depriving, oppressing, stigmatizing, and devaluing people who are not heterosexual (Herek 2004; Szymanski et al. 2008).

The belief that heterosexuality is superior to non-heterosexuality results in prejudice and discrimination against gays, lesbians, and bisexuals. **Prejudice** refers to negative attitudes, whereas **discrimination** refers to behavior that denies individuals or groups equality of treatment (see Chapter 9). In turn, this often leads non-heterosexual individuals to question the legitimacy of their same-sex attractions (Harper et al. 2004). Before reading further, you may wish to complete this chapter's *Self and Society* feature, which assesses your knowledge of and attitudes toward lesbians, gays, and bisexual individuals.

master status The status that is considered the most significant in a person's social identity.

internalized homophobia (or **internalized heterosexism**) The internalization of negative messages about homosexuality by lesbian, gay, and bisexual individuals as a result of direct or indirect social rejection and stigmatization.

oppression The use of power to create inequality and limit access to resources, which impedes the physical and/or emotional well-being of individuals or groups of people.

privilege When a group has a special advantage or benefits as a result of cultural, economic, societal, legal, and political factors.

heterosexism A form of oppression that refers to a belief system that gives power and privilege to heterosexuals, while depriving, oppressing, stigmatizing, and devaluing people who are not heterosexual.

prejudice Negative attitudes and feelings toward or about an entire category of people.

discrimination Actions or practices that result in differential treatment of categories of individuals.

Please respond to the items in the following scale. For each item, write the number that indicates the extent to which each statement is characteristic or uncharacteristic of you or your views. Please try to respond to every item.

1	2	3	4	5	6

VERY UNCHARACTER-ISTIC OF ME OR MY VIEWS

VERY CHARACTER-ISTIC OF ME OR MY VIEWS

Please consider the ENTIRE statement when making your rating, as some statements contain two parts.

1. I feel qualified to educate others about how to be affirmative regarding LGB issues. ___

2. I have conflicting attitudes or beliefs about LGB people. ___

3. I can accept LGB people even though I condemn their behavior. ___

4. It is important to me to avoid LGB individuals. ___

5. I could educate others about the history and symbolism behind the "pink triangle." ___

6. I have close friends who are LGB. ___

7. I have difficulty reconciling my religious views with my interest in being accepting of LGB people. ___

8. I would be unsure what to do or say if I met someone who is openly lesbian, gay, or bisexual. ___

9. Hearing about a hate crime against a LGB person would not bother me. ___

10. I am knowledgeable about the significance of the Stonewall riots to the gay liberation movement. ___

11. I think marriage should be legal for same-sex couples. ___

12. I keep my religious views to myself in order to accept LGB people. ___

13. I conceal my negative views toward LGB people when I am with someone who doesn't share my views. ___

14. I sometimes think about being violent toward LGB people. ___

15. Feeling attracted to another person of the same sex would not make me uncomfortable. ___

16. I am familiar with the work of the National Gay and Lesbian Task Force. ___

17. I would display a symbol of gay pride (pink triangle, rainbow, etc.) to show my support of the LGB community. ___

18. I would feel self-conscious greeting a known LGB person in a public place. ___

19. I have had sexual fantasies about members of my same sex. ___

20. I am knowledgeable about the history and mission of the PFLAG organization. ___

21. I would attend a demonstration to promote LGB civil rights. ___

22. I try not to let my negative beliefs about LGB people harm my relationships with lesbian, gay, or bisexual individuals. ___

23. Hospitals should acknowledge same-sex partners equally to any other next of kin. ___

24. LGB people deserve the hatred they receive. ___

25. It is important to teach children positive attitudes toward LGB people. ___

26. I conceal my positive attitudes toward LGB people when I am with someone who is homophobic. ___

27. Health benefits should be available equally to same-sex partners as to any other couple. ___

28. It is wrong for courts to make child custody decisions based on a parent's sexual orientation. ___

Scoring

HATE = 4, 24, 8, 14, 9, 18
KNOWLEDGE = 20, 10, 16, 5, 1
CIVIL RIGHTS = 27, 23, 11, 28, 25
RELIGIOUS CONFLICT = 26, 12, 22, 7, 3, 13, 2
INTERNALIZED AFFIRMATIVENESS = 19, 15, 17, 6, 21

To calculate your subscale scores, add up all the values of your responses on the items in a subscale (e.g., HATE) and divide by the number of items summed. A higher score on each subscale is associated with a *stronger belief* on that subscale. For example, a high score on the Hate subscale would indicate a high level of hate. A high score on the Internalized Affirmativeness subscale would indicate a high level of internal affirmativeness.

NOTE: LGB = Lesbian, Gay, or Bisexual

Source: Worthington, Roger L., Frank R. Dillon, and Ann M. Becker-Shutte. "Development, Reliability, and Validity of the Lesbian, Gay, and Bisexual Knowledge and Attitudes Scale for Heterosexuals (LGB-KASH)." *Journal of Counseling Psychology* 52:104–118. Copyright © 2005 by the American Psychological Association. Reprinted by permission.

Homophobia and Biphobia

The term **homophobia** is commonly used to refer to negative or hostile attitudes directed toward same-sex sexual behavior, a non-heterosexually identified individual, and communities of non-heterosexuals. Homophobia is made possible by heterosexism. Homophobia is not necessarily a clinical phobia, i.e., one involving a compelling desire to avoid the feared object despite recognizing that the fear is unreasonable (e.g., fear of flying). Rather, research indicates that disgust and anger, rather than fear and anxiety, are more involved in heterosexuals' negative attitudes and emotional responses to LGB individuals (Herek 2004). When prejudice is directed toward bisexual individuals, this is called **biphobia** (Rust 2002). Both heterosexuals and

homophobia Negative or hostile attitudes directed toward non-heterosexual sexual behavior, a non-heterosexually identified individual, and communities of non-heterosexuals.

biphobia When prejudice is directed toward bisexual individuals.

The term *homophobia* is commonly used to refer to negative or hostile attitudes directed toward same-sex sexual behavior, a non-heterosexually identified individual, and communities of non-heterosexuals.

non-heterosexuals often reject bisexuals; thus, bisexual men and women can experience "double discrimination." Other terms that refer to negative attitudes, beliefs, and emotions toward non-heterosexuals include *homonegativity, bi-negativity, anti-gay bias,* and *sexual prejudice* (Eliason 2001; Herek 2000a; Szymanski et al. 2008).

The decriminalization of homosexuality in the United States is one indicator of the growing acceptance LGB individuals. Further, a recent survey indicates that 59 percent of Americans believe that gay and lesbian relations are morally acceptable (Newport & Himelfarb 2013). Another measure of acceptance is the election of non-heterosexual officials (Page 2011). All 50 states have been served by openly LGB elected politicians in some capacity, and at least 41 states have elected openly LGB politicians to one or both houses of their state legislatures (Gay and Lesbian Victory Fund 2013).

Attitudes toward homosexuality and bisexuality, although narrowing, vary by gender (Pew Research Center 2013a). In general, men are more likely than women to have negative attitudes toward gay individuals. Although limited in number and scope, research that assesses attitudes toward gay men versus gay women has found that heterosexual women and men hold similar attitudes toward lesbians, but that men are more negative toward gay men (Falomir-Pichastor & Mugny 2009; Herek 2000b, 2002). Generational differences in attitudes toward same-sex marriage have also been observed. As shown in Figure 11.2, younger cohorts of Americans are more likely to have favorable attitudes toward same-sex marriage than older cohorts (Pew Research Center 2013b).

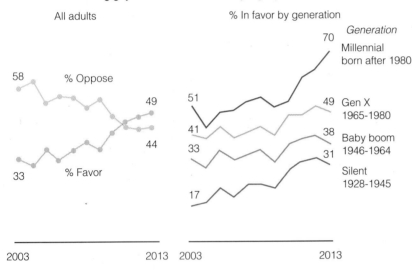

Figure 11.2 Support for Same-Sex Marriage by Age Cohorts, 2003–2013
Source: Pew Research Center 2013a.

Cultural Origins of Anti-LGB Bias

Nearly two-thirds of Americans agree that discrimination against gays and lesbians is a "very" or "somewhat serious" problem in the United States. This is particularly true among LGBT-identified Americans, 88 percent of whom agree that discrimination against gays and lesbians is a "very" or "somewhat serious" problem in the United States (Jones 2012). These findings suggest that not only are attitudes shifting toward a more positive view of LGBT individuals, but there is also still much headway to be made in the fight for equality and non-discrimination. Anti-LGB bias has its roots in various aspects of American culture including religion, rigid gender roles, and myths and negative stereotypes about non-heterosexuals.

Religion. Organized religion has been both a source of comfort and distress for many lesbian, gay, and bisexual Americans. Countless LGB individuals have been forced to leave their faith communities due to the condemnation embedded in doctrine and practice. Research has found that higher levels of religiosity (APA 2009; Brown & Henriquez 2008; Shackelford & Besser 2007), religious fundamentalism (Summers 2010), and more conservative political beliefs (APA 2009; Brown & Henriquez 2008; Pew Research Center 2013b; Shackelford & Besser 2007) are consistently associated with negative attitudes toward homosexuality.

At the same time, there exists variability within the Judeo-Christian faiths regarding attitudes toward LGB individuals. As early as the 1970s, the United Church of Christ became the first major Christian church to ordain an openly gay minister, and in 2005, became the largest Christian denomination to endorse same-sex marriages (Fone 2000). More recently, an increasing number of organized religious groups have issued official statements welcoming LGB members and are supportive of such LGB issues as freedom from discrimination, the affirmation of same-sex marriages, and the ordination of openly gay clergy. Reformed Judaism ordains openly lesbian, gay, and bisexual individuals as rabbis (HRC 2011a).

Other Christian denominations, with various levels of inclusiveness, include Lutherans, Episcopalians, Methodists, and Presbyterians (Fone 2000; Goodstein 2010; HRC 2011a). The Association of Welcoming and Affirming Baptists is a network of more than 89 churches and organizations that advocate for the inclusion of LGB individuals within the Baptist community of faith (AWAB 2013). With some exceptions, the majority of religious groups in the United States have been silent with respect to transgender individuals.

Rigid Gender Roles. Disapproval of homosexuality also stems from rigid gender roles. Kimmel (2011), a sociologist who specializes in research on masculinity, writes about the relationship between perceived masculinity and bullying:

> Why are some students targeted? Because they're gay or even "seem" gay—which may be just as disastrous for a teenage boy. After all, the most common put-down in American high schools today is "that's so gay," or calling someone a "fag."
> It refers to anything and everything: what kind of sneakers you have on, what you're eating for lunch, some comment you made in class, who your friends are or what sports team you like. . . . Calling someone gay or a fag has become so universal that it's become synonymous with dumb, stupid, or wrong. (p. 10)

From a conflict perspective, heterosexual men's subordination and devaluation of gay men reinforces gender inequality. "By devaluing gay men . . . heterosexual men devalue the feminine and anything associated with it" (Price & Dalecki 1998, pp. 155–156). Negative views toward lesbians also reinforce the patriarchal system of male dominance. Social disapproval of lesbians is a form of punishment for women who relinquish traditional female sexual and economic dependence on men. Not surprisingly, research findings suggest that individuals with traditional gender role attitudes tend to hold more negative views toward homosexuality (Brown & Henriquez 2008; Louderback & Whitley 1997).

Myths and Negative Stereotypes. The stigma associated with homosexuality and bisexuality can also stem from myths and negative stereotypes. One negative myth about non-heterosexuals is that they are sexually promiscuous and lack "family values," such as monogamy and commitment to a relationship. Although some gay men and lesbians do engage in casual sex, as do some heterosexuals, many same-sex couples develop and maintain long-term committed relationships. In a recent survey of a sample of LGBT respondents, 66 percent of gay women, 40 percent of gay men, 68 percent of bisexual women, and 40 percent of bisexual men reported that they were in a committed relationship (Pew Research Center 2013c).

Another myth is that non-heterosexuals, as a group, are a threat to children—most notably in regard to child molestation. In other words, people confuse homosexuality with pedophilia, which refers to the perpetration by an adult against a youth who has not yet reached puberty or has just achieved puberty. Having a non-heterosexual orientation is unrelated to pedophilia. Research has *not* demonstrated a connection between an adult's same-sex attraction and an increased likelihood of molesting children and teenagers. In fact, a pedophile does not have an adult sexual orientation because they are fixated on children and teenagers (Herek 2009).

What Do You Think? On May 23, 2013, the Boy Scouts of America decided to allow all openly gay youths to be members; however, openly gay adult Scout leaders continue to be banned from the organization. Do you agree or disagree with the Boy Scouts' decisions?

To the contrary, Bergman et al. (2010) found that gay men who became parents had heightened self-esteem as a result of becoming parents and raising children. Thus, the simple act of becoming a father had a very positive outcome on gay men's sense of self-worth. Research has also shown that gay men have been stigmatized by social service employees, such as social workers and managers, as being "perverts" as well as being viewed as presenting "problematic" models of **gender expression**, that is, being too feminine (Hicks 2006). Most often, child abusers are the father, stepfather, or a heterosexual relative or family friend (Herek 2009).

Discrimination Against Lesbians, Gays, and Bisexuals

In June 2003, a Supreme Court decision in *Lawrence v. Texas* invalidated state laws that criminalize sodomy—oral and anal sexual acts. This historic decision overruled a 1986 Supreme Court case (*Bowers v. Hardwick*), which upheld a Georgia sodomy law as constitutional. The 2003 ruling, which found that sodomy laws were discriminatory and unconstitutional, removed the legal stigma and criminal branding that sodomy laws have long placed on LGB individuals. Nonetheless, sodomy or "unnatural acts" are still illegal in 14 states, although are primarily used against gay men and lesbians (Liebelson 2013).

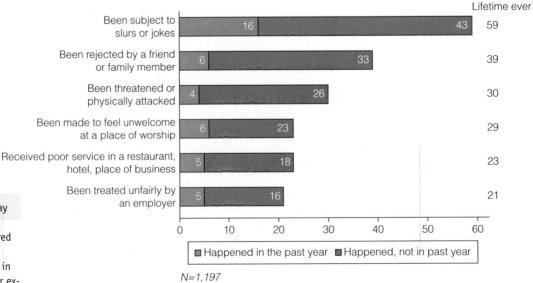

N=1,197

gender expression The way in which a person presents her- or himself as a gendered individual (i.e., masculine, feminine, or androgynous) in society. A person could, for example, have a gender identity as male but nonetheless present their gender as female.

Figure 11.3 **Percent of Respondents Who Experienced Discrimination and/or Exclusion Because of LGBT Status, 2013**
Source: Pew Research Center 2013c.

Like other minority groups in American society, gays, lesbians, and bisexual individuals experience various forms of discrimination (see Figure 11.3). Next, we look at discrimination in the workplace, in marriage and parenting, in the violent expressions of hate, and in treatment by police.

Workplace Discrimination and Harassment

Most U.S. adults, 77 percent, agree that non-heterosexuals should have health insurance and other employee benefits for their gay and lesbian domestic partners or spouses (Gallup Organization 2012). Yet, it is still legal in 29 states to fire, decline to hire or promote, or otherwise discriminate against employees because of their sexual orientation (HRC 2013f). A survey conducted in Utah found that 43 percent of the LGB respondents reported being ". . . fired, denied a job, denied a promotion, or experiencing other forms of discrimination in employment at some point in their lives" (Rosky et al. 2011, p. 8).

As discussed later in this chapter, many workplaces have non-discrimination policies that cover LGBT employees. But gay-affirmative policies do not ensure friendly attitudes and behaviors from coworkers. In a national probability survey representative of the United States, just one-third of LGBT employees were "out" to someone at work, and only 25 percent were "out" to everyone at work. Employees who are open about their LGB status are more likely to be harassed, lose their job, and, in general, be discriminated against when compared to employees who are not "out" at work (Sears & Mallory 2011).

Nearly half of LGBT employees report that they hear coworkers express negative views concerning LGBT issues "at least once in a while," and 61 percent report hearing coworkers tell jokes about LGBT people "at least once in a while." More than one in five respondents said they looked for a new job in the last year because of the uncomfortable working environment in their current job (Fidas 2009). A gay police officer who felt compelled to hide his sexual orientation from his coworkers describes his experience:

> Can you imagine going to work every day and avoiding any conversations about with whom you had a date . . . or a great weekend . . . or an argument—basically not sharing any part of your personal life for fear of reprisal or being ostracized. I did this in a career that prides itself on integrity, honesty, and professional-ism—and where a bond with one's colleagues and partner is critical in danger-ous and potentially deadly situations. (Carney 2007)

For many Americans, the workplace is the military. Since President Clinton signed a bill instituting a "don't ask, don't tell" (DADT) policy, more than 13,000 servicemen and women have been discharged from the military on the basis of their sexual orientation (Stone 2011), and countless others have decided not to enlist. DADT did not prohibit lesbians, gays, and bisexuals from serving in the military; it prohibited recruiters from asking enlistees their sexual orientation and prohibited LGB servicemen and women from revealing their sexual orientation.

In 2010, not long after a poll revealed that the majority of Americans supported the repeal of DADT (Morales 2010), the law was revoked (O'Keefe 2010). In July 2011, the President, Secretary of Defense, and chairman of the Joint Chiefs of Staff certified that the Department of Defense was ready to enact the change. Lesbian, gay, and bisexual troops began to serve openly 60 days later (Singer & Belkin 2012). Contrary to the fears of some, a report by U.S. military school professors concluded that the repeal of DADT had no overall negative impact on military readiness or its "component dimensions," such as cohesion, re-cruitment, retention, assaults, harassment, or morale (Belkin et al. 2012). In 2013, the Pentagon extended military benefits (e.g., health care, housing allow-ances, retirement benefits) to spouses of gay military personnel (Ackerman 2013; see Chapter 15).

. . . [A] report by U.S. military school professors concluded that the repeal of DADT had no overall negative impact on military readiness or its "component dimensions," such as cohesion, recruitment, retention, assaults, harassment, or morale.

Before the 2003 Massachusetts Supreme Court ruling in *Goodridge v. Department of Public Health,* no state had declared that same-sex couples have a constitutional right to be legally married. In response to growing efforts to secure legal recognition of same-sex couples, opponents of same-sex marriage have prompted antigay marriage legislation. For example, in 1996, Congress passed and President Clinton signed the **Defense of Marriage Act (DOMA)**, which (1) states that marriage is a "legal union between one man and one woman"; (2) denies federal recognition of same-sex marriage; and (3) allows states to either recognize or not recognize same-sex marriages performed in other states.

During June 2013, the Supreme Court issued two key rulings affecting same-sex marriage in the United States. In one ruling, the parts of DOMA that denied equal benefits and recognition to legally married same-sex spouses that married heterosexual spouses have were deemed unconstitutional. As a result of the ruling, at the federal level, couples married in a state that provide legal same-sex marriages can now receive over a thousand federal rights, benefits, and responsibilities of marriage heretofore reserved for heterosexual married couples.

For the 37 states that do not allow same-sex marriage, the effect of the ruling remains unclear. For example, what will the federal government do when a couple has a valid marriage license in Vermont and they later move to Texas? The Supreme Court justices also cleared the way for same-sex marriages to resume in California after rejecting an appeal on a ruling that struck down the state's Proposition 8 (Savage 2013).

Over 70,000 same-sex couples have been legally married in the United States, although that is likely an underestimate (DeSilver 2013). One survey found that over half of gay men and gay women say that they would like to get married. Contrary to stereotypes, gay men and gay women were similarly likely to respond affirmatively to the question, "If you could, would you like to get married someday?" (DeSilver 2013). As conflict theorists note, approval of same-sex marriages is economically beneficial to states that have marriage equality. For example, it is projected that if marriage rights where extended to same-sex couples in Minnesota, there would be an increase of over $40 million in the wedding and tourism business within the first three years (Kastanis & Badgett 2013).

Arguments against Same-Sex Marriage. Some opponents of same-sex marriage view homosexuality as sick, unnatural, or immoral. They argue that granting legal status to same-sex unions would convey social acceptance of homosexuality and would thus teach youth to view homosexuality as an acceptable lifestyle. Opponents are also concerned that if same-sex marriages are legalized, schools would be pressured to treat LGB individuals as any other minority group resulting in, for example, classes on gay history, gay literature, and the like.

Opponents of same-sex marriages also commonly argue that such marriages would subvert the stability and integrity of the heterosexual family. However, as Sullivan (1997) notes in a now-classic statement, lesbians, gays, and bisexuals are already part of heterosexual families:

> [Homosexuals] are sons and daughters, brothers and sisters, even mothers and fathers, of heterosexuals. The distinction between "families" and "homosexuals" is, to begin with, empirically false; and the stability of existing families is closely linked to how homosexuals are treated within them. (p. 147)

Many opponents of same-sex marriage base their opposition on religious grounds. When asked about the reasons against legal same-sex marriage, people are most likely to cite religious views as the basis of their position (Newport 2012b). In one poll, 76 percent of Americans who claim no religious affiliation favor legal same-sex marriage, compared to 60 percent support among Catholics, and 38 percent support among Protestants (Saad 2013). Research has demonstrated the importance of religion when predicting attitudes toward same-sex marriage and civil unions (see Figure 11.4), as well as the differences within religions (Olson et al. 2006; Whitehead 2010) (see this chapter's *The Human Side* feature). For example, for evangelical Protestants, the question of same-sex marriage elicits a strong, unfavorable response, perhaps because of the belief in the religious sanctity of marriage between a man and a woman (Whitehead 2010).

Defense of Marriage Act (DOMA) Federal legislation that states that marriage is a legal union between one man and one woman and denies federal recognition of same-sex marriage.

The following is from a blog written by Dannika Nash, a junior studying English and theology at the University of Sioux Falls in South Dakota. Dannika, a straight ally, was inspired to write the letter after attending a Mackelmore concert at which the singer performed his marriage equality anthem, "Same Love." Dannika's letter, which garnered nearly 4,000 comments since its release, has gained national attention. Although Dannika was widely praised, she experienced some backlash for posting the blog, including being fired from her summer job as a camp counselor.

Church,

. . . My point in writing this isn't to protect gay people. Things are changing—the world is becoming a safer place for my gay friends. They're going to get equal rights. I'm writing this because I'm worried about the safety of the Church. The Church keeps scratching its head, wondering why 70 percent of 23- to 30-year-olds who were brought up in church leave. I'm going to offer a pretty candid answer, and it's going to make some people upset, but I care about the Church too much to be quiet. We're scared of change. We always have been. When scientists proposed that the Earth could be moving through space, church bishops condemned the teaching. . . .

But the scientific theory continued, and the Church still exists. I'm saying this: We cannot keep pitting the church against humanity, or progress. DON'T hear me saying that we can't fight culture on anything. Lots of things in culture are absolutely contradictory to love and equality and we *should* be battling those things. The way culture treats women, or pornography? Get AT that, Church. I'll be right there with you. But my generation, the generation that can smell bullshit, especially holy bullshit, from a mile away, will not stick around to see the church fight gay marriage against our better judgment. It's my generation who is overwhelmingly supporting marriage equality, and Church, as a young person and as a theologian, it is not in your best interest to give them that ultimatum.

My whole life, I've been told again and again that Christianity is not conducive with homosexuality. It just doesn't work out. I was forced to choose between the love I had for my gay friends and so-called biblical authority. I chose gay people, and I'm willing to wager I'm not the only one. I said, "If the Bible *really* says this about gay people, I'm not too keen on trusting what it says about God." And I left my church. It has only been lately that I have seen evidence that the Bible could be saying something completely different about love and equality.

. . . Christians can be *all about* gay people, it's possible. People do it every day with a clear biblical conscience. Find out if you think there's truth in that view before you sweep us under the rug. You CAN have a conservative view on gay marriage, or gay ordination. You can. But I want you to have some serious conversations with God, your friends that disagree with you, and maybe even some gay people, Christians or not, before you decide that *this one view* is worth marginalizing my generation. Weigh those politics against what you're giving up: us. We want to stay in your churches . . . but it's hard to hear about love from a God who doesn't love our gay friends (and we all have gay friends). Help us find love in the church before we look for it outside.

Love,

A College Kid Who Misses You

Source: From Dannika Nash, "An Open Letter to the Church from My Generation," Posted on April 7, 2013.

What Do You Think? The U.S. Constitution and, specifically, the Second Amendment has been interpreted as creating a "wall" between the government on the one hand and organized religion on the other. If same-sex marriages are denied on the basis of religious beliefs, is the prohibition in violation of the Second Amendment? What do you think?

Arguments in Favor of Same-Sex Marriage. Advocates of same-sex marriage argue that banning same-sex marriages or refusing to recognize same-sex marriages granted in other states is a violation of civil rights that denies same-sex couples the countless legal and financial benefits that are granted to heterosexual married couples. DOMA prevented many of the over 1,100 federal rights, benefits, and responsibilities of marriage from being afforded to legally married same-sex couples. For example, married couples have the right to inherit from a spouse who dies without a will, to avoid inheritance taxes between spouses, to make crucial medical decisions for a partner, and to take family leave to care for a partner in the event of the partner's critical injury or illness (Badgett et al. 2011). A 2012 poll indicated that the majority of Americans (78 percent) believe that there should be inheritance rights for gay and lesbian domestic partners or spouses (Newport 2012a).

Spouses can receive Social Security survivor benefits and include their partner on health insurance coverage. The majority of Americans (77 percent) believe that health

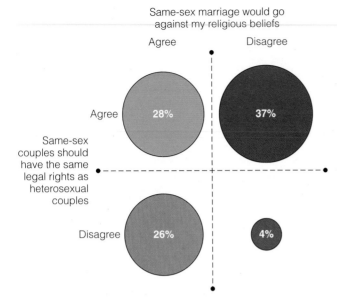

Same-sex marriage would go
against my religious beliefs

Agree   Disagree

Agree   28%   37%

Same-sex
couples should
have the same
legal rights as
heterosexual
couples

Disagree   26%   4%

Figure 11.4 Gay Marriage, Religious Beliefs, and Legal Rights for Gay Couples Source: Pew Research Center 2013a.

Ironically, the same pro-marriage groups that stress that children are better off in married-couple families disregard the benefits of same-sex marriage to children.

"Gays and lesbians aren't a threat to the sanctity of my marriage. It's all the straight women who sleep with my husband."

insurance and other employee benefits should be granted to gay and lesbian domestic partners or spouses (Newport 2012b). Other rights bestowed married partners include assumption of a spouse's pension, bereavement leave, burial determination, domestic violence protection, divorce protections, and immunity from testifying against a spouse. Finally, unlike other countries that recognize same-sex couples for immigration purposes, the United States does not recognize same-sex couples in granting immigration status (Caldwell 2011).

Another argument for same-sex marriage is that it would promote relationship stability among gay and lesbian couples. "To the extent that marriage provides status, institutional support, and legitimacy, gay and lesbian couples, if allowed to marry, would likely experience greater relationship stability" (Amato 2004, p. 963). Greater relationship stability benefits not only same-sex couples, but their children as well. Children in same-sex families would gain a range of securities and benefits, including the right to get health insurance coverage and Social Security survivor benefits from a nonbiological parent, and the right to continue living with a nonbiological parent should their biological mother or father die (Tobias & Cahill 2003). Ironically, the same pro-marriage groups that stress that children are better off in married-couple families disregard the benefits of same-sex marriage to children.

Finally, a cross-cultural and historical view of marriage and family suggests that marriage is a social construction that comes in many forms. In response to supporters of a constitutional amendment banning gay marriage as a threat to civilization, the American Anthropological Association (AAA 2004) released the following statement:

> The results of more than a century of anthropological research on households, kinship relationships, and families, across cultures and through time, provide no support whatsoever for the view that either civilization or viable social orders depend upon marriage as an exclusively heterosexual institution. Rather, anthropological research supports the conclusion that a vast array of family types, including families built upon same-sex partnerships, can contribute to stable and humane societies.

Children and Parental Rights

Several respected national organizations—including the American Academy of Family Physicians, American Academy of Pediatrics, American Bar Association, American Medical Association, American Psychological Association, Child Welfare League of America, National Adoption Center, and the National Association of Social Workers—have taken the position that a parent's sexual orientation has nothing to do with his or her ability to be a good parent (AMA 2011; HRC 2011b; U.S. Court of Appeals 2010).

Over 25 years of scholarly research on the children of non-heterosexual parents was thought to provide clear and consistent evidence that irrespective of whether children are from divorced lesbian and gay parents or born to lesbian or gay parents, children are at least as well adjusted as children from opposite-sex parents (Patterson 2009). However, in 2012, a large-scale research study found significant differences in, for example, the incomes of adult children raised in heterosexual households compared to non-heterosexual households (Regnerus 2012). As the author himself acknowledges in his response to the many critiques of his work (compare to Amato 2013), the results should be interpreted with caution. The data do not lend themselves to establishing causal relationships, that is, it would be incorrect to say that being raised in a non-heterosexual household lowers the income of adult children.

Spencer Grant/PhotoEdit

Laws concerning same-sex adoption vary considerably by state. Here, a multiracial same-sex couple eat dinner with their two adopted Hispanic sons in their kitchen at home in Long Beach, California.

What Do You Think? Studies have shown that not only are adolescents reared by lesbian parents just as psychologically healthy as their peers from opposite-sex parents, but also that stigmatization can have a negative impact on an adolescent's mental health (Bos et al. 2013). What do you think can be done to help mitigate this connection between being discriminated against for having same-sex parents and psychological adjustment issues such as, say, anxiety?

Despite the weight of supportive research, the legal contexts for lesbian and gay parents and their children vary significantly from one jurisdiction to another. In Massachusetts, the law recognizes same-sex marriages, joint adoptions, and second-parent adoptions, and a parent's sexual orientation is considered irrelevant when it concerns foster care, child custody, and visitation proceedings. Yet, in Mississippi, the law does not recognize same-sex marriages, and lesbian and gay parents are discriminated against in custody and visitation proceedings. Same-sex couples are prohibited from adopting in Mississippi, and this includes joint adoptions and second-parent adoptions (HRC 2013c, 2013d).

Violence, Hate, and Criminal Victimization

On October 6, 1998, Matthew Shepard, a 21-year-old student at the University of Wyoming, was abducted and brutally beaten. Two motorcyclists who initially thought he was a scarecrow found him tied to a wooden ranch fence. His skull had been smashed, and his head and face had been slashed. The only apparent reason for the attack: Matthew Shepard was gay. On October 12, Matthew died of his injuries. Media coverage of his brutal attack and subsequent death focused nationwide attention on hate crimes against non-heterosexuals. It can be argued that Matthew Shepard's death put a face on anti-LGBT hate crimes in a way that previously had not been done.

Anti-LGBT hate crimes are crimes against individuals or their property that are based on bias against the victim because of perceived sexual orientation or gender identity. Such crimes include verbal threats and intimidation, vandalism, sexual assault and rape, physical assault, and murder. According to the Federal Bureau of Investigation

(FBI), 20.8 percent of reported hate crimes in 2011 were motivated by sexual orientation involving 1,572 victims. Over half of the incidents were motivated by anti-gay male bias, 12.7 percent were motivated by anti-lesbian bias, and 3.0 percent were classified as anti-bisexual bias (FBI 2012). But, as discussed in Chapter 9, FBI hate crime statistics underestimate the incidence of hate crimes. The National Coalition of Anti-Violence Programs (NCAVP) documented 2,016 survivors and victims of violence due to perceived non-heterosexuality and gender non-conformity in 2012, including 25 murders. The report found that 73.1 percent of all anti-LGBT homicide victims were people of color; 53.8 percent of the victims were transgender women, a considerable increase from 2011.

Anti-LGB Hate and Harassment in Schools and on Campuses. On September 19, 2010, Seth Walsh of Tehachapi, California, was found unconscious following an attempt to hang himself. Seth was on life support until his death on September 28. He was 13 years old, openly gay, and tormented by years of unrelenting bullying based on his sexual orientation. Two days later, Tyler Clementi, an 18-year-old Rutgers University student, jumped from the George Washington Bridge to his death just days after learning that his roommate and another student broadcast a romantic encounter involving Tyler and another male on the Internet (McKinley 2010). These and other suicides by LGBT youth provoked a nationwide dialogue about bullying, sexual orientation, and gender expression.

Hostile school environments, characterized by anti-LGBT attitudes, remarks, and actions of students and teachers, have been documented to exist as early as elementary school. One study examined hostile school environments using a national sample of 1,065 elementary school students in 3rd through 6th grade and 1,099 elementary school teachers of kindergarten through 6th grade (GLSEN & Harris Interactive 2012). One quarter of the students and teachers reported hearing students make comments like "fag" or "lesbo" at least "sometimes." Twenty-three percent of students attributed bullying and name-calling of a boy to acting or looking "too much like a girl" and of a girl to acting or looking "too much like a boy."

A survey of over 8,500 LGBT middle and high school students documents the prevalence of anti-gay sentiment during adolescence. According to the National School Climate Survey (Kosciw et al. 2012), in the 2011 school year,

- 81.9 percent of LGBT students reported being verbally harassed at school, 38.3 percent reported being physically harassed at school, and 18.3 percent reported being physically assaulted at school because of their sexual orientation.
- 55.2 percent of LGBT students experienced electronic harassment (e.g., via text messages or Facebook posts) because of their sexual orientation.
- 71.3 percent of LGBT students heard homophobic remarks, such as "faggot" or "dyke," frequently or often at school.
- nearly two-thirds of LGBT students, 63.5 percent, reported that they felt unsafe in school because of their sexual orientation.
- 31.8 percent of LGBT students missed at least one day of school in the previous month because of safety concerns or feeling uncomfortable.

LGBT bullying is also common among college students. For example, findings from the College Climate Survey conducted by the Iowa Pride Network indicate that LGBT students face more physical and cyber-harassment than their heterosexual peers. The results also indicate that over 80 percent of students have heard racist, sexist, homophobic, or negative comments about gender expression from students on campus (Gardner & Roemerman 2011). Law enforcement agencies report that nearly 10.0 percent of hate crimes based on sexual orientation occur in schools or on college campuses (FBI 2012).

The Effects of Antigay Harassment on Teenagers and Young Adults. LGBT youth in middle and high school settings undergo harsh treatment by their peers and may feel uncomfortable seeking the support of teachers, school administrators, and family members. Thus, it is not surprising that the reported grade point averages of students who were frequently harassed because of their sexual orientation or gender expression are lower than for students who were less often harassed (Kosciw et al. 2012).

Beyond performance in school, LGBT youth may suffer from mental health problems as a result of minority stress. The **minority stress theory** (Meyer 2003) explains that when an individual experiences the social environment as emotionally or physically threatening due to social stigma, the result is an increased risk for mental health problems. If someone is exposed to considerable hostility, or is aware that it happens to similar others, she or he may develop expectations that discrimination or harassment will happen to them. Internalized homophobia, biphobia, and/or transphobia may also develop.

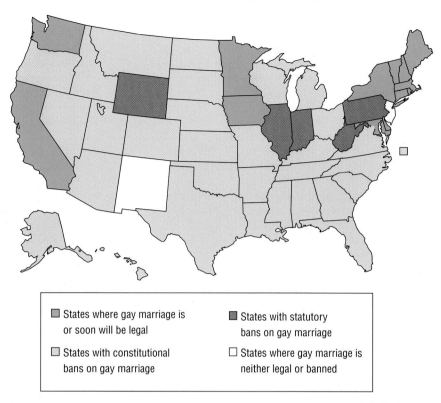

States where gay marriage is or soon will be legal

States with constitutional bans on gay marriage

States with statutory bans on gay marriage

States where gay marriage is neither legal or banned

Figure 11.5 **U.S. States and Marriage Equality Legal Status, 2013**
Source: The Pew Forum 2013b.

Toomey et al. (2010) found that victimization due to LGBT status was associated with negative adjustment variables, such as depression and lower levels of life satisfaction, which carried over into adulthood (see this chapter's *Social Problems Research Up Close* feature). Johnson et al. (2011) report that non-heterosexual female high school students were more likely to report bullying and experience more depression than their heterosexual peers. Minority stress has also been linked to increased risk for substance abuse problems among sexual minority adolescents. One meta-analysis found that the best predictors of substance abuse among LGB youth were victimization, a lack of social support, psychological stress, bad coming-out experiences, other problem behaviors, and housing status (Goldbach et al. 2013).

Police Mistreatment

Due to fear of further victimization, many cases of LGBT violence are not reported to the police (Ciarlante & Fountain 2010; NCAVP 2013). In 2012, among LGBT victims who reported violence to the police, nearly half reported police misconduct. Of those who were victims of violence and interacted with the police, 26.8 percent reported police hostility, an increase of 18 percent from the previous year. It is worth noting that the police violence and misconduct directed toward LGBT individuals reflects the disproportionate number of similar incidents experienced by people of color and transgender individuals (NCAVP 2013).

As with heterosexuals, intimate partner violence and sexual assault occur in the lives of LGBT individuals. Many cases of LGBT violence are *not* reported to the police out of fear of further victimization and trepidation in formally acknowledging the nature of the relationship (Ciarlante & Fountain 2010; NCAVP 2013). Further, failure of the police to define LGBT partner violence as *criminal* also contributes to underreporting. Many officers, assuming intimate partner violence is between a man and a woman and that it is the woman who is the victim, simply assign the label of "mutual abuse" and arrest both parties. It is estimated that police are 10 to 15 times more likely to make a dual arrest in cases of same-sex domestic violence than in heterosexual domestic violence (Ciarlante & Fountain 2010).

minority stress theory
Explains that when an individual experiences the social environment as emotionally or physically threatening due to social stigma, the result is an increased risk for mental health problems.

Adolescents encounter an array of challenges in their transition to adulthood; however, LGB adolescents face additional challenges that are related to the social stigma of their sexual orientation. For some lesbian, gay, and bisexual adolescents, this stigma may induce psychosocial stress, leading to increased health risk behaviors and poorer health outcomes (Meyer 2003). The following research study sought to determine whether the social environment surrounding lesbian, gay, and bisexual youth may contribute to their higher rates of suicide attempts, controlling for individual-level risk factors.

Sample and Methods

Nearly 32,000 11th grade students in Oregon completed the Oregon Healthy Teens survey over a two-year period. Students were identified as lesbian, gay, or bisexual (LGB) by asking respondents to indicate "which of the following best describes you"—"heterosexual (straight)," "gay or lesbian," "bisexual," or "not sure." Of the total sample respondents, 90.3 percent identified as heterosexual, 0.9 percent identified as gay or lesbian, and 3.3 percent identified as bisexual. The 1.9 percent of respondents who reported that they were unsure of their sexual orientation was excluded from the analysis.

The extent to which a social environment is supportive, the independent variable, was measured by five indicators across the 34-county area: (1) the percentage of same-sex couples living in a county, (2) the percentage of registered Democrats living in a county, (3) the percentage of schools with gay-straight alliances (GSAs), (4) the percentage of schools with non-discrimination policies that specifically protect LGB students, and (5) the percentage of schools with anti-bullying policies that specifically protect LGB

students. Hatzenbuehler (2011) hypothesizes that the higher each of these indicators and, thus, the higher the social environment measure, the lower LGB student suicide attempts.

The dependent variable was the number of times a student reported attempting suicide in the previous 12 months. The number of suicide attempts was then recoded into zero, no suicide attempts, and one, the presence of one or more suicide attempt(s). Respondents were also asked about other variables empirically documented to be associated with suicide attempts including depression, binge drinking, harassment at school by other students, and whether or not they had ever been physically abused by an adult. Each of these variables was measured in a "yes" or "no" format.

Findings and Discussion

The analysis took place in several phases, the first answering the question whether or not LGB youth when compared to heterosexual youth are more likely to attempt suicide. Based upon the data collected, the answer to this question is yes—LGB youth are more likely to attempt suicide than their heterosexual counterparts, approximately five times more likely. Further, LGB youth have significantly higher rates of depression, binge drinking, peer harassment at school, and adult lifetime physical abuse.

The second question answered is whether or not the supportiveness of the social environment in which a student lives is associated with the presence or absence of a suicide attempt. The answer, once again, is yes. The more supportive the social environment, the lower the odds that a student will attempt suicide. Further, when a supportive social environment is divided into a positive social environment and a negative social

environment, students residing in negative social environments are 20 percent more likely to attempt suicide. There are also differences in suicide attempts by heterosexuals based on the supportiveness of their social environment. However, the increased risk of a suicide attempt by a heterosexual living in a negative social environment is only 9 percent.

The third question answered is whether or not the relationship between LGB status and social environment is a function of individual-level risk factors. Thus, Hatzenbuehler (2011) examined the relationship between LGB status and social environment adjusting for the presence or absence of depression, binge drinking, peer harassment, and lifetime adult physical abuse. Although LGB status remains a significant predictor of student suicide attempts, its predictive power was reduced in the presence of individual-level risk factors. This means that part of the relationship between LGB status and social environment is explained by individual-level risk variables.

The author concludes that the results of this study can be used in reducing LGB youth suicide attempts. For example, only a 5 percent increase in the supportiveness of the social environment would be associated with a 10 percent decrease in suicide attempts. Hatzenbuehler (2011) is quick to note, however, that the data were collected at one point in time, and therefore, a causal relationship between social environment and suicide attempts cannot be established. Further, in terms of limitations of the study, the sample by virtue of being collected in a school did not include homeless youth, runaways, dropouts, and students absent on the day the survey was administered.

Source: Hatzenbuehler 2011.

Strategies for Action: Toward Equality for All

As discussed in this chapter, attitudes toward homosexuality in the United States have become more accepting over the years, and support for protecting civil rights of gays and lesbians is increasing. Many of the efforts to change policies and attitudes regarding non-heterosexuals and gender non-conforming individuals have been spearheaded by organizations that specifically advocate for LGBT rights including the Human Rights Campaign, the National Gay and Lesbian Task Force (NGLTF), the Gay and Lesbian Alliance Against Defamation (GLAAD), the International Lesbian and Gay Association (ILGA), and Amnesty International. But the effort to achieve sexual orientation equality is not a "gay agenda"; it is a human rights agenda that many heterosexuals and mainstream organizations support.

Gays, Lesbians, and the Media

Inspired by the fight for equal rights and taking on the often homophobic lyrics of other rap artists, Mackelmore's song "Same Love," featuring Mary Lambert, is an example of the impact that media can have on social issues. Lyrics such as "Kids are walking around the hallway—plagued by pain in their heart—a world so hateful—some would rather die than be who they are" calls attention to LGBT youth suicides and the distress caused by bullying. As of 2013, over 53 million people have watched the single's video on YouTube (McKinley 2013).

The media has been instrumental in the lives of LGBT individuals for several reasons. First, it has provided LGBT individuals, most importantly youth, role models for "**coming out**," a phrase referring to the ongoing process whereby a lesbian, gay, or bisexual individual becomes aware of his or her sexuality, accepts and incorporates it into his or her overall sense of self, and shares that information with others such as family, friends, and coworkers (APA 2008). The phrase "coming out" also applies to transgender individuals who become aware of their gender identity and share this with others (Herman 2009). Coming out happens because heterosexuality and gender conformity are considered normative in society.

National Coming Out Day, October 11, is recognized in many countries as a day to raise awareness of the LGBT population and foster discussion of LGBT rights issues. National Coming Out Day signifies the recognition that coming out is not only an important step in the lives of LGBT individuals, but is also a critical component of the fight for equality in the United States. Based on a survey of over 1,100 LGBT adults, research indicates that youth for the first time consider that they may be "something other than straight or heterosexual" at a young age but do not know "for sure" until later in life. The median age for the former was 12 years of age, and for the latter, 17 years of age (Lopez 2013).

Coming out can have positive outcomes in terms of strengthening relationships, instilling hope in LGBTs who conceal their identities and, in turn, their emotional well-being. However, when LGBT individuals come out, they also risk rejection from friends, coworkers, and family. In a study of fathers coming out to their sons and daughters, several of the children rejected their fathers and, at the time of the research, remained hostile (Tasker et al. 2010). Further, publically acknowledging your LGBT status may be met with criminal victimization.

As increasingly more LGBT Americans "come out" to their family, friends, and coworkers, heterosexuals and gender-conforming individuals have more personal contact with LGBT individuals. Psychologist Gordon Allport (1954) asserted that contact between groups is necessary for the reduction of prejudice—an idea known as the **contact hypothesis**. Research has shown that heterosexuals have more favorable attitudes toward gay men and lesbian women if they have had prior contact with or know someone who is gay or lesbian (Bonds-Raacke et al. 2007; Mohipp & Morry 2004). The most common reason for switching from an antigay marriage stance to a pro-gay marriage stance is knowing someone who is a homosexual (Pew Research Center 2013b).

Second, in addition to providing role models, LGBT visibility in the media counteracts stereotypes of LGBT individuals (Wilcox & Wolpert 2000) and allows non-LGBT individuals to see them not as abstractions, but as real people. In 1998, Ellen DeGeneres came out on her sitcom *Ellen*. After complaints about the episode, the show was canceled. Today, she is an Emmy-winning daytime talk show host. Since Ellen's groundbreaking "coming out"

coming out The ongoing process whereby a lesbian, gay, or bisexual individual becomes aware of his or her sexuality, accepts and incorporates it into his or her overall sense of self, and shares that information with others such as family, friends, and coworkers.

National Coming Out Day Celebrated on October 11, this day is recognized in many countries as a day to raise awareness of the LGBT population and foster discussion of gay rights issues.

contact hypothesis The idea that contact between groups is necessary for the reduction of prejudice.

> . . . Coming out not only is an important step in the lives of LGBT individuals, but also a critical component of the fight for equality in the United States.

episode, many television viewers have watched shows depicting LGBT characters in more realistic ways and that take a supportive stance on LGBT rights issues, including but not limited to *Buffy the Vampire Slayer*, *Glee*, *Queer Eye for the Straight Guy*, the *Big Gay Sketch Show*, *Chelsea Lately*, *True Blood*, *Will & Grace*, *Modern Family*, *Orange is the New Black*, and *The New Normal*. "Honest, non-stereotyped, and diverse portrayals of gays and lesbians in prime time can offer youth a realistic representation of the gay community . . . [and] can offer positive role models for gay and lesbian youth" (Miller et al. 2002, p. 21). Importantly, one study found that college students reported lower levels of antigay prejudice after watching television shows with prominent gay characters (Schiappa et al. 2005).

The visibility of famous gay, lesbian, and bisexual individuals has also had a societal impact. Many music and television celebrities, as well as politicians, have "come out" over the years, including Anderson Cooper, Wanda Sykes, Lance Bass, Neil Patrick Harris, Melissa Etheridge, former U.S. Representative Barney Frank, Jane Lynch, Rosie O'Donnell, David Hyde Pierce, Elton John, Suze Orman, Adam Lambert, and Ricky Martin.

Pictured is Jason Collins, NBA basketball player, marching in the 43rd annual Boston Pride March. In 2013, Mr. Collins became the first major league basketball player to announce that he is gay.

John Tlumacki/The Boston Globe/Getty Images

Advocates from the World of Sports. In recent years, various notable athletes and athletic organizations have taken a stand against homophobia. For example, former New York Giants defensive end and current Fox NFL analyst Michael Strahan filmed a public service announcement (PSA) supporting the legalization of same-sex marriage in the state of New York. Phoenix Suns point guard Steve Nash and New York Rangers forward Sean Avery have also filmed PSAs (Coogan 2011). At the *Straight for Equality Awards* ceremony on April 4, 2013, Baltimore Ravens player Brendon Ayanbadejo and Minnesota Vikings player Chris Kluwe were presented with awards by former National Football League (NFL) commissioner Paul Tagliabue honoring their support of equality for the LGB community (Battista 2013).

Athletes who reveal that they are gay have also contributed to the sports community's support of LGBT issues. In 2013, the number one draft pick of the Women's National Basketball Association (WNBA), Brittney Griner, announced she was gay (Morgan 2013). One month later, soccer player Robbie Rogers and basketball player Jason Collins revealed their sexual orientations, becoming the first athlete in their respective major league franchise to openly discuss their homosexuality (Collins & Lindz 2013; Mitz 2013).

The media has also provided examples of the increasing social disapproval of LGB prejudice. In 2011, some notable celebrities experienced personal and professional consequences as a result of their use of antigay slurs. Los Angeles Lakers player Kobe Bryant, angered by a technical foul penalty, hurled an offensive antigay epithet at a referee. This was followed by Bryant's swift apology; he also paid a $100,000 fine. Additionally, Atlanta Braves pitching coach Roger McDowell attracted attention by verbalizing antigay slurs and making offensive gestures at a group of fans. He was briefly suspended from playing, and later apologized at a news conference (Mungin 2011). Finally, comedian Tracy Morgan, from *Saturday Night Live* and *30 Rock*, also received a great deal of criticism following a "homophobic rant" during a stand-up routine in Nashville, Tennessee. Morgan subsequently apologized to the public, yet many are not satisfied with Morgan's apology (Brown 2011).

Finally, social media has been important in addressing the needs of LGBT youth. Following several gay teenage suicides that appeared in the news during September 2010, Dan Savage and his partner created the It Gets Better Project. The It Gets Better Project has turned into a viral cyber-movement, inspiring thousands of user-created videos that instill messages of hope and support for LGBT youth who have been bullied, feel that they must hide in shame, or who experience rejection from family members. The project has received submissions from celebrities, organizations, activists, politicians, and media personalities, ranging widely from President Barack Obama and Adam Lambert to staff members of Google and Pixar (It Gets Better Project 2011).

Ending Workplace Discrimination

Presently, federal law only protects against discrimination on the bases of race, religion, national origin, sex, age, and disability. When LGBT respondents were asked to prioritize specific policies that affect the LGBT population as "a top priority," "very important but not a top priority," "a somewhat important priority," or "not a priority at all," equal employment rights was the most commonly listed top priority followed by legally sanctioned marriages for non-heterosexual couples (Pew Research Center 2013b, p. 104).

The **Employment Non-Discrimination Act (ENDA)**, a proposed federal bill that would protect LGBT individuals from workplace discrimination, has been debated in Congress since 1994, but has never been signed into law. It would offer individuals basic protections against workplace discrimination on the basis of sexual orientation or gender identity. It was reintroduced in the U.S. House of Representatives and the U.S. Senate in 2013. To the surprise of some, ENDA passed the U.S. Senate on November 7, 2013, and is now in the House of Representatives where many predict it will not be passed.

With the absence of federal legislation prohibiting discrimination based on sexual orientation, some state and local governments, as well as private corporations, prohibit employment discrimination based on sexual orientation and extend domestic partner benefits to same-sex couples. Twenty-one states and the District of Columbia have laws banning sexual orientation discrimination in the workplace (HRC 2013f; Sears & Mallory 2011). Additionally, 26 states and the District of Columbia offer benefits for their government employees and their same-sex spouses. The Domestic Partnership Benefits and Obligations Act, currently pending in Congress, would extend domestic partnership benefits to all federal employees and their partners (HRC 2013a). A record 96.6 percent of Fortune 500 companies, a total of 483, include sexual orientation in their equal employment opportunity or non-discrimination policies, and more than half offer same-sex domestic partner health benefits to their employees (Equality Forum 2013).

Law and Public Policy

Marriage Equality. Thirteen states and Washington, DC, issue marriage licenses to same-sex couples. Several other states allow same-sex couples many of the same legal rights and responsibilities as married heterosexual couples. For example, in some states, same-sex couples can apply for a civil union license. A **civil union** is a legal status parallel to civil marriage under state law and entitles same-sex couples to nearly all of the rights and responsibilities available under state law to opposite-sex married couples.

In the 13 states that permit same-sex marriage, federal law now extends the same rights that opposite-sex couples have had to same-sex couples; however, federal law does not recognize the rights of partners in same-sex civil unions, so they do not have the more than 1,000 federal protections that go along with civil marriage. Some states, counties, cities, and workplaces allow unmarried couples, including gay couples, to register as **domestic partners** (or "reciprocal beneficiaries" in Hawaii). The rights and responsibilities granted to domestic partners vary from place to place but may include coverage under a partner's health and pension plan, rights of inheritance and community property, tax benefits, access to married student housing, child custody and child and spousal support obligations, and mutual responsibility for debts.

Employment Non-Discrimination Act (ENDA) This proposed federal bill that would protect LGBTs from workplace discrimination has been up for congressional debate on a number of occasions since 1994, but has never been signed into law.

civil union A legal status that entitles same-sex couples who apply for and receive a civil union certificate to nearly all of the benefits available to married couples.

domestic partners Unmarried couples (same- or opposite-sex) who have been granted domestic partnership status—a status which conveys various legal rights and responsibilities.

Bill Aron/PhotoEdit

Two gay Jewish men wearing yarmulkes stand under a chuppah during their outdoor wedding ceremony. The reformed Jewish community was an early supporter of LGBT equality.

Respect for Marriage Act (RMA) A bill that, if passed, would overturn DOMA and grant federal recognition to same-sex marriages, regardless of the state laws in which they reside.

Every Child Deserves a Family Act This federal legislation would remove obstacles to non-heterosexual (as well as transgender) individuals from providing loving homes for adoption or foster care.

Matthew Shepard and James Byrd, Jr. Hate Crimes Prevention Act (HCPA) This law expands the original 1969 federal hate crimes law to cover hate crimes based on actual or perceived sexual orientation, gender, gender identity, and disability.

In 2011, the White House formally endorsed the **Respect for Marriage Act (RMA)**, which was first introduced into Congress in 2009 (ACLU 2011; Savage & Stolberg 2011). After the U.S. Supreme Court's decisions on DOMA and Proposition 8, the RMA was again introduced into the U.S. Senate and House of Representatives (Esselink 2013). If passed, the RMA would overturn DOMA and grant federal recognition to same-sex marriages, regardless of the state laws in which a couple resides. The bill would *not* require states to recognize same-sex marriages performed in other states. The legal recognition of same-sex marriages by the federal government would mean that same-sex couples would realize all of the benefits of their opposite-sex counterparts (Badgett et al. 2011).

Parental Rights. In the United States, the **Every Child Deserves a Family Act (ECDFA)** has been introduced into Congress for several years, most recently in 2013. But, to date, the ECDFA has yet to receive legislative approval. If passed, the act would remove obstacles to non-heterosexual individuals providing homes for adoption or foster care by prohibiting public child welfare agencies from discriminating on the basis of sexual orientation, gender identity, or marital status. Agencies that do discriminate would be in danger of losing federal financial assistance (HRC 2013b; Moulton 2011).

Laws regulating adoption by LGB individuals not only vary by state, as previously discussed, but also vary across countries well. Argentina and Uruguay, for example, allow adoption by same-sex married couples. However, in Cuba, adoption by a same-sex couple or a partner of a gay man or woman is illegal. One man expresses his frustrations over his lack of rights in caring for his partner's biological son. He goes on to say, "I've held him in my arms since his birth . . . [but] if the baby has to go to the hospital . . . I have no legal authority to decide anything about his illness" (Gonzalez 2013, p. 1).

Hate Crimes Legislation. In 2009, President Obama signed into law the **Matthew Shepard and James Byrd, Jr. Hate Crimes Prevention Act (HCPA)**. This law expands the original 1969 federal hate crimes law to include hate crimes based on actual or perceived sexual orientation, gender, gender identity, and disability. The law was named after Matthew Shepard, a gay Wyoming teenager who died after being severely beaten, and James Byrd, Jr., an African American man who was attacked, chained to a vehicle, and dragged to his death in Texas (CNN 2009; FBI 2011). Hate crime laws call for tougher sentencing when prosecutors can prove that the crime committed was a hate crime. According to the most recent data available, 30 states and the District of Columbia have hate crime laws that include sexual orientation, 15 states have hate crime laws that do not include sexual orientation, and 5 states have no hate crime statutes (HRC 2013e; National Gay and Lesbian Task Force 2013).

Educational Strategies and Activism

Educational institutions bear the responsibility of promoting the health and well-being of all students. Thus, they must address the needs and promote acceptance of LGBT youth through certain policies and programs. The strategies for attaining these goals are to include LGBT issues in sex education, include LGBT-affirmative classroom curricula, and promote tolerance in learning environments through policies, education, and activism.

Sex Education and LGBT-Affirmative Classroom Curricula. The censorship of LGBT current issues, historical figures and events, and sexual health in both the classroom and in school libraries is a heated debate across the nation (Casey 2011). Whether LGBT themes can be brought into public school classrooms varies considerably between and within states. Several years ago, a Maryland county school system permitted lessons on non-heterosexuality in the classroom after consulting with the American Academy of Pediatrics on the issue, a bold move at the time (Schemo 2007). More recently, the California Department of Education's recommended reading list for students in grades K–12 gained considerable media attention for its inclusion of so-called "gay literature." The list of more than 7,800 books includes 32 publications with LGBT themes (Malmsheimer 2013).

California requires schools to teach about the contributions of women, people of color, and other historically underrepresented groups. California Governor Jerry Brown signed legislation in 2011 "making California the first state to require that textbooks and history lessons include the contributions of gay, lesbian, bisexual, and transgender Americans" (McGreevy 2011, p. 1). In 2012, the guidelines of California's FAIR Education Act were revised in order to prohibit schools from adopting learning materials that portray people with disabilities, lesbians, and/or gay men in a negative light in history and social studies lessons. The act also encourages inclusiveness. For example, when teaching about the Holocaust, teachers are encouraged to mention that in addition to the over 6 million Jews who were executed, the Nazis targeted and killed people based on sexual orientation and disability status ("FAIR Education Act" 2013).

In advocating for the end to censorship of LGBT resources in school libraries, Lambda Legal (2011b) points to the multiple benefits of providing education to students in a way that normalizes and affirms LGBT experiences. This includes promoting greater awareness of human diversity, the reduction of bullying and harassment, and the provision of support, reassurance, and information that non-heterosexual and gender non-conforming students need.

Opposition to LGBT curricula topics has garnered momentum in some areas of the country. In 2011, legislation dubbed by opponents as the "Don't Say Gay" Bill would have prohibited Tennessee teachers from discussing homosexuality with students from kindergarten through the eighth grade (Humphrey 2011). In 2013, the bill was reintroduced by state representative John Ragan after lawmakers abandoned it the previous year. As most recently proposed, the legislation would prohibit elementary and middle school teachers from discussing sexual activity that is not related to "natural human reproduction" or even acknowledges that homosexuality *exists*. Further, new language would require school officials to tell parents when students are—or might be—gay. In June 2013, 11-year-old Marcel Neergaard, an openly gay Tennessee boy, successfully garnered enough signatures in a petition to revoke representative Ragan's award of educational "Reformer of the Year" (Sieczkowski 2013).

Promotion of Tolerance in Learning Environments. Concerns over harassment and/or bullying of students have become the focus of many states. Sixteen states and Washington, DC, have laws that protect LGBT students from harassment and/or bullying, and 26 states prohibit bullying of all students including lesbian, gay, and transgender students.

Some individual school districts, regardless of state law, have taken measures to prevent LGBT bullying and harassment. In 2011, the Miami-Dade County Public School District added language to its anti-bullying and harassment policy that specifically protects students and teachers from harm based on sexual orientation and gender identity (Rapado & Campbell 2011). This is particularly notable given that Florida prohibits bullying in general, but does not identify protected classes.

For a number of years, U.S. Congresswoman Linda Sanchez has spearheaded the effort to pass the Safe Schools Improvement Act (SSIA). This bill, if passed, would require school districts to adopt policies prohibiting bullying and harassment based on, among other things, race, gender, religion, sexual orientation, and gender identity. Reintroduced in 2013 by Ms. Sanchez and others, the bill is presently pending in Congress (Kennedy & Temkin 2013).

A number of programs exist that aim to create a "harassment-free" climate and promote understanding and acceptance of sexual orientation and gender diversity in the K–12 school setting. The Gay, Lesbian, and Straight Education Network (GLSEN) is a national organization that collaborates with educators, policy makers, community leaders, and students to (1) protect LGBT students from bullying and harassment, (2) advance comprehensive safe schools laws and policies, (3) empower principals to make their schools safer, and (4) build the skills of educators to teach respect for all people. GLSEN sponsors such initiatives as the **National Day of Silence** in April, the No Name-Calling Week in January, and the "thinkb4youspeak" website.

GLSEN also provides support for **gay-straight alliances (GSAs)**, which are school-sponsored clubs that address antigay harassment and promote respect for all middle and high school students (GLSEN 2011b). Research has demonstrated that the presence of GSAs at schools has a positive impact. In general, students in schools with GSAs report hearing fewer homophobic remarks than students in schools that do not have GSAs. Further, LGBT students in schools with GSAs report feeling safer than LGBT students in schools without GSAs (GLSEN 2007).

What Do You Think? Pennsylvania does not have an LGBT-inclusive anti-bullying law, and it is up to local school districts in that state to protect the welfare of its LGBT students. One Pennsylvania school board denied students' request to form a GSA club at their school, while numerous other student clubs had been approved (Loviglio 2013). Do you think that the board should face legal action for their decision? If so, why?

Campus Programs. Many LGBT-affirmative and educational initiatives of middle and high schools also occur on college campuses. Student groups in colleges and universities have been active in the gay liberation movement since the 1960s. In addition to university-wide non-discrimination policies, other measures to support the LGBT college student population include gay and lesbian studies programs, social centers, and support groups, as well as campus events and activities that celebrate diversity. Some campuses have a "Lavender Graduation" ceremony in which LGBT graduates are honored and receive rainbow tassels for their caps (Dudash 2013). Many campuses also have Safe Zone programs designed to visibly identify students, staff, and faculty who support the LGBT population. Safe Zone programs require a training session that provides a foundation of knowledge needed to be an effective ally to LGBT students and those questioning their sexuality (University of Alabama 2011).

National Day of Silence A day during which students do not speak in recognition of the daily harassment that LBGT students endure.

gay-straight alliances (GSAs) School-sponsored clubs led by middle or high schools that strive to address anti-LGBT name-calling and promote respect for all students.

What Do You Think? Some schools, like Pepperdine University, have denied student requests to create support groups for LGBT students. Pepperdine, which is affiliated with the Church of Christ, made its decision based on religious ideology. The University of Notre Dame, which is affiliated with the Catholic Church, in a change from previous policies, agreed that a LGBT student group could be formed (Basu 2012). Do you think that colleges and universities that prohibit LGBT-affirmative groups are, essentially, putting their students at risk for negative experiences of bias and prejudice?

Understanding Sexual Orientation and the Struggle for Equality

In recent years, a growing acceptance of lesbians, gay men, and bisexuals, as well as increased legal protection and recognition of these marginalized populations has occurred. The advancements in LGB rights are notable and include progress with respect to marriage equality, the repeal of DADT, the growing public support of LGB civil rights, the endorsement of LGB civil rights by mental health and medical organizations, and the growing adoption of non-discrimination policies that include sexual orientation.

But winning these battles in no way signifies that LGBT individuals have secured equal rights. As evidenced by the years of court maneuvering in the struggle against Proposition 8, gay rights previously won can be taken away. LGB individuals employed at workplaces with anti-discrimination policies still experience harassment and rejection from their coworkers, and students in schools with policies against bullying are still subjected to antigay taunts. Although many countries worldwide are increasing legal protections for LGBT individuals, homosexuality is formally condemned in some countries, with penalties ranging from fines to imprisonment to death.

Many of the advancements in gay rights have been the result of political action and legislation. Barney Frank (1997), an openly gay former U.S. Congressman, emphasized the importance of political participation in influencing social outcomes. He noted that demonstrative and cultural expressions of gay activism, such as "**gay pride**" celebrations, marches, demonstrations, or other cultural activities promoting gay rights, are important in organizing gay activists, but cannot be used as substitutes for "conventional, boring, but essential" political participation (p. xi).

As both structural functionalists and conflict theorists note, non-heterosexuality challenges traditional definitions of family, child rearing, and gender roles. Every victory in achieving legal protection and social recognition for non-heterosexuals fuels the backlash against them by groups who are determined to maintain traditional notions of family and gender. Often, this determination is rooted in and derives its strength from religious ideology.

As symbolic interactionists note, the meanings associated with homosexuality are learned. Powerful individuals and groups opposed to gay rights focus their efforts on maintaining the negative meanings of homosexuality to keep the gay, lesbian, and bisexual population marginalized.

But political efforts to undermine gay rights and recognition must realize that prejudice and discrimination against individuals based on statuses over which research suggests they have no control hurts everyone. Using gay epithets as a way of questioning a boy's or man's masculinity threatens heterosexual as well as non-heterosexual males and the community as a whole. Antigay harassment has been identified as a precipitating factor in several school shootings (Pollack 2000a, 2000b).

There is the loss of human capital, i.e., the potential contributions LGBT individuals would have made if not for the prevalence of antigay sentiment. For example, fear of reprisal keeps many LGBT employees from sharing information about their personal lives at work. Employees who are "closeted" at work are less satisfied with their jobs, less trusting of their employers, and are more likely to leave their position than their "out" counterparts (Hewlett & Sumberg 2011).

States that do not legally recognize unmarried couples in committed relationships equally ignore the rights of heterosexual and same-sex couples. Heterosexuals who "look" or "act" gay may be the victims of hate crimes. Heterosexual mothers and fathers live in fear that their children will be victimized—harassed, fired, or even killed—by antigay prejudice and discrimination.

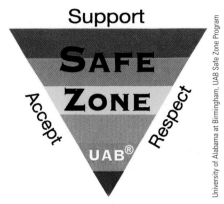

University of Alabama at Birmingham, UAB Safe Zone Program

Safe Zone programs provide a visible network of volunteers for lesbian, gay, bisexual, transgender, and other students, staff, and faculty seeking information and assistance regarding sexual orientation, gender identity, and expression.

> But political efforts to undermine gay rights and recognition must realize that prejudice and discrimination against individuals based on statuses over which research suggests they have no control hurts everyone.

gay pride Demonstrative and cultural expressions of gay activism that include celebrations, marches, demonstrations, or other cultural activities promoting gay rights.

True, the American public is becoming increasingly supportive of gay rights, and LGBT individuals have been granted legal rights in several states. But, as one scholar notes, "The new confidence and social visibility of homosexuals in American life have by no means conquered homophobia. Indeed it stands as the last acceptable prejudice" (Fone 2000, p. 411).

CHAPTER REVIEW

- **Are there any countries in which homosexuality is illegal?**
Seventy-six countries worldwide continue to criminalize same-sex relations. Legal penalties vary for violating laws that prohibit homosexual acts. In some countries, homosexuality is punishable by prison sentences and/or corporal punishment, such as whipping or lashing, and in five nations—Iran, Mauritania, Saudi Arabia, Yemen, and Sudan—people found guilty of engaging in same-sex sexual behavior may receive the death penalty.

- **Is there any country where same-sex couples can be legally married?**
Yes. In, 2000 the Netherlands became the first country in the world to offer full legal marriage to same-sex couples. In 2003, Belgium became the second country to legalize same-sex marriages. Since then, 13 more countries have legalized same-sex marriage: Canada (2005), Spain (2005), South Africa (2006), Norway (2009), Sweden (2009), Portugal (2010), Iceland (2010), Argentina (2010), Denmark (2012), Uruguay (2013), New Zealand (2013), France (2013), and Brazil (2013). In some countries, such as the United States (2003), Mexico (2009), and Brazil (2011), same-sex marriage is legal only in certain jurisdictions.

- **In what ways is the classification of individuals into sexual orientation categories problematic?**
Classifying individuals into sexual orientation categories is problematic for a number of reasons: (1) Distinctions among sexual orientation categories are not clear-cut and may better be represented by a continuum; (2) research with same-sex populations has tended to define sexual orientation based on one of three components whereas sexual orientation is complex and multidimensional; and (3) social stigma associated with non-heterosexuality leads people to conceal or falsely portray their sexuality.

- **What is the relationship between beliefs about what "causes" homosexuality and attitudes toward homosexuality?**
Individuals who believe that homosexuality or bisexuality is biologically based or inborn tend to be more accepting of LGB individuals. In contrast, individuals who believe that lesbian, gay, and bisexual people choose their sexual orientation are less tolerant of LGB individuals.

- **What is the official position of numerous respected professional organizations regarding sexual orientation change efforts (SOCE) for gays and lesbians?**
Many professional organizations agree that sexual orientation cannot be changed and that efforts to change sexual

orientation (using conversion, reparative, or reorientation therapy) do not work and may, in fact, be harmful.

- **What three cultural changes have influenced the worldwide increase in liberalized national policies on same-sex relations and the gay rights movement?**
The worldwide increase in liberalized national policies on same-sex relations and the gay rights social movement have been influenced by (1) the rise of individualism, (2) increasing gender equality, and (3) the emergence of a global society in which nations are influenced by international pressures.

- **How does social stigma impact the mental health of marginalized individuals, such as LGBT individuals?**
The minority stress theory explains that when an individual experiences the social environment as emotionally or physically threatening due to social stigma, the result is an increased risk for mental health problems. If someone is exposed to considerable hostility, or is aware that it happens to similar others, he or she may develop expectations that discrimination or harassment will happen to them. Internalized homophobia, biphobia, and transphobia may also develop.

- **Is employment discrimination based on sexual orientation illegal in all 50 states?**
It is still legal in 29 states to fire, decline to hire or promote, or otherwise discriminate against employees because of their sexual orientation, and in 33 states, it remains legal to discriminate against an employee for being transgender. As of this writing, a federal bill called the Employment Non-discrimination Act (ENDA), which would ban employment discrimination against individuals on the basis of sexual orientation, is pending before Congress.

- **What is the Defense of Marriage Act (DOMA)?**
The Defense of Marriage Act is federal legislation that states that marriage is a legal union between one man and one woman and denies federal recognition of same-sex marriage. In a June 2013 5–4 ruling in *United States v. Windsor*, the Supreme Court struck down a provision of the 17-year-old act that denies federal benefits to same-sex couples legally married.

- **Same-sex couples want the same legal and financial benefits that are granted to heterosexual married couples. What are some of these benefits?**
Married couples have the right to inherit from a spouse who dies without a will, to avoid inheritance taxes between spouses, to make crucial medical decisions for a partner

and to take family leave to care for a partner in the event of the partner's critical injury or illness, to receive Social Security survivor benefits, and to include a partner in health insurance coverage. Other rights bestowed on married (or once-married) partners include assumption of a spouse's pension, bereavement leave, burial determination, domestic violence protection, reduced rate memberships, divorce protections (such as equitable division of assets and visitation of partner's children), filing joint tax returns, automatic housing lease transfer, immunity from testifying against a spouse, and spousal immigration eligibility.

- **What are some reasons for the underreporting of LGBT intimate partner violence and sexual assault to the police?**
Many cases of LGBT violence are not reported to the police for fear of further victimization by police, including indifference or some form of abuse. Additionally, the failure of police to identify such incidents as occurring in the context of an intimate partnership often erroneously leads to the arrest of both the victim and the perpetrator.

- **Why is the media important in the advancement of LGBT civil rights?**
LGBT visibility in the media counteracts stereotypes of LGBT individuals and allows non-LGBT individuals to see them not as an abstraction, but as real people. This is consistent with Gordon Allport's contact hypothesis that suggests more contact with or exposure to a group results in the reduction of prejudice. The media has also provided LGBT individuals, most importantly youth, with role models for "coming out."

- **What are Safe Zone programs?**
Safe Zone programs are designed to visibly identify students, staff, and faculty who support the LGBT population. Participants in Safe Zone programs display a sign or placard outside their office or residence hall room that identifies them as individuals who are willing to provide a safe haven and support for LGBT people and those struggling with sexual orientation issues.

TEST YOURSELF

1. Research has indicated that many individuals are not exclusively heterosexual or homosexual and that sexual orientation can be represented on a continuum.
   a. True
   b. False
2. A national study of U.S. college students found that what percentage identified as gay, lesbian, or bisexual?
   a. 1.2   c. 6.4
   b. 3.0   d. 10.1
3. According to the 2010 Census, an estimated 30 percent of Americans live with a same-sex partner or spouse.
   a. True
   b. False
4. NARTH has claimed all but the following:
   a. They are not "antigay."
   b. Collectively, they have a success rate of 75 percent.
   c. They are not a religious group.
   d. They assist those who are distressed by unwanted sexual attractions.
5. In 2013, ministers of the European Union instructed their diplomats around the world to do what?
   a. Repeal same-sex marriage laws
   b. Add constitutional amendments prohibiting marriage equality
   c. Defend the rights of LGBTI people
   d. Prohibit adoptions by transgender individuals
6. Exodus International
   a. is a SOCE organization
   b. is expanding their pro-gay ministry
   c. no longer exists
   d. Both a and c are correct

7. Nearly half of all Americans describe discrimination against gays and lesbians as a "very" or "somewhat serious" problem in the United States.
   a. True
   b. False
8. Why was the It Gets Better Project started?
   a. Because same-sex couples continue to encounter obstacles in second-parent and joint adoptions
   b. Because LGBT youth continue to be bullied and harassed by their peers, resulting in shame, isolation, and even suicide
   c. Because LGB servicemen and women have been waiting a long time to serve openly in the military
   d. Because American society seems to be growing in its acceptance of the legal recognition of same-sex couples
9. The "thinkb4youspeak" website created by GLSEN uses public service announcements made by celebrities to help combat biased language (e.g., "You're so gay," etc.).
   a. True
   b. False
10. The first country to legalize same-sex marriage was
   a. Norway
   b. England
   c. The Netherlands
   d. Argentina

Answers: 1. A; 2. C; 3. B; 4. B; 5. C; 6. D; 7. A; 8. B; 9. A; 10. C.

Population Growth and Aging

"Population may be the key to all the issues that will shape the future: economic growth; environmental security; and the health and well-being of countries, communities, and families."

—Nafis Sadik, former executive director,
United Nations Population Fund

In this stainless steel bio-cremation container, heat and lye transform human remains into a non-toxic liquid.

AS PEOPLE AGE, one of the issues they think about is what they would like done with their body after they die. Although traditions, customs, and laws concerning what to do with human remains vary across cultures and across time, some form of cremation and/or burial is common practice around the world. Have you ever driven past a cemetery and wondered how many generations can continue to be buried before we run out of land? Several cemeteries in New York City have already run out of land and have stopped selling burial plots (Santora 2010).

Japan offers an innovative answer to the problem of not enough grave space: a high-tech graveyard in a multistoried building where the ashes of the dead are stored in urns on shelves. Visitors use plastic swipe cards and a touch screen to identify the remains they wish to "visit" and a robotic arm collects the requested urn and delivers it to a "mourning room" where visitors pay tribute to their deceased loved ones (Here and Now 2009). To maximize grave space, some countries allow the practice of vertical burials in which the deceased are wrapped in a biodegradable shroud and buried in a vertical, standing-up position (Dunn 2008). In another alternative to burials, the dead can be poured down the drain if the remains are put through a process called alkaline hydrolysis, in which heat and lye are used to transform the body into a nontoxic liquid, leaving behind a small amount of dry bone residue for the family to scatter (Frynes-Clinton 2011). Also known as "liquid cremation" and "bio cremation," this alternative to cremation or burial is legal in at least eight states, and legalization is pending or anticipated in another 20 states.

Running out of land for grave sites is only one of many concerns that stem from the growth in human population over the last century or so. The world's population is not only growing, but it also is aging. In this chapter, we focus on social problems associated with population growth and aging.

The Global Context: A Worldview of Population Growth and Aging

The well-known scientist Jane Goodall remarked that, "It's our population growth that underlies just about every single one of the problems that we've inflicted on the planet. If there were just a few of us, then the nasty things we do wouldn't really matter and Mother Nature would take care of it—but there are so many of us" (quoted in Koch 2010, n.p.). In the following sections, we describe how the size and life span of human populations have increased over time.

World Population: History, Current Trends, and Future Projections

Humans have existed on this planet for at least 200,000 years. For 99 percent of human history, population growth was restricted by disease and limited food supplies. Around 8000 B.C., the development of agriculture and the domestication of animals led to increased food supplies and population growth, but even then, harsh living conditions and disease still put limits on the rate of growth. This pattern continued until the mid-18th century, when the Industrial Revolution improved the standard of living for much of the world's population. The improvements included better food, cleaner drinking water, improved housing and sanitation, and advances in medical technology, such as antibiotics and vaccinations against infectious diseases—all of which contributed to rapid increases in population.

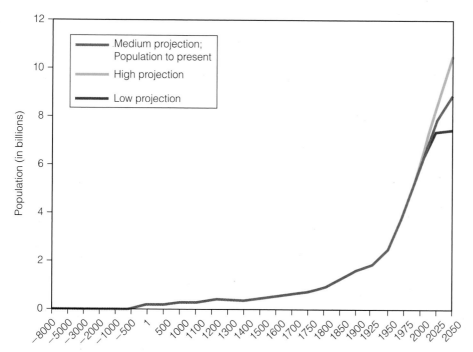

Note: Time is not to scale.

Figure 12.1 **World Population through History**
Source: Weeks 2012.

Population **doubling time** is the time required for a population to double from a given base year if the current rate of growth continues. It took several thousand years for the world's population to double to a size of 14 million, but then took only a thousand years to nearly double to 27 million and another thousand to reach 50 million. From there, it took only 500 years to double from 50 million to 100 million, and 400 years for the next doubling to occur. When the Industrial Revolution began around 1750, population growth exploded, taking only 100 years to double. The most recent doubling—from three billion in 1960, to six billion in 1999—took only about 40 years (Weeks 2012). Although world population will continue to grow in the coming decades, it will probably not double in size again. However, the population of the least developed countries is projected to double from 898 million in 2013 to 1.8 billion in 2050 (United Nations 2013).

Although thousands of years passed before the world's population reached 1 billion around the year 1800, the population exploded from 1 billion to 6 billion in less than 300 years (see Figure 12.1). World population was 1.6 billion when we entered the 20th century, and 6.1 billion when we entered the 21st century. World population is projected to grow from 7.2 billion in 2013, to 8.1 billion in 2025, 9.6 billion in 2050, and 10.9 billion by 2100 (United Nations 2013).

Most of the world's population lives in less developed countries, primarily in Asia and Africa (see Figure 12.2). The most populated country in the world today is China, where about one in five people on this planet live. By 2050, India will become the most populated country (see Figure 12.3).

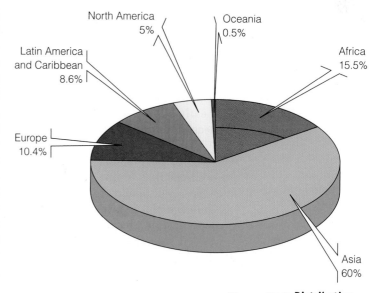

Figure 12.2 **Distribution of World Population by Region: 2013**
Source: United Nations 2013.

doubling time The time required for a population to double in size from a given base year if the current rate of growth continues.

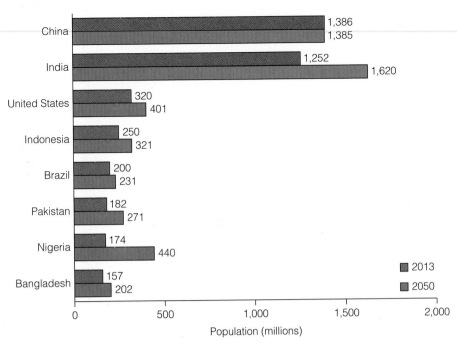

Figure 12.3 **World's Eight Largest Countries in Population, 2013 and 2050 (Projected)**
Source: United Nations 2013.

Nearly all of world population growth from now until 2100 will occur in less developed countries, particularly the least developed countries of the world, and mostly in Africa and Asia. The population of developing countries is expected to increase from 5.9 billion in 2013 to 9.6 billion in 2100. In contrast, the population of more developed countries is expected to grow minimally, from 1.25 billion in 2013 to 1.28 billion in 2100, and would decline if it were not for the migration from developing countries to developed countries (United Nations 2013). Higher population growth in developing countries is largely due to the higher **total fertility rates**—the average lifetime number of births per woman in a population. Although total fertility rates worldwide have declined significantly from nearly 5 in the 1950s to 2.53 between 2005 and 2010, they continue to be highest in the least developed countries of the world.

Will there be an end to the rapid population growth that has occurred in recent decades? Will the population of the world stabilize? Although some predict that population will stabilize around the middle of the 21st century, no one knows for sure. Despite the overall decline in fertility rates, there are still about 30 countries with high fertility (average of 5 or more children per woman) (United Nations 2013). To reach population stabilization, fertility rates throughout the world would need to achieve what is called "replacement level," whereby births would replace, but not outnumber, deaths. **Replacement-level fertility** is 2.1 births per woman, that is, slightly more than 2 because not all female children will live long enough to reach their reproductive years. Between 2005 and 2010, 75 countries had below-replacement fertility (United Nations 2013). In some of these countries, population will continue to grow for several decades because of **population momentum**—continued population growth as a result of past high fertility rates that have resulted in a large number of young women who are currently entering their childbearing years. But 43 countries or areas have populations that are projected to decrease between 2013 and 2050 (United Nations 2013). Although the U.S. population has a total fertility rate slightly lower than the replacement level, the U.S. population is expected to continue to increase through 2050 because of immigration.

In sum, two population size trends are occurring simultaneously that appear to be contradictory: (1) The total number of people on this planet is rising and is expected to continue to increase over the coming decades; and (2) fertility rates are so low in some countries that the countries' populations are likely to decline over the coming years. As we discuss later in this chapter, each of these trends presents a set of problems and challenges.

total fertility rates The average lifetime number of births per woman in a population.

replacement-level fertility The level of fertility at which a population exactly replaces itself from one generation to the next; currently, the number is 2.1 births per woman (slightly more than 2 because not all female children will live long enough to reach their reproductive years).

population momentum Continued population growth as a result of past high fertility rates that have resulted in a large number of young women who are currently entering their childbearing years.

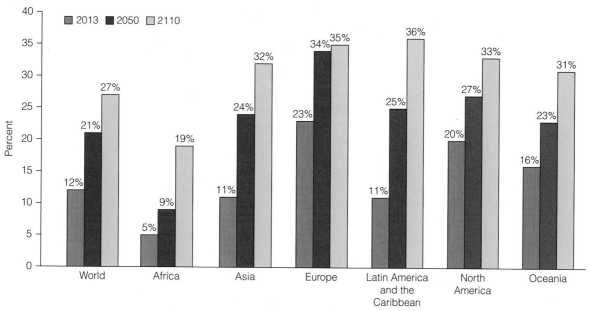

Note: Percentages are rounded to nearest whole percentage.

Figure 12.4 **Percent of Population Aged 60+, by Region: 2013–2110 (projected)**
Source: United Nations 2013.

The Aging of the World's Population

Another demographic trend that presents its own set of challenges is the increasing number and proportion of older individuals in the total population. Globally, the population aged 60 and older is the fastest growing age group (United Nations 2013). Between 2013 and 2050, the percentage of older individuals (ages 60 and over) in the world population is expected to nearly double (see Figure 12.4). In 2050, the global population of people aged 60 years or over will outnumber the population of children (0 to 14 years) for the first time in human history (United Nations 2012). The 80 and older population is also growing throughout the world, increasing from nearly 2 percent of world population in 2013 to 4 percent in 2050 to nearly 8 percent in 2100 (United Nations 2013). In the United States, the 60 and older and 80 and older populations are also growing (see Figure 12.5).

The aging of the population is the result of both increased longevity and declining fertility rates. Global life expectancy increased from 47 years between 1950 and 1955 to 69 years between 2005 and 2010, and is expected to increase to 76 years between 2045 and 2050 and to 82 years between 2095 and 2100 (United Nations 2013). In the United States, population aging is also occurring as the **baby boomers**—the generation of Americans born during a period of high birthrates between 1946 and 1964—are entering their older years. The first of the baby boomer generation turned 65 in 2011.

Fertility rates globally have fallen, from 4.44 in 1970 to 2.53 in 2010. In most developed countries, including the United States, fertility rates have fallen under 2 children per woman, and have been below this replacement level for two to three decades. This means that as more of the population is living longer into older ages, fewer children are being born, so an increased percentage of the population is older.

Population aging increases pressure on a society's ability to support its elderly members because as the proportion of older people increases in a population, there are fewer

> In 2050, the global population of people aged 60 years or over will outnumber the population of children (0 to 14 years) for the first time in human history.

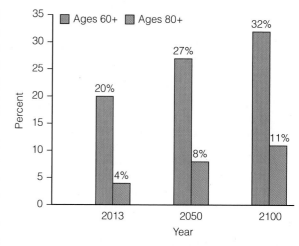

Note: Percentages are rounded to nearest whole percentage.

Figure 12.5 **Percent of U.S. Population Age 60+ and 80+, 2013–2100 (projected)**
Source: United Nations 2013.

baby boomers The generation of Americans born between 1946 and 1964, a period of high birthrates.

working-age adults to support the elderly population. A commonly used indicator of this pressure is the **elderly support ratio**, calculated as the number of working-age people divided by the number of people 65 or older. Globally, "working age" is considered to be ages 15 to 64; in the United States, "working age" is considered to be ages 20 to 64. Since 1950, the global elderly support ratio has been declining. In 1950, there were 12 working-age people for every older person; in 2012, there were 8 working-age people for every older person, and the elderly support ratio is projected to further decrease to 4:1 by 2050 (United Nations 2012). In developed countries, the elderly support ratio is much lower than in less developed regions, meaning that there are fewer working-age adults to support each older person in developed countries. Germany, Italy, and Japan have an elderly support ratio of 3:1—the lowest in the world (Population Reference Bureau 2010). By 2050, Japan will have only one working-age adult for every elderly person; Germany and Italy will each have two.

The decrease in the elderly support ratio raises concerns about whether there will be enough workers to take care of the older population. Population aging raises other concerns as well: How will societies provide housing, medical care, transportation, and other needs of the increasing elderly population? Later in this chapter, we look at the problems and challenges of meeting the needs of a growing elderly population.

Sociological Theories of Population Growth and Aging

The three main sociological perspectives—structural functionalism, conflict theory, and symbolic interactionism—can be applied to the study of population and aging.

Structural-Functionalist Perspective

Structural functionalism focuses on how changes in one aspect of the social system affect other aspects of society. For example, the **demographic transition theory** of population describes how industrialization and economic development affect population growth by influencing birth and death rates. According to this theory, traditional agricultural societies have both high birth rates and high death rates. As a society becomes industrialized and urbanized, improved sanitation, health, and education lead to a decline in mortality. The increased survival rate of infants and children along with the declining economic value of children leads to a decline in birthrates. The demographic transition theory is a generalized model that does not apply to all countries, and does not account for population change due to HIV/AIDS, war, migration, and changes in gender roles and equality. Many countries with low fertility rates have entered what is known as a "second demographic transition," in which fertility falls below the two-child replacement level. This second demographic transition has been linked to greater educational and job opportunities for women, increased availability of effective contraception, and the rise of individualism and materialism (Population Reference Bureau 2004).

The growing elderly population is a social change that has led to a number of other social changes, such as increased pressure on the workforce and on federal programs to support the aging population (e.g., Medicare and Social Security). Increased longevity also affects families, who often bear the brunt of elder care. The problems and challenges associated with caring for the aging population are discussed later in this chapter.

The structural-functionalist perspective also focuses attention on the unintended, or latent, consequences of social behavior. Although the intended, or manifest, function of modern contraception is to control and limit childbearing, there have also been far-reaching unintended effects of contraception on the social and economic status of women. The development of modern contraceptives—particularly the birth control pill—has led to fewer births to high school- and college-aged women, increased age at first marriage, and increased participation by women in the workforce. "The advent of the pill allowed women greater freedom in career decisions, by allowing them to invest in higher education and a career with far less risk of an unplanned pregnancy" (Sonfield 2011, p. 9). Another example of an unintended consequence can be found in

elderly support ratio The ratio of working-age adults (15 to 64) to adults ages 65 and older in a population.

demographic transition theory A theory that attributes population growth patterns to changes in birth rates and death rates associated with the process of industrialization.

the findings of several studies that link social security programs in various countries and lowered fertility (Kornblau 2009). One reason that people have children is to provide security in old age. Social security reduces the need to have children to secure care in old age.

Conflict Perspective

The conflict perspective focuses on how wealth and power, or the lack thereof, affect population problems. In 1798, Thomas Malthus predicted that the population would grow faster than the food supply and that masses of people were destined to be poor and hungry. According to Malthusian theory, food shortages would lead to war, disease, and starvation, which would eventually slow population growth. However, conflict theorists argue that food shortages result primarily from inequitable distribution of power and resources (Livernash & Rodenburg 1998).

Conflict theorists also note that population growth results from pervasive poverty and the subordinate position of women in many less developed countries. Poor countries have high infant and child mortality rates. Hence, women in many poor countries feel compelled to have many children to increase the chances that some will survive into adulthood. Their subordinate position prevents many women from limiting their fertility. In many developing countries, a woman must get her husband's consent before she can receive any contraceptive services. Thus, according to conflict theorists, population problems result from continued economic and gender inequality.

Some conflict theorists view the elderly population as a special-interest group that competes with younger populations for scarce resources. Debates about funding programs for the elderly (such as Social Security and Medicare) versus funding for youth programs (such as public schools and child health programs) largely represent conflicting interests of the young versus the old. A growing elderly population means that the elderly have increased power in political issues. In the United States, adults ages 65 to 74 have the highest rate of voting of any age group (File & Crissey 2010).

> In the United States, adults ages 65 to 74 have the highest rate of voting of any age group.

Symbolic Interactionist Perspective

The symbolic interactionist perspective focuses on how meanings, labels, and definitions learned through interaction affect population problems. For example, many societies are characterized by **pronatalism**—a cultural value that promotes having children. Throughout history, many religions have worshiped fertility and recognized it as being necessary for the continuation of the human race. In many countries, religions prohibit or discourage birth control, contraceptives, and abortion. Women in some pronatalistic societies learn through interaction with others that deliberate control of fertility is socially unacceptable. Women who use contraception in communities in which family planning is not socially accepted face ostracism by their community, disdain from relatives and friends, and even divorce and abandonment by their husbands (Women's Studies Project 2003). However, once some women learn new definitions of fertility control, they become role models and influence the attitudes and behaviors of others in their personal networks (Bongaarts & Watkins 1996).

The symbolic interactionist perspective emphasizes the importance of examining social meanings and definitions associated with aging. "Old age" is largely a social construct; there is no biological marker that indicates when a person is "old." Rather, old age is a matter of social definition. In the United States, and much of the world, a person is considered to be "old" (or a "senior citizen") when they reach 65, as this is the age that company pension plans and Social Security have used to define when a person retires and collects benefits. Due to changes in the Social Security system, the retirement age for receiving full Social Security benefits has been increased to 67, yet we continue to use 65 to define "elderly."

pronatalism A cultural value that promotes having children.

We also learn both positive and negative meanings associated with old age, such as "wise" and "experienced" as well as "frail" and "impaired" (Kornadt & Rothermund 2010). In U.S. culture, negative labels of older people, such as "crone," "old geezer," and "old biddy" are predominant, and reflect ageism—a topic we discuss later in this chapter.

Finally, the symbolic interactionist perspective reminds us that understanding the quality of life among the elderly requires consideration of how the elderly subjectively define their experiences. For example, to assess social isolation among the elderly, researchers often use objective measures such as the frequency of social contacts among the elderly. But it is also important to consider how the elderly subjectively experience their level of social contact, as some older adults have a preference for engaging in solitary activities and experience being alone as enjoyable (Cloutier-Fisher et al. 2011).

Social Problems Related to Population Growth and Aging

Social problems related to population growth include environmental problems; poverty, unemployment, and global insecurity; and poor maternal and infant health. This chapter's *Animals and Society* feature looks at the problem of pet overpopulation and the importance of controlling the fertility of companion animals.

Problems related to the aging of the population include ageism—prejudice and discrimination against older individuals—employment and retirement concerns of older Americans, and the challenge of meeting the various needs of the elderly population, particularly retirement income and health care. Health issues and Medicare are discussed in Chapter 2, and elder abuse is discussed in Chapter 5.

Environmental Problems and Resource Scarcity

According to a survey of faculty at the State University of New York College of Environmental Science and Forestry (2009), overpopulation is the world's top environmental problem, followed closely by climate change and the need to replace fossil fuels with renewable energy sources. As we discuss in Chapter 13, population growth places increased demands on natural resources, such as forests, water, cropland, and oil, and results in increased waste and pollution.

According to the United Nations Environment Programme (UNEP), half of the planet's forests have been cleared for human land use and, in 2025, two-thirds of the world's population will be living in countries with water scarcity or stress, and the world's fisheries will be depleted by the middle of this century (Engelman 2011). The countries that suffer most from shortages of water, farmland, and food are developing countries with the highest population growth rates. However, countries with the largest populations do not necessarily have the largest impact on the environment. This is because the demands that humanity makes on the earth's natural resources—each person's **environmental footprint**—is determined by the patterns of production and consumption in that person's culture. The environmental footprint of an average person in a high-income country is much larger than that of someone in a low-income country. Hence, although population growth is a contributing factor in environmental problems, patterns of production and consumption are at least as important in influencing the effects of population on the environment.

environmental footprint The demands that humanity makes on the earth's natural resources.

Poverty, Unemployment, and Global Insecurity

Poverty and unemployment are problems that plague countries with high population growth. Less developed, poor countries with high birthrates do not have enough jobs for a rapidly growing population, and land for subsistence farming becomes increasingly scarce as populations grow. In some ways, poverty leads to high fertility, because poor women are less likely to have access to contraception and are more likely to have large families in the hope that some children will survive to adulthood and support them in old age. But high fertility also exacerbates poverty, because families have more children to support, and national budgets for education and health care are stretched thin.

A Population Institute report titled *Breeding Insecurity: Global Security Implications of Rapid Population Growth* warns that rapid population growth is a contributing factor to global insecurity, including civil unrest, war, and terrorism (Weiland 2005). Although world population is, overall, aging, some countries in Africa and the Middle East are experiencing a "youth bulge"—a high proportion of 15- to 29-year-olds relative to the adult population. Youth bulges result from high fertility rates and declining infant mortality rates, a common pattern in developing countries today. The combination of a youth bulge with other characteristics of rapidly growing populations, such as resource scarcity, high unemployment rates, poverty, and rapid urbanization, sets the stage for political unrest. "Large groups of unemployed young people, combined with overcrowded cities and lack of access to farmland and water create a population that is angry and frustrated with the status quo and thus is more likely to resort to violence to bring about change" (Weiland 2005, p. 3).

Poor Maternal, Infant, and Child Health

As noted in Chapter 2, maternal deaths (deaths related to pregnancy and childbirth) are the leading cause of mortality for reproductive-age women in the developing world. Having several children at short intervals increases the chances of premature birth, infectious disease, and death for the mother or the baby. Childbearing at young ages (teens) also increases the risks of health problems and death for both women and infants (United Nations Population Division 2009). In developing countries, one in four children is born unwanted, increasing the risk of neglect and abuse. In addition, the more children a woman has, the fewer the parental resources (parental income, time, and maternal nutrition) and social resources (health care and education) available to each child.

Ageism: Prejudice and Discrimination toward the Elderly

Ageism refers to negative stereotyping, prejudice, and discrimination based on a person's or group's perceived chronological age. Ageism is reflected in negative stereotypes of the elderly, such as that they are slow, they don't change their ways, they are grumpy, they are poor drivers, they cannot/don't want to learn new things, they are incompetent, or they are physically and/or cognitively impaired (Nelson 2011). Ageism also occurs when older individuals are treated differently because of their age, such as when they are spoken to loudly in simple language, assuming they cannot understand normal speech, or when they are denied employment due to their age. Another form of ageism—**ageism by invisibility**—occurs when older adults are not included in advertising and educational materials. Before reading further, you may want to complete the "Ageism Survey" in this chapter's *Self and Society* feature on page 389.

ageism Negative stereotyping, prejudice, and discrimination based on a person's or group's perceived chronological age.

ageism by invisibility Occurs when older adults are not included in advertising and educational materials.

More than a third of U.S. households include dogs or cats as pets, also referred to as companion animals (Coate & Knight 2010). Many pets live out their lives in homes where they are loved and well cared for. But each year, about 8 million stray and unwanted cats and dogs are taken to animal shelters across the United States, and about half of these animals end up being euthanized, most commonly through intravenous injection (American Humane Association 2011; Humane Society 2009). In fact, shelter euthanasia is the number one cause of death for both dogs and cats in the United States (American Humane Association 2011).

According to the National Council on Pet Population Study and Policy (2009), the top 10 reasons for taking a companion dog or cat to an animal shelter include moving, cost of pet maintenance, and allergies (see the table within this feature). Regarding cost, many people do not realize that, aside from the initial purchase price of a dog or cat, the cost of maintaining a pet over a 12-year lifetime can cost $22,000 or more for food, treats, vet visits, grooming, toys, medications, training, and other expenses (Greenfield 2011). City dwellers that hire dog walkers can spend an additional $5,000 a year just for this service.

During the recent housing crisis, animal shelters and pet rescue groups saw a surge in abandoned pets as a result of the increase in foreclosures (Brenoff 2011). Although it is a crime in all 50 states to abandon a pet, some families evicted from their homes leave their pets in the yard or in the house to fend for themselves. Animal shelters in college towns see an increase in abandoned pets at the end of the school year when many students move away and cannot take their pets with them (Kidd 2009).

Not all abandoned pets are taken to shelters; some are left in the street or are dropped off in parking lots or other public places. Abandoned pets left out in the wild struggle with starvation and disease, and pose a threat to both humans and other animals, such as endangered birds. In the wild, unsterilized cats and dogs can multiply quickly. Consider that if an unsterilized cat gives birth to two litters a year, that can result in 400,000 cats over seven years (Shikina 2011).

The millions of unwanted and feral dogs and cats is part of a larger problem of pet overpopulation. The American Society for the Prevention of Cruelty to Animals (ASPCA) suggests that the best method for reducing pet overpopulation is sterilization, and recommends that (1) all companion dogs and cats, except those that are a part of a responsible breeder's program, be spayed or neutered at an early age (2 months); and (2) all communities have accessible and affordable spay/neuter programs (ASPCA 2011). Many vets and animal welfare agencies offer low-cost or free spaying and neutering, and some organizations have mobile clinics that travel to low-income or rural areas where people are not able to transport a pet to a vet.

Despite the problem of pet overpopulation, about a third of pet owners in the United States do not spay or neuter their pet (American Humane Association 2011). Some pet owners want to breed their pets, either for profit or because they think it would be fun to have a litter of puppies or kittens. Others mistakenly believe that their pet won't become accidentally pregnant.

Some states and municipalities have laws to try to curb pet overpopulation. Rhode Island mandates that dogs and cats must be spayed or neutered before being released from

Old age is stereotypically viewed as a negative time during which older individuals suffer a loss of identity (retirement from job), loss of respect from society, and increasing dependence on others. Although stereotypes of old people may accurately describe a number of old people, these stereotypes do not apply for many older individuals. Contrary to negative views of aging, people in their later years can be productive and fulfilled (see Table 12.1).

Ageism is different from the other "isms" in that everyone is vulnerable to ageism if they live long enough.

Ageism is embedded in our culture and is much more widely accepted than other "isms" such as racism, sexism, and heterosexism. Nelson (2011) states that "there is no other group like the elderly about which we feel free to openly express stereotypes and even subtle hostility" (p. 40). According to Margaret Gullette (2011), "ageism is to the twenty-first century what sexism, racism, homophobia, and ableism were earlier in the twentieth—entrenched and implicit systems of discrimination, without adequate movements of resistance to oppose them" (p. 15). Ageism is different from the other "isms" in that everyone is vulnerable to ageism if they live long enough.

a shelter; a Los Angeles ordinance requires pet owners to sterilize their dogs or cats by age 4 months; and some cities, including Austin, Texas, Albuquerque, New Mexico, and West Hollywood, California, completely ban the retail sales of dogs and cats (Humane Society 2010; Shikina 2011). Hawaii is considering a bill that would require pet retailers to sterilize all cats and dogs before selling them (Shikina 2011). And some major pet stores, including Petco and PetSmart, refuse to sell puppies and kittens at their stores (although many major pet stores allow local animal and shelter groups to display dogs and cats for adoption).

Finally, people wanting to acquire a dog or cat are encouraged to adopt from an animal shelter or nonprofit animal rescue organization. Each year, an estimated 17 million Americans acquire a pet dog or cat, but only about 20 percent adopt their pet from a shelter (American Humane Association 2011). Dogs and cats that are not adopted or rescued from a shelter are, sadly, destroyed. In some shelters, more than 90 percent of animals are put to death.

The "no kill" movement seeks to end the killing of animals in shelters, and to date at least 160 U.S. cities and towns are designated as "no kill" communities in which shelters save from 90 to 99 percent of their animals (Sandberg 2013). Nathan Winograd, a leader in the no kill movement, views pet overpopulation as a myth that is used to justify

Top Ten Reasons for Pet Relinquishment to U.S. Shelters

Dogs	Cats
1. Moving	1. Too many in house
2. Landlord issues	2. Allergies
3. Cost of pet maintenance	3. Moving
4. No time for pet	4. Cost of pet maintenance
5. Inadequate facilities	5. Landlord issues
6. Too many pets in home	6. No homes for littermates
7. Pet illness	7. House soiling
8. Personal problems	8. Personal problems
9. Biting	9. Inadequate facilities
10. No homes for littermates	10. Doesn't get along with other pets

Source: National Council on Pet Population Study and Policy 2009.

the unnecessary mass killing of animals. He argues that 3 to 4 million dogs and cats are killed every year in shelters, not because there are too few homes to adopt them but because shelter directors fail to find adoptive homes for the animals. No kill shelters achieve high-volume adoptions by working with volunteers, foster families, and rescue groups and individuals. They treat medical and behavior problems and neuter and release, rather than kill, feral cats. Although costs associated with pet ownership lead some pet owners to abandon an animal, Winograd argues that encouraging more adoptions from

shelters not only saves animals' lives, but it also contributes to the local economy:

> Each new pet owner will spend an average of $1,100 a year on his or her pet, and that means more tax revenue for the community. It's an economic boost to local pet stores, groomers, veterinarians, and boarding kennels. No kill is also consistent with public health and safety, because owning pets improves the quality of people's lives and their interactions with one another. We win on so many issues beyond saving the lives of dogs and cats. (quoted in Sandberg 2013, p. 12)

What Do You Think? Ageism is perpetuated in a variety of ways that we commonly accept as harmless fun. For example, birthday greeting cards for aging adults often communicate the message, "Sorry to hear you are another year older." Nelson (2011) remarks, "think about the outrage that would ensue if there was a section of cards that communicated the message 'sorry to hear you're Black' or 'ha ha ha too bad you're Jewish'—yeah, it wouldn't go over so well. So why does society allow, and even condone, the same message directed against older persons?" (p. 41). We give adult birthday gag gifts that make fun of old people with the theme of "over the hill," and we tell or forward jokes that make fun of the elderly, thinking the jokes are humorous rather than offensive. When people are forgetful, they say they are "having a senior moment" without considering that this statement reflects ageism. After considering these points, do you think you will view such acts of making fun of the elderly as harmless fun, or as examples of ageism?

TABLE 12.1 Accomplishments of Famous Older Individuals

At 70, fitness guru Jack LaLanne towed 70 boats, carrying a total of 70 people, a mile and a half through Long Beach Harbor while handcuffed and shackled.

At 72, feminist author Betty Friedan published *The Fountain of Age*, where she debunks misconceptions about aging.

At age 75, Nelson Mandela was elected as the first President of a democratic South Africa.

At 81, Benjamin Franklin facilitated the compromise that led to the adoption of the U.S. Constitution.

At 82, Winston Churchill wrote *A History of the English-Speaking Peoples*.

At 85, actress Mae West starred in the film *Sextette*.

At 85, Coco Chanel was the head of a fashion design firm.

At 87, Mary Baker Eddy created the newspaper *Christian Science Monitor*.

At 88, actress Betty White became the oldest person to ever host *Saturday Night Live*.

At 89, Doris Haddock, also known as "Granny D," began a 3,200-mile walk from Los Angeles to Washington, DC, to raise awareness for the issue of campaign finance reform. She walked 10 miles a day for 14 months, skiing 100 miles when snowfall made walking impossible, and completed her cross-country walk at age 90.

At 90, Pablo Picasso was producing drawings and engravings.

At 93, George Bernard Shaw wrote the play *Farfetched Fables*.

At 94, philosopher Bertrand Russell was active in promoting peace in the Middle East.

At 100, Grandma Moses, noted for her rural American landscapes, was painting. She only started painting at age 78, but by the time she died at 101, she had created over 1,500 works of art.

Nelson Mandela was elected President of South Africa at age 75.

Another indicator of ageism in our society is the negative view of wrinkles, gray hair, and other physical signs of aging. Millions of Americans purchase products or treatments to make them look younger, spending substantial sums of money and undergoing unnecessary and often risky medical procedures.

Unlike more traditional societies that honor and respect their elders, ageism is widespread in modern societies. With the invention of the printing press, elders lost their special status as the keepers of a culture's stories and knowledge (Nelson 2011). Ageism also stems partly from fear and anxiety surrounding aging and death. Many people are uncomfortable around the topic of death and don't like to acknowledge death as a natural part of the life cycle. We use euphemisms to avoid talking about death: We say someone "passed away," "has gone to a better place," "is resting in peace," "kicked the bucket," "has departed," and so on. Old people are reminders of our mortality, and as such, take on negative social meanings.

Employment Age Discrimination. Three in four U.S. workers plan to work past retirement age, either because they will want to (40 percent) or because they will have to (35 percent) (Saad 2013). One of the obstacles older workers face in finding and keeping employment is age discrimination. In one experimental study, younger job seekers were 40 percent more likely to be offered a job interview than older job seekers with similar resumes (Lahey 2008).

Older workers may be more vulnerable to being "let go" because, although they have seniority on the job, they may also have higher salaries, and businesses that need to cut payroll expenses can save more money by letting higher-salaried personnel go. Prospective employers may view older job applicants as "overqualified" for entry-level positions, less productive than younger workers, and/or more likely to have health problems that could affect not only their productivity but also the cost of employer-based group insurance premiums. Employers may also be concerned about the ability of older workers to learn new skills and adapt to new technology.

Family Caregiving for Our Elders

Many adults have or will provide care and/or financial support for aging spouses, parents, grandparents, and in-laws. Adults who care for their aging parents while also taking care of their own children are referred to as members of the **sandwich generation**—they are "sandwiched" in between taking care of both parents and children. Some adults are caring for not two but *three* generations of family members. The **club sandwich generation** includes adults (often in their 50s or 60s) who are caring for aging parents, adult children, and grandchildren, or adults (usually in their 30s and 40s) who are caring for their own young children, aging parents, and grandparents.

For thousands of years, caring for older adults has been an important function of the family. For example, Chinese culture embraces the tenet of *filial piety*, which entails respecting, obeying, pleasing, and offering support and care to parents. Due to social, economic, and cultural changes, it has become more difficult for families throughout the world to care for aging parents. This chapter's *Social Problems Research Up Close* feature examines the challenges of elder care in U.S. families.

Countries around the world have used various methods to encourage parental support. The United States and Taiwan offer tax deductions and credits to adult children caring for elderly parents. In China, which has the largest aging population in the world, some parents have taken their adult children to court for failing to support them. Millions

How much money do you spend on products that promise to keep you looking young?

sandwich generation A generation of people who care for their aging parents while also taking care of their own children.

club sandwich generation Includes adults (often in their 50s or 60s) who are caring for aging parents, adult children, and grandchildren, or adults (usually in their 30s and 40s) who are caring for their own young children, aging parents, and grandparents.

In this feature, we briefly describe a national research study entitled, *The Elder Care Study: Everyday Realities and Wishes for Change* (Aumann et al. 2010). This research provides both quantitative and qualitative data about the experience of providing care to elderly family members.

Sample and Methods

The 2008 National Study of the Changing Workforce (NSCW) gathered data through telephone interviews with a nationally representative sample of 3,502 employed people using a random-digit dial procedure. The response rate was 54.6 percent. From this initial study, 1,589 caregivers—both those who were currently providing care or who had provided care to someone who had died within the past five years—were asked to participate in a telephone interview about their experiences in providing care to an elderly family member. A subsample of 421 family caregivers agreed to participate in a follow-up interview exploring their experiences caring for an elderly relative or in-law, and of these, 140 were successfully contacted and interviewed.

Selected Quantitative Findings

In response to the question, "Within the past five years, have you provided special attention or care for a relative or in-law 65 years old or older—helping with things that were difficult or impossible for them to do themselves?", nearly half—42 percent—responded yes. Three-quarters of the elderly relatives cared for by participants in the qualitative study were age 75 and older.

The types of care family members provided to elders encompass a wide range of tasks and responsibilities, including direct, in-person care (e.g., preparing meals, performing housework, providing transportation to doctor appointments, bathing, etc.) and indirect care (e.g., shopping, arranging for doctor appointments and other services, managing finances, etc.). Most elders in this study (67 percent) lived in their own homes. Nearly one in four caregivers live with their elderly relative or in-law, either in the caregiver's home (18 percent) or in the elderly person's home (6 percent), and 52 percent live 20 minutes or less from the person for whom they are providing care.

Three-quarters (76 percent) of caregivers relied solely on themselves and their families to care for their elderly relative, with no paid assistance. Elder care providers in this study provided elder care for an average of 4.1 years; one in four provided care for 5 or more years. Women (20 percent) and men (22 percent) were equally likely to provide care for elders, although on average, women spent more time than men providing care (9.1 hours a week for women versus 5.7 hours for men). Many of these caregivers are in the "sandwich generation"; 46 percent of women caregivers and 40 percent of men caregivers also have children under the age of 18 at home. The majority of caregivers report not having enough time for their children (71 percent), their spouse/partner (63 percent), and themselves (63 percent).

Selected Qualitative Findings

Family caregivers expressed the wish for more support from workplaces in the form of greater job flexibility and more time off for elder care, especially paid time off without having to use vacation time. Caregivers also expressed wanting more active involvement and help from other family members. But despite the challenges and frustration involved in the caregiving role, many caregivers expressed appreciation for the opportunity to spend time with their elderly relatives and to establish a closer relationship. Interestingly, former caregivers (whose relatives had died) were much more likely to report positive

of Chinese families have signed a **Family Support Agreement**—a voluntary contract between older parents and adult children that specifies the details of how the adult children will provide parental care (Chou 2011).

What Do You Think? In Singapore, the Maintenance of Parents Act of 1995 makes supporting parents a legal obligation, and parents can sue children who fail to support them. And under French law, children are obligated to honor and respect their parents, pay them an allowance, provide or fund a home for them, and to stay informed of their parents' state of health and intervene if there are medical problems. France also has a law that obligates parents to leave their estates to their children. What do you think about such policies? Do you think adult children should be legally responsible for providing support for their aging parents? And should parents be legally obligated to leave their estates to their children?

Family Support Agreement In China, a voluntary contract between older parents and adult children that specifies the details of how the adult children will provide parental care.

changes in their relationship with their elderly relative during caregiving than were current caregivers. The researchers explained:

It appears that the death of the elder alters caregivers' perspective on the caregiving experience and its impact on their relationship. It is possible that the demands and challenges of family caregiving may negatively impact the caregiver's perception of the relationship with the elder during the caregiving experience. Quite possibly, caregivers do not have enough time or mental resources to reflect on the caregiving experience and the relationship with the care recipient until after the caregiving experience is over. The grieving process can thus be seen as a healing process. (p. 24)

Many caregivers reported deriving satisfaction from helping their elder avoid being placed in a nursing home. One caregiver said:

Knowing that she is not in a nursing home, knowing that there is no one harming her, no one taking advantage of her. . . . It is awful. I work in that industry, and the stories are horrifying, so knowing that she is safe in her own home—those are

the comforts. And knowing that I can do everything I can for her. (p. 22)

Many caregivers also described learning valuable lessons from their caregiving experience, including the importance of planning ahead for one's own aging and elder care. When asked what they hoped for in their own aging, caregivers expressed the desire to (1) not "burden" others, especially their children; (2) not burden themselves or others with unaffordable expenses; and (3) not end up in a nursing home.

Finally, researchers noted that

[a]n alarming theme that emerged from our interviews is that family caregivers overwhelmingly seem to view aging and receiving elder care as profoundly negative, depressing processes to be avoided if at all possible. People seem to dread the idea of aging and needing care so much they say they would rather be killed in some other way or even commit suicide. (p. 43)

As one caregiver expressed:

I don't even want to think about it. I want to pass in my sleep of old age. It's an ugly time of life—the last few years of suffering. I would rather die in a car wreck

than put anyone through what I had to go through taking care of my mother. (p. 41)

The findings of "The Elder Care Study" point to the need for better models of elder care that involve more support from the workplace, family/friends, and the health care system. The researchers make this point at the conclusion of their report, where they quote a doctor, who, when standing in a hospital next to an elderly person, said the following:

Look at that bed and imagine how many more people are going to be in beds just like this in the coming years. You and I will be in those beds someday! We *have* to make things better than they are now. (p. 44)

Source: Aumann, Kerstin, Ellen Galinsky, Kelly Sakai, Melissa Brown, and James T. Bond. 2010. *The Elder Care Study: Everyday Realities and Wishes for Change*. Families and Work Institute. Available at www.familiesandwork.org

Retirement Concerns of Older Americans and the Role of Social Security

One of the concerns as we age is financial planning for retirement. About half of U.S. workers are "not at all confident" (28 percent) or "not too confident" (21 percent) that they will have enough money to live comfortably in retirement (Helman et al. 2013). These concerns are not unfounded: The Center for Retirement Research estimates that more than half of U.S. households will not have enough retirement income to maintain their preretirement standard of living (Munnell et al. 2013).

Types of Retirement Plans. Retirement plans include (1) traditional pensions, which are "defined benefit" plans, in which retirees receive a specified annual amount until their death; and (2) "defined contribution" plans in which workers contribute money to 401k plans or individual retirement accounts (IRAs), without any guarantee of what their future benefits will be (see Figure 12.6).

Pensions, 401ks, and IRAs provide limited financial security for old age. Fewer private employers are offering guaranteed pensions, and budget shortfalls threaten the pensions

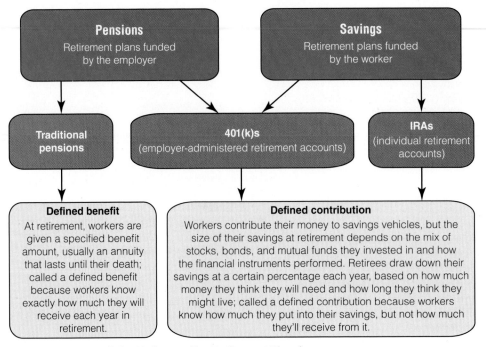

Figure 12.6 Framework for Understanding Retirement Planning
Source: From Edwards, Kathryn A., Anna Turner, and Alexander Hertel-Fernandez. *A Young Person's Guide To Social Security*. Copyright © 2012 Economic Policy Institute. Reprinted with permission.

and other retirement benefits for government workers. Defined contribution retirement plans (401ks and IRAs) are risky because they involve investments in the market and their value fluctuates. In the recent economic crisis, the stock market took a huge hit, producing losses that decimated the IRAs and 401ks of older Americans. Many workers who were planning to retire could no longer afford to stop working. Others who had recently retired and then lost a lot of money in the market felt compelled to reenter the labor force. Although some older workers want to retire but cannot afford to, other older workers want to continue working but are forced to retire due to health problems/disability, job cuts or displacement, or the need to care for parents or spouses (Szinovacz 2011).

The Role of Social Security in Retirement. In addition to retirement plans and savings, most workers are eligible to receive Social Security retirement benefits when they reach retirement age. **Social Security**, actually titled "Old Age, Survivors, Disability, and Health Insurance," is a federal insurance program established in 1935 that protects against loss of income due to retirement, disability, or death. Most (70 percent) Social Security benefits paid in 2011 were retirement benefits, 19 percent were disability benefits, and 11 percent were survivor benefits (to dependent spouses and children) (Edwards et al. 2012). The amount people receive from Social Security is based on how much they earned during their working history—higher lifetime earnings result in higher benefits. Benefit payments also depend on the age at which a person retires. The minimum age for receiving full benefits was 65 for many years, but in 1983, Congress phased in a gradual increase in the full retirement age from 65 to 67. People born in 1960 and later are subject to the new retirement age of 67. Retirees can claim reduced benefits as early as age 62; they receive a larger benefit if they wait until age 70 to claim benefits. In the United States, spouses are entitled to one-half of their partners' benefits regardless of their own work histories and Social Security contributions.

> Without Social Security income, nearly half of all seniors would be living in poverty.

Social Security Also called "Old Age, Survivors, Disability, and Health Insurance," a federal program that protects against loss of income due to retirement, disability, or death.

In July 2013, the average monthly Social Security benefit to retired workers was $1,269.38, which totals about $15,000 a year (Social Security Administration 2013). When Social Security was established in 1935, it was not intended to be a person's sole economic support in old age; rather, it was meant to supplement other savings and assets. But Social Security is a major source of family income for most older Americans: For more than

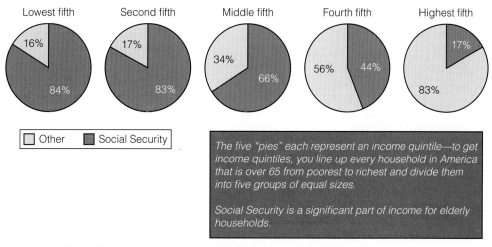

Lowest fifth   Second fifth   Middle fifth   Fourth fifth   Highest fifth

16% / 84%   17% / 83%   34% / 66%   56% / 44%   17% / 83%

☐ Other   ■ Social Security

The five "pies" each represent an income quintile—to get income quintiles, you line up every household in America that is over 65 from poorest to richest and divide them into five groups of equal sizes.

Social Security is a significant part of income for elderly households.

Figure 12.7 **Share of Income from Social Security of Households 65 or Over by Income Quintile, 2010**
Source: From Edwards, Kathryn A., Anna Turner, and Alexander Hertel-Fernandez. *A Young Person's Guide To Social Security*. Copyright © 2012 Economic Policy Institute. Reprinted with permission.

half of Americans over age 65, Social Security provides more than half of their income, and without Social Security income, nearly half of all seniors would be living in poverty (see Figure 12.7). Instead, as noted in Chapter 6, poverty rates for U.S. adults ages 65 and older are lower than for any other age group. Because Social Security payments are based on the number of years of paid work and preretirement earnings, women and minorities, who often earn less during their employment years, receive less in retirement benefits.

How Is Social Security Funded? Social Security is funded by workers through a payroll tax called the Federal Insurance Contributions Act (FICA) that comprises 12.4 percent of a worker's wages (6.2 percent is deducted from the worker's paycheck and 6.2 percent is paid by the employer). Self-employed workers pay the entire FICA tax. FICA taxes are paid on wages up to a certain level, or tax cap, which changes each year based on average U.S. wages. In 2012, the tax cap was $110,100, meaning that people whose wages were more than this amount paid FICA tax only on the first $110,100 of their earnings.

Another source of funding for Social Security is a tax on higher-income beneficiaries. For most recipients, Social Security benefits are not taxed, but for recipients who have income from other sources that is above a certain amount, a portion of Social Security benefits is taxed to help finance Social Security.

Finally, Social Security funds are held in a trust fund and invested in securities guaranteed as to both principal and interest by the federal government. Social Security benefits are paid out of this trust fund, and each year a Board of Trustees issues a report on the financial status of the trust fund.

Is Social Security in Crisis? In 2012, Social Security's total income ($840 billion, which included $731 billion from taxes paid into Social Security and $109 billion in interest earnings) exceeded its total expenditures ($786 billion) (Social Security Trustees 2013). However, a number of factors threaten the long-term ability of Social Security to meet its financial obligations to future retirees, including the retirement of the baby boomers, increasing longevity, the declining elderly support ratio (fewer workers per beneficiary), widening wage inequality (which means that more income is not subject to Social Security taxes), high rates of unemployment, and wage stagnation.

Since 1984, surpluses have been accumulating in the trust fund, creating significant reserves for the baby boomers' retirement. According to a report of the Social Security Trustees (2013), Social Security's total income, including interest earnings on trust fund assets, will be sufficient to cover annual costs until 2033. The bottom line seems to be that, in the short term, Social Security is not "broke" and there is no immediate crisis. But long-term changes to the Social Security system will be needed to ensure that the program is able to meet its financial obligations to future retirees. Options for Social Security reform are discussed in the following "Strategies for Action" section of this chapter.

Strategies for Action: Responding to Problems of Population Growth and Aging

Next we look at efforts to curb population growth as well as efforts to increase population in countries experiencing population decline. We end the chapter with a discussion of strategies that address social problems associated with aging.

TABLE 12.2 Total Fertility Rates by Region: 1970–1975 to 2005–2010	1970–1975	2005–2010
World	4.44	2.53
Africa	6.66	4.88
Asia	4.99	2.25
Europe	2.17	1.54
Latin America and Caribbean	5.02	2.30
North America	2.01	2.02
Oceania	3.23	2.47

Source: United Nations 2013.

Recognizing that men play a crucial role in family planning decisions, family planning programs are making efforts to include men in family planning education and services.

Efforts to Curb Population Growth: Reducing Fertility

Although worldwide fertility rates have fallen significantly since the 1970s, they are still high in many less developed regions (see Table 12.2). Approaches to reducing fertility in high-fertility regions include family planning, economic development, improving the status of women, providing access to safe abortion, government population control policies such as China's one-child policy, and voluntary childlessness.

Family Planning and Contraception. Since the 1950s, governments and nongovernmental organizations such as the International Planned Parenthood Federation have sought to lower fertility through family planning programs that provide reproductive health services and access to contraceptive information and methods. Yet there are still 222 million women in developing countries who want to delay or stop childbearing but are not using contraception (World Health Organization 2013). Nearly one-quarter of reproductive-age women in developing countries have an "unmet need" for family planning services (United Nations 2013).

Population expert William Ryerson (2011) claims that, in most countries, lack of access to family planning is a very minor reason for not using contraception. In Nigeria, less than one percent of nonusers who don't want to be pregnant cite lack of access as the reason. Other reasons for not using modern contraception include (1) the belief that modern methods of contraception are dangerous; (2) male partners are opposed to using modern contraception; (3) the belief that one's religion opposes the use of contraception; and (4) the belief that God should decide how many children a woman has (Ryerson 2011). Encouraging women to use modern contraception requires providing access to affordable contraceptive methods, but providing access is not enough. Women and men need education to dispel myths about the dangers of using contraception and to understand the health and economic benefits of delayed, spaced, and limited childbearing. Involving men in family planning services is important because, although men play a central role in family planning decisions, they often do not have access to information and services that would empower them to make informed decisions about contraceptive use (Women's Studies Project 2003).

Economic Development. Although fertility reduction can be achieved without industrialization, economic development may play an important role in slowing population growth. Families in poor countries often rely on having many children to provide enough labor and income to support the family. Economic development decreases the economic value of children and is also associated with more education for women and greater gender equality.

Economic development tends to result in improved health status of populations. Reductions in infant and child mortality are important for fertility decline, because couples no longer need to have many pregnancies to ensure that some children survive into adulthood. Finally, the more developed a country is, the more likely women are to be exposed to meanings and values that promote fertility control through their interaction in educational settings and through media and information technologies (Bongaarts & Watkins 1996).

> Families in poor countries often rely on having many children to provide enough labor and income to support the family.

The Status of Women: The Importance of Education and Employment.

Throughout the developing world, the primary status of women is that of wife and mother. Women in developing countries traditionally have not been encouraged to seek education or employment; rather, they are encouraged to marry early and have children.

Improving the status of women by providing educational and occupational opportunities is vital to curbing population growth. Educated women are more likely to marry later, want smaller families, and use contraception. Data from many countries have shown that women with at least a secondary-level education eventually give birth to between one-third and one-half as many children as women with no formal education (Population Reference Bureau 2007).

Compared with women with less education, women with more education tend to delay marriage and exercise more control over their reproductive lives, including decisions about childbearing. In addition, "education can result in smaller family size when the education provides access to a job that offers a promising alternative to early marriage and childbearing" (Population Reference Bureau 2004, p. 18). Providing employment opportunities for women is also important to slow population growth, because high levels of female labor force participation and higher wages for women are associated with smaller family size. Primary school enrollment of both young girls and boys is also related to declines in fertility rates. In countries where primary school enrollment is widespread or nearly universal, fertility declines more rapidly because (1) schools help spread attitudes about the benefits of family planning, and (2) universal education increases the cost of having children, because parents sometimes are required to pay school fees for each child and because they lose potential labor that children could provide (Population Reference Bureau 2004). However, providing access to contraception and increasing girls' education are unlikely to slow population growth without changes in male attitudes toward family planning. Indeed, attempts to provide free primary education may increase African men's desire for more children because the costs are decreased (Frost & Dodoo 2009).

Another important component of family planning and reproductive health programs involves changing traditional male attitudes toward women. According to traditional male gender attitudes, (1) a woman's most important role is being a wife and mother, (2) it is a husband's right to have sex with his wife at his demand, and (3) it is a husband's right to refuse to use condoms and to forbid his wife to use any other form of

> Women with more education tend to delay marriage and exercise more control over their reproductive lives, including decisions about childbearing.

contraception. In parts of Africa, men express their reluctance to use condoms by arguing that "you wouldn't eat a banana with the peel on or candy with the wrapper on" (Frost & Dodoo 2009). A number of programs around the world work with groups of boys and young men to change such traditional male gender attitudes (Schueller 2005).

Access to Safe Abortion.

Abortion is a sensitive and controversial issue that has religious, moral, cultural, legal, political, and health implications (see also Chapter 14). Worldwide, one in five pregnancies ends in abortion, and one in ten pregnancies ends in unsafe abortion, including those performed in unhygienic conditions by unskilled providers (such as traditional or religious healers and herbalists) and those that are self-induced by a woman inserting a foreign object into her uterus, ingesting toxic substances, or inflicting trauma to the abdomen (Mesce & Clifton 2011). Of the estimated 42 million abortions each year, about 47,000 girls and women die. Almost all unsafe abortions take place in developing countries (Mesce & Clifton 2011).

More than one-quarter of the world's population lives in countries where abortion is prohibited or allowed only to save the life of the woman. Where abortion is illegal, it is very often unsafe, and where abortion is legal, and widely accessible through formal health systems, it is highly safe (Barot 2011). Because the abortion rate is similar in regions with legal abortion and in regions with restrictive abortion laws, Barot (2011) notes that "legal restrictions on abortion largely do not affect whether women will get an abortion, but they can have a major impact on whether abortion takes place under safe or unsafe conditions and, therefore, whether it jeopardizes women's health and lives" (p. 25).

Some couples in China are eligible for an exemption from the one-child policy and are encouraged to have a second child.

AP Images/Imaginechina

China's One-Child Policy. In 1979, China initiated a national family planning policy that encourages families to have only one child by imposing a monetary fine on couples that have more than one child. The implementation and enforcement of this policy varies from one province to another, and there are a number of exemptions that allow some couples to have two or even more children. Nevertheless, the one-child policy has been effective in reducing population growth in China, which has the highest rate of modern contraceptive use in the world. Of married women ages 15 to 49 in China, 86 percent use modern contraceptives (Population Reference Bureau 2010).

China has been criticized for using extreme measures to enforce its one-child policy, including steep fines, seizure of property, and forced sterilizations and abortions. In addition, because of a traditional preference for male heirs, many Chinese couples have aborted female fetuses with the hope of having a boy in a subsequent pregnancy. This has led to an imbalanced sex ratio, with many more Chinese males than females.

Another problem with the one-child policy is that, as older Chinese retire, there are fewer workers to take their place and to support pensions for the elderly. Largely due to concerns of these imbalances in the population structure, China is considering adopting a two-child policy (Branigan 2011).

Voluntary Childlessness. In the United States, as in other countries of the world, the cultural norm is for women and couples to, sooner or later, want to have children. However, a small but growing segment of U.S. women and men does not want children and chooses to be childfree. In the United States, of the 19 percent of women ages 40 to 44 who are childless, half are childless by choice (Notkin 2013). Other women who are childless are either unable to have children or are waiting for love and the right relationship before motherhood. In general, childfree couples are more educated, have higher incomes, live in urban areas, are less religious, and do not adhere to traditional gender ideology (Parks 2005). In a study of 121 childless-by-choice women, the top reason women gave for not wanting children is that they simply love their life as it is (Scott 2009). Other reasons included valuing freedom and independence and not wanting to take on the responsibility. Three-quarters of the women said they "had no desire to have a child, no maternal/paternal instinct." Another reason some individuals choose not to have children is concern for overpopulation and a deep caring for the health of the planet. Yet, ironically, voluntarily childless individuals are often criticized as being selfish and individualistic, as well as less well adjusted and less nurturing (Parks 2005). In this chapter's *The Human Side* feature, one man explains why he chose to have a vasectomy and not have children.

Matt Leonard lives in California, where he works on climate justice and energy issues with the organizations Greenpeace and Rising Tide North America. In this The Human Side *feature, Matt Leonard explains why he has chosen to have a vasectomy—a simple surgical procedure that makes men unable to cause pregnancy.*

Courtesy of Matt Leonard

Last year, I . . . had a vasectomy. While it's actually a very common procedure (nearly 500,000 are performed every year in the United States), it raises eyebrows—and a lot of questions.

The first one is always simply: *Why?*

Although this was a very personal decision for me, it was also a choice I made out of larger societal, political, and environmental motivations. I consider the environmental ones paramount. In an economic system that demands infinite growth with finite resources, not doubling my own consumption is one small stone in a big river.

More importantly, I live in the United States, and any child I had would have been raised here and would consume (despite my best efforts) far more resources than I am comfortable accepting. Living even a modest lifestyle in the United States comes as a direct result of the oppression, domination, and deaths of many unseen people, not to mention the exploitation of natural resources at rates that threaten the ability of our planet to sustain life. These facts shouldn't be cause for guilt or shame; instead, they should spur us to organize to confront the systems and institutions that have created these problems. On a personal level, contributing another person to the system that I have spent my adult life fighting is just not something I'm willing to do. . . .

The next question is usually: *But what if you change your mind?*

I view my decision as permanent. As I see it, I already made the decision years ago not to have children, based on sound, rational reasons. If I change my mind in the future, I believe that change would be fundamentally selfish, and I am comfortable committing myself to rational reasons now.

People typically follow up with: *Aren't there other forms of birth control?*

Yes, of course, and most of us here in the United States are lucky to be able to choose the form that is best for our lifestyles, our preferences, and our relationships. A vasectomy fit my needs best. I guess there's always abstinence, but that's no fun, right? I suppose the rhythm method is an option, but almost everyone knows how [in]effective that is. Condoms are fine and dandy in many situations, but they have their downsides as well, and can seem pointless if you are in a monogamous relationship.

All the other common birth control methods have one aspect in common: They place the onus on women. Not only does our society expect women to deal with the logistics of birth control, but these methods also have severe physiological drawbacks, from roller-coaster hormonal changes to intensifying menstruation cycles to weight and skin changes. Although these methods have come a long way in a few decades, they still burden women and their bodies. Is it any coincidence that in a male-dominated society, the medical establishment has thus far focused on birth control methods that leave the burden solely on women?

For men, vasectomies are simple. There are almost no side effects and no long-term impacts; it's a quick, low-cost, outpatient procedure. Having decided that I want to take an active role in birth control, a vasectomy is fair, easy, and it confronts my privilege on this issue.

What if you decide you want children in the future? people ask.

Many of my friends whom I deeply respect have chosen to have children or will do so in the future. Some people do feel that there is something special and important about having a blood-related child. I just don't share that feeling.

There are thousands of beautiful children all over the world who need parents, and if I ever decide that being a father is something I want in my life, I would be remiss to ignore the existing children needing support and love. For me, adoption is the best option. We need more parents in this world, not more kids.

Finally: *But don't we need the smart, progressive people to reproduce?*

I'm of the nurture-over-nature camp. I think the whole "passing on genes" obsession can sometimes border on eugenics. I'm fairly confident there is no gene that instructs your child to fight for justice, peace, and sustainability. That comes from living those values and instilling them in the communities we are a part of. That's what I want to prioritize in my life—and I feel I can share those things more effectively without a child.

And besides—I've got messed-up teeth, I'm legally blind, bald, and have a history of heart disease. Let Matt Damon pass on his genes instead.

Efforts to Increase Population

Whereas some countries are struggling to slow population growth, others are challenged with maintaining or even increasing their populations. In some countries with below-replacement fertility levels, population strategies have focused on *increasing* rather than decreasing the population. For example, in 2013, Singapore announced that it would offer a package of incentives to increase population, including government-paid time off for adoption and paternity leave, funding for fertility treatment, and state-funded assistance with medical care for children. In Australia and Japan, the government has paid women a monetary bonus for having babies. Aside from monetary rewards, many countries encourage childbearing by implementing policies designed to help women combine child rearing with employment. For example, as noted in Chapters 5 and 7, many European countries have generous family leave policies and universal child care. Another way to increase population is to increase immigration. A number of European countries, for example, have eased restrictions on immigration as a way to gain population.

What Do You Think? One strategy for encouraging childbearing in European countries with low fertility rates is to provide work–family supports to make it easier for women to combine childbearing with employment. If the United States offered more generous work–family benefits, such as paid parenting leave and government-supported child care, would the U.S. birthrate increase? Would such policies affect the number of children you would want to have?

Combating Ageism and Age Discrimination in the Workplace

One strategy to combat ageism involves incorporating positive views of aging in educational lessons, beginning in preschool, and in other forms of media. Such lessons should teach about aging as a normal part of life, rather than something to fear or be embarrassed about. Younger people should have opportunities to learn from the wisdom, experience, and life perspective of older individuals.

In 1967, Congress passed the Age Discrimination in Employment Act (ADEA), which was designed to ensure continued employment for people between the ages of 40 and 65. In 1986, the upper age limit was removed, making mandatory retirement illegal in most occupations. Under ADEA, it is illegal to discriminate against people because of their age with respect to hiring, firing, promotion, layoff, compensation, benefits, job assignments, and training. Age discrimination is difficult to prove. Nevertheless, thousands of age discrimination cases are filed annually with the Equal Employment Opportunity Commission (EEOC).

What Do You Think? The Age Discrimination in Employment Act prohibits imposing mandatory retirement on employees, but there are exceptions to this rule. In the private sector, certain executives such as partners in accounting firms can be subject to mandatory retirement. Firefighters, law enforcement officers, military personnel, pilots, air traffic controllers, and in some states, judges are also subject to mandatory retirement. Mandatory retirement is justified by the argument that certain occupations require high levels of physical and/or mental ability, and that age-related decline could jeopardize the safety or well-being of the public and/or the worker. Because mandatory retirement policies are based on a fixed age (varies according to the profession and by state for judges) and not on an evaluation of the worker's abilities, some say that mandatory retirement is a form of age discrimination. Do you think mandatory retirement is justified, or is it age discrimination?

Social Security is the most solvent part of the U.S. government. It is funded through its own separate tax and interest from its trust fund to pay beneficiaries. No other government program or agency is fully funded. Social Security is also a very efficient program. Although it collects taxes from more than 90 percent of the workforce and sends benefits to more than 50 million Americans, Social Security spends less than one cent of every dollar on administration.

Nevertheless, as discussed earlier in this chapter, due to a number of factors, unless changes are made in years to come, Social Security won't have enough funds to provide payouts to future retirees. Options for reforming Social Security include cutting Social Security benefits, increasing Social Security revenue, and expanding Social Security benefits. The following outlines these options (Edwards et al. 2012; Morrissey 2011).

Senior citizens generally oppose proposals to cut Social Security benefits.

Cut Social Security Benefits. Some reform proposals call for cuts in Social Security benefits, which would create a significant financial burden on households that depend on Social Security. Aside from simply reducing the amount of benefits paid to recipients, another way to cut benefits is to increase the retirement age. In 1983, the retirement age (the age at which a worker could collect full Social Security retirement benefits) was raised from 65 to 67 to be phased in over 23 years. Some policy makers suggest raising the retirement age even further to 68, 69, or even 70. However, due to the link between higher socioeconomic status and longer life expectancy, raising retirement age imposes the greatest burden on low earners who have lower life expectancies and therefore fewer years to collect Social Security. In addition, older workers already face employment discrimination, and finding or keeping employment is difficult for seniors. Finally, the older workers become, the more likely they will experience health problems, and hence file disability claims.

Increase Social Security Revenue. Instead of cutting benefits, many advocates for the elderly suggest raising revenues. One option for increasing Social Security revenue is simply to raise the tax that funds Social Security (the payroll or FICA tax) from its current rate of 12.4 percent (6.2 percent paid by employer; 6.2 percent paid by employee) to a higher rate. To offset gains in life expectancy, Social Security taxes increased 19 times, from 1 percent between 1937 and 1949 to 6.2 percent in 1990 (paid by both employer and employee). As of this writing (August 2013), Social Security taxes have not increased since 1990—the longest period without an increase to the payroll tax. In 2011 and 2012, the workers' share was cut to 4.2 percent under a "payroll tax holiday" economic stimulus measure. The downside to raising the payroll tax is that it could result in fewer jobs, as employers would have a higher financial burden. Paying a higher payroll tax would also disproportionately burden lower-wage earners.

Another option is to raise or even eliminate the tax cap, so that more earnings are taxed, providing more funds to the Social Security program. This would affect higher-income earners, who currently earn more than the tax cap, but only pay payroll taxes on the first $110,100 (the tax cap in 2012) of their earnings.

Finally, policies that increase employment and wages of all workers would lead to more revenue for Social Security. Because health insurance costs are deducted from taxable income, policies to reduce health care costs would also indirectly increase Social Security funding.

Expand Social Security Benefits. In this current economic climate, people need more, not less, economic support. Options for increasing Social Security benefits include raising the minimum benefit amount, offering unemployed parents who are taking care of their children wage credits, and increasing the benefits for the very old (85 years and older). Another proposal involves restoring a student benefit (which existed between 1965 and 1985) so that children of the retired, deceased, or disabled can continue to receive benefits until age 22 if attending college or vocational school.

Not surprisingly, proposals to cut Social Security benefits or raise retirement age for receiving benefits are vehemently opposed by seniors and advocacy groups for older adults. Raising the tax cap would go far to ensure the future financial viability of Social Security, but this option is generally not supported by the wealthy segment of the population who would bear the burden of higher taxes. Until legislators enact changes to prevent the long-term Social Security deficit, there is little chance of seeing Social Security benefits increased.

Understanding Problems of Population Growth and Aging

What can we conclude from our analysis of population growth and aging? First, although fertility rates have declined significantly in recent years and although many countries are experiencing a decline in their fertility rates, world population will continue to grow for several decades. This growth will occur largely in developing regions. Finance columnist Paul Farrell (2009) says that population growth is "the key variable in every economic equation . . . impacting every other major issue facing world economies . . . from peak oil to global warming . . . from foreign policy to nuclear threats . . . from religion to science . . . everything." Given the problems associated with population growth, such as environmental problems and resource depletion, global insecurity, poverty and unemployment, and poor maternal and infant health, most governments recognize the value of controlling population size and supporting family planning programs. However, efforts to control population must go beyond providing safe, effective, and affordable methods of birth control. Slowing population growth necessitates interventions that change the cultural and structural bases for high fertility rates. Reducing fertility necessitates improving the status of women so that women have more power to control their choices regarding contraception and reproduction and have more life options other than being a wife and mother. Addressing problems associated with population growth also requires the willingness of wealthier countries to commit funds to providing reproductive health care to women, improving the health of populations, and providing universal education for people throughout the world.

As fertility rates decline and life expectancy increases, the United States and other countries are experiencing population aging. Many areas of social life are affected by the aging of the population, and more attention is being given to concerns related to older people. What it means to be old is culturally defined. Unfortunately, many cultural meanings associated with the aged are negative stereotypes. Prejudice toward and discrimination against older people constitutes ageism, but unlike other "isms"—racism, sexism, heterosexism—ageism is more socially acceptable and will affect all of us who reach a certain age.

Although older individuals struggle against ageism, families and governments face challenges of meeting the needs of seniors. Caring for elders has traditionally been an important function of the family, but social, cultural, and economic changes have made it more challenging for families to fulfill this function. Social Security plays an important role in supporting many older retirees, but due to projections of a future shortfall, some reforms to Social Security are necessary. What type of reform measures legislators enact continues to be hotly debated. Until such reform measures are in place, the future of Social Security and the well-being of older Americans is uncertain.

Like other social problems, slowing population growth and meeting the needs of aging populations require political will and leadership. Given all the pressing concerns in the world, control of population may not seem like a priority. But as finance columnist Paul Farrell (2009) warns,

> [p]opulation is the core problem that, unless confronted and dealt with, will render all solutions to all other problems irrelevant. Population is the one variable in an economic equation that impacts, aggravates, irritates, and accelerates all other problems.

And if taking care of older populations is not a priority now, those of us who are not yet "old" will suffer the consequences in the future, if, or when, we enter our "golden years."

- **How long did it take for the world's population to reach 1 billion? How long did it take for it to reach 6 billion?**
It took thousands of years for the world's population to reach 1 billion, and just another 300 years for the population to grow from 1 billion to 6 billion.

- **Is the world population still increasing? Is it decreasing? Or is it remaining stable?**
World population is still growing. It is projected to grow from 7.2 billion in 2013, to 8.1 billion in 2025, 9.6 billion in 2050, and 10.9 billion by 2100.

- **Where is most of the world's population growth occurring?**
Most world population growth is in developing countries, primarily in Africa and Asia.

- **Between 2013 and 2050, how is the percentage of older individuals (ages 60 and over) expected to change globally?**
Globally, the population aged 60 and older is the fastest growing age group. Between 2013 and 2050, the percentage of older individuals (ages 60 and over) in the world population is expected to nearly double from 12 percent to 21 percent.

- **What factors are contributing to population aging?**
Population aging is a function of lowered fertility as well as increased longevity. In the United States, population aging is also occurring because the baby boomers are reaching their senior years.

- **What is the demographic transition theory?**
The demographic transition theory of population describes how industrialization and economic development affect population growth by influencing birth and death rates. According to this theory, traditional agricultural societies have both high birth rates and high death rates. As a society becomes industrialized and urbanized, improved sanitation, health, and education lead to a decline in mortality. The increased survival rate of infants and children along with the declining economic value of children leads to a decline in birth rates. The demographic transition theory is a generalized model that does not apply to all countries, and does not account for population change due to HIV/AIDS, war, migration, and changes in gender roles and equality.

- **Many countries are experiencing below-replacement fertility (fewer than 2.1 children born to each woman). Why are some countries concerned about their low fertility?**
In countries with below-replacement fertility, there are or will be fewer workers to support a growing number of elderly retirees and to maintain a productive economy.

- **What kinds of environmental problems are associated with population growth?**
Population growth places increased demands on natural resources, such as forests, water, cropland, and oil, and results in increased waste and pollution. Although population growth is a contributing factor in environmental problems, patterns of production and consumption are at least as important in influencing the effects of population on the environment.

- **Why is population growth considered a threat to global security?**
In developing countries, rapid population growth results in a "youth bulge"—a high proportion of 15- to 29-year-olds relative to the adult population. The combination of a youth bulge with other characteristics of rapidly growing populations, such as resource scarcity, high unemployment rates, poverty, and rapid urbanization, sets the stage for civil unrest, war, and terrorism, because large groups of unemployed young people resort to violence in an attempt to improve their living conditions.

- **How is ageism different from other "isms" (racism, sexism, heterosexism)?**
Ageism is more widely accepted than other "isms" and, unlike other "isms," everyone is vulnerable to experiencing ageism if they live long enough.

- **How important is Social Security income for senior citizens in the United States?**
For more than half of Americans over age 65, Social Security provides more than half of their income, and without Social Security income, nearly half of all seniors would be living in poverty.

- **Is Social Security in crisis?**
In the short term, Social Security is not "broke" and there is no immediate crisis. But long-term changes to the Social Security system will be needed to ensure that the program is able to meet its financial obligations to future retirees.

- **What is the "sandwich generation?" and the "club sandwich generation"?**
The "sandwich generation" refers to adults who care for their aging parents while also taking care of their own children—they are "sandwiched" in between taking care of both parents and children. The "club sandwich generation" includes adults (often in their 50s or 60s) who are caring for aging parents, adult children, and grandchildren, or adults (usually in their 30s and 40s) who are caring for their own young children, aging parents, and grandparents.

- **Efforts to curb population growth include what strategies?**
Efforts to curb population growth include strategies to reduce fertility by providing access to family planning services, involving men in family planning, implementing a one-child policy as in China, and improving the status of women by providing educational and employment opportunities. Achievements in economic development and health are also associated with reductions in fertility.

- **What are three general options for reforming Social Security?**
Options for reforming Social Security include strategies for increasing Social Security revenue, cutting benefits, and expanding Social Security benefits.

1. In 2013, world population was 7.2 billion. In 2050, world population is projected to be ___ billion.
   a. 6.5
   b. 7.4
   c. 8.8
   d. 9.6

2. How many countries have achieved below-replacement fertility rates?
   a. None
   b. 5
   c. 28
   d. More than 70

3. What would happen if every country in the world achieved below-replacement fertility rates?
   a. Population growth would stop and world population would remain stable.
   b. World population would immediately begin to decline.
   c. World population would continue to grow for several decades.
   d. World population would decline, but then go up again.

4. For more than half of Americans over age 65, Social Security provides more than half of their income.
   a. True
   b. False

5. Conflict theorists argue that food shortages result primarily from overpopulation of the planet.
   a. True
   b. False

6. Pronatalism is a cultural value that promotes which of the following?
   a. Car ownership
   b. Abstaining from sex until one is married
   c. Having children
   d. Urban living

7. According to the "Ageism Survey," jokes and birthday cards that poke fun at the elderly represent a form of ageism.
   a. True
   b. False

8. What is the number one cause of death for dogs and cats in the United States?
   a. Getting hit by a car
   b. Shelter euthanasia
   c. Cancer
   d. Eating something poisonous

9. U.S. women with advanced education are more likely than women with less education to voluntarily choose to have no children.
   a. True
   b. False

10. In 1983, the retirement age (the age at which a worker could collect full Social Security retirement benefits) was raised from 65 to 70 to be phased in over 23 years.
   a. True
   b. False

Answers: 1. D; 2. D; 3. C; 4. A; 5. B; 6. C; 7. A; 8. B; 9. A; 10. B

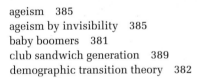

Environmental Problems

"The world will no longer be divided by the ideologies of 'left' and 'right,' but by those who accept ecological limits and those who don't."

—Wolfgang Sachs

The Global Context: Globalization and the Environment

Sociological Theories of Environmental Problems

Social Problems Research Up Close: **The Seven Sins of Greenwashing**

Environmental Problems: An Overview

Photo Essay: **Deepwater Horizon Oil Rig Explosion and Fukushima Nuclear Power Plant Accident**

The Human Side: **A Casualty of Water Contamination at Camp Lejeune**

Social Causes of Environmental Problems

Strategies for Action: Responding to Environmental Problems

Self and Society: **Attitudes toward Government Interventions to Reduce Global Warming**

Understanding Environmental Problems

Chapter Review

The first act of civil disobedience authorized by the Sierra Club in its 120-year history was a protest at the White House to voice opposition to the Keystone XL Pipeline.

IN EARLY 2013, four dozen environmental activists, including actress Daryl Hannah, were arrested for participating in peaceful protests at the White House to voice opposition to the proposed Keystone XL oil pipeline. The activists were charged with failure to disperse and obey lawful orders. The protest, which was organized by the Sierra Club, was the first act of civil disobedience authorized by the Sierra Club in its 120-year history (Broder 2013). The proposed Keystone XL pipeline would carry 900,000 barrels a day of tar sands oil from Canada to the Gulf, crossing Montana, South Dakota, Nebraska, Kansas, Oklahoma, and Texas. **Tar sands** are large, naturally occurring deposits of sand, clay, water, and a dense form of petroleum (that looks like tar). **Tar sands oil** has been referred to as the world's dirtiest oil. Converting tar sands into liquid fuel requires energy and generates high levels of greenhouse gases (that cause global warming and climate change), and also leaves behind large amounts of toxic waste. The mining of tar sands requires large amounts of water and involves destruction of forests and wetlands, which disrupts wildlife habitats. And transporting the tar sands oil through a pipeline involves risk of leakage that could affect the safety of drinking water (Swift et al. 2011). The protest against the proposed Keystone XL pipeline is one of the largest civil disobedience actions in decades. NASA scientist James Hansen said if the Canadian tar sands are fully developed, it is "essentially game over for the climate" (quoted in Elk 2011).

In this chapter, we focus on environmental problems that threaten the lives and well-being of people, plants, and animals all over the world—today and in future generations. After examining how globalization affects environmental problems, we view environmental issues through the lens of structural functionalism, conflict theory, and symbolic interactionism. We then present an overview of major environmental problems, examining their social causes and exploring strategies to reduce or alleviate them.

The Global Context: Globalization and the Environment

Two aspects of globalization that have affected the environment are (1) the permeability of international borders to pollution and environmental problems, and (2) the growth of free trade and transnational corporations.

Permeability of International Borders

tar sands Large, naturally occurring deposits of sand, clay, water, and a dense form of petroleum (that looks like tar).

tar sands oil Oil that results from converting tar sands into liquid fuel. It is known as the world's dirtiest oil because producing it requires energy and generates high levels of greenhouse gases (that cause global warming and climate change), and also leaves behind large amounts of toxic waste.

The most urgent environmental problem we face today—global warming and climate change—impacts the global civilization. No country can escape global warming's effects, such as loss of biodiversity and extreme weather events (we discuss these effects in detail later in this chapter).

Other environmental problems also can extend far beyond their source to affect distant regions and even the entire planet. For example, toxic chemicals (such as polychlorinated biphenyls [PCBs]) from the Southern Hemisphere have been found in the Arctic. In as few as five days, chemicals from the tropics can evaporate from the soil, ride the winds thousands of miles north, condense in the cold air, and fall on the Arctic in the form of toxic snow or rain (French 2000). This phenomenon was discovered in the mid-1980s, when scientists found high levels of PCBs in the breast milk of Inuit women in the Canadian Arctic region. Another example involves the leaking of radioactive contaminated water from Japan's crippled Fukushima Daiichi nuclear power plant into the Pacific Ocean (see this Chapter's *Photo Essay*). Radioactive seawater originating in Japan could reach the west coast of the United States in three to five years (Kiger 2013).

Another environmental problem involving permeability of borders is **bioinvasion**: the intentional or accidental introduction of organisms in regions where they are not native. Bioinvasion is largely a product of the growth of global trade and tourism (Chafe 2005). Invasive species compete with native species for food, start an epidemic, or prey on native species. Nonnative wood-boring insects, which are among the 450 nonnative insects that are established in the United States, cost an estimated $1.7 billion in local government expenses and $830 million in lost residential property values every year (Aukema et al. 2011). Other nonnative insects are red fire ants, which traveled from Paraguay and Brazil on shiploads of lumber to Mobile, Alabama in 1957, and have since spread throughout the southern states, causing damage to gardens and yards, invading the food supplies (seeds, young plants, and insects) of other animals, and harming humans with their painful sting (Hilgenkamp 2005). You might be surprised to learn that the domestic cat is considered among the world's 100 worst invasive species. Native to northeast Africa, cats have spread to every part of the world and are responsible for the decline and extinction of many species of birds (Global Invasive Species Database 2013).

Toxic chemicals travel thousands of miles from the Southern Hemisphere to the Arctic, where they have been found in the breast milk of Inuit women.

The Growth of Transnational Corporations and Free Trade Agreements

As discussed in Chapter 7, the world's economy is dominated by transnational corporations, many of which have established factories and other operations in developing countries where labor and environmental laws are lax. Transnational corporations have been implicated in environmentally destructive activities—from mining and cutting timber to dumping toxic waste.

The World Trade Organization (WTO) and free trade agreements such as the North American Free Trade Agreement (NAFTA) and the Free Trade Area of the Americas (FTAA) allow transnational corporations to pursue profits, expand markets, use natural resources, and exploit cheap labor in developing countries while weakening the ability of governments to protect natural resources or to implement environmental legislation (Bruno & Karliner 2002).

Under NAFTA's Chapter 11 provisions, corporations can challenge local and state environmental policies, federal-controlled substances regulations, and court rulings if such regulatory measures and government actions negatively affect the corporation's profits. Any country that decides, for example, to ban the export of raw logs as a means of conserving its forests or, as another example, to ban the use of carcinogenic pesticides, can be charged under the WTO by member states on behalf of their corporations for obstructing the free flow of trade and investment. A secret tribunal of trade officials would then decide whether these laws were "trade restrictive" under the WTO rules and should therefore be struck down. Once the secret tribunal issues its edict, no appeal is possible. The convicted country is obligated to change its laws or face the prospect of perpetual trade sanctions (Clarke 2002, p. 44). For example, in the late 1990s, Ethyl, a U.S. chemical company, used NAFTA rules to challenge Canada's decision to ban the gasoline additive methylcyclopentadienyl manganese tricarbonyl (MMT), which is believed to have harmful effects on human health. Ethyl won the suit, and Canada paid $13 million in damages and legal fees to Ethyl and reversed the ban on MMT (Public Citizen 2005).

Sociological Theories of Environmental Problems

The three main sociological theories—structural functionalism, conflict theory, and symbolic interactionism—provide insights into social causes of and responses to environmental problems.

bioinvasion The intentional or accidental introduction of plant, animal, insect, and other species in regions where they are not native.

Structural-Functionalist Perspective

Structural functionalism views social systems (e.g., families, workplaces, societies) as composed of different parts that work together to keep the whole system functioning. Likewise, humans are part of a larger **ecosystem**—which consists of all the organisms living in a particular area, as well as all the nonliving, physical components of the environment, like air, water, soil, and sunlight, that interact to keep the whole ecosystem functioning. Each living and nonliving part of the ecosystem plays a vital role in maintaining the whole; disrupt or eliminate one element of the ecosystem, and every other part could be affected.

> The capitalistic pursuit of profit encourages making money from industry regardless of the damage done to the environment.

Structural functionalism focuses on how changes in one aspect of the social system affect other aspects of society. For example, agriculture, forestry, and fishing provide 50 percent of all jobs worldwide and 70 percent of jobs in sub-Saharan Africa, East Asia, and the Pacific (World Resources Institute 2000). As croplands become scarce or degraded, as forests shrink, and as marine life dwindles, millions of people who make their living from these natural resources must find alternative livelihoods. By 2020, an estimated 50 million people globally will be **environmental refugees**—individuals who have migrated because they were forced from their homelands and/or who can no longer secure a livelihood as a result of environmental problems (Zelman 2011). As individuals lose their source of income, so do nations. In a quarter of the world's nations, crops, timber, and fish contribute more to the nation's economy than industrial goods do (World Resources Institute 2000).

The structural-functionalist perspective raises our awareness of latent dysfunctions—negative consequences of social actions that are unintended and not widely recognized. For example, the expanding production of biofuels made from corn reduces reliance on fossil fuels, but also has the unintended consequence of raising food prices.

Conflict Perspective

The conflict perspective focuses on the role that wealth, power, and the pursuit of profit plays in environmental problems and solutions. The capitalistic pursuit of profit encourages making money from industry regardless of the damage done to the environment. To maximize sales, manufacturers design products intended to become obsolete. As a result of this **planned obsolescence**, consumers continually get rid of used products and purchase replacements. **Perceived obsolescence**—the *perception* that a product is obsolete—is a marketing tool used to convince consumers to replace certain items even though the items are still functional. Fashion is a prime example as consumers are encouraged to buy the latest trends in clothing style every season, even though their current clothing may still be in good condition. Both planned and perceived obsolescence benefit industry profits, but at the expense of the environment, which must sustain the constant production and absorb ever-increasing amounts of waste.

Industries use their power and wealth to influence politicians' environmental and energy policies as well as the public's beliefs about environmental issues. For example, some utility companies, whose profits are being threatened by the growing solar energy industry, are lobbying to reduce or eliminate solar energy tax incentives and are waging publicity campaigns that criticize solar energy companies for spending "hard-earned tax dollars to subsidize their wealthy customers" (Gunther 2013). Other utility companies are joining forces with the solar industry, trying to make money off of solar power as opposed to fighting against the solar energy industry.

In the 2011–2012 election cycle, the oil and gas industry was among the top 10 interest groups contributing money to members of Congress. Of the more than $20 million the oil/gas industry gave to members of the House and Senate, 86 percent went to Republicans (OpenSecrets.org 2013). Government policies that support ethanol have been linked to political financial contributions by major players in the ethanol industry, such as Archer Daniels Midland, the nation's largest ethanol producer (Food & Water Watch and Network for New Energy Choices 2007). President Obama, who called nuclear power an "important part" of his energy agenda—even after the 2011 Fukushima nuclear power

ecosystem A biological environment consisting of all the organisms living in a particular area, as well as all the nonliving, physical components of the environment such as air, water, soil, and sunlight, that interact to keep the whole ecosystem functioning.

environmental refugees Individuals who have migrated because they can no longer secure a livelihood as a result of deforestation, desertification, soil erosion, and other environmental problems.

planned obsolescence The manufacturing of products that are intended to become inoperative or outdated in a fairly short period of time.

perceived obsolescence The perception that a product is obsolete; used as a marketing tool to convince consumers to replace certain items even though the items are still functional.

plant disaster, had received generous campaign contributions from Exelon Corporation—the company that operates all 11 of Illinois's nuclear reactors (McCormick 2011).

> **What Do You Think?** About one in five members of Congress own stocks in oil or gas companies (Beckel 2010). Do you think that members of Congress, who make and vote on energy policies, should be allowed to have investments in energy-related industries? Would you feel the same way about members of Congress who invested in renewable energy industries, such as solar and wind power?

The conflict perspective is also concerned with **environmental injustice** (also known as *environmental racism*)—the tendency for marginalized populations and communities to disproportionately experience adversity due to environmental problems. Although environmental pollution and degradation and depletion of natural resources affect us all, some groups are more affected than others. For example, although developing countries have emitted far more greenhouse gases (that cause global warming and climate change), the effects of global climate change are expected to be felt most severely by poor, developing nations (Miranda et al. 2011).

In the United States, polluting industries and industrial and waste facilities are often located in minority communities. More than half (56 percent) of people living within 1.8 miles of a commercial hazardous waste site are people of color (Bullard et al. 2007). Rates of poverty are also higher among households located near hazardous waste facilities. However, "racial disparities are more prevalent and extensive than socioeconomic disparities, suggesting that race has more to do with the current distribution of the nation's hazardous waste facilities than poverty" (Bullard et al. 2007, p. 60). In North Carolina, hog industries—and the associated environmental and health risks associated with hog waste—tend to be located in communities with large black populations, low voter registration, and low incomes (Edwards & Driscoll 2009).

The recycling of "e-waste"—waste from discarded items such as computers, cell phones, and televisions—is done in developing countries under conditions that expose recycling workers and local residents to toxic metals from burning the waste. Although the Basel Convention on the Control of Transboundary Movements of Hazardous Wastes and Their Disposal bans the exchange of hazardous waste (including e-waste) between developed and developing countries, the United States, which is the largest generator of e-waste worldwide, remains the only industrialized nation that has not ratified the Basel Convention (McAllister 2013).

Symbolic Interactionist Perspective

The symbolic interactionist perspective focuses on how meanings, labels, and definitions learned through interaction and through the media affect environmental problems. Whether an individual recycles, drives a sport-utility vehicle (SUV), or joins an environmental activist group is influenced by the meanings and definitions of these behaviors that the individual learns through interaction with others.

Large corporations and industries commonly use marketing and public relations strategies to construct favorable meanings of their corporation or industry. The term **greenwashing** refers to the way in which environmentally and socially damaging companies portray their corporate image, products, and services as being "environmentally friendly" or socially responsible. Greenwashing is commonly used by public relations firms that specialize in damage control for clients whose reputations and profits have been hurt by poor environmental practices. For example, coal is associated with the devastation of communities through the mining practice of mountaintop removal, and burning coal is the biggest contributor to pollution that causes global warming. The coal industry has spent enormous sums to convince the public that coal is clean. The "clean coal" campaign has invited widespread criticism from environmentalists: "Saying coal is clean is like talking about healthy cigarettes. There's no such thing as clean coal" (Beinecke 2009).

environmental injustice Also known as *environmental racism,* the tendency for marginalized populations and communities to disproportionately experience adversity due to environmental problems.

greenwashing The way in which environmentally and socially damaging companies portray their corporate image and products as being "environmentally friendly" or socially responsible.

Go into any large grocery or "big box" store today and take notice of how many products are advertised as "green," "all natural," or "earth-friendly." Marketing companies know that "green" sells, as consumers are becoming more eco-minded. But how valid are the environmental claims made on the labels of the products we buy? Are the claims trustworthy? Or do marketers deliberately mislead consumers regarding the environmental practices of a company or the environmental benefits of a product or service—a practice known as greenwashing? In this *Social Problems Research Up Close* feature, we present a study that attempts to answer these questions.

Sample and Methods

In 2008 and 2009, TerraChoice, an environmental marketing company, sent researchers into leading "big box" retailers in the United States, Canada, the United Kingdom, and Australia with instructions to record every product making an environmental claim. For each product, the researchers recorded product details, claim(s) details, any supporting information, and any explanatory detail or offers of additional information or support. In the United States and Canada, a total of 2,219 products making 4,996 green claims were recorded. These claims were tested against guidelines provided by the U.S. Federal Trade Commission, Competition Bureau of Canada, Australian Competition and Consumer Commission, and the International Organization for Standardization (ISO) 14021 standard for environmental labeling. The researchers wanted to answer the following questions about each green claim they identified: (1) Is the claim truthful? (2) Does the company offer validation for its claim from an independent and trusted third party? (3) Is the claim specific, using terms that have agreed-upon definitions, not vague ones like "natural" or "nontoxic"? (4) Is the claim relevant to the product it accompanies? (5) Does the claim address the product's principal environmental impact(s) or does it distract consumers from the product's real problems?

Selected Findings and Conclusions

This study found that green claims are most common for products related to children (toys and baby products), cosmetics, and cleaning products. Of the 2,219 North American products surveyed, over 98 percent committed at least one of the following seven "sins" of greenwashing:

1. *Sin of the hidden trade-off,* committed by suggesting a product is "green" based on an unreasonably narrow set of attributes without attention to other important environmental issues. Paper, for example, is not necessarily environmentally preferable just because it comes from a sustainably harvested forest. Other important environmental issues in the papermaking process, including energy, greenhouse gas emissions, and water and air pollution, may be equally or more significant.
2. *Sin of no proof,* committed by an environmental claim that cannot be substantiated by easily accessible supporting information or by a reliable third-party certification. Common examples are facial or toilet tissue products that claim various percentages of postconsumer recycled content without providing any evidence.
3. *Sin of vagueness,* committed by every claim that is so poorly defined or broad that consumers are likely to misunderstand its real meaning. "All natural" is an example. Arsenic, uranium, mercury, and formaldehyde are all naturally occurring, and poisonous. "All natural" isn't necessarily "green."
4. *Sin of irrelevance,* committed by making an environmental claim that may be truthful but is unimportant or unhelpful for consumers seeking environmentally preferable products. "CFC-free" is a common example, because it is a frequent claim despite the fact that CFCs are banned by law.
5. *Sin of lesser of two evils,* committed by claims that may be true within the product category, but that may distract the consumer from the greater environmental impacts of the category as a whole. Organic cigarettes are an example of this category, as are fuel-efficient sport-utility vehicles.
6. *Sin of fibbing,* the least frequent sin, is committed by making environmental claims that are simply false. The most common examples were products falsely claiming to be Energy Star–certified or Energy Star–registered.
7. *Sin of worshiping false labels* is committed by a product that, through either words or images, gives the impression of third-party endorsement where no such endorsement actually exists.

Although many companies are making meaningful efforts to minimize their impact on the environment, this study alerts consumers to the widespread practice of greenwashing and suggests that there is a need for more accountability and transparency in marketing products as "green."

Source: TerraChoice Group Inc. 2009.

Although greenwashing involves manipulation of public perception to maximize profits, many corporations make genuine and legitimate efforts to improve their operations, packaging, or overall sense of corporate responsibility toward the environment. For example, in 1990, McDonald's announced that it was phasing out foam packaging and switching to a new, paper-based packaging that is partially degradable. But many environmentalists are not satisfied with what they see as token environmentalism, or as Peter Dykstra of Greenpeace suggested, 5 percent of environmental virtue to mask 95 percent of environmental vice (Hager & Burton 2000).

Corporations also engage in **pinkwashing**—the practice of using the color pink and pink ribbons and other marketing strategies that suggest a company is helping to fight breast cancer, even when the company may be manufacturing or selling chemicals linked to cancer. "Through the pink ribbon, corporate America has embraced cause-related

pinkwashing The practice of using the color pink and pink ribbons to indicate a company is helping to fight breast cancer, even when the company may be using chemicals linked to cancer.

TABLE 13.1 Critical Questions to Ask before You Buy Pink
1. Does any money from this purchase go to support breast cancer programs? How much?
2. What organization will get the money? What will they do with the funds, and how do these programs turn the tide of the breast cancer epidemic?
3. Is there a "cap" on the amount the company will donate? Has this maximum donation already been met? Can you tell?
4. Does this purchase put you or someone you love at risk for exposure to toxins linked to breast cancer? What is the company doing to ensure that its products are not contributing to the breast cancer epidemic?

Source: Breast Cancer Action. San Francisco CA. Used by permission.

marketing—reframing shopping as a way to fight disease" (Lunden 2013). The cosmetic company Estée Lauder was the first company to use the pink ribbon symbol of fighting breast cancer in its marketing, followed by Avon and Revlon. Yet all three cosmetics companies produce and sell products with ingredients that include hormone disruptors and other suspected carcinogens. Avon, Revlon, and Estée Lauder also declined to sign the Compact for Safe Cosmetics, a pledge to produce personal care products that are free of chemicals known or strongly suspected of causing cancer, mutation, or birth defects. And, through their trade association, the Personal Care Products Council, all three companies opposed a California bill that would require cosmetics manufacturers to disclose their use of chemicals linked to cancer or birth defects (Lunden 2013).

What Do You Think? A wide range of companies has adopted the pink ribbon marketing strategy, including Clorox, Dansko, Evian, Ford, and American Airlines. Does the presence of a pink ribbon on a product or service influence your purchasing behavior? Why or why not? *Think Before You Pink*, a project of the Breast Cancer Action, encourages consumers to ask critical questions before they buy products with the pink ribbon symbol (see Table 13.1).

Environmental Problems: An Overview

Over the past 50 years, humans have altered ecosystems more rapidly and extensively than in any other comparable period of time in history (Millennium Ecosystem Assessment 2005). As a result, humans have created environmental problems, including depletion of natural resources; air, land, and water pollution; global warming and climate change; environmental illness; threats to biodiversity; and light pollution. Because many of these environmental problems are related to the ways that humans produce and consume energy, we begin this section with an overview of global energy use.

Energy Use Worldwide

When Hurricane Sandy hit the Northeast United States in 2012, 8.5 million people from Indiana to Maine experienced power outage. Until we experience a prolonged power outage, most of us take the availability of electricity for granted, and don't think about how dependent we are on energy. Being mindful of environmental problems means seeing the connections between energy use and our daily lives:

Everything we consume or use—our homes, their contents, our cars and the roads we travel, the clothes we wear, and the food we eat—requires energy to produce and package, to distribute to shops or front doors, to operate, and then to get rid of. We rarely consider where this energy comes from or how much of it we use—or how much we truly need. (Sawin 2004, p. 25)

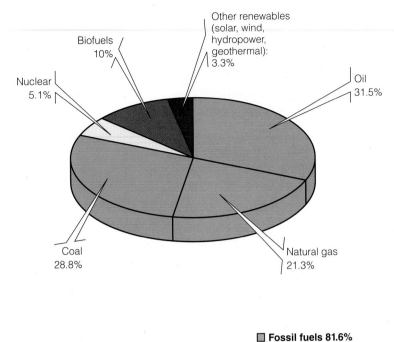

Figure 13.1 **World Energy Production by Source, 2011** Source: International Energy Agency 2013.

Biofuels 10%

Other renewables (solar, wind, hydropower, geothermal): 3.3%

Nuclear 5.1%

Oil 31.5%

Coal 28.8%

Natural gas 21.3%

☐ Fossil fuels 81.6%

Most of the world's energy comes from fossil fuels, which include petroleum (or oil), coal, and natural gas (see Figure 13.1). As you continue reading this chapter, notice that the major environmental problems facing the world today—air, land, and water pollution, destruction of habitats, biodiversity loss, global warming, and environmental illness—are linked to the production and use of fossil fuels.

Renewable energy includes hydroelectric power, which involves generating electricity from moving water. As water passes through a dam into a river below, a turbine in the dam produces energy. Although hydroelectric power is nonpolluting and inexpensive and is considered to be a clean and renewable form of energy, it is criticized for affecting natural habitats. For example, dams make certain fish unable to swim upstream to reproduce. Other forms of clean, renewable energy include solar, wind, geothermal (heat from the earth), and **biomass**, which refers to material derived from plants and animals (such as dung, wood, crop residues, and charcoal).

Nuclear power is associated with a number of problems related to radioactive nuclear waste (discussed later in this chapter). As of 2013, there were 432 operable nuclear power reactors in the world, with another 68 under construction (World Nuclear Association 2013). The United States has more nuclear reactors than any other country. There are 104 operable U.S. nuclear reactors at 65 nuclear power plants, most of which are located east of the Mississippi River (see Figure 13.2).

The safety of nuclear power is highly questionable, as equipment failures and natural disasters can result in leaks of harmful radiation (see this chapter's *Photo Essay*). The worst nuclear power plant accident in the United States occurred in 1979, when the cooling system at the Three Mile Island reactor failed, causing a partial meltdown. A small amount of radioactive gas was vented from the building to prevent an explosion. Cleanup of the site cost $1 billion and took 14 years (Clemmitt 2011). In 2013, former Nuclear Regulatory Commission Chairman Gregory B. Jaczko said he believes that all 104 U.S. nuclear power reactors have safety problems that cannot be fixed and that they should all be phased out of operation (Wald 2013). The Indian Point nuclear power plant, located 30 miles from New York City, has had numerous safety problems and is situated near an earthquake fault line. A nuclear disaster at Indian Point would threaten the entire population of New York City and its surrounding metropolitan area. Emergency evacuation would be impossible (Nader 2013).

Depletion of Natural Resources: Our Growing Environmental Footprint

Humans have used more of the earth's natural resources since 1950 than in the million years preceding 1950 (Lamm 2006). In 1961, humanity used only about two-thirds of earth's natural resources; in the early 70s, human demand for resources began exceeding what the planet could renewably produce (Global Footprint Network 2013). The demands that humanity makes on the earth's natural resources are known as the **environmental footprint**. Currently, the environmental footprint exceeds the earth's biocapacity—the area of land and oceans available to produce renewable resources and absorb CO_2 emissions—by more than 50 percent, meaning that we currently use 1.5 planet earths to support our consumption. Rich countries that consume more goods and produce more carbon dioxide have a much larger environmental footprint than poor countries. If every person in the world lived like an average resident of the United States, a total of four earths would be required to support humanity's annual demand on nature. If the current

biomass Material derived from plants and animals, such as dung, wood, crop residues, and charcoal; used as cooking and heating fuel.

environmental footprint The demands that humanity makes on the earth's natural resources.

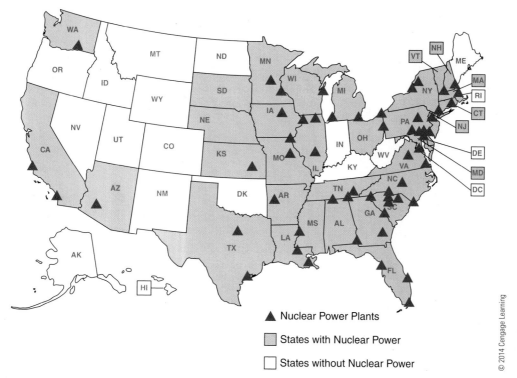

Nuclear Power Plants

States with Nuclear Power

States without Nuclear Power

© 2014 Cengage Learning

Figure 13.2 **Location of Nuclear Power Plants in the United States**

global patterns of consumption continue, we would need the equivalent of 2.9 planet earths to support us by 2050 (WWF 2012).

Every year the Global Footprint Network identifies **Earth Overshoot Day**—the approximate date on which humanity's annual demand on the planet's resources exceeds what our planet can renew in a year. In 2013, Earth Overshoot Day was August 20, meaning that in less than eight months, we used as much natural resources as our planet can renew in a year (Global Footprint Network 2013).

Population growth and consumption patterns are depleting natural resources such as forests, water, minerals, and fossil fuels (see also Chapter 12). About 1.2 billion people—nearly one-fifth of the world's population—live in areas of physical water scarcity, which occurs when there is not enough water to meet demand (Kumar 2013). Water supplies around the world are dwindling, while the demand for water continues to increase because of population growth, industrialization, rising living standards, and changing diets that include more food products that require larger amounts of water to produce: milk, eggs, chicken, and beef. With 70 percent of freshwater use going to agriculture, water shortages threaten food production and supply.

The world's forests are also being depleted due to the expansion of agricultural land, human settlements, wood harvesting, and road building. The result is **deforestation**—the conversion of forestland to nonforestland. Global forest cover has been reduced by half of what it was 8,000 years ago (Gardner 2005). Between 2000 and 2010, the world's forests shrank by an area roughly the size of France (Normander 2011). Deforestation displaces people and wild species from their habitats; soil erosion caused by deforestation can cause severe flooding; and, as we explain later in this chapter, deforestation contributes to global warming. Deforestation also contributes to **desertification**—the degradation of semiarid land, which results in the expansion of desert land that is unusable for agriculture. As more land turns into desert, populations can no longer sustain a livelihood on the land, and so they migrate to urban areas or other countries, contributing to social and political instability.

> If every person in the world lived like an average resident of the United States, a total of four earths would be required to support humanity's annual demand on nature.

Earth Overshoot Day The approximate date on which humanity's annual demand on the planet's resources exceeds what our planet can renew in a year.

deforestation The conversion of forestland to nonforestland.

desertification The degradation of semiarid land, which results in the expansion of desert land that is unusable for agriculture.

Transportation vehicles, fuel combustion, industrial processes (such as burning coal and processing minerals from mining), and solid waste disposal have contributed to the growing levels of air pollutants, including carbon monoxide, sulfur dioxide, arsenic, nitrogen dioxide, mercury, dioxins, and lead. Leaded aviation gasoline is one of the few fuels in the United States to still contain lead, and it's the single largest source of lead emissions in the country (Kessler 2013). Air pollution, which is linked to heart disease, lung cancer, emphysema, chronic bronchitis, and asthma, kills about 3 million people a year (Pimentel et al. 2007). In the United States, 42 percent of the population lives in areas where they are exposed to unhealthy levels of air pollution (ozone or particulate pollution) (American Lung Association 2013).

Indoor pollution is a serious problem in developing countries. As this woman cooks food for her family, she is exposed to harmful air contaminants from the fumes.

Ted Spiegel/Documentary Value/Corbis

Indoor Air Pollution. Around 3 billion people, mostly in low- and middle-income countries, use biomass fuels—wood, charcoal, crop residues, and dung—for cooking and heating their homes (World Health Organization 2012). Biomass is typically burned on open fires or stoves without chimneys, creating smoke and indoor air pollution. Exposure to this indoor smoke increases risk of pneumonia, chronic respiratory disease, asthma, cataracts, tuberculosis, and lung cancer, and is responsible for nearly 2 million deaths a year (World Health Organization 2012). Exposure is particularly high among women and children, who spend the most time near the domestic hearth or stove.

Even in affluent countries, much air pollution is invisible to the eye and exists where we least expect it—in our homes, schools, workplaces, and public buildings. Sources of indoor air pollution include lead dust (from old lead-based paint); secondhand tobacco smoke; by-products of combustion (e.g., carbon monoxide) from stoves, furnaces, fireplaces, heaters, and dryers; and other common household, personal, and commercial products. Some of the most common indoor pollutants include carpeting (which emits more than a dozen toxic chemicals); mattresses, sofas, and pillows (which emit formaldehyde and fire retardants); pressed wood found in kitchen cabinets and furniture (which emits formaldehyde); and dry-cleaned clothing (which emits perchloroethylene). Air fresheners, deodorizers, and disinfectants emit the pesticide paradichlorobenzene. Potentially harmful organic solvents are present in numerous office supplies, including glue, correction fluid, printing ink, carbonless paper, and felt-tip markers. Many homes today contain a cocktail of toxic chemicals: "Styrene (from plastics), benzene (from plastics and rubber), toluene and xylene, trichloroethylene, dichloromethane, trimethylbenzene, hexanes, phenols, pentanes, and much more outgas from our everyday furnishings, construction materials, and appliances" (Rogers 2002).

Destruction of the Ozone Layer. The ozone layer of the earth's atmosphere protects life on earth from the sun's harmful ultraviolet rays. Yet the ozone layer has been weakened by the use of certain chemicals, particularly chlorofluorocarbons (CFCs), used in refrigerators, air conditioners, and spray cans. The ozone hole over the Antarctic in 2012 was, at its peak, slightly smaller than the area of North America (Blunden & Arndt 2013). The depletion of the ozone layer allows hazardous levels of ultraviolet rays to reach the earth's surface and is linked to increases in skin cancer and cataracts, weakened immune systems, reduced crop yields, damage to ocean ecosystems and reduced fishing yields, and adverse effects on animals. Despite measures that have ended production of

CFCs, the ozone is not expected to recover significantly for about another decade because CFCs already in the atmosphere remain for 40 to 100 years.

Acid Rain. Air pollutants, such as sulfur dioxide and nitrogen oxide, mix with precipitation to form **acid rain**. Polluted rain, snow, and fog contaminate crops, forests, lakes, and rivers. As a result of the effects of acid rain, all the fish have died in a third of the lakes in New York's Adirondack Mountains (Blatt 2005). Because winds carry pollutants in the air, industrial pollution in the Midwest falls back to earth as acid rain on southeast Canada and the northeast New England states. In China, most of the electricity comes from burning coal, which creates sulfur dioxide pollution and acid rain that falls on one-third of China, damaging lakes, forests, and crops (Woodward 2007). Acid rain also deteriorates the surfaces of buildings and statues. "The Parthenon, Taj Mahal, and Michelangelo's statues are dissolving under the onslaught of the acid pouring out of the skies" (Blatt 2005, p. 161).

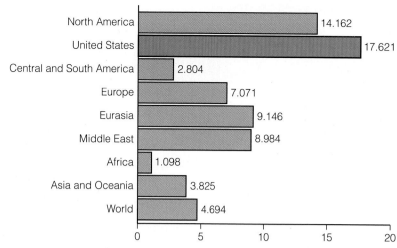

Note: Unit of measurement 5 metric tons.

Figure 13.3 Per Capita Carbon Dioxide Emissions, by Region, 2011
Source: Energy Information Administration 2013.

Global Warming and Climate Change

Global warming refers to the increasing average global temperature of earth's atmosphere, water, and land, caused mainly by the accumulation of various gases (greenhouse gases) that collect in the atmosphere. According to the Intergovernmental Panel on Climate Change (IPCC)—an international team of scientists from countries around the world—"Warming of the climate system is unequivocal. . . .The atmosphere and ocean have warmed, the amounts of snow and ice have diminished, sea level has risen, and the concentrations of greenhouse gases have increased" (2013, p. SPM-3). In the United States, 2012 was the warmest year since records began in 1895 (Blunden & Arndt 2013).

Causes of Global Warming. The prevailing scientific view is that **greenhouse gases**—primarily carbon dioxide (CO_2), methane, and nitrous oxide—accumulate in the atmosphere and act like the glass in a greenhouse, holding heat from the sun close to the earth. Most scientists believe that global warming has resulted from the marked increase in global atmospheric concentrations of greenhouse gases since industrialization began. Global increases in carbon dioxide concentration are due primarily to the actions of humankind, particularly the use of fossil fuels.

Deforestation also contributes to increasing levels of carbon dioxide in the atmosphere. Trees and other plant life use carbon dioxide and release oxygen into the air. As forests are cut down or are burned, fewer trees are available to absorb the carbon dioxide.

The growth of greenhouse gas emissions is strongest in developing countries, particularly China, which emits more carbon dioxide than any other nation. In 2010, China consumed nearly half of all coal worldwide and surpassed the United States as the world's largest consumer of energy (BP 2011). However, the United States has the highest per capita emissions of carbon dioxide (see Figure 13.3).

Even if greenhouse gases are stabilized, global air temperature and sea level are expected to continue to rise for hundreds of years. That is because global warming that has already occurred contributes to further warming of the planet—a process known as a *positive feedback loop*. For example, the melting of Siberia's frozen peat bog could release billions of tons of methane, a potent greenhouse gas, into the atmosphere (Pearce 2005).

> Even if greenhouse gases are stabilized, global air temperature and sea level are expected to continue to rise for hundreds of years.

acid rain The mixture of precipitation with air pollutants, such as sulfur dioxide and nitrogen oxide.

global warming The increasing average temperature of earth's atmosphere, water, and land, caused mainly by the accumulation of various gases (greenhouse gases) that collect in the atmosphere.

greenhouse gases Gases (primarily carbon dioxide, methane, and nitrous oxide) that accumulate in the atmosphere and act like the glass in a greenhouse, holding heat from the sun close to the earth.

The price we pay for energy includes the human, environmental, and financial costs associated with unexpected accidents in energy-production industries. This photo essay looks at two major energy-related disasters: the 2010 British Petroleum (BP) Deepwater Horizon oil rig explosion and the 2011 Fukushima Daiichi nuclear power plant accident.

AP Images/Anonymous/US Coast Guard

◄ In April 2010, the Deepwater Horizon oil drilling rig, owned by Transocean and leased to BP, exploded over the Macondo oil well, killing 11, injuring 17 oil rig workers, and opening a gusher that released oil and methane gas into the Gulf of Mexico. By the time the oil well was sealed months later, over 4 million barrels of oil had spilled into the Gulf, creating what may be the worst environmental disaster in U.S. history. In the decade prior to the Deepwater Horizon accident, there had been 948 fires and explosions in the Gulf of Mexico oil industry. A presidential commission on the 2010 Gulf oil spill concluded that the disaster was caused by a series of preventable human and engineering mistakes made by BP, Halliburton, and Transocean, and by lack of regulatory oversight by the government (National Commission on the BP Deepwater Horizon Oil Spill and Offshore Drilling 2011).

Dead birds, sea turtles, dolphins, and whales were found, covered in oil, along the 650 miles of coastline that were affected by the BP oil spill. Although much of the oil in the Deepwater Horizon disaster was recovered by burning and skimming, 156 million gallons were left in the environment. In an attempt to lessen the impact of the oil slick on the ecosystem, BP used 1.84 million gallons of chemical dispersants—detergent-like compounds that break up spilled oil into tiny droplets that would mix with water. But using dispersants may have added new risks to the harm already done. Some of the 57 chemicals found in dispersants are associated with cancer, skin and eye irritation, respiratory problems, kidney problems, and toxicity to aquatic life (Foster 2011). The long-term toll of the oil spill and the use of dispersants on the Gulf water and coastline ecosystems is unknown and could last for decades (NRDC 2011). ▶

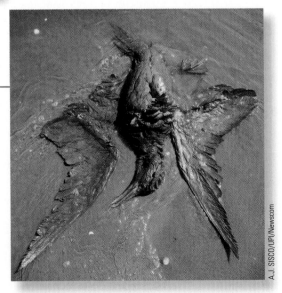

A.J. SISCO/UPI/Newscom

AFLO/Newscom

◄ Following the accident at the Fukushima nuclear power plant, a mandatory evacuation was issued for people living within 12 miles of the plant; those living outside this radius were advised to stay indoors, close doors and windows, turn off the air conditioner, cover their mouths with masks, and avoid drinking tap water. In the weeks following the accident, the evacuation zone expanded to 18 miles or more, resulting in a total of about 200,000 people being forced to leave their homes.

The amount of radiation released over a period of about five months following the Fukushima nuclear disaster is the equivalent to more than 29 Hiroshima-type atomic bombs and the amount of uranium released is equivalent to 20 Hiroshima bombs (Jamail 2011). The areas surrounding the Fukushima nuclear power plant could remain uninhabitable for decades due to high radiation.

jeremy sutton-hibbert/Alamy

Following the Fukushima nuclear power plant accident, there were many protests against nuclear power in Japan, as well as in other countries. ►

AP Images/Anonymous

◄ Following a 9.0 magnitude earthquake and tsunami on March 11, 2011, a series of equipment failures and nuclear meltdowns occurred at the Fukushima Daiichi nuclear power plant in Japan. The accident was assessed as Level 7 on the International Nuclear Event Scale— the maximum scale value that is defined as "[A] major release of radioactive material with widespread health and environmental effects requiring implementation of planned and extended countermeasures" (Jamail 2011). The only other nuclear accident rated at Level 7 prior to Fukushima was the 1986 Chernobyl nuclear power plant accident in Ukraine.

In August 2013, more than two years after this nuclear accident, news reports confirmed that several tanks and pipes at the Fukushima Daiichi plant were leaking massive amounts of radioactive water into the Pacific Ocean and into the ground surrounding the plant. The level of radiation around one tank was reported to be enough to kill an exposed person in four hours (Kiger 2013). The total amount of radioactivity released by the Fukushima accident may not be known for years.

LEE CELANO/Reuters/Landov

◄ Fishing and tourist industries suffered economic losses in the tens of billions of dollars due to the Gulf oil spill in 2010.

And the melting of ice and snow—another result of global warming—exposes more land and ocean area, which absorbs more heat than ice and snow, further warming the planet.

For more than 20 years, the fossil fuel industry and its allies have launched an aggressive misinformation campaign attacking and discrediting climate science, scientists, and scientific institutions (Greenpeace USA 2013). This well-funded "**climate denial machine**" has been effective in swaying public views of climate change: Despite the overwhelming scientific consensus that human activity causes global warming (Cook et al. 2013), more than half (57 percent) of U.S. adults believe that global warming is due more to natural changes in the environment (Saad 2013).

Effects of Global Warming and Climate Change. Climate change kills an estimated 30,000 people per year, mostly in the developing world (Global Humanitarian Forum 2009). The majority of these deaths are attributed to crop failure leading to malnutrition and water problems such as flooding and drought. The effects of global warming and climate change also include the following:

Shrinking Arctic ice threatens the survival of the polar bear.

Melting Ice and Sea-Level Rise. Between 1901 and 2010, average global sea level rose by about 7.5 inches (Intergovernmental Panel on Climate Change 2013). Some forecasts predict that sea-level rise could reach 3 to 6.5 feet over the 21st century (Mulrow & Ochs 2011). The two major factors that are causing a rise in the sea level are (1) thermal expansion caused by the warming of the oceans (water expands as it warms), and (2) the melting of glaciers and the Greenland and polar ice sheets. In 2012, sea ice extent (ocean area covered by ice) was at a record low, and in mid-summer, 97 percent of the Greenland ice sheet was melting (UNEP 2013). Scientists say the Arctic Ocean in summer could be ice-free by the end of the century (Leitzell 2011). Rising sea levels pose a threat to 10 percent of the world's population that live in coastal areas, and 13 of the world's 20 largest cities that are located in coastal areas (Mulrow & Ochs 2011). As sea levels rise, some island countries, as well as some barrier islands off the U.S. coast, are likely to disappear, and low-lying coastal areas will become increasingly vulnerable to storm surges and flooding.

Hans Strand/CORBIS

Flooding and Spread of Disease. Increased heavy rains and flooding caused by global warming contribute to increases in drownings and increases in the number of people exposed to insect- and water-related diseases, such as malaria and cholera. Flooding, for example, provides fertile breeding grounds for mosquitoes that carry a variety of diseases including encephalitis, dengue fever, yellow fever, West Nile virus, and malaria (Knoell 2007). With the warming of the planet, mosquitoes are now living in areas in which they previously were not found, placing more people at risk of acquiring one of the diseases carried by the insect.

Threat of Species Extinction. At least 19 species extinctions have been attributed to climate change (Staudinger et al. 2012). Scientists have predicted that, in certain areas of the world, global warming will lead to the extinction of up to 43 percent of plant and animal species, representing the potential loss of 56,000 plant species and 3,700 vertebrate species (Malcolm et al. 2006). The U.S. Geological Survey (2007) predicts that, due to the effects of climate change, the entire polar bear population of Alaska may be extinct in the next 43 years.

Extreme Weather: Hurricanes, Droughts, and Heat Waves. Rising temperatures are causing drought in some parts of the world and too much rain in other parts. Warmer tropical ocean temperatures can cause more intense hurricanes (Chafe 2006). With rising

climate denial machine A well-funded and aggressive misinformation campaign run by the fossil fuel industry and its allies that involves attacking and discrediting climate science, scientists, and scientific institutions.

temperatures, an increase in the number, intensity, and duration of heat waves is expected, with the accompanying adverse health effects (Intergovernmental Panel on Climate Change 2007). Droughts, as well as floods, can be devastating to crops and food supplies.

Forest Fires. Another effect of global warming is an increase in the number and size of forest fires (Westerling et al. 2006). For every degree Celsius warming in the Western states, scientists project a two- to sixfold increase in area burned by wildfire (Staudinger et al. 2012). Warmer temperatures dry out brush and trees, creating ideal conditions for fires to spread. Warmer weather also allows bark beetles to breed more frequently, which leads to more trees dying from beetle infestation (Staudinger et al. 2012). Dead trees become dry and increase risk of fire. Global warming also means that spring comes earlier, making the fire season longer.

The increase in the number and size of forest fires in recent years has been linked to global warming.

Effects on Recreation. Winter sports and recreation, such as skiing and snowboarding, are threatened by decreased and unreliable snowfall, causing high economic losses for winter recreation businesses, not to speak of frustration for winter sports enthusiasts. In coastal areas, beach recreation is also projected to suffer due to coastal erosion caused by sea-level rise and increased storms association with climate change (Staudinger et al. 2012).

What Do You Think? A Pew Research (2013) poll of people in 39 countries found that concern about global climate change is particularly prevalent in Latin America, Europe, sub-Saharan Africa, and the Asian/Pacific region, and majorities in Lebanon, Tunisia, and Canada also say climate change is a major threat to their countries. In contrast, only four in ten Americans view climate change as a major threat, making Americans among the least concerned about global climate change. Why do you think this is so?

Land Pollution

About 30 percent of the world's surface is land, which provides soil to grow the food we eat. Increasingly, humans are polluting the land with nuclear waste, solid waste, and pesticides. In 2013, 1,320 hazardous waste sites in the United States (also called Superfund sites) were on the National Priorities List (EPA 2013a).

Nuclear Waste. Nuclear waste, resulting from both nuclear weapons production and nuclear reactors or power plants, contains radioactive plutonium, a substance linked to cancer and genetic defects. Radioactive wastes and contaminated materials from nuclear power remain potentially harmful to human and other life for 250,000 years (Nader 2013).

In the United States, nuclear waste is being stored temporarily in 121 aboveground sites in 39 states. The first planned U.S. repository for nuclear waste was in Yucca Mountain, 100 miles northwest of Las Vegas. However, President Obama has rejected this plan, saying there are too many questions about whether nuclear waste storage at Yucca Mountain would be safe. The question remains about how to safely dispose of nuclear waste.

Nuclear plants have tons of radioactive spent fuel, with some of it sealed in casks (Vedantam 2005b). Because of inadequate oversight and gaps in safety procedures, radioactive spent fuel is missing or unaccounted for at some U.S. nuclear power plants, which raises serious safety concerns (Vedantam 2005a). Accidents at nuclear power plants, such as the 2011 accident at Fukushima, Japan, and the potential for nuclear reactors to be targeted by terrorists add to the actual and potential dangers of nuclear power plants. Political activist Ralph Nader (2013) commented,

> With all the technological advancements in energy efficiency, solar, wind, and other renewable energy sources, surely there are better and more efficient ways to meet our electricity needs without burdening future generations with deadly waste products and risking the radioactive contamination of entire regions should anything go wrong. . . . It is telling that Wall Street, which rarely considers the consequences of gambling on a risk, will not finance the construction of a nuclear plant without a full loan guarantee from the U.S. government. Nuclear power is also uninsurable in the private insurance market. The Price-Anderson Act of 1957 requires taxpayers to cover almost all the cost if a meltdown should occur. (n.p.)

Recognizing the hazards of nuclear power plants and their waste, Germany became the first country to order all of its 19 nuclear power plants shut down by 2020 ("Nukes Rebuked" 2000). Belgium is also phasing out nuclear reactors, and Austria, Denmark, Italy, and Iceland have prohibitions against nuclear energy.

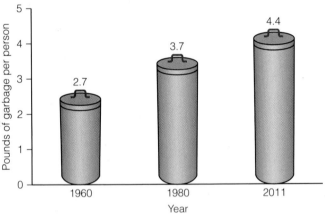

Figure 13.4 Pounds of Garbage per Person, per Day, United States, 1960–2011
Source: EPA 2013b.

Solid Waste. In 1960, each U.S. citizen generated 2.7 pounds of garbage on average every day. This figure increased to 4.4 pounds in 2011 (see Figure 13.4) (EPA 2013b). This figure does not include mining, agricultural, and industrial waste; demolition and construction wastes; junked autos; or obsolete equipment wastes. Just over half of this waste is dumped in landfills; the rest is recycled or composted. The availability of landfill space is limited, however. Some states have passed laws that limit the amount of solid waste that can be disposed of; instead, they require that bottles and cans be returned for a deposit or that lawn clippings be used in community composting programs.

What Do You Think? Plastic bags, commonly used in grocery and retail stores, are associated with a number of environmental problems. In the United States, most plastic shopping bags, which contain toxic chemicals, end up in landfills where it takes 1,000 years for them to degrade (Cheeseman 2007). Plastic bags also end up in oceans, where marine life can choke or starve after swallowing them. Some countries, such as South Africa, China, Thailand, and Bangladesh, have banned plastic shopping bags. In 2007, San Francisco became the first U.S. city to ban plastic bags from supermarkets and chain pharmacies. Since then, a number of plastic bag bans or taxes have been implemented across the country. Would you support a ban, or a tax, on plastic bags in your community?

Solid waste includes discarded electrical appliances and electronic equipment, known as **e-waste**. Ever think about where your discarded computer, cell phone, CD player, television, or other electronic product ends up when you replace it with a newer model? Most discarded electronics end up in landfills, incinerators, or hazardous waste exports. The main concern about dumping e-waste in landfills is that hazardous

e-waste Discarded electrical appliances and electronic equipment.

substances, such as lead, cadmium, barium, mercury, PCBs, and polyvinyl chloride, can leach out of e-waste and contaminate the soil and groundwater.

Pesticides. Pesticides are used worldwide for crops and gardens; outdoor mosquito control; the care of lawns, parks, and golf courses; and indoor pest control. Pesticides contaminate food, water, and air and can be absorbed through the skin, swallowed, or inhaled. Many common pesticides are considered potential carcinogens and neurotoxins (Blatt 2005). Even when a pesticide is found to be hazardous and is banned in the United States, other countries from which we import food may continue to use it. In an analysis of more than 5,000 food samples, pesticide residues were detected in 43 percent of the domestic samples and 31 percent of the imported samples (Food and Drug Administration 2013). Pesticides also contaminate our groundwater supplies.

Water Pollution

Our water is being polluted by a number of harmful substances, including plastics, pesticides, vehicle exhaust, acid rain, oil spills, sewage, and industrial, military, and agricultural waste. Water pollution is most severe in developing countries, where more than 1 billion people lack access to clean water. In developing nations, more than 80 percent of untreated sewage is dumped directly into rivers, lakes, and seas that are also used for drinking and bathing (World Water Assessment Program 2009).

In the United States, one indicator of water pollution is the thousands of fish advisories issued by the U.S. Environmental Protection Agency (EPA) that warn against the consumption of certain fish caught in local waters because of contamination with pollutants such as mercury and dioxin. The EPA advises women who may become pregnant, pregnant women, nursing mothers, and young children to avoid eating certain fish altogether (swordfish, shark, king mackerel, and tilefish) because of the high levels of mercury (EPA 2004).

Pollutants in drinking water can cause serious health problems and even death. At Camp Lejeune—a Marine Corps base in Onslow County, North Carolina—as many as 1 million people were exposed to water contaminated with trichloroethylene (TCE), an industrial degreasing solvent, and perchloroethylene (PCE), a dry-cleaning agent from 1957 until 1987 (Sinks 2007). Exposure to these chemicals has been linked to a number of health problems, including kidney, liver, and lung damage, as well as cancer, childhood leukemia, and birth defects. In this chapter's *The Human Side* feature, a retired Marine tells the story of his daughter's illness and death that he believes resulted from the contaminated water at Camp Lejeune.

Water pollution also affects the health and survival of fish and other marine life. In the Gulf of Mexico, as well as in the Chesapeake Bay and Lake Erie, there are areas known as "dead zones" that—due to pollution runoff from agricultural uses of fertilizer—have oxygen levels so low they cannot support life (Scavia 2011).

In recent years, there has been increasing public concern about the effects of hydraulic fracturing, or "**fracking**"—a process used in natural gas production that involves injecting at high pressure a mixture of water, sand, and chemicals into deep underground wells to break apart shale rock and release gas. Opponents of fracking cite a number of concerns about the damaging impacts to the environment and to human health, including the production of toxic wastewater and contamination of drinking water, air pollution, land damage, and global warming emissions (Ridlington & Rumpler 2013).

Another growing concern surrounds the increasing amount of plastic pollution found in the world's oceans: There is not a single cubic meter of ocean water that does not contain some plastic. Much of this plastic is difficult to see because of its small size. Microplastics, which are fragments of plastic that measure less than 5 mm, come from the degradation of plastic products and from small pellets that are used to make plastic products such as bottles, bags, and packaging. Some of these pellets are accidentally spilled into the environment and have been found on beaches and in ocean water around the world (Takada 2013). These plastic pellets and other plastic debris contain high concentrations of hazardous chemicals that can have adverse effects on marine life and humans that consume seafood.

fracking Hydraulic fracturing, commonly referred to as "fracking," involves injecting a mixture of water, sand, and chemicals into drilled wells to crack shale rock and release natural gas into the well.

About 3 million tons of toxic chemicals are released into the environment each year (Pimentel et al. 2007). Chemicals in the environment enter our bodies via the food and water we consume, the air we breathe, and the substances with which we come in contact. For example, bisphenol A (BPA), a chemical that disrupts endocrine function, is commonly found in food packaging, including cans, plastic wraps, and food storage containers (Betts 2011). During a 2004 World Health Organization convention, 44 hazardous chemicals were found in the bloodstreams of top European Union officials (Schapiro 2004). And in a study of umbilical cord blood of 10 newborns, researchers found an average of 200 industrial chemicals, pesticides, and other pollutants (Environmental Working Group 2005).

This mother puts sunscreen on her child to protect against sunburn. But sunscreen, like other personal care products, may contain harmful chemicals. Learn what chemicals are in the personal care products you use at the Environmental Working Group's website Skin Deep at www.ewg.org/skindeep.

The *12th Report on Carcinogens* (U.S. Department of Health and Human Services 2011) lists 240 chemical substances that are "known to be human carcinogens" or "reasonably anticipated to be human carcinogens," meaning that they are linked to cancer. These may constitute only a fraction of actual human carcinogens. When the Toxic Substances Control Act of 1976 was enacted, it "grandfathered" in the 62,000 chemicals then on the market. Since then, the EPA has restricted the uses of only 5 of the 80,000 chemicals used in the United States; 95 percent of the chemicals in use have not been tested for safety (Lunden 2013).

Most cancer researchers believe that the environment in which we live and work may be a major contributor to the development of cancer—a disease that develops in half of U.S. men and a third of U.S. women (U.S. Department of Health and Human Services 2011). A review of 152 research studies of environmental pollution and breast cancer found that evidence supports a link between breast cancer and a number of environmental pollutants (Brody et al. 2007). Many of the chemicals we are exposed to in our daily lives can cause not only cancer but also other health problems, such as infertility, birth defects, and a number of childhood developmental and learning problems (Fisher 1999; Kaplan & Morris 2000; McGinn 2000; Schapiro 2007). Chemicals found in common household, personal, and commercial products can result in a variety of temporary acute symptoms, such as drowsiness, disorientation, headache, dizziness, nausea, fatigue, shortness of breath, cramps, diarrhea, and irritation of the eyes, nose, throat, and lungs. Long-term exposure can affect the nervous system, reproductive system, liver, kidneys, heart, and blood. Fragrances, which are found in perfumes and colognes, shampoos, deodorants, laundry detergents, tampons, air "fresheners," and a host of other consumer products, are known to be respiratory irritants. Some of the 4,000 ingredients that are used in fragrance manufacturing have been linked with cancer, birth defects, neurotoxic effects, and endocrine disruption (Bradshaw 2010).

What Do You Think? Some businesses, universities, hospitals, and local governments are voluntarily limiting fragrances to accommodate employees and consumers who experience ill effects from them. What do you think about banning fragrances in the workplace or other public places? If your college or university were considering instituting a ban on fragrances on campus, would you support the ban? Why or why not?

Children are more vulnerable than adults to the harmful effects of most pollutants for a number of reasons. For instance, children drink more fluids, eat more food, and inhale more air per unit of body weight than do adults; in addition, crawling and a tendency to put their hands and other things in their mouths provide more opportunities for children to ingest chemical or heavy metal residues.

Jerry Ensminger holds a portrait of his daughter Janey, whose death from leukemia is believed to have been caused by contaminated water at Camp Lejeune.

Retired Marine Master Sergeant Jerry Ensminger is one of about 900 individuals who have filed administrative claims worth $4 billion against Camp Lejeune in Onslow County, North Carolina, for damages that allegedly resulted from drinking and bathing in water contaminated with the dry-cleaning agent perchloroethylene (PCE) and the industrial degreasing solvent trichloroethylene (TCE) (Hefling 2007).

Here, we present Jerry Ensminger's account of his daughter, Janey, who died of leukemia at age 9 in 1985. Janey's illness and death is believed to have been caused by contaminated water at Camp Lejeune that Janey's mother was exposed to during her pregnancy with Janey. If passed, a bill before Congress—the Honoring America's Veterans and Caring for Camp Lejeune Families Act—would provide health care to veterans and their families who have health problems caused by exposure to contaminated water at Camp Lejeune. Ensminger's story is excerpted from the book *Poisoned Nation* (2007):

My little girl died in my arms and 15 years later, I found out that the people I had faithfully served for almost 25 years knew she was being poisoned by the water all along. . . . It was 1997 before I finally found out why my daughter died.

Even then, it was just by chance. A local TV station had picked up the story. The evening news was turned on in the living room, and I was carrying my dinner in on a plate from the kitchen. All of a sudden I heard the newscaster saying that the water at Camp Lejeune had been highly contaminated from 1968 to 1985, and that the chemicals it contained had been linked to childhood leukemia. When I heard that, I just dropped my plate right on the floor and began shaking. The next day . . . I started reading everything I could find and making contacts with everyone I knew. There are stages you go through when you lose a child to a catastrophic illness. First you go into shock; then you start wondering why it happened to your child. So, years ago, I checked my family history and her mother's and found there was nothing on either side. But that nagging question of why Janey got leukemia had stayed with me throughout her illness, her death, and for 14 and a half years after it. And in that moment, that one moment, when I heard the newscast and dropped the plate of food right out of my hands, it all became clear. I suddenly knew why my little girl died.

I signed up with the Marines to serve my country, but I never signed anything that gave them the right to kill my child, to knowingly poison her. . . . I've said that publicly and the Marine Corps has never refuted anything I've said, including that the contamination went back to the 1950s. I'm sure they know that from the geological studies. . . .

The day she died . . . Janey was in a lot of pain, so they suggested that she take morphine. She didn't want to, because she had tried it before and it made her so tired. But this time she just couldn't handle the pain. Janey went through hell for nearly two and a half years, and I went through hell with her. . . . I was there with her every step of the way. Her mother couldn't handle it. Every time Janey went to the hospital, I was the parent who went with her. Sometimes, she was screaming in my ear,

"Daddy, don't let them hurt me." Like when she had the bone marrow transplant and the spinal taps. The last time she went into the hospital was the last day of July, just before her ninth birthday. She didn't come out until September 20, and that was in a casket.

About a week before she died, the head of hematology had come in talking about a new form of therapy. He said it would cause severe burns and ulcers, and they didn't recommend it, but Janey looked at them, blinking to control her tears, and said, "This is my life you are talking about, and I'm not giving up. Let's try it."

The ulcers were all over her mouth, her legs, inside her nose and her vagina. The day she died, she was in such intense pain she could hardly speak, but finally she managed to whisper, "I want to die peacefully." When Janey said that, I started sobbing. She hugged me and said, "Stop it, stop crying." I said, "I can't help it, I love you." "I know you do," she replied. "I love you too. But, Daddy, I hurt so bad." "Do you want some morphine?" I asked. She was already being given methadone. "Yes, Daddy," she said, "I'm ready."

When the nurse heard that Janey wanted morphine, she knew the time had come. Then, just as they started to give it to her, she said, "Wait. Stop. I want some for my daddy, too."

"This is a very powerful pain medicine. I can't give it to your daddy," the nurse said.

"But my daddy hurts, too," Janey answered. You see, I always took a little of whatever she took to show her I was with her. The morphine killed her. It's a respiratory depressant. . . .

The organization I served faithfully for 24 and a half years knew about this all along, and they never said anything, well, shame on them.

Source: From Hefling 2007; Sinks 2007; Schwartz-Nobel 2007.

Multiple Chemical Sensitivity Disorder. **Multiple chemical sensitivity** (MCS), also known as environmental illness, is a condition whereby individuals experience adverse reactions when exposed to low levels of chemicals found in everyday substances (vehicle exhaust, fresh paint, housecleaning products, perfume and other fragrances, synthetic building materials, and numerous other petrochemical-based products). Symptoms of MCS

multiple chemical sensitivity Also known as "environmental illness," a condition whereby individuals experience adverse reactions when exposed to low levels of chemicals found in everyday substances.

TABLE 13.2 Threatened Species Worldwide, 2013

Category	Number of Threatened* Species in 2013
Mammals	1,140
Birds	1,313
Amphibians	1,948
Reptiles	847
Fishes (bony)	1,914
Insects	835
Crustaceans	723
Snails, etc.	1,707
Plants	9,829
Other	671
Total	20,927

*Threatened species include those classified as critically endangered, endangered, and vulnerable.

Source: IUCN 2013.

include headache, burning eyes, difficulty breathing, stomach distress or nausea, loss of mental concentration, and dizziness. The onset of MCS is often linked to acute exposure to a high level of chemicals or to chronic long-term exposure. Individuals with MCS often avoid public places and/or wear a protective breathing filter to avoid inhaling the many chemical substances in the environment. Some individuals with MCS build houses made from materials that do not contain the chemicals that are typically found in building materials.

In a national study on the prevalence of multiple chemical sensitivity, 11.2 percent of U.S. adults reported an unusual hypersensitivity to common chemical products such as perfume, fresh paint, and household cleaning products, and 2.5 percent said they had been diagnosed with MCS (Caress & Steinemann 2004). Two-thirds of those with hypersensitivity described their symptoms as either severe or moderately severe. More than a third (39.5 percent) of the sample reported having trouble shopping in public places due to chemical sensitivity.

Threats to Biodiversity

There are an estimated 8.7 million species of life on earth (some scientists believe the number is much higher), 1.4 million of which have been named and cataloged (Staudinger et al. 2012). This enormous diversity of life, known as **biodiversity,** provides food, medicines, fibers, and fuel; purifies air and freshwater; pollinates crops and vegetation; and makes soils fertile.

One species of life on earth goes extinct every three hours (Leahy 2009). As shown in Table 13.2, 20,927 species worldwide are threatened with extinction. Most extinctions today result from habitat loss caused by deforestation and urban sprawl, overharvesting, global warming, and bioinvasion—environmental problems caused by human activity (Cincotta & Engelman 2000; Leahy 2009). For example, the ocean's absorption of carbon dioxide—a greenhouse gas by-product of fossil fuel use—has resulted in an increase in ocean acidity, which poses a serious threat to reef-forming corals, crustaceans, and mollusks (The International Programme on the State of the Ocean 2013).

> **One species of life on earth goes extinct every three hours.**

Light Pollution

The United States, like much of the rest of the world, has become increasingly "lit up" with artificial light. Artificial light has negative impacts on the well-being of both humans and wildlife (Bogard 2013). **Light pollution** refers to artificial lighting that is annoying, unnecessary, and/or harmful to life forms on earth:

> The 24-hour day/night cycle, known as the circadian clock, affects physiologic processes in almost all organisms. These processes include brain wave patterns, hormone production, cell regulation, and other biologic activities. Disruption of the circadian clock is linked to several medical disorders in humans, including depression, insomnia, cardiovascular disease, and cancer. (Chepesiuk 2009, p. A22)

About 30 percent of vertebrates and 60 percent of invertebrates are nocturnal and vulnerable to light pollution disrupting their patterns of mating, migration, feeding, and pollination (Bogard 2013). For example, artificial lighting around beaches where leatherback turtles nest threatens the survival of hatchlings and is a major cause of declining leatherback turtle populations. Hatchlings, which instinctually follow the

biodiversity The diversity of living organisms on earth.

light pollution Artificial lighting that is annoying, unnecessary, and/or harmful to life forms on earth.

reflected light of the stars and moon from the beach to the ocean, instead follow the light of hotels and streetlights, with the result that they die of dehydration, are eaten by predators, or run over by cars.

Social Causes of Environmental Problems

Various structural and cultural factors have contributed to environmental problems. These include population growth, industrialization and economic development, and cultural values and attitudes such as individualism, consumerism, and militarism.

Population Growth

The world's population is growing, exceeding 7 billion in 2013 and projected to grow to more than 9 billion in 2050 (see Chapter 12). Population growth places increased demands on natural resources and results in increased waste. As Hunter (2001) explained:

> Global population size is inherently connected to land, air, and water environments because each and every individual uses environmental resources and contributes to environmental pollution. While the scale of resource use and the level of wastes produced vary across individuals and across cultural contexts, the fact remains that land, water, and air are necessary for human survival. (p. 12)

However, population growth itself is not as critical as the ways in which populations produce, distribute, and consume goods and services. As shown in Figure 13.5, regions with the highest populations have the lowest environmental footprint.

Industrialization and Economic Development

Many of the environmental problems confronting the world are associated with industrialization, in large part because industrialized countries consume more energy and natural resources and contribute more pollution to the environment than poor countries. The relationship between level of economic development and environmental pollution is curvilinear rather than linear. Industrial emissions are minimal in regions with low

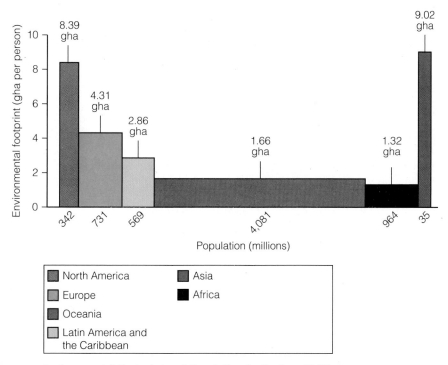

Figure 13.5 **Environmental Footprint and Population by Region, 2007**
Source: Ewing et al. 2010.

levels of economic development and are high in the middle-development range as developing countries move through the early stages of industrialization. However, at more advanced stages of industrialization, industrial emissions ease because heavy-polluting manufacturing industries decline, "cleaner" service industries increase, and because rising incomes are associated with a greater demand for environmental quality and cleaner technologies.

Cultural Values and Attitudes

Cultural values and attitudes that contribute to environmental problems include individualism, consumerism, and militarism.

Individualism. Individualism, which is a characteristic of U.S. culture, puts individual interests over collective welfare. Even though recycling is good for our collective environment, many individuals do not recycle because of the personal inconvenience involved in washing and sorting recyclable items. Similarly, individuals often indulge in countless behaviors that provide enjoyment and convenience for themselves at the expense of the environment: long showers, recreational boating, frequent meat eating, the use of air conditioning, and driving large, gas-guzzling SUVs, to name just a few.

Consumerism. Consumerism—the belief that personal happiness depends on the purchasing of material possessions—also encourages individuals to continually purchase new items and throw away old ones. The media bombard us daily with advertisements that tell us life will be better if we purchase a particular product. Consumerism contributes to pollution and environmental degradation by supporting polluting and resource-depleting industries and by contributing to waste.

Militarism. The cultural value of militarism also contributes to environmental degradation (see also Chapter 15). "It is generally agreed that the number one polluter in the United States is the American military. It is responsible each year for the generation of more than one-third of the nation's toxic waste . . . an amount greater than the five largest international chemical companies combined" (Blatt 2005, p. 25). Toxic substances from military vehicles, weapons materials, and munitions pollute the air, land, and groundwater in and around military bases and training areas. The Pentagon has asked Congress to loosen environmental laws for the military, and the EPA is forbidden to investigate or sue the military (Blatt 2005; Janofsky 2005).

Strategies for Action: Responding to Environmental Problems

Responses to environmental problems include environmental activism, environmental education, the use of green energy, modifications in consumer products and behavior, efforts to slow population growth, and government regulations and legislation. Sustainable economic development, international cooperation and assistance, and institutions of higher education also play important roles in alleviating environmental problems.

Environmental Activism

In 1962, Rachel Carson, a scientist, ecologist, and writer, published *Silent Spring*—a book that warned the public about the long-term effects of pesticides and argued that if humankind poisoned nature, nature would in turn poison humankind (Griswold 2012). Rachel Carson and *Silent Spring*, which has sold more than 2 million copies, are credited with igniting the U.S. environmental movement.

With more than 6,500 national and 20,000 local environmental organizations with a combined membership of between 20 million and 30 million, the U.S. environmental movement may be the largest single social

> The U.S. environmental movement may be the largest single social movement in the United States.

movement in the United States (Brulle 2009). A Gallup survey found that about one in six (17 percent) U.S. adults reports being an active participant in the environmental movement (Gallup Organization 2013) (see Table 13.3).

What Do You Think? Gallup Poll results indicate that women are more likely than men to worry about the environment, to be active in or sympathetic toward the environmental movement, and to give precedence to the environment over economic and energy concerns. Why do think this gender difference exists?

Environmental organizations exert pressure on government and private industry to initiate or intensify actions related to environmental protection. Environmentalist groups also design and implement their own projects and disseminate information to the public about environmental issues. Environmental organizations use the Internet and e-mail to send e-mail action alerts to members, informing them when Congress and other decision makers threaten the health of the environment. These members can then send e-mails and faxes to Congress, the president, and business leaders, urging them to support policies that protect the environment.

TABLE 13.3 Involvement in the Environmental Movement*	
Involvement	Percentage of U.S. Adults
Active participant	17
Sympathetic but not active	42
Neutral	29
Unsympathetic	10
No opinion	2

*In a 2013 Gallup survey, a national sample of U.S. adults was asked: "Do you think of yourself as an active participant in the environmental movement, sympathetic toward the movement, but not active, neutral, or unsympathetic toward the movement?"

Source: Gallup Organization 2013.

Religious Environmentalism. From a religious perspective, environmental degradation can be viewed as sacrilegious, sinful, and an offense against God (Gottlieb 2003a). "The world's dominant religions—as well as many people who identify with the 'spiritual' rather than with established faiths—have come to see that the environmental crisis involves much more than assaults on human health, leisure, or convenience. Rather, humanity's war on nature is at the same time a deep affront to one of the essentially divine aspects of existence" (Gottlieb 2003b, p. 489). This view has compelled religious groups to take an active role in environmental activism.

For example, the National Association of Evangelicals, an umbrella group of 51 church denominations, adopted a platform called For the Health of the Nation: An Evangelical Call to Civic Responsibility. This platform, which has been signed by nearly 100 evangelical leaders, calls on the government to "protect its citizens from the effects of environmental degradation" (Goodstein 2005). Larry Schweiger, president of the National Wildlife Federation, welcomes evangelicals as allies and explains that conservative lawmakers who might not pay attention to what environmental groups say may be more likely to pay attention to what the faith community is saying.

Radical Environmentalism. The **radical environmental movement** is a grassroots movement of individuals and groups that employs unconventional and often illegal means of protecting wildlife or the environment. Radical environmentalists believe in what is known as **deep ecology**—the view that maintaining the earth's natural systems should take precedence over human needs, that nature has a value independent of human existence, and that humans have no right to dominate the earth and its living inhabitants (Brulle 2009). The best-known radical environmental groups are the Earth Liberation Front (ELF) and the Animal Liberation Front (ALF), which are international underground movements consisting of autonomous individuals and small groups who engage in "direct action" to (1) inflict economic damage on those profiting from the destruction and exploitation of the natural environment, (2) save animals from places of abuse (e.g., laboratories, factory farms, and fur farms), and (3) reveal information and educate the public on atrocities committed against the earth and all the species that populate it.

radical environmental movement A grassroots movement of individuals and groups that employs unconventional and often illegal means of protecting wildlife or the environment.

deep ecology The view that maintaining the earth's natural systems should take precedence over human needs, that nature has a value independent of human existence, and that humans have no right to dominate the earth and its living inhabitants.

This Greenpeace activist is climbing the 630-foot chimney at Kingsnorth coal power plant.

In 2007, six Greenpeace environmental activists climbed a 630-foot chimney at Kingsnorth coal power plant in England, intending to shut down the plant by occupying the chimney. They planned to write the words, "Gordon, bin it" on the chimney to pressure Prime Minister Gordon Brown to stop the building of new coal power plants, but after writing "Gordon," the six activists were served with an injunction and came down from the chimney. The "Kingsnorth Six," as they are called, were criminally charged for property damage; it cost 35,000 euros (equal to US $53,000 at the time) to remove the graffiti. Jurors in the case found the Kingsnorth Six "not guilty," accepting the defense arguments that the six activists had a "lawful excuse" to damage property at the Kingsnorth power station to prevent even greater damage caused by global warming (McCarthy 2008). The Criminal Damage Act of 1971 allows individuals to damage property to prevent even greater damage—such as breaking down the door of a burning house to put out a fire. James Hansen, a top NASA climate scientist, testified for the defense, and told the jury that carbon dioxide emissions from the Kingsnorth power plant would contribute to climate change.

The Kingsnorth Six Greenpeace activists were acquitted of criminal charges of property damage as jurors agreed that the defendants had a "lawful excuse" for their actions. But other radical environmentalists have been prosecuted as terrorists. **Ecoterrorism** is defined as any crime intended to protect wildlife or the environment that is violent, puts human life at risk, or results in damages of $10,000 or more (Denson 2000). Many environmentalists question whether "terrorist" is an appropriate label and argue that the real terrorists are corporations that plunder the earth.

ecoterrorism Any crime intended to protect wildlife or the environment that is violent, puts human life at risk, or results in damages of $10,000 or more.

What Do You Think? Should motives be considered in imposing penalties on individuals who are convicted of acts of ecoterrorism? For example, should a person who sets fire to a business to protest that business's environmentally destructive activities receive the same penalty as a person who sets fire to a business for some other reason?

The Role of Corporations in the Environmental Movement. Corporations are major contributors to environmental problems and often fight against environmental efforts that threaten their profits. However, some corporations are joining the environmental movement for a variety of reasons, including pressure from consumers and environmental groups, the desire to improve their public image, genuine concern for the environment, and/or concern for maximizing current or future profits.

In 1994, out of concern for public and environmental health, Ray Anderson, founder and chairman of Interface carpet company, set a goal of being a sustainable company by 2020—"a company that will grow by cleaning up the world, not by polluting or degrading it" (McDaniel 2005, p. 33). Anderson envisioned recycling all the materials used, not releasing any toxins into the environment, and using solar energy to power all production. The company has made significant progress toward these goals, reducing use of fossil fuels by 45 percent, and reducing water and landfill use by as much as 80 percent.

Rather than hope that industry voluntarily engages in eco-friendly practices, corporate attorney Robert Hinkley suggested that corporate law be changed to mandate socially responsible behavior. Hinkley explained that corporations pursue profit at the expense of the public good, including the environment, because corporate executives are bound by corporate law to try to make a profit for shareholders (Cooper 2004). Hinkley suggested that corporate law should include a Code for Corporate Citizenship that would say the following: "The duty of directors henceforth shall be to make money for shareholders but not at the expense of the environment, human rights, public health and safety, dignity of employees, and the welfare of the communities in which the company operates" (quoted by Cooper 2004, p. 6).

Environmental Education

One goal of environmental organizations and activists is to educate the public about environmental issues and the seriousness of environmental problems. Being informed about environmental issues is important because people who have higher levels of environmental knowledge tend to engage in higher levels of pro-environment behavior. For example, environmentally knowledgeable people are more likely to save energy in the home, recycle, conserve water, purchase environmentally safe products, avoid using chemicals in yard care, and donate funds to conservation (Coyle 2005).

A main source of information about environmental issues for most Americans is the media. However, because corporations and wealthy individuals with corporate ties own the media, unbiased information about environmental impacts of corporate activities may not readily be found in mainstream media channels. Indeed, the public must consider the source in interpreting information about environmental issues. Propaganda by corporations sometimes comes packaged as "environmental education." The Heartland Institute, a leader in the climate denial campaign, has mailed books and other "educational" material that dismisses climate change to teachers across the United States (Greenpeace USA 2013).

The American Legislative Exchange Council (ALEC) developed a bill, the Environmental Literacy Improvement Act, that requires a "balance" in deciding which textbooks and teaching materials to use to teach students about climate change (Greenpeace USA 2013). The bill requires materials to be approved by a council of people who are not allowed to have any credentials in environmental science. At least four states—Louisiana, Texas, South Dakota, and Tennessee—have passed laws that weaken teachers' ability to accurately present the science of climate change to their K–12 students. The scientific consensus that human activity is causing global warming and climate change is now, under law, only a "controversial theory among other theories" (Horn 2012).

> Being informed about environmental issues is important because people who have higher levels of environmental knowledge tend to engage in higher levels of pro-environment behavior.

"Green" Energy

Increasing the use of **green energy**—energy that is renewable and nonpolluting—can help alleviate environmental problems associated with fossil fuels. Also known as clean energy, green energy sources include solar power, wind power, biofuel, and hydrogen.

> **What Do You Think?** The World Bank defines "clean energy" as energy that does not produce carbon dioxide when generated. Under this definition, nuclear energy is considered "clean energy." When Obama, in 2011, announced he wanted 80 percent of the nation's electricity to come from clean energy sources by 2035, his definition of "clean" included nuclear power and even natural gas and "clean coal." Do you think that nuclear power, natural gas, and coal should be labeled as clean energy?"

Morton Beebe/Encyclopedia/Corbis

Wind energy is harnessed by turbines such as those pictured in this photo of a wind farm in Altamont Pass, California.

Solar and Wind Energy. Solar power involves converting sunlight to electricity through the use of photovoltaic cells. Other forms of solar power include the use of solar thermal collectors, which capture the sun's warmth to heat building space and water, and "concentrating solar power plants," which use the sun's heat to make steam to turn electricity-producing turbines.

Wind turbines, which turn wind energy into electricity, are operating in 82 countries. The United States is the world's leading generator of wind energy (Kitasei 2011). One disadvantage of wind power is that wind turbines have been known to result in bird mortality. However, this problem has been mitigated in recent years through the use of painted blades, slower rotational speeds, and careful placement of wind turbines.

Biofuel. Biofuels are fuels derived from agricultural crops. Two types of biofuels are ethanol and biodiesel.

Ethanol is an alcohol-based fuel that is produced by fermenting and distilling corn or sugar. Ethanol is blended with gasoline to create E85 (85 percent ethanol and 15 percent gasoline). Vehicles that run on E85, called flexible fuel vehicles, have been used by the government and in private fleets for years and have just recently become available to consumers. Most ethanol is produced in the United States and Brazil (Shrank 2011).

A problem associated with ethanol fuel is that increased demand for corn, which is used to make most ethanol in the United States, has driven up the price of corn, resulting in higher food prices (many processed food items contain corn, and animal feed is largely corn). And as corn prices rise, so too do those of rice and wheat because the crops compete for land. Rising food prices threaten the survival of the world's poorest 2 billion people who depend on grain to survive. The grain it takes to fill a 25-gallon tank with ethanol would feed one person for an entire year (Brown 2007).

Increased corn and/or sugar cane production to meet the demand for ethanol also has adverse environmental effects, including increased use and runoff of fertilizers, pesticides, and herbicides; depletion of water resources; and soil erosion. In addition, tropical forests are being clear-cut to make room for "energy crops," leaving less land for conservation and wildlife (Price 2006); biofuel refineries commonly run on coal and natural gas (which emit greenhouse gases); farm equipment and fertilizer production

green energy Also known as clean energy, energy that is nonpolluting and/or renewable, such as solar power, wind power, biofuel, and hydrogen.

require fossil fuels; and the use of ethanol involves emissions of several pollutants. Finally, even if 100 percent of the U.S. corn crop were used to produce ethanol, it would only displace less than 15 percent of U.S. gasoline use (Food & Water Watch and Network for New Energy Choices 2007).

Biodiesel fuel is a cleaner-burning diesel fuel made from vegetable oils and/or animal fats, including recycled cooking oil. Some individuals who make their own biodiesel fuel obtain used cooking oil from restaurants at no charge.

Hydrogen Power. Hydrogen, the most plentiful element on earth, is a clean-burning fuel that can be used for electricity production, heating, cooling, and transportation. Many see a movement to a hydrogen economy as a long-term solution to the environmental and political problems associated with fossil fuels. Further research is needed, however, to develop nonpolluting and cost-effective ways to extract and transport hydrogen.

Carbon Capture and Storage. Coal-fired power plants emit more carbon dioxide than any other source. One proposal to reduce CO_2 emissions is carbon capture and storage (CCS)—a process of removing CO_2 from the smokestacks of coal-burning plants and storing it deep underground. The technology required for this process, which is still in the development stage, is expensive and requires large inputs of energy. The development of carbon capture and storage technology also promotes continued use of coal, and diverts or reduces investments in renewable energy such as solar, wind, and geothermal energy. Finally, scientists are concerned that stored carbon dioxide could leak out into the environment and cause sudden and drastic climate change (Miller & Spoolman 2009).

Modifications in Consumer Behavior

In the United States and other industrialized countries, many consumers are making "green" choices in their behavior and purchases that reflect concern for the environment. In some cases, these choices carry a price tag, such as paying more for organically grown food or for clothing made from organic cotton. Consumers are also motivated to make green purchases that save money. For example, rising gas prices have led to increased sales in more fuel-efficient cars, such as hybrids. Consumers often consider their utility bill when they choose energy-efficient appliances and electrical equipment. Although some eco-minded individuals choose "green" products and services, others choose to reduce their overall consumption and "buy nothing" rather than "buy green." For example, many consumers are choosing not to buy bottled water and to drink tap water instead. The switch from bottled to tap is partly fueled by the need to cut down on unnecessary spending in hard economic times, but environmental concerns are also a factor. The production and transportation of bottled water uses fossil fuels, and the disposal of plastic water bottles adds to our already overburdened landfills.

Tiny houses have less impact on the environment.

Although the average size of new housing in the United States has increased considerably, some homeowners are choosing to downsize their housing. For some, the driving force behind housing downsizing is economic, but others are moving into smaller dwellings out of concern for the environment.

Table 13.4 presents tips for how consumers can reduce the amount of carbon dioxide each of us produces.

Green Building

The U.S. Green Building Council developed green building standards known as Leadership in Energy and Environmental Design (LEED). These standards consist of 69 criteria to be met by builders in six areas, including energy use and emissions, water use, materials and resource use, and sustainability of the building site. LEED buildings include the Pentagon Athletic Center, the Detroit Lions' football training facility, and the David L. Lawrence Convention Center in Pittsburgh.

TABLE 13.4 Top Ten Things You Can Do to Fight Global Warming

National Geographic's Green Guide (www.thegreenguide.com) lists the following tips for consumers to reduce the amount of carbon dioxide they produce each year. All of the following CO_2 reductions listed are on an annual basis:

1. Replace five incandescent lightbulbs in your home with compact fluorescent bulbs (CFLs): Swapping those 75-watt incandescent bulbs with 19-watt CFLs can cut 275 pounds of CO_2.

2. Instead of short-haul flights of 500 miles or so, take the train and bypass 310 pounds of CO_2.

3. Sure it may be hot, but get a fan, set your thermostat at 75°F, and blow away 363 pounds of CO_2.

4. Replace refrigerators more than 10 years old with today's more energy-efficient Energy Star models and save more than 500 pounds of CO_2.

5. Shave your eight-minute shower to five minutes for a savings of 513 pounds.

6. Caulk, weatherize, and insulate your home. If you rely on natural gas heating, you'll stop 639 pounds of CO_2 from entering the atmosphere (472 pounds for electric heating). And this summer, you'll save 226 pounds from air conditioner use.

7. Whenever possible, dry your clothes on a line outside or a rack indoors. If you air-dry half your loads, you'll dispense with 723 pounds of CO_2.

8. Trim down on the red meat. Because it takes more fossil fuels to produce red meat than fish, eggs, and poultry, switching to these foods will slim your CO_2 emissions by 950 pounds.

9. Leave the car at home and take public transportation to work. Taking the average U.S. commute of 12 miles by light-rail will leave you 1,366 pounds of CO_2 lighter than driving. The standard, diesel-powered city bus can save 804 pounds, and heavy rail subway users save 288.

10. Finally, support the creation of wind, solar, and other renewable energy facilities by choosing green power if offered by your utility. To find a green power program in your state, call your local utility or visit the U.S. Department of Energy's Green Power Network page at http://apps3.eere.energy.gov/greenpower/

With the use of air conditioning increasing rapidly in developing countries, architects are exploring strategies to keep buildings cool in ways that do not harm the environment. Air conditioning poses a threat to the environment as it draws on electricity (largely from the use of fossil fuel–fired power plants) and uses refrigerants that also produce greenhouse gases. One building strategy involves using passive cooling systems that transfer heat from inside a building to the outside air, earth, and/or water through special designs in the building. Pablo LaRoche, a professor of architecture at California State Polytechnic University, explains that by providing pathways to carry heat from the interior of the building to the outdoors, the building itself becomes the air conditioner, using little or no energy at all (Dahl 2013).

Slow Population Growth

As discussed in Chapter 12, slowing population growth is an important component of efforts to protect the environment. One study concluded that providing the more than 200 million women worldwide who want access to contraception but currently don't have it would reduce projected world population in 2050 by a half billion people, and would prevent the emission of at least 34 gigatons of carbon dioxide (a gigaton equals one billion tons) (Wire 2009).

Although Americans who are concerned about the environment may think about how their home energy use, travel, food choices, and other lifestyle behaviors affect the environment, they rarely consider the environmental impact of their reproductive choices. Researchers at Oregon State University estimate that each child born in the United States, and the children that child has as an adult, adds 9,441 metric tons of carbon dioxide to the environment—the equivalent of burning 972,160 gallons of gas (Murtaugh & Schlax 2009).

As shown in Table 13.5, reducing the number of children a woman has by one saves far more carbon dioxide from being released in the environment than do many other environmentally friendly lifestyle choices.

Government Policies, Programs, and Regulations

Government policies and regulations can play an important role in protecting and restoring the environment. Before reading further, assess your attitudes toward government interventions to reduce global warming in this chapter's *Self and Society* feature.

Cap and Trade Programs. Cap and trade programs are a free-market approach used to control pollution by providing economic incentives to power plants and other industries for achieving reductions in the emissions of pollutants. In a cap and trade system, a limit is set on the amount of carbon dioxide that can be released into the air. Polluters buy credits allowing them to emit a limited amount of carbon dioxide. They can sell leftover credits to other polluters, creating a monetary incentive to reduce emissions. Twenty-three U.S. states have joined regional agreements to lower carbon emissions through the cap and trade system (Pew Center on the States 2009). In 2010, California adopted the nation's most stringent rules to curb greenhouse gas emissions; the rules reward industries that cut emissions by allowing them to sell carbon credits to other industries. Critics of the cap and trade approach argue that it fails to achieve the lowest possible emissions because it does not require all plants to use the best available technology to reduce emissions. By allowing some plants to have higher emissions, it also exposes populations living near these high-emissions plants to excessive air pollution.

In 2009, the U.S. House passed the American Clean Energy and Security Act, which sought to establish a federal cap and trade system modeled after the European Union Emission Trading Scheme. The bill was defeated in the Senate and continues to be debated in Congress.

Policies and Regulations on Energy. In 2004, more than 20 countries committed to specific targets for the renewable share of total energy use (UNEP 2007). A number of states have set goals of producing a minimum percentage of electricity from wind power, solar power, or other renewable sources (Prah 2007). In addition, more than 70 mayors and other local leaders from around the world signed the Urban Environmental Accords, pledging to obtain 10 percent of energy from renewable resources by 2012, and to reduce greenhouse gases 25 percent by 2030 (Stoll 2005). In 2013, nine northeastern states organized in the Regional Greenhouse Gas Initiative (RGGI) agreed that power plants in 2014 must cut their greenhouse gas emissions by 45 percent, and after 2014, the limit will drop another 2.5 percent each year until 2020 (Malewitz 2013).

Taxes. Some environmentalists propose that governments use taxes to discourage environmentally damaging practices and products (Brown & Mitchell 1998). In the 1990s, a number of European governments increased taxes on environmentally harmful activities and products (such as gasoline, diesel, and motor vehicles) and decreased taxes on income and labor (Renner 2004). As a result of high gasoline taxes in Europe, gas there costs as much as $10 a gallon, which has increased consumer demand for small, fuel-efficient cars. Raising gasoline taxes in the United States is highly unpopular with voters and consumers. On the other hand, tax incentives and credits are used for renewable energy, hybrid and electric cars, and energy efficiency.

Fuel Efficiency Standards. In 2012, President Obama issued a new average fuel-efficiency standard for cars and light trucks of 54.5 mpg by 2025, a significant increase from the previous 35.5 mpg. This new standard is expected to save consumers more

TABLE 13.5 Carbon Reduction Figures for Various Lifestyle Behaviors*

Behavior	Amount of CO_2 Not Released into the Environment
Replace windows with energy-efficient windows	12 metric tons
Recycle newspapers, magazines, glass, plastic, aluminum, and steel cans	17 metric tons
Replace 10 75-watt incandescent bulbs with 25-watt energy-efficient lights	36 metric tons
Increase auto gas mileage from 20 mpg to 30 mpg	148 metric tons
Reduce number of children by one	9,441 metric tons

*Calculated over an 80-year period.
Source: Murtaugh & Schlax 2009.

Answer each of the following questions:

1. Some people believe that the United States government should limit the amount of greenhouse gases thought to cause global warming that U.S. businesses can produce. Other people believe that the government should not limit the amount of greenhouse gases that U.S. businesses put out. What do you think?

2. Do you favor or oppose each of the following as a way for the federal government to try to reduce future global warming?
   a. Increased taxes on electricity so people use less of it
   b. Increased taxes on gasoline so people either drive less, or buy cars that use less gas
   c. Tax breaks for companies to produce more electricity from water, wind, and solar power

3. For each of the following, do you think the government should require by law, encourage with tax breaks but not require, or stay out of entirely?
   a. Building cars that use less gasoline
   b. Building air conditioners, refrigerators, and other appliances that use less electricity
   c. Building new homes and offices that use less energy for heating and cooling

4. Do you think that U.S. actions to reduce global warming in the future would hurt the U.S. economy, help the economy, or have no effect on the U.S. economy?

5. Do you think the United States should take action on global warming only if other major industrial countries such as China and India agree to do equally effective things, that the United States should take action even if these other countries do less, or that the United States should not take action on this at all?

6. Do you think most U.S. business leaders *want* the federal government to do things to stop global warming, or do you think most U.S. business leaders *do not want* the federal government to do things to stop global warming?

7. How important is the issue of global warming to you personally—extremely important, very important, somewhat important, not too important, or not at all important?

Comparison Data

Telephone interviews with a sample of 1,000 U.S. adults from across the nation found that, overall, Americans want government to be involved in reducing global warming:

1. Government should limit greenhouse gases from U.S. businesses (76%)
   Government should not limit greenhouse gases from U.S. businesses (20%)
   Don't know (3 percent)

2. To reduce future global warming, government should . . .

	Favor	Oppose
a. Increase taxes on electricity	22%	78%
b. Increase taxes on gasoline	28%	71%
c. Give companies tax breaks to produce more electricity from water, wind, and solar power	84%	15%

3. Should the government require by law, encourage with tax breaks but not require, or stay out of entirely?

	Require	Encourage	Stay out of
a. Building cars that use less gasoline	31%	50%	19%
b. Building air conditioners, refrigerators, and other appliances that use less electricity	29%	51%	20%
c. Building new homes and offices that use less energy for heating and cooling	24%	56%	20%

4. U.S. actions to reduce global warming in the future would:
   Hurt U.S. economy (20%)
   Help U.S. economy (56%)
   Not affect economy (23%)

5. The United States should take action on global warming:
   Only if other countries do (14%)
   Even if other countries do less (68%)
   Not take action at all (18%)

6. Most U.S. business leaders ___ the government to do things to stop global warming.
   do want (25%) do not want (72%) don't know (3%)

7. How important is the issue of global warming to you personally?
   Extremely or very important (46%)
   Somewhat important (30%)
   Not too important (12%)
   Not at all important (12%)

Source: Adapted from Jon Krosnick. 2010. *Global Warming Poll*. Stanford University. Available at woods.stanford.edu. Used by permission.

than 1.7 trillion in gasoline and reduce U.S. oil consumption by 12 billion barrels a year (White House 2012).

Policies on Chemical Safety. In 2003, the European Union drafted legislation known as Registration, Evaluation, and Authorization and Restriction of Chemicals (REACH) that requires chemical companies to conduct safety and environmental tests to prove that the chemicals they are producing are safe. If they cannot prove that a chemical is safe, it will be banned from the market (Rifkin 2004). The European Union has become a world leader in environmental stewardship by placing the "precautionary principle" at the center of EU regulatory policy. The precautionary principle requires industry to prove that their products are safe. In contrast, in the United States, chemicals are assumed to be safe unless proven otherwise, and the burden is put on the consumer, the public, or the government to prove that a chemical causes harm.

International Cooperation and Assistance

Global environmental concerns call for global solutions forged through international cooperation and assistance. For example, the 1987 Montreal Protocol on Substances That Deplete the Ozone Layer forged an agreement made by 70 nations to curb the production of CFCs (which contribute to ozone depletion and global warming).

In 1997, delegates from 160 nations met in Kyoto, Japan, and forged the **Kyoto Protocol**—the first international agreement to place legally binding limits on greenhouse gas emissions from developed countries. The United States, the world's largest producer of greenhouse gas emissions, rejected the Kyoto Protocol in 2001. As of September 2013, 191 countries and 1 region had signed and ratified the Kyoto Protocol; the United States had not ratified the Kyoto Protocol.

Avoiding dangerous climate change will require rich countries to cut carbon emissions at least 80 percent by the end of the 21st century, with cuts of 30 percent by 2020—significantly more than the cuts required under the Kyoto Protocol (United Nations Development Programme 2007). In 2009, world leaders met at a climate summit meeting in Copenhagen, where the United States, China, and dozens of other countries signed on to the Copenhagen Accord—a voluntary agreement to curb climate change by cutting greenhouse gas emissions. But the Copenhagen Accord has no provisions for enforcement of the agreement, and even if each country actually meets its pledge, the pledges are "drastically inadequate" to meet the cuts that climate scientists say are needed (Brecher 2011).

At the 2010 United Nations Climate Change Conference in Cancun, Mexico, the international community reached a set of agreements to address climate change. The Cancun Agreements included plans to reduce greenhouse gas emissions, and to help developing nations protect themselves from climate impacts and build their own sustainable futures.

Public support of international agreements to address global climate change is not universal or unanimous. A study of citizens in four countries (France, Germany, the United Kingdom, and the United States) found that public support is higher for global climate agreements that involve lower cost, that include a higher number of participating countries, are monitored by an independent third party, and include a low sanction for countries failing to meet their emission reduction target (Bechtel & Scheve 2013).

Sustainable Economic and Human Development

Achieving global cooperation on environmental issues is difficult, in part, because developed countries (primarily in the Northern Hemisphere) have different economic agendas from those of developing countries (primarily in the Southern Hemisphere). The northern agenda emphasizes preserving wealth and affluent lifestyles, whereas the southern agenda focuses on overcoming mass poverty and achieving a higher quality of life. Southern countries are concerned that northern industrialized countries—having already achieved economic wealth—will impose international environmental policies that restrict the economic growth of developing countries.

As discussed in Chapter 6, development involves more than economic growth and the alleviation of poverty. The human development approach views the well-being of populations in terms of not only their income, but their access to education, and their

Kyoto Protocol The first international agreement to place legally binding limits on greenhouse gas emissions from developed countries.

ability to lead long, healthy lives in societies that respect and value everyone. The 2013 *Human Development Report* adds, "To sustain progress in human development, far more attention needs to be paid to the impact human beings are having on the environment. The goal is high human development and a low ecological footprint per capita" (UNDP 2013, p. 94).

The 2013 *Human Development Report* presents data on each country's Human Development Index (based on measures of health and longevity, education, and income) as well as each country's environmental footprint. Although it might seem that carbon productivity—GDP per unit of carbon dioxide emission—would increase with human development, the correlation is weak. This finding supports the idea that progress in human development can be sustainable. **Sustainable development** is development that enables human populations to have fulfilling lives without degrading the planet. "The aim here is for those alive today to meet their own needs without making it impossible for future generations to meet theirs. . . . This in turn calls for an economic structure within which we consume only as much as the natural environment can produce, and make only as much waste as it can absorb" (McMichael et al. 2000, p. 1067).

The development and use of clean, renewable energy technologies plays an important role in sustainable human development. Renewable energy projects in developing countries have demonstrated that providing affordable access to green energy helps to alleviate poverty by providing energy for creating business and jobs and by providing power for refrigerating medicine, sterilizing medical equipment, and supplying fresh water and sewer services needed to reduce infectious disease and improve health (Flavin & Aeck 2005).

The Role of Institutions of Higher Education

Colleges and universities can play an important role in efforts to protect the environment by encouraging use of bicycles on campus, using hybrid and electric vehicles, establishing recycling programs, using local and renewable building materials for new buildings, involving students in organic gardening to provide food for the campus, using clean energy, and incorporating environmental education into the curricula. A growing number of colleges and universities are establishing **Green Revolving Funds (GRFs)**, which are funds dedicated to financing cost-saving energy-efficiency upgrades and other projects that decrease resource use and minimize environmental impact (Sustainable Endowments Institute 2011). The resulting savings in operating expenses are returned to the fund and then reinvested in additional projects. In some colleges and universities, students play a role in deciding how Green Revolving Funds are spent and in implementing the various energy upgrades and projects. Nevertheless, David Newport, director of the environmental center at the University of Colorado at Boulder, believes that institutions of higher education are not doing enough to promote sustainability. "We're supposed to be on the leading edge, and we're behind the curve. . . . There are, what 4,500 colleges in the United States, and how many of them are really doing something? Less than 100 or 200?" (quoted by Carlson 2006, p. A10).

Understanding Environmental Problems

sustainable development
Occurs when human populations can have fulfilling lives without degrading the planet.

Green Revolving Funds (GRFs) College and university funds that are dedicated to financing cost-saving energy-efficiency upgrades and other projects that decrease resource use and minimize environmental impacts.

Environmental problems are linked to corporate globalization, rapid and dramatic population growth, expanding world industrialization, patterns of excessive consumption, and reliance on fossil fuels for energy. The Global Footprint Network (2010) offers the following analysis of environmental problems:

> Climate change is not the problem. Water shortages, overgrazing, erosion, desertification, and the rapid extinction of species are not the problem. Deforestation, reduced cropland productivity, and the collapse of fisheries are not the problem. Each of these crises, though alarming, is a symptom of a single, overriding issue. Humanity is simply demanding more than the earth can provide.

Whether we understand environmental problems as resulting from a complex set of causes, or from one, simple underlying cause such as consuming more than the earth can provide, we cannot afford to ignore the growing evidence of the irreversible effects of global warming and loss of biodiversity, and the adverse health effects of toxic waste and other forms of pollution. Although civilizations throughout history have collapsed, these civilizations were limited to a particular region. Never before has civilization on a global scale been threatened, as it is now, by destruction of our ecosystems. To avoid the collapse of the global civilization, it is essential to reduce greenhouse gas emissions to half of current levels by 2050 (Ehrlich & Ehrlich 2013). This would require a major shift away from the use of fossil fuels—a shift the fossil fuel industry cannot support without jeopardizing its own profits. "Because the ethics of some businesses include knowingly continuing lethal but profitable activities . . . it is hardly surprising that interests with large financial stakes in fossil fuel burning have launched a gigantic and largely successful disinformation campaign in the USA to confuse people about climate disruption . . . and block attempts to deal with it" (Ehrlich & Ehrlich 2013, p. 3).

Many Americans believe in a "technological fix" for the environment—that science and technology will solve environmental problems. Paradoxically, the same environmental problems that have been caused by technological progress may be solved by technological innovations designed to clean up pollution, preserve natural resources and habitats, and provide clean forms of energy. But leaders of government and industry must have the will to finance, develop, and use technologies that do not pollute or deplete the environment. When asked how companies can produce products without polluting the environment, Robert Hinkley suggested that, first, it must become a goal to do so:

> I don't have the technological answers for how it can be done, but neither did President John F. Kennedy when he announced a national goal to land a man on the moon by the end of the 1960s. The point is that, to eliminate pollution, we first have to make it our goal. Once we've done that, we will devote the resources necessary to make it happen. We will develop technologies that we never thought possible. But if we don't make it our goal, then we will never devote the resources, never develop the technology, and never solve the problem. (quoted by Cooper 2004, p. 11)

But the direction of technical innovation is largely in the hands of big corporations that place profits over environmental protection. Unless the global community challenges the power of transnational corporations to pursue profits at the expense of environmental and human health, corporate behavior will continue to take a heavy toll on the health of the planet and its inhabitants. Because oil has been implicated in political and military conflicts involving the Middle East (see Chapter 15), such conflicts are likely to continue as long as oil plays the lead role in providing the world's energy.

Global cooperation is also vital to resolving environmental concerns but is difficult to achieve because rich and poor countries have different economic development agendas: Developing poor countries struggle to survive and provide for the basic needs of their citizens; developed wealthy countries struggle to maintain their wealth and relatively high standard of living. Can both agendas be achieved without further pollution and destruction of the environment? Is sustainable economic development an attainable goal? With mounting concern about climate change, the health impacts of air pollution, rising oil prices, and the need to ensure energy access to all, governments worldwide have strengthened their commitment to sustainable, renewable energy policies and projects (UNEP 2007).

In the United States, there is significant opposition to governmental regulations designed to address environmental problems. Opponents of regulation argue that rules and restrictions are harmful to the economy and destroy jobs. Yet, an analysis of the costs and benefits of the Clean Air Act Amendments of 1990 found that value of the benefits ($1.3 trillion) was 25 times the cost ($53 billion). In 2010 alone, the Clean Air Act Amendments of 1990 saved an estimated 160,000 lives (Shapiro & Irons 2011). Lax regulation was a contributing factor in the BP Deepwater Horizon oil spill of 2010—the largest oil spill in U.S. history, which had devastating effects on jobs and the economy in the Gulf states. The claim that environmental regulation hurts jobs is not supported by the evidence, and the costs of regulations are far outweighed by the benefits to health, safety, and well-being (Shapiro & Irons 2011). Protecting the environment does not mean that jobs must be sacrificed in the process. Indeed, transitioning to a low-carbon economy can simultaneously protect the environment, create national energy independence, lower energy prices, *and* create jobs (Brecher 2011).

Our collective response to the precarious state of the environment constitutes a test that we cannot afford to fail. As environmentalist Bill McKibben (2008) notes,

> The next few years are a kind of final exam for the human species. Does that big brain really work or not? It gave us the power to build coal-fired power plants and SUVs and thereby destabilize the working of the earth. But does it give us the power to back away from those sources of power, to build a world that isn't bent on destruction? Can we think, and feel, our way out of this, or are we simply doomed to keep acting out the same set of desires for MORE that got us into this fix?

In 2004, the Nobel Peace Prize was given to Wangari Maathai for leading a grassroots environmental campaign to plant 30 million trees across Kenya. This was the first time ever that the Nobel Peace Prize was awarded to someone for accomplishments in restoring the environment. In her acceptance speech, Maathai explained, "A degraded environment leads to a scramble for scarce resources and may culminate in poverty and even conflict" (quoted by Little 2005, p. 2). With ongoing conflict around the globe, it is time for world leaders to recognize the importance of a healthy environment for world peace and to prioritize environmental protection in their political agendas.

CHAPTER REVIEW

- **How do free trade agreements pose a threat to environmental protection?**
Free trade agreements such as NAFTA and the FTAA provide transnational corporations with privileges to pursue profits, expand markets, use natural resources, and exploit cheap labor in developing countries while weakening the ability of governments to protect natural resources or to implement environmental legislation.

- **What are environmental refugees?**
Environmental refugees are individuals who have migrated because they are forced from their homelands and/or can no longer secure a livelihood as a result of deforestation, desertification, soil erosion, and other environmental problems.

- **What is greenwashing?**
Greenwashing refers to the ways in which environmentally and socially damaging companies portray their corporate image and products as being "environmentally friendly" or socially responsible.

- **Where does most of the world's energy come from?**
Most of the world's energy comes from fossil fuels, which include oil, coal, and natural gas. This is significant because many of the serious environmental problems in the world today, including global warming and climate change, biodiversity loss, and pollution, stem from the use of fossil fuels.

- **What are the major causes and effects of deforestation?**
The major causes of deforestation are the expansion of agricultural land, human settlements, wood harvesting, and road building. Deforestation displaces people and wild species from their habitats, contributes to global warming, and contributes to desertification, which results in the expansion of desert land that is unusable for agriculture. Soil erosion caused by deforestation can cause severe flooding.

- **What are the effects of air pollution on human health?**
Air pollution, which is linked to heart disease, lung cancer, and respiratory ailments, such as emphysema, chronic bronchitis, and asthma kills about 3 million people a year.

- **What are some examples of common household, personal, and commercial products that contribute to indoor pollution?**
Some common indoor air pollutants include carpeting, mattresses, drain cleaners, oven cleaners, spot removers, shoe polish, dry-cleaned clothes, paints, varnishes, furniture polish, potpourri, mothballs, fabric softener, caulking compounds, air fresheners, deodorizers, disinfectants, glue, correction fluid, printing ink, carbonless paper, and felt-tip markers.

- **What is the primary cause of global warming?**
The prevailing view on what causes global warming is that greenhouse gases—primarily carbon dioxide, methane, and nitrous oxide—accumulate in the atmosphere and act like the glass in a greenhouse, holding heat from the sun close to the earth. The primary greenhouse gas is carbon dioxide, which is released into the atmosphere by burning fossil fuels.

- **How does global warming contribute to further global warming?**
As global warming melts ice and snow, it exposes more land and ocean area, which absorbs more heat than ice and

snow, further warming the planet. The melting of Siberia's frozen peat bog—a result of global warming—could release billions of tons of methane, a potent greenhouse gas, into the atmosphere and cause further global warming. This process, whereby the effects of global warming cause further global warming, is known as a positive feedback loop.

- **What is the relationship between level of economic development and environmental pollution?**
There is a curvilinear relationship between level of economic development and environmental pollution. In regions with low levels of economic development, industrial emissions are minimal, but emissions rise in countries that are in the middle economic development range as they move through the early stages of industrialization. However, at more advanced stages of industrialization, industrial emissions ease because heavy-polluting manufacturing industries decline, "cleaner" service industries increase, and because rising incomes are associated with a greater demand for environmental quality and cleaner technologies.

- **What does the term *environmental injustice* refer to?**
Environmental injustice, also called *environmental racism,* refers to the tendency for marginalized populations and communities to disproportionately experience adversity due to environmental problems. For example, in the United States, polluting industries, industrial and waste facilities, and transportation arteries (which generate vehicle emissions pollution) are often located in minority communities.

- **What are some of the concerns about nuclear energy?**
Nuclear waste contains radioactive plutonium, a substance linked to cancer and genetic defects. Nuclear waste in the environment remains potentially harmful to human and other life for thousands of years, and disposing of nuclear waste is problematic. Accidents at nuclear power plants, such as the 2011 Fukushima disaster, and the potential for nuclear reactors to be targeted by terrorists add to the actual and potential dangers of nuclear power plants.

- **What is the "climate denial machine"?**
The "climate denial machine" is a well-funded misinformation campaign by the fossil fuel industry and its allies that involves attacking and discrediting climate science,

scientists, and scientific institutions. This campaign has been effective in swaying public views of climate change: Despite the overwhelming scientific consensus that human activity causes global warming, a 2013 Gallup Poll found that more than half (57 percent) of U.S. adults believe that global warming is due more to natural changes in the environment.

- **Why are women who may become pregnant, pregnant women, nursing mothers, and young children advised against eating certain types of fish?**
The U.S. Environmental Protection Agency advises women who may become pregnant, pregnant women, nursing mothers, and young children to avoid eating certain fish altogether (swordfish, shark, king mackerel, and tilefish) because of the high levels of mercury.

- **How often does a species of life on earth become extinct?**
One species of life on earth goes extinct every three hours.

- **What social and cultural factors contribute to environmental problems?**
Social and cultural factors that contribute to environmental problems include population growth, industrialization and economic development, and cultural values and attitudes such as individualism, consumerism, and militarism.

- **What are some of the strategies for alleviating environmental problems?**
Strategies for alleviating environmental problems include efforts to lower fertility rates and slow population growth, environmental activism, environmental education, the use of "green" energy, modifications in consumer products and behavior, and government regulations and legislation. Sustainable economic development and international cooperation and assistance also play important roles in alleviating environmental problems.

- **According to 2004 Nobel Peace Prize winner Wangari Maathai, why is environmental protection important for national and international security?**
In her acceptance speech for the 2004 Nobel Peace Prize, Wangari Maathai explained that "a degraded environment leads to a scramble for scarce resources and may culminate in poverty and even conflict."

TEST YOURSELF

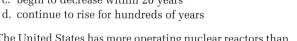

1. Tar sands oil is known as the world's ___ oil.
   a. most expensive
   b. least expensive
   c. cleanest
   d. dirtiest
2. Red fire ants are an example of
   a. a bioinvasion
   b. an extinct species
   c. a threatened species
   d. an alternative fuel
3. Which innovation has resulted in increased food prices?
   a. Solar power
   b. Bottled water

   c. Wind turbines
   d. Biofuels

4. If greenhouse gases were to be stabilized today, global air temperature and sea level would be expected to
   a. remain at their current level
   b. decrease immediately
   c. begin to decrease within 20 years
   d. continue to rise for hundreds of years

5. The United States has more operating nuclear reactors than any other country.
   a. True
   b. False

6. Most solid waste in the United States is recycled.
   a. True
   b. False

7. In 2007, which U.S. city became the first to ban plastic bags from supermarkets and chain pharmacies?
   a. Miami
   b. San Francisco
   c. Atlanta
   d. Portland

8. In the United States, the EPA has required testing on all of the more than 80,000 chemicals that have been on the market since 1976.
   a. True
   b. False

9. Which of the following is the number one polluter in the United States?
   a. The military
   b. Dow Chemical
   c. Archer Daniels Midland
   d. ExxonMobil

10. A growing number of ___ are establishing Green Revolving Funds (GRFs), which are funds dedicated to financing cost-saving energy-efficiency upgrades and other projects that decrease resource use and minimize environmental impact.
    a. state governments
    b. colleges and universities
    c. manufacturing industries
    d. countries

Answers: 1. D; 2. A; 3. D; 4. D; 5. A; 6. B; 7. B; 8. B; 9. A; 10. B.

KEY TERMS

acid rain 413
biodiversity 422
bioinvasion 405
biomass 410
climate denial machine 416
deep ecology 425
deforestation 411
desertification 411
e-waste 418
Earth Overshoot Day 411
ecosystem 406

ecoterrorism 426
environmental footprint 410
environmental injustice 407
environmental refugees 406
fracking 419
global warming 413
green energy 428
greenhouse gases 413
Green Revolving Funds (GRFs) 434
greenwashing 407
Kyoto Protocol 433

light pollution 422
multiple chemical sensitivity 421
perceived obsolescence 406
pinkwashing 408
planned obsolescence 406
radical environmental movement 425
sustainable development 434
tar sands 404
tar sands oil 404

Dan McCoy/Rainbow/Getty Images

Science and Technology

"We have arranged things so that almost no one understands science and technology. This is a prescription for disaster. We might get away with it for a while, but sooner or later this combustible mixture of ignorance and power is going to blow up in our faces."

—Carl Sagan, astronomer and astrobiologist

439

Pictured is Neda Agha-Soltan, a 26 year old college student who was shot dead while watching a demonstration protesting the re-election of Iranian President Mahmoud Ahmadinejad. The use of the wrong Facebook picture led to Professor Neda Soltani, who looked very much like the young student, fleeing her homeland leaving family and career behind. Professor Soltani was granted asylum in Germany.

STUDENT ACTIVIST NEDA AGHA-SOLTAN, shot in 2009 by a government sniper, became a martyr in the Iranian pro-democracy movement following the reelection of President Mahmoud Ahmadinejad (Ghafour 2012). Her photo was placed on placards and carried in the streets, and her ". . . serene smile went viral as every news organization on the planet broadcast her image in their coverage of the election's violent aftermath" (p. 1). But the woman in the pictures was not Neda Agha-Soltan but Professor Neda Soltani, an English professor at Azad University in Tehran. Professor Soltani's picture had mistakenly been taken from her Facebook page and she was now "caught in the terrifying currents of social media, where every tweet, Facebook post, Flickr image reconfirmed the false story" (p. 1). After being interrogated by Iranian officials, Professor Soltani was asked to make a video stating that she was the student activist and that the stories of her death were simply Western propaganda. When she refused, she was threatened with death and soon after fled Iran, leaving behind a successful career, her fiancé, family, and friends.

technological dualism The tendency for technology to have both positive and negative consequences.

science The process of discovering, explaining, and predicting natural or social phenomena.

technology Activities that apply the principles of science and mechanics to the solutions of a specific problem.

mechanization Dominant in an agricultural society, the use of tools to accomplish tasks previously done by hand.

automation The replacement of human labor with machinery and equipment.

cybernation Dominant in postindustrial societies, the use of machines to control other machines.

Technological dualism is the tendency for technology to have both positive and negative consequences. Clearly, Facebook has allowed friends and family to remain in contact with one another in a way that was simply unheard of before the advent of social networking sites. On the other hand, in this case of mistaken identity, Facebook nearly cost Professor Soltani her life. Like social media, the 3-D printer is another example of technological dualism. Although efficient and cost-effective for manufacturers, 3-D printers may lead to higher rates of unemployment, particularly among assembly-line laborers (see Chapter 7). Further, 3-D printers can be used to make a robotic hand for a child with amniotic band syndrome or to make a semiautomatic gun used by a mass murderer.

Science and technology go hand in hand. **Science** is the process of discovering, explaining, and predicting natural or social phenomena. A scientific approach to understanding acquired immunodeficiency syndrome (AIDS), for example, might include investigating the molecular structure of the virus, the means by which it is transmitted, and public attitudes about AIDS. **Technology**, as a form of human cultural activity that applies the principles of science and mechanics to the solution of problems, is intended to accomplish a specific task—in this case, the development of an AIDS vaccine.

Societies differ in their level of technological sophistication and development. In agricultural societies, which emphasize the production of raw materials, the use of tools to accomplish tasks previously done by hand, or **mechanization**, dominates. As societies move toward industrialization and become more concerned with the mass production of goods, automation prevails. **Automation** involves the use of self-operating machines, as in an automated factory where autonomous robots assemble automobiles. Finally, as a society moves toward postindustrialization, it emphasizes service and information professions (Bell 1973). At this stage, technology shifts toward **cybernation**, whereby machines control machines—making production decisions, programming robots, and monitoring assembly performance.

What are the effects of science and technology on humans and their social world? How do science and technology help to remedy social problems, and how do they contribute to social problems? Is technology, as author Neil Postman (1992) suggested, both a friend and a foe to humankind? We address each of these questions in this chapter.

The Global Context: The Technological Revolution

Less than 50 years ago, traveling across state lines was an arduous task; a long-distance phone call was a memorable event; and mail carriers brought belated news of friends and relatives from far away. Today, travelers journey between continents in a matter of hours, and for many, e-mail, faxes, instant messaging, texting, and cell phones have replaced previously conventional means of communication.

The world is a much smaller place than it used to be, and it will become even smaller as the technological revolution continues. In 2012, the Internet had 2.4 billion users in more than 200 countries, with 245 million users in the United States (Internet Statistics 2013). Of all Internet users, the highest proportion come from Asia (44.8 percent), followed by Europe (21.5 percent), North America (11.4 percent), Latin America and the Caribbean (10.4 percent), Africa (7.0 percent), the Middle East (3.7 percent), and Oceania/Australia (1.0 percent) (Internet Statistics 2013).

Although the **penetration rate**, that is, the percentage of people who have access to and use the Internet in a particular area, is higher in industrialized countries, there is some movement toward the Internet becoming a truly global medium as Africans, Middle Easterners, and Latin Americans increasingly "get online." For example, Internet use in the United States grew 152 percent between 2000 and 2010; the number of Internet users in Nigeria increased by 21,891 percent during the same time period (Internet Statistics 2011). Further, Google is actively working to connect 1 billion or more new wireless Internet users in such areas as sub-Saharan Africa and Southeast Asia (Efrati 2013).

baobao ou/Getty Images

Forty-five percent of all Internet users live in Asia. That translates to over 1 billion people, nearly half of whom live in China. The sheer number of people online in China is a function of China's population. However, China's penetration rate is only 40 percent compared to, for example, Japan that has 80 percent of its population online.

The movement toward globalization of technology is, of course, not limited to the use and expansion of the Internet. The world robot market and the U.S. share of it continues to expand; Microsoft's Internet platform and support products are sold all over the world; scientists collect skin and blood samples from remote islanders for genetic research; a global treaty regulating trade of genetically altered products has been signed by more than 100 nations; and Intel's central processing units (CPUs) power an estimated 80 percent of the world's personal computers (PCs) (Robinson-Avila 2013).

To achieve such scientific and technological innovations, sometimes called research and development (R&D), countries need material and economic resources. *Research* entails the pursuit of knowledge; *development* refers to the production of materials, systems, processes, or devices directed to the solution of practical problems. According to the National Science Foundation (NSF), the United States spends over $400 billion a year in research and development, accounting for about 31 percent of the global total—the largest single performing country in the world (NSF 2013). As in most other countries, U.S. funding sources are primarily from private industry, 62 percent of the total, followed by the federal government and nonprofit organizations such as research institutes at colleges and universities (NSF 2013).

The United States leads the world in science and technology, although there is some evidence that we are falling behind (Dutta & Mia 2011; ITIF 2012; Price 2008). For example,

penetration rate The percentage of people who have access to and use the Internet in a particular area.

a report by the World Economic Forum compares information and communication technologies (ICTs) across countries using a networked readiness index (NRI). The NRI is composed of four subsections: (1) the quantity and quality of the *environment* for ICTs (e.g., political environment, regulatory environment), (2) ICTs, *readiness* (e.g., affordability), (3) ICTs, *usage* (e.g., individual, business), and (4) the *impact* of ICTs (e.g., social, economic) (WEF 2013). In 2013, the NRI results indicated two dominant groups—Nordic countries (Finland, Sweden, Norway, and Denmark all in the top 10 rankings) and Asian countries (Singapore, Taiwan/China, Republic of Korea, and Hong Kong, all in the top 14 rankings). The United States ranked number nine on the overall index, but scored higher on the readiness subsection and lower on the remaining three subsections.

> In the past, the United States granted the highest proportion of science and engineering PhD degrees in the world but in recent years has been outpaced by China.

The decline of U.S. supremacy in science and technology is likely to be the result of several interacting forces (ITIF 2009; Lemonick 2006; Price 2008; World Bank 2009). First, the federal government has been scaling back its investment in research and development in response to fiscal deficits. Second, corporations, the largest contributors to research and development, have begun to focus on short-term products and higher profits as pressure from stockholders mounts. Third, developing countries, most notably China and India, are expanding their scientific and technological capabilities at a faster rate than the United States. Although the United States is ranked 27th in its change score for business research and development investments between 1999 and 2008, China is ranked second (ITIF 2011).

Fourth, there has been concern over science and math education in the United States, both in terms of quality and quantity. In the past, the United States granted the highest proportion of science and engineering PhD degrees in the world but in recent years has been outpaced by China (NSF 2013). A recent report by a presidential advisory board concludes that over the next decade, more than 1 million more college students in science, technology, engineering, and mathematics (**STEM**), will be needed to maintain global competitiveness (PCAT 2012).

Finally, Mooney and Kirshenbaum (2009) document "unscientific America"—the tremendous disconnect between the citizenry, media, politicians, religious leaders, education, and the entertainment industry (e.g., *CSI, The Big Bang Theory, Grey's Anatomy*) on the one hand, and science and scientists on the other. Post-World War II America, in part because of the Cold War, invested in R&D, leading to such scientific and technological advances as the space program, the development of the Internet, and the decoding of the genome. Yet, despite these significant contributions and the recognition of the significance of STEM disciplines, most Americans know very little about science (NSF 2013) (see this chapter's *Self and Society*). Figure 14.1 displays the primary sources of science and technology information and events as measured by a recent survey.

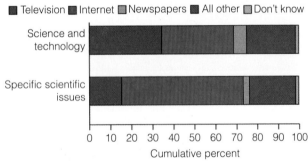

NOTE: "All other" includes radio, magazines, books, government agencies, family, and friends/colleagues.

Figure 14.1 **Sources of Science and Technology Information and Events**
Source: NSF 2013.

What Do You Think? Scientific discoveries and technological developments require the support of a country's citizens and political leaders. For example, although abortion has been technically possible for years, millions of the world's citizens live in countries where abortion is either prohibited or permitted only when the life of the mother is in danger. Can you name other scientific discoveries or technological developments that are technically possible, but likely to be rejected by large segments of the population?

STEM An acronym for science, technology, engineering, and mathematics.

Answer each of the following questions and then calculate the number that you answered correctly.

1. All radioactivity is man-made.
   a. True
   b. False
2. Electrons are smaller than atoms.
   a. True
   b. False
3. Lasers work by focusing sound waves.
   a. True
   b. False
4. The continents on which we live have been moving their location for millions of years and will continue to move in the future.
   a. True
   b. False
5. Which of the following types of solar radiation does sunscreen protect the skin from?
   a. X-rays
   b. infrared
   c. ultraviolet
   d. microwaves
6. Nanotechnology deals with things that are extremely . . .
   a. small
   b. large
   c. cold
   d. hot

7. Which gas makes up most of the earth's atmosphere?
   a. hydrogen
   b. nitrogen
   c. carbon dioxide
   d. oxygen
8. What is the main function of red blood cells?
   a. fight disease in the body
   b. carry oxygen to all parts of the body
   c. help the blood to clot
9. Which of these is a major concern about the overuse of antibiotics?
   a. it can lead to antibiotic-resistant bacteria
   b. antibiotics are very expensive
   c. people will become addicted to antibiotics
10. Which is an example of a chemical reaction?
    a. water boiling
    b. sugar dissolving
    c. nails rusting
11. Which is the better way to determine whether a new drug is effective in treating diseases? If a scientist has a group of 1,000 volunteers with the disease to study, should she
    a. give the drug to all of them and see how many get better
    b. give the drug to half of them but not to the other half, and compare how many in each group get better

12. Which gas do most scientists believe causes temperatures in the atmosphere to rise?
    a. carbon dioxide
    b. hydrogen
    c. helium
    d. radon
13. Which natural resource is extracted in a process known as "fracking"?
    a. oil
    b. diamonds
    c. natural gas
    d. silicon

Answers: 1. b; 2. a; 3. b; 4. a; 5. c; 6. a; 7. b; 8. b; 9. a; 10. c; 11. b; 12. a; 13. c.

DIRECTIONS: Compare the percentage of questions you answered correctly to the following distribution. For example, if you answered all the questions correctly, you scored better than 93 percent of the public (the sum of all percents below you) and the same as 7 percent of the public. If you answered 11 of the questions correctly, you scored better than 75 percent of the public (the sum of all percents below you), the same as 10 percent of the public, and worse than 15 percent of the public (the sum of all percents above you).

Source: Pew Research Center 2013.

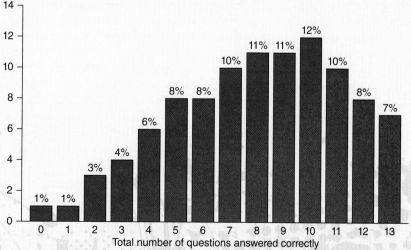

Distribution of the percent of respondents answering 0–11 questions correctly (N=1006)*

*Based on a sample of 1006 randomly selected adults who took part in a national survey in March, 2013.

Many Americans believe that social problems can be resolved through a **technological fix** (Weinberg 1966) rather than through social engineering. For example, a social engineer might approach the problem of water shortages by persuading people to change their lifestyle: use less water, take shorter showers, and wear clothes more than once before washing. A technologist would avoid the challenge of changing people's habits and motivations and instead concentrate on the development of new technologies that would increase the water supply.

Social problems can be tackled through both social engineering and a technological fix. In recent years, for example, social engineering efforts to reduce drunk driving have included imposing stiffer penalties for drunk driving and disseminating public service announcements, such as "Friends don't let friends drive drunk." An example of a technological fix for the same problem is the development of car airbags, which reduce injuries and deaths resulting from car accidents.

Not all individuals, however, agree that science and technology are good for society. **Postmodernism** is an emerging worldview which holds that rational thinking and the scientific perspective have fallen short in providing the "truths" they were once presumed to hold. During the industrial era, science, rationality, and technological innovations were thought to hold the promises of a better, safer, and more humane world. Today, postmodernists question the validity of the scientific enterprise, often pointing to the unforeseen and unwanted consequences of resulting technologies. Automobiles, for example, began to be mass-produced in the 1930s, in response to consumer demands. But the proliferation of automobiles has also led to increased air pollution and the deterioration of cities as suburbs developed, and, today, traffic fatalities are the number one cause of accident-related deaths.

Sociological Theories of Science and Technology

Each of the three major sociological frameworks helps us to better understand the nature of science and technology in society.

Structural-Functionalist Perspective

Structural functionalists view science and technology as emerging in response to societal needs—that "science was born indicates that society needed it" (Durkheim 1973/1925). As societies become more complex and heterogeneous, finding a common and agreed-on knowledge base becomes more difficult. Science fulfills the need for an assumed objective measure of "truth" and provides a basis for making intelligent and rational decisions. In this regard, science and the resulting technologies are functional for society.

If society changes too rapidly as a result of science and technology, however, problems may emerge. When the material part of culture (i.e., its physical elements) changes at a faster rate than the nonmaterial part (i.e., its beliefs and values), a **cultural lag** may develop (Ogburn 1957). For example, the typewriter, the conveyor belt, and the computer expanded opportunities for women to work outside the home. With the potential for economic independence, women were able to remain single or to leave unsatisfactory relationships and/or establish careers. But although new technologies have created new opportunities for women, beliefs about women's roles, expectations of female behavior, and values concerning equality, marriage, and divorce have lagged behind.

Robert Merton (1973), a structural functionalist and founder of the subdiscipline sociology of science, also argued that scientific discoveries or technological innovations may be dysfunctional for society and may create instability in the social system. For example, the development of time-saving machines increases production, but it also displaces workers and contributes to higher rates of employee alienation. Defective technology can have disastrous effects on society. For example, the use of in vitro fertilization (IVF) is associated with an increased risk of birth defects (Davies et al. 2012).

technological fix The use of scientific principles and technology to solve social problems.

postmodernism A worldview that questions the validity of rational thinking and the scientific enterprise.

cultural lag A condition in which the material part of culture changes at a faster rate than the nonmaterial part.

The world was made a smaller place by the Pony Express in the late 1800s, and today a number of technological feats make the world even smaller. Pictured is Apple's iPhone 5C, a less expensive model of the iPhone 5S. The computer giant hopes to expand their mobile phone market into developing countries.

Conflict Perspective

Conflict theorists, in general, argue that science and technology benefit a select few. For some conflict theorists, technological advances occur primarily as a response to capitalist needs for increased efficiency and productivity and thus are motivated by profit. As McDermott (1993) noted, most decisions to increase technology are made by "the immediate practitioners of technology, their managerial cronies, and for the profits accruing to their corporations" (p. 93). In the United States, private industry spends more money on research and development than the federal government does. The Dalkon Shield (IUD) and silicone breast implants are examples of technological advances that promised millions of dollars in profits for their developers. However, the rush to market took precedence over thorough testing of the products' safety. Subsequent lawsuits filed by consumers resulted in large damage awards for the plaintiffs.

Science and technology also further the interests of dominant groups to the detriment of others. The need for scientific research on AIDS was evident in the early 1980s, but the required large-scale funding was not made available so long as the virus was thought to be specific to homosexuals and intravenous drug users. Only when the virus became a threat to mainstream Americans were millions of dollars allocated to AIDS research. Hence, conflict theorists argue that granting agencies act as gatekeepers to scientific discoveries and technological innovations. These agencies are influenced by powerful interest groups and the marketability of the product rather than by the needs of society.

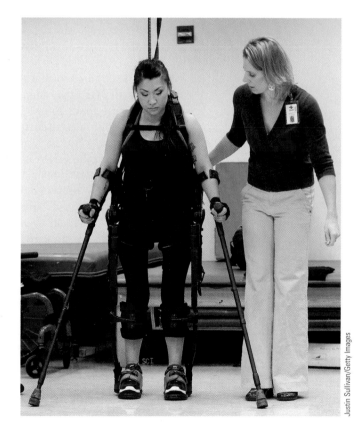

Robots are not only for the factory floor. Here, paralysis victim Stephanie Sablan walks with the use of battery-powered robotic legs called eLEGS.

When the dominant group feels threatened, it may use technology as a means of social control. For example, the use of the Internet is growing dramatically in China, the world's largest Internet market. Censorship has been consolidated under the State Council Information Office, also known as the "great firewall of China" (Chen 2011). A study by Harvard Law School researchers indicates that, of the 204,000 websites accessed, nearly 20,000 were inaccessible. Top Google search results for such words as *Tibet, equality, Taiwan,* and *democracy China* were consistently blocked (Associated Press 2010). China is not alone, however. The OpenNet Initiative, a collaborative effort of three academic institutions, reports that Ethiopia, Iran, Turkmenistan, Afghanistan, Syria, Uzbekistan, and Vietnam also have pervasive political censorship of the Internet (OpenNet 2012).

Finally, conflict theorists as well as feminists argue that technology is an extension of the patriarchal nature of society that promotes the interests of men and ignores the needs and interests of women. As in other aspects of life, women play a subordinate role in reference to technology in terms of both its creation and its use. For example, washing machines, although time-saving devices, disrupted the communal telling of stories and the resulting friendships among women who gathered together to do their chores. Bush (1993) observed that, in a "society characterized by a sex-role division of labor, any tool or technique . . . will have dramatically different effects on men than on women" (p. 204).

Symbolic Interactionist Perspective

Knowledge is relative. It changes over time, over circumstances, and between societies. We no longer believe that the world is flat or that the earth is the center of the universe, but such beliefs once determined behavior because individuals responded to what they thought to be true. The scientific process is a social process in that "truths"—socially constructed truths—result from the interactions between scientists, researchers, and the lay public.

Kuhn (1973) argued that the process of scientific discovery begins with assumptions about a particular phenomenon (e.g., the world is flat). Because unanswered questions always remain about a topic (e.g., why don't the oceans drain?), science works to fill these gaps. When new information suggests that the initial assumptions were incorrect (e.g., the world is not flat), a new set of assumptions or framework emerges to replace the old one (e.g., the world is round). It then becomes the dominant belief or paradigm.

Symbolic interactionists emphasize the importance of this process and the effect that social forces have on it. Lynch et al. (2008) describe the media's contribution in framing societal beliefs about racial discrimination, racism, and genetic determinism. Social forces also affect technological innovations, and their success depends, in part, on the social meaning assigned to any particular product. As social constructionists argue, individuals socially construct reality as they interpret the social world around them, including the meaning assigned to various technologies. If claims makers can successfully define a product as impractical, cumbersome, inefficient, or immoral, the product is unlikely to gain public acceptance. Such is the case with RU-486, an oral contraceptive known as the "abortion pill" that is widely used in France, Great Britain, China, and the United States, but is opposed by many Americans (NARAL 2013).

Not only are technological innovations subject to social meaning, but who becomes involved in what aspects of science and technology is also socially defined. Men, for example, far outnumber women in earning computer science degrees, as many as 10 to 1 at some schools. Although women make up 47.2 percent of the general workforce, they make up just 30.5 percent of computer scientists and 22 percent of computer programmers (U.S. Census Bureau 2013b). Societal definitions of men as being rational, mathematical, and scientifically minded and as having greater mechanical aptitude than women are, in part, responsible for these differences. This chapter's *Social Problems Research Up Close* feature highlights one of the consequences of the masculinization of technology, as well as the ways in which computer hacker identities and communities are socially constructed.

Cyberstalking, pornography on the Internet, and identity theft are crimes that were unheard of before the computer revolution and the enormous growth of the Internet. One such "high-tech" crime, computer hacking, ranges from childish pranks to deadly viruses that shut down corporations. In this classic study, Jordan and Taylor (1998) enter the world of hackers, analyzing the nature of this illegal activity, hackers' motivations, and the social construction of the "hacking community."

Sample and Methods

Jordan and Taylor (1998) researched computer hackers and the hacking community through 80 semistructured interviews, 200 questionnaires, and an examination of existing data on the topic. As is often the case in crime, illicit drug use, and other similarly difficult research areas, a random sample of hackers was not possible. Snowball sampling is often the preferred method in these cases; that is, one respondent refers the researcher to another respondent, who then refers the researcher to another respondent, and so forth. Through their analysis, the investigators provide insight into this increasingly costly social problem and the symbolic interactionist notion of "social construction"—in this case, of an online community.

Findings and Conclusions

Computer hacking, or "unauthorized computer intrusion," is an increasingly serious problem, particularly in a society dominated by information technologies. Unlawful entry into computer networks or databases can be achieved by several means, including (1) guessing someone's password, (2) tricking a computer about the identity of another computer (called "IP spoofing"), or (3) "social

engineering," a slang term referring to getting important access information by stealing documents, looking over someone's shoulder, going through their garbage, and so on.

Hacking carries with it certain norms and values, because, according to Jordan and Taylor (1998), the hacking community can be thought of as a culture within a culture. The two researchers identified six elements of this socially constructed community:

- *Technology.* The core of the hacking community is the technology that allows it to occur. As one professor who was interviewed stated, the young today have "lived with computers virtually from the cradle, and therefore have no trace of fear, not even a trace of reverence."
- *Secrecy.* The hacking community must, on the one hand, commit secret acts because their "hacks" are illegal. On the other hand, much of the motivation for hacking requires publicity to achieve the notoriety often sought. In addition, hacking is often a group activity that bonds members together. As one hacker stated, hacking "can give you a real kick sometimes. But it can give you a lot more satisfaction and recognition if you share your experiences with others."
- *Anonymity.* Whereas secrecy refers to the hacking act, anonymity refers to the importance of the hacker's identity remaining unknown. Thus, for example, hackers and hacking groups take on names such as Legion of Doom, the Inner Circle I, Mercury, and Kaos, Inc.
- *Membership fluidity.* Membership is fluid rather than static, often characterized by high turnover rates, in part, as a response to law enforcement pressures. Unlike more

structured organizations, there are no formal rules or regulations.

- *Male dominance.* Hacking is defined as a male activity; consequently, there are few female hackers. After recounting an incident of sexual harassment, Jordan and Taylor (1998) also note that "the collective identity hackers share and construct . . . is in part misogynist" (p. 768).
- *Motivation.* Contributing to the articulation of the hacking communities' boundaries are the agreed-upon definitions of acceptable hacking motivations, including (1) addiction to computers, (2) curiosity, (3) excitement, (4) power, (5) acceptance and recognition, and (6) community service through the identification of security risks.

Finally, Jordan and Taylor (1998, p. 770) note that hackers also maintain group boundaries by distinguishing between their community and other social groups, including "an antagonistic bond to the computer security industry (CSI)." Ironically, hackers admit a desire to be hired by the CSI, which would not only legitimize their activities but also give them a steady income.

Jordan and Taylor conclude that the general fear of computers and of those who understand them underlies the common, although inaccurate, portrayal of hackers as pathological, obsessed computer "geeks." When journalist Jon Littman asked hacker Kevin Mitnick if he was being demonized because of increased dependence on and fear of information technologies, Mitnick replied, "Yeah. . . . That's why they're instilling fear of the unknown. That's why they're scared of me. Not because of what I've done, but because I have the capability to wreak havoc" (Jordan & Taylor 1998, p. 776).

Technology and the Transformation of Society

A number of modern technologies are considerably more sophisticated than technological innovations of the past. Nevertheless, older technologies have influenced the nature of work as profoundly as the most mind-boggling modern inventions. Postman (1992) described how the clock—a relatively simple innovation that is taken for granted in today's world—profoundly influenced the workplace and with it the larger economic institution:

> The clock had its origin in the Benedictine monasteries of the twelfth and thirteenth centuries. The impetus behind the invention was to provide a more or

less precise regularity to the routines of the monasteries, which required, among other things, seven periods of devotion during the course of the day. The bells of the monastery were to be rung to signal the canonical hours; the mechanical clock was the technology that could provide precision to these rituals of devotion.... What the monks did not foresee was that the clock is a means not merely of keeping track of the hours but also of synchronizing and controlling the actions of men. And thus, by the middle of the fourteenth century, the clock had moved outside the walls of the monastery, and brought a new and precise regularity to the life of the workman and the merchant. . . . In short, without the clock, capitalism would have been quite impossible. The paradox . . . is that the clock was invented by men who wanted to devote themselves more rigorously to God; it ended as the technology of greatest use to men who wished to devote themselves to the accumulation of money. (pp. 14–15)

> . . . [T]echnology continues to have far-reaching effects not only on the economy but also on every aspect of social life.

Today, technology continues to have far-reaching effects not only on the economy but also on every aspect of social life. In the following section, we discuss societal transformations resulting from various modern technologies including workplace technology, computers, the Internet, and science and biotechnology.

Technology and the Workplace

All workplaces—from government offices to factories and from supermarkets to real estate agencies—have felt the impact of technology. Some technology lessens the need for supervisors and makes control by employers easier. For example, employees in the Department of Design and Construction in New York City must scan their hands each time they enter or leave the workplace. The use of identifying characteristics such as hands, fingers, and eyes is part of a technology called *biometrics*. Union leaders "called the use of biometrics degrading, intrusive and unnecessary and said that experimenting with the technology could set the stage for a wider use of biometrics to keep tabs on all elements of the workday" (Chan 2007, p. 1).

Technology can also make workers more accountable by gathering information about their performance. Further, through time-saving devices such as personal digital assistants (PDAs) and battery-powered store-shelf labels, technology can enhance workers' efficiency. Medical software marketed by Wal-Mart at a cost of $25,000 a year for one physician and $10,000 a year for each additional physician in a practice will not only save time and save money on costly record keeping but is also likely to improve patient care (Lohr 2009). Worldwide over 50,000 employees in such companies as Walgreens, Cold Stone, and Verizon are held accountable by a software program called *When I Work* used to schedule, monitor, and communicate with workers (WIW 2013).

However, technology can also contribute to worker error. In a study of a popular hospital computer system, researchers found several ways that the computerized drug-ordering program endangered the health of patients. For example, the software program warned a doctor of a patient's drug allergy only *after* the drug was ordered and, rather than showing the usual dose of a particular drug, the program showed the dosage available in the hospital pharmacy (DeNoon 2005).

Technology is also changing the location of work (see Chapter 7). Nearly a quarter of employed Americans report telecommuting, i.e., regularly completing work tasks at home (Noonan & Glass 2012). Research on telecommuting indicates several benefits including higher productivity, lower absenteeism rates, and higher retention. Parents are no more likely to telecommute than the general population, and mothers are no more likely to telecommute than fathers. Unlike structural functionalists who would argue that telecommuting provides a flexible work option for working parents, Noonan and Glass (2012), consistent with a conflict perspective, suggest that ". . . telecommuting appears . . . to have become instrumental in the general expansion of work hours, facilitating workers' needs for additional work time beyond the standard workweek and/or the ability of employers to increase or intensify work demand among their salaried employees." (p. 38).

Telepresencing, a much more technologically sophisticated version of teleconferencing, allows life-sized participants in the virtual presence of one another to realistically communicate through broadcast quality sound and images (Houlahan 2006; Sharkey 2009; Whitlock 2013). The telepresence industry includes such giants as Microsoft, Cisco Systems, and AT&T—all of which have invested millions of dollars into the R&D of this new technology.

What Do You Think? Telepresencing is no longer confined to the boardroom. Segway-like robots that allow real-time face-to-face conversations for anyone who can afford the price tag are now available. Telepresence robots on a stand range from $35,000 for a service robot to $200 for an android "Botiful" (Olson 2013). Using a combination of sensors, cameras, speakers, and microphones, telepresence robots move smoothly and quietly, and are easy to move about—all you need is a computer and a browser. How would you feel if you signed up for a face-to-face class and your professor turned out to be a telepresence robot?

Robotic technology has also revolutionized work. Although the economic downturn of 2009 and 2010 led to a reduction in the sales of robotic equipment, the record sales and growth of 2011 is predicted to continue through 2015 (IFR 2013). Ninety percent of robots work in factories, and more than half of these are used in heavy industry, such as automobile manufacturing. Robots are most commonly used for materials handling and spot welding, although there has recently been a trend toward integrating them more fully into the manufacturing process (Pethokoukis 2004).

Technology has also changed the nature of work. Federal Express not only created a FedEx intranet for its employees, they allow customers to enter their package-tracking database, saving the company millions of dollars a year. The used car industry has been revolutionized by the disabler—a remote device wired to a car's ignition system. When a customer fails to make a car payment on time, the device prevents the car from starting. Because the device has made it less likely that borrowers will default on their loans, there is some evidence that dealerships have become more willing to qualify low-income, low-credit customers (Welsh 2009).

Adam Lubroth/Riser/Getty Images

Automation means that machines can now perform the labor originally provided by human workers, such as robots performing tasks on automobile assembly lines.

The Computer Revolution

Early computers were much larger than the small machines we have today and were thought to have only esoteric uses among members of the scientific and military communities. In 1951, only about a half dozen computers existed (Ceruzzi 1993). The development of the silicon chip and sophisticated microelectronic technology allowed tens of thousands of components to be imprinted on a single chip smaller than a dime. The silicon chip led to the development of laptop computers, cellular phones, digital cameras, the iPad, and portable DVDs. The silicon chip also made computers affordable. Although the first PC was developed only 30 years ago, today over 75 percent of U.S. households report having a computer in the home compared to 61.8 percent just a decade ago (File 2013).

telepresencing A sophisticated technology that allows life-sized participants in the virtual presence of one another to realistically communicate through broadcast quality sound and images.

Over 80 percent of Asian and white, non-Hispanic Americans report having at least one computer at home, compared to just 68 percent of black and Hispanic Americans.

Computer use is associated with several demographic variables. Individuals between the ages of 35 and 45 are the most likely to have a home with at least one computer, and those 65 years of age and older are the least likely (File 2013). Over 80 percent of Asian and white, non-Hispanic Americans report having at least one computer at home, compared to just 68 percent of black and Hispanic Americans. As educational attainment and household income increases, the likelihood of having a computer in the home increases. There are no significant gender differences in home computer ownership.

According to a report by Forrester Research, by 2016, there will be an estimated 2 billion PCs and 760 million tablets (e.g., iPads, Fire) in use worldwide (Lunden 2012). Globally, Israel has the highest rate of computer ownership (122 computers for every 100 people), and Honduras has one of the lowest, with just 2.5 computers for every 100 people (*The Economist* 2008; U.S. Census Bureau 2011). The most common computer activities include accessing the Internet, sending e-mail, using a search engine, getting the news or e-mail, followed by word processing, working with spreadsheets or databases, and accessing or updating calendars or schedules.

Not surprisingly, computer education has also mushroomed in the last two decades. In 1998, 38,027 U.S. college students were enrolled in graduate degree programs in computer science; by 2010, that number had increased to 51,516 (Digest of Education Statistics 2013). Universities increasingly require their students to have laptop computers, provide wireless campus corridors, and spend millions of dollars on hardware and software.

Computers are big business, and the United States is one of the most successful producers of computer technology in the world, boasting several of the top companies. Hewlett-Packard (HP) has the highest global PC market share followed by Dell, Acer, Apple, and Lenovo (Gartner Research 2013). U.S. computer sales revenues exceeded $85.5 billion in 2011; in the same year, global sales are estimated to have been $328 billion.

Computer software is also big business and, in some cases, too big. In 1999, a federal judge found that Microsoft Corporation was in violation of antitrust laws, laws that prohibit unreasonable restraint of trade. At issue were Microsoft's Windows operating system and the vast array of Windows-based applications (e.g., spreadsheets, word processors, tax software)—applications that *only* work with Windows. As a result of the appellate process and other delays, Microsoft's compliance with the final judgment remains incomplete and continues to be monitored by the courts (U.S. Department of Justice 2011). Further, in 2013, the European Union (EU) fined Microsoft $733 million for failing to offer consumers a choice of Internet browsers when installing their Windows operation system (Sterling 2013).

Microsoft is not the only software giant under legal attack. In 2012, a South Korean court held that Samsung and Apple had violated each other's patents and banned the sales of several Apple (e.g., iPhone 4) and Samsung products (e.g., Galaxy S2) (Botelho 2013). In the same year, a U.S. court found that Samsung was guilty of intentionally violating Apple's patents and was ordered to pay $1 billion to the software company, making it—according to a Stanford law professor—one of the largest patent awards in history.

Information and Communication Technology and the Internet

Information and communication technology, or ICT, refers to any technology that carries information. Most information technologies were developed within a 100-year span: taking pictures and telegraphy (1830s), rotary power printing (1840s), the typewriter (1860s), transatlantic cable (1866), the telephone (1876), motion pictures (1894), wireless telegraphy (1895), magnetic tape recording (1899), radio (1906), and television (1923) (Beniger 1993). The concept of an "information society" dates back to the 1950s, when an economist identified a work sector he called "the production and distribution of knowledge." In 1958, 31 percent of the labor force was employed in this sector—today, more than 50 percent is. When this figure is combined with those in service occupations, more than 75 percent of the labor force is involved in the information society.

Internet An international information infrastructure available through universities, research institutes, government agencies, libraries, and businesses.

The **Internet** is an international infrastructure—a network of networks—available in universities, research institutions, government agencies, libraries, and businesses. In 2011, 70 percent of all Americans used the Internet from some location, although home Internet access is generally lower (see Table 14.1). Those who access the Internet are equally likely to be male or female, and most likely to be between the ages of 18 and 34, white non-Hispanic, college educated, employed, and with incomes over $150,000 annually (File 2013). Among those who do not go online, about half report that the main reason they don't use the Internet is because they don't think it is relevant to their lives, i.e., they're not interested in it (Zickuhr & Smith 2012). Other reasons include "don't have a computer," "too expensive," "too difficult," and "it's a waste of time."

The Internet, or the "Web" as it is most commonly known, has evolved to what is now called **Web 2.0**—a platform for millions of users to express themselves online in the common areas of cyberspace (Grossman 2006; Pew 2007). The development of Web 2.0 is a story about collaboration, about user-generated information on social network sites, blogs, and wikis and, ultimately, about *synergy* as the resulting content (e.g., Wikipedia) and products (e.g., YouTube) are greater than the sum of their parts (e.g., each entry).

Wireless access to the Internet is also altering Internet use. The growth of mobile Internet capabilities has led to "always-present" connectivity as users working on laptops, tablets, and smartphones access the Internet from coffee shops, classrooms, and shopping malls. Teens are particularly likely to access the "net" from mobile devices (Madden et al. 2013). Nearly 75 percent of 12- to 17-year-olds say that they access the Internet using their cell phones, tablets, and/or other mobile devices, at least occasionally, and a quarter say that they access the Internet via their cell phones daily.

High-speed broadband use has increased over time, from 4 percent of American households in 2001 to 62 percent of American households in 2011. The demographics of broadband access is similar to those of Internet use, i.e., those who are younger, more educated, and wealthier are more likely to have home broadband (Zickuhr & Smith 2012).

TABLE 14.1 Home Internet Use and/or Smartphone Use by Selected Demographic Variables, 2011

Selected Characteristics	Home Internet Users Percent	Smartphone Users[1] Percent	Either[2] Percent
Total 15 years and older[3]	**67.2**	**48.2**	**75.9**
Age[4]			
Under 25 years	70.5	67.8	87.5
25–34 years	74.5	67.4	86.2
35–44 years	77.1	58.9	85.2
45–54 years	71.2	45.1	77.3
55 years and over	54.0	23.3	58.4
Race and Hispanic origin			
White alone	68.9	48.0	76.8
White non-Hispanic alone	72.5	48.6	79.2
Black alone	53.8	47.3	67.9
Asian alone	78.3	51.6	83.0
Hispanic (of any race)	51.2	45.4	65.5
Sex of householder			
Male	68.5	48.6	76.8
Female	65.9	47.8	75.0
Employment status			
Employed	75.8	58.4	85.1
Unemployed	63.5	49.4	75.9
Not in labor force	52.2	29.3	59.1
Total 25 years and older	**66.5**	**44.1**	**73.5**
Educational Attainment			
Less than high school graduate	25.3	20.5	35.2
High school graduate or GED	52.8	32.1	61.5
Some college or associate degree	73.5	48.7	81.3
Bachelor's degree or higher	87.3	59.2	91.0

[1] Smartphone use includes anyone who reported using their phone to browse the Web, e-mail, use maps, play games, access social networking sites, download apps, listen to music, or take photos and videos.
[2] This includes the number and percentage of individuals who either use the Internet at home, use a Smartphone, or both.
[3] Data in this table are from questions asked only of household respondents and then weighted to reflect the total population.
[4] Because household respondents tended to be older, the data for those below the age of 25 had more variability than for older respondents. The estimates in this section for those under 25 should therefore be interpreted with caution.

Source: U.S. Census Bureau 2013a.

Web 2.0 A platform for millions of users to express themselves online in the common areas of cyberspace.

The growth of mobile Internet capabilities has led to "always-present" connectivity as users working on laptops, tablets, and smartphones access the Internet from coffee shops, classrooms, and shopping malls.

E-Commerce. **E-commerce** is the buying and selling of goods and services over the Internet, and primarily includes online shopping and online banking. Online sales represent nearly half of all U.S. sales with books and magazines followed by clothing, computer hardware, and computer software being the most common online purchases (U.S. Census Bureau 2013b). Motivations to purchase products online include convenience, the ability to compare prices, and less expensive products in some states where there is no Internet sales tax. The Marketplace Fairness Act, presently being debated in Congress, would ". . . grant states the authority to compel online and catalog retailers . . . no matter where they are located, to collect sales tax at the time of the transaction . . ." (MFA 2013, p. 1).

As Table 14.2 indicates, 61 percent of survey respondents use the Internet for banking, making it almost as popular of an online activity as social networking (Zickuhr &

TABLE 14.2 Users within Each Demographic Group Engaging in Online Activities, 2011 (N = 2260)*

	Search	Email	Buy a Product	Use Social Network Sites	Bank Online
All adults	**92%**	**91%**	**71%**	**64%**	**61%**
Men	93	89	69	63	65
Women	91	93	74	66	57
Race/ethnicity					
White, Non-Hispanic	93	92	73	63	62
Black, Non-Hispanic	91	88	74	70	67
Hispanic (English- and Spanish-speaking	87	86	59	67	52
Age					
18–29	96	91	70	87	61
30–49	91	93	73	68	68
50–64	91	90	76	49	59
65+	87	86	56	29	44
Household income					
Less than $30,000/yr	90	85	51	68	42
$30,000–$49,999	91	93	77	65	65
$50,000–$74,999	93	94	80	61	74
$75,000+	98	97	90	66	80
Educational attainment					
No high school diploma	81	69	33	63	32
High school graduate	88	87	59	60	47
Some college	94	95	74	73	66
College graduate +	96	97	87	63	74

*18 years old and older.

Source: Zickuhr & Smith 2012.

e-commerce The buying and selling of goods and services over the Internet.

Smith 2012). Online banking as well as mobile banking, that is, the use of a mobile phone to access banking information, has increased significantly over the last decade (Fox 2013). For example, in 2011, 18 percent of cell phone owners used their mobile device to bank online compared to 35 percent in 2013.

What Do You Think? The advent and proliferation of Wi-Fi (i.e., wireless access to the Internet) has facilitated a variety of Internet software and hardware innovations such as smartphones, the iPad and other tablets, and thousands of downloadable applications ("apps"). It has even led to institutional shifts; for example, e-commerce now includes m-commerce or the ability to make financial transactions (e.g., mobile banking) from a mobile device such as a smartphone. Do you think that the advantages of 24/7 access to the Internet outweigh the disadvantages that accompany "always-present" technologies?

Health and Digital Medicine. Many Americans turn to the Internet to address health issues and concerns. In a recent survey on the Internet and health, over a third of U.S. adults reported going online to research a health condition (Fox & Duggan 2013). Women, youth, those with annual household incomes over a $75,000, and the college educated were the most likely to seek an online diagnosis.

Many other Americans, 72 percent, sought general health or medical information on the Internet, and half of those searches were on behalf of friends or family members. Interestingly, sociologically, there is a "social life" to health information on the Internet. It is not uncommon for people to share their health stories, post symptoms and diagnoses, and otherwise provide peer-to-peer health support (Fox & Duggan 2013).

Besides providing medical information, there is considerable evidence that online medical records help improve medical care:

> A paper record is a passive, historical document. An electronic health record can be a vibrant tool that reminds and advises doctors. It can hold information on a patient's visits, treatments, and conditions, going back years, even decades. It can be summoned with a mouse click, not hidden in a file drawer in a remote location and thus useless in medical emergencies. (Lohr 2008, p. 1)

Digital patient records are thus the first step in creating **learning health systems**, whereby physicians, looking across patient populations, can identify successful treatments or detect harmful interactions (Lohr 2011). Further, mobile devices such as smartphones and tablets ". . . have the computing capability, display, and battery power to become powerful medical devices that measure vital signs and provide intelligent interpretation or immediate transmission of information" (p. 883).

Lastly, there is evidence that technology can help mediate the soaring cost of health care. Much of the savings comes from "increases in efficiency, such as shorter hospital stays because of better coordination, better productivity for nurses, and more efficient drug utilization" (Atkinson & Castro 2008, p. 27). Other important sources of cost effectiveness include electronic claims processing and reducing medical errors through more effective diagnostic and treatment interventions. As of July 2012, the federal government spent over $6.6 billion in health information technology incentives (Castro 2013).

The Search for Knowledge and Information. The Internet, perhaps more than any other technology, is the foundation of the information society. Whether reading an online book, taking a MOOC (massive open online course) (see this chapter's *The Human Side* for a light look at online education; also see Chapter 8), mapping directions, visiting the Louvre virtually, or accessing Wikipedia, the Internet provides millions of surfers with instant answers to questions previously requiring a trip to the library.

learning health systems The result of electronic records whereby physicians can look across patient populations and identify successful treatments or detect harmful interactions.

There is concern, however, that the very way in which the "Google generation" reads, thinks, and approaches problems has been altered by the new technology. In 2008, an article asked, "Is Google making us stupid?" which was eventually expanded into a book entitled *The Shallows: What the Internet Is Doing to Our Brains* (Carr 2010). The author argues that the use of the Internet promotes a chronic state of distraction that we are often unaware of:

> Our focus on a medium's content can blind us to the deep effects. We're too busy being dazzled or disturbed by the programming to notice what's going on inside our heads. In the end, we come to pretend that the technology itself doesn't matter. It's how we use it that matters, we tell ourselves. The implication, comforting in its hubris, is that we're in control. The technology is a tool, inert until we pick it up and inert again once we set it aside. (p. 3)

Finally, with the lightning growth of the Internet, there are concerns that information is often outdated, difficult to find, and limited in scope (Mateescu 2010). Some are already looking ahead to a time when search engines look for information not syntactically (i.e., based on *combinations* of words and phrases) but semantically (i.e., based on the *meaning* of words and phrases). The **Semantic Web**, sometimes referred to as Web 3.0, entails not only pages of information but also pages that describe the interrelationship between the pages of information resulting in "smart media" (Semantic Media 2011).

Games and Entertainment. Over half of all Americans play video games, and 51 percent of all U.S. households have a dedicated home game console. About a third of "gamers" are under the age of 18, a third between the ages of 18 and 35, and a third over 36 years old (ESA 2013). Gamers and purchasers of video games are more likely to be male, although women over the age of 18 comprise a larger proportion of the game-playing population (31 percent) than boys 17 years old or younger (19 percent). In 2012, consumers spent over $20 billion on video and computer games (ESA 2013).

Video and computer games are a multibillion-dollar industry comparable to movies, television, and print media and, like those mediums, there is the concern over consumption outcomes. Dietrich (2013), after investigating online and off-line role-playing games, reports that the ability to create avatars is significantly limited to white characters in terms of skin color, facial features, and hair, leading the researcher to conclude that "character creation options reflect and reinforce a sense of normative whiteness in terms of the anesthetic presentation of race" (p. 96). Further, in a content analysis of 399 teen- or mature-rated video game boxes, Near (2013) reports that the presence of a noncentral hypersexualized female is associated with increased sales; alternatively, the presence of a female central character or female-only characters is associated with reduced sales.

Semantic Web Sometimes called Web 3.0, a version of the Internet in which pages not only contain information but also describe the interrelationship between pages; sometimes called smart media.

Finally, competition for Internet entertainment dollars extends beyond video and computer games. YouTube, Hulu, and Netflix provide access to online movies and television shows, and fantasy sports would have been nearly impossible prior to the advent and popularization of information and communication technologies. Fantasy sports or other "vicarious management" entertainment initiatives, according to one researcher, are rooted in the structure of white supremacy, whereby predominantly white males evaluate, purchase, and trade prized commodities—disproportionately, the black athlete (Oates 2009).

Politics and e-Government. Technology is changing the world of politics. In 2010, approximately 73 percent of U.S. adult Internet users went online to find news or information about the 2010 midterm elections, or to send or receive political messages through e-mail, instant messaging, Twitter, and the like (Smith 2011).

When then Iranian President-elect Mahmoud Ahmadinejad banned foreign media from the streets of Tehran, protestors contesting the election used Twitter, social networking sites, Flickr, and YouTube to stay in contact with one another and to transmit messages and images around the world. More recently, Facebook played an essential role in several of the Middle East uprisings (Preston 2011). Elliot Schrage, vice president for global communications, public policy, and marketing of Facebook, said, "We've witnessed brave people of all ages coming together to effect a profound change in their country. Certainly, technology was a vital tool in their efforts but we believe their bravery and determination mattered most" (quoted in Preston 2011, p. 1).

The United States is just one of over 100 countries worldwide that hosts a government website. In 2012, the United Nations ranked e-government sites on a variety of criteria including information delivery, ease of obtaining information, delivery of public services, and citizen–government interaction. The overall ranking, as indicated by the E-Government Development Index (EGDI), reveals that the Republic of Korea is ranked first in the world followed by the Netherlands, the United Kingdom, and Denmark (United Nations 2012). In general, developed countries' e-government capabilities are greater than those in developing countries, with all of the countries ranked in the top 20 being high-income nations. The number of countries with a government web page as well as the topic areas available, has increased in recent years (see Figure 14.2).

Social Networking and Blogging. Social networks (e.g., Facebook, Twitter) and blogs comprise a sector of the Internet called **membership communities**. Membership communities have changed in three substantively significant ways in recent years. First, the *number of people* who visit membership communities has increased. In 2012, 850 million people used Facebook monthly (Honigman 2013). Second, the *amount of time* members spend at a membership community site has grown dramatically. Globally, nearly a quarter of Facebook users check their account five or more times a day.

Finally, *who joins* membership communities is changing. Although people under the age of 30 still comprise the largest proportion of social media users, of those who use the Internet, 57 percent of 50- to 64-year-olds and 35 percent of those 65 years and older report using Facebook (Duggan & Brenner 2013). African Americans, Latinos, and women are more likely to use Instagram, Pinterest, and Twitter—the micro-blogging site that asks "What are you doing?" There are an estimated 175 million "tweets" sent every day (Honigman 2013).

Video games, a billion-dollar industry, most often have male lead characters. When females are present, they are often portrayed as either victim or prize, and are hypersexualized.

What Do You Think? Professor Laurence Thomas of Syracuse University walked out of his classroom after a student, in the front row of a large lecture hall, was seen sending a text (Jaschik 2008). Dr. Thomas's behavior has been both praised and criticized. What do you think? What should a professor do if students are texting in class?

membership communities
Internet sites where participation requires membership and members regularly communicate with one another for personal and/or professional reasons.

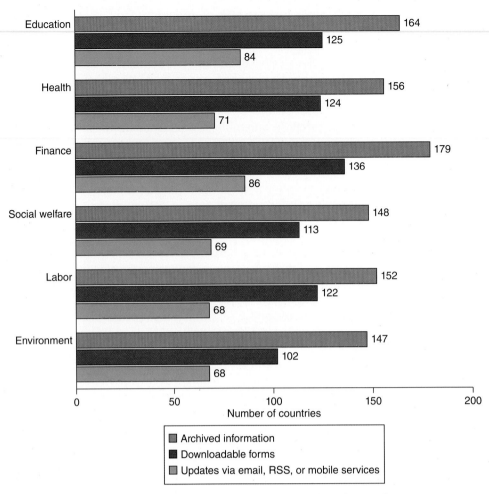

Figure 14.2 Availability of E-Government Service Areas by Number of Countries, 2012
Source: United Nations 2012.

Science and Biotechnology

Although recent computer innovations and the establishment of the Internet have led to significant cultural and structural changes, science and its resulting biotechnologies have produced not only dramatic changes but also hotly contested issues with public policy implications. In this section, we look at some of the issues raised by developments in genetics, food and biotechnology, and reproductive technologies.

Genetics. Molecular biology has led to a greater understanding of the genetic material found in all cells—DNA (deoxyribonucleic acid)—and with it the ability for **genetic testing.** Genetic testing ". . . involves examining your DNA, the chemical database that carries instructions for your body's functions," and using that information to identify abnormalities or alterations that may lead to disease or illness (Mayo Clinic 2013, p. 1).

Currently, researchers are trying to complete genetic maps that will link DNA to particular traits. There is some evidence that personality characteristics are inherited; other evidence links certain conditions previously thought of as psychological in nature (e.g., addiction, anorexia, and autism) as, at least in part, genetically induced (Harmon 2006). Already, specific strands of DNA have been identified as carrying physical traits such as eye color and height, as well as such diseases as sickle-cell disease, breast cancer, cystic fibrosis, prostate cancer, depression, Alzheimer's, and mitochondrial disorders (ORNL 2011; Picard & Turnbull 2013).

genetic testing Examination of DNA in order to identify abnormalities or alterations that may lead to disease or illness.

The U.S. Human Genome Project (HGP), a 13-year effort to decode human DNA, was completed in 2003. Conclusion of the project is transforming medicine:

> All diseases have a genetic component whether inherited or resulting from the body's response to environmental stresses like viruses or toxins. The successes of the HGP have . . . enabled researchers to pinpoint errors in genes—the smallest units of heredity—that cause or contribute to disease. The ultimate goal is to use this information to develop new ways to treat, cure, or even prevent the thousands of diseases that afflict humankind. (Human Genome Project 2007, p. 1)

The hope is that, if a defective or missing gene can be identified, possibly a healthy duplicate can be acquired and transplanted into the affected cell. This is known as **gene therapy**. Alternatively, viruses have their own genes that can be targeted for removal. Experiments are now under way to accomplish these biotechnological feats.

Food and Biotechnology. **Genetic engineering** is the ability to manipulate the genes of an organism in such a way that the natural outcome is altered. Genetically modified (GM) food, also known as genetically engineered food, and genetically modified organisms involve this process of DNA recombination—scientists transferring genes from one plant into the genetic code of another plant.

In the United States, genetically modified organisms (GMOs) are in an estimated 80 percent of all packaged food sold in the United States and Canada (Spaeth 2011). Yet, a national survey of adults found that one-third of Americans believe that genetically modified foods are unsafe to eat, 93 percent believe the government should require labels that identify food as "genetically modified" or "bio-engineered," and over half report they would use the labels to avoid GMO foods (Langer 2013).

> Human health concerns [of genetically modified foods] include possible toxicity, carcinogenicity, food intolerance, antibiotic resistance buildup, decreased nutritional value, and food allergens. . . .

Biotechnology companies and other supporters of GM foods commonly cite the alleviation of hunger and malnutrition as a main benefit, claiming that this technology can enable farmers to produce crops with higher yields. Critics of GM foods argue that the world already produces enough food for all people to have a healthy diet. According to the World Hunger Education Service (2011), if all the food produced worldwide were distributed equally, every person would be able to consume 2,720 calories a day. Biotechnology, critics argue, will not alter the fundamental causes of hunger, which are poverty and lack of access to food and to land on which to grow food.

Biotechnology companies claim that GM foods approved by the Food and Drug Administration are safe for human consumption, and they even cite potential health benefits such as the use of genetic modification to remove harmful allergens from foods or to improve nutritional benefits (Kaplan 2009; ORNL 2011). But critics claim that research on the effects of GM crops and foods on human health is inadequate, especially concerning long-term effects (Olster 2013). Human health concerns include possible toxicity, carcinogenicity, food intolerance, antibiotic resistance buildup, decreased nutritional value, and food allergens in GM foods.

Biotechnology skeptics are also concerned about the environmental effects of GM crops. Biotechnology companies claim that crops that are genetically designed to repel insects negate the need for chemical (pesticide) control and thus reduce pesticide poisoning of land, water, animals, foods, and farmworkers. However, critics are concerned that insect populations can build up resistance to GM plants with insect-repelling traits, which would necessitate increased rather than decreased use of pesticides. Similarly, the use of Roundup, a top-selling weed killer, has led to the development of "superweeds" requiring the use of more rather than less toxic herbicides (Neuman & Pollack 2010).

GM seed contamination is also of concern (Associated Press 2013). First, organic farmers fear that cross-pollination will contaminate neighboring crops, reducing demand and thus sales for organically grown produce (Pollack 2011). Second, to maintain control over

gene therapy The transplantation of a healthy gene to replace a defective or missing gene.

genetic engineering The manipulation of an organism's genes in such a way that the natural outcome is altered.

Activists protesting the agribusiness giant Monsanto and genetically modified organisms (GMOs) wear anonymous masks in front of the White House in 2013.

their products, some biotechnology companies have developed "terminator" seeds, which cause the plant to produce sterile seeds. For example, Monsanto initially agreed not to market its terminator technology but later adopted a positive stance on genetic seed sterilization, suggesting that the commercialization of terminator technology might occur in the future (ETC Group 2003). If the seed sterility trait in terminator crops inadvertently contaminates traditional crops and plant life, the ramifications would be devastating to life on earth. In 2013, thousands of protesters in 52 countries and 436 cities participated in the "March against Monsanto," carrying such signs as "Label GMOs" and "Real Food 4 Real People" (Associated Press 2013; Occupy Monsanto 2013).

Biotechnology critics also raise concerns about insufficient safeguards and regulatory mechanisms. In 2000, worldwide concern about the safety of GM crops resulted in 160 nations signing the landmark Biosafety Protocol, which requires producers of a GM food to demonstrate that it is safe before it is widely used. The Biosafety Protocol also allows countries to ban the importation of GM crops based on suspected health, ecological, or social risks. In 2011, officials from 15 countries and the European Union signed a supplementary agreement that addressed liability and regress from damages resulting from living modified organisms (UNEP 2011).

Reproductive Technologies. The evolution of "reproductive science" has been furthered by scientific developments in biology, medicine, and agriculture. At the same time, however, its development has been hindered by the stigma associated with sexuality and reproduction, its link with unpopular social movements (e.g., contraception), and the feeling that such innovations challenge the natural order (Clarke 1990). Nevertheless, new reproductive technologies have been and continue to be developed.

Although there are other reproductive technologies (e.g., in vitro fertilization), abortion more than any other biotechnology epitomizes the potentially explosive consequences of technological innovation (see also Chapter 13). **Abortion** is the removal of an embryo or fetus from a woman's uterus before it can survive on its own. Globally, according to Sedgh et al. (2012):

- The proportion of all abortions that take place in the developing world has increased over time.
- Between 2003 and 2008, the number of abortions decreased in the developed world, but increased in the developing world.
- Regionally, Latin America has the highest abortion rate followed by Africa, Asia, and Europe.
- Subregionally, as a result of low contraception use and high failure rates of the types of contraception used (e.g., withdrawal, rhythm method), eastern Europe has the highest abortion rate and western Europe the lowest abortion rate.
- China's abortion rate has risen since 2003 as a result of increased premarital sex and decreased access to contraception.
- Although varying by region, nearly half of all abortions are unsafe.

In countries where abortion laws are restrictive, abortion rates are higher. For example, abortion rates are higher in Africa and Latin America, two regions where abortion is illegal in the majority of countries, but abortion rates are low in western Europe where abortion is generally permitted (Sedgh et al. 2012). In 2008, 97 percent of abortions in Africa and 95 percent of abortions in Latin America were unsafe, i.e., were not performed in medically safe environments by individuals with the necessary skill set. Worldwide, unsafe abortions are estimated to be responsible for 5 million women being disabled and 47,000 deaths annually (WHO 2012).

abortion The intentional termination of a pregnancy.

In the United States, since the U.S. Supreme Court's ruling in *Roe v. Wade* in 1973, abortion has been legal. However, several Supreme Court decisions have limited the scope of the *Roe v. Wade* decision (e.g., *Planned Parenthood of Southeastern Pennsylvania v. Casey*). Further, several recent legislative initiatives in Congress are intended to restrict or prohibit a woman's access to a lawful abortion. For example, the U.S. House of Representatives is considering the Life at Conception Act, and the Pain-Capable Unborn Child Protection Act is being debated in the Senate (Government Track 2013). In 2013, the U.S. House of Representatives, led by Republicans, passed an abortion bill that would ban abortions after 20 weeks of pregnancy.

Historically, abortions are banned when the fetus is considered viable, usually around 22 to 26 weeks from conception. Recently, however, state laws have made legal abortions more restrictive. With an exception for the life or health of the mother, abortions after 20 weeks are prohibited in Alabama, Arkansas, Indiana, Kansas, Louisiana, Nebraska, North Carolina, and North Dakota (Guttmacher Institute 2013). Other recently enacted state restrictions include (1) that abortion must take place in a hospital and be performed by a licensed physician, (2) gestational limits, (3) restrictions on state spending for abortions, (4) the right of individual or institutional refusal to perform an abortion, (5) state-mandated counseling, (6) waiting periods, and (7) parental involvement if involving a minor (Guttmacher Institute 2013).

Nineteen states prohibit intact dilation and extraction (D & E) abortions, which often take place in the second trimester of pregnancy. Opponents refer to such abortions as **partial-birth abortions** because the limbs and the torso are typically delivered before the fetus has expired. However, former National Organization for Women president Kim Gandy states, "Try as you might, you won't find the term 'partial birth abortion' in any medical dictionary. That's because it doesn't exist in the medical world—it's a fabrication of the anti-choice machine" (U.S. Newswire 2003, p. 1). D & E abortions are performed because the fetus has a serious defect, the woman's health is jeopardized by the pregnancy, or both. In 2003, a federal ban on partial-birth abortions was signed into law and, in 2007, after constitutional challenges, the ban was upheld by the U.S. Supreme Court (Greenhouse 2007; White House 2003). The significance of this case lies in the fact that it was the first time the U.S. Supreme Court upheld a ban on any type of abortion procedure.

> Abortion is . . . a complex issue for societies, which must respond to the pressures of conflicting attitudes toward abortion and the reality of high rates of unintended and unwanted pregnancy.

partial-birth abortions Also called an intact dilation and extraction (D & E) abortion, the procedure may entail delivering the limbs and the torso of the fetus before it has expired.

TABLE 14.3 Survey Responses to Question "Would you like to see the Supreme Court completely overturn its *Roe v. Wade* decision, or not?" (N = 1502)	Overturn Decision %	Not Overturn %	Don't Know/ Refused %
Total	29	63	7=100
Men	29	63	9=100
Women	30	64	6=100
18–29	27	68	5=100
30–49	31	61	8=100
50–64	26	69	6=100
65+	36	52	12=100
White	29	66	6=100
Black	29	67	4=100
College graduate +	22	73	4=100
Post-graduate	13	82	5=100
College graduate	27	69	4=100
Some college	27	67	6=100
HS grad or less	36	53	11=100
Republican	46	48	6=100
Democrat	20	74	6=100
Independent	28	64	8=100
Protestant	35	58	7=100
White evangelical	54	42	4=100
White mainline	17	76	7=100
Black protestant	29	65	5=100
Catholic	38	55	7=100
White Catholic	33	63	4=100
Unaffiliated	9	82	9=100
Attend religious services			
Weekly or more	50	44	7=100
Less often	17	76	7=100

Source: Pew 2013.

Abortion is a complex issue for everyone, but especially for women, whose lives are most affected by pregnancy and childbearing. Women who have abortions are disproportionately poor, unmarried minorities who say that they intend to have children in the future. Abortion is also a complex issue for societies, which must respond to the pressures of conflicting attitudes toward abortion and the reality of high rates of unintended and unwanted pregnancy. The debate over abortion has also complicated health care reform as conservatives, citing a 30-year ban on using taxpayer's money to pay for elective abortions, battle the Obama administration and abortion rights supporters.

Attitudes toward abortion tend to be polarized between two opposing groups of abortion activists—pro-choice and pro-life. As recently as 2013, public opinion surveys indicated that the majority of Americans, 63 percent, would not like the Supreme Court to overturn *Roe v. Wade* (Pew 2013) (see Table 14.3). Advocates of the pro-choice movement hold that freedom of choice is a central human value, that procreation choices must be free of government interference, and that because the woman must bear the burden of moral choices, she should have the right to make such decisions. Alternatively, pro-lifers hold that an unborn fetus has a right to live and be protected, that abortion is immoral, and that alternative means of resolving an unwanted pregnancy should be found.

Cloning, Therapeutic Cloning, and Stem Cells. In July 1996, scientist Ian Wilmut of Scotland successfully cloned an adult sheep named Dolly. To date, cattle, goats, mice, pigs, cats, rabbits, and horses have also been cloned. Although not illegal in the United States, worldwide concerns over human cloning have led to its prohibition in over 60 countries. There are also several international agreements banning the procedure (Rovner 2013).

One argument in favor of developing human cloning technology is its medical value; it may potentially allow everyone to have "their own reserve of therapeutic cells that would increase their chance of being cured of various diseases, such as cancer, degenerative disorders, and viral or inflammatory diseases" (Kahn 1997, p. 54). Human cloning could also provide an alternative reproductive route for couples who are infertile and for those in which one partner is at risk for transmitting a genetic disease.

Arguments against cloning are largely based on moral and ethical considerations. Critics of human cloning suggest that, whether used for medical therapeutic purposes or as a means of reproduction, human cloning is a threat to human dignity (Human Cloning Prohibition Act of 2007). Cloned humans would be deprived of their individuality, and as Kahn (1997, p. 119) pointed out, "creating human life for the sole purpose of preparing therapeutic material would clearly not be for the dignity of the life created."

Therapeutic cloning uses stem cells from human embryos. **Stem cells** can produce any type of cell in the human body and thus can be "modeled into replacement parts for people suffering from spinal cord injuries or regenerative diseases, including Parkinson's and diabetes" (Eilperin & Weiss 2003, p. A6). In 2009, the U.S. Food and Drug Administration approved the first trials of embryonic stem cell therapy for paralyzed patients with spinal cord injuries (Park 2009).

To bypass the use of embryonic stem cells altogether, some scientists are using induced pluripotent stem (IPS) cells that mimic many of the scientific properties of embryonic stem cells (Erickson 2011). IPS cells "offer the possibility of a renewable source of replacement cells and tissues to treat a myriad of diseases, conditions, and disabilities including Parkinson's disease, amyotrophic lateral sclerosis, spinal cord injury, burns, heart disease, diabetes, and arthritis" (NIH 2012, p. 1).

In 2009, President Obama lifted the ban on federal funding of embryonic stem cell research and directed the National Institutes of Health (NIH) to develop research support guidelines. Abortion opponents criticized the executive order for publicly financing research that destroys embryos, something conservatives argue is morally wrong. Although 70 percent of Americans would support the use of embryonic stem cells in the treatment of a serious disease for themselves or a family member (NSF 2013), the recent cloning of human embryos is likely to reignite the debate (Stein & Doucleff 2013).

Despite what appears to be a universal race to the future and the indisputable benefits of scientific discoveries such as the workings of DNA and the technology of IVF and stem cells, some people are concerned about the duality of science and technology. Science and the resulting technological innovations are often life assisting and life giving; they are also potentially destructive and life threatening. The same scientific knowledge that led to the discovery of nuclear fission, for example, led to the development of both nuclear power plants and the potential for nuclear destruction. Thus, we now turn our attention to the problems associated with science and technology.

Societal Consequences of Science and Technology

Scientific discoveries and technological innovations have implications for all social actors and social groups. As such, they also have consequences for society as a whole. Figure 14.3 displays the public's relative assessment of the costs and benefits of scientific research. Although most Americans believe that the benefits of scientific research outweigh the potential for harmful results, there is no denying that science and the resulting technologies have some negative consequences.

Social Relationships, Social Networking, and Social Interaction

Technology affects social relationships and the nature of social interaction. The development of telephones has led to fewer visits with friends and relatives; with the advent of DVRs, cable television, and video streaming, the number of places where social life occurs (e.g., movie theaters) has declined. Even the nature of dating has changed as computer networks facilitate instant messaging, cyberdates, and "private" chat rooms. As technology increases, social relationships and human interaction are transformed.

Technology also makes it easier for individuals to live in a cocoon—to be self-sufficient in terms of finances (e.g., Quicken), entertainment (e.g., Hulu), work (e.g., telecommuting), news (e.g., Twitter), recreation (e.g., Wii), shopping (e.g., Amazon), networking (e.g. LinkedIn), communication (e.g., e-mail, texting), family conferences (e.g., Skype), and many other aspects of social life. The popularity of social networking sites has led to fears over the privacy and security of information posted (Della Cava 2010), as well as concerns over their impact on social relationships (Klotz 2004).

Children who use a home computer "spend much less time on sports and outdoor activities than non-computer users" (Attewell et al. 2003, p. 277). Members of social networking sites are less likely to socialize with their neighbors or to rely on them for care

therapeutic cloning Use of stem cells to produce body cells that can be used to grow needed organs or tissues; regenerative cloning.

stem cells Undifferentiated cells that can produce any type of cell in the human body.

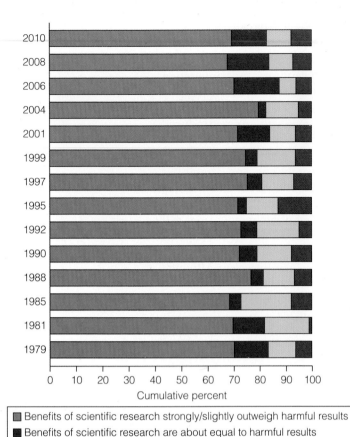

Benefits of scientific research strongly/slightly outweigh harmful results

Benefits of scientific research are about equal to harmful results

Harmful results of scientific research strongly/slightly outweigh benefits

Don't know

Figure 14.3 Public Assessment of Scientific Research, 1979–2010
Source: NSF 2013.

and assistance (Pew 2009). A survey researcher at the University of Denver reports that 40 percent of people say they would avoid someone in real life who had "unfriended" them on Facebook (Kelly 2013). Finally, there is some evidence that suggests that Internet recommender systems, those systems that personalize search results, contribute to polarization between groups in society by echoing already existing biases (Dandekar et al. 2013).

Loss of Privacy and Security

In 2012, identity theft was the number one complaint filed with the Federal Trade Commission for the 13th year in a row (FTC 2013; see also Chapter 4). Through computers, individuals can obtain access to someone's phone bills, tax returns, medical reports, credit histories, bank account balances, and driving records. Several security breaches are notable:

- The software giant Adobe announced that hackers had accessed the names, credit or debit card numbers, and card expiration dates of as many as 29 million customers (Konrad 2013).
- Chinese hackers infiltrated the computer systems of *The New York Times* over four months, securing passwords for its reporters and other employees (Perlroth 2013).
- A quarter of a million Twitter account holders had their security breached (Jones 2013).
- An unencrypted external hard drive containing personal information on over 25 million veterans was stolen from the home of a U.S. government employee (Wait 2012).
- In one of the largest incidents of hacking a state government site, South Carolina's Department of Revenue was compromised when 3.8 million tax records and nearly 400,000 credit card numbers were breached (Presti 2012).

What Do You Think? Wikileaks is a nonprofit organization whose stated mission is "bringing important news and information to the public" (Wikileaks 2011, p. 1). In 2010 and 2011, classified documents regarding the wars in Iraq and Afghanistan, as well as 250,000 confidential U.S. Embassy cables, were released to the public. Do you think Julian Assange, the founder of Wikileaks, is a hero or a villain?

Although just inconvenient for some, unauthorized disclosure of, for example, medical records, is potentially devastating for others. If a person's medical records indicate that he or she is human immunodeficiency virus (HIV)-positive, that person could be in danger of losing his or her job or health benefits. If DNA testing of hair, blood, or skin samples reveals a condition that could make the person a liability in an insurer's or employer's opinion, the individual could be denied insurance benefits, medical care, or even employment. In response to such fears, the **Genetic Information Nondiscrimination Act of 2008** (GINA) was passed. GINA is a federal law that prohibits discrimination in health coverage or employment based on genetic information (Department of Health and Human Services 2009).

Technology has created threats not only to the privacy of individuals but also to the security of entire nations (also see Chapter 15). As revealed by Edward Snowden,

Genetic Information Nondiscrimination Act of 2008
A federal law that prohibits discrimination in health coverage or employment based on genetic information.

the National Security Administration (NSA) computer program XKeyscore allows ". . . analysts to search with no prior authorization through vast databases containing e-mails, online chats, and the browsing histories of millions of individuals . . ." (Greenwald 2013, p. 1). Although the extent to which Snowden is telling the truth is unknown, allegations that the United States has used surveillance techniques on other countries have resulted in strained foreign relations. For example, as result of the U.S. espionage program, the president of Brazil "postponed" her state trip to Washington (Colitt & Galvao 2013).

Unemployment, Immigration, and Outsourcing

Some technologies replace human workers and many of those jobs are unlikely to return to the labor market:

> The global economy is being reshaped by machines that generate and analyze vast amounts of data; by devices such as smartphones and tablet computers that let people work just about anywhere, even when they're on the move; by smarter, nimbler robots; and by services that let businesses rent computer power when they need it, instead of installing expensive equipment and hiring IT staffs to run it. Whole employment categories, from secretaries to travel agents, are starting to disappear. (Condon & Wiseman 2013, p. 1)

After analyzing employment records in 20 countries, examining hiring patterns, job losses and gains, and interviewing economic and technology experts, unemployed workers, and corporate executives, researchers conclude that:

- Over the last 30 years, technology has reduced the number of jobs in both the manufacturing and service sectors.
- The most vulnerable workers are those that engage in repetitive tasks that can be replaced by a computer program.
- Technology is replacing workers in all types of organizations, from schools to hospitals to corporations to the military.
- As a result of technology, the total number of workers in 2012 decreased while corporate profits increased. (Condon & Wiseman 2013)

In addition to and in agreement with the previous information, Brynjolfsson and McAfee (2011) conclude that the growth of technology is also responsible for declining U.S. incomes and greater economic inequality.

Unemployment rates can also increase when companies **outsource** (sometimes called off-shore) jobs to lower-wage countries. For example, in the United States, "offshore outsourcing in information technology, finance, and other back-office functions such as human resources has nixed 1.1 million jobs since 2008 and will result in another 1.3 million positions lost by 2014 . . ." (Dignan 2010). It should be noted, however, that economic trends impact receiving countries as well; in other words, jobs lost in one country impact outsourcing in another.

Finally, there is some concern about the number of immigrant employees who are in the United States on H-1B visas. H-1B visas permit employers to temporarily hire foreign workers in certain specialty occupations including high-tech industries. The need for immigrant high-tech employees is a consequence of the lack of American counterparts in STEM occupations, that is, occupations in science, technology, engineering, and mathematics (Price 2008). The number of H-1B visas the federal government issues each year is limited to 65,000 (U.S. Citizenship and Immigration Services 2013). However, critics of the H-1B visa limitations are pressuring the federal government to increase the number of visas available annually.

The H-1B visa program is not, however, without problems for immigrant employees. Workers' visas are temporary—valid for a maximum of six years—at which time visa holders must leave the United States unless permanent residency has been granted. Employees with H-1B visas are often paid less than American employees are, and cannot voluntarily quit their jobs for fear of deportation (Price 2008; U.S. Citizenship and Immigration Services 2013).

outsource A practice in which a business subcontracts with a third party to provide business services.

The Digital Divide

One of the most significant social problems associated with science and technology is the increased division between the classes. In a now oft-quoted statement, Welter (1997) notes:

> It is a fundamental truth that people who ultimately gain access to, and who can manipulate, the prevalent technology are enfranchised and flourish. Those individuals (or cultures) that are denied access to the new technologies, or cannot master and pass them on to the largest number of their offspring, suffer and perish. (p. 2)

The fear that technology will produce a "virtual elite" is not uncommon. Several theorists hypothesize that, as technology displaces workers—most notably the unskilled and uneducated—certain classes of people will be irreparably disadvantaged—the poor, minorities, and women. There is even concern that biotechnologies will lead to a "genetic stratification," whereby genetic testing, gene therapy, and other types of genetic enhancements are available only to the rich.

Globally, the digital divide reflects the economic and social conditions of a country. In general, wealthier and more educated countries have more technology than poorer countries with a less educated populace, and the gap is growing (see Figure 14.4) (Wakefield 2010; White et al. 2011). For example, 97.1 percent of the population of Iceland is online compared to less than 2 percent of the populations in Myanmar, Chad, Congo, and Ethiopia (Internet Statistics 2013).

Similarly, in the United States, the wealthier the family, the more likely the family is to have Internet access. Of households with annual incomes of $150,000 or more, 51.3 percent have connectivity at multiple locations for multiple devices. However, only 11.6 percent of households with income levels below $25,000 a year have connectivity at multiple locations for multiple devices (File 2013). Internet use and broadband home access follow a similar pattern.

> Several theorists hypothesize that, as technology displaces workers—most notably the unskilled and uneducated—certain classes of people will be irreparably disadvantaged—the poor, minorities, and women.

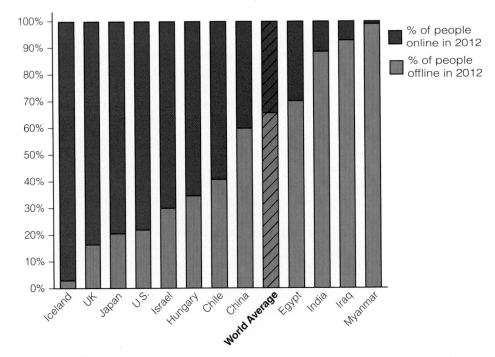

Figure 14.4 Percent of People Online in 2012
Source: Internet Statistics 2013.

Income, however, cannot explain all of the variation in Internet use and broadband home access. Some of the differences in Internet use and broadband home access are a function of housing patterns. Inner-city neighborhoods are disproportionately populated by racial and ethnic minorities and are simply less likely to be "wired," that is, to have the telecommunications hardware necessary for access to online services. In fact, cable and telephone companies are less likely to lay fiber optic cables in these areas—a practice called "information apartheid" or "electronic redlining."

Racial and ethnic minorities' lack of access to computers and the Internet, although signaling a type of digital divide, may be less common than what researchers are now calling the **participation gap**. Black and Latino children, for example, are more likely to access the Internet via mobile devices, use Twitter, use the Internet to play games, participate in social networking, and watch video games than their advantaged, white counterparts (McCollum 2011). Further, some research indicates that having a home computer, although increasing computer skills, actually lowers academic achievement rather than increasing it, as one might expect (Malamud & Pop-Eleches 2010).

There are few gender disparities in computer use and access in developed countries such as the United States and Japan. However, in developing counties, women play a subordinate role in information communication technologies (ICT), which affects their employability:

> The perception of women being passive consumers of ICT rather than producers extends to their work-related use as well, where one continues to see a feminization of lower-level ICT jobs. . . . Women continue to be concentrated in tedious, repetitive tasks as when they were during the first wave of industrialization, in manufacturing sectors such as textiles, clothing, and electronics. The lower-skilled ICT jobs that women typically find themselves in are word-processing and data entry. (Thas et al. 2007, p. 10)

Concern over accessibility to broadband connectivity has led to a debate over net neutrality. **Net neutrality** advocates hold that Internet users should be able to visit any website and access any content without Internet service providers (ISP) (e.g., cable or telephone companies) acting as gatekeepers by controlling, for example, the speed of downloads. Why would an ISP do that? Hypothetically, if Internet service provider company X signs an agreement with search engine Y, then it's in the best interest of Internet service provider X to slow down all other search engines' performances so that you will switch to search engine Y. Internet service providers argue that Internet users, be they individuals or corporations, who use more than their "fair share" of the Internet should pay more. Why should you pay the same monthly fee as your neighbor who nightly downloads full-length movie files? Others fear any government regulation of the Internet and/or prefer a strictly market model.

In response to the debate, the Federal Communications Commission (FCC) ordered that a free and open Internet be preserved (FCC 2011). To promote a free and open Internet, the regulatory agency adopted three principles. *Transparency* requires that broadband providers disclose their policies, terms, and conditions to users. The principle of *no blocking* prohibits broadband providers from blocking ". . . lawful content, applications, services, or nonharmful devices . . ." (p. 59,192). Lastly, broadband providers may *not unreasonably discriminate* in the transmission of lawful network traffic (FCC 2011).

Problems of Mental and Physical Health

Youth between the ages of 8 and 18 spend an average of 7 hours and 38 minutes a day consuming some type of media, and because more than one medium can be consumed at a time, the actual total exposure to media per day is 10 hours and 45 minutes (Kaiser Family Foundation 2010). Media consumption of all types, except for reading, have increased over the last decade, and heavier media consumption is associated with lower levels of reported personal contentment, more boredom, having fewer friends, and not being happy at school.

participation gap The tendency for racial and ethnic minorities to participate in information and communication technologies (e.g., using smartphones to access the Internet rather than a computer) that place them in a disadvantaged position (e.g., difficult to research a term paper on a smartphone).

net neutrality A principle that holds that Internet users should be able to visit any website and access any content without Internet service provider interference.

What Do You Think? Slade (2012) observes that "Instead of visiting, we phone; instead of phoning, we text; instead of texting, we post 'updates' for our 'friends' on Facebook's wall, and when they don't like these updates, we 'unfriend' them. In public, we catch up on our smartphones with people we no longer have time to visit, or we steal a few moments during our commute to listen to a playlist while reading an e-book on a Kindle. The luminal spaces of our cities are full of people experiencing and practicing uncomfortable 'elevator silences'" (p. 11). Does technology lead to loneliness and separation? How would separating individuals from one another financially benefit corporations?

Padilla-Walker et al. (2012), using longitudinal data, examined the relationship between media use and family connections between mothers, fathers, and their 13- to 16-year-old adolescent children. Parent–child connection was measured in terms of perceived warmth and support. Although frequency of cell phone use and watching television and movies together was positively associated with family connections, communication through social network sites was negatively associated with parent–child connections.

Technology changes what we do, and when and how we do it. A study on the relationship between sleep and technology use found that 95 percent of Americans use some kind of technology the hour before going to sleep (National Sleep Foundation 2011). Watching television, surfing on the net, texting, or playing video games are disruptive to sleep patterns and may be responsible for the millions of Americans who suffer from insomnia, restlessness, and general "sleepiness" during the week.

The multitasking that is associated with technology is linked to distraction, a false sense of urgency, and the inability to focus. Stanford researchers constructed experiments to test the reactions of media multitaskers compared to non–media multitaskers. Over various trials, media multitaskers were unable to ignore irrelevant information, had poorer memories when asked to remember a sequence of alphabetical letters, and were unable to perform primary tasks if distracted (Ophir et al. 2009). Such distractions may be responsible for the relationship between cell phone use and traffic accidents.

Some technologies have unknown risks. Biotechnology, for example, has promised and, to some extent, has delivered everything from life-saving drugs to hardier pest-free tomatoes. Limbs are being replaced by bionic devices controlled by the recipient's thoughts (Martinez 2013), and micro-sized telescopes are being implanted in the eyes of people with damaged retinas (Eisenberg 2009). However, biotechnologies have also created **technology-induced diseases**. According to the American Academy of Environmental Medicine, genetically modified foods have "more than a casual association" with "adverse health effects" (AAEM 2009, p. 1).

Malicious Use of the Internet

technology-induced diseases Diseases that result from the use of technological devices, products, and/or chemicals.

Internet piracy Illegally downloading or distributing copyrighted material (e.g., music, games, software).

malware A general term that includes any spyware, viruses, and adware that is installed on an owner's computer without their knowledge.

Some Internet users access the Internet for malicious purposes, including but not limited to cybercrime and prostitution (see Chapter 4), hacking, piracy, electronic aggression (e.g., cyber-bullying), and "questionable content" sites. **Internet piracy** entails illegally downloading or distributing copyrighted material (e.g., music, games, or software). Court cases indicate that trade organizations (e.g., the Recording Industry Association of American) are pursuing criminal cases against violators, and courts are imposing strict penalties. In 2013, Spain passed legislation that would allow sentences for up to six years in jail for owners of websites that link to pirated content (*Reuters* 2013).

Malware is a general term that includes any spyware, crimeware, worms, viruses, and adware that is installed on owners' computers without their knowledge. Anderson et al. (2012) estimated that, globally, online bank fraud through the use of password-capturing malware costs consumers upward of $70 million and businesses over $300 million in 2010. In 2012, computer owners around the world spent over $10 billion to "disinfect" or replace their PCs, laptops, and tablets due to malware.

Electronic aggression is defined as any kind of aggression that takes place with the use of technology (David-Ferdon & Hertz 2009). For example, **cyber-bullying** refers to the use of electronic communication (e.g., websites, e-mail, instant messaging, or text messaging) to send or post negative or hurtful messages or images about an individual or a group (Kharfen 2006). Cyber-bullying differs from traditional bullying in several significant ways including the potential for a larger audience, anonymity, the inability to respond directly and immediately to the bully, and reduced levels of adult or peer supervision (Sticca & Perren 2013).

Estimates of the frequency of involvement in cyber-bullying either as victim, perpetrator, or both range dramatically, although electronic aggression researchers generally agree that texting is the most common means of cyber-bullying (David-Ferdon & Hertz 2009). Because cyber-bullying is capable of reaching wider audiences, many states and school districts have begun creating cyber-bullying disciplinary policies. As of 2013, 18 states have anti-bullying laws that include cyber-bullying, and in 47 states, anti-bullying laws also prohibit electronic harassment (Hinduja & Patchin 2013).

The Challenge to Traditional Values and Beliefs

Technological innovations and scientific discoveries often challenge traditionally held values and beliefs, in part because they enable people to achieve goals that were previously unobtainable. Before recent advances in reproductive technology, for example, women could not conceive and give birth after menopause. Technology, that allows postmenopausal women to give birth, challenges societal beliefs about childbearing and the role of older women. The techniques of egg retrieval, in vitro fertilization, and gamete intrafallopian transfer make it possible for two different women to each make a biological contribution to the creation of a new life. Such technology requires society to reexamine its beliefs about what a family is and what a mother is. Should family be defined by custom, law, or the intentions of the parties involved?

Medical technologies that sustain life lead us to rethink the issue of when life should end. The increasing use of computers throughout society challenges the traditional value of privacy. New weapons systems make questionable the traditional idea of war as something that can be survived and even won. And cloning causes us to wonder about our traditional notions of family, parenthood, and individuality. Toffler (1970) coined the term **future shock** to describe the confusion resulting from rapid scientific and technological changes that unravel our traditional values and beliefs.

What Do You Think? Perhaps no technologies are as futuristic and as controversial as nanotechnologies. Nanotechnologies are a classification of different technologies, all of which have one thing in common—they are small, very, very small. A nanometer is one-billionth of a meter. To put this in perspective, one sheet of paper is 100,000 nanometers thick (PCAST 2012). That nanotechnologies will change our lives and the social world we live in is indisputable. Would you wear a T-shirt that, through nanotechnology, converts the energy your movements generate to electricity that can then power your laptop? Would you be comfortable with paint that turns colors because you told it to? How about nano-sized medical robots that are injected into your blood stream to look for and fight off diseases? Do any of these innovations, all presently proposed by nanotechnology researchers, make you feel a little uneasy? Why or why not?

Strategies for Action: Controlling Science and Technology

As technology increases, so does the need for social responsibility. Nuclear power, genetic engineering, cloning, and computer surveillance all increase the need for social responsibility: "Technological change has the effect of enhancing the importance of public decision making in society, because technology is continually creating new possibilities

cyber-bullying The use of electronic communication (e.g., websites, e-mail, instant messaging, text messaging) to send or post negative or hurtful messages or images about an individual or a group.

future shock The state of confusion resulting from rapid scientific and technological changes that unravel our traditional values and beliefs.

Annually, "tens of millions of animals are used for biomedical research, chemical testing, and training" (Bowman 2011, p. 1). An estimated 1 million are dogs and cats, monkeys and apes, hamsters and guinea pigs, rabbits, pigs, and sheep, and other farm animals. All else, including birds, fish, rats and mice, frogs, and lizards, number somewhere between 80 to 100 million. Each has or will be subject to **vivisection,** the practice of cutting into or otherwise harming living, nonhuman animals for the purpose of scientific research.

The use of animals for research purposes is not a new phenomenon, but the century-old debate over the morality of using animals in research has recently picked up momentum as animal rights groups such as PETA (People for the Ethical Treatment of Animals) and the Animal Liberation Front make inroads into the American collective consciousness. Additionally, there is evidence that pets are assuming an increasingly important role in our emotional lives (Power 2008). In a national survey of pet owners, 57 percent of respondents stated that if they were stranded on a desert island and could have only one companion, they would choose their pet (Cohen 2002).

The issues surrounding the vivisection of nonhuman animals are not easily resolved. Should humans' rights trump those of non-human animals? Do nonhuman animals, as human animals, have a right to a clean, safe, and pain-free environment? Should nonhuman

animals die in the hopes of gathering data that might prolong or save human life?

Opinion polls indicate that the number of Americans who believe that medical testing on animals is "morally acceptable" has steadily declined over the years (NSF 2013). Over half of Americans, 52 percent, are in favor of using animals in research although there are striking gender differences. Sixty-two percent of men are in favor of using animals in research compared to 42 percent of women. There are also differences based on the type of animals used—for example, mice versus chimpanzees or dogs (NSF 2013).

The arguments both for (at one end of the continuum) and against (at the other end of the continuum) the use of animals in research generally fall into three camps—yes, no, and sometimes. Advocates argue that any distinction that differentiates dogs and chimpanzees, for example, from birds and fish, is an artificial one. The only substantively significant distinction to be made is between human animals and nonhuman animals.

That said, if your daughter were ill and you had the choice of giving her drug A (which killed all of the dogs and cats it was tested on), drug B (which killed some of the dogs and cats it was tested on), or drug C (which killed none of the dogs and cats it was tested on), which would you choose? Obviously, the rational answer is drug C, proving that, or so many scientists would argue, using animals in research provides us with

important information in making life and death decisions. Test on animals or your child dies—your choice. In a recent survey of bio-medical scientists, 90 percent responded that the use of animals in research is "essential" (*Nature News* 2011).

Finally, the Food and Drug Administration *requires* that drugs and procedures be tested on nonhuman animals before they can be consumed by or used on human animals. And to opponents who are quick to point out that human and nonhuman animals are too different to result in any meaningful findings, Holland (2010) responds:

> Pigs are clearly not human, but heart valves in both species are nearly interchangeable. Some viruses that attack rodents, cats, and primates can do the same damage to humans. And diseases in some species have highly similar "first cousins" that affect people. The key issue for researchers is that the biological mechanisms for many human and animal viruses are highly similar, if not the same. So understanding how to stop one form of virus in humans can provide the key to doing the same for animals and vice versa.

Hanson (2010), however, convincingly argues that such justifications are an ethical "bait and switch." Few of us could even bear to look at pictures of "monkeys, with their electrode-implanted brains and bolted heads, being put through their paces in a

for social action as well as new problems that have to be dealt with" (Mesthene 1993, p. 85). In the following sections, we address various aspects of the public debate, including science, ethics and the law, the role of corporate America, and government policy.

Science, Ethics, and the Law

genetic exception laws Laws that require that genetic information be handled separately from other medical information.

vivisection The practice of cutting into or otherwise harming living, nonhuman animals for the purpose of scientific research.

Science and its resulting technologies alter the culture of society through the challenging of traditional values. Public debate and ethical controversies, however, have led to structural alterations in society as the legal system responds to calls for action. For example, several states now have what are called genetic exception laws. **Genetic exception laws** require that genetic information be handled separately from other medical information, leading to what is sometimes called patient shadow files (Legay 2001). The logic of such laws rests with the potentially devastating effects of genetic information being revealed to insurance companies, other family members, employers, and the like. Their importance has also grown—there are now genetic tests for over 1,000 diseases, and hundreds more are being researched. Examples of protected tests include the BRCA1 and BRAC2 tests for desperate attempt to get a life-sustaining sip of water" (p. 3). Yet, researchers argue that such testing techniques may lead to finding the cause or cure for Alzheimer's disease. Hanson, a renowned neuroscientist and PETA member, continues noting that "these experiments . . . are so thoroughly unrelated to the neuropathology of Alzheimer's . . . that in more than 28 years of research in the neuroscience of the disease, I have never come across a single reference to them in any scientific literature on neurodegenerative diseases" (p. 3).

Moderates acknowledge that the use of animals in scientific research may be necessary but that the issue is multifaceted. Henry and Pulcino's (2009) study of the characteristics that influence attitudes toward the use of animals in research is a case in point. In a pen and pencil experiment, the researchers varied three independent variables: (1) type of animal used (chimpanzees, dogs, or mice), (2) level of harm to the animal (no harm, serious injury, death), and (3) severity of the disease being studied (eczema, rheumatoid arthritis, cancer). In addition to individual differences (for example, females were less tolerant of the use of animals in research than males), all of the independent variables were statistically significant in predicting what a given subject would recommend be done. Subjects were more likely to be in favor of using animals for testing if mice were being used, there was no harm, and the disease under investigation

was cancer. Although the results are not particularly surprising, the lack of significant interactions is. Respondents considered each of the three research variables independently of one another rather than simultaneously.

There are laws that protect the use of animals in scientific research. The Laboratory Animal Welfare Act was passed in 1966, after an article appeared in *Life* magazine entitled, "Concentration Camps for Lost and Stolen Pets." A later amendment required that oversight "care and use" committees be established at research institutions, but most are often staffed by animal use advocates (Hanson 2010). Further, in 2009, Japan, Canada, the United States, and the European Union signed an international agreement that would reduce the number of animals used in product safety testing (NIEHS 2009). The new agreement, however, has little impact on the use of animals in biotechnology research.

There are also norms within the culture of "doing science," some of which have become formalized: minimize pain and suffering, use as few animals as possible, hire caretakers (e.g., veterinarians), look for alternative means of accomplishing the same end, and, given a choice, use lower-level animals rather than higher-level animals (Research Animal Resources 2003). The problem is that there is overwhelming evidence that all animals suffer, and even "lower-level" animals, such as mice, grimace when experiencing pain and laugh when being tickled; in other words,

they have emotions (Ferdowsian 2010; Flecknell 2010).

Ironically, the same thing that led to the use of animals in research, the pursuit of scientific and technological know-how, may lead to ending the use of animals in scientific research. For example, a "surrogate in-vitro human immune system . . . has been developed [in a test tube] to help predict an individual's immune response system" (Ferdowsian 2010). Further, the ICCVAM (Interagency Coordinating Committee on the Validation of Alternative Methods), established in 2000 under the National Institute of Health, is designed to

promote the development, validation, and regulatory acceptance of new, revised, and alternative regulatory safety testing methods. Emphasis is on alternative methods that will reduce, refine (less pain and distress), and replace the use of animals in testing while maintaining and promoting scientific quality and the protection of human health, animal health, and the environment. (ICCVAM 2011)

The issue is complex, and there are no easy answers. But asking the right questions is a step in the right direction. Perhaps as philosopher Jeremy Bentham said over 200 years ago, the right question is not, "Can they speak?" nor "Can they reason?" but "Do they suffer?"

breast cancer and carrier screening for such conditions as sickle-cell anemia and cystic fibrosis (Genetics and Public Policy Center 2010).

Are such regulations necessary? In a society characterized by rapid technological and thus social change—a society in which custody of frozen embryos is part of a divorce agreement—many would say yes. Cloning, for example, is one of the most hotly debated technologies in recent years. Bioethicists and the public vehemently debate the various costs and benefits of this scientific technique. Despite such controversy, however, the chairman of the National Bioethics Advisory Commission warned nearly 15 years ago that human cloning will be "very difficult to stop" (McFarling 1998). At least 15 states have laws pertaining to human cloning, with some prohibiting cloning for reproductive purposes, some prohibiting therapeutic cloning, and still others prohibiting both (NCSL 2008; Rovner 2013).

Should the choices that we make as a society be dependent on what we can do or what we should do? Whereas scientists and the agencies and corporations that fund them often determine what we *can* do, who should determine what we *should* do (see this chapter's *Animals and Society* feature)? Although such decisions are likely to have a strong legal component—that is, they must be consistent with the rule of law and the constitutional right of scientific inquiry—legality or the lack thereof often fails to answer the question, "What should be done?" *Roe v. Wade* (1973) did little to quash the public debate over abortion and, more specifically, the question of when life begins. Thus, it is likely that the issues surrounding the most controversial of technologies will continue throughout the 21st century with no easy answers.

Technology and Corporate America

As philosopher Jean-Francois Lyotard noted, knowledge is increasingly produced to be sold. The development of GMOs, the commodification of women as egg donors, direct-to-consumer genetic testing, and the harvesting of regenerated organ tissues are all examples of market-driven technologies. Like the corporate pursuit of computer technology, profit-motivated biotechnology creates several concerns.

First is the concern that only the rich will have access to life-saving technologies such as genetic testing and cloned organs. Such fears may be justified. Myriad Genetics patented breast and ovarian cancer genes (Bollier 2009). However, because of the resulting **gene monopolies** and the associated astronomical patient costs for genetic testing and treatment, in 2009, the American Civil Liberties Union and several plaintiffs filed a lawsuit against Myriad Genetics, alleging that the patents are "invalid and unconstitutional" (Genomics Law Report 2011). Appeals of several lower court decisions took the case to the U.S. Supreme Court, which ruled in favor of the plaintiffs—human genes cannot be patented (ACLU 2013).

The commercialization of technology causes several other concerns, including issues of quality control and the tendency for discoveries to remain closely guarded secrets rather than collaborative efforts (Crichton 2007; Lemonick & Thompson 1999; Mayer 2002; Rabino 1998). In addition, industry involvement has made government control more difficult because researchers depend less and less on federal funding. More than 62 percent of research and development in the United States is supported by private industry using their own company funds (NSF 2013).

Runaway Science and Government Policy

Science and technology raise many public policy issues. Policy decisions, for example, address concerns about the safety of nuclear power plants, the privacy of e-mail, the hazards of chemical warfare, and the ethics of cloning. In creating science and technology, have we created a monster that has begun to control us rather than the reverse? What controls, if any, should be placed on science and technology? And are such controls consistent with existing law? Consider the use of the file-sharing network BitTorrent to download music and movie files (the question of intellectual property rights and copyright infringement); laws limiting children's access to material on the Internet (free speech issues); and Acxiom, the "cookie"-collecting company that helps corporations customize advertising on websites by tracing clicks and keystrokes (Fourth Amendment privacy issues).

> Policy decisions . . . address concerns about the safety of nuclear power plants, the privacy of e-mail, the hazards of chemical warfare, and the ethics of cloning.

Concerns over "runaway science" are not uniquely American. Although genetic data may be collected legally, in the course of an arrest, for example, there is growing evidence that governments around the world are maintaining large genetic databases. In Great Britain, police collected samples of nearly 7 million people, the equivalent of 10 percent of the population. The database was significantly reduced when the European Court of Human Rights ruled that such "blanket and indiscriminate" storage of DNA material was a privacy intrusion (Lawless 2013). In the United States, the Federal Bureau of Investigation's DNA database contains genetic information on over 11 million suspected or convicted criminals.

And yet it is the government, often through Congress, regulatory agencies, or departments, that is responsible for controlling technology, prohibiting some (e.g., assisted-suicide devices) and requiring others (e.g., seat belts). A good example is the Stem Cell Research Advancement Act of 2013 that was introduced in the U.S. House of Representatives. As

gene monopolies Exclusive control over a particular gene as a result of government patents.

proposed, the act (1) supports the use of embryonic stem cells, including human embryonic stems cells, (2) defines the types of human embryonic stem cells eligible for use in research (e.g., donated from in vitro fertilization clinics), (3) mandates that the Department of Health and Human Services maintain, review, and update guidelines in support of human stem cell research, and (4) prohibits public funds from being used for human cloning (Stem Cell Research Advancement Act 2013).

Of late, two areas of concern have required increased government scrutiny—the National Security Agency (NSA) surveillance programs and international and domestic cyber-threats. One of the most disturbing revelations about the NSA is their 10-year effort to subvert encryption technology that included collusion with encryption software companies (Ball et al. 2013). As a result of their successes, hundreds of millions of e-mails, medical records, and online transactions were breached. Despite the agency's claim that decryption techniques are a necessary evil in the fight against terrorism, 19 bills designed to restrict NSA's surveillance capabilities are currently in Congress (Richardson & Greene 2013).

There are also privacy concerns about the passage of the Cyber Intelligence Sharing and Protection Act (CISPA). The act would allow private companies (e.g., Facebook and Verizon) that become aware of cyber-threats to share information with the U.S. government, even if personal user information is revealed, without fear of reprisal (Whittaker 2013).

As the risks of cyber-attacks from individuals, groups, or governments increase, there are universal calls for stricter regulation. Despite then Secretary of Defense Panetta's warning that we are on the eve of a "cyber-Pearl Harbor," the Cybersecurity Act of 2012 failed to pass the U.S. Senate, leading to heavy criticism (Bumiller & Shanker 2012). In 2013, more than a dozen cyber-security bills were introduced into either the U.S. House of Representatives or the U.S. Senate (Fischer 2013).

One way to deter cyber-attacks is to develop a profile of those most likely to engage in the behavior. Holt and Kilger (2012) investigated politically motivated cyber-attacks against a home or hypothetical country by presenting a sample of undergraduate and graduate students with different scenarios, and asking them under what conditions they would be willing to participate in off-line or online attacks against "critical infrastructures." Results indicate that nationalism and patriotism are unrelated to the likelihood of physical or virtual protest against a home or foreign government. Respondents who identified with the United States as their homeland were more likely to engage in physical protests compared to international students who had higher likelihoods of virtual protests. Finally, the researchers report that respondents who say they would engage in violent off-line actions were also more likely to say they would commit an online act of violence.

Lastly, the government has several science and technology boards and initiatives, including the National Science and Technology Council, the Office of Science and Technology Policy, the President's Council of Advisors on Science and Technology, and the U.S. National Nanotechnology Initiative. These agencies advise the president on matters of science and technology, including research and development, implementation, national policy, and coordination of different initiatives.

What Do You Think? Although still in its infancy, direct-to-consumer (DTC) genetic testing is available now. How accurate it is in predicting disease depends on the disease in question. In some cases, the presence of a gene indicates with absolute certainty that you will get the disease; in others, your risk may be much lower. As availability of DTC genetic testing increases and cost decreases, would you consider genetic testing for health purposes?

Understanding Science and Technology

What are we to understand about science and technology from this chapter? As structural functionalists argue, science and technology evolve as a social process and are a natural part of the evolution of society. As society's needs change, scientific discoveries and technological innovations emerge to meet these needs, thereby serving the functions of the whole. Consistent with conflict theory, however, science and technology also meet the needs of select groups and are characterized by political components. As Winner (1993) noted, the structure of science and technology conveys political messages, including "power is centralized," "there are barriers between social classes," "the world is hierarchically structured," and "the good things are distributed unequally" (p. 288).

The scientific discoveries and technological innovations that society does or does not embrace are socially determined. Research indicates that science and the resulting technologies have both negative and positive consequences—a *technological dualism*. Technology saves lives, time, and money; it also leads to death, unemployment, alienation, and estrangement. Weighing the costs and benefits of technology poses ethical dilemmas, as does science itself. Ethics, however, "is not only concerned with individual choices and acts. It is also and, perhaps, above all concerned with the cultural shifts and trends of which acts are but the symptoms" (McCormick & Richard 1994, p. 16).

Thus, society makes a choice by the very direction it follows. These choices should be made on the basis of guiding principles that are both fair and just, such as those listed here:

1. Science and technology should be prudent. Adequate testing, safeguards, and impact studies are essential. Impact assessment should include an evaluation of the social, political, environmental, and economic factors.
2. No technology should be developed unless all groups, and particularly those who will be most affected by the technology, have at least some representation "at a very early stage in defining what that technology will be" (Winner 1993, p. 291). Traditionally, the structure of the scientific process and the development of technologies have been centralized (i.e., decisions have been made by a few scientists and engineers); decentralization of the process would increase representation.
3. Means should not exist without ends. Each new innovation should be directed to fulfilling a societal need rather than the more typical pattern in which a technology is developed first (e.g., high-definition television) and then a market is created (e.g., "You'll never watch a regular TV again!"). Indeed, from the space program to research on artificial intelligence, the vested interests of scientists and engineers, whose discoveries and innovations build careers, should be tempered by the demands of society. (Buchanan et al. 2000; Eibert 1998; Goodman 1993; Murphie & Potts 2003; Winner 1993)

What the 21st century will hold, as the technological transformation continues, may be beyond the imagination of most of society's members. Technology empowers; it increases efficiency and productivity, extends life, controls the environment, and expands individual capabilities. According to a National Intelligence Council report, "Life in 2015 will be revolutionized by the growing effort of multidisciplinary technology across all dimensions of life: social, economic, political, and personal" (NIC 2003, p. 1).

As we proceed further into the first computational millennium, one of the great concerns of civilization will be the attempt to reorder society, culture, and government in a manner that exploits the digital bonanza yet prevents it from running roughshod over the checks and balances so delicately constructed in those simpler pre-computer years.

CHAPTER REVIEW

• **What are the three types of technology?**
The three types of technology, escalating in sophistication, are mechanization, automation, and cybernation. Mechanization is the use of tools to accomplish tasks previously done by hand. Automation involves the use of self-operating machines, and cybernation is the use of machines to control machines.

• **What are some of the reasons the United States may be "losing its edge" in scientific and technological innovations?**
The decline of U.S. supremacy in science and technology is likely to be the result of five interacting social forces: First, the federal government has been scaling back its investment in research and development. Second, corporations

have begun to focus on short-term products and higher profits. Third, there has been a drop in science and math education in U.S. schools in terms of both quality and quantity. Fourth, developing countries, most notably China and India, are expanding their scientific and technological capabilities at a faster rate than the United States. Finally, as documented in the book *Unscientific American,* there is a disconnect between American society and the principles of science.

• **What are some Internet global trends?**
In 2012, the Internet had 2.4 billion users in more than 200 countries with 245 million users in the United States (Internet Statistics 2013). Of all Internet users, the highest proportion come from Asia (44.8 percent), followed by Europe (21.5 percent), North America (11.4 percent), Latin America and the Caribbean (10.4 percent), Africa (7.0 percent), the Middle East (3.7 percent), and Oceania/Australia (1.0 percent) (Internet Statistics 2013). Penetration rates vary but, in general, are higher in wealthier rather than poorer nations.

• **According to Kuhn, what is the scientific process?**
Kuhn describes the process of scientific discovery as occurring in three steps. First are assumptions about a particular phenomenon. Next, because unanswered questions always remain about a topic, science works to start filling in the gaps. Then, when new information suggests that the initial assumptions were incorrect, a new set of assumptions or framework emerges to replace the old one. It then becomes the dominant belief or paradigm until it is questioned and the process repeats.

• **What is meant by the computer revolution?**
The silicon chip made computers affordable. Today, over 75 percent of U.S. households report having a computer in the home compared to 61.8 percent just a decade ago (File 2013).

• **What is the Human Genome Project?**
The U.S. Human Genome Project is an effort to decode human DNA. The 13-year-old project is now complete, allowing scientists to "transform medicine" through early diagnosis and treatment as well as possibly preventing disease through gene therapy. Gene therapy entails identifying a defective or missing gene and then replacing it with a healthy duplicate that is transplanted to the affected area.

• **What is the legal status of abortion in the United States?**
In the United States, since the U.S. Supreme Court's ruling in *Roe v. Wade* in 1973, abortion has been legal. However, recent Supreme Court and state court decisions have limited the scope of *Roe v. Wade.*

• **How are some of the problems of the Industrial Revolution similar to the problems of the technological revolution?**
The most obvious example is unemployment. Just as the Industrial Revolution replaced many jobs with technological innovations, so too has the technological revolution. Furthermore, research indicates that many of the jobs created by the Industrial Revolution, such as working on a factory assembly line, were characterized by high rates of alienation. Rising rates of alienation are also a consequence of increased estrangement as high-tech employees work in "white-collar factories."

• **What is meant by outsourcing, and why is it important?**
Outsourcing is the practice of a business subcontracting with a third party, often in low-wage countries such as China and India, for services. The problem with outsourcing is that it tends to lead to higher rates of unemployment in the export countries.

• **What is the digital divide?**
The digital divide is the tendency for technology to be most accessible to the wealthiest and most educated. For example, some fear that there will be "genetic stratification," whereby the benefits of genetic testing, gene therapy, and other genetic enhancements will be available to only the richest segments of society.

• **What is meant by the commercialization of technology?**
The commercialization of technology refers to profit-motivated technological innovations. Whether in regard to the isolation of a particular gene, genetically modified organisms, or the regeneration of organ tissues, where there is a possibility for profit, private enterprise will be there.

TEST YOURSELF

1. Which of the following technologies is associated with industrialization?
   a. Mechanization
   b. Cybernation
   c. Hibernation
   d. Automation
2. Analysis of data over time suggests that the science and technology gap between poorer and richer nations is getting greater rather than shrinking.
   a. True
   b. False
3. The U.S. government, as part of the technological revolution, spends more money on research and development than educational institutions and corporations combined.
   a. True
   b. False
4. Which theory argues that technology is often used as a means of social control?
   a. Structural functionalism
   b. Social disorganization
   c. Conflict theory
   d. Symbolic interactionism
5. The global legalization of abortion has all but eliminated unsafe abortion procedures.
   a. True
   b. False
6. The ability to manipulate the genes of an organism to alter the natural outcome is called
   a. Gene therapy
   b. Gene splicing
   c. Genetic engineering
   d. Genetic testing

7. Genetically modified foods have been documented as harmless to humans by the Food and Drug Administration.
   a. True
   b. False
8. In 2007, the U.S. Supreme Court upheld the partial-birth abortion ban in a 5 to 4 decision.
   a. True
   b. False
9. The practice of outsourcing entails
   a. creating high-tech jobs in the United States for immigrants
   b. hiring temporary workers to cover for employees who are absent
   c. allowing company workers to work from home
   d. subcontracting jobs often to workers in low-wage countries
10. The Genetic Information Nondiscrimination Act (GINA)
   a. makes human cloning illegal in the United States
   b. establishes a criminal penalty for human cloning in the United States
   c. is a federal law that prohibits discrimination in health coverage or employment based on genetic information
   d. all of the above

Answers: 1. D; 2. A; 3. B; 4. C; 5. B; 6. C; 7. B; 8. A; 9. D; 10. C.

KEY TERMS

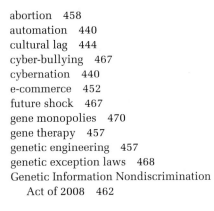

abortion 458
automation 440
cultural lag 444
cyber-bullying 467
cybernation 440
e-commerce 452
future shock 467
gene monopolies 470
gene therapy 457
genetic engineering 457
genetic exception laws 468
Genetic Information Nondiscrimination Act of 2008 462

genetic testing 456
Internet 450
Internet piracy 466
learning health systems 453
malware 466
mechanization 440
membership communities 455
net neutrality 465
outsource 463
partial-birth abortions 459
participation gap 465
penetration rate 441
postmodernism 444

science 440
Semantic Web 454
STEM 442
stem cells 461
technological dualism 440
technological fix 444
technology 440
technology-induced diseases 466
telepresencing 449
therapeutic cloning 461
vivisection 468
Web 2.0 451

Conflict, War, and Terrorism

15

"Every gun that is made, every warship launched, every rocket fired signifies, in the final sense, a theft from those who hunger and are not fed, those who are cold and not clothed."

General Dwight D. Eisenhower,
former U.S. president
and military leader

In this photo taken Thursday, July 29, 2010, Staff Sergeant Melinda Miller hugs Gina after a workout on an obstacle course at Peterson Air Force Base in Colorado Springs, Colorado. Gina was a playful 2-year-old German shepherd when she went to Iraq as a highly trained bomb-sniffing dog with the U.S. military, but months of door-to-door searches and noisy explosions left her cowering and fearful. After she came home to Peterson Air Force Base in June 2009, a military veterinarian diagnosed her with post-traumatic stress disorder.

THINGS WERE STARTING to turn around for the 28-year-old Marine vet Clay Hunt. After two tours of duty, one in Iraq and one in Afghanistan, he suffered from PTSD (post-traumatic stress disorder) and the guilt of surviving what his buddies hadn't, but he—unlike so many other vets—had done everything right (Hefling 2011; Wise 2011). He moved closer to his family, took his medications and sought counseling, and volunteered in both Haiti and Chile after the earthquakes. He even made public service announcements warning other vets of the dangers of PTSD—reaching out to them to get the help he had. Nonetheless, when Clay got a construction job, bought a truck, and started dating again, his friends and relatives were relieved. After all, he had made plans for a big reunion with his fellow marines the following weekend. But Clay never made it to the reunion. On March 31, 2011, this young man with movie-star good looks and a friendly, easy manner bolted himself inside his Houston apartment and shot himself. Ironically, the Purple Heart recipient had a tattoo on his arm from J.R.R. Tolkien's *Lord of the Rings*: "Not all those who wander are lost." In 2013, CBS's *60 Minutes* presented a mini-documentary on "The Life and Death of Clay Hunt."

War is one of the great paradoxes of human history. It both protects and annihilates. It creates and defends nations but may also destroy them. **War**, the most violent form of conflict, refers to organized armed violence aimed at a social group in pursuit of an objective. Wars have existed throughout human history and continue in the contemporary world. Whether war is just or unjust, defensive or offensive, it involves the most horrendous atrocities known to humankind. This is especially true in the 21st century, when nearly all wars are fought in populated areas rather than on remote battlefields, having deadly consequences for civilians. Thus, war is not only a social problem in and of itself, but it also contributes to a host of other social problems—death, disease, and disability, crime and immorality, psychological terror, loss of economic resources, and environmental devastation. In this chapter, we discuss each of these issues within the context of conflict, war, and terrorism, the most threatening of all social problems.

The Global Context: Conflict in a Changing World

As societies have evolved and changed throughout history, the nature of war has also changed. Before industrialization and the sophisticated technology that resulted, war occurred primarily between neighboring groups on a relatively small scale. In the modern world, war can be waged between nations that are separated by thousands of miles as well as between neighboring nations. Increasingly, war is a phenomenon internal to states, involving fighting between the government and rebel groups or among rival contenders for state power. Indeed, Figure 15.1 documents that wars between states, that is, interstate wars, recently made up the smallest percentage of armed conflicts. In the following sections, we examine how war has changed our social world and how our changing social world has affected the nature of war in the industrial and postindustrial information age.

war Organized armed violence aimed at a social group in pursuit of an objective.

War and Social Change

The very act that now threatens modern civilization—war—is largely responsible for creating the advanced civilization in which we live. Before large political states existed, people lived in small groups and villages. War broke the barriers of autonomy between local groups and permitted small villages to be incorporated into larger political units known as chiefdoms. Centuries of warfare between chiefdoms culminated in the development of the state. The **state** is "an apparatus of power, a set of institutions—the central government, the armed forces, the regulatory and police agencies—whose most important functions involve the use of force, the control of territory, and the maintenance of internal order" (Porter 1994, pp. 5–6). Social historian Charles Tilly famously said, "war makes states, and states make war" (1992). The creation of the state in turn led to other profound social and cultural changes:

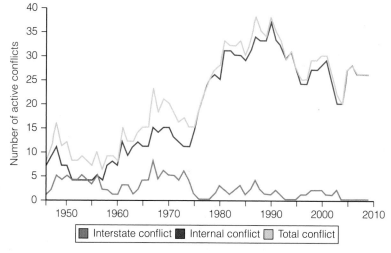

Figure 15.1 Global Trends in Violent Conflict, 1946–2009
Source: Hewitt et al. 2012.

> And once the state emerged, the gates were flung open to enormous cultural advances, advances undreamed of during—and impossible under—a regimen of small autonomous villages. . . . Only in large political units, far removed in structure from the small autonomous communities from which they sprang, was it possible for great advances to be made in the arts and sciences, in economics and technology, and indeed in every field of culture central to the great industrial civilizations of the world. (Carneiro 1994, pp. 14–15)

Industrialization and technology could not have developed in the small social groups that existed before military action consolidated them into larger states. Thus, war contributed indirectly to the industrialization and technological sophistication that characterize the modern world. Industrialization, in turn, has had two major influences on war. Cohen (1986) calculated the number of wars fought per decade in industrial and preindustrial nations and concluded, "as societies become more industrialized, their proneness to warfare decreases" (p. 265). Thus, for example, in 2012, there were 32 ongoing armed conflicts, all of which were taking place in less developed countries in the Middle East, South Asia, Africa, and South America; over 90 percent of the combatants involved in these conflicts represented less developed countries (UCDP 2013).

Although industrialization may decrease a society's propensity to war, it also increases the potential destruction of war. With industrialization, military technology became more sophisticated and more lethal. Rifles and cannons replaced the clubs, arrows, and swords used in more primitive warfare and, in turn, were replaced by tanks, bombers, and nuclear warheads. Today, the use of new technologies such as high-performance sensors, information processors, directed energy technologies, precision-guided munitions, and computer worms and viruses, has changed the very nature of conflict, war, and terrorism. The use of unoccupied aerial vehicles (UAVs), also known as drones, has raised national security and ethical concerns. The Obama administration has defended the use of drones as an essential form of defense in the global war on terror in which military ground operations are not always possible to catch suspected terrorists and prevent future violence. Critics argue that the use of drones hurts the national security interests of the United States because the strikes typically violate the sovereign airspace of the countries where the attacks

> The use of new technologies such as high-performance sensors, information processors, directed energy technologies, precision-guided munitions, and computer worms and viruses, has changed the very nature of conflict, war, and terrorism.

state The organization of the central government and government agencies such as the military, police, and regulatory agencies.

take place, angering the governments of other countries and alienating potential allies. The use of drones in another country's airspace without their permission violates international laws, but defenders suggest such actions are legal when done in self-defense. Critics have also raised ethical concerns. Strikes in Pakistan, Yemen, and Somalia have resulted in the deaths of nearly 1,000 unintentional civilian targets, including hundreds of children. Defenders of the U.S. drone policy argue, however, that unintentional deaths of civilians are much higher in conventional ground wars and the use of drones is ethically preferable to a large-scale ground war (Matthews 2013).

What Do You Think? Reports indicate that "lone wolf" terrorist attacks in Boston and London in 2013 were motivated by the use of drones and other military actions by the U.S. and British forces in the Middle East (Greenwald 2013). Does the use of drones prevent violence by killing terrorist leaders, or spur more violence by inspiring others to carry out terrorist attacks? What do you think?

United States Air Force

In the 20th century, industrialization spurred technological innovations that transformed warfare more rapidly than at any other time in human history. Here, a drone—an unoccupied aerial vehicle—is being prepared for a mission. In 2013, Niger agreed to house U.S. surveillance drones used for watching the actions of Islamic militants in the area. Drones also have nonmilitary uses including aerial photography, weather observations, law enforcement, and fighting wildfires.

In the postindustrial information age, computer technology has not only revolutionized the nature of warfare, it has made societies more vulnerable to external attacks. In 2013, the U.S. Department of Defense publicly accused the Chinese government of mounting numerous cyberattacks against U.S. government computer systems. The Pentagon's report claimed that the government and military of China were engaged in a widespread strategy of cyber-espionage with the intent of "building a picture of U.S. defense networks, logistics, and related military capabilities that could be exploited during a crisis" (U.S. Department of Defense 2013).

The Economics of Military Spending

The increasing sophistication of military technology has commanded a large share of resources; world military expenditures in 2012 totaled $1.76 trillion, or about $249 for each person in the world (SIPRI 2013). This total expenditure represents a decline of about 0.4 percent from 2011, the first decline in global military expenditure since 1998. Although the United States has the highest military expenditure of any country in the world, part of this decline can be attributed to the drawdown of American troops from Iraq and Afghanistan. Other global economic and social factors also played a role. The global economic crisis that began in 2008 led many European countries to enact austerity measures that drastically reduced military spending, limiting the ability of the North Atlantic Treaty Organization (NATO) forces to project power or carry out protracted irregular warfare operations (Larrabee et al. 2012). Concerns over the budget deficit in the United States also sparked a national debate about military funding. Between 2011 and 2012, the United States reduced its military spending by 5.6 percent while western and central European countries decreased military spending by about 2 percent. At the same time, countries in East and Southeast Asia have increased military spending by 5.0 to 6.0 percent (SIPRI 2013).

The United States spends far more than any other country on its military. The **Cold War**, the state of political tension, economic competition, and military rivalry that existed between the United States and the former Soviet Union for nearly 50 years, provided justification for large expenditures for military preparedness. However, the end of the Cold War, along with the rising national debt, resulted in cutbacks in the U.S. military

Cold War The state of military tension and political rivalry that existed between the United States and the former Soviet Union from the 1950s through the late 1980s.

budget in the 1990s. Today, military spending has nearly returned to the levels during the Cold War. In 2012, the United States spent $685.3 billion on its military, which accounts for over 40 percent of the world's military spending and is more than the combined military expenditures of the next 13 highest military spending nations, including China ($106.4 billion) and Russia ($30 billion) (SIPRI 2013).

The U.S. government not only spends more money than other countries on its own military and defense, but it also sells military equipment to other countries, either directly or by helping U.S. companies sell weapons abroad. Although the purchasing countries may use these weapons to defend themselves from hostile attack, foreign military sales may pose a threat to the United States by arming potential antagonists. For example, the United States, the world's leading arms-exporting nation, supplied weapons to Iraq to use against Iran during the war between 1980 and 1988. These same weapons were then used against Americans in the 1991 Gulf War and 2003 Iraq War (Silverstein 2007). Similarly, after the Soviet invasion of Afghanistan in 1979, the United States funded Afghan rebel groups. Years after the Soviets left Afghanistan, rebels continued to fight for control of the country. Using weapons supplied by the United States, the Taliban took over much of Afghanistan and sheltered al Qaeda and Osama bin Laden—also a former recipient of U.S. support—as they planned the attacks on September 11 (Bergen 2002; Rashid 2000).

The United States regularly transfers arms to countries in active conflict. For example, in 2006 and 2007, the United States sold about $9.8 billion in arms to allies for use in war zones in Pakistan, Iraq, Israel, Afghanistan, and Colombia (Berrigan 2009). A 2005 report titled *U.S. Weapons at War: Promoting Freedom or Fueling Conflict?* concluded that, far "from serving as a force for security and stability, U.S. weapons sales frequently serve to empower unstable, undemocratic regimes to the detriment of U.S. and global security" (Berrigan & Hartung 2005). For example, during the 2011 so-called "Arab Spring"—a series of popular uprisings against authoritarian regimes in the Middle East—reports surfaced that authorities in Egypt and Bahrain used tear gas and weapons obtained through authorized arms agreements with the United States against unarmed civilians (Braun 2011; Wali & Sami 2011).

What Do You Think? In June 2013, the Obama administration announced it would send military aid, including weapons and ammunition, to support Syrian rebels in their attempts to overthrow the regime of Syrian President al-Assad (White House 2013). Some critics argued that, given the high levels of civilian casualties, much more military aid is needed than the United States plans to provide, while others argued that no military aid should be provided given the likelihood that these weapons could fall into the hands of enemies of the United States. Under what conditions do you think the United States should send arms to support fighters in other countries?

Historically, wars have been associated with economic growth and technological innovation. During World War II, for example, a surge in government spending and investment in public works projects led to spikes in employment rates and gross domestic product (GDP) that had been at historically low levels during the Great Depression of the 1930s. At the same time, the government raised taxes to fund the war, and U.S. consumption declined as Americans on the home front sacrificed personal comforts as part of the war effort. The wars in Iraq and Afghanistan, however, represented the first time in modern history that the United States has gone to war without seeing an increase in GDP. It is also the first time that the government has chosen to fund a war through increased deficits rather than increased taxes (IEP 2013). The experience of these wars on the home front was also very different from previous eras. During WWII, the U.S. military filled its ranks largely through the use of the draft, whereas the military today is all-volunteer; during WWII, approximately 9 percent of the U.S. population went to war, and during the wars in Iraq and Afghanistan, fewer than half of one percent of Americans went to war (Clever & Segal 2012).

OPERATION	2001 AND 2002	2003	2004	2005	2006	2007	2008	2009	2010	2001–2010	2012 Request 2011	2012
Iraq	0	53	75.9	85.5	101.6	131.2	142.1	95.5	71.3	756.2	49.3	17.7
Afghanistan	20.8	14.7	14.5	20	19	36.9	42.1	59.5	93.8	324.4	118.6	113
Domestic security	13	8	3.7	2.1	0.8	0.5	0.1	0.1	0.1	28.5	0.1	0.1
Unallocated	0	5.5	0	0	0	0	0	0	0	5.5	0	0
TOTAL*	33.8	81.1	94.1	107.6	121.4	171	183.3	155.1	165.3	1283.3	168.1	131.7

*Totals may not add due to rounding.
Source: Belasco 2011.

The costs of war are difficult to estimate because it is nearly impossible to disentangle direct and indirect costs, as well as the costs associated with the loss of human life, productivity, and infrastructure. The budgetary costs alone for the wars in Iraq and Afghanistan are estimated at approximately $1 trillion (IEP 2013). In a 2008 book, Stiglitz and Bilmes estimated that when calculating the budgetary costs as well as the cost of future resources, the impact of oil prices on the economy, and numerous other indirect factors, the cost of the Iraq War alone was $3 trillion. At the end of the war in 2010, they updated their estimate to a total of $5 trillion, including long-term costs such as lifetime care for the wounded, equipment replacement, and interest on debt payments (Stiglitz & Bilmes 2010). Veterans' health care and retirement benefits as well as other veterans' services are not included in the defense budget and represented an additional $120 billion in spending in 2012 (NCVAS 2013).

The wars in Iraq and Afghanistan have also taken a major toll within these societies, including the loss of life and the disruption to social and economic development. One study estimated the death toll in Iraq, including civilians and fighters, at the midpoint of the war in 2006 at 650,000 deaths (Burnham et al. 2006). A 2013 study used information from the WikiLeaks documents to check repetition in the reporting of deaths, and found that the total death toll, particularly for civilians, was likely much higher than earlier estimates (Carpenter et al. 2013). The United Nations High Commissioner for Refugees (UNHCR), the United Nations agency that aids refugees, estimates that the "population of concern" originating from Iraq, which includes refugees, asylum seekers, and other displaced people, is more than 3 million people (UNHCR 2013). It is impossible to calculate the long-term economic costs that the disruption and devastation of war causes for millions of people.

Sociological Theories of War

Sociological perspectives can help us understand various aspects of war. In this section, we describe how structural functionalism, conflict theory, and symbolic interactionism can be applied to the study of war.

Structural-Functionalist Perspective

Structural functionalism focuses on the functions that war serves and suggests that war would not exist unless it had positive outcomes for society. We have already noted that war has served to consolidate small autonomous social groups into larger political states. An estimated 600,000 autonomous political units existed in the world at about 1000 B.C. Today, that number has dwindled to fewer than 200.

Another major function of war is that it produces social cohesion and unity among societal members by giving them a "common cause" and a common enemy. For example, in

2005, *Newsweek* began a new feature about everyday American heroes called "Red, White, and Proud." Unless a war is extremely unpopular, military conflict also promotes economic and political cooperation. Internal domestic conflicts between political parties, minority groups, and special interest groups often dissolve as they unite to fight the common enemy. During World War II, U.S. citizens worked together as a nation to defeat Germany and Japan.

In the short term, war may also increase employment and stimulate the economy. The increased production needed to fight World War II helped pull the United States out of the Great Depression. The investments in the manufacturing sector during World War II also had a long-term impact on the U.S. economy. Hooks and Bloomquist (1992) studied the effect of the war on the U.S. economy between 1947 and 1972, and concluded that the U.S. government "directed, and in large measure, paid for a 65 percent expansion of the total investment in plants and equipment" (p. 304). War can also have the opposite effect, however. In a 2005 restructuring of the military, the Pentagon, seeking a "meaner, leaner fighting machine," recommended shutting down or reconfiguring nearly 180 military installations, "ranging from tiny Army reserve centers to sprawling Air Force bases that have been the economic anchors of their communities for generations" (Schmitt 2005), at a cost of thousands of civilian jobs.

Wars also function to inspire scientific and technological developments that are useful to civilians. For example, innovations in battlefield surgery during World War II and the Korean War resulted in instruments and procedures that later became common practice in civilian hospital emergency wards (Zoroya 2006). Research on laser-based defense systems led to laser surgery, research in nuclear fission facilitated the development of nuclear power, and the Internet evolved from a U.S. Department of Defense research project. In the U.S. airline industry, which owes much of its technology to the development of air power by the U.S. Department of Defense, the distinction between military and civilian technology is important because different government agencies regulate its exports. The U.S. Department of Commerce regulates the export of parts produced for use on commercial airlines whereas the Department of State imposes stricter controls on parts produced for military aircraft to prevent sales to countries at odds with U.S. foreign policy objectives (Millman 2008). Today, **dual-use technologies**, a term referring to defense-funded innovations that also have commercial and civilian applications, are quite common. For example, "almost all information technology is dual use. We both use the same operating systems, the same networking protocols, the same applications, and even the same security software" (Schneier 2008, p. 1).

War also serves to encourage social reform. After a major war, members of society have a sense of shared sacrifice and a desire to heal wounds and rebuild normal patterns of life. They put political pressure on the state to care for war victims, improve social and political conditions, and reward those who have sacrificed lives, family members, and property in battle. As Porter (1994) explained, "Since . . . the lower economic strata usually contribute more of their blood in battle than the wealthier classes, war often gives impetus to social welfare reforms" (p. 19). For example, much of today's social welfare benefits for impoverished mothers and children grew out of the benefits programs enacted for the widows and wives of deceased and injured soldiers of the U.S. Civil War (Skocpol 1992).

When members of groups who are discriminated against in society are perceived as sacrificing for the benefit of the dominant group, military policy typically precedes the expansion of equal rights laws in the society at large. Thus, the bravery and sacrifice of African American troops in segregated units during WWII led to the racial integration of the military in the early 1950s, more than a decade before the passage of civil rights legislation in U.S. law. In 2011, as states debated the legal rights of lesbian, gay, and other sexual minorities, the U.S. military formally ended its "Don't Ask, Don't Tell" policy that

> Wars . . . function to inspire scientific and technological developments. . . . Research on laser-based defense systems led to laser surgery, research in nuclear fission facilitated the development of nuclear power, and the Internet evolved from a U.S. Department of Defense research project.

dual-use technologies
Defense-funded technological innovations with commercial and civilian use.

barred LGBT service members from serving openly and living with their partners. Less than a month after the Supreme Court struck down the "Defense of Marriage Act" that barred federal recognition of same-sex marriages authorized by states in 2013, the military announced that the spouses of military personnel and civilian defense employees would be eligible to receive military benefits, regardless of whether they are stationed in a state in which same-sex marriage is outlawed (Hicks 2013) (see Chapter 11).

Finally, the U.S. military has historically provided an alternative for the advancement of poor or disadvantaged groups who otherwise face discrimination or limited opportunities in the formal economy. The military's specialized training, tuition assistance programs for college education, and preferential hiring practices improve the prospects of veterans to find a decent job or career after their service (Military 2007).

Conflict Perspective

Conflict theorists emphasize that the roots of war are often antagonisms that emerge whenever two or more ethnic groups (e.g., Bosnians and Serbs), countries (United States and Vietnam), or regions within countries (the U.S. North and South) struggle for control of resources or have different political, economic, or religious ideologies. In addition, conflict theory suggests that war benefits the corporate, military, and political elites. Corporate elites benefit because war often results in the victor taking control of the raw materials of the losing nations, thereby creating a bigger supply of raw materials for its own industries.

Jacob Silberberg/Getty Images News/Getty Images

Conflict theorists draw attention to the ways in which war benefits elites and is tied to the economic system. Antiwar demonstrations were common occurrences in the 1960s and 1970s, as students demonstrated against the war in Vietnam (1959 to 1975). They erupted again in opposition to the Gulf War (1990 to 1991), the war in Iraq (2013 to 2011), and the war in Afghanistan (2001 to present). There are those who have protested any U.S. intervention in the Syrian Civil War.

Indeed, many corporations profit from defense spending. Under the Pentagon's bid and proposal program, for example, corporations can charge the cost of preparing proposals for new weapons as overhead on their Department of Defense contracts. Also, Pentagon contracts often guarantee a profit to the developing corporations. Even if the project's cost exceeds initial estimates, called a cost overrun, the corporation still receives the agreed-on profit. In the late 1950s, President Dwight D. Eisenhower referred to this close association between the military and the defense industry as the **military-industrial complex**.

Contemporary examples of the military-industrial complex include the former Bush administration's direct link to defense spending. Even as the U.S. government was deciding on the war in Iraq, "many former Republican officials and political associates of the Bush administration [were] associated with the Carlyle Group, an equity investment firm with billions of dollars in military and aerospace assets" (Knickerbocker 2002, p. 2). Further, news media have come to rely heavily on retired military professionals with ties to the defense establishment and military contractors for interpretation of the Iraq and Afghanistan wars. This close "intersection of network news and wartime commerce" blurs the line between security policy and private commercial interests (Barstow 2008, p. 1).

The military elite benefit because war and the preparations for it provide prestige and employment for military officials. For example, Military Professional Resources Inc. (MPRI), an organization staffed by former military, defense, law enforcement, and other professionals, operates in more than 40 countries with U.S. and other government contracts involving military and police training, democracy and governance support, disaster management, and other operations in "post-conflict and transitional environments"

military-industrial complex
A term first used by Dwight D. Eisenhower to connote the close association between the military and defense industries.

(MPRI 2011). According to some estimates, private contractors such as MPRI and many others contributed more than 180,000 civilians to the occupation of Iraq, about 20,000 more than the U.S. military and government employees deployed in country (Miller 2007). Throughout the wars in Iraq and Afghanistan, contractor personnel outnumbered uniformed military personnel (Schwartz & Swain 2011); in 2013, the U.S. military employed over 100,000 civilian contractors in Iraq and Afghanistan (CENTCOM 2013).

Although thousands of these employees perform security roles in combat zones, the vast majority is employed in noncombat support roles such as "KP" or kitchen patrol. In total, 18,366, or approximately 16 percent of all combat zone contractors were security personnel, while 91,224 worked in nonsecurity roles such as interpreters, construction workers, truck drivers, and office administration. U.S. citizens make up less than one-third of these civilian contractors; the majority comes from the local area or other countries (CENTCOM 2013). Although most private contractors work in noncombat roles, many are still at risk of injury and death; at least 430 American contractors were reported killed in Afghanistan in 2011, compared with 418 American soldiers. Because private contractors, particularly those companies based outside of the United States, are not subject to the same strict reporting requirements as the military, the number of contractors killed and injured is likely far higher. The largest American contracting agency is L-3 Communications, which operates MPRI; L-3 has the third highest number of personnel fatalities in Iraq and Afghanistan, after the United States and Britain. The high number of contractor deaths led one law professor to remark, "by continuing to outsource high-risk jobs that were previously performed by soldiers, the military, in effect, is privatizing the ultimate sacrifice" (Nordland 2012).

One private company that provided security personnel to the U.S. military is the North Carolina-based firm Xe—pronounced "z"—which until 2009 was known as Blackwater Worldwide. The company received widespread attention in 2004, when four of its employees in Iraq were killed by a Sunni mob in Fallujah, where their charred corpses were hung along public streets. In September 2007, Blackwater personnel guarding a U.S. diplomatic convoy opened fire at a traffic circle in Baghdad, killing 17 and wounding 24 Iraqi civilians. Company officials initially claimed that their contractors responded proportionately to a nearby attack. In response, and after many years of lodging complaints about alleged indiscriminate firings by private security contractors, the Iraqi government revoked Blackwater's license to operate in Iraq (Tavernise 2007). An FBI investigation of the incident found that most of the killings were in violation of the rules for use of deadly force (Johnston & Broder 2007). The U.S. Department of Justice charged five former Blackwater employees with voluntary manslaughter in January 2009, but the case was dismissed on procedural grounds in December 2009 (CNN 2009). An appeal overruled the original judge's decision that the evidence was tainted, and new charges were brought against four of the guards in October 2013 (Horwitz 2013).

War also benefits the political elite by giving government officials more power. Porter (1994) observed that "throughout modern history, war has been the level by which ... governments have imposed increasingly larger tax burdens on increasingly broader segments of society, thus enabling ever-higher levels of spending to be sustained, even in peacetime" (p. 14). Political leaders who lead their country to a military victory also benefit from the prestige and hero status conferred on them.

Finally, feminists and many other analysts often note the overwhelming association between war and gender. By and large, active combat has historically been carried out by men. Nature-based arguments about gender—i.e., that men are innately aggressive or violent and women inherently peaceful—are not generally supported by social science research and do not adequately explain why men are more likely to kill than women. Feminists emphasize the social construction of aggressive masculine identities and their manipulation by elites as important reasons for the association between masculinity and militarized violence (Alexander & Hawkesworth 2008).

Although some feminists view women's participation in the military as a matter of equal rights, others object because they see war as an extension of patriarchy and the subordination of women in male-dominated societies. Ironically, because protection of women is perceived as a feature of masculine identity, feminists also point out that war and other conflicts are often justified using "the language of feminism" (Viner 2002). For example, former President Bush used respect for women's rights and protection of women subjugated under the Taliban as a partial justification for the attack on Afghanistan in 2001 (Viner 2002).

The recent entry of women into the U.S. armed forces is changing how women's roles in combat are perceived and even how the military conducts operations in war. During the wars in Iraq and Afghanistan, "tens of thousands of American military women have lived, worked, and fought with men for prolonged periods" (Myers 2009, p. 1). Under the 1994 Ground Combat Exclusion Policy, women were barred from serving in units whose primary missions involved ground combat operations. This policy excluded women from approximately 200,000 military positions. The unconventional nature of the wars in Iraq and Afghanistan, however, exposed thousands of men and women in occupations defined by the military as noncombat roles into harm's way. Particularly in Iraq, the use of improvised explosive devices (IEDs) turned roads into the front lines of the war, posing new dangers to traditionally noncombat roles such as truck drivers. According to the Department of Defense, 155 women were killed and 965 were injured in the wars in Iraq and Afghanistan (DCAS 2013).

In both Iraq and Afghanistan, female linguists, intelligence specialists, and military police were needed in combat units to speak with and search Muslim women who traditionally avoid contact with men outside of their families. The realities of nonconventional warfare and the operational needs of the military put increasing pressure on the military to revise its policy that excluded women from combat roles. In January 2013, the Department of Defense announced its plans to rescind the female combat exclusion policy and set a deadline in 2014 for each service to develop a plan for the full integration of women into all military roles, including developing "gender-neutral" standards for each occupation, or to petition for an exemption from the policy (Burrelli 2013).

Despite the many advancements women have achieved in the military, many barriers remain. Women make up approximately 15 percent of the military, but only 7 percent of generals and admirals. Because combat experience is a major factor in military promotion decisions, the integration of women into combat roles may help, over time, to bring more women into the highest leadership positions. The lack of women in high levels of leadership and the male-dominated culture of the military have been blamed for the high rate of sexual assault among women in the military, which is about twice the rate among civilian women. In January 2013, Chairman of the Joint Chiefs of Staff General Dempsey stated that the integration of women into combat units could help to address the sexual assault crisis in the military because "the more we treat people equally, the more they will treat each other equally" (quoted in Portero 2013). Although some high-ranking military leaders have linked the sexual assault crisis to the military culture itself, Americans are more likely to view this problem as stemming from individual deviance. In a nationally representative poll, 54 percent of Americans said that reports of sexual assaults by military personnel represent individual acts of misconduct, while 40 percent said these reports represent underlying problems with military culture (see Table 15.2).

> The realities of nonconventional warfare and the operational needs of the military put increasing pressure on the military to revise its policy that excluded women from combat roles.

Tammy Duckworth, a double-amputee veteran of the Iraq War, was elected to Congress in 2013. Duckworth lost her legs in 2004, when insurgents shot down the helicopter she was piloting. Hundreds of women were killed or injured in the wars in Iraq and Afghanistan, despite being formally barred from serving in combat units.

Andrew Harrer/Bloomberg/Getty Images

What Do You Think? General Dempsey stated that one major cause of the sexual assault crisis in the military is a culture of inequality that has stemmed from the military defining men and women as "separate classes of military personnel" (quoted in Portero 2013). In what ways do you think inequality contributes to a "culture of rape," both in the military and in civilian society (see Chapters 4 and 10)?

Symbolic Interactionist Perspective

The symbolic interactionist perspective focuses on how meanings and definitions influence attitudes and behaviors regarding conflict and war. The development of attitudes and behaviors that support war begins in childhood. American children learn to glorify and celebrate the Revolutionary War, which created our nation. Movies romanticize war, children play war games with toy weapons, and various video and computer games glorify heroes conquering villains.

Symbolic interactionism helps to explain how military recruits and civilians develop a mind-set for war by defining war and its consequences as acceptable and necessary. The word *war* has achieved a positive connotation through its use in various popular public policies—the war on drugs, the war on poverty, and the war on crime. Positive labels and favorable definitions of military personnel facilitate military recruitment and public support of armed forces. In 2005, the Army National Guard launched a $38 million marketing campaign targeting young men and women with advertisements, showing "troops with weapons drawn, helicopters streaking and tanks rolling," all "in an attempt to remind people what the Guard has been about since Colonial Days: fighting wars and protecting the homeland." The new slogan? "The most important weapon in the war on terrorism. You" (Davenport 2005, p. 1).

TABLE 15.2 Do Reports of Sexual Assaults by Military Personnel Indicate Individual Acts of Misconduct, or Underlying Problems with Military Culture?

	Individual acts of misconduct %	Underlying problems with military culture %	Don't know %
Total	54	40	6
Men	56	38	6
Women	51	43	6
18–29	59	37	4
30–49	54	41	5
50–64	53	40	7
65+	47	44	10
Republican	69	25	6
Democrat	44	49	6
Independent	56	41	3
Military household			
Yes	62	33	6
No	51	43	6
Following story			
Very closely	42	52	6
Less closely	56	38	6

*Totals may not add to 100 because of rounding. Poll conducted June 6–9, 2013.
Source: Pew Research Center 2013.

Many government and military officials convince the masses that the way to ensure world peace is to be prepared for war. Patriotism is a popular sentiment in American society. For example, 75 percent of Americans say they display a U.S. flag at home, at the office, on their car, or on their clothing (Pew Research Center 2011a).

Governments may use propaganda and appeals to patriotism to generate support for war efforts and to motivate individuals to join armed forces. Salladay (2003), for example, notes that those in favor of the war in Iraq have commandeered the language of patriotism, making it difficult but necessary for peace activists to use the same symbols or phrases. In their study of the U.S. peace movement, Woehrle et al. (2008) observed that the government and supporters of U.S. wars often framed the issue as "supporting our boys" or "supporting the troops." This made public discussions about whether a particular war is effective or justifiable appear to be betrayals of soldiers. By analyzing public statements from leading peace movement groups that opposed the first Gulf War (1991) and the Iraq War, the researchers documented how opponents of these wars developed a counternarrative that "peace is patriotic" and reframed the issue around how war itself endangered the troops, how government policies failed to provide for the welfare of troops, and how war negatively affected the well-being of civilians.

Governments may use propaganda and appeals to patriotism to generate support for war efforts and to motivate individuals to join armed forces.

To legitimize war, the act of killing in war is not regarded as "murder." Deaths that result from war are referred to as casualties. Bombing military and civilian targets appears more acceptable when nuclear missiles are "peacekeepers" that are equipped with multiple "peace heads." Killing the enemy is more acceptable when derogatory and dehumanizing labels such as Gook, Jap, Chink, Kraut, and Haji convey the attitude that the enemy is less than human.

Such labels are socially constructed as images, often through the media, and are presented to the public. Social constructionists, like symbolic interactionists in general, emphasize the social aspects of "knowing." Thus, Li and Izard (2003) used content analysis to analyze newspaper and television coverage of the World Trade Center and Pentagon attacks on September 11. The researchers examined the first eight hours of coverage of the attacks presented on CNN, ABC, CBS, NBC, and Fox, as well as in eight major U.S. newspapers (including the *Los Angeles Times, The New York Times,* and *The Washington Post*). Results of the analysis indicated that newspaper articles tended to have a "human interest" emphasis, whereas television coverage was more often "guiding and consoling." Other results suggested that both media relied most heavily on government sources, that newspapers and the networks were equally factual, and that networks were more homogeneous in their presentation than newspapers. One indication of the importance of the media lies in former President George W. Bush's creation of the Office of Global Communications—"a huge production company, issuing daily scripts on the Iraq War to U.S. spokesmen around the world, auditioning generals to give media briefings, and booking administration stars on foreign news shows" (Kemper 2003, p. 1).

Causes of War

The causes of war are numerous and complex. Most wars involve more than one cause. The immediate cause of a war may be a border dispute, for example, but religious tensions that have existed between the two combatant countries for decades may also contribute to the war. The following section reviews various causes of war.

Conflict over Land and Other Natural Resources

Nations often go to war in an attempt to acquire or maintain control over natural resources, such as land, water, and oil. Michael Klare, author of *Resource Wars: The New Landscape of Global Conflict* (2001), predicted that wars would increasingly be fought over resources as supplies of the most needed resources diminish. Disputed borders have been common motives for war. Conflicts are most likely to arise when borders are physically easy to cross and are not clearly delineated by natural boundaries, such as major rivers, oceans, or mountain ranges.

In the modern era, oil has been a major resource at the center of many conflicts. Not only do the oil-rich countries in the Middle East present a tempting target in themselves, but war in the region can also threaten other nations that are dependent on Middle Eastern oil. Thus, when Iraq seized Kuwait and threatened the supply of oil from the Persian Gulf, the United States and many other nations reacted militarily in the Gulf War. In the digital era, rare earth minerals needed to make cell phones, computers, and other advanced technologies have come into increasing global demand, leading to fears that control over these resources could be the next major source of conflict. In 2012, the United States, European Union, and Japan, the three biggest consumers of digital products, filed trade grievances with the UN against China—the largest exporter of rare earth minerals—over its decision to restrict exports of these minerals to produce more digital products domestically (Geitner 2012).

Water is another valuable resource that has led to wars. Unlike other resources, water is universally required for survival. At various times, the empires of Egypt, Mesopotamia, India, and China all went to war over irrigation rights. In 1998, five years after Eritrea gained independence from Ethiopia, forces clashed over control of the port city Assab, and with it, access to the Red Sea. In a document prepared for the Center for Strategic and International Studies, Starr and Stoll (1989) warned that soon "water, not oil, will be the dominant resource issue of the Middle East" (p. 1). Despite such predictions, tensions in the Middle East have erupted into fighting repeatedly in recent years—but not over water. In July 2006, Israel and Lebanon fought a border war that killed a thousand people and displaced a million Lebanese. Civil war erupted in the Palestinian territories between rival parties when Hamas seized control of the Gaza Strip in response to Fatah's refusal to hand over the government after Hamas won legislative elections (BBC 2007). Beginning in 2011, pro-democracy and anticorruption protests across the Middle East collectively known as the Arab Spring led to the outbreak of major civil wars in Egypt, Yemen, Libya, and Syria.

Conflict over Values and Ideologies

Many countries initiate war not over resources but over beliefs. World War II was largely a war over differing political ideologies: democracy versus fascism. The Cold War involved the clash of opposing economic ideologies: capitalism versus communism. Conflicts over values or ideologies are not easily resolved. They are less likely to end in compromise or negotiation because they are fueled by people's convictions. For example, when a representative sample of American Jews was asked, "Do you agree or disagree with the following statement? 'The goal of Arabs is not the return of occupied territories but rather the destruction of Israel,'" 76 percent agreed, 19 percent disagreed, and 5 percent were unsure (American Jewish Committee 2011).

If ideological differences can contribute to war, do ideological similarities discourage war? The answer seems to be yes; in general, countries with similar ideologies are less likely to engage in war with each other than countries with differing ideological values (Dixon 1994). Referred to as the **democratic peace theory**, research has shown that democratic nations are particularly disinclined to wage war against one another (Brown et al. 1996; Rasler & Thompson 2005).

Racial, Ethnic, and Religious Hostilities

Racial, ethnic, and religious groups vary in their cultural beliefs, values, and traditions. Thus, conflicts between racial, ethnic, and religious groups often stem from conflicting values and ideologies. Such hostilities are also fueled by competition over land and other scarce natural and economic resources. Gioseffi (1993) noted that "experts agree that the depleted world economy, wasted on war efforts, is in great measure the reason for renewed ethnic and religious strife. 'Haves' fight with 'have-nots' for the smaller piece of the pie that must go around" (p. xviii). Racial, ethnic, and religious hostilities are sometimes perpetuated by a wealthy minority to divert attention away from their exploitations and to maintain their own position of power. Such **constructivist explanations** of ethnic conflict—those that emphasize the role of leaders of ethnic groups in stirring up intercommunal hostility—differ sharply from **primordial explanations**, or those that emphasize the existence of "ancient hatreds" rooted in deep psychological or cultural differences between ethnic groups.

As described by Paul (1998), sociologist Daniel Chirot argued that the recent worldwide increase in ethnic hostilities is a consequence of "retribalization," that is, the tendency for groups, lost in a globalized culture, to seek solace in the "extended family of an ethnic group" (p. 56). Chirot identified five levels of ethnic conflict: (1) multiethnic societies without serious conflict (e.g., Switzerland), (2) multiethnic societies with controlled conflict (e.g., United States and Canada), (3) societies with ethnic conflict that has been resolved (e.g., South Africa), (4) societies with serious ethnic conflict leading to warfare (e.g., Sri Lanka), and (5) societies with genocidal ethnic conflict, including "ethnic cleansing" (e.g., Darfur).

Religious differences as a source of conflict have recently come to the forefront. An Islamic jihad, or holy war, has been blamed for the September 11 attacks on the World Trade Center and Pentagon as well as for bombings in Kashmir, Sudan, the Philippines, Indonesia, Kenya, Tanzania, Saudi Arabia, Spain, and Great Britain. Some claim that Islamic beliefs in and of themselves have led to recent conflicts (Feder 2003). Others contend that religious fanatics, not the religion itself, are responsible for violent

democratic peace theory
A prevalent theory in international relations suggesting that the ideological similarities between democratic nations makes it unlikely that such countries will go to war against each other.

constructivist explanations
Those explanations that emphasize the role of leaders of ethnic groups in stirring up hatred toward others external to one's group.

primordial explanations
Those explanations that emphasize the existence of "ancient hatreds" rooted in deep psychological or cultural differences between ethnic groups, often involving a history of grievance and victimization, real or imagined, by the enemy group.

confrontations and emphasize that misunderstandings between cultural groups can further fuel these tensions. For example, most Americans understand the term *jihad* as radical Muslims have used it to justify violent conflict as a holy war, but many moderate Muslims point out that the more conventional understanding of the term is as a faith-based internal struggle to achieve a life of peace (Bonner 2006).

Conflicts between different sects of the same religion can also lead to long-lasting and devastating wars. The civil war that erupted in Syria in 2011 was driven in large part by the perceived injustice of Sunni's representing the largest segment of the population yet holding a minority of leadership positions within the Assad regime. The conflict between these two sects has deep historical roots, and the conflict in Syria has reignited tensions between Sunni and Shiite populations in Iraq, Iran, and Libya (Arango & Barnard 2013). Wars over differing religious beliefs have led to some of the worst episodes of bloodshed in history, in part, because some religions lend themselves to martyrdom—the idea that dying for one's beliefs leads to eternal salvation. For example, Islamic leader Osama bin Laden claimed that unjust U.S.–Middle East policies are responsible for "dividing the whole world into two sides—the side of believers and the side of infidels" (Williams 2003, p. 18).

> Wars over differing religious beliefs have led to some of the worst episodes of bloodshed in history, in part, because some religions lend themselves to martyrdom—the idea that dying for one's beliefs leads to eternal salvation.

Defense against Hostile Attacks

The threat or fear of being attacked may cause leaders of a country to declare war on the nation that poses the threat. This is an example of what experts in international relations refer to as the **security dilemma**: "[A]ctions to increase one's security may only decrease the security of others and lead them to respond in ways that decrease one's own security" (Levy 2001, p. 7). Such situations may lead to war inadvertently. The threat may come from a foreign country or from a group within the country. After Germany invaded Poland in 1939, Britain and France declared war on Germany out of fear that they would be Germany's next victims. Germany attacked Russia in World War I, in part out of fear that Russia had entered the arms race and would use its weapons against Germany. Japan bombed Pearl Harbor, hoping to avoid a later confrontation with the U.S. Pacific fleet, which posed a threat to the Japanese military.

In 2001, a U.S.-led coalition bombed Afghanistan in response to the September 11 terrorist attacks. Moreover, in March 2003, the United States, Great Britain, and a loosely coupled "coalition of the willing" invaded Iraq in response to perceived threats of weapons of mass destruction and the reported failure of Saddam Hussein to cooperate with United Nations' weapons inspectors. Yet, in 2005, a presidential commission concluded that the attack on Iraq was based on faulty intelligence and that, in fact, "America's spy agencies were 'dead wrong' in most of their judgments about Iraq's weapons of mass destruction" (Shrader 2005, p. 1). As a result, by 2007, many Americans, more than 60 percent, favored a partial or complete withdrawal from Iraq (CNN/Opinion Research Corporation Poll 2007). Many believe that the public's desire for withdrawal from Iraq was a key factor in the outcome of the 2008 U.S. presidential election. The Obama administration withdrew the last remaining combat brigade from Iraq in August 2010.

security dilemma A characteristic of the international state system that gives rise to unstable relations between states; as State A secures its borders and interests, its behavior may decrease the security of other states and cause them to engage in behavior that decreases A's security.

Revolutions and Civil Wars

Revolutions and civil wars involve citizens warring against their own government and often result in significant political, economic, and social change. The difference between a revolution and a civil war is not always easy to determine. Scholars generally agree that revolutions involve sweeping changes that fundamentally alter the distribution of power in society (Skocpol 1994). The American Revolution resulted from colonists revolting against British control. Eventually, they succeeded and established a republic where none existed before. The Russian Revolution involved a revolt against a corrupt, autocratic, and out-of-touch ruler, Czar Nicholas II. Among other changes, the revolution led to wide-scale seizure of land by peasants who formerly were economically dependent on large landowners. More recently, after a long history of civil war and semiautonomous rule, and as part of a 2005 peace deal with northern Sudan, southern Sudan held a referendum and declared independence from northern Sudan. The Republic of South Sudan became the newest member state of the United Nations on July 14, 2011 (Worsnip 2011).

Civil wars may result in a different government or a new set of leaders but do not necessarily lead to such large-scale social change. Because the distinction between a revolution and a civil war depends upon the outcome of the struggle, it may take many years after the fighting before observers agree on how to classify it. Revolutions and civil wars are more likely to occur when a government is weak or divided, when it is not responsive to the concerns and demands of its citizens, and when strong leaders are willing to mount opposition to the government (Barkan & Snowden 2001; Renner 2000).

One of the world's longest-running civil wars came to an end in May 2009. Since 1983, the government of Sri Lanka fought an insurgency led by the Liberation Tigers of Tamil Eelam (LTTE). Also known as the Tamil Tigers, the LTTE were separatist militants who sought to carve an independent state out of the northern and eastern portions of this island country. The war resulted in more than 68,000 deaths (Gardner 2007). The Sri Lankan Army defeated the last remnants of the LTTE and killed their leader in May 2009 (Buncombe 2009). Like many civil wars, the war in Sri Lanka was also a struggle between a majority community (in this case, Sinhalese Buddhists) and a relatively poor and disadvantaged minority community (Hindu Tamils). Despite the end of the war, a political settlement addressing the minority community's grievances and the question of power sharing has not been achieved. Civil wars have also erupted in newly independent republics created by the collapse of communism in eastern Europe, as well as in Rwanda, Sierra Leone, Chile, Uganda, Liberia, and Sudan.

The civil war in Syria has quickly become one of the deadliest in recent decades. Protests against government corruption and the arrests of political dissenters in March 2011 met with a police crackdown that left several protestors dead. Continued protests spurred the Syrian government, led by President Bashar al-Assad, to deploy military force against the protestors. As military and government officials began to defect from the regime and join the side of the rebels, under the banner of the Free Syrian Army, the conflict moved from one of political dissent to a full-scale civil war between two militaries fighting for control of the government. In the summer of 2012, the United Nations formally accused the Assad regime of war crimes, and one year later, the United States pledged military aid to the Free Syrian Army after evidence emerged that the Assad regime was using chemical weapons against civilians (Kaphle 2013).

Nationalism

Some countries engage in war in an effort to maintain or restore their national pride. For example, Scheff (1994) argued that "Hitler's rise to power was laid by the treatment Germany received at the end of World War I at the hands of the victors" (p. 121). Excluded from the League of Nations, punished by the Treaty of Versailles, and ostracized by the world community, Germany turned to nationalism as a reaction to material and symbolic exclusion.

In the late 1970s, Iranian militants seized the U.S. Embassy in Tehran and held its occupants hostage for more than a year. President Carter's attempt to use military forces to free the hostages was not successful. That failure intensified doubts about the United States' ability to use military power effectively to achieve its goals. The hostages in Iran were eventually released after President Reagan took office, but doubts about the strength and effectiveness of the U.S. military still called into question the United States' status as a world power. Subsequently, U.S. military forces invaded the small island of Grenada because the government of Grenada was building an airfield large enough to accommodate major military armaments. U.S. officials feared that this airfield would be used by countries in hostile attacks on the United States. From one point of view, the large-scale "successful" attack on Grenada functioned to restore faith in the power and effectiveness of the U.S. military.

Terrorism

Terrorism is the premeditated use, or threatened use, of violence against civilians by an individual or group to gain a political or social objective (Barkan & Snowden 2001; Brauer 2003; Goodwin 2006). Terrorism may be used to publicize a cause, promote an ideology, achieve religious freedom, attain the release of a political prisoner, or rebel against a government. Terrorists use a variety of tactics, including assassinations, skyjackings, suicide bombings, armed attacks, kidnapping and hostage taking, threats, and various forms of bombings. Through such tactics, terrorists struggle to induce fear within a population, create pressure to change policies, or undermine the authority of a government they consider objectionable. Most analysts agree that, unlike war—where a clear winner is more likely—terrorism is unlikely to be completely defeated:

> There can be no final victory in the fight against terrorism, for terrorism (rather than full-scale war) is the contemporary manifestation of conflict, and conflict will not disappear from earth as far as one can look ahead and human nature has not undergone a basic change. But it will be in our power to make life for terrorists and potential terrorists much more difficult. (Laqueur 2006, p. 173)

Types of Terrorism

Terrorism can be either transnational or domestic. **Transnational terrorism** occurs when a terrorist act in one country involves victims, targets, institutions, governments, or citizens of another country. The 1988 bombing of Pan Am Flight 103 over Lockerbie, Scotland, exemplifies transnational terrorism. The incident took the lives of 270 people, including 35 Syracuse University undergraduates returning from an overseas studies program in London. After a 10-year investigation, Abdel Basset Ali al-Megrahi, a Libyan intelligence agent (CNN 2001), was sentenced to life imprisonment in Scotland for his role in preparing the bomb that brought down the plane. In 2003, the Libyan government agreed to pay $2.7 billion in compensation to the victims' families (Smith 2004). After a diagnosis of terminal cancer, the Scottish government released Megrahi "on compassionate grounds" to return to Libya where he received a hero's welcome organized by the Libyan government (Cowell & Sulzberger 2009).

The 2001 attacks on the World Trade Center, the Pentagon, and Flight 93—the most devastating in U.S. history—are also the deadliest examples of transnational terrorism. Al Qaeda, a global alliance of militant Sunni Islamic groups advocating violence against the Western targets, was also responsible for attacks on U.S. embassies in Kenya and Tanzania (1998) and the bombing of a naval ship, the USS *Cole*, moored in Aden Harbor, Yemen (2000). Al Qaeda has since been linked to deadly bombings in Bali, Indonesia (2002), Madrid (2004), and London (2005). After nearly 20 years of effort to track him down, President Obama announced on May 2, 2011, that Osama bin Laden, the leader of al Qaeda, was dead, killed by Navy Seals and CIA operatives during a raid on a private residential compound in Abbottabad, Pakistan, about 30 miles northeast of Islamabad (Baker et al. 2011). Then Secretary of Defense Leon Panetta later announced that the U.S. government was "within reach" of defeating al Qaeda (Burns 2011). Two years after bin Laden's death, 40 percent of Americans polled responded that they were very or somewhat worried that they or a family member would be a victim of terrorism (see Figure 15.2).

Many groups other than al Qaeda use terrorism to further their own social and political goals. In fact, the U.S. Department of State identifies 51 "foreign terrorist organizations . . . [that] threaten the security of U.S. nationals or the national security (national defense, foreign relations, or the economic interests) of the United States" (Office of the Coordinator for Counterterrorism 2013). In 2012, the Haqqani network, a militant organization

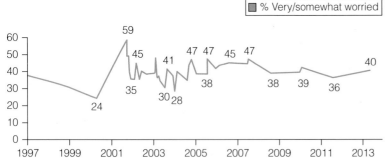

□ % Very/somewhat worried

How worried are you that you or someone in your family will become a victim of terrorism—very worried, somewhat worried, not too worried, or not worried at all?

Figure 15.2 Concern about Terrorism, U.S. Adults Aged 18 and Older, 1997–2013
Source: Gallup Poll 2013.

terrorism The premeditated use or threatened use of violence by an individual or group to gain a political objective.

transnational terrorism Terrorism that occurs when a terrorist act in one country involves victims, targets, institutions, governments, or citizens of another country.

affiliated with the Taliban and operating around the Afghanistan–Pakistan border, was added to this list after a series of attacks on U.S.- and NATO-affiliated targets in Afghanistan. In one attack in September 2011, militants laid siege to the American Embassy and NATO headquarters in Kabul, leading to a battle that lasted 19 hours and resulted in the deaths of 5 Afghan police officers, 6 children, and 5 other Afghan civilians (Healy & Rubin 2011). Members of the Haqqani network, which is based out of the Waziristan region of Pakistan, have been the targets of repeated drone attacks by the United States, one of which killed 16 members of the group in July 2013 (Masood & Mehsud 2013).

Bill Greene/The Boston Globe/Getty Images

The Somali-based group al-Shabaab is another example of a transnational terrorist organization. Added to the State Department's terrorist organization list in 2008, the group, comprised of several thousand members from all over the world, is believed to have at least 50 American-born members who were recruited from Somali-American communities in Minnesota (Ferran et al. 2013). In September 2013, dozens of al-Shabaab militants stormed an upscale shopping mall in Nairobi, Kenya. The mall was frequented by Kenyan elite and many Americans and Europeans visiting or working in Nairobi. Over the course of four days, the militants held hundreds of shoppers hostage as they engaged in a tense standoff with Kenyan military and police forces. By the time the Kenyan forces had retaken the mall, at least 67 civilians had been killed, and 200 injured. The group has sought to establish Shari'a law in Somalia, and claimed that Kenyan interference in Somalian governance motivated the attack (Tharoor 2013).

Domestic terrorism, sometimes called insurgent terrorism (Barkan & Snowden 2001), is exemplified by the 1995 truck bombing of a nine-story federal office building in Oklahoma City, resulting in 168 deaths and the injury of more than 200 people. Gulf War veteran Timothy McVeigh and Terry Nichols were convicted of the crime. McVeigh is reported to have been a member of a paramilitary group that opposes the U.S. government. In 1997, McVeigh was sentenced to death for his actions, and he was executed in 2001 (Barnes 2004). More recently, in 2010, Major Nidal Malik Hasan, an Arab American and a U.S. army psychiatrist, was convicted of killing 13 people in a Fort Hood, Texas and, in 2013, was convicted and sentenced to death. In the same year, two homemade bombs were detonated near the finish line of the Boston Marathon, killing 3 people and injuring 24. Surveillance footage implicated brothers Dzhokhar and Tamerlan Tsarnaev, American citizens who were born in Dagestan, Russia, of carrying out the attack. After a high-speed chase and gun fight with police, Tamerlan Tsarnaev was killed, and 19-year old Dzhokhar was ultimately taken into police custody and charged with carrying out the attacks; he has pleaded not guilty and awaits trial (Zaremba et al. 2013).

The 2004 bombing of a Russian school by Chechen militants, which killed 324 people—nearly half of them children—is also an act of domestic terrorism, as Chechen rebels continue to fight for an independent state. Since 1968, the Basque separatist group ETA has used bombings and assassinations to fight for the political independence of the Basque population from Spain. In January 2011, weakened by hundreds of arrests in recent years, the group declared a "permanent and general" cease-fire (Tremlatt 2011).

Patterns of Global Terrorism

A report by the National Counterterrorism Center (2012) described patterns of terrorism around the world (see Figure 15.3). In 2011:

- There were over 10,000 domestic and international terrorist attacks recorded in 70 countries around the world.
- Over 12,500 people lost their lives as a result of these attacks.

In April 2013, after bombs exploded during the Boston Marathon, SWAT teams searched house to house in Boston looking for accused bomber Dzhokhar Tsarnaev. The manhunt led to a "shelter in place" order for the entire city of Boston, forcing the closure of businesses and schools for nearly 24 hours. The lockdown sparked a debate about the boundaries between public safety and American liberty in the wake of an act of domestic terrorism. Representative Dutch Ruppersberger (D-MD) suggested the lockdown plays into the fears that terrorists hope to foster: "We have to stand up as Americans to this. . . . We've got to continue to go to baseball games, continue to go to events. We can't allow these people to shut us down."

domestic terrorism Domestic terrorism, sometimes called insurgent terrorism, occurs when the terrorist act involves victims, targets, institutions, governments, or citizens from one country.

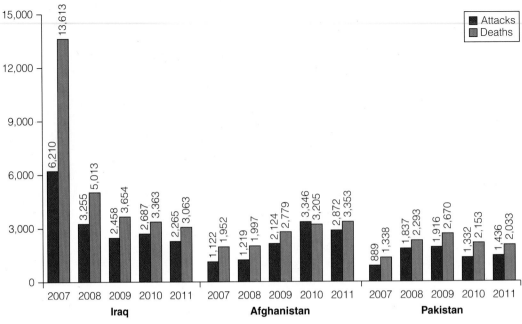

Figure 15.3 Trends in Terrorism Attacks and Deaths in Iraq, Afghanistan, and Pakistan
Source: National Counterterrorism Center 2012.

- More than 75 percent of the world's terrorist attacks and resulting deaths occurred in the Near East or South Asia.
- Africa and the Western Hemisphere experienced five-year highs in the number of attacks.
- Attacks in Afghanistan, Iraq, and Pakistan accounted for 85 percent of terrorist attacks in the Near East and South Asian region and 64 percent of attacks worldwide.
- Although attacks in Afghanistan and Iraq decreased by 14 to 16 percent between 2010 and 2011, attacks in Pakistan increased by 8 percent.

> . . . Americans report that they are more concerned about international terrorism than any other international security matter, including the wars in Afghanistan and Iraq, Iran's and North Korea's nuclear capabilities, and the Israeli-Palestinian conflict.

A poll of randomly selected U.S. adults in the weeks after the Boston Marathon bombing showed that half believed another terrorist attack was very or somewhat likely to occur in the next few weeks, up from 38 percent in 2011 (Saad 2013). In an earlier poll, Americans reported that they were more concerned about international terrorism than any other international security matter, including the wars in Afghanistan and Iraq, Iran's and North Korea's nuclear capabilities, and the Israeli-Palestinian conflict (Morales 2009).

The Roots of Terrorism

In 2003, a panel of terrorist experts came together in Oslo, Norway, to address the causes of terrorism (Bjorgo 2003). Although not an exhaustive list, several causes emerged from the conference:

- A failed or weak state, which is unable to control terrorist operations
- Rapid modernization, when, for example, a country's sudden wealth leads to rapid social change
- Extreme ideologies—religious or secular
- A history of political violence, civil wars, and revolutions
- Repression by a foreign occupation (i.e., invaders to the inhabitants)
- Large-scale racial or ethnic discrimination
- The presence of a charismatic leader

It's Not Right	It's Not Fair	It's Your Fault	You're Evil

Social and Economic Deprivation → Inequality and Resentment → Blame/Attribution → Generalizing/Stereotyping / Dehumanizing/Demonizing the Enemy (Cause)

| Context | Comparison | Attribution | Reaction |

Figure 15.4 The Process of Ideological Development
Source: Borum 2003.

Note that Iraq has several of the characteristics listed here, including rapid modernization (e.g., oil reserves), extreme ideologies (e.g., Islamic fundamentalism), a history of violence (e.g., invasion of Kuwait), large-scale ethnic discrimination (e.g., persecution of Kurdish minority), and a weak state that is unable to control terrorist operations (e.g., the newly elected Iraqi government).

The causes of terrorism listed here, however, are macro in nature. What about social-psychological variables? (See Figure 15.4.) How do individuals choose to join terrorist organizations or use terrorist tactics? Borum (2011) reviews several models of how people move from radicalization into violent extremism (RVE) from academic studies, the military, and law enforcement, and suggests that "understanding the mind-set" of a terrorist can help in the fight against terrorism. The RVE process is very complex, Borum argues, involving social relationship networks, grievances, perceived rewards, experiences with incarceration, and psychosocial vulnerabilities. Borum points out that the vast majority of people with militant extremist beliefs don't engage in violent action, and that counterviolent extremist (CVE) efforts need to focus on more than the "battle of ideas." Rather, CVE efforts should take a multifaceted approach to understanding and attempting to mitigate the grievances, psychosocial vulnerabilities, and social network influences that interact to produce violent extremism.

America's Response to Terrorism

A government can use both defensive and offensive strategies to fight terrorism. Defensive strategies include using metal detectors and X-ray machines at airports and strengthening security at potential targets, such as embassies and military command posts. The Department of Homeland Security (DHS) coordinates such defensive tactics for the U.S. government. DHS was created in 2002, from 22 domestic agencies (e.g., the U.S. Coast Guard, the Immigration and Naturalization Service, and the Secret Service) and, as of 2011, has more than 240,000 employees. With a budget of $53.9 billion in 2012, the mission of DHS is as follows: "We will lead efforts to achieve a safe, secure, and resilient homeland. We will counter terrorism and enhance our security; secure and manage our borders; enforce and administer our immigration laws; protect cyber networks and critical infrastructure; and ensure resilience from disasters . . ." (U.S. Department of Homeland Security 2013, p. 1).

The government's capacity to investigate and counteract terrorism, domestically and internationally, is much larger than the Department of Homeland Security (see this chapter's *Animals and Society* feature). A *Washington Post* investigation of the U.S. counterterrorism industry concluded that nearly 1,300 government organizations and 2,000 private companies are involved in intelligence and counterterrorism (Priest & Arkin 2010). The industry publishes 50,000 intelligence reports each year, and over 854,000 of their employees hold top-secret security clearances. The *Post* report concluded that "The government has built a national security and intelligence system so big, so complex, and so hard to manage, no one really knows if it's fulfilling its most important purpose: keeping its citizens safe" (*Washington Post* 2010, p. 1).

There are monuments to them all over the world but none quite as spectacular as the Animals in War Memorial in England. Opened in 2004, the memorial honors the millions of animals that valiantly "suffered and served":

> . . . to the mules that were silenced for the Burmese jungle by having their vocal cords severed; the donkeys that collapsed under panniers of ammunition; and the dogs that ripped their paws raw digging for survivors, or had half their faces blown off searching for mines but carried on to find more—they are all remembered; the camels and canaries; the elephants and oxen; the messenger pigeons that flew home on one wing; and even the glow worms, by whose gentle light the soldiers read their maps in the first world war. (Price 2004)

Analogous to the Victoria Cross, the most prestigious military award in England, the Dickin Medal honors animals that have shown "conspicuous gallantry or devotion to duty while serving in military conflict" (Treo 2010, p. 1). In 2010, after 62 previous winners—including 32 WWII messenger pigeons, 26 dogs, 3 horses, and a cat—an 8-year-old black Lab named Treo was honored. Treo saved countless military and civilian lives in Afghanistan by locating Taliban-hidden improvised explosive devices (IEDs), many of them in what are called daisy chains—a series of IEDs connected together.

Since 9/11, animals have increasingly been used for guarding U.S. borders and protecting against terrorism. U.S. military forces commonly use dogs because of their keen sense of smell, loyalty, and intelligence. Worldwide, there are presently over 1,300 working K-9 teams, the majority of which are involved in bomb and drug detection (Animals at Arms 2010). However, when the military speaks of "chem dogs," they are referring to dogs that locate weapons of mass destruction. These dogs have such a sophisticated sense of smell that they can detect chemical weapons, or components of chemical weapons, while they are still concealed and before they are released into the atmosphere. Labrador retrievers, Belgian Malinois, and German shepherds, among others, are trained at schools such as the Canine Enforcement Training Center, where each year about 100 certified chemical detector dogs graduate, many of which will become employees of the U.S. Customs and Border Protection (Langley n.d.).

The U.S. Navy, however, is responsible for training two unlikely additions to the military lineup—bottlenose dolphins and California sea lions. The Navy's Marine Mammal Program, relatively unknown until the 1990s, trains nearly 80 dolphins and 40 sea lions for a variety of missions (Larsen 2011; Oliver 2011). Both animals have skills that set them apart from other mammals and from each other. When given a cue by a handler, dolphins use *echolocation* or biosonar to search for and find a particular target. The dolphin then uses a series of clicks that bounce off the target and echo back to the dolphin's jawbone, which then transmits the information to the brain, creating a "mental image of the object" (Animals at Arms 2010; Simon 2003; U.S. Navy 2010). The dolphin then returns to the handler and conveys information about the target to him or her through a series of noises and movements.

Because of this ability, dolphins are also used to detect mines in harbors and other waterways, thereby protecting U.S. citizens and military forces from any number of threats

Here, Brady Rusk, 12, hugs Eli, his older brother's military working dog, at an early retirement and adoption ceremony at Lackland Air Force Base, Texas. Private First Class Colton Rusk's family adopted the black Lab after their son was killed by a sniper in Afghanistan.

U.S. Air Force photo/Tech. Sgt. Bennie J. Davis III

Two recent incidents of top-secret information being publicly leaked have raised questions about the U.S. government's response to both domestic and international terrorism. In 2010, Private Bradley Manning was arrested for leaking videos and over 700,000 pages of documents related to the conduct of the U.S. military in Iraq and Afghanistan to the website WikiLeaks. One video showed a U.S. helicopter crew laughing as they launched an airstrike that killed a dozen civilians (Walker 2013). Manning claimed his objective in leaking the information was to "spark a domestic debate as to the role of the military and foreign policy in general" (PBS NewsHour 2013). Similarly, Edward Snowden, one of the over 850,000 civilian contractors with top-secret clearance, leaked information about the National Security Agency's (NSA) domestic surveillance program, sparking a public debate about the balance between the rights of American citizens to privacy and the government's obligation to ensure public safety. In 2013, Private First Class Manning was convicted of 17 of the 22 charges that were brought against him, although he was cleared of the most serious charge, that of "aiding the enemy" (Pilkington 2013b). Snowden fled to Hong Kong and ultimately Russia where he seeks asylum to avoid U.S. charges of espionage and theft of government property (Reuters 2013).

(Larsen 2011; NPR 2009; Oliver 2011). After locating a mine, dolphins are trained to attach a line to the mine, which is then attached to a buoy that floats to the top of the water, thereby informing military personnel where the mine is located. Dolphins not only detect mines that are tethered off the bottom of the ocean, but they can also locate mines buried deep under the ocean's floor (Animals at Arms 2010, p. 1).

Dolphins also protect ships, submarines, harbors, nuclear power plants, and U.S. borders against any kind of waterborne threat. After the attack on the USS *Cole*, there were fears that enemy swimmers might try to attach explosive devices to other U.S. ships. In response to the attack, the U.S. Navy created the anti-swimmer dolphin system (Frey 2003).

Anti-swimmer dolphins are able to detect and mark combat swimmers with "an illuminated floating beacon" and, if necessary, use their rock-hard bottlenoses to delay the combat swimmer's movements until U.S. military personnel arrive (German 2011, p. 1; Larsen 2011). Dolphins, who comfortably work in a variety of environments, have served "tours of duty" in Vietnam, the Gulf War, Bahrain, and Iraq, and can be deployed anywhere in the world within a 72-hour period (Frey 2003; Larsen 2011). Recently, sea mammals have been used as sentries around Trident submarines in Kings Bay, Georgia, and Bangor, Washington (German 2011).

Sea lions complete tasks similar to those completed by dolphins but do so in a slightly different manner because of their unique skills.

Sea lions can see in the dark as well as underwater, and have underwater directional hearing. Unlike the sheer force that dolphins use to intercept combat swimmers, sea lions use a little finesse:

> The sea lions are trained to detect swimmers or divers approaching military ships or piers. The animals carry a clamp in their mouths. They approach the swimmer quietly from behind and attach the clamp, which is connected to a rope, to the swimmer's leg. With the person restrained, sailors aboard ships can pull the swimmer out of the water. . . . Navy officials say the sea lions, part of the Shallow Water Intruder Detection System program, are so well-trained that the clamp is on the swimmer before he is aware of it. (Leinwand 2003, p. 1)

Sea lions—and to a lesser extent, dolphins—can also be used to retrieve lost materials but, unlike dolphins, sea lions can go ashore. Since 1962, it is estimated that sea lions in the Marine Mammal Program have "recovered millions of dollars of U.S. Naval torpedoes and instrumentation dropped on the sea floor" (Larsen 2011, p. 1). Sea lions are also useful in retrieving expensive military gear dropped during practice exercises. The sea lions locate an item by listening for the acoustic beacon inside the dropped item and then dive for it, holding a metal plate in their teeth. The metal plate is then attached to the dropped equipment and hauled to the water's surface (German 2011).

Dolphins and sea lions are clearly the first choice for the U.S. Navy, but "man's best friend" recently edged out the two sea mammals. In 2011, Operation Neptune Spear, the mission to kill bin Laden, used 79 U.S. Navy SEALS (of the human variety) and a dog to enter Abbottabad, Pakistan. However, only 24 commandos slid down the ropes from the hovering Black Hawk helicopter into bin Laden's compound, and only one had a dog strapped to his back (Frankel 2011a; Jeon 2011). Although details about the dog are as top secret as the identity of the Navy SEALS, it is rumored that, soon after the mission, a Belgian Malinois named Cairo, a member of the Navy SEALS, had a closed-door session with President Obama and the rest of the team (Frankel 2011b).

Like their fellow dolphin and sea lion warriors, dogs form strong relationships with their animal handlers, and vice versa. When Private First Class Colton Rusk, a 20-year-old marine, was killed by a sniper, his black Labrador retriever, Eli, crawled on top of him to protect him from further harm. In the obituary, Eli was listed first as one of the surviving members of the grieving family, and the 3-year-old Lab, after retiring early from the military, was adopted by the Rusks. When Eli arrived at his new home, said mother Kathy Rusk, he ran to Colton's bed, sniffed around, and jumped right in (Bumiller 2011; Roughton 2011).

No machine, no technology, no sophisticated military equipment can do what an animal does or feel what an animal feels.

What Do You Think? Both Bradley Manning and Edward Snowden claimed patriotic motivations for leaking documents about questionable actions of the U.S. government in the fight against terrorism. Some argue that these leaks are acts of whistle-blowing that publicize government corruption and abuse, while others suggest that these leakers are traitors whose actions harm the U.S. fight against terror and endanger the lives of service members. Are the actions of Manning and Snowden those of traitors or heroes? What do you think?

Offensive strategies include retaliatory raids, such as the U.S. bombing of terrorist facilities in Afghanistan, group infiltration, and preemptive strikes. New legislation facilitates such offensive tactics. In October 2001, the USA PATRIOT Act (Uniting and Strengthening America by Providing Appropriate Tools Required to Intercept and Obstruct Terrorism) was signed into law. The act increases police powers both domestically and abroad. Advocates of the PATRIOT Act argue that, during war, some restrictions of civil liberties are necessary. Moreover, the legislation is not "a substantive shift in policy but a mere revitalization of already established precedents" (Smith 2003, p. 25). Critics hold that the act poses a danger to civil liberties. For example, the original act provides for the *indefinite* detention of immigrants if the immigrant group is defined as a "danger to national security" (Romero 2003). In 2007, Congress revised the act to address some of these concerns, particularly to limit the government's authority to conduct wiretaps. Unlike the earlier PATRIOT Act, which had "sunset" clauses mandating review by Congress before renewal, most of the provisions in the new act are now permanent features of the law.

Among the most controversial U.S. policies in the war on terrorism is the indefinite detention of "enemy combatants" at a military prison and interrogation camp in Guantanamo, Cuba. This is the primary detention center for the Taliban or their allies captured in Afghanistan, as well as suspected terrorists, including al Qaeda members from other regions. Since 2002, as many as 779 detainees have been held at "Gitmo" (*New York Times* 2013). The Bush administration argued that because these detainees were not members of a state's army, they were not covered by the **Geneva Conventions**, the principal international treaties governing the laws of war and in particular the treatment of prisoners of war and civilians during wartime. In 2006, the U.S. Supreme Court rejected this argument, ruling that the detainees were subject to minimal protections under the Conventions. Furthermore, in 2008, the Court ruled that the constitution guarantees the right of detainees to challenge their detention in a federal court (Greenburg & de Vogue 2008).

On his second day in office, President Obama ordered the Pentagon to close Guantanamo Bay and all other detention facilities by January 2010. Concerns over security have led many to oppose bringing the detainees to the U.S. mainland for trial and possible incarceration. Nonetheless, in November 2009, the Obama administration announced that Khalid Shaikh Mohammed, the reputed mastermind of the September 11 attacks, would stand trial in a federal courtroom in Manhattan, a few blocks away from the site of the World Trade Center bombings (Savage 2009). In December 2009, the administration announced plans to convert an Illinois state prison to a federal detention center that would house dozens of Guantanamo terrorism suspects (Slevin 2009). In 2011, Congress prohibited the transfer of Guantanamo detainees to U.S. soil. This thwarted the Obama administration's plans to put Khalid Shaikh Mohammed and the other detainees on trial in U.S. courts (Ryan & Khan 2011).

In protest over their treatment and indefinite detention, over 100 prisoners began a hunger strike in February 2013. After months of media attention over treatment of the prisoners, especially the practice of force-feeding striking prisoners, the Obama administration announced in May it would renew efforts to close the prison and transfer low-risk prisoners back to their countries of origin. In July 2013, two Algerian prisoners were approved for repatriation to Algeria (Nakamura & Kenber 2013). Out of the 166 men imprisoned at Guantanamo Bay in 2013, the government listed 46 as "indefinite detainees" i.e., those who are considered too dangerous to release or move to other prisons, and whose cases are considered impossible to try in either military or civilian courts (Pilkington 2013a).

The treatment of detainees at Guantanamo and at secret detention centers (so-called "black sites") around the world has sparked public debate about interrogation techniques used on suspected terrorists. According to documents released by the Department of Justice in 2009, between 2002 and 2005, interrogators with the CIA and U.S. military intelligence used sleep deprivation, extended periods of standing, prolonged exposure to cold and noise, sexual degradation, and "waterboarding" (i.e., simulated drowning), among many other aggressive techniques (Mazzetti & Shane 2009a). The Bush administration and other supporters argued that such "enhanced interrogation techniques" were necessary to extract useful information that protected society from future terrorist attacks. Opponents, including many officials and security experts, claim that these practices: (1) did not elicit useful information, (2) were counterproductive politically, (3) were banned by international treaties to which the United States is obligated, and (4) constituted torture. Congress and the White House have since banned these practices. In August 2009, U.S. Attorney General Eric Holder appointed a federal prosecutor to investigate alleged abuses by the CIA and to consider whether or not a full criminal investigation was warranted (Mazzetti & Shane 2009b). The probe ended in 2011 without widespread criminal investigations.

Geneva Conventions A set of international treaties that govern the behavior of states during wartime, including the treatment of prisoners of war.

Public opinion on this topic is mixed. In response to a question about whether "torture to gain important information from suspected terrorists is justified," 48 percent of American adults said that they believed torture is "often" or "sometimes" justified, whereas 47 percent believed that torture is "rarely" or "never" justified (Pew Research Center 2009). Republicans were twice as likely as Democrats to say that torture was "sometimes" justified (49 percent versus 24 percent), and Democrats were nearly three times more likely than Republicans to say that torture was "never" justified (38 percent versus 14 percent).

Combating terrorism is difficult, and recent trends will make it increasingly problematic (Strobel et al. 2001; Zakaria 2000). First, hackers who illegally gain access to classified information can easily acquire data stored on computers. For example, interlopers obtained the fueling and docking schedules of the USS *Cole*. Second, the Internet permits groups with similar interests, once separated by geography, to share plans, fund-raising efforts, recruitment strategies, and other coordinated efforts. Worldwide, thousands of terrorists keep in touch through e-mail, and virtually all terrorist groups maintain websites for recruitment, fund-raising, internal communication, and propagandizing (Weimann 2006). Third, globalization contributes to terrorism by providing international markets where the tools of terrorism—explosives, guns, electronic equipment, and the like—can be purchased. Finally, fighting terrorism under guerrilla warfare-like conditions is increasingly a concern. Unlike terrorist activity, which targets civilians and may be committed by lone individuals, **guerrilla warfare** is committed by organized groups opposing a domestic or foreign government and its military forces.

The possibility of terrorists using weapons of mass destruction is the most frightening scenario of all and, as stated earlier, the motivation for the 2003 war with Iraq. **Weapons of mass destruction (WMD)** include chemical, biological, and nuclear weapons. Anthrax, for example, although usually associated with diseases in animals, is a highly deadly disease in humans and, although preventable by vaccine, has a "lethal lag time." In a hypothetical city of 100,000 people, delaying a vaccination program one day would result in 5,000 deaths; a delay of six days would result in 35,000 deaths. In 2001, trace amounts of anthrax were found in several letters sent to media and political figures, resulting in five deaths and the inspection and closure of several postal facilities (Baliunas 2004). Despite widespread speculation that al Qaeda or Saddam Hussein was responsible for the attacks, investigators soon began to suspect that the source was domestic. In 2008, shortly after the FBI informed Bruce Ivins, a microbiologist at a U.S. Army laboratory in Fort Detrick, Maryland, that they intended to charge him with the crime, the scientist committed suicide (Associated Press 2008). In 2004, the poison ricin was detected on a mail-opening machine in Senate Majority Leader Bill Frist's Washington, DC, office (Associated Press 2005).

Despite the understandable concerns that Americans have about the potential of terrorist attacks, it is important to remember that death by terrorism is an extraordinarily rare occurrence worldwide, and especially in American society.

The use of chemical weapons in the Syrian civil war led to tensions and near military conflict between the United States and the Russian-supported Syrian regime in September 2013. After evidence emerged that the Syrian government had used chemical weapons banned by international law against civilians in the summer of 2013, the Obama administration threatened military strikes. After weeks of tense negotiations, the Russian government brokered a deal between the Obama administration and the Syrian government to eradicate all chemical weapons stockpiled by the Syrian government in order to prevent a strike by the United States military. By October, international mediators assured world leaders that the Syrian stockpile was being eradicated (Rapoza 2013). Although many praised the diplomatic effort between the United States and Russia for averting military action, others criticized both countries for demanding the eradication of chemical weapons in Syria when both countries have large stockpiles of chemical weapons of their own. Although each signed a treaty in 1994 to destroy their chemical weapons arsenal by 2012; as of 2013, Russia has an estimated 16,000 metric tons of chemical weapons, while the United States holds an estimated 3,000 metric tons

guerrilla warfare Warfare in which organized groups oppose domestic or foreign governments and their military forces; often involves small groups of individuals who use camouflage and underground tunnels to hide until they are ready to execute a surprise attack.

weapons of mass destruction (WMD) Chemical, biological, and nuclear weapons that have the capacity to kill large numbers of people indiscriminately.

TABLE 15.3 Overall Combating WMD Desired End States

The U.S. Armed Forces, in concert with other elements of U.S. national power, deter WMD use.

The U.S. Armed Forces are prepared to defeat an adversary threatening to use WMD and prepared to deter follow-up use.

Existing worldwide WMD are secure, and the U.S. Armed Forces contribute, as appropriate, to secure, reduce, reverse, or eliminate them.

Current or potential adversaries are dissuaded from producing WMD.

Current or potential adversaries' WMD are detected and characterized, and elimination is sought.

Proliferation of WMD and related materials to current and/or potential adversaries is dissuaded, prevented, defeated, or reversed.

If WMD are used against the United States or its interests, the U.S. Armed Forces are capable of minimizing the effects in order to continue operations in a WMD environment and assist U.S. civil authorities, allies, and partners.

The U.S. Armed Forces assist in attributing the source of an attack, respond decisively, and/or deter future attacks.

Allies, partners, and U.S. civilian agencies are capable partners in combating WMD.

Source: Moroney et al. 2009. From Moroney, Jennifer D.P. et al. "Building Partner Capacity to combat Weapons of Mass Destruction." Copyright 2009 Rand National Defense Institute. Reprinted with permission.

(Taylor 2013). Table 15.3 lists the U.S. government's military goals in combating WMD, including stopping their proliferation, securing existing stockpiles, and working with allies to prevent the use of WMD by potential adversaries (Moroney et al. 2009).

Despite the understandable concerns that Americans have about the potential of terrorist attacks, it is important to remember that death by terrorism is an extraordinarily rare occurrence worldwide, and especially in American society. There has been no major terrorist event in the United States since the attacks of 2001, and as one analyst observed, except for 2001, more Americans have died struck by lightning than by international terrorists (Mueller 2006, p. 13). Nonetheless, fears about terrorist attacks continue to be widespread. For example, in a poll taken the day after the announcement of Osama bin Laden's death during a U.S. raid on his compound, 62 percent said they thought "an act of terrorism is either 'very' or 'somewhat likely' to occur in the U.S. in the next few weeks" (Saad 2011, p. 1).

Social Problems Associated with Conflict, War, and Terrorism

Social problems associated with conflict, war, and terrorism include death and disability; rape, forced prostitution, and displacement of women and children; social-psychological costs; diversion of economic resources; and destruction of the environment.

Death and Disability

Many American lives have been lost in wars, including 53,000 in World War I, 292,000 in World War II, 34,000 in Korea, and 47,000 in Vietnam (Leland & Oboroceanu 2010). As of July 2013, 6,675 U.S. military personnel have been killed in the wars in Iraq and Afghanistan, and 50,958 have been wounded. Many civilians and enemy combatants also die or are injured in war. Despite over 100,000 Iraqi civilian deaths, many Americans are unaware of this tremendous loss of life. In a program the Pentagon developed, American reporters were "embedded" into U.S. military units to provide "journalists with a detailed understanding of military culture and life on the frontlines" (Lindner 2009, p. 21). One of the by-products of this program, however, was that 90 percent of the stories by embedded journalists were written from the perspective of American soldiers, focusing "on the horrors facing the troops, rather than upon the thousands of Iraqis who died" (p. 45).

The impact of war and terrorism extends far beyond those who are killed. Many of those who survive war incur disabling injuries or contract diseases. For example, in South Sudan alone, over 4,200 people were killed or wounded by land mines following the end of civil war with northern Sudan in January 2005 (UNMAO 2011). In 1997, the Mine Ban Treaty, which requires that governments destroy stockpiles within 4 years and clear land mine fields within 10 years, became international law. To date, 156 countries have signed the agreement; 36 countries remain, including China, India, Israel, Russia, and the United States (ICBL 2013). War-related deaths and disabilities also deplete the labor force, create orphans and single-parent families, and burden taxpayers who must pay for the care of orphans and disabled war veterans (see Chapter 2 for a discussion of military health care).

The killing of unarmed civilians is also likely to undermine the credibility of armed forces and make their goals more difficult to defend. In Iraq, for example, the events in Haditha, a city in western Iraq, became international news, outraged Iraqis, and led to intense condemnation of the U.S. mission. In November 2005, after a roadside bomb killed one marine and wounded two others, "Marines shot five Iraqis standing by a car and went house to house looking for insurgents, using grenades and machine guns to clear houses" (Watkins 2007, p. 1). Twenty-four Iraqis were killed, many of them women and children, "shot in the chest and head from close range" (McGirk 2006, p. 3). After *Time* magazine broke the story in March 2006 (McGirk 2006), the military investigated and reversed its claim that the civilians died as a result of the roadside bomb. Eight marines were accused of wrongdoing during the incident; charges against six of them were dismissed, and one was cleared of all charges (Puckett & Faraj 2009; Reuters 2008). The remaining defendant was acquitted in 2012 (Savage & Bumiller 2012).

Lastly, individuals who participate in experiments for military research may also suffer physical harm. U.S. representative Edward Markey of Massachusetts identified 31 experiments dating back to 1945, in which U.S. citizens were subjected to harm from participation in military experiments. Markey charged that many of the experiments used human subjects whom were captive audiences or populations considered "expendable," such as elderly individuals, prisoners, and hospital patients. Eda Charlton of New York was injected with plutonium in 1945. She and 17 other patients did not learn of their poisoning until 30 years later. Her son, Fred Schultz, said of his deceased mother:

> I was over there fighting the Germans who were conducting these horrific medical experiments . . . at the same time my own country was conducting them on my own mother. (Miller 1993, p. 17)

Rape, Forced Prostitution, and the Displacement of Women and Children

Half a century ago, the Geneva Convention prohibited rape and forced prostitution in war. Nevertheless, both continue to occur in modern conflicts.

Before and during World War II, Japanese officials forced between 100,000 and 200,000 women and teenage girls into prostitution as military "comfort women." These women were forced to have sex with dozens of soldiers every day in "comfort stations." Many of the women died as a result of untreated sexually transmitted diseases, harsh punishment, or indiscriminate acts of torture.

Since 1998, Congolese government forces have fought Ugandan and Rwandan rebels. Women have paid a high price for this civil war, in which gang rape is "so violent, so systematic, so common . . . that thousands of women are suffering from vaginal fistula, leaving them unable to control bodily functions and enduring ostracism and the threat of debilitating health problems" (Wax 2003, p. 1). Though much less common than violence against women, aid workers also see increasing incidents of rape and sexual violence against men as "yet another way for armed groups to humiliate and demoralize Congolese communities into submission" (Gettleman 2009, p. 1). United Nations officials called the situation in Congo "the worst sexual violence in the world" (Gettleman 2008, p. 1).

Feminist analyses of wartime rape emphasize that the practice reflects not only a military strategy but also ethnic and gender dominance. For example, Refugees International, a humanitarian aid group, reports that rape is "a systematic weapon of ethnic cleansing" against Darfuris and is "linked to the destruction of their communities"

Safin Hamed/AFP/Getty Images

Nearly 2 million Syrians have become refugees of the ongoing civil war, many of them fleeing to Jordan, Egypt, and Turkey seeking safety. Some refugee camps become as large as major cities; the Za'atari camp in Jordan sprouted up in July 2012, when 450 Syrians crossed the border to escape the civil war. By July 2013, Za'atari was home to 120,000 Syrians (UNHCR 2013).

A child soldier in Liberia points his gun at a cameraman while carting a teddy bear on his back. Although reliable figures are hard to obtain, the UN estimates that about 300,000 child soldiers are fighting in wars worldwide.

(Boustany 2007, p. 9). Under Darfur's traditional law, prosecution of rapists is nearly impossible: Four male witnesses are required to accuse a rapist in court, and single women risk severe corporal punishment for having sex outside of marriage.

War and terrorism also force women and children to flee to other countries or other regions of their homeland. For example, as a result of the civil war in Syria, millions of children have been "... killed, maimed, and denied access to food and medicine" (Uenuma 2013, p. 1). Refugee women and female children are particularly vulnerable to sexual abuse and exploitation by locals, members of security forces, border guards, or other refugees. Wars are also dangerous for the very young—"Nine out of ten countries with the highest under-5 mortality rates are experiencing, or emerging from, armed conflict." These include Sierra Leone, Angola, Afghanistan, Niger, Liberia, Somalia, Mali, Chad, and the Democratic Republic of the Congo (Save the Children 2007, p. 13). Despite the *Convention on the Rights of the Child* (UNICEF 2011), which was enacted to protect children from the effects of armed conflict, it is estimated that, worldwide, 40 million children of primary school age—nearly one in three of all children in war zones—are prevented from attending school because of armed conflict (Save the Children 2009).

Social-Psychological Costs

Terrorism, war, and living under the threat of war disrupt social-psychological well-being and family functioning. For example, Myers-Brown et al. (2000) report that Yugoslavian children suffered from depression, anxiety, and fear as a response to conflicts in that region, emotional responses not unlike those Americans experienced after the events of 9/11 (NASP 2003). More recently, as a result of the war, there is evidence that many Iraqi children suffer from everything from "nightmares and bedwetting to withdrawal, muteness, panic attacks, and violence towards other children, sometimes even to their own parents" (Howard 2007, p. 1). Further, a study of children in postwar Sierra Leone found that over 70 percent of boys and girls whose parents had been killed were at "serious risk" of suicide (Morgan & Behrendt 2009).

Guerrilla warfare is particularly costly in terms of its psychological toll on soldiers. In Iraq, soldiers were repeatedly traumatized as "guerrilla insurgents attack[ed] with impunity," and death was as likely to come from "hand grenades thrown by children, [as] earth-rattling bombs in suicide trucks, or snipers hidden in bombed-out buildings" (Waters 2005, p. 1). In 2008, the U.S. Army estimated that for every 100,000 soldiers deployed to Iraq, 20.2 committed suicide compared to 12 suicides per 100,000 soldiers who were not deployed to Iraq (Alvarez 2009; U.S. Army Medical Command 2007). This was the first time since the Vietnam War that the suicide rate for U.S. soldiers surpassed that for civilians. The suicide rate reached "epidemic" proportions by 2013, with approximately 22 veteran suicides a day (Haiken 2013). When reservists are included in the totals with active-duty personnel, more military members committed suicide than died in combat in Iraq and Afghanistan (Donnelly 2011) (see this chapter's *Social Problems Research Up Close*).

The suicide rate reached "epidemic" proportions by 2013, with approximately 22 veteran suicides a day. When reservists are included in the totals with active-duty personnel, more military members committed suicide than died in combat in Iraq and Afghanistan.

Not long after the invasion of Iraq in 2003, the suicide rate among military personnel and veterans began to increase dramatically. The wars in Iraq and Afghanistan required the military to deploy its troops to war zones more often and for longer periods of time than ever before. In addition, the nature of the counterinsurgency operations in these war zones introduced new sources of stress, as military personnel had to balance humanitarian outreach with war fighting. The suicide rate among military personnel increased from 2005 to 2009, plateaued in 2010 and 2011, and spiked again in 2012. The unconventional nature of the wars in Iraq and Afghanistan, and the strain associated with repeated and prolonged deployments led many to speculate that the growing suicide rate was linked with the stresses of deployment and combat.

This troubling pattern led Dr. Cynthia LeardMann and colleagues, research scientists at the Naval Health Research Center (NHRC), to launch a long-term study with the goal of understanding whether or not there is a link between combat experience and risk of suicide. The information gained from this research could help military leaders and policy makers understand what kind of organizational changes and outreach efforts might help prevent future suicides.

Sample and Methods

Beginning in July 2001, the NHRC launched the Millennium Cohort Study (MCS), a longitudinal study of over 150,000 randomly selected military personnel. Every three years, the study participants completed a questionnaire detailing their physical, mental, and functional health. These surveys were collected among both active-duty personnel and those who had left the service since the study

began. Using the names and Social Security numbers of the study participants over an 11-year period, the researchers matched the MCS records with records from the National Death Index and the Defense Medical Mortality Registry; they found that 83 of the 151,560 study participants had committed suicide over the previous decade. The researchers also matched the survey records of the study participants with their DOD service records, indicating the number and length of their deployments and whether they served in combat areas. The MCS survey also asked respondents to report whether they had ever witnessed or been exposed to specific traumatic events, such as witnessing a death, torture, or seeing maimed or decomposing bodies. By matching these records together and employing a statistical hazard ratio model, the researchers were able to determine which factors were associated with risk of suicide.

Findings and Conclusions

Contrary to expectations, the researchers did not find that deployments, combat, or traumatic experiences in war were risk factors for suicide. Instead, the risk factors for suicide among the military population were identical to those among the civilian population: Those who were male, reported alcohol abuse, and/or had a history of mental illness, such as depression or manic-depressive disorder, were at a higher risk of suicide. The frequency, length, and nature of deployment experiences were not associated with suicide. The research team concludes that:

> The findings from this study do not support an association between deployment or combat with suicide; rather they are consistent with previous research indicating that men-

tal health problems increase suicide risk. Therefore, knowing the psychiatric history, screening for mental and substance use disorders, and early recognition of associated suicidal behaviors combined with high-quality treatment are likely to provide the best potential for mitigating suicide risk.

The results of this research help to focus the attention of military leaders on areas where they can help prevent future suicides: by improving substance abuse and mental health resources for soldiers and veterans. However, some questions remain. If the risk of suicide is not linked to deployments and combat, then why did the rate of suicide increase so dramatically during the most intense years of the wars in Iraq and Afghanistan? One of the study investigators, Dr. Nancy Crum-Cianflone suggests that "perhaps it's not being deployed so much as being in a war during a high-stress time period." Other experts also emphasize the link between wartime stressors and mental health problems; Kim Ruocco, the director for TAPS—the military's Tragedy Assistance Program for Survivors—says, "We so often just link military suicide to combat trauma. But there are many others: long hours, separation from support systems, sleeplessness. All are stressors. All add to increases in mental health issues" (Dao 2013).

The findings of the study indicate that the increased suicide rate was a result of an increase in mental health and substance abuse problems among military members during the war years, and point to the necessity of improving mental health and substance abuse resources for all service members and veterans, regardless of whether they have served in combat.

Source: LeardMann et al. 2013.

Military personnel who engage in combat and civilians who are victimized by war may experience a form of psychological distress known as **post-traumatic stress disorder (PTSD)**, a clinical term referring to a set of symptoms that can result from any traumatic experience, including crime victimization, rape, or war. Symptoms of PTSD include sleep disturbances, recurring nightmares, flashbacks, and poor concentration (NCPSD 2007). For example, Canadian Lieutenant-General Roméo Dallaire, head of the United Nations peacekeeping mission in Rwanda, witnessed horrific acts of genocide. For years after his return, he continued to have images of "being in a valley at sunset, waist deep in bodies, covered in blood" (quoted in Rosenberg 2000, p. 14). PTSD is also associated with other personal problems, such as alcoholism, family violence, divorce, and suicide.

post-traumatic stress disorder (PTSD) A set of symptoms that may result from any traumatic experience, including crime victimization, war, natural disasters, or abuses.

Estimates of PTSD vary widely, although they are consistently higher among combat versus noncombat veterans. In a telephone study of 1,965 military personnel who had been deployed to Iraq and Afghanistan, 14 percent reported current symptoms of PTSD, 14 percent reported current symptoms of major depression, and 9 percent reported symptoms consistent with both conditions (Tanielian et al. 2008). The Department of Veterans Affairs estimates that PTSD afflicts 31 percent of Vietnam veterans, 11 percent of Afghan war veterans, and 20 percent of Iraqi war veterans (National Institute of Health 2009). According to one PTSD survivor:

> Post-traumatic stress disorder (PTSD) isn't something you just get over. You don't go back to being who you were. It's more like a snow globe. War shakes you up, and suddenly all those pieces of your life—muscles, bones, thoughts, beliefs, relationships, even your dreams—are floating in the air out of your grip. They'll come down. I'm here to tell you that, with hard work, you'll recover. But they'll never come down where they once were. You're a changed person after combat. Not better or worse, just different. (Montalván 2011, p. 50)

Figure 15.5 describes the types of traumas to which Afghanistan and Iraq war veterans were exposed.

The rate of PTSD among soldiers is difficult to measure for several reasons. First, there is a lag, often of several years, between the time of exposure to trauma and the manifestation of symptoms. Second, many are reluctant to seek help because of the social stigma associated with being labeled as having mental illness (Mittal et al. 2013). When they do seek help, army officials say the Department of Veterans Affairs (VA) is slow to respond (Dao 2009). Troubled by VA policies that required veterans to prove a specific stressor event that caused PTSD symptoms, in 2009, VA Secretary Eric Shinseki announced expanded eligibility for VA benefits for veterans suffering from PTSD symptoms (Friedman 2013). See this chapter's *The Human Side* for a touching description of one man's efforts to deal with PTSD.

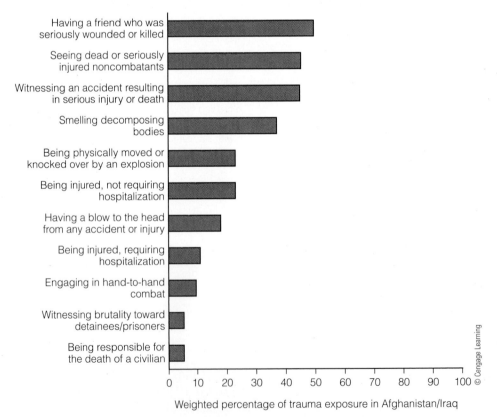

Weighted percentage of trauma exposure in Afghanistan/Iraq

Figure 15.5 Trauma Exposures Reported by Afghanistan/Iraq Service Members

Luis Montalván, a 17-year veteran and former captain in the U.S. Army, had courage to spare. He had won a Purple Heart, two Bronze Stars, and a Combat Action Badge, and yet his toughest challenge was yet to come. Returning from Iraq with PTSD and traumatic brain injury, he was unable to function and turned to the bottle for help, isolating himself from the world. But then he met Tuesday, a beautiful, sensitive golden retriever who had been trained as a service dog but who had suffered his own losses. In this excerpt, Luis and Tuesday go on a date with a beautiful woman:

I let my date step up first . . . then stepped into the small entryway with Tuesday.

"No dogs," the bus driver barked.

"Oh, this is my service dog," I said with a smile, expecting her to let me pass. . . .

She looked Tuesday over, her lips pursed. "That's not a service dog. . . ."

"Yes, Tuesday is my service dog. See his vest. See my cane."

"Service dog don't wear a vest like that. Service dog has a big handle you hold onto."

"That's a guide dog for the blind," I said, trying to hold myself together. "This is a service dog for the disabled."

"Sir, I know a service dog when I see a service dog, and that ain't no service dog."

I pulled out my cell phone. "Then call the cops," I said angrily, "because I am not getting off this bus."

I looked the bus driver straight in the eye. . . .

"Please," I said quietly. "I'm on a date. Please let me on."

"No sir," she said loudly, trying to embarrass me.

"Then I'm calling the police," I said angrily, "because you are violating my rights. I hope you are ready to explain to your boss why you wouldn't let a disabled person on your bus."

She gave me a nasty look, waiting for thirty seconds to see if I would back down, then let me pass with a grunt. . . .

"Are you all right?"

I took a deep breath and petted the back of Tuesday's head. . . .

"I'm sorry."

"Don't be," I said. I looked at her. Smart, beautiful, understanding. She smiled, patted me on the arm, and . . .

"That ain't no service dog."

I looked up. It was the bus driver. She was talking to a woman in the first seat, presumably a friend, but she was intentionally talking loud enough for the whole bus to hear.

Keep it together, Luis. "I think you'll like this restaurant. . . ."

"I've been driving this bus a long time," the bus driver continued, clearly trying to embarrass me. "I know service dogs."

My mind was crumbling. "I think you'll um, I think you'll like . . ."

"Service dog's got a handle."

She was like the voice of PTSD always playing inside my head, always bringing up betrayals.

"Ain't no service dog. I know a service dog."

She was harassing me and wouldn't stop.

"He thinks I don't know a service dog. I know a service dog."

"I'm not deaf," I said in a raised voice. "That's not my disability."

Some of the other passengers laughed. Tuesday turned and nuzzled me with his snout. I grabbed him around the neck, and he leaned into my chest. . . .

"Sorry about the dog," she said sarcastically to the people at the next stop. "Man says it's a service dog."

I went inside myself. I held Tuesday and tried to beat down the anger. I could feel a migraine coming, but I pushed it away. *Just a few hours*, I thought. *A few hours and it will be over.* (pp. 179–181)

Source: Montalván 2011.

Diversion of Economic Resources

As discussed earlier, maintaining the military and engaging in warfare require enormous financial capital and human support. In 2012, world military expenditures totaled $1.76 trillion, or about $249 for each person in the world (SIPRI 2013).

The decision to spend $567 million, equal to the operating cost of the Smithsonian Institution (Center for Defense Information 2003), for one Trident II D-5 missile is a political choice. Similarly, allocating $2.3 billion for a "Virginia" attack submarine while our schools continue to deteriorate is also a political choice. Between 2001 and 2011, taxpayers paid $1.28 trillion for the cost of the wars in Iraq and Afghanistan, the equivalent of providing (1) one year of low-income health care for 259 million people, or (2) a year's salary for over 19 million elementary school teachers, or (3) one year of Head Start for 165 million children, or (4) 285 million households with renewable electricity (solar photovoltaic), or (5) one year of scholarships for 159 million university students (National Priorities Project 2011). Proposed expenditures for the 2014 fiscal year include more money for national defense than for justice, transportation, veterans' benefits, and natural resources and the environment combined (Office of Management and Budget 2013a). (See Figure 15.6.)

Destruction of the Environment

The environmental damage that occurs during war devastates human populations long after war ends. As the casings for land mines erode, poisonous substances—often carcinogenic—leak into the ground (UNAUSA 2004). In 1991, during the Gulf War, Iraqi

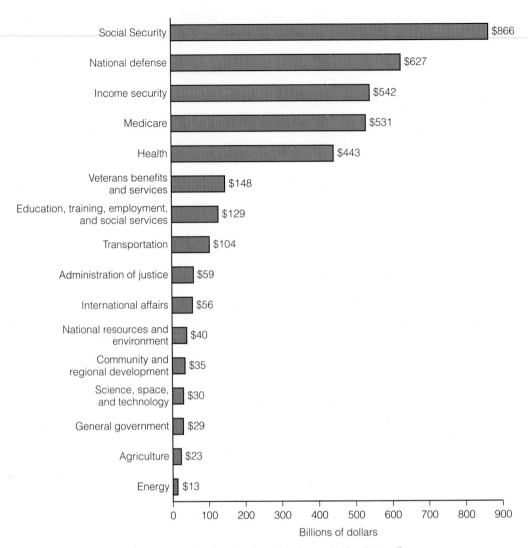

Figure 15.6 Selected Federal U.S. Outlays by Function for 2014 (estimated)
Source: Office of Management and Budget 2013b.

troops set 650 oil wells on fire, releasing oil, which covered the surface of the Kuwaiti desert and continues to seep into the ground, threatening underground water supplies. The smoke from the fires that hung over the Gulf region for eight months contained soot, sulfur dioxide, and nitrogen oxides—the major components of acid rain—and a variety of toxic and potentially carcinogenic chemicals and heavy metals. The U.S. Environmental Protection Agency estimates that, in March 1991, about 10 times as much air pollution was being emitted in Kuwait as by all U.S. industrial and power-generating plants combined (Environmental Media Services 2002; Funke 1994; Renner 1993).

Combatants often intentionally exploit natural resources to fuel their efforts. The local elephant population was heavily depleted during the civil war in southern Angola. The rebel group UNITA killed the animals to trade ivory for money to buy weapons. In addition, many elephants were killed or fatally crippled by land mines planted by the guerrillas, though scientists report that, remarkably, the animals seemed to have learned to avoid mined areas (Marshall 2007). Others have been fitted with prosthetic devices for lost limbs. Between 1992 and 1997, UNITA also earned $3.7 billion to support its fighting from the sale of "conflict" or "blood" diamonds (GreenKarat 2007). Depending upon the location of diamonds, mining can be highly destructive to riverbed ecosystems or to areas surrounding open-pit mines.

The ultimate environmental catastrophe facing the planet is thermonuclear war. Aside from the immediate human casualties, poisoned air, poisoned crops, and radioactive rain, many scientists agree that the dust storms and concentrations of particles created by a massive exchange of nuclear weapons would block vital sunlight and lower temperatures in the northern hemisphere, creating a **nuclear winter**. In the event of large-scale nuclear war, most living things on earth would die. For example, a nuclear blast and the resulting blast wave create overpressure—the amount of pressure in excess of ordinary atmospheric levels as measured by pounds per square inch (psi). As described in a report sponsored by the U.S. Air Force:

- At 20 psi of overpressure, even reinforced-concrete buildings are destroyed.
- Ten psi will collapse most factories and commercial buildings, as well as wood-frame and brick houses.
- Five psi flattens most houses and lightly constructed commercial and industrial structures.
- Three psi suffices to blow away the walls of steel-frame buildings.
- Even 1 psi will produce flying glass and debris sufficient to injure large numbers of people. (Ochmanek & Schwartz 2008, p. 6)

The fear of nuclear war has greatly contributed to the military and arms buildup, which, ironically, also causes environmental destruction even in times of peace. For example, in practicing military maneuvers, the armed forces demolish natural vegetation, disturb wildlife habitats, erode soil, silt up streams, and cause flooding. Bombs exploded during peacetime leak radiation into the atmosphere and groundwater. From 1945 to 1990, 1,908 bombs were tested—that is, exploded—at more than 35 sites around the world. Although underground testing has reduced radiation, some radioactive material still escapes into the atmosphere and is suspected of seeping into groundwater.

Finally, although arms control and disarmament treaties of the last decade have called for the disposal of huge stockpiles of weapons, no completely safe means of disposing of weapons and ammunition exist. Many activist groups have called for placing weapons in storage until safe disposal methods are found. Unfortunately, the longer the weapons are stored, the more they deteriorate, increasing the likelihood of dangerous leakage. In 2003, a federal judge gave permission, despite objections by environmentalists, to incinerate 2,000 tons of nerve agents and mustard gas left from the Cold War era. Although the army said that it is safe to dispose of the weapons, it issued protective gear in case of an "accident" to the nearly 20,000 residents who lived nearby (CNN 2003a).

AP Images/Michel Lipchitz

Oil smoke from the 650 burning oil wells left in the wake of the Gulf War contains soot, sulfur dioxide, and nitrogen oxides, the major components of acid rain, along with a variety of toxic and potentially carcinogenic chemicals and heavy metals.

Strategies for Action: In Search of Global Peace

Various strategies and policies are aimed at creating and maintaining global peace. These include the redistribution of economic resources, the creation of a world government, peacekeeping activities of the United Nations, mediation and arbitration, and arms control.

Redistribution of Economic Resources

Inequality in economic resources contributes to conflict and war because the increasing disparity in wealth and resources between rich and poor nations fuels hostilities and resentment. Therefore, any measures that result in a more equal distribution of economic resources are likely to prevent conflict. John J. Shanahan (1995), retired U.S. Navy vice admiral and former director of the Center for Defense Information, suggested that wealthy nations can help reduce the social and economic roots of conflict by providing economic assistance to poorer countries. Nevertheless, U.S. military expenditures for national defense far outweigh U.S. economic assistance to foreign countries. For instance, the Obama administration's 2014 budget request included $627 billion for national security, but just $56 billion for international affairs (see Figure 15.6).

nuclear winter The predicted result of a thermonuclear war whereby thick clouds of radioactive dust and particles would block out vital sunlight, lower temperature in the northern hemisphere, and lead to the death of most living things on earth.

As discussed in Chapter 12, strategies that reduce population growth are likely to result in higher levels of economic well-being. Funke (1994) explained that "rapidly increasing populations in poorer countries will lead to environmental overload and resource depletion in the next century, which will most likely result in political upheaval and violence as well as mass starvation" (p. 326). Although achieving worldwide economic well-being is important for minimizing global conflict, it is important that economic development does not occur at the expense of the environment.

Finally, former United Nations Secretary General Kofi Annan, in an address to the United Nations, observed that it is not poverty per se that leads to conflict but rather the "inequality among domestic social groups" (Deen 2000). Referencing a research report completed by the Tokyo-based United Nations University, Annan argued that "inequality . . . based on ethnicity, religion, national identity, or economic class . . . tends to be reflected in unequal access to political power that too often forecloses paths to peaceful change" (Deen 2000).

The United Nations

Founded in 1945 after the devastation of World War II, the United Nations (UN) today includes 193 member states and is the principal organ of world governance. In its early years, the UN's main mission was the elimination of war from society. In fact, the UN charter begins, "We the people of the United Nations—determined to save succeeding generations from the scourge of war." During the past 65 years, the UN has developed major institutions and initiatives in support of international law, economic development, human rights, education, health, and other forms of social progress. The Security Council is the most powerful branch of the United Nations. Comprised of 15 member states, it has the power to impose economic sanctions against states that violate international law. It can also use force, when necessary, to restore international peace and security.

The United Nations has engaged in 68 peacekeeping operations since 1948 (see Table 15.4) (United Nations 2013):

United Nations peacekeepers—military personnel in their distinctive blue helmets or blue berets, civilian police and a range of other civilians—help implement peace agreements, monitor cease-fires, create buffer zones, or support complex military and civilian functions essential to maintain peace and begin reconstruction and institution building in societies devastated by war. (United Nations 2003, p. 1)

In 2013, the over 100,000 UN personnel were involved in overseeing 15 multinational peacekeeping forces in Afghanistan, the Sahara, Mali, Haiti, the Congo, Darfur, Syria, Cyprus, Lebanon, the Sudan, Côte d'Ivoire, Kosovo, India, and Pakistan (United Nations 2013).

TABLE 15.4 United Nations Peacekeeping Operations: Summary Data, 2013

Military personnel and civilian police serving in peacekeeping operations	92,099
Countries contributing military personnel and civilian police	116
International civilian personnel	5,107
Local civilian personnel	11,724
UN volunteers	2,088
Total number of fatalities in peacekeeping operations since 1948	3,108
Approved budgets for July 1, 2012, to June 30, 2013	$7.33 billion

Source: United Nations 2013.

In the last few years, the UN has come under heavy criticism. First, in recent missions, developing nations have supplied more than 75 percent of the troops while developed countries—United States, Japan, and Europe—have contributed 85 percent of the finances. As one UN official commented, "You can't have a situation where some nations contribute blood and others only money" (quoted by Vesely 2001, p. 8). Second, a review of UN peacekeeping operations noted several failed missions, including an intervention in Somalia in which 44 U.S. marines were killed (Lamont 2001). Third, as typified by the debate over the disarming of Iraq, the UN cannot take sides but must wait for a consensus of its members that, if not forthcoming, undermines the strength of the organization (Goure 2003). Even if a consensus emerges, without a standing army, the UN relies on troop and equipment contributions from member states, and there can be significant delays and logistical problems in assembling a force for intervention. The consequences of delays can be staggering. For example, under the Genocide Convention of 1948, the UN is obligated to prevent instances of genocide, defined as "acts committed with the intent to destroy, in whole or in part, a national, racial, ethnical, or religious group" (United Nations 1948). In 2000, the UN Security Council formally acknowledged its failure to prevent the 1994 genocide in Rwanda (BBC 2000). After the death of 10 Belgian soldiers in the days leading up to the genocide, and without a consensus for action among the members, the Security Council ignored warnings from the mission's commander about impending disaster and withdrew its 2,500 peacekeepers.

Finally, the concept of the UN is that its members represent individual nations, not a region or the world. And because nations tend to act in their own best economic and security interests, UN actions performed in the name of world peace may be motivated by nations acting in their own interests.

As a result of such criticisms, outgoing UN Secretary General Kofi Annan called on the member states of the UN to approve the most far-reaching changes in the 60-year history of the organization (Lederer 2005). One of the most controversial recommendations concerns the composition of the Security Council, the most important decision-making body of the organization. Annan's recommendation that the 15 members of the Security Council—a body dominated by the United States, Great Britain, France, Russia, and China—be changed to include a more representative number of nations could, if approved, shift the global balance of power. Ban Ki-moon, former foreign minister of South Korea, was elected as the eighth secretary general in October 2006 (MacAskill et al. 2006), after campaigning for the position on a platform that included support for the UN reforms. So far, the United Nations has not been able to reach agreement about whether and how to expand the membership of the Security Council (Macfarquhar 2010).

What Do You Think? On March 17, 2011, after a monthlong civil war between opposition forces and forces loyal to Libyan head of state Muammar Gaddafi, the United Nations Security Council passed a resolution authorizing international military intervention in Libya. The resolution empowered UN and NATO forces "to take all necessary measures" short of an occupying force in order "to protect civilians . . . under threat of attack" (United Nations Security Council 2011, p. 3). Near the end of 2011, opposition forces occupied Tripoli, the capital of Libya, and Gaddafi was killed (Gamel & Keath 2011). Given the number of other civil wars in countries around the world, including Syria, the Sudan, and Mali, on what basis should the UN make decisions about intervention? What do you think?

Mediation and Arbitration

Most conflicts are resolved through nonviolent means. Mediation and arbitration are just two of the nonviolent strategies used to resolve conflicts and to stop or prevent war. In mediation, a neutral third party intervenes and facilitates negotiation between representatives or leaders of conflicting groups. Good mediators do not impose solutions but rather help disputing parties generate options for resolving the conflict (Conflict Research Consortium 2003). Ideally, a mediated resolution to a conflict meets at least some of the concerns and interests of each party to the conflict. In other words, mediation attempts to find "win–win" solutions in which each side is satisfied with the solution.

Although mediation is used to resolve conflict between individuals, it is also a valuable tool for resolving international conflicts. For example, former U.S. Senator George Mitchell successfully mediated talks between parties to the conflict in Northern Ireland in 1998. The resulting political agreement continues to hold today. Also, in May 2008, the government of Qatar mediated talks between Lebanon's political parties that resulted in an agreement that averted civil war after an 18-month political crisis. Using mediation as a means of resolving international conflict is often difficult, given the complexity of the issues. However, research by Bercovitch (2003) shows that there are more mediators available and more instances of mediation in international affairs than ever before. For example, Bercovitch identified more than 250 separate mediation attempts during the Balkan Wars of the early and mid-1990s.

Arbitration also involves a neutral third party who listens to evidence and arguments presented by conflicting groups. Unlike mediation, however, the neutral third party in arbitration arrives at a decision that the two conflicting parties agree in advance to accept. For instance, the Permanent Court of Arbitration—an intergovernmental organization based in The Hague since 1899—arbitrates disputes about territory, treaty compliance, human rights, commerce, and investment among any of its 107 member states who signed and ratified either of its two founding legal conventions. Recent cases include a dispute between France, Britain, and Northern Ireland about the Eurotunnel, a dispute between Pakistan and India about the construction of a hydroelectric power project, and boundary disputes between Eritrea and Ethiopia, as well as between the government of Sudan and the Sudan People's Liberation Army (Permanent Court of Arbitration 2011).

What Do You Think? International talks were held between Iran and the United States in 2013, in the hopes of resolving the nuclear standoff between the two countries. In return for removing economic sanctions against Iran, the United States hopes that Iran will freeze their nuclear enrichment program (Smith-Spark et al. 2013). However, the proposed U.S. concessions are opposed by many in Congress and have strained U.S.–Israeli relations (Weisman 2013). What do you think? Should the United States negotiate with Iran regarding their nuclear program, or impose tougher sanctions?

Arms Control and Disarmament

In the 1960s, the United States and the Soviet Union led the world in a nuclear arms race, with each competing to build a larger and more destructive arsenal of nuclear weapons than its adversary. If either superpower were to initiate a full-scale war, the retaliatory powers of the other nation would result in the destruction of both nations as well as much of the planet. Thus, the principle of **mutually assured destruction (MAD)** that developed from nuclear weapons capabilities transformed war from a win–lose proposition to a lose–lose scenario. If both sides would lose in a war, the theory suggested, neither side would initiate war. At its peak year in 1966, the U.S. stockpile of nuclear weapons included more than 32,000 warheads and bombs (see this chapter's *Self and Society* feature).

As their arsenals continued to grow at an astronomical cost, both sides recognized the necessity for nuclear arms control, including the reduction of defense spending, weapons production and deployment, and armed forces. Throughout the Cold War and

arbitration Dispute settlement in which a neutral third party listens to evidence and arguments presented by conflicting groups and arrives at a decision that the parties have agreed in advance to accept.

mutually assured destruction (MAD) A Cold War doctrine referring to the capacity of two nuclear states to destroy each other, thus reducing the risk that either state will initiate war.

1. The first nuclear weapon ever used by a country was used by _____ against _____ .
   a. Russia, China
   b. The United States, Japan
   c. Germany, England

2. When the nucleus of an atom splits into two pieces it is called:
   a. beta decay
   b. nuclear fusion
   c. nuclear fission

3. The first energy-releasing nuclear chain reaction was developed in:
   a. The United States
   b. Germany
   c. England

4. Uranium, a key element used in the development of nuclear weapons, takes _____ years to naturally decay.
   a. 700,000,000
   b. 1,000,000
   c. 500,000

5. What was the name of the bomb dropped on Nagasaki during WWII?
   a. U-235
   b. Little Boy
   c. Fat Man

6. Historically, nuclear weapons were carried by aircraft (e.g. B-29 bombers). Today, nuclear weapon delivery systems are more likely to be:
   a. drones
   b. ballistic missiles
   c. satellites

7. What is the center of a nuclear bomb blast called?
   a. fallout radius
   b. ground zero
   c. blast zone

8. How high do temperatures at the center of an atomic blast reach?
   a. Up to 300 million degrees Fahrenheit
   b. Up to 400 million degrees Fahrenheit
   c. Up to 500 million degrees Fahrenheit

9. Which of the following countries does not have nuclear weapons capabilities?
   a. Israel
   b. France
   c. Canada

10. In the event of multiple nuclear explosions, what may cause a "nuclear winter"?
    a. Rising dust and radioactive material blocking the sun's light
    b. Plunging surface temperatures as the polar ice caps melt
    c. Destruction of heat sources as wood, coal, gas, and oil are contaminated

Answers: 1. b; 2. c; 3. a; 4. a; 5. c; 6. b; 7. b; 8. c; 9. c; 10. a.

Scoring: Give yourself one point for every correct answer, and then assess how well you did using the following scale.

Number of Correct Answers

10	The Pentagon needs you!
9	Congratulations, professor.
8	You're the bomb!
7	Better than most . . .
6	Take a physics course!
5	Time to turn in your lab coat.
4 or below	Don't quit your day job.

Based on: William Harris, Craig Freudenrich, and John Fuller. "How do Nuclear Bombs Work." Available at http://science.howstuffworks.com/nuclear-bomb.htm

even today, much of the behavior of the United States and the Soviet Union has been governed by major arms control initiatives. These initiatives include:

- The Limited Test Ban Treaty that prohibited testing of nuclear weapons in the atmosphere, underwater, and in outer space
- The Strategic Arms Limitation Treaties (SALT I and II) that limited the development of nuclear missiles and defensive antiballistic missiles
- The Strategic Arms Reduction Treaties (START I and II) that significantly reduced the number of nuclear missiles, warheads, and bombs
- The Strategic Offensive Reduction Treaty (SORT) that required that the United States and the Russian Federation each reduce their number of strategic nuclear warheads to between 1,700 and 2,200 by 2012. This goal was met, with the United States currently holding 1,950 strategic warheads and the Russian Federation holding 1,800 (Center for Arms Control and Non-Proliferation 2013).
- The New Strategic Arms Reduction Treaty (new START) that further reduces the number of deployed strategic nuclear warheads to 1,550 for each country and significantly reduces the number of strategic nuclear missile launchers allowed (Atomic Archive 2011)

With the end of the Cold War came the growing realization that, even as Russia and the United States greatly reduced their arsenals, other countries were poised to acquire nuclear weapons or expand their existing arsenals. Thus, the focus on arms control shifted toward **nuclear nonproliferation**, i.e., the prevention of the spread of nuclear technology to nonnuclear states.

Nuclear Nonproliferation Treaty. The Nuclear Nonproliferation Treaty (NPT) was signed in 1970, and was the first treaty governing the spread of nuclear weapons

nuclear nonproliferation
Efforts to prevent the spread of nuclear weapons, or the materials and technology necessary for the production of nuclear weapons.

technology from the original nuclear weapons states (i.e., the United States, the Soviet Union, the United Kingdom, France, and China) to nonnuclear countries. The NPT was renewed in 2000, and is subject to review every five years. Currently, 189 countries have adopted it. The NPT holds that countries without nuclear weapons will not try to get them; in exchange, the countries with nuclear weapons agree that they will not provide nuclear weapons to countries that do not have them. Signatory states without nuclear weapons also agree to allow the International Atomic Energy Agency to verify compliance with the treaty through on-site inspections (Atomic Archive 2011). Only India, Israel, and Pakistan have not signed the agreement, although each of these states is known to possess a nuclear arsenal. Further, many experts suspect that Iran and Syria—both signatories to the NPT—are developing nuclear weapons programs. Both countries claim that their nuclear reactors are for peaceful purposes (e.g., domestic power consumption), not to develop a nuclear weapons arsenal. However, a report by the International Atomic Energy Agency tentatively concluded that Iran has "sufficient information to be able to design and produce a workable" atomic bomb (Broad & Sanger 2009, p. 1).

In January 2003, North Korea, under suspicion of secretly producing nuclear weapons, announced that it was withdrawing from the treaty, effective immediately (Arms Control Association 2003). In July 2005, after "more than a year of stalemate, North Korea agreed . . . to return to disarmament talks . . . and pledged to discuss eliminating its nuclear-weapons program" (Brinkley & Sanger 2005, p. 1). However, the North Korean government is believed to have tested plutonium bombs in 2006, 2009, and 2013 (Dahl 2013).

Even if military superpowers honor agreements to limit arms, the availability of black-market nuclear weapons and materials presents a threat to global security. One of the most successful nuclear weapons brokers was Pakistan's Abdul Qadeer Khan. Khan, the "father" of Pakistan's nuclear weapons program, sold the technology and equipment required to make nuclear weapons to rogue states such as Iran and North Korea (Powell & McGirk 2005). Further nuclear tests took place after the new Korean leader Kim Jong-un came to power in December 2011. However, after the public failure of several missile tests and intense international pressure, the North Korean leader announced his intentions to renew talks on ending his country's nuclear program (Sang-Hun & Buckley 2013).

> Even if military superpowers honor agreements to limit arms, the availability of black-market nuclear weapons and materials presents a threat to global security.

In 2003, the United States accused Iran of operating a covert nuclear weapons program. Although Iranian officials responded that their nuclear program was solely for the purpose of generating electricity, concerns escalated with Iran's successful test of a 1,200-mile-range missile in 2009 (Dareini 2009). Hostile public statements about Israel by Iran's president Mahmoud Ahmadinejad deepen already grave concerns about Iran's possible acquisition of nuclear weapons. Analysts fear the development of nuclear weapons by Iran would spur a nuclear competition with Israel, the only other state in the Middle East to possess nuclear weapons, and provoke other states in the region to acquire them (Salem 2007). After years of tensions with the United States over its nuclear program, both the Obama administration and Iran's newly elected president, Hassan Rouhani, have indicated a desire for direct and open talks about Iran's nuclear capabilities in 2013 (Gordon 2013).

Many observers consider South Asia the world's most dangerous nuclear rivalry. India and Pakistan are the only two nuclear powers that share a border and have repeatedly fired upon each other's armies while in possession of nuclear weapons (Stimson Center 2007). India first detonated nuclear weapons in 1974. Pakistan detonated six weapons in 1998, a few weeks after India's second round of nuclear tests. Although precise figures are hard to come by, experts estimate that India has about 50 nuclear bombs, and Pakistan has about 60 nuclear bombs (Carnegie Endowment for International Peace 2009). Pakistan, however, is aggressively expanding its nuclear weapons program (Shanker & Sanger 2009). Many fear that a conventional military confrontation between these two countries may someday escalate to an exchange of nuclear weapons. Tensions have further escalated in the wake of the U.S. drawdown from Afghanistan as India and Pakistan compete for influence in the region, leading some analysts to speculate that the threat of direct, and possibly nuclear, confrontation will increase over the next decade (Dalrymple 2013).

As states that want to obtain nuclear weapons are quick to point out, nuclear states that advocate for nonproliferation possess well over 25,000 weapons, a huge reduction in the world's arsenal from Cold War days but still a massive potential threat to the earth. "Do as I say not as I do" is a weak bargaining position. Recognizing this situation and with concern about the possibility of new arms races, many high-level experts have begun to advocate a more comprehensive approach to banning nuclear weapons—with the United States leading the

AP Images/Bullit Marquez

nuclear powers toward complete nuclear disarmament. In January 2007, three former U.S. Secretaries of State (George Shultz, William Perry, and Henry Kissinger) and former chairman of the Senate Armed Services Committee Sam Nunn published an editorial in the *Wall Street Journal* advocating that the United States "take the world to the next stage" of disarmament to "a world free of nuclear weapons" (Schultz et al. 2007). As a start, their proposal advocated the elimination of all short-range nuclear missiles, a complete halt in the production of weapons-grade uranium, and continued reduction of nuclear forces. President Obama visited Russia in July 2009, to sign the New Strategic Arms Reduction Treaty that reduces the strategic nuclear arsenals of the United States and the Russian Federation by at least 25 percent. The Senate ratified the treaty in December 2010 (Sheridan & Branigin 2010). This was reportedly "a first step in a broader effort intended to reduce the threat of such weapons drastically and to prevent their further spread to unstable regions" (Levy & Baker 2009).

Although the United States is the world's top arms exporter, it is also the world's leader in the destruction of conventional weapons. In 2010, the Office of Weapons Removal and Abatement contributed over $160 million to 43 countries to destroy conventional weapons. Since 2001, the United States has helped destroy 90,000 tons of ammunition and 1.5 million weapons (U.S. Department of State 2011).

The Problem of Small Arms

Although the devastation caused by even one nuclear war could affect millions, the easy availability of conventional weapons fuels many active wars around the world. Small arms and light weapons include handguns, submachine guns and automatic weapons, grenades, mortars, land mines, and light missiles. The Small Arms Survey estimated that, in 2013, 875 million firearms were in circulation worldwide, with national armed forces and law enforcement agencies possessing less than one-quarter of these weapons and nonstate armed groups and gangs possessing only 1.3 percent; civilians held the remaining 75 percent of small arms. Half a million people die each year as a result of small arms use, 200,000 in homicides and suicides, and the rest during wars and other armed conflicts (Geneva Graduate Institute for International Studies 2013).

Unlike control of weapons of mass destruction such as chemical and biological weapons, controlling the flow of small arms—especially firearms—is not easy because they have many legitimate uses by the military, by law enforcement officials, and for recreational or sporting activities (Schroeder 2007). Small arms are easy to afford, to use, to conceal, and to transport illegally. The small arms trade is also lucrative. According to official records, the top five exporters of small arms in 2010 were the United States ($821 million), Germany ($495 million), Italy ($473 million), Brazil ($326 million), and Switzerland ($215 million) (Geneva Graduate Institute for International Studies 2013).

The U.S. Department of State's Office of Weapons Removal and Abatement (WRA) administers a program that has supported the destruction of over 1.6 million small arms and light weapons and 90,000 tons of ammunition in 38 countries since its founding in 2001 (Office of Weapons Removal and Abatement 2013). The availability of these weapons fuels terrorist groups and undermines efforts to promote peace after wars have formally concluded. "If not expeditiously destroyed or secured, stocks of arms and ammunition left over after the cessation of hostilities frequently recirculate into neighboring regions, exacerbating conflict and crime" (U.S. Department of State 2007, p. 1).

Understanding Conflict, War, and Terrorism

As we come to the close of this chapter, how might we have an informed understanding of conflict, war, and terrorism? Each of the three theoretical positions discussed in this chapter reflects the realities of global conflict. As structural functionalists argue, war offers societal benefits—social cohesion, economic prosperity, scientific and technological developments, and social change. Furthermore, as conflict theorists contend, wars often occur for economic reasons because corporate elites and political leaders benefit from the spoils of war—land and water resources and raw materials. The symbolic interactionist perspective emphasizes the role that meanings, labels, and definitions play in creating conflict and contributing to acts of war.

The September 11 attacks on the World Trade Center and the Pentagon and the aftermath—the battle against terrorism, the wars in Iraq and Afghanistan—changed the world Americans live in. For some theorists, these events were inevitable. Political scientist Samuel P. Huntington argued that such conflict represents a **clash of civilizations**. In *The Clash of Civilizations and the Remaking of World Order* (1996), Huntington argued that in the new world order,

> the most pervasive, important and dangerous conflicts will not be between social classes, rich and poor, or economically defined groups, but between people belonging to different cultural entities . . . the most dangerous cultural conflicts are those along the fault lines between civilizations . . . the line separating peoples of Western Christianity, on the one hand, from Muslim and Orthodox peoples on the other. (p. 28)

Some public opinion surveys seem to support Huntington's view. In an interview of almost 10,000 people from nine Muslim states representing half of all Muslims worldwide, only 22 percent had favorable opinions toward the United States (CNN 2003b). Even more significantly, 67 percent saw the September 11 attacks as "morally justified," and the majority of respondents found the United States to be overly materialistic and secular and having a corrupting influence on other nations. Moreover, according to a poll of world public opinion taken every year since 2002, positive attitudes about the United States are consistently lowest in predominantly Muslim countries (Pew Research Center 2011b).

Conversely, according to a poll, 38 percent of Americans have unfavorable views of Islam, and 35 percent of the respondents believe that, compared to other religions, Islam is more likely to encourage violence (Pew Research Center for the People & the Press 2010). Moreover, a recent Gallup Poll indicates that barely half (53 percent) of Americans believe that Muslims living in the United States are supportive of the United States, 36 percent believe that Muslims living in the United States are "too extreme in their religious beliefs," and 28 percent believe that U.S. Muslims are "sympathetic to the al Qaeda terrorist organization" (Newport 2011, p. 1).

The clash of civilizations perspective has been vehemently criticized, however, by many scholars who see this view as divisive, overly simplistic, and historically inaccurate. David Brooks summarizes this critique by suggesting that Huntington committed a "Fundamental Attribution Error. That is, he ascribed to traits qualities that are actually determined by context" (2011). Huntington suggested that Arab societies are intrinsically opposed to democracy and not nationalistic, but the Arab Spring revolutions highlighted how certain political regimes can effectively, although never permanently, suppress

clash of civilizations A hypothesis that the primary source of conflict in the 21st century has shifted away from social class and economic issues and toward conflict between religious and cultural groups, especially those between large-scale civilizations such as the peoples of Western Christianity and Muslim and Orthodox peoples.

national patriotism and the intrinsic human desire for liberty. Brooks also suggests that Huntington fundamentally misunderstood the nature of culture for, despite our intrinsic and cultural differences, we are all alike:

> Huntington minimized the power of universal political values and exaggerated the influence of distinct cultural values . . . underneath cultural differences there are universal aspirations for dignity, for political systems that listen to, respond to and respect the will of the people.

Ultimately, we are all members of one community—earth—and have a vested interest in staying alive and protecting the resources of our environment for our own and future generations. But, as we have seen, conflict between groups is a feature of social life that is not likely to disappear. What is at stake—human lives and the ability of our planet to sustain life—merits serious attention. World leaders have traditionally followed the advice of philosopher Carl von Clausewitz: "If you want peace, prepare for war." Thus, nations have sought to protect themselves by maintaining large military forces and massive weapons systems. These strategies are associated with serious costs, particularly in hard economic times. In diverting resources away from other social concerns, defense spending undermines a society's ability to improve the overall security and well-being of its citizens. Conversely, defense-spending cutbacks, although unlikely in the present climate, could potentially free up resources for other social agendas, including lowering taxes, reducing the national debt, addressing environmental concerns, eradicating hunger and poverty, improving health care, upgrading educational services, and improving housing and transportation. Therein lies the promise of a "peace dividend." The hope is that future dialogue on the problems of war and terrorism will redefine national and international security to encompass social, economic, and environmental well-being.

CHAPTER REVIEW

- **What is the relationship between war and industrialization?**
  War indirectly affects industrialization and technological sophistication because military research and development advances civilian-used technologies. Industrialization, in turn, has had two major influences on war: The more industrialized a country is, the lower the rate of conflict, and if conflict occurs, the higher the rate of destruction.

- **What are the latest trends in armed conflicts?**
  Since World War II, wars between two or more states make up the smallest percentage of armed conflicts. In the contemporary era, the majority of armed conflicts have occurred between groups in a single state, who compete for the power to control the resources of the state or to break away and form their own state.

- **In general, how do feminists view war?**
  Feminists are quick to note that wars are part of the patriarchy of society. Although women and children may be used to justify a conflict (e.g., improving women's lives by removing the repressive Taliban in Afghanistan), the basic principles of male dominance and control are realized through war. Feminists also emphasize the social construction of aggressive masculine identities and their manipulation by elites as important reasons for the association between masculinity and militarized violence.

- **What are some of the causes of war?**
  The causes of war are numerous and complex. Most wars involve more than one cause. Some of the causes of war are conflict over land and natural resources; values or ideologies; racial, ethnic, and religious hostilities; defense against hostile attacks; revolution; and nationalism.

- **What is terrorism, and what are the different types of terrorism?**
  Terrorism is the premeditated use, or threatened use, of violence by an individual or group to gain a political or social objective. Terrorism can be either transnational or domestic. Transnational terrorism occurs when a terrorist act in one country involves victims, targets, institutions, governments, or citizens of another country. Domestic terrorism involves only one nation, such as the 1995 truck bombing of a nine-story federal office building in Oklahoma City.

- **What are some of the macro-level "roots" of terrorism?**
  Although not an exhaustive list, some of the macro-level "roots" of terrorism include (1) a failed or weak state, (2) rapid modernization, (3) extreme ideologies, (4) a history of violence, (5) repression by a foreign occupation, (6) large-scale racial or ethnic discrimination, and (7) the presence of a charismatic leader.

- **How has the United States responded to the threat of terrorism?**

The United States has used both defensive and offensive strategies to fight terrorism. Defensive strategies include using metal detectors and X-ray machines at airports and strengthening security at potential targets, such as embassies and military command posts. The Department of Homeland Security coordinates such defensive tactics. Offensive strategies include retaliatory raids such as the U.S. bombing of terrorist facilities in Afghanistan, group infiltration, and preemptive strikes.

- **What is meant by "diversion of economic resources"?**

Worldwide, the billions of dollars used on defense could be channeled into social programs dealing with, for example, education, health, and poverty. Thus, defense monies are economic resources diverted from other needy projects.

- **What are some of the criticisms of the United Nations?**

First, in recent missions, developing nations have supplied more than 75 percent of the troops. Second, several recent UN peacekeeping operations have failed. Third, the UN cannot take sides but must wait for a consensus of its members that, if not forthcoming, undermines the strength of the organization. Fourth, the concept of the UN is that its members represent individual nations, not a region or the world. Because nations tend to act in their own best economic and security interests, UN actions performed in the name of world peace may be motivated by nations acting in their own interests. Finally, the Security Council limits power to a small number of states.

- **What problems do small arms pose?**

Even after a conflict ends, these weapons circulate in society, making crime worse or falling into the hands of terrorists. Trade in small arms is legal because they have many legitimate uses—for example, by the military, police, and hunters. Because they are small and simple to handle, these weapons are easily concealed and transported, making it difficult to control them.

TEST YOURSELF

1. War between states is still the most common form of warfare.
   a. True
   b. False
2. The rise of the modern state is most directly a result of
   a. industrialization and the creation of national markets
   b. innovations in communications technology
   c. the development of armies to control territory
   d. the development of police to control a population
3. Structural-functionalist explanations about war emphasize
   a. that war is a biological necessity
   b. that despite its destructive power, war persists because it fulfills social needs
   c. that war is an anachronism that will eventually disappear
   d. that war is necessary because it benefits political and military elites
4. On the whole, conflicts over values and ideologies are more difficult to resolve than those over material resources.
   a. True
   b. False
5. All wars are a result of unequal distribution of wealth.
   a. True
   b. False
6. Next to defense spending, transfers to foreign governments are the most expensive item in the U.S. government budget.
   a. True
   b. False
7. Which of the following factors is a likely cause of revolutions or civil wars?
   a. A weak or failed state
   b. An authoritarian government that ignores major demands from citizens
   c. The availability of strong opposition leaders
   d. All of the above
8. Primordial explanations of ethnic conflict suggest that
   a. ethnic leaders instigate hostilities to serve their own interests
   b. people become hostile when they blame their frustration with economic hardship on competing ethnic groups
   c. ancient hatreds compel ethnic groups to continue fighting
   d. none of the above
9. The consequences of war and the military on the environment
   a. are prevalent only during wartime
   b. persist in peacetime or for many years after a war is over
   c. are negligible
   d. have been mostly reduced by technological innovations
10. Advocates of nuclear nonproliferation seek to
    a. ban all nuclear weapons
    b. prevent construction of nuclear power plants
    c. prevent new states from acquiring nuclear weapons
    d. none of the above

Answers: 1. B; 2. C; 3. B; 4. A; 5. B; 6. B; 7. D; 8. C; 9. B; 10. C.

KEY TERMS

arbitration 508

clash of civilizations 512

Cold War 478

constructivist explanations 487

democratic peace theory 487

domestic terrorism 491

dual-use technologies 481

Geneva Conventions 496

guerrilla warfare 497

military-industrial complex 482

mutually assured destruction
(MAD) 508

nuclear nonproliferation 509

nuclear winter 505

post-traumatic stress disorder
(PTSD) 501

primordial explanations 487

security dilemma 488

state 477

terrorism 490

transnational terrorism 490

war 476

weapons of mass destruction
(WMD) 497

Appendix

Methods of Data Analysis

Description, Correlation, Causation, Reliability and Validity, and Ethical Guidelines in Social Problems Research

There are three levels of data analysis: description, correlation, and causation. Data analysis also involves assessing reliability and validity.

Description

Qualitative research involves verbal descriptions of social phenomena. Having a homeless and single pregnant teenager describe her situation is an example of qualitative research.

Quantitative research often involves numerical descriptions of social phenomena. Quantitative descriptive analysis may involve computing the following: (1) means (averages), (2) frequencies, (3) mode (the most frequently occurring observation in the data), (4) median (the middle point in the data; half of the data points are above the median and half are below), and (5) range (the highest and lowest values in a set of data).

Correlation

Researchers are often interested in the relationship between variables. *Correlation* refers to a relationship between or among two or more variables. The following are examples of correlational research questions: What is the relationship between poverty and educational achievement? What is the relationship between race and crime victimization? What is the relationship between religious affiliation and divorce?

If there is a correlation or relationship between two variables, then a change in one variable is associated with a change in the other variable. When both variables change in the same direction, the correlation is positive. For example, in general, the more sexual partners a person has, the greater the risk of contracting a sexually transmissible disease (STD). As variable A (number of sexual partners) increases, variable B (chance of contracting an STD) also increases. Similarly, as the number of sexual partners decreases, the chance of contracting an STD decreases. Notice that in both cases, the variables change in the same direction, suggesting a positive correlation (see Figure A.1).

When two variables change in opposite directions, the correlation is negative. For example, there is a negative correlation between condom use and contracting STDs. In other words, as condom use increases, the chance of contracting an STD decreases (see Figure A.2).

The relationship between two variables may also be curvilinear, which means that it varies in both the same and opposite directions. For example, suppose a researcher finds that after drinking one alcoholic beverage, research participants are more prone to violent behavior. After two drinks, violent behavior is even more likely, and this trend continues for three and four drinks. So far, the correlation between alcohol consumption and violent behavior is positive. After the research participants have five alcoholic drinks, however, they become less prone to violent behavior. After six and seven drinks, the likelihood of engaging in violent behavior decreases further. Now the correlation between alcohol consumption and violent behavior is negative. Because the correlation

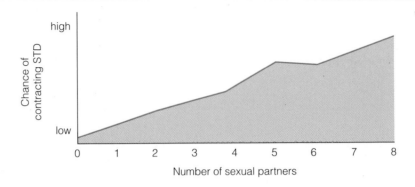

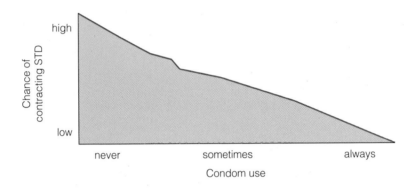

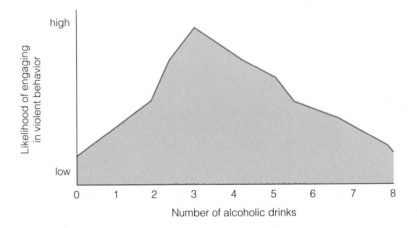

changed from positive to negative, we say that the correlation is curvilinear (the correlation may also change from negative to positive) (see Figure A.3).

A fourth type of correlation is called a spurious correlation. Such a correlation exists when two variables appear to be related, but the apparent relationship occurs only because each variable is related to a third variable. When the third variable is controlled through a statistical method in which the variable is held constant, the apparent relationship between the first two variables disappears. For example, blacks have a lower average life expectancy than whites do. Thus, race and life expectancy appear to be related. This apparent correlation exists, however, because both race and life expectancy are related to socioeconomic status. Because blacks are more likely than whites to be impoverished, they are less likely to have adequate nutrition and medical care.

Causation

If the data analysis reveals that two variables are correlated, we know only that a change in one variable is associated with a change in another variable. We cannot assume, however, that a change in one variable causes a change in the other variable unless our data collection and analysis are specifically designed to assess causation. The research method that best assesses causality is the experimental method (discussed in Chapter 1).

To demonstrate causality, three conditions must be met. First, the data analysis must demonstrate that variable A is correlated with variable B. Second, the data analysis must demonstrate that the observed correlation is not spurious. Third, the analysis must demonstrate that the presumed cause (variable A) occurs or changes before the presumed effect (variable B). In other words, the cause must precede the effect.

It is extremely difficult to establish causality in social science research. Therefore, much social research is descriptive or correlative, rather than causative. Nevertheless, many people make the mistake of interpreting a correlation as a statement of causation. As you read correlative research findings, remember the following adage: "Correlation does not equal causation."

Reliability and Validity

Assessing reliability and validity is an important aspect of data analysis. *Reliability* refers to the consistency of the measuring instrument or technique; that is, the degree to which the way information is obtained produces the same results if repeated. Measures of reliability are made on scales and indexes (such as those in the *Self and Society* features in this text) and on information-gathering techniques, such as the survey methods described in Chapter 1.

Various statistical methods are used to determine reliability. A frequently used method is called the "test-retest method." The researcher gathers data on the same sample of people twice (usually 1 or 2 weeks apart) using a particular instrument or method and then correlates the results. To the degree that the results of the two tests are the same (or highly correlated), the instrument or method is considered reliable.

Measures that are perfectly reliable may be absolutely useless unless they also have a high validity. *Validity* refers to the extent to which an instrument or device measures what it intends to measure. For example, police officers administer "Breathalyzer" tests to determine the level of alcohol in a person's system. The Breathalyzer is a valid test for measuring alcohol consumption.

Validity measures are important in research that uses scales or indexes as measuring instruments. Validity measures are also important in assessing the accuracy of self-report data that are obtained in survey research. For example, survey research on high-risk sexual behaviors associated with the spread of HIV relies heavily on self-report data on topics such as number of sexual partners, types of sexual activities, and condom use. Yet how valid are these data? Do survey respondents underreport the number of their sexual partners? Do people who say they use a condom every time they engage in intercourse really use a condom every time? Because of the difficulties in validating self-reports of number of sexual partners and condom use, we may not be able to answer these questions.

Ethical Guidelines in Social Problems Research

Social scientists are responsible for following ethical standards designed to protect the dignity and welfare of people who participate in research. These ethical guidelines include the following:

1. Freedom from coercion to participate. Research participants have the right to decline to participate in a research study or to discontinue participation at any time during the study. For example, professors who are conducting research using college students should not require their students to participate in their research.
2. Informed consent. Researchers are required to inform potential participants of any aspect of the research that might influence a subject's willingness to participate.

After informing potential participants about the nature of the research, researchers typically ask participants to sign a consent form indicating that the participants are informed about the research and agree to participate in it.

3. Deception and debriefing. Sometimes the researcher must disguise the purpose of the research to obtain valid data. Researchers may deceive participants as to the purpose or nature of a study only if there is no other way to study the problem. When deceit is used, participants should be informed of this deception (debriefed) as soon as possible. Participants should be given a complete and honest description of the study and why deception was necessary.

4. Protection from harm. Researchers must protect participants from any physical and psychological harm that might result from participating in a research study. This is both a moral and a legal obligation. It would not be ethical, for example, for a researcher studying drinking and driving behavior to observe an intoxicated individual leaving a bar, getting into the driver's seat of a car, and driving away.

     Researchers are also obligated to respect the privacy rights of research participants. If anonymity is promised, it should be kept. Anonymity is maintained in mail surveys by identifying questionnaires with a number coding system rather than with the participants' names. When such anonymity is not possible, as is the case with face-to-face interviews, researchers should tell participants that the information they provide will be treated as confidential. Although interviews may be summarized and excerpts quoted in published material, the identity of the individual participants is not revealed. If a research participant experiences either physical or psychological harm as a result of participation in a research study, the researcher is ethically obligated to provide remediation for the harm.

5. Reporting of research. Ethical guidelines also govern the reporting of research results. Researchers must make research reports freely available to the public. In these reports, a researcher should fully describe all evidence obtained in the study, regardless of whether the evidence supports the researcher's hypothesis. The raw data collected by the researcher should be made available to other researchers who might request it for purposes of analysis. Finally, published research reports should include a description of the sponsorship of the research study, its purpose, and all sources of financial support.

Glossary

3-D printing A revolutionary manufacturing technology that involves downloading a digital file containing a design for a product. A printer reads the file and then shoots out the product (made of specialized plastic or other raw materials) through a heated nozzle.

abortion The intentional termination of a pregnancy.

absolute poverty The lack of resources necessary for material well-being—most importantly, food and water, but also housing, sanitation, education, and health care.

acculturation The process of adopting the culture of a group different from the one in which a person was originally raised.

achieved status A status that society assigns to an individual on the basis of factors over which the individual has some control.

acid rain The mixture of precipitation with air pollutants, such as sulfur dioxide and nitrogen oxide.

acquaintance rape Rape committed by someone known to the victim.

adaptive discrimination Discrimination that is based on the prejudice of others.

affirmative action A broad range of policies and practices in the workplace and educational institutions to promote equal opportunity as well as diversity.

Affordable Care Act (ACA) Health care reform legislation that Obama signed into law in 2010, with the goal of expanding health insurance coverage to more Americans; also known as the Patient Protection and Affordable Care Act, or "Obamacare."

ageism Negative stereotyping, prejudice, and discrimination based on a person's or group's perceived chronological age.

ageism by invisibility Occurs when older adults are not included in advertising and educational materials.

alienation A sense of powerlessness and meaninglessness in people's lives.

allopathic medicine The conventional or mainstream practice of medicine; also known as Western medicine.

alternative certification programs Programs whereby college graduates with degrees in fields other than education can become certified if they have "life experience" in industry, the military, or other relevant jobs.

androgyny Having both traditionally defined feminine and masculine characteristics.

anomie A state of normlessness in which norms and values are weak or unclear.

anti-miscegenation laws Laws banning interracial marriage until 1967, when the Supreme Court (in *Loving v. Virginia*) declared these laws unconstitutional.

arbitration Dispute settlement in which a neutral third party listens to evidence and arguments presented by conflicting groups and arrives at a decision that the parties have agreed in advance to accept.

ascribed status A status that society assigns to an individual on the basis of factors over which the individual has no control.

assimilation The process by which formerly distinct and separate groups merge and become integrated as one.

automation The replacement of human labor with machinery and equipment.

aversive racism A subtle form of prejudice that involves feelings of discomfort, uneasiness, disgust, fear, and pro-white attitudes.

baby boomers The generation of Americans born between 1946 and 1964, a period of high birthrates.

Basic Economic Security Tables Index (BEST) A measure of the basic needs and income workers require for economic security.

behavior-based safety programs A strategy used by business management that attributes health and safety problems in the workplace to workers' behavior, rather than to work processes and conditions.

beliefs Definitions and explanations about what is assumed to be true.

bigamy The criminal offense in the United States of marrying one person while still legally married to another.

bilingual education In the United States, teaching children in both English and their non-English native language.

binge drinking As defined by the U.S. Department of Health and Human Services, drinking five or more drinks on the same occasion on at least one day in the past 30 days prior to the National Survey on Drug Use and Health.

biodiversity The diversity of living organisms on earth.

bioinvasion The intentional or accidental introduction of plant, animal, insect, and other species in regions where they are not native.

biomass Material derived from plants and animals, such as dung, wood, crop residues, and charcoal; used as cooking and heating fuel.

biphobia When prejudice is directed toward bisexual individuals.

bisexuality The emotional, cognitive, and sexual attraction to members of both sexes.

boy code A set of societal expectations that discourages males from expressing emotion, weakness, or vulnerability, or asking for help.

breeding ground hypothesis A hypothesis that argues that incarceration serves to increase criminal behavior through the transmission of criminal skills, techniques, and motivations.

bullying Bullying "entails an imbalance of power that exists over a long period of time in which the more powerful intimidate or belittle others" (Hurst 2005, p. 1).

capital punishment The state (the federal government or a state) takes the life of a person as punishment for a crime.

capitalism An economic system characterized by private ownership of the means of production and distribution of goods and services for profit in a competitive market.

character education Education that emphasizes the moral and ethical aspects of an individual.

charter schools Schools that originate in contracts, or charters, which articulate a plan of instruction that local or state authorities must approve.

chattel slavery A form of slavery in which slaves are considered property that can be bought and sold.

chemical dependency A condition in which drug use is compulsive and users are unable to stop because of physical and/or psychological dependency.

child abuse The physical or mental injury, sexual abuse, negligent treatment, or maltreatment of a child younger than age 18 by a person who is responsible for the child's welfare.

child labor Involves a child performing work that is hazardous, that interferes with a child's education, or that harms a child's health or physical, mental, social, or moral development.

civil union A legal status that entitles same-sex couples who apply for and receive a civil union certificate to nearly all of the benefits available to married couples.

clash of civilizations A hypothesis that the primary source of conflict in the 21st century has shifted away from social class and economic issues and toward conflict between religious and cultural groups, especially those between large-scale civilizations such as the peoples of Western Christianity and Muslim and Orthodox peoples.

classic rape Rape committed by a stranger, with the use of a weapon, resulting in serious bodily injury to the victim.

clearance rate The percentage of crimes in which an arrest and official charge have been made and the case has been turned over to the courts.

climate denial machine A well-funded and aggressive misinformation campaign run by the fossil fuel industry and its allies that involves attacking and discrediting climate science, scientists, and scientific institutions.

club sandwich generation Includes adults (often in their 50s or 60s) who are caring for aging parents, adult children, and grandchildren, or adults (usually in their 30s and 40s) who are caring for their own young children, aging parents, and grandparents.

Cold War The state of military tension and political rivalry that existed between the United States and the former Soviet Union from the 1950s through the late 1980s.

color-blind racism A form of racism that is based on the idea that overcoming racism means ignoring race, but color-blindness is, in itself, a form of racism because it prevents acknowledgment of privilege and disadvantage associated with race, and therefore allows the continuation of institutional forms of racial bias.

coming out The ongoing process whereby a lesbian, gay, or bisexual individual becomes aware of his or her sexuality, accepts and incorporates it into his or her overall sense of self, and shares that information with others such as family, friends, and coworkers.

complementary and alternative medicine (CAM) Refers to a broad range of health care approaches, practices, and products that are not considered part of conventional medicine.

comprehensive primary health care An approach to health care that focuses on the broader social determinants of health, such as poverty and economic inequality, gender inequality, environment, and community development.

compressed workweek A work arrangement that allows employees to condense their work into fewer days (e.g., four 10-hour days each week).

computer crime Any violation of the law in which a computer is the target or means of criminal activity.

constructivist explanations Those explanations that emphasize the role of leaders of ethnic groups in stirring up hatred toward others external to one's group.

contact hypothesis The idea that contact between groups is necessary for the reduction of prejudice.

corporal punishment The intentional infliction of pain intended to change or control behavior.

corporate violence The production of unsafe products and the failure of corporations to provide a safe working environment for their employees.

corporate welfare Laws and policies that benefit corporations.

corporatocracy A system of government that serves the interests of corporations and that involves ties between government and business.

covenant marriage A type of marriage (offered in a few states) that requires premarital counseling and that permits divorce only under condition of fault or after a marital separation of more than two years.

crack A crystallized illegal drug product produced by boiling a mixture of baking soda, water, and cocaine.

crime An act, or the omission of an act, that is a violation of a federal, state, or local criminal law for which the state can apply sanctions.

crime rate The number of crimes committed per 100,000 population.

cultural imperialism The indoctrination into the dominant culture of a society.

cultural lag A condition in which the material part of culture changes at a faster rate than the nonmaterial part.

cultural sexism The ways in which the culture of society perpetuates the subordination of an individual or group based on the sex classification of that individual or group.

culture The meanings and ways of life that characterize a society, including beliefs, values, norms, sanctions, and symbols.

cyber-bullying The use of electronic communication (e.g., websites, e-mail, instant messaging, text messaging) to send or post negative or hurtful messages or images about an individual or a group.

cybernation Dominant in postindustrial societies, the use of machines to control other machines.

cycle of abuse A pattern of abuse in which a violent or abusive episode is followed by a makeup period when the abuser expresses sorrow and asks for forgiveness and "one more chance," before another instance of abuse occurs.

decriminalization Entails removing state penalties for certain drugs, and promotes a medical rather than criminal approach to drug use that encourages users to seek treatment and adopt preventive practices.

deep ecology The view that maintaining the earth's natural systems should take precedence over human needs, that nature has a value independent of human existence, and that humans have no right to dominate the earth and its living inhabitants.

Defense of Marriage Act (DOMA) Federal legislation that states that marriage is a legal union between one man and one woman and denies federal recognition of same-sex marriage.

deforestation The conversion of forestland to nonforestland.

deinstitutionalization In the U.S. model for psychiatric care, the transformation from long-term inpatient care in institutions to drug therapy and community-based mental health centers.

demand reduction One of two strategies in the U.S. war on drugs (the other is supply reduction), demand reduction focuses on reducing the demand for drugs through treatment, prevention, and research.

democratic peace theory A prevalent theory in international relations suggesting that the ideological similarities between democratic nations makes it unlikely that such countries will go to war against each other.

demographic transition theory A theory that attributes population growth patterns to changes in birth rates and death rates associated with the process of industrialization.

dependent variable The variable that the researcher wants to explain; the variable of interest.

deregulation The reduction of government control over, for example, certain drugs.

desertification The degradation of semiarid land, which results in the expansion of desert land that is unusable for agriculture.

deterrence The use of harm or the threat of harm to prevent unwanted behaviors.

devaluation hypothesis The hypothesis that women are paid less because the work they perform is socially defined as less valuable than the work men perform.

developed countries Countries that have relatively high gross national income per capita, also known as high-income countries.

developing countries Countries that have relatively low gross national income per capita, also known as less-developed or middle-income countries.

discrimination Actions or practices that result in differential treatment of categories of individuals.

divorce mediation A process in which divorcing couples meet with a neutral third party (mediator) who assists the individuals in resolving issues such as property division, child custody, child support, and spousal support in a way that minimizes conflict and encourages cooperation.

domestic partners Unmarried couples (same- or opposite-sex) who have been granted domestic partnership status—a status which conveys various legal rights and responsibilities.

domestic partnership A status that some states, counties, cities, and workplaces grant to unmarried couples, including gay and lesbian couples, which conveys various rights and responsibilities.

domestic terrorism Domestic terrorism, sometimes called insurgent terrorism, occurs when the terrorist act involves victims, targets, institutions, governments, or citizens from one country.

doubling time The time required for a population to double in size from a given base year if the current rate of growth continues.

drug Any substance other than food that alters the structure or functioning of a living organism when it enters the bloodstream.

drug abuse The violation of social standards of acceptable drug use, resulting in adverse physiological, psychological, and/or social consequences.

drug courts Special courts that divert drug offenders to treatment programs in lieu of probation or incarceration.

dual-use technologies Defense-funded technological innovations with commercial and civilian use.

earned income tax credit (EITC) A refundable tax credit based on a working family's income and number of children.

earnings premium The benefits of having a college degree far outweighs the cost of getting one.

Earth Overshoot Day The approximate date on which humanity's annual demand on the planet's resources exceeds what our planet can renew in a year.

e-commerce The buying and selling of goods and services over the Internet.

economic institution The structure and means by which a society produces, distributes, and consumes goods and services.

ecosystem A biological environment consisting of all the organisms living in a particular area, as well as all the nonliving, physical components of the environment such as air, water, soil, and sunlight, that interact to keep the whole ecosystem functioning.

ecoterrorism Any crime intended to protect wildlife or the environment that is violent, puts human life at risk, or results in damages of $10,000 or more.

education dividend The additional benefits of universal education for women including the reduction of the death rate of children under 5 years of age.

elder abuse The physical or psychological abuse, financial exploitation, or medical abuse or neglect of the elderly.

elderly support ratio The ratio of working-age adults (15 to 64) to adults ages 65 and older in a population.

e-learning Learning in which, by time or place, the learner is separated from the teacher.

emotion work Work that involves caring for, negotiating, and empathizing with people.

Employment Non-Discrimination Act (ENDA) This proposed federal bill that would protect LGBTs from workplace discrimination has been up for congressional debate on a number of occasions since 1994, but has never been signed into law.

environmental footprint The demands that humanity makes on the Earth's natural resources.

environmental injustice Also known as **environmental racism**, the tendency for marginalized populations and communities to disproportionately experience adversity due to environmental problems.

environmental racism See *environmental injustice*.

environmental refugees Individuals who have migrated because they can no longer secure a livelihood as a result of deforestation, desertification, soil erosion, and other environmental problems.

Equal Employment Opportunity Commission (EEOC) A U.S. federal agency charged with ending employment discrimination in the United States and responsible for enforcing laws against discrimination, including Title VII of the 1964 Civil Rights Act that prohibits employment discrimination on the basis of race, color, religion, sex, or national origin.

equal rights amendment (ERA) The proposed 28th amendment to the Constitution, which states that "equality of rights under the law shall not be denied or abridged by the United States, or by any state, on account of sex."

ethnicity A shared cultural heritage or nationality.

Every Child Deserves a Family Act This piece of federal legislation would remove obstacles to non-heterosexual (as well as transgender) individuals providing loving homes for adoption or foster care.

e-waste Discarded electrical appliances and electronic equipment.

experiments Research methods that involve manipulating the independent variable to determine how it affects the dependent variable.

expressive roles Roles into which women are traditionally socialized (i.e., nurturing and emotionally supportive roles).

expulsion Occurs when a dominant group forces a subordinate group to leave the country or to live only in designated areas of the country.

extreme poverty Living on less than $1.25 a day.

family A kinship system of all relatives living together or recognized as a social unit, including adopted people.

Family and Medical Leave Act (FMLA) A federal law that requires public agencies and companies with 50 or more employees to provide eligible workers with up to 12 weeks of job-protected, unpaid leave so that they can care for an ill child, spouse, or parent; stay home to care for their newborn, newly adopted, or newly placed child; or take time off when they are seriously ill, and up to 26 weeks of unpaid leave to care for a seriously ill or injured family member who is in the armed forces, including the National Guard or Reserves.

Family Support Agreement In China, a voluntary contract between older parents and adult children that specifies the details of how the adult children will provide parental care.

feminism The belief that men and women should have equal rights and responsibilities.

feminist criminology An approach that focuses on how the subordinate position of women in society affects their criminal behavior and victimization.

feminization of poverty The disproportionate distribution of poverty among women.

fetal alcohol syndrome A syndrome characterized by serious physical and mental handicaps as a result of maternal drinking during pregnancy.

field research Research that involves observing and studying social behavior in settings in which it occurs naturally.

flextime A work arrangement that allows employees to begin and end the workday at different times so long as 40 hours per week are maintained.

forced labor Also known as slavery, any work that is performed under the threat of punishment and is undertaken involuntarily.

fracking Hydraulic fracturing, commonly referred to as "fracking," involves injecting a mixture of water, sand, and chemicals into drilled wells to crack shale rock and release natural gas into the well.

free trade agreement (FTA) A pact between two or more countries that makes it easier to trade goods across national boundaries and that protects intellectual property rights.

future shock The state of confusion resulting from rapid scientific and technological changes that unravel our traditional values and beliefs.

gateway drug A drug (e.g., marijuana) that is believed to lead to the use of other drugs (e.g., cocaine).

gay A term that can refer to either women or men who are attracted to same-sex partners.

gay pride Demonstrative and cultural expressions of gay activism that include celebrations, marches, demonstrations, or other cultural activities promoting gay rights.

gay-straight alliances (GSAs) School-sponsored clubs led by middle or high schools, that strive to address anti-LGBT name-calling and promote respect for all students.

gender The social definitions and expectations associated with being female or male.

gender deviance hypothesis The tendency to overconform to gender norms after an act(s) of gender deviance; a method of neutralization.

gender expression The way in which a person presents her- or himself as a gendered individual (i.e., masculine, feminine, or androgynous) in society. A person could, for example, have a gender identity as male but nonetheless present their gender as female.

gender non-conforming Often used synonymously with transgender, gender non-conforming (sometimes called gender variant) refers to displays of gender that are inconsistent with society's expectations.

gender roles Patterns of socially defined behaviors and expectations associated with being female or male.

gene monopolies Exclusive control over a particular gene as a result of government patents.

gene therapy The transplantation of a healthy gene to replace a defective or missing gene.

genetic engineering The manipulation of an organism's genes in such a way that the natural outcome is altered.

genetic exception laws Laws that require that genetic information be handled separately from other medical information.

Genetic Information Nondiscrimination Act of 2008 A federal law that prohibits discrimination in health coverage or employment based on genetic information.

genetic testing Examination of DNA in order to identify abnormalities or alterations than may lead to disease or illness.

Geneva Conventions A set of international treaties that govern the behavior of states during wartime, including the treatment of prisoners of war.

genocide The deliberate, systematic annihilation of an entire nation or people.

glass ceiling An invisible barrier that prevents women and other minorities from moving into top corporate positions.

glass escalator effect The tendency for men seeking or working in traditionally female occupations to benefit from their minority status.

GLBT See *LGBT*.

global economy An interconnected network of economic activity that transcends national borders and spans the world.

global warming The increasing average global temperature of Earth's atmosphere, water, and land, caused mainly by the accumulation of various gases (greenhouse gases) that collect in the atmosphere.

globalization The growing economic, political, and social interconnectedness among societies throughout the world.

globesity The high prevalence of obesity around the world.

green energy Also known as clean energy, energy that is nonpolluting and/or renewable, such as solar power, wind power, biofuel, and hydrogen.

greenhouse gases Gases (primarily carbon dioxide, methane, and nitrous oxide) that accumulate in the atmosphere and act like the glass in a greenhouse, holding heat from the sun close to the earth.

Green Revolving Funds (GRFs) College and university funds that are dedicated to financing cost-saving energy-efficiency upgrades and other projects that decrease resource use and minimize environmental impacts.

greenwashing The ways in which environmentally and socially damaging companies portray their corporate image and products as being "environmentally friendly" or socially responsible.

guerrilla warfare Warfare in which organized groups oppose domestic or foreign governments and their military forces; often involves small groups of individuals who use camouflage and underground tunnels to hide until they are ready to execute a surprise attack.

harm reduction A public health position that advocates reducing the harmful consequences of drug use for the user as well as for society as a whole.

hate crime An unlawful act of violence motivated by prejudice or bias.

Head Start Begun in 1965 to help preschool children from the most disadvantaged homes, Head Start provides an integrated program of health care, parental involvement, education, and social services for qualifying children.

health According to the World Health Organization, "a state of complete physical, mental, and social well-being."

heavy drinking As defined by the U.S. Department of Health and Human Services, five or more drinks on the same occasion on each of five or more days in the past 30 days prior to the National Survey on Drug Use and Health.

heterosexism A form of oppression that refers to a belief system that gives power and privilege to heterosexuals, while depriving, oppressing, stigmatizing, and devaluing people who are not heterosexual.

heterosexuality The predominance of emotional, cognitive, and sexual attraction to individuals of the other sex.

homophobia Negative or hostile attitudes directed toward non-heterosexual sexual behavior, a non-heterosexually identified individual, and communities of non-heterosexuals.

homosexuality The predominance of emotional, cognitive, and sexual attraction to individuals of the same sex.

honor killings Murders, often public, as a result of a female dishonoring, or being perceived to have dishonored, her family or community.

human capital hypothesis The hypothesis that pay differences between females and males are a function of differences in women's and men's levels of education, skills, training, and work experience.

hypothesis A prediction or educated guess about how one variable is related to another variable.

identity theft The use of someone else's identification (e.g., Social Security number, birth date) to obtain credit or other economic rewards.

incapacitation A criminal justice philosophy that argues that recidivism can be reduced by placing offenders in prison so that they are unable to commit further crimes against the general public.

independent variable The variable that is expected to explain change in the dependent variable.

index offenses Crimes identified by the FBI as the most serious, including personal or violent crimes (homicide, assault, rape, and robbery) and property crimes (larceny, motor vehicle theft, burglary, and arson).

individual discrimination Occurs when individuals treat other individuals unfairly or unequally due to their group membership.

individualism The tendency to focus on one's individual self-interests and personal happiness rather than on the interests of one's family and community.

industrialization The replacement of hand tools, human labor, and animal labor with machines run by steam, gasoline, and electric power.

infant mortality Deaths of live-born infants under 1 year of age.

insider trading The use of privileged (i.e., nonpublic) information by an employee of an organization that gives that employee an unfair advantage in buying, selling, and trading stocks or other securities.

institution An established and enduring pattern of social relationships.

institutional discrimination Discrimination in which institutional policies and procedures result in unequal treatment of and opportunities for minorities.

institutional racism The systematic distribution of power, resources, and opportunity in ways that benefit whites and disadvantage minorities.

instrumental roles Roles into which men are traditionally socialized (i.e., task-oriented roles).

integration hypothesis A theory that the only way to achieve quality education for all racial and ethnic groups is to desegregate the schools.

intergenerational poverty Poverty that is transmitted from one generation to the next.

internalized homophobia (or internalized heterosexism) The internalization of negative messages about homosexuality by lesbian, gay, and bisexual individuals as a result of direct or indirect social rejection and stigmatization.

Internet An international information infrastructure available through universities, research institutes, government agencies, libraries, and businesses.

Internet piracy Illegally downloading or distributing copyrighted material (e.g., music, games, software).

Interpol The largest international police organization in the world.

intimate partner violence (IPV) Actual or threatened violent crimes committed against individuals by their current or former spouses, cohabiting partners, boyfriends, or girlfriends.

Islamophobia Anti-Muslim and anti-Islam bias.

job burnout Prolonged job stress that can cause or contribute to high blood pressure, ulcers, headaches, anxiety, depression, and other health problems.

job exportation The relocation of jobs to other countries where products can be produced more cheaply.

Kyoto Protocol The first international agreement to place legally binding limits on greenhouse gas emissions from developed countries.

labor unions Worker advocacy organizations that developed to protect workers and represent them at negotiations between management and labor.

larceny Larceny is simple theft; it does not entail force or the use of force, or breaking and entering.

latent functions Consequences that are unintended and often hidden.

learning health systems The result of electronic records whereby physicians can look across patient populations and identify successful treatments or detect harmful interactions.

least developed countries The poorest countries of the world.

legalization Making prohibited behaviors legal; for example, legalizing drug use or prostitution.

lesbian A woman who is attracted to same-sex partners.

LGBT, LGBTQ, and LGBTQI Terms used to refer collectively to lesbian, gay, bisexual, transgender, questioning or "queer," and/or intersexed individuals.

life expectancy The average number of years that individuals born in a given year can expect to live.

light pollution Artificial lighting that is annoying, unnecessary, and/or harmful to life forms on earth.

living apart together (LAT) relationships An emerging family form in which couples—married or unmarried—live apart in separate residences.

living wage laws Laws that require state or municipal contractors, recipients of public subsidies or tax breaks, or, in some cases, all businesses to pay employees wages that are significantly above the federal minimum, enabling families to live above the poverty line.

long-term unemployment rate The share of the unemployed who have been out of work for 27 weeks or more.

malware A general term that includes any spyware, viruses, and adware that is installed on an owner's computer without their knowledge.

managed care Any medical insurance plan that controls costs through monitoring and controlling the decisions of health care providers.

manifest functions Consequences that are intended and commonly recognized.

marital decline perspective A pessimistic view of the current state of marriage that includes the beliefs that (1) personal happiness has become more important than marital commitment and family obligations, and (2) the decline in lifelong marriage and the increase in single-parent families have contributed to a variety of social problems.

marital resiliency perspective A view of the current state of marriage that includes the beliefs that (1) poverty, unemployment, poorly funded schools, discrimination, and the lack of basic services (such as health insurance and child care) represent more serious threats to the well-being of children and adults than does the decline in married two-parent families, and (2) divorce provides a second chance for happiness for adults and an escape from dysfunctional environments.

masculine overcompensation thesis The thesis that men have a tendency to act out in an exaggerated male role when believing their masculinity to be threatened.

master status The status that is considered the most significant in a person's social identity.

maternal mortality Deaths that result from complications associated with pregnancy, childbirth, and unsafe abortion.

Matthew Shepard and James Byrd, Jr. Hate Crimes Prevention Act (HCPA) This new law expands the original 1969 federal hate crimes law to cover hate crimes based on actual or perceived sexual orientation, gender, gender identity, and disability.

McDonaldization The process by which principles of the fast-food industry (efficiency, calculability, predictability, and control through technology) are being applied to more sectors of society, particularly the workplace.

means-tested programs Assistance programs that have eligibility requirements based on income and/or assets.

mechanization Dominant in an agricultural society, the use of tools to accomplish tasks previously done by hand.

Medicaid A public health insurance program, jointly funded by the federal and state governments, that provides health insurance coverage for the poor who meet eligibility requirements.

medicalization Defining or labeling behaviors and conditions as medical problems.

medical tourism A global industry that involves traveling, primarily across international borders, for the purpose of obtaining medical care.

Medicare A federally funded program that provides health insurance benefits to the elderly, disabled, and those with advanced kidney disease.

membership communities Internet sites where participation requires membership and members regularly communicate with one another for personal and/or professional reasons.

mental health The successful performance of mental function, resulting in productive activities, fulfilling relationships with other people, and the ability to adapt to change and to cope with adversity.

mental illness Refers collectively to all mental disorders, which are characterized by sustained patterns of abnormal thinking, mood (emotion), or behaviors that are accompanied by significant distress and/or impairment in daily functioning.

meritocracy A social system in which individuals get ahead and earn rewards based on their individual efforts and abilities.

meta-analysis Meta-analysis combines the results of several studies addressing a research question; i.e., it is the analysis of analyses.

microcredit programs The provision of loans to people who are generally excluded from traditional credit services because of their low socioeconomic status.

military-industrial complex A term first used by Dwight D. Eisenhower to connote the close association between the military and defense industries.

Millennium Development Goals Eight goals that comprise an international agenda for reducing poverty and improving lives.

minority group A category of people who have unequal access to positions of power, prestige, and wealth in a society and who tend to be targets of prejudice and discrimination.

minority stress theory Explains that when an individual experiences the social environment as emotionally or physically threatening due to social stigma, the result is an increased risk for mental health problems.

modern racism A subtle form of racism that involves the belief that serious discrimination in the United States no longer exists, that any continuing racial inequality is the fault of minority

group members, and that the demands for affirmative action for minorities are unfair and unjustified.

monogamy Marriage between two partners; the only legal form of marriage in the United States.

mortality Death.

motherhood penalty The tendency for women with children, particularly young children, to be disadvantaged in hiring, wages, and the like compared to women without children.

multicultural education Education that includes all racial and ethnic groups in the school curriculum, thereby promoting awareness and appreciation for cultural diversity.

Multidimensional Poverty Index A measure of serious deprivation in the dimensions of health, education, and living standards that combines the number of deprived and the intensity of their deprivation.

multiple chemical sensitivity Also known as "environmental illness," a condition whereby individuals experience adverse reactions when exposed to low levels of chemicals found in everyday substances.

mutually assured destruction (MAD) A Cold War doctrine referring to the capacity of two nuclear states to destroy each other, thus reducing the risk that either state will initiate war.

National Coming Out Day Celebrated on October 11, this day is recognized in many countries as a day to raise awareness of the LGBT population and foster discussion of gay rights issues.

National Day of Silence A day during which students do not speak in recognition of the daily harassment that LBGT students endure.

nativist extremist groups Organizations that not only advocate restrictive immigration policy, but also encourage their members to use vigilante tactics to confront or harass suspected undocumented immigrants.

naturalized citizens Immigrants who apply for and meet the requirements for U.S. citizenship.

neglect A form of abuse involving the failure to provide adequate attention, supervision, nutrition, hygiene, health care, and a safe and clean living environment for a minor child or a dependent elderly individual.

neonatal abstinence syndrome (NAS) A condition in which a child, at birth, goes through withdrawal as a consequence of maternal drug use.

net neutrality A principle that holds that Internet users should be able to visit any website and access any content without Internet service provider interference.

no-fault divorce A divorce that is granted based on the claim that there are irreconcilable differences within a marriage (as opposed to one spouse being legally at fault for the marital breakup).

norms Socially defined rules of behavior, including folkways, mores, and laws.

nuclear nonproliferation Efforts to prevent the spread of nuclear weapons, or the materials and technology necessary for the production of nuclear weapons.

nuclear winter The predicted result of a thermonuclear war whereby thick clouds of radioactive dust and particles would block out vital sunlight, lower temperature in the northern hemisphere, and lead to the death of most living things on earth.

objective element of a social problem Awareness of social conditions through one's own life experiences and through reports in the media.

occupational sex segregation The concentration of women in certain occupations and men in other occupations.

Occupy Wall Street A protest movement that began in 2011, and is concerned with economic inequality, greed, corruption, and the influence of corporations on government.

offshoring The relocation of jobs to other countries.

oppression The use of power to create inequality and limit access to resources, which impedes the physical and/or emotional well-being of individuals or groups of people.

organized crime Criminal activity conducted by members of a hierarchically arranged structure devoted primarily to making money through illegal means.

outsource See *outsourcing*.

outsourcing A practice in which a business subcontracts with a third party to provide business services.

overt discrimination Discrimination that occurs because of an individual's own prejudicial attitudes.

parental alienation The intentional efforts of one parent to turn a child against the other parent and essentially destroy any positive relationship a child has with the other parent.

parent trigger laws State legislation that allows parents to intervene in their children's education and schools.

parity In health care, a concept requiring equality between mental health care insurance coverage and other health care coverage.

parole Release from prison, for a specific time period and subject to certain conditions, before an inmate's sentence is finished.

partial-birth abortions Also called an intact dilation and extraction (D & E) abortion, the procedure may entail delivering the limbs and the torso of the fetus before it has expired.

participation gap The tendency for racial and ethnic minorities to participate in information and communication technologies (e.g., using smartphones to access the Internet rather than a computer) that place them in a disadvantaged position (e.g., difficult to research a term paper on a smartphone).

patriarchy A male-dominated family system that is reflected in the tradition of wives taking their husband's last name and children taking their father's name.

penetration rate The percentage of people who have access to and use the Internet in a particular area.

perceived obsolescence The perception that a product is obsolete; used as a marketing tool to convince consumers to replace certain items even though the items are still functional.

pink-collar jobs Jobs that offer few benefits, often have low prestige, and are disproportionately held by women.

pinkwashing The practice of using the color pink and pink ribbons to indicate a company is helping to fight breast cancer, even when the company may be using chemicals linked to cancer.

planned obsolescence The manufacturing of products that are intended to become inoperative or outdated in a fairly short period of time.

pluralism A state in which racial and ethnic groups maintain their distinctness but respect each other and have equal access to social resources.

political alienation A rejection of or estrangement from the political system accompanied by a sense of powerlessness in influencing government.

polyandry The concurrent marriage of one woman to two or more men.

polygamy A form of marriage in which one person may have two or more spouses.

polygyny A form of marriage in which one husband has more than one wife.

population momentum Continued population growth as a result of past high fertility rates that have resulted in a large number of young women who are currently entering their childbearing years.

postindustrialization The shift from an industrial economy dominated by manufacturing jobs to an economy dominated by service-oriented, information-intensive occupations.

postmodernism A worldview that questions the validity of rational thinking and the scientific enterprise.

post-traumatic stress disorder (PTSD) A set of symptoms that may result from any traumatic experience, including crime victimization, war, natural disasters, or abuses.

prejudice Negative attitudes and feelings toward or about an entire category of people.

primary deviance Deviant behavior committed before a person is caught and labeled an offender.

primary groups Usually small numbers of individuals characterized by intimate and informal interaction.

primordial explanations Those explanations that emphasize the existence of "ancient hatreds" rooted in deep psychological or cultural differences between ethnic groups, often involving a history of grievance and victimization, real or imagined, by the enemy group.

privilege When a group has a special advantage or benefits as a result of cultural, economic, societal, legal, and political factors.

probation The conditional release of an offender who, for a specific time period and subject to certain conditions, remains under court supervision in the community.

progressive taxes Taxes in which the tax rate increases as income increases, so that those who have higher incomes are taxed at higher rates.

pronatalism A cultural value that promotes having children.

psychotherapeutic drugs The nonmedical use of any prescription pain reliever, stimulant, sedative, or tranquilizer.

public housing Federally subsidized housing that is owned and operated by local public housing authorities (PHAs).

race A category of people who are perceived to share distinct physical characteristics that are deemed socially significant.

racial profiling The law enforcement practice of targeting suspects on the basis of race.

racism The belief that race accounts for differences in human character and ability and that a particular race is superior to others.

radical environmental movement A grassroots movement of individuals and groups that employs unconventional and often illegal means of protecting wildlife or the environment.

recession A significant decline in economic activity spread across the economy and lasting for at least six months.

recidivism A return to criminal behavior by a former inmate, most often measured by rearrest, reconviction, or reincarceration.

refined divorce rate The number of divorces per 1,000 married women.

registered partnerships Federally recognized relationships that convey most but not all the rights of marriage.

rehabilitation A criminal justice philosophy that argues that recidivism can be reduced by changing the criminal through such programs as substance abuse counseling, job training, education, and so on.

relative poverty The lack of material and economic resources compared with some other population.

replacement-level fertility The level of fertility at which a population exactly replaces itself from one generation to the next; currently, the number is 2.1 births per woman (slightly more than 2 because not all female children will live long enough to reach their reproductive years).

Respect for Marriage Act (RMA) A bill that, if passed, would overturn DOMA and grant federal recognition to same-sex marriages, regardless of the state laws in which they reside.

restorative justice A philosophy primarily concerned with reconciling conflict between the victim, the offender, and the community.

roles The set of rights, obligations, and expectations associated with a status.

sample A portion of the population, selected to be representative so that the information from the sample can be generalized to a larger population.

sanctions Social consequences for conforming to or violating norms.

sandwich generation A generation of people who care for their aging parents while also taking care of their own children.

school-to-prison pipeline The established relationship between severe disciplinary practices, increased rates of dropping out of school, lowered academic achievement, and court or juvenile detention involvement.

school vouchers Tax credits that are transferred to the public or private school that parents select for their child.

science The process of discovering, explaining, and predicting natural or social phenomena.

second shift The household work and child care that employed parents (usually women) do when they return home from their jobs.

secondary deviance Deviant behavior that results from being caught and labeled as an offender.

secondary groups Involving small or large numbers of individuals, groups that are task-oriented and are characterized by impersonal and formal interaction.

Section 8 housing A housing assistance program in which federal rent subsidies are provided either to tenants (in the form of certificates and vouchers) or to private landlords.

security dilemma A characteristic of the international state system that gives rise to unstable relations between states; as State A secures its borders and interests, its behavior may decrease the security of other states and cause them to engage in behavior that decreases A's security.

segregation The physical separation of two groups in residence, workplace, and social functions.

selective primary health care An approach to health care that focuses on using specific interventions to target specific health problems.

self-fulfilling prophecy A concept referring to the tendency for people to act in a manner consistent with the expectations of others.

Semantic Web Sometimes called Web 3.0, a version of the Internet in which pages not only contain information but also describe the interrelationship between pages; sometimes called smart media.

serial monogamy A succession of marriages in which a person has more than one spouse over a lifetime but is legally married to only one person at a time.

sex A person's biological classification as male or female.

sexism The belief that innate psychological, behavioral, and/or intellectual differences exist between women and men and that these differences connote the superiority of one group and the inferiority of the other.

sexual harassment In reference to workplace harassment, when an employer requires sexual favors in exchange for a promotion, salary increase, or any other employee benefit and/or the existence of a hostile environment that unreasonably interferes with job performance.

sexual orientation A person's emotional and sexual attractions, relationships, self-identity, and behavior.

sexual orientation change efforts (SOCE) Collectively refers to reparative, conversion, and reorientation therapies, according to the APA.

shaken baby syndrome A form of child abuse whereby a caretaker shakes a baby to the point of causing the child to experience brain or retinal hemorrhage.

single-payer health care A health care system in which a single tax-financed public insurance program replaces private insurance companies.

slums Concentrated areas of poverty and poor housing in urban areas.

social group Two or more people who have a common identity, interact, and form a social relationship.

social movement An organized group of individuals with a common purpose to either promote or resist social change through collective action.

social problem A social condition that a segment of society views as harmful to members of society and in need of remedy.

Social Security Also called "Old Age, Survivors, Disability, and Health Insurance," a federal program that protects against loss of income due to retirement, disability, or death.

socialism An economic system characterized by state ownership of the means of production and distribution of goods and services.

socioeconomic status A person's position in society based on the level of educational attainment, occupation, and income of that person or that person's household.

sociological imagination The ability to see the connections between our personal lives and the social world in which we live.

state The organization of the central government and government agencies such as the military, police, and regulatory agencies.

State Children's Health Insurance Program (SCHIP) A public health insurance program, jointly funded by the federal and state governments, that provides health insurance coverage for children whose families meet income eligibility standards.

status A position that a person occupies within a social group.

STEM An acronym for science, technology, engineering, and mathematics.

stem cells Undifferentiated cells that can produce any type of cell in the human body.

stereotype threat The tendency of minorities and women to perform poorly on high-stakes tests because of the anxiety created by the fear that a negative performance will validate societal stereotypes about one's member group.

stereotypes Exaggerations or generalizations about the characteristics and behavior of a particular group.

stigma A discrediting label that affects an individual's self-concept and disqualifies that person from full social acceptance.

structural sexism The ways in which the organization of society, and specifically its institutions, subordinate individuals and groups based on their sex classification.

structure The way society is organized including institutions, social groups, statuses, and roles.

structured choice Choices that are limited by the structure of society.

subjective element of a social problem The belief that a particular social condition is harmful to society, or to a segment of society, and that it should and can be changed.

subprime mortgages High-interest or adjustable-rate mortgages that require little money down and are issued to borrowers with poor credit ratings or limited credit history.

Supplemental Nutrition Assistance Program (SNAP) The largest U.S. food assistance program.

supply reduction One of two strategies in the U.S. war on drugs (the other is demand reduction), supply reduction concentrates on reducing the supply of drugs available on the streets through international efforts, interdiction, and domestic law enforcement.

survey research A research method that involves eliciting information from respondents through questions.

sustainable development Occurs when human populations can have fulfilling lives without degrading the planet.

sweatshops Work environments that are characterized by less-than-minimum wage pay, excessively long hours of work (often without overtime pay), unsafe or inhumane working conditions, abusive treatment of workers by employers, and/or the lack of worker organizations aimed to negotiate better working conditions.

symbol Something that represents something else.

synthetic drugs A category of drugs that are "designed" in laboratories rather than naturally occurring in plant material.

tar sands Large, naturally occurring deposits of sand, clay, water, and a dense form of petroleum (that looks like tar).

tar sands oil Oil that results from converting tar sands into liquid fuel. It is known as the world's dirtiest oil because producing it requires energy and generates high levels of greenhouse gases (that cause global warming and climate change), and also leaves behind large amounts of toxic waste.

technological dualism The tendency for technology to have both positive and negative consequences.

technological fix The use of scientific principles and technology to solve social problems.

technology Activities that apply the principles of science and mechanics to the solutions of a specific problem.

technology-induced diseases Diseases that result from the use of technological devices, products, and/or chemicals.

telecommuting A work arrangement involving the use of information technology that allows employees to work part or full time at home or at a satellite office.

telepresencing A sophisticated technology that allows life-sized participants in the virtual presence of one another to realistically communicate through broadcast quality sound and images.

Temporary Assistance for Needy Families (TANF) A federal cash welfare program that involves work requirements and a five-year lifetime limit.

terrorism The premeditated use or threatened use of violence by an individual or group to gain a political objective.

theory A set of interrelated propositions or principles designed to answer a question or explain a particular phenomenon.

therapeutic cloning Use of stem cells to produce body cells that can be used to grow needed organs or tissues; regenerative cloning.

therapeutic communities Organizations in which approximately 35 to 500 individuals reside for up to 15 months to abstain from drugs, develop marketable skills, and receive counseling.

total fertility rates The average lifetime number of births per woman in a population.

toxic workplace A work environment in which employees are subjected to co-workers and/or bosses who engage in a variety of negative, stress-inducing behaviors such as intimidation and workplace bullying, gossiping, and "backstabbing."

transgender individual A transgender individual is a person whose sense of gender identity—masculine or feminine—is inconsistent with their birth (sometimes called chromosomal) sex (male or female).

transnational corporations Also known as multinational corporations, corporations that have their home base in one country and branches, or affiliates, in other countries.

transnational crime Criminal activity that occurs across one or more national borders.

transnational terrorism Terrorism that occurs when a terrorist act in one country involves victims, targets, institutions, governments, or citizens of another country.

under-5 mortality Deaths of children under age 5.

underemployment Unemployed workers as well as (1) those working part-time but who

wish to work full-time, (2) those who want to work but have been discouraged from searching by their lack of success, and (3) others who are neither working nor seeking work but who want and are available to work and have looked for employment in the last year. Also refers to the employment of workers with high skills and/or educational attainment working in low-skill or low-wage jobs.

unemployment To be currently without employment, actively seeking employment, and available for employment, according to U.S. measures of unemployment.

union density The percentage of workers who belong to unions.

universal health care A system of health care, typically financed by the government, that ensures health care coverage for all citizens.

value-added measurement (VAM). VAM is the use of student achievement data to assess teacher effectiveness.

values Social agreements about what is considered good and bad, right and wrong, desirable and undesirable.

variable Any measurable event, characteristic, or property that varies or is subject to change.

victimless crimes Illegal activities that have no complaining participant(s) and are often thought of as crimes against morality, such as prostitution.

vivisection The practice of cutting into or otherwise harming living, nonhuman animals for the purpose of scientific research.

war Organized armed violence aimed at a social group in pursuit of an objective.

wealth The total assets of an individual or household minus liabilities.

wealthfare Laws and policies that benefit the rich.

weapons of mass destruction (WMD) Chemical, biological, and nuclear weapons that have the capacity to kill large numbers of people indiscriminately.

Web 2.0 A platform for millions of users to express themselves online in the common areas of cyberspace.

white-collar crime Includes both *occupational crime*, in which individuals commit crimes in the course of their employment, and *corporate crime*, in which corporations violate the law in the interest of maximizing profit.

worker cooperatives Democratic business organizations controlled by their members, who actively participate in setting their policies and making decisions; also known as *workers' self-directed enterprises*.

workers' compensation Also known as workers' comp, an insurance program that provides medical workers' compensation and living expenses for people with work-related injuries or illnesses.

workers' self-directed enterprises See *worker cooperatives*.

Workforce Investment Act Legislation passed in 1998 that provides a wide array of programs and services designed to assist individuals to prepare for and find employment.

working poor Individuals who spend at least 27 weeks per year in the labor force (working or looking for work) but whose income falls below the official poverty level.

References

Chapter 1

Associated Press. 2006 (September 7). "Florida Appeals Court Upholds Ban of Veil in Driver's License Photo." *Religious News.* Pew Forum on Religion and Public Life. Available at http://pewforum.org/

Berger, Peter. 1963. *Invitation to Sociology.* Garden City, N.J.: Doubleday.

Blumer, Herbert. 1971. "Social Problems as Collective Behavior." *Social Problems* 8(3):298–306.

Byard, Eliza. 2013 (March 26). "The 81%. How Young Adults are Reshaping the Marriage Equality Debate." Available at http://tv.msnbc.com

Canedy, Dana. 2002. "Lifting Veil for Photo ID Goes Too Far, Driver Says." *New York Times,* June 27. Available at www.nytimes.com

Career Cast. 2013. "Jobs Rated 2013: Ranking 200 Jobs from Best to Worse." Available at www.careercast.com

Centers for Disease Control and Prevention. 2008. (August 1). Trends In HIV- and STD-Related Risk Behaviors among High School Students—United States, 1991–2007. *Morbidity and Mortality Weekly Report* 57(30):817–822.

Centers for Disease Control and Prevention. 2010 (June 4). Youth Risk Behavior Surveillance, United States, 2009. *Morbidity and Mortality Weekly Report* 59 No. SS-5.

Centers for Disease Control and Prevention. 2012 (July 27). Trends in HIV-Related Risk Behaviors among High School Students—United States, 1991–2011. *Morbidity and Mortality Weekly Report* 61(29):556–560.

The Eleanor Roosevelt Papers. 2008. "Teaching Eleanor Roosevelt: American Youth Congress." Available at www.nps.gov

Farrell, Dan and James C. Petersen. 2010. "The Growth of Internet Research Methods and the Reluctant Sociologist." *Sociological Inquiry* 80(1):114–125.

File, Thom. 2013. "Computer and Internet Use in the United States." U.S. Department of Commerce. U.S. Census Bureau, P20-569.

Finn, Kathy. 2013. "Ruling against BP Clears Way for Appeal of Spill Payouts." April 5. Reuters News Service. Available at www.reuters.com

Fleming, Zachary. 2003. "The Thrill of It All." In *In Their Own Words,* ed. Paul Cromwell, 99–107. Los Angeles, CA: Roxbury.

Gallup Poll. 2013a. "Most Important Problem." May 7. Available at www.gallup.com

Gallup Poll. 2013b. "Satisfaction with the United States." May 14. Available at www.gallup.com

George, Nimala. 2012. "Outrage Grows in India over Gang Rape on Bus." *The Christian Science Monitor,* December 19. Available at www.csmonitor.com

Gottipati, Sruthi, Anjani Trivedi, and Saritha Rai. 2012. "Protests across India Over Death of Gang Rape Victim." *The New York Times,* December 29. Available at http://india.blogs.nytimes.com

Herd, D. 2011. "Voices from the Field: The Social Construction of Alcohol Problems in Inner-City Communities." *Contemporary Drug Problems,* 38(1):7–39.

Indiana State Government. 2013. "General Information." Department of Toxicology. Available at www.in.gov/isdt/2340.htm

Kmec, Julie A. 2003. "Minority Job Concentration and Wages." *Social Problems* 50:38–59.

Matza, David. 1990. *Delinquency and Drift.* Brunswick, N.J: Transaction Publishers.

Merton, Robert K. 1968. *Social Theory and Social Structure.* New York: Free Press.

Mills, C. Wright. 1959. *The Sociological Imagination.* London: Oxford University Press.

Mooney, Linda A. 2015. *An Author's "Human Side."* Personal Essay.

National Highway Traffic Safety Administration. 2013 (April). "Using Electronic Devices While Driving is a Serious Safety Problem." *Safety in Numbers* 1(1):all. Available at www.distraction.gov

Newport, Frank. 2012 (November 29). "Democrats, Republicans Diverge on Capitalism, Federal Gov't." Available at www.gallup.com

Obama, Barack. 2013. "Inaugural Address by President Barack Obama." January 21. Available at www.whitehouse.gov

Palacios, Wilson R., and Melissa E. Fenwick. 2003. "'E' Is for Ecstasy." In *In Their Own Words,* ed. Paul Cromwell, 277–283. Los Angeles, CA: Roxbury.

Park, Madison. 2013. "Anger, Frustration Over Rapes in India: 'Mindset Hasn't Changed.'" April 23. Available at www.cnn.com

Pope, Carl. 2011. "No, BP Won't Make it Right." *Huffington Post,* February 25. Available at www.huffingtonpost.com

Pryor, J. H., K. Eagan, L. Paluki Blake, S. Hurtado, J. Berdan, and M. Case. 2012. *The American Freshman: National Norms 2012.* Los Angeles: Higher Education Research Institute, UCLA.

Recovery.gov. 2013. "Overview of Funding." Available at www.recovery.gov

Reiman, Jeffrey, and Paul Leighton. 2013. *The Rich Get Richer and the Poor Get Prison,* 10th ed. Pearson.

Rifkind, Hugo. 2009. "Student Activism Is Back." *Times Online,* February 16. Available at http://women.timesonline.co.uk

Schlosser, Jim. 2000. "Activist Recalls 'Catalyst for Civil Rights.'" *Greensboro News and Record,* February 2. Available at www.sitins.com

The Scranton Report. 1971. *The Report of the President's Commission on Campus Unrest.* Washington, DC: U.S. Government Printing Office.

Schoenberg,Tom, and Phil Mattingly. 2013. "Peanut Corp. of America Officials Charged Over Salmonella." February 21. *Bloomberg News Service.* Available at www.bloomberg.com

Simi, Pete, and Robert Futrell. 2009. "Negotiating White Power." *Social Problems* 56(1):98–110.

Smith, Jacquelyn. 2013. "The Best and Worst Jobs for 2013." *Forbes.* April 23. Available at www.forbes.com.

Sykes, Marvin. 1960. "Negro College Students Sit at Woolworth Lunch Counter." *Greensboro Record,* February 2. Available at www.sitins.com

Thomas, W. I. 1931/1966. "The Relation of Research to the Social Process." In *W. I. Thomas on Social Organization and Social Personality,* ed. Morris Janowitz, 289–305. Chicago: University of Chicago Press.

U.S. Department of State. 2013. " USA Jobs." Available at www.usajobs.gov

The Wall Street Journal (WSJ). 2013. "Best and Worst Jobs of 2013." April 22. Available at www.wsj.com

Weir, Sara, and Constance Faulkner. 2004. *Voices of a New Generation: A Feminist Anthology.* Boston, MA: Pearson Education Inc.

Wilhelm, Ian. 2012 (May 20). "Northern Arizona U. Overhauls Curriculum to Focus on 'Global

*The authors and Cengage Learning acknowledge that some of the Internet sources may have become unstable; that is, they are no longer hot links to the intended reference. In that case, the reader may want to access the article through the search engine or archives of the homepage cited (e.g., fbi.gov, cbsnews.com).

Competence.'" *Chronicle of Higher Education.* Available at http://chronicle.com

Wilson, John. 1983. *Social Theory.* Englewood Cliffs, NJ: Prentice Hall.

Chapter 2

Allen, P. L. 2000. *The Wages of Sin: Sex and Disease, Past and Present.* Chicago: University of Chicago Press.

Alvarez, Lizette, and Erik Eckholm. 2009. "Purple Heart Is Ruled Out for Traumatic Stress." *New York Times,* January 27, p. A1.

American College Health Association. 2012. *American College Health Association National College Health Assessment II: Reference Group Executive Summary Spring 2012.* Hanover MD: American College Health Association.

Angell, Marcia. 2003. "Statement of Dr. Marcia Angell Introducing the U.S. National Health Insurance Act." Physicians for a National Health Program. Available at www.pnhp.org

Arehart-Treichel, Joan. 2008. "Psychiatrists and Farmers: Alliance in the Making?" *Psychiatric News* 43(10):15.

Bafana, Busani. 2013 (January 11). "Morphine Kills Pain but Its Price Kills Patients." InterPress Service. Available at http://ipsnews.net

Barker, Kristin. 2002. "Self-Help Literature and the Making of an Illness Identity: The Case of Fibromyalgia Syndrome (FMS)." *Social Problems* 49(3):279–300.

Barnes, Brooks. 2012. "Promoting Nutrition, Disney to Restrict Junk Food Ads." *New York Times,* June 5. Available at www.nytimes.com

Beronio, K., R. Po, L. Skopec, and S. Glied. 2013 (February 20). "Affordable Care Act Expands Mental Health and Substance Use Disorder Benefits and Federal Parity Protections for 62 Million Americans." *ASPE Issue Brief.* U.S. Department of Health and Human Services. Available at aspe.hhs.gov

Biset, Blain. 2013. "No Woman Should Die Giving Life." Inter Press Service, January 31. Available at www.ipsnews.net

Borgelt, L. M., K. L. Franson, A. M. Nussbaum, and G. S. Wang. 2013. "The Pharmacologic and Clinical Effects of Medical Cannabis." *Pharmacotherapy* 33(2):195–209.

Braine, Theresa. 2011. "Race against Time to Develop New Antibiotics." *Bulletin of the World Health Organization* 89:88–89.

Brill, Steven (2013, March 4). "Bitter Pill," *Time Magazine* 181(8):16–55.

Carolla, Bob. 2013 (March). "Entry on Mental Illness Added to AP Stylebook." *NAMI NOW.* Available at www.nami.org

Center for Science in the Public Interest. 2012. "Petition to Ensure the Safe Use of Added Sugars." Available at www.cspi.org

Centers for Disease Control and Prevention. 2013a. "Making Health Care Safer: Stop Infections from Lethal CRE Germs Now." *Vital Signs* (March). Available at www.cdc.gov

Centers for Disease Control and Prevention. 2013b. "Deaths: Final Data for 2010." *National Vital Statistics Report* 61(4): Table 7.

Centers for Disease Control and Prevention. 2011 (January 14). *CDC Health Disparities and Inequalities Report—United States, 2011.* MMWR 60:all.

Chandler, C. K. 2005. *Animal Assisted Therapy in Counseling.* New York: Routledge.

Clemmitt, Marcia. 2012 (September 21). "Assessing the New Health Care Law." *CQ Researcher* 22(3):791–811.

Cockerham, William C. 2007. *Medical Sociology,* 10th ed. Upper Saddle River, NJ: Prentice Hall.

Cohen, R. A., R. M. Gindi, and W. K. Kirzinger. 2012. "Burden of Medical Care Cost: Early Release of Estimates from the National Health Interview Survey, January–June 2011." National Center for Health Statistics. Available at www.cdc.gov/nchs

Crudo, Dana. 2013 (January). "New Semester, New NAMI on Campus Clubs." *NAMI Now.* Available at www.nami.org

Davis, Karen, Cathy Schoen, Stephen C. Schoenbaum, Michelle M. Doty, Alyssa L. Holmgren, Jennifer L. Kriss, and Katherine K. Shea. 2007 (May). *Mirror, Mirror on the Wall: An International Update on the Comparative Performance of American Health Care.* Commonwealth Fund. Available at http://commonwealthfund.org

Davis, Mathew A., Brook I. Martin, Ian D. Coulter, and William B. Weeks. 2013. "U.S. Spending on Complementary and Alternative Medicine during 2002–2008 Plateaued, Suggesting Role in Reformed Health System." *Health Affairs* 32(1):45–52.

Delany, Bill. 2012 (June 26). "New Hampshire Native: Allow Compassionate Use in the Granite State." *Medical Cannabis: Voices from the Frontlines. Blog Archive.* Safe Access Now. Available at http://safeaccessnow.org

DeNavas-Walt, Carmen, Bernadette D. Proctor, and Jessica C. Smith. 2013. Income, Poverty, and Health Insurance Coverage in the United States: 2012. *Current Population Reports* P60–245. U.S. Census Bureau. Available at www.census.gov

Devries, K. M., J. Y. T. Mak, C. García-Moreno, M. Petzold, J. C. Child, G. Falder, S. Lim, L. J. Bacchus, R. E. Engell, L. Rosenfeld, C. Pallitto, T. Vos, N. Abrahams, and C. H. Watts. 2013. "The Global Prevalence of Intimate Partner Violence against Women." *Science* (June 28):1527–1528.

Dingfelder, Sadie F. 2009 (June). "The Military's War on Stigma." *Monitor on Psychology* 40(6):52.

Families USA. 2007 (January 9). "No Bargain: Medicare Drug Plans Deliver Higher Prices." Available at www.familiesusa.org

Farmer, Paul, Julio Frenk, Felicia M. Knaul, Lawrence N. Shulman, George Alleyne, Lance Armstrong, Rifat Atun, Douglas Blayney, Lincoln Chen, Richard Feachem, Mary Gospodarowicz, Julie Gralow, Sanjay Gupta, Ana Langer, Julian Lob-Levyt, Claire Neal, Anthony Mbewu, Dina Mired, Peter Piot, K. Srinath Reddy, Jeffrey D. Sachs, Mahmoud Sarhan, and John R. Seffrin. 2010 (October 2). "Expansion of Cancer Care and Control in Countries of Low and Middle Income: A Call to Action." *Lancet* 376(9747):1186–1193.

Federal Trade Commission. 2012. "A Review of Food Marketing to Children and Adolescents." Available at www.ftc.gov

Fine, Aubrey H. 2010. "Forward." In *Handbook on Animal-Assisted Therapy,* 3rd ed., Aubrey H. Fine, ed. (pp. xix–xxi). Burlington MA: Academic Press.

Fischman, J. 2010 (September 12). "The Pressure of Race." *The Chronicle of Higher Education.* Available at http://chronicle.com

Goldstein, Michael S. 1999. "The Origins of the Health Movement." In *Health, Illness, and Healing: Society, Social Context, and Self,* Kathy Charmaz and Debora A. Paterniti, eds. (pp. 31–41). Los Angeles, CA: Roxbury.

Goode, Erica, and Jack Healy. 2013 (January 31). "Focus on Mental Health Laws to Curb Violence is Unfair, Some Say." *New York Times.* Available at www.nytimes.com

Grossman, Amy. 2009. "A Birth Pill." *New York Times,* May 9. Available at www.nytimes.com

Hallal, P. C., L. B. Andersen, F. C. Bull, R. Guthold, W. Haskell, and U. Ekelund. 2012. "Global Physical Activity Levels: Surveillance, Progress, Pitfalls, and Prospects." *The Lancet* 380:247–257.

Harrison, Joel A. 2008. "How Much Is the Sick U.S. Health Care System Costing You?" *Dollars and Sense,* May/June. Available at dollarsandsense.com

Harvard School of Public Health. 2013. "The Obesity Prevention Source. Obesity Causes; Globalization." Available at www.hsph.harvard.edu

Hawkes, Corinna. 2006. "Uneven Dietary Development: Linking the Policies and Processes of Globalization with the Nutrition Transition, Obesity, and Diet-Related Chronic Diseases." *Globalization and Health* 2:4.

Himmelstein, D. U., D. Thorne, E. Warren, and S. Woolhandler. 2009. "Medical Bankruptcy in the United States, 2007: Results of a National Study." *The American Journal of Medicine* 122(8):741–46.

Hingle, Melanie, Mimi Nichter, Melanie Medeiros, and Samantha Grace. 2013. "Texting for Health: The Use of Participatory Methods to Develop Healthy Lifestyle Messages for Teens." *Journal of Nutrition Education and Behavior* 45(1): 12–19.

Huffington Post. 2013 (March 12). "NYC Soda Ban Rejected: Judge Strikes Down Limit in Large Sugary Drinks as 'Arbitrary, Capricious.'" Available at www.hufffingtonpost.com

Hummer, Robert A., and Elaine M. Hernandez. 2013. "The Effects of Educational Attainment on Adult Mortality in the United States." *Population Bulletin* 68(1): all.

Kaiser, Chris. 2013. "Sugary Drinks Add to Global Death Rates." *MedPage Today,* March 20. Available at www.medpagetoday.com

Kaiser Commission on Medicaid and the Uninsured. 2010 (September). *The Uninsured and the Difference Health Insurance Makes.* Washington, DC: Kaiser Family Foundation.

Kaiser Family Foundation. 2013 (March). *Kaiser Health Tracking Poll: Public Opinion of Health Care Issues.* Available at www.kff.org

Kaiser Family Foundation. 2009 (June). "Health Tracking Poll." Available at www.kff.org

Kaiser Family Foundation. 2007. "How Changes in Medical Technology Affect Health Care Costs." Available at www.kff.org

Kaiser Family Foundation/HRET. 2012. "Employer Health Benefits 2012 Annual Survey." Available at www.kff.org

Katel, Peter. 2010. "Food Safety." *CQ Researcher* 20(44): all.

Kindig, David A., and Erika R. Cheng. 2013. "Even as Mortality Fell in Most U.S. Counties, Female Mortality Nonetheless Rose in 42.8 Percent of Counties from 1992 to 2006." *Health Affairs* 32(3):451–458.

Kolappa, K., D. C. Henderson, and S. P. Kishore. 2013. "No Physical Health Without Mental Health: Lessons Unlearned?" *Bulletin of the World Health Organization* 91(3):3–3A.

Kruger, K. A., and J. A. Serpell. 2010. "Animal-Assisted Interventions in Mental Health: Definitions and Theoretical Foundations." In *Handbook on Animal-Assisted Therapy*, 3rd ed., Aubrey H. Fine ed. (pp. 33–48). San Diego: Academic Press.

Lee, Kelley. 2003. "Introduction." In *Health Impacts of Globalization*, ed. Kelley Lee, 1–10. New York: Palgrave MacMillan.

Light, Donald W., and Rebecca Warburton. 2011 (February 7). "Demythologizing the High Costs of Pharmaceutical Research." *Biosocieties*, pp. 1–17.

Link, Bruce G., and Jo Phelan. 2001. "Social Conditions as Fundamental Causes of Disease." In *Readings in Medical Sociology*, 2nd ed., William C. Cockerham, Michael Glasser, and Linda S. Heuser, eds. (pp. 3–17). Upper Saddle River, NJ: Prentice Hall.

Mahar, Maggie. 2006. *Money-Driven Medicine.* New York: Harper-Collins.

Mannix, Jeff. 2009. "The Patients' Perspective: Locals Find Relief with Medical Marijuana." *Durango Telegraph,* November 19. Available at www.durangotelegraph.com

Mayer, Lindsay Renick. 2009. "Insurers Fight Public Health Plan." Capitol Eye Blog, June 18. Center for Responsive Politics. Available at www.opensecrets.org

Morain, Stephanie, and Michelle M. Mello. 2013. "Survey Finds Public Support for Legal Interventions Directed at Health Behavior to Fight Noncommunicable Disease." *Health Affairs* 32(3):486–496.

Musumeci, Mary Beth. 2012 (July). "A Guide to the Supreme Court's Affordable Care Act Decision." Kaiser Family Foundation. Available at www.kff.org

Nader, Ralph. 2009 (July 25). "Health Care Hypocrisy." *Common Dreams.* Available at www.commondreams.org

National Alliance on Mental Illness. 2012. *College Students Speak: A Survey Report on Mental Health.* Available at www.nami.org

National Center for Complementary and Alternative Medicine (NCCAM). 2012. "What is CAM?" Available at http://nccam.nih.gov

National Center for Health Statistics. 2012. *Health, United States, 2011 with Special Feature on Socioeconomic Status and Health.* Hyattsville, MD: U.S. Government Printing Office.

National Coalition on Health Care. 2009. "Health Insurance Costs." Available at www.nchc.org

National Conference of State Legislators. 2010. "Tanning Restrictions for Minors—A State-by-State Comparison." Available at www.ncsl.org

National Research Council and Institute of Medicine. 2013. *U.S. Health in International Perspective: Shorter Lives, Poorer Health.* Washington D.C.: The National Academies Press.

Park, Madison. 2009 (September 18). "45,000 American Deaths Associated with Lack of Insurance." CNN. Available at http://articles.cnn.com

Peters, Sharon. 2011. "Animals Can Assist in Psychotherapy." *USA Today,* January 17. Available at www.usatoday.com

Pew Health Initiatives. 2013 (February 6). "Record-High Antibiotic Sales for Meat and Poultry Production." Available at www.pewhealth.org

Potter, Wendell. 2010. *Deadly Spin: An Insurance Company Insider Speaks Out on How Corporate PR is Killing Health Care and Deceiving Americans.* New York: Bloomsbury Press.

ProCon.org. 2013 (February). "Medical Marijuana. 18 Legal Medical Marijuana States and D.C." Available at http://medicalmarijuanaprocon.org

Quadagno, Jill. 2004. "Why the United States Has No National Health Insurance: Stakeholder Mobilization against the Welfare State 1945–1996." *Journal of Health and Social Behavior* 45:25–44.

Reinberg, Steven. 2009. "Tanning Beds Get Highest Carcinogen Rating." *U.S. News & World Report,* July 28. Available at http://health.usnews.com

Ruiz, John M., Patrick Steffen, and Timothy B. Smith. 2013. "Hispanic Mortality Paradox: A Systematic Review and Meta-Analysis of the Longitudinal Literature." *American Journal of Public Health* 103(3):e52–e60.

Sanders, David, and Mickey Chopra. 2003. "Globalization and the Challenge of Health for All: A View from Sub-Saharan Africa." In *Health Impacts of Globalization*, ed. Kelley Lee (pp. 105–119). New York: Palgrave Macmillan.

Scal, Peter, and Robert Town. 2007. "Losing Insurance and Using the Emergency Department: Critical Effect of Transition to Adulthood for Youth with Chronic Conditions." *Journal of Adolescent Health* 40 (2 Suppl. 1):S4.

Sered, Susan Starr, and Rushika Fernandopulle. 2005. *Uninsured in America: Life and Death in the Land of Opportunity.* Berkeley and Los Angeles, CA: University of California Press.

Shally-Jensen, Michael. 2013. "Introduction." In *Mental Health Care Issues in America: An Encyclopedia*, M. Shally-Jensen, ed. (pp. i–xxix). Santa Barbara CA: ABC-CLIO, LLC.

Shern, David, and Wane Lindstrom. 2013. "After Newtown: Mental Illness and Violence." *Health Affairs* 32(3):447–450.

Sidel, Victor W., and Barry S. Levy. 2002. "The Health and Social Consequences of Diversion of Economic Resources to War and Preparation for War." In *War or Health: A Reader*, Ilkka Taipale, P. Helena Makela, Kati Juva, and Vappu Taipale, eds. (pp. 208–221). New York: Palgrave MacMillan.

Smith-McDowell, Keiana. 2013 (March). "Executive Order Calls for New Recommendations on Mental Health Issues." *NAMI Now.* Available at www.nami.org

Squires, David A. 2012. "Explaining High Health Care Spending in the United States: An International Comparison of Supply, Utilization, Prices, and Quality." *Issues in International Health Policy* (May). The Commonwealth Fund. Available at www.commonwealthfund.org

Substance Abuse and Mental Health Services Administration. 2012. *Results from the 2011 National Survey on Drug Use and Health: Mental Health Findings.* Rockville, MD: Substance Abuse and Mental Health Services Administration.

Szasz, Thomas. 1961/1970. *The Myth of Mental Illness: Foundations of a Theory of Personal Conduct.* New York: Harper & Row.

Tavernise, S. 2012. "Life Span Shrink for Least-Educated Whites in the U.S." *New York Times,* September 20. Available at www.nytimes.com

"Testimonials." 2013. NORML. Available at http://norml.org/about/item/testimonials

Thomson, George, and Nick Wilson. 2005. "Policy Lessons from Comparing Mortality from Two Global Forces: International Terrorism and Tobacco." *Globalization and Health* 1:18.

Trust for America's Health. 2012. "F as in Fat: How Obesity Threatens America's Future." Available at http://healthyamericans.org

Turner, Leigh, and Jill R. Hodges. 2012. "Introduction: Health Care Goes Global." In J. R. Hodges, L. Turner, and A. M. Kimball (Eds.), *Risks and Challenges in Medical Tourism*, pp. 1–18. Santa Barbara: Praeger.

UNICEF. 2012a. *State of the World's Children.* Available at www.unicef.org

UNICEF. 2012b. *Progress for Children: A Report Card on Adolescents.* Available at www.childinfo.org

Urichuk, L. with D. Anderson. 2003. *Improving Mental Health through Animal-Assisted Therapy.* Edmonton, Alberta: The Chimo Project.

U.S. Department of Health and Human Services. 2001. *Mental Health: Culture, Race, and Ethnicity—A Supplement to Mental Health: A Report of the Surgeon General.* Rockville, MD: U.S. Government Printing Office.

Weitz, Rose. 2013. *The Sociology of Health, Illness, and Health Care: A Critical Approach.* 6th ed. Belmont, CA: Wadsworth/Cengage.

White, Frank. 2003. "Can International Public Health Law Help to Prevent War?" *Bulletin of the World Health Organization* 81(3):228.

Williams, David R. 2012. "Miles to Go before We Sleep: Racial Inequities in Health." *Journal of Health and Social Behavior* 53(3):279–295.

Williams, David R. 2003. "The Health of Men: Structured Inequalities and Opportunities." *American Journal of Public Health* 93(5): 724–731.

Williams, D. R., M. B. McClellan, and A. M. Rivlin. 2010 (August). "Beyond the Affordable Care Act: Achieving Real Improvements in Americans' Health." *Health Affairs* 29(8): 1481–1488.

Wisner, K. L., D. K. Y. Sit, M. C. McShea, et al. 2013. "Onset Timing, Thoughts of Self-Harm, and Diagnoses in Postpartum Women with Screen-Positive Depression Findings." *JAMA Psychiatry* (Online March 13). Available at http://archpsych.jamanetwork.com

Wootan, Margo. 2012 (December 21). "Little Improvement Seen in Food Marketing." www.cspinet.org

World Health Organization. 2013. *World Health Statistics 2013.* Available at www.who.int

World Health Organization. 2012a. *World Health Statistics 2012.* Available at www.who.int

World Health Organization. 2012b. "The Top 10 Causes of Death." *Fact Sheet No. 310.* Available at www.who.int

World Health Organization. 2010. "Mental Health and Development: Targeting People with Mental Health Conditions as a Vulnerable Group." Available at www.who.org

World Health Organization. 2000. The World Health Report 2000. Available at http://www.who.int

World Health Organization. 1946. *Constitution of the World Health Organization.* New York: World Health Organization Interim Commission.

Chapter 3

Abadinsky, Howard. 2014. *Drug Use and Abuse: A Comprehensive Introduction.* Belmont, CA: Wadsworth.

Alcoholics Anonymous. 2013 (April 8). "Estimates of A.A. Groups and Members as of January 1, 2013." *Service Material from the General Service Office.* Available at www.aa.org

Alderman, Lisa. 2013 (June 12). "E-Cigarettes are in Vogue and at a Crossroads." Available at www.nytimes.com

Alfonsi, Sharyn, and Hanna Siegel. 2010. "Heroin Use in Suburbs on the Rise." ABC News, March 29. Available at abcnews.go.com

American College Health Association. 2013. *American College Health Association National College Health Assessment II.* Reference Group Executive Summary Fall 2012. Hanover, MD: American College Health Association.

Allsop, D. J., J. Copeland, M. M. Norberg, S. Fu, A. Molnar, J. Lewis, and A. J. Budney. 2012. "Quantifying the Clinical Significance of Cannabis Withdrawal." *Public Library of Science* 7(9):44864–44864.

American College Health Association. 2013. *American College Health Association National College Health Assessment II.* Reference Group Executive Summary Fall 2012. Hanover, MD: American College Health Association.

Balsa, A. I., J. F. Homer, and M. T. French. 2009. "The Health Effects of Parental Problem Drinking on Adult Children." *Journal of Mental Health Policy and Economics* 12(2):55–66.

Becker, H. S. 1966. *Outsiders: Studies in the Sociology of Deviance.* New York: Free Press.

Behrendt, S., H.-U. Wittchen, M. Höfler, R. Lieb, and K. Beesdo. 2009. "Transitions from First Substance Use to Substance Use Disorders in Adolescence: Is Early Onset Associated with a Rapid Escalation?" *Drug and Alcohol Dependence* 99:68–78.

Bellum, Sara, 2013. "Bath Salts: An Emerging Danger." *NIDA for Teens,* February 5. Available at http://teens.drugabuse.gov

Branson, Richard. 2012 (December 7). "War on Drugs Trillion-Dollar Failure." Available at www.cnn.com

BMA (British Medical Association). 2013. *Drugs of Dependence: The Role of Medical Professionals.*

BMA Board of Science. January. Available at http://bma.org.uk

Campaign for Tobacco Free Kids. 2013 (March 13). *Not Your Grandfather's Cigars.* Available at www.tobaccofreekids.org

CASA (National Center on Addiction and Substance Abuse). 2012. "National Survey of American Attitudes on Substance Abuse XVII: Teens." Available at www.casacolumbia.org

CASA. 2009. "The Impact of Substance Abuse on Federal, State, and Local Budgets." New York: Columbia University.

Cattan, Nacha. 2010. "How Mexican Drug Gangs Use YouTube against Rival Groups." *Christian Science Monitor,* November 5. Available at www.csmonitor.com

CDC (Centers for Disease Control). 2013a. "Smoking in the Movies." Available at www.cdc.gov

CDC. 2013b. "Binge Drinking: A Serious, Underrecognized Problem among Women and Girls." *Vital Signs,* January. Available at www.cdc.gov

CDC. 2013c (April 17). "Impaired Driving: Get the Facts." *Injury Prevention and Control: Motor Vehicle Safety.* Available at www.cdc.gov

CDC. 2012a (January 10). "Health Effects of Cigarette Smoking." *Smoking and Tobacco Use.* Available at www.cdc.gov

CDC. 2012b. "Frequently Asked Questions." *Alcohol and Public Health,* November 7. Available at www.cdc.gov

CDC. 2012c. "Vital Signs: Drinking and Driving among High School Students Aged ≥16 Years — United States, 1991–2011." *Morbidity and Mortality Weekly Report (MMWR),* October 12. Available at www.cdc.gov/mmwr/

CDC. 2011a. "Excessive Drinking Costs U.S. $223.5 B." *CDC Features,* October 17. Available at www.cdc.gov/alcohol

CDC. 2011b (September 22). "Fact about FASDs." *Fetal Alcohol Spectrum Disorders (FASDs).* Available at: www.cdc.gov

Champion, Katrina E., Nicola C. Newton, Emma L. Barrett, and Maree Teesson. 2013. "A Systematic Review of School-Based Alcohol and Other Drug Prevention Programs Facilitated by Computers or the Internet." *Drug and Alcohol Review* 32(2):115–123.

Chauvin, Chantel D. 2012. "Social Norms and Motivations Associated with College Binge Drinking." *Sociological Inquiry* 82(2):257–281.

Chouvy, Piere-Arnaud. 2013. "A Typology of Unintgended Consequences of Drug Crop Reduction." *Journal of Drug Issues* 43(2):216–230.

Copes, Heigh, Andy Hochstetler, and J. Patrick Williams. 2008. "'We Weren't Like No Regular Dope Fiends': Negotiating Hustler and Crackhead Identities." *Social Problems* 55(2):254–270.

CTC (Count the Costs). 2013. "The Seven Costs." Available at www.countthecosts.org

Crisp, Elizabeth. 2012. "Mo. Legislature Approves Change in Crack Cocaine Sentencing." *The St. Louis Post-Dispatch,* May 18.

Csomor, Marina. 2012. "There's Something (Potentially Dangerous) about Molly." *CNN Health,* August 16. Available at www.cnn.com

DEA. 2010. "Fiction: Drug Production Does Not Damage the Environment." Just Think Twice: Facts and Fiction. Available at www.justthinktwice.com

Degenhardt, Louisa, Wai-Tat Chiu, Nancy Sampson, Ronald C. Kessler, James C. Anthony, Matthias Angermeyer, Ronny Bruffaerts, Giovanni de Girolamo, Oye Gureje, Yueqin Huang, Aimee Karam, Stanislav Kostyuchenko, Jean Pierre Lepine, Maria Elena Medina Mora, Yehuda Neumark, J. Hans Ormel, Alejandra Pinto-Meza, José Posada-Villa, Dan J. Stein, Tadashi Takeshima, and J. Elisabeth Wells. 2008. "Toward a Global View of Alcohol, Tobacco, Cannabis, and Cocaine Use: Findings from the WHO World Mental Health Surveys." *PLoS Medicine* 5(1):1053–1077.

Dekel, Rachel, Rami Benbenishty, and Yair Amram. 2004. "Therapeutic Communities for Drug Addicts: Prediction of Long-Term Outcomes." *Addictive Behaviors* 29(9):1833–1837.

Dinan, Stephen, and Ben Conery. 2009. "DEA Pot Raids Go On; Obama Opposes." *The Washington Times,* February 5. Available at www.washingtontimes.com

Dinno, Alexis, and Stanton Glantz. 2009. "Tobacco Control Policies are Egalitarian: A Vulnerabilities Perspective on Clean Indoor Air Laws, Cigarette Prices, and Tobacco Use Disparities." *Social Science & Medicine* 68:1439–1447.

DPA (Drug Policy Alliance). 2013. *Fact Sheet: Women, Prison, and the Drug War.* Available at www.drugpolicy.org/

Doward, Jamie. 2013. "Western Leaders Study 'Gamechanging' Report on Global Drugs Trade." *The Guardian,* May 18. Available at www.guardian.co.uk

Earhart, James, Shelley R. Hart, Shane R. Jimerson, Tyler Renshaw, Elina Saeki, and Renee D. Singh. 2011. "A Summary and Synthesis of Contemporary Empirical Evidence Regarding the Effects of the Drug Abuse Resistance Education program (D.A.R.E.)." *Contemporary School Psychology* 15(January):93–102.

The Economist. 2013 (February 23). "Winding Down the War on Drugs towards a Ceasefire." Available at: www.economist.com

The Economist. 2009 (March 7). "A Toker's Guide." Available at www.economist.com

EMCDDA (European Monitoring Centre for Drugs and Drug Addiction). 2013 (June 6). "Perspectives on Drugs: The New EU Drugs Strategy (2013–2020)." Available at www.emcdda.org

Eriksen, M., J. Mackay, and H. Ross. 2012. *The Tobacco Atlas,* 4th ed. Atlanta, GA: American Cancer Society; New York, NY: World Lung Foundation. Available at www.TobaccoAtlas.org

Feagin, Joe R., and C. B. Feagin. 1994. *Social Problems.* Englewood Cliffs, NJ: Prentice Hall.

Feilding, Amanda. 2013. "At Last, the Edifice of Drugs Prohibition Starts to Crumble." *The Guardian,* June 14. Available at www.guardian.co.uk

Flanzer, Jerry P. 2005. "Alcohol and Other Drugs Are Key Causal Agents of Violence." In *Current Controversies on Family Violence,* 2nd ed., Donileen R. Loseke, Richard J. Gelles, and Mary M. Cavanaugh, eds. (pp. 163–174).

FDA (Food and Drug Administration). 2011. "Menthol Cigarettes and Public Health: Review of the Scientific Evidence and Recommendations." Tobacco Products Scientific Advisory Committee. Available at www.fda.gov

Foundation for a Drug Free World. 2013. "Real Life Stories about Drug Abuse." Available at www.drugfreeworld.org

Freeman, Dan, Merrie Brucks, and Melanie Wallendorf. 2005. "Young Children's Understanding of Cigarette Smoking." *Addiction* 100(10):1537–1545.

Friedman-Rudovsky, Jean. 2009. "Red Bull's New Cola: A Kick from Cocaine?" *Time/CNN*, May 25. Available at www.time.com

Gallup Poll. 2013. "Illegal Drugs: Gallup Historical Trends." Available at www.gallup.com

Gallahue, Patrick, Ricky Gunawan, Fifa Rahman, Karim El Mufti, Najam U Din, and Rita Felten. 2012. "The Death Penalty for Drug Offences: Global Overview 2012, Tipping the Scales for Abolition." International Harm Reduction Association. London. Available at www.ihra.net

GAO (Government Accounting Office). 2013 (January). "State Approaches Taken to Control Access to Key Methamphetamine Ingredient Show Varied Impact on Domestic Drug Labs." GAO-13-204. Available at www.gao.gov

Gilbert, R., C. S. Widom, K. Browne, D. Fergusson, E. Webb, and S. Janson. 2009. "Burden and Consequence of Child Maltreatment in High-Income Countries." *The Lancet* 73(9657):68–81.

Goodnough, Abby, and Katie Zezima, 2011. "An Alarming New Stimulant, Legal in Many States." *The New York Times,* July 16. Available at www.nytimes.com

Greenwald, Glenn. 2009 (April 2). "Drug Decriminalization in Portugal: Lessons for Creating Fair and Successful Drug Policies." CATO Institute: Washington, DC. Available at www.cato.org

Gusfield, Joseph. 1963. *Symbolic Crusade: Status Politics and the American Temperance Movement.* Urbana, IL: University of Illinois Press.

Haasnoot, Shirley. 2013. "Dutch Drug Policy, Pragmatic as Ever." *The Guardian,* January 3. Available at www.guardian.co.uk

Hastings, Deborah. 2013. "Bodies Pile Up as Mexican Drug Cartels Kill and Dismember Journalists." *N.Y. Daily News,* May 6. Available at www.nydailynews.com

Heinrich, Henry. 2009. "Obama Drug Policy to Do More to Ease Health Risks." Reuters, March 16. Available at: www.reuters.com

Hingson, Ralph W., Timothy Heeren, and Michael R. Winter. 2006. "Age at Drinking Onset and Alcohol Dependence." *Archives of Pediatrics & Adolescent Medicine* 160:739–746.

Human Rights Watch. 2007. "Reforming the Rockefeller Drug Laws." Available at www.hrw.org /campaigns/drugs

Ingold, John. 2013. "Colorado Court Upholds Firing for Off-the-Job Medical Marijuana Use." *The Denver Post,* April 25.

Jargin, Sergei V. 2012. "Social Aspects of Alcohol Consumption in Russia." *South African Medical Journal* 102(9):719.

Jennings, Ashley. 2013 (January 3). "Idaho Inmates Sue Beer, Wine Companies for $1B." Available at: http://abcnews.go.com

Jervis, Rick. 2009. "YouTube Riddled with Drug Cartel Videos, Messages." *USA Today,* April 9. Available at www.usatoday.com

Johnson, K. Z. Pan, L. Young, J. Vanderhoff, S. Shamblen, T. Browne, K. Linfield, and G. Suresh. 2008. "Therapeutic Community Drug Treatment Success in Peru: A Follow-Up Outcome Study." *Substance Abuse Treatment and Prevention* 3(December 3):26.

Kearns-Bodkin, J. N., and K. E. Leonard. 2008. "Relationship Functioning among Adult Children of Alcoholics." *Journal of Studies on Alcohol and Drugs* 69(6):941–950.

Kelly, J. F., and B. B. Hoeppner. 2013. "Does Alcoholics Anonymous Work Differently for Men and Women? A Moderated Multiple-Mediation Analysis in a Large Clinical Sample." *Drug and Alcohol Dependence* 130(1–3):186–193.

Kelly, Adrian B., Gary C. K. Chan, John W. Toumbourou, Martin O'Flaherty, Ross Homel, George C. Patton, and Joanne Williams. 2012. "Very Young Adolescents and Alcohol: Evidence of a Unique Susceptibility to Peer Alcohol Use." *Addiction* 37(4):414–419.

King, Ryan S., and Jill Pasquarella. 2009. "Drug Courts: A Review of the Evidence." The Sentencing Project, April. Available at www .sentencingproject.org

LaBrie, Joseph W., Phillip J. Ehret, and Justin F. Hummer. 2013. "Are They All the Same? An Exploratory, Categorical Analysis of Drinking Game Types." *Addictive Behaviors* 38(5):2133–2139.

Lee, Yon, and M. Abdel-Ghany. 2004. "American Youth Consumption of Licit and Illicit Substances." *International Journal of Consumer Studies* 28(5):454–465.

Liberty Mutual and SADD (Students Against Destructive Decisions). 2012 (February 22). "Promoting Responsible Teen Behavior." Available at www.libertymutual.com

MacCoun, Robert J. 2011. "What Can We Learn from the Dutch Cannabis Coffeeshop System?" *Addiction* 106(11):1899–1910.

Margolis, Robert D., and Joan E. Zweben. 2011. *Treating Patients with Alcohol and Other Drug Problems: An Integrated Approach.* Chapter 3: "Models and Theories of Addiction." Washington DC: American Psychological Association.

Marine-Street, Natalie. 2012. "Stanford Researchers' Cigarette Ad Collection Reveals How Big Tobacco Targets Women and Adolescent Girls." *Gender News,* April 26. Available at http:// gender.stanford.edu

Martin, David. S. 2012. "Vets Feel Abandoned after Secret Drug Experiments." *CNN Health,* March 1. Available at www.cnn.com

Mauer, Marc. 2009. "The Changing Racial Dynamics of the War on Drugs." The Sentencing Project, April 2009. Washington, DC. Available at www.sentencingproject.org

McLaughlin, Michael. 2012. "Bath Salt Incidents Down since DEA Banned Synthetic Drug." Huffington Post, September 4. www.huffingtonpost.com

McMillen, Matt. 2011. "'Bath Salts' Drug Trend: Expert Q&A." WebMD: Mental Health. Available at www.webmd.com

Mears, Bill. 2012 (November 26). "Tobacco Companies Ordered to Publicly Admit Deception on Smoking Dangers." Available at www.cnn.com

Merolla, David. 2008. "The War on Drugs and the Gender Gap in Arrests: A Critical Perspective." *Critical Sociology* 34 (March):255–270.

MTF (Monitoring the Future). 2011. *National Results on Adolescent Drug Use.* The University of Michigan, Institute for Social Research. Ann Arbor, Michigan. Available at monitoringthefuture.org

Morgan, Patricia A. 1978. "The Legislation of Drug Law: Economic Crisis and Social Control." *Journal of Drug Issues* 8:53–62.

MADD (Mothers Against Drunk Driving). 2011. "Why 21? Addressing Underage Drinking." Available at www.madd.org

Mulvey, Edward P. 2011. "Highlights from Pathways to Desistance: A Longitudinal Study of Serious Adolescent Offenders." U.S. Department of Justice, March 11. Available at http:// ncjrs.gov

Murray, Rheana. 2012. "Heroin Use among Suburban Teens Skyrockets: Experts Say Prescription Pills are the New Gateway Drug." *New York Daily,* June 20. Available at www.nydailynews .com

Myers, Matthew. 2011 (April 25). "FDA Acts to Protect Public Health by Extending Authority over Tobacco Products, Including E-Cigarettes." Available at www.tobaccofreekids.org

Myers, Matthew L. 2009. "U.S. Court of Appeals Affirms 2006 Lower Court Ruling That Tobacco Companies Committed Fraud for Five Decades and Lied about the Dangers of Smoking." Press Office Release, May 22. Available at www .tobaccofreekids.org

National Institute of Justice. 2013. "Drug Courts: Background." Available at www.ncjrs.gov

National Institute on Drug Abuse (NIDA). 2013. "Monitoring the Future 2012 Survey Results." Available at www.drugabuse.gov

Nebehay, Stephanie. 2011. "Alcohol Kills More than AIDS, TB, or Violence: WHO." Reuters, February 11. Available at www.reuters.com

NIDA. 2012a. "Commonly Abused Drugs." Available at www.drugabuse.gov

NIDA. 2012b. "Is Drug Addiction Treatment worth its Cost?" *Principles of Drug Addiction Treatment.* 3rd edition. Available at www.drugabuse .gov

NIDA. 2006. (National Institute of Drug Abuse). *Principles of Drug Addiction Treatment: A Research- Based Guide.* Available at http:// drugabuse.gov

NSDUH (National Survey on Drug Use and Health). 2013. "Results from the 2012: Volume I. Summary of National Findings." Substance Abuse and Mental Health Services Administration. Office of Applied Statistics, NSDUH Series H-44, HHS Publication No. (SMA) 12-4713. Rockville, MD.

NSDUH (National Survey on Drug Use and Health). 2004 (February 13). "Alcohol Dependence or Abuse among Parents with Children Living at Home." Office of Applied Studies, Substance Abuse and Mental Health Services Administration. Available at www.oas.samhsa .gov

OAS (Organization of American States). 2013 (May 17). "Highlights." *OAS Report on the Drug Problem in the Americas.* Available at www.oas .org/documents/eng/press/highlights.pdf

ONDCP (Office of National Drug Control Policy). 2013 (April). "FY 2014 Funding Highlights." *National Drug Control Budget.* Available at www.whitehouse.gov

ONDCP. 2009. *National Youth Anti-Drug Media Campaign.* Retrieved from www.theantidrug .com

ONDCP. 2006. *Methamphetamine.* Available at www.whitehousedrugpolicy.gov/drugfact/methamphetamine

Pereltsvaig, Aysa. 2013 (March 18). "Global Alcohol Consumption: World Map." Available at http://geocurrents.info

PATS (Partnership Attitude Tracking Study). 2013 (April 23). *PATS Key Findings: Partnership Attitude Tracking Study.* MetLife Foundation in conjunction with the Partnership at Drugfree.org. Available at www.drugfree.org

Peralta, Robert L. Jennifer L. Steele, Stacey Nofziger, and Michael Rickles. 2010. "The Impact of Gender on Binge Drinking Behavior among U.S. College Students Attending a Midwestern University: An Analysis of Two Gender Measures." *Feminist Criminology* 10(5):355–379.

Peters, Jeremy W. 2009. "Albany Takes Step to Repeal '70s-Era Drug Laws." *The New York Times*, March 5. Available at www.nytimes.com

Phoenix House. 2013. "About Phoenix House." Available at www.phoenixhouse.org

Porter, Eduardo. 2012. "Numbers Tell of Failure in Drug War." *New York Times*, July 3. Available at www.nytimes.com

Primack, Brian A., James E. Bost, Stephanie R. Land, and Michael J. Fine. 2007. "Volume of Tobacco Advertising in African American Markets: Systematic Review and Meta-Analysis." *Public Health Reports* 122(5):607–615.

Rasmussen. 2012. "7% Think U.S. is Winning War on Drugs." *Rasmussen Reports*, November 13. Available at www.rasmussenreports.com

Rabinowitz, Mikaela, and Arthur Lurigio. 2009. "A Century of Losing Battles: The Costly and Ill-Advised War on Drugs." Conference Paper. American Sociological Association. San Francisco, August 8–11.

Rorabaugh, W. J. 1979. *The Alcoholic Republic: An American Tradition.* New York: Oxford University Press.

Rose-Jacobs, Ruth, Marilyn Augustyn, Marjorie Beeghly, Brett Martin, Howard J. Cabral, Timothy C. Heeren, Mark A. Richardson, and Deborah A. Frank. 2012. "Intrauterine Substance Exposures and Wechsler Individual Achievement Test-II Scores at 11 Years of Age." *Vulnerable Children and Youth Studies* 7(2):186–197.

Saad, Lydia. 2012. "Majority in U.S. Drink Alcohol, Averaging Four Drinks a Week." *Gallup Poll*, August 17. Available at www.gallup.com

Saloner, B., and B. LeCook. 2013. "Blacks and Hispanics Are Less Likely Than Whites to Complete Addiction Treatment, Largely Due to Socioeconomic Factors." *Health Affairs*, January 7. Available at www.rwjf.org

SAMHSA (Substance Abuse and Mental Health Services Administration). 2012. "More than 7 Million Children Live with a Parent with Alcohol Problems." *Data Spotlight*, February 16.

SAMHSA. 2009. "Children Living with Substance-Dependent or Substance-Abusing Parents: 2002–2007." *The NSDUH Report*, April 16. Available at http://oas.samhsa.gov

SAMHSA. 2007. "Parental Substance Abuse Raises Children's Risk." *Practice What You Preach*, February 20. Retrieved from www.family.samhsa.gov

Sánchez-Moreno, Maria McFarland. 2012. "A Discussion about Drug Policy Is Long Overdue." Human Rights Watch, September 4. Available at www.hrw.org

Schmidt, Lorna. 2013 (June 18). "Tobacco Company Marketing to Kids." *Campaign for Tobacco Free Kids.* Available at www.tobaccofreekids.org

Sevigny, Eric L., Harold A. Pollack, and Peter Reuter. 2013. "Can Drug Courts Help to Reduce Prison and Jail Populations?" *Academy of Political and Social Science* 647(1):190–212.

Sifferlin, Alexandra. 2013. "FDA Approves New Cigarettes in First Use of New Regulatory Power Over Tobacco." *Time Magazine*, June 26. Available at: http://healthland.time.com

Sohn, Emily. 2010. "Side Effects of Drugs in Water Still Murky." Discovery News, September 28. Available at: news.discovery.com

Szalavitz, Maia. 2009. "Drugs in Portugal: Did Decriminalization Work?" *Time*, April 26. Available at www.time.com

Taifia, Nkechi. 2006 (May). "The 'Crack/Powder Disparity': Can the International Race Convention Provide a Basis for Relief?" American Constitution Society for Law and Policy White Paper. Available at acslaw.org

Tarter, Ralph E., Michael Vanyukov, Levent Kirisci, Maureen Reynolds, and Duncan B. Clark. 2006. "Predictors of Marijuana Use in Adolescents Before and After Licit Drug Use: Examination of the Gateway Hypothesis." *American Journal of Psychiatry* 163:2134–2140.

Terry-McElrath, Yvonne, Glen Szczypka, and Lloyd D. Johnston. 2011. "Potential Exposure to Anti-Drug Advertising and Drug-Related Attitudes, Beliefs, and Behaviors among United States Youth, 1995–2006." *Addictive Behaviors* 36(1/2):116–124.

Thio, Alex. 2007. *Deviant Behavior.* Boston: Allyn and Bacon.

Timeline. 2001. "Timeline of Tobacco Litigation." Fox News, March 8. Available at www.foxnews.com

U.S. Department of Health and Human Services. 2010. *Ending the Tobacco Epidemic: A Tobacco Control Strategic Action Plan for the U.S. Department of Health and Human Services.* Washington, DC: Office of the Assistant Secretary for Health.

U.S. Department of Justice. 2010. "The Impact of Drugs on Society." *National Drug Threat Assessment 2010.* Available at www.justice.gov

Van Dyck, C., and R. Byck. 1982. "Cocaine." *Scientific American* 246:128–141.

Volpp, K. G., A. B. Troxel, M. V. Pauly, H. A. Glick, A. Puig, D. A. Asch, R. Galvin, J. Zhu, F. Wan, J. DeGuzman, E. Corbett, J. Weiner, J. Audrain-McGovern. 2009. "A Randomized, Controlled Trial of Financial Incentives for Smoking Cessation." *New England Journal of Medicine* 360(7):699–709.

Wadley, Jared. 2012. "American Teens Are Less Likely than European Teens to Use Cigarettes and Alcohol, but More Likely to Use Illicit Drugs." News Release. University of Michigan News Service, June 1. Available at www.ns.umich.edu

WDR (World Drug Report). 2013. *World Drug Report 2013.* United Nations Office on Drugs and Crime (UNODC). New York: United Nations.

Wechsler, William, and Toben F. Nelson. 2008. "What We Have Learned from the Harvard School of Public Health College Alcohol Study: Focusing Attention on College Student Alcohol Consumption and the Environmental Conditions that Promote It." *Journal of Alcohol Studies* (July):1–9.

Weiss, Debra Cassens 2013. "Drug-Abusing Pregnant Women May Be Prosecuted under Endangerment Laws, Ala. Supreme Court Says." *ABA Journal*, January 14. Available at www.abajournal.com

West, Steven L., and Keri K. O'Neal. 2004. "Project D.A.R.E. Outcome Effectiveness Revisited." *American Journal of Public Health* 94(6):1027–1029.

Williams, Jenny, Frank J. Chaloupka, and Henry Wechsler. 2005. "Are There Differential Effects of Price and Policy on College Students' Drinking Intensity?" *Contemporary Economic Policy* 23(1):78–90.

Willing, Richard. 2002. "Study Shows Alcohol Is Main Problem for Addicts." *USA Today*, October 3, p. B4.

Wilson, Joy Johnson. 1999. "Summary of the Attorneys General Master Tobacco Settlement Agreement." National Conference of State Legislators—AFI Health Committee, March. Available at http://academic.udayton.edu/health

Witters, Weldon, Peter Venturelli, and Glen Hanson. 1992. *Drugs and Society*, 3rd ed. Boston: Jones & Bartlett.

WHO (World Health Organization). 2013. "Tobacco." Geneva: World Health Organization.

WHO (World Health Organization). 2011a. "Data and Statistics." Geneva: World Health Organization.

WHO. 2011b (February). "Tobacco." Geneva: World Health Organization. Available at www.who.int

WHO. 2011c. *Global Status Report on Alcohol and Health 2011.* Geneva: World Health Organization. Available at www.who.int

WHO. 2008. *WHO Report on the Global Tobacco Epidemic, 2008: The MPOWER Package.* Geneva: World Health Organization, 2008.

Wu, Li-Tzy, George E. Woody, Chongming Yang, Jeng-Jong, and Dan G. Blazer. 2011. "Racial/Ethnic Variations in Substance-Related Disorders among Adolescents in the United States." *Archives of General Psychiatry* 68(11):1176–1185.

Wysong, Earl, Richard Aniskiewicz, and David Wright. 1994. "Truth and Dare: Tracking Drug Education to Graduation and as Symbolic Politics." *Social Problems* 41:448–468.

Zickler, Patrick. 2003. "Study Demonstrates that Marijuana Smokers Experience Significant Withdrawal." *NIDA Notes* 17:7, 0.

Chapter 4

Afterschool Alliance. 2013. "After School Programs: Making a Difference in America's Communities by Improving Academic Achievement, Keeping Kids Safe and Healthy Working Families." Available at www.afterschoolalliance.org

Albanese, Jay S. 2012. "Deciphering the Linkages between Organized Crime and Transnational Crime."*Journal of International Affairs* 66(1):1–11.

ABA (American Board of Anesthesiology). 2011. "Anesthesiologists and Capital Punishment." Available at www.deathpenaltyinfo.org

ACLU (American Civil Liberties Union). 2013 (June). *The War on Marijuana in Black and White.* Available at www.aclu.org

Amnesty International. 2013.*Death Sentences and Executions 2012.*Available atwww.amnesty.org

Anderson, David A. 2012. "The Cost of Crime." *Foundations and Trends in Microeconomics* 7(3):209–265.

Arnold, Tim. 2013. "The Real Weapons of Mass Destruction: America's 300 Million Guns." *The Huffington Post*, May 17. Available at www .huffingtonpost.com

Ascani, Nathaniel. 2012. "Labeling Theory and the Effects of Sanctioning on a Staff Delinquent Peer Association: A New Approach to Sentencing Juveniles."*Perspectives* (Spring):80–84.

Austin, Andrew D. 2013 (April 25). "Discretionary Budget Authority by Subfunction: An Overview." Washington, DC: Congressional Research Service.

Barkan, Steven. 2006. *Criminology*. Upper Saddle River, NJ: Prentice Hall.

Barrett, Ted, and Tom Cohen.2013 (April 18). "Senate Rejects Expanded Gun Background Checks." Available at www.cnn.com

Bartunek, Robert-Jan. 2013. "Robbers Pull Off Huge Diamond Heist at Brussels Airport." Reuters News Service, February 19. Available at www.reuters.com

Baze v. Rees, 553 U.S. 35, 2008. U.S. Supreme Court. Available at www.supremecourtus.gov

Becker, Howard S. 1963. *Outsiders: Studies in the Sociology of Deviance*. New York: Free Press.

Bell, Kerryn E. 2009. "Gender and Gangs: A Quantitative Comparison."*Crime and Delinquency* 55(3):363–387.

Berlow, Alan, and Gordon Witkin. 2013 (May 1). "Gun Lobby's Money and Power Still Holds Sway over Congress." Available at www .publicintegrity.org

Blackden, Richard. 2013. "US Investigates Claims Microsoft Bribes Chinese Officials."*The Telegraph,* March 19. Available at www.telegraph.co.uk

Brown, Patricia Leigh.2013. "Opening Up, Students Transform a Vicious Circle." *The New York Times*, April 3. Available at www.nytimes.com

BOP (Bureau of Prisons). 2013 (March 18). "Annual Determination of Average Cost of Incarceration." Available at www.gpo.gov

BJS (Bureau of Justice Statistics). 2013a (June 28). "NCVS Redesign." Available at www.bjs.gov

BJS. 2013b (March 7). "Female Victims Of Sexual Violence, 1994–2010." Available at www.bjs.gov

BJS. 2013c (September 24). "Study Finds Some Racial Differences In Perceptions of Police Behavior During Contact With Police." Available at www.bjs.gov

BJS. 2012a (November). *Identity Theft Reported by Households, 2005–2010.* NCJ23624. Available at www.bjs.gov

BJS. 2012b (November 29). "One in 34 US Adults Under Correctional Supervision in 2011, Lowest Rate since 2000." Available at www.bjs.gov

BJS. 2011a. "Correctional Populations in the United States, 2011." NCJ239972. Available at www.bjs.gov

BJS. 2011b. "Prisoners in 2011." NCJ239808. Available at www.bjs.gov

BJS. 2011c. "Probation and Parole in the United States, 2011." NCJ239686. Available at www .bjs.gov

Carlson, Joseph. 2009. "Prison Nurseries: A Pathway to Crime-Free Futures." *Corrections Compendium* 34(1):17–24.

Carter, Zach, and Lauren Berlin. 2012. "State of the Union: Pres. Obama's Financial Fraud Team Tied to Banks." The Huffington Post, January 24. Available at www.huffingtonpost.com

CBS. 2013 (May 29). "Shaniya Davis Update: Death Penalty for Mario Andrette McNeill, N.C. Man Found Guilty of Killing Five-Year-Old." Available at www.cbsnews.com

CBS. 2012. "The Cost of a Nation of Incarceration."*Face the Nation*, April 22. Available at www.cbsnews.com

CDCP (Center for Disease Control and Prevention). 2008. "Youth Risk Behavior Surveillance." *Morbidity and Mortality Weekly Report*, June 6, pp.1–131. Available at www.cdc.gov

Chen, Elsa Y. 2008. "The Liberation Hypothesis and Racial and Ethnic Disparities in the Application of California's Three Strikes Law."*Journal of Ethnicity in Criminal Justice* 6 (2):83–102.

Chesney-Lind, Meda, and Randall G. Shelden. 2004. *Girls, Delinquency, and Juvenile Justice*. Belmont, CA: Wadsworth.

Cho, Seo-Young, Axel Dreher, and Eric Neumayer. 2013. "Does Legalized Prostitution Increase Human Trafficking?" *World Development* 41:67–82.

Civitas. 2012. "Comparisons of Crime in OECD Countries."*Crime Briefing*, January. Available at www.civitas.org.uk

Clark, Maggie. 2013. "George Zimmerman Verdict Renews Stand Your Ground Law Debate.*" The Huffington Post*, July 16.Available at www. huffingtonpost.com

Cohen, Mark A., and Alex R. Piquero. 2009. "New Evidence on the Monetary Value of Saving a High Risk Youth."*Journal of Quantitative Criminology* 25:25–49.

Conklin, John E. 2007. *Criminology*, 9th ed. Boston: Allyn and Bacon.

COPS (Community Oriented Policing Services). 2012. *Success Stories.* Available at www.cops .usdoj.gov

COPS (Community Oriented Policing Services). 2009. *Community Policing Defined.* Available at www.cop.usdoj.gov

Correctional Association. 2009. Education from the Inside Out: The Multiple Benefits of College Programs in Prison. The Correctional Association of New York. Available at www .correctionalassociation.org

D'Alessio, David, and Lisa Stolzenberg. 2002. "A Multilevel Analysis of the Relationship between Labor Surplus and Pretrial Incarceration."*Social Problems* 49:178–193.

Daugherty, Scott. 2013. "Somali Pirates Receive Life Sentences from Federal Jury." *The Virginian-Pilot*, August 3. Available at hamptonroads.com

Davey, Monica. 2010. "Safety is Issue as Budget Cuts Free Prisoners." *The New York Times*, March 4. Available at www.nytimes.org

Diamond, Milton, Eva Jozifkova, and Petr Weiss. 2011. "Pornography and Sex Crimes in the Czech Republic." *Archives of Sexual Behavior* 40:1037–1043.

Donald, Brooke. 2013. "Stanford Law's Three Strikes Project Works for Fair Implementation of New Statute."*Stanford Report*, June 6. Available at http://news.stanford.edu

Drakulich, Kevin M. 2013. "Strangers, Neighbors, and Race: A Content Model of Stereotypes and Racial Anxieties about Crime."*Race and Justice* 2(4):322–355.

Eisen, Lauren-Brooke. 2013 (May 22). "Should Judges Consider the Cost of Sentences?" Brennan Center for Justice. New York University School of Law. Available at www.brennancenter .org

Eng, James. 2012. "Judge Rules Race Tainted North Carolina Death Penalty Case; Inmate Marcus Robinson Spared From Death Row." *NBC New.* Available athttp://usnews.nbcnews.com

Erikson, Kai T. 1966. *Wayward Puritans*. New York: Wiley.

Europol. 2011. "Frequently Asked Questions." Available at www.europol.europa.eu

Fact Sheet. 2012 (April). "Major Federal Legislation Concerned with Child Protection, Child Welfare, and Adoption." *Child Welfare Information Gateway*. Available at www.childwelfare .gov

FBI (Federal Bureau of Investigation). 2013. "Crime in the United States, 2012." *Annual Uniform Crime Report*. Washington, DC: U.S. Government Printing Office.

FBI. 2009 (April). N*ational Incident-Based Reporting System*. Available at www.fbi.gov

FTC (Federal Trade Commission). 2013 (March). "Consumer Sentinel Network Data Book." Available at www.ftc.gov

Fight Crime: Invest in Kids. 2013. About Us. Available at www.fightcrime.org

Finkelhor, David, Heather Turner, and Sherry Hamby. 2011(October). "Questions and Answers about the National Survey of Children's Exposure to Violence."Office of Juvenile Justice and Delinquency Prevention. Available at www .ncjrs.gov

Florida Criminal Code. 2009. Available at www .leg.state.fl.us/statutes

Ford, Jason. 2005. "Substance Use, Social Bond, and Delinquency."*Sociological Inquiry* 75(1):109–128.

Fox, James Alan, and Jack Levin. 2011. *Extreme Killing: Understanding Serial and Mass Murder.* Washington, DC: Sage.

Frank, Ted. 2011. "Refutation of Toyota Sudden Acceleration Hysteria Doesn't Stop Toyota Sudden Acceleration Litigation."*Forum*, May 2. Available at www.pointoflaw.com

Gabriel, Trip. 2013. "Girl's Death by Gunshot Rejected as Symbol." *The New York Times,* May 5. Available at www.nytimes.com

Gallup Poll. 2007 "Gallup's Pulse of Democracy: Crime." Available at www.galluppoll.com

Gault-Sherman, Martha. 2012. It's a Two-Way Street: The Bi-Directional Relationship Between Parenting and Delinquency." *Journal of Youth & Adolescence* 41(2):121–145.

Goldenberg, Suzanne. 2013. "The U.S.Government Assessment of BP Oil Spill Will Not Account for Damage."*The Guardian,* July 20. Available at www.guardian.co.uk

Greenblatt, Alan. 2008. "Second Chance Programs Quietly Gain Acceptance."*Congressional Quarterly Weekly,* September 15. Available at www.cq.com

Greene, Richard Allen. 2012. "Norway Massacre Could Have Been Avoided, Report Finds." CNN News, August 13. Available at www.cnn.com

Hartney, Christopher, and Linh Vuong. 2009. *Created Equal: Racial and Ethnic Disparities in the U.S. Criminal Justice System.* Oakland, CA: National Council on Crime and Delinquency.

Heimer, Karen, Stacy Wittrock, and Halime Unal. 2005. "Economic Marginalization and the Gender Gap in Crime." In *Gender and Crime: Patterns of Victimization and Offending,* Karen Heimer and Candace Kruttschnitt, eds. (pp. 115–136). New York University Press.

Henrichson, Christian, and Ruth Delaney. 2012 (July 20). *The Price of Prisons: What Incarceration Costs Taxpayers.* Vera Institute of Justice. Available at www.vera.org

The Herald Sun. 2008 (January 17). "Jessica's Victim Impact Statement." Available at www.news.com.au/heraldsun

Hilzenrath, David. 2011. "Goldman Sachs Subpoenaed."*The Washington Post,* June 2. Available at www.washingtonpost.com

Hirschi, Travis. 1969. *Causes of Delinquency.* Berkeley, CA: University of California Press.

Holtfreter, Kristy, Shanna Van Slyke, Jason Bratton, and Marc Gertz. 2008. "Public Perceptions of White-Collar Crime and Punishment."*Bureau of Criminal Justice* 36:50–60.

HR 446. 2013. "Text of 'National Criminal Justice Commission Act of 2013.'" Available at www.govtrack.us

ICC (International Chamber of Commerce). 2013. "IMB Piracy Report." Available at www.icc-ccs.org

ICCC (Internet Crime Complaint Center). 2013. "2012 Internet Crime Report." Available at www.ic3.gov

ICPC (International Centre for the Prevention of Crime). 2013. "Missions and Activities." Available at www.crime-prevention-intl.org/

The Innocence Project. 2013. "Understand the Causes." Available at www.innocenceproject.org

Infographic. 2012. "Just How Many Guns Do Americans Own?" *The Huffington Post,* August 9. Available at www.huffingtonpost.com

Interpol. 2013. "Overview." Available at www.interpol.int

Interpol. 2011. "Trafficking in Human Beings." Available at www.interpol.int

Jacobs, David, Zhenchao Qian, Jason Carmichael, and Stephanie Kent. 2007. "Who Survives on Death Row? An Individual and Contextual Analysis."*American Sociological Review* 72:610–632.

Jones, Karen. 2013 (April 10). "Background Checks on Gun Sales: How Do They Work?" Available at www.cnn.com

Jones, Jeffrey M. 2013a (April 5). "Minneapolis-St. Paul Area Residents Most Likely to Feel Safe." Available at www.gallup.com

Jones, Jeffrey M. 2013b (February 1). "Men, Married, Southerners, Most Likely to be Gun Owners." Available at www.gallup.com

Jones, Jeffrey M. 2013c (January 29). "Party Views Diverge Most on U.S. Gun Policies." Available at www.gallup.com

Katz, Rebecca. 2012. "Environmental Pollution: Corporate Crime and Cancer Mortality." *Contemporary Justice Review: Issues in Criminal, Social, and Restorative Justice* 15(1):97–125.

Kerley, Kent, Michael Benson, Matthew Lee, and Francis Cullen. 2004. "Race, Criminal Justice Contact, and Adult Position in the Social Stratification System."*Social Problems* 51(4):549–568.

Kirkham, Chris. 2013. "Lake Erie Prison Plagued by Violence and Drugs after Corporate Takeover." Huffington Post, March 22. Available at www.huffingtonpost.com

Kitamura, Makiko, and Adi Narayan. 2013. "Your Pushes to Keep Lethal Injection Drugs from US Prisons." *Business Week,* February 7. Available at www.businessweek.com

Kohm, Stephen A., Courtney Wade-Lindberg, Michael Weinrath, Tara O'Connor Shelley, and Rhonda R. Dobbs. 2012. "The Impact of Media on Fear of Crime among University Students: A Cross National Comparison." *Canadian Journal of Criminology and Criminal Justice* 1:67–100.

Kubrin, Charis E. 2005. "Gangsters, Thugs, and Hustlas: Identity and the Code of the Street in Rap Music."*Social Problems* 52(3):360–378.

Kubrin, Charis, and Ronald Weitzer. 2003. "Retaliatory Homicide: Concentrated Disadvantage and Neighborhood Culture."*Social Problems* 50:157–180.

Langston, Lynn. 2012 (November 8). "Firearms Stolen during Household Burglaries and Other Property Crimes, 2005–2010." NCJ 239436. Bureau of Justice Statistics. Available at www.bjs.gov

Langston, Lynn, Marcus Berzofsky, Christopher Krebs, and Hope Smiley–McDonald. 2012. *Victimization Is Not Reported to the Police, 2006–2010.* Capital NCJ 238536. August. US Department of Justice. Available at www.bjs.gov

Leverentz, Andrea. 2012. "Narratives of Crime and Criminals: How Places Socially Construct the Crime Problem." *Sociological Forum* 27(2):348–371.

Lichtblau, Eric, David Johnston, and Ron Nixon. 2008. "F.B.I. Struggles to Handle Financial Fraud Cases." *The New York Times*, October 19. Available at www.nytimes.com

Lott, John R., Jr. 2003. "Guns Are an Effective Means of Self-Defense." In *Gun Control,* Helen Cothran, ed. (pp. 86–93). Farmington Hills, MI: Greenhaven Press.

MAD DADS. 2013."Who are Mad Dads?"Available at www.maddads.com

Male Survivor. 2012."Sentencing of Sandusky is Beginning of the Next Chapter, Not End of the Story." Available at www.malesurvivor.org

Marks, Alexandria. 2006. "Prosecutions Drop for US White Collar Crime."*The Christian Science Monitor*, August 31. Available at www.csmonitor.com

Maynard, Micheline. 2010. "Toyota Cited $100 Million Savings after Limiting Recall."*The New York Times,* February 21. Available at www.nytimes.com

Marsh, Julia. 2013. "'Smoking' Gun nets 60G Fine." *New York Post*, May 3. Available at www.nypost.com

Martinez, Michael. 2013. "California to Challenge Court Order to Release 10,000 Inmates by Year's End." *CNN News.* Available at www.cnn.com

McCurry, Justin. 2012. "Yakuza Gangs Face Fight for Survival as Japan Cracks Down on Organized Crime." *The Guardian,* January 5.Available at www.guardian.co.uk

Merton, Robert. 1957. *Social Theory and Social Structure.* Glencoe, IL: Free Press.

Mertz, Janice. 2013. "Collaboration to Recover U.S. Exploited Youth: The FBI's Innocence Lost National Initiative." *The Police Chief* 80:24–25.

Mouilso, Emily R., and Karen S. Calhoun. 2013. "The Role of Rape Myth Acceptance and Psychopathy in Sexual Assault Perpetration." *Journal of Aggression, Maltreatment and Trauma* 22(2):159–174.

Myers, Laura L. 2012. "Green River Killer Case: More Remains Tied to Gary Ridgway Identified." *The Huffington Post,* June 18.

Murphy, Paul.2013. "Feds Say They Rescued Five Women from Sex Trafficking Ring in Quarter." *Crime News,* January 31. Available at www.wwltv.com

NATW (National Association of Town Watch). 2013. "About Us." Available at www.natw.org

NCMEC (National Center on Missing and Exploited Children). 2013. *Child Sexual Exploitation.* Available at www.missingkids.com

NCPC (National Crime Prevention Council). 2012 (April 11). "The Impact of Gang Violence on Business and Communities." Available at http://ncpc.typepad.com

NCPC. 2005. "Preventing Crime Saves Money." Available at www.ncpc.org

NCSL (National Conference of State Legislatures). 2013 (July 2). "2013 Legislation Regarding Internet Gambling and Lotteries." Available at www.ncsl.org

NCVC (National Center for Victims of Crime). 2013 (January). *New Challenges, New Solutions.* U.S. Department of Justice. Available at http://ovc.ncjrs.gov

NGC (National Gang Center). 2012 (April). "National Youth Gang Survey Analysis." Available at www.nationalgangcenter.gov

NGIC (National Gang Intelligence Center). 2012. *2011 National Gang Threat Assessment.* Available at www.fbi.gov

NWCCC (National White Collar Crime Center). 2010. "National Public Survey on White Collar Crime." Available at http://crimesurvey.nw3c.org

National Research Council. 1994. *Violence in Urban America: Mobilizing a Response.* Washington, DC: National Academy Press.

National Security Council. 2011. Strategy to Combat Transnational Organized Crime. July 19. Available at www.whitehouse.gov

Newport, Frank. 2010. "Americans Want BP to Pay All Losses, No Matter the Cost." Gallup, June 15. Available at www.gallup.com

Obama, Barack. 2013. "Statement to the Press Corps: Remarks about Trayvon Martin." Available at www.washingtonpost.com

OJJDP (Office of Juvenile Justice and Delinquency Prevention). 2012. "Program Summary." *Internet Crimes Against Children Task Force Program.* Available at www.ojjdp.gov

O'Toole, James. 2013. "Smith and Wesson Will Record Sales as Gun Debate Raged." *CNN Money,* June 13. Available at http://money.cnn.com

Ouimet, Marc. 2012. "The Effect of Economic Development, Income Inequality, and Excess Infant Mortality on the Homicide Rate for 165 Countries in 2010." *Homicide Studies* 16(3):238–258.

PEP (Prison Entrepreneurship Program). 2012. *Prison Entrepreneurship Program, 2011 Annual Report.* Available at www.pep.org/

Pertossi, Mayra. 2000 (September 27). *Analysis: Argentine Crime Rate Soars.* Available at http://news.excite.com

Pew. 2013. (March 8). "U.S. Prison Count Continues to Drop." Pew Charitable Trust. Available at http://pewstates.org

Pew. 2012. (June 6). "Time Served: The High Cost, Low Return of Longer Prison Terms." Pew Charitable Trust. Available at http://pewstates.org

Pew. 2011 (April). "State of Recidivism: The Revolving Door of America's Prisons." Pew Charitable Trust. Available at http://pewresearch.org

Planty, Michael, and Jennifer L Truman. 2013. *Firearm Violence, 1993–2011.* Bureau of Justice Statistics (May). NCJ 241730 Available at www.bjs.gov

Polaris Project. 2013. "Sex Trafficking in the U.S." Available at www.polarisproject.org

Prichard, Jeremy, Caroline Spiranovic, Paul Watters, and Christopher Lueg. 2013. "Young People, Child Pornography, and Subcultural Norms on the Internet." *Journal of the American Society for Information Science and Technology* 65(5):992–1000.

Pridemore, William Alex, and Sang-Weon Kim. 2007. "Socioeconomic Change and Homicide in a Transitional Society." *Sociological Quarterly* 48:229–251.

Protess, Ben, and Azam Ahmed. 2012 (August 9). "S.E.C. and Justice Department End Mortgage Investigations into Goldman." Available at http://dealbook.nytimes.com

PUP (Prison University Project). 2013. *Academics.* Available at www.prisonuniversityproject.org

Rankin, Jennifer. 2013. "Japan Carmakers Recall 3 Million Vehicles over Air Bag Vault." *The Guardian,* April 11. Available at www.guardian.co.uk

Reiman, Jeffrey, and Paul Leighton. 2012. *The Rich Get Richer and the Poor Get Prison.* Boston: Allyn and Bacon.

Resource Center (Racial Profiling Data Collection Resource Center). 2013. "Legislation and Litigation." Northeastern University. Available at http://racialprofilinganalysis.neu.edu

Rogers, Simon. 2012. "The Gun Ownership and Gun Homicides Murder Map of the World." *The Guardian.* July 22. Available at http://www.thequardian.com

Rubin, Paul H. 2002. "The Death Penalty and Deterrence." *Forum,* Winter, pp. 10–12.

Saad, Lydia. 2011. "Americans Expressed Mixed Confidence in Criminal Justice System." *Gallup Poll,* July 11. Available at www.gallup.com

Saad, Lydia. 2010. "Nearly 4 in 10 Americans Still Fear Walking Alone at Night." Gallup Poll, November 5. Available at www.gallup.com

Salant, Jonathan D. 2013. "Snowden Seen as Whistleblower by Majority in Poll." Bloomberg News Service, July 10. Available at www.bloomberg.com

Sampson, Robert J., Jeffrey D. Morenoff, and Stephen W. Raudenbush. 2005. "Social Anatomy of Racial and Ethnic Disparities in Violence." *American Journal of Public Health* 95(2):224–232.

Santora, Marc. 2013. "In Hours, Thieves Took $45 Million in an ATM Scheme." *The New York Times,* May 10. Available at www.nytimes.com

Schelzig, Erik. 2007 (September 19). "Court Ruling Halts Tennessee Executions." Available at www.wral.com

Schiesel, Seth. 2011. "Supreme Court Has Ruled: Now Games Have a Duty." *New York Times,* June 28. Available at www.nytimes.com

Schweinhart, Lawrence J. 2007. "Crime Prevention by the High/Scope Perry Preschool Program." *Victims and Offenders* 2:141–160.

Seelye, Katherine Q. 2013. "Bulger Guilty of Gangland Crimes, Including Murder." *The New York Times,* August 12.

Severson, Kim. 2013. "North Carolina Repeals Law Allowing Racial Bias Claim in Death Penalty Challenges." *The New York Times,* June 5. Available at www.nytimes.com

Shapland, Joanna, and Matthew Hall. 2007. "What Do We Know about the Effects of Crime on Victims?" University of Sheffield: Great Britain. *International Review of Victimology* 14:175–217.

Shelley, Louise. 2007. "Terrorism, Transnational Crime and Corruption Center." American University. Available at traccc.gmu.edu/

Sherman, Lawrence. 2003. "Reasons for Emotions." *Criminology* 42:1–37.

Siegel, Larry. 2006. C*riminology,* 9th ed. Belmont, CA: Wadsworth.

Steinhauer, Jennifer. 2009. "To Cut Costs, States Relax Prison Policies." *The New York Times,* March 25. Available at www.nytimes.com

Surgeon General. 2002. "Cost-Effectiveness." In *Youth Violence: A Report of the Surgeon General.* Available at www.ncbi.nlm.nih.gov/books/NBK44295/#A13029

Sutherland, Edwin H. 1939. *Criminology.* Philadelphia: Lippincott.

Sweeten, Gary, Alex R. Piquero, and Laurence Steinberg. 2013. "Age and the Explanation of Crime, Revisited." *Journal of Youth and Adolescence* 42:921–938.

Takei, Carl. 2013 (July 9). "Anonymous Exposes U.S.'s Biggest Private Prison Company as a Bad Financial Investment." Available at www.aclu.org

The Sentencing Project. 2012. "Incarcerated Women." Available at http://sentencing project.org

Truman, Jennifer, and Michael Planty. 2012. *Criminal Victimization, 2011.* October. U.S. Department of Justice, Bureau of Justice Statistics. Available at www.bjs.gov

Truman, Jennifer L., and Erica L. Smith. 2012. "Prevalence of Violent Crime among Households with Children, 1993–2010." U.S. Department of Justice, Office of Justice Programs. Bureau of Justice Statistics, NCJ 238799.

Turner, Wendy. 2007. "Experiences of Offenders in Prison Canine Program." *Federal Probation* 71(1):38–43.

UNODC (United Nations Office on Drug and Crimes). 2012. *Global Study on Homicide: Friends, Contacts, Data.* Available at www.unodc.org

U.S. Census Bureau. 2013 (May 2). "National Crime Victimization Survey." Washington, DC: U.S. Government Printing Office.

U.S. Census Bureau. 2012. *Statistical Abstract of the United States,* 132nd edition. Washington, DC: U.S. Government Printing Office.

U.S. Congress. 2013–2014. "Bill Summary and Status." Available at http://thomas.loc.gov

U.S. Department of Commerce. 2011. *Service Annual Survey: 2009.* Washington, D.C. Government Printing Office.

U.S. Department of Health and Human Services. 2012. "Fact Sheet: Child Victims of Human Trafficking." August 8. Available at www.acf.hhs.gov

U.S. Department of Justice. 2012a. "Transnational Organized Crime." Available at justice.gov

U.S. Department of Justice. 2012b (August 9). "Nearly 3.4 Million Violent Crimes Per Year Went Unreported to Police from 2006 to 2010." Bureau of Justice Statistics. Available at www.ojp.usdoj.gov

U.S. Department of Justice. 2012c. "The Prostitution of Children in the United States." *Child Exploitation and Obscenity Section.* Available at www.justice.gov

U.S. Department of Justice. 2008. "Serial Murder: Multi-Disciplinary Perspectives for Investigators." Washington, DC: Behavioral Analysis Unit, National Center for the Analysis of Violent Crime.

U.S. Department of State. 2013. *Trafficking in Persons Report.* June. Available at www.state.gov

Verizon. 2013. *2013 Data Breach Investigations Report.* April. Available at www.verizonenterprise.com

Victim Statements. 2009. *U.S. v. Bernard L. Madoff.* 2009. U.S. Department of Justice. Available at www.pbs.org

VORP (Victim-Offender Reconciliation Program). 2012. *About Victim-Offender Mediation and Reconciliation.* Available at www.vorp.com

Vu, Pauline. 2007. "Executions Halted as Doctors Balk." *Stateline,* March 21. Available at www.stateline.org

Walmsley, Roy. 2012. *World Prison Population List,* 9th edition. International Centre for Prison Studies. Available at www.prisonstudies.org

Welsh-Huggins, Andrew. 2013. "Hundreds of New Charges Filed in US Kidnap Case." *Associated Press,* July 13. Available at apnews.com

Williams, Linda. 1984. "The Classic Rape: When Do Victims Report?" *Social Problems* 31:459–467.

Winslow, Robert W., and Sheldon Zhang. 2008. *Criminology: A Global Perspective.* Prentice Hall.

Wright, Darlene, and Kevin Fitzpatrick. 2006. "Violence and Minority Youth: The Effects of Risk and Asset among African-American Children and Adolescents." *Adolescence* 41(162):251–263.

Chapter 5

Ahrons, C. 2004. *We're Still Family: What Grown Children Have to Say about Their Parents' Divorce.* New York: HarperCollins.

Allendorf, Keera. 2013. "Schemas of Marital Change: From Arranged Marriages to Eloping for Love." *Journal of Marriage and Family* (April): 453–464.

Amato, Paul. 2004. "Tension between Institutional and Individual Views of Marriage." *Journal of Marriage and Family* 66:959–965.

Amato, Paul. 2003. "The Consequences of Divorce for Adults and Children." In *Family in Transition*, 12th ed., Arlene S. Skolnick and Jerome H. Skolnick, eds. (pp. 190–213). Boston: Allyn and Bacon.

Amato, Paul. 1999. "The Postdivorce Society: How Divorce Is Shaping the Family and Other Forms of Social Organization." In *The Postdivorce Family: Children, Parenting, and Society*, R. A. Thompson and P. R. Amato, eds. (pp. 161–190). Thousand Oaks, CA: Sage.

Amato, P. R., and J. Cheadle. 2005. "The Long Reach of Divorce: Divorce and Child Well-Being across Three Generations." *Journal of Marriage and the Family* 67:191–206.

Amato, P. R., A. Booth, D. R. Johnson, and S. J. Rogers. 2007. *Alone Together: How Marriage in America Is Changing*. Cambridge MA: Harvard University Press.

American Humane Association. 2010 (April 20). "Orange County Animal Services and Harbor House Create First Pets and Women's Shelter (PAWS) Program in Central Florida." News Release. Available at www.americanhumane.org

Anderson, Kristin L. 2013. "Why Do We Fail to Ask: 'Why' about Gender and Intimate Partner Violence?" *Journal of Marriage and Family* 75(April):314–318.

Anderson, Kristin L. 1997. "Gender, Status, and Domestic Violence: An Integration of Feminist and Family Violence Approaches." *Journal of Marriage and the Family* 59:655–669.

Applewhite, Ashton. 2003. "Covenant Marriage Would Not Benefit the Family." In *The Family: Opposing Viewpoints*, Auriana Ojeda, ed. (pp. 189–195). Farmington Hill, MI: Greenhaven Press.

Arroyo, J., Payne, K. K., Brown, S. L., and Manning, W. D. 2013. "Crossover in Median Age at First Marriage and First Birth: Thirty Years of Change." National Center for Family & Marriage Research. Available at http://ncfmr.bgsu.edu

Ascione, F. R. 2007. "Emerging Research on Animal Abuse as a Risk Factor for Intimate Partner Violence." In *Intimate Partner Violence*, K. Kendall-Tackett and S. Giacomoni, eds. (pp. 3.1–3.17). Kingston, NJ: Civic Research Institute.

Ascione, Frank R., and Kenneth Shapiro. 2009. "People and Animals, Kindness and Cruelty: Research Directions and Policy Implications." *Journal of Social Issues* 65(3):569–587.

Baker, Amy J. L. 2007. *Adult Children of Parental Alienation Syndrome: Breaking the Ties that Bind*. New York: W. W. Norton & Co.

Baker, Amy J. L. 2006. "The Power of Stories/Stories about Power: Why Therapists and Clients Should Read Stories about Parental Alienation Syndrome." *The American Journal of Family Therapy* 34:191–203.

Baker, Amy J. L., and Jaclyn Chambers. 2011. "Adult Recall of Childhood Exposure to Parental Conflict: Unpacking the Black Box of Parental Alienation." *Journal of Divorce & Remarriage* 52(1):55–76.

Bernet, William, and Amy J. L. Baker. 2013. "Parental Alienation, DSM-5, and ICD-11: Response to Critics." *Journal of the American Academy of Psychiatric Law* 41(1):98–104.

Bernstein, Nina. 2007. "Polygamy, Practiced in Secrecy, Follows Africans to New York." *New York Times*, March 23, p. A1.

Bonach, Kathryn. 2009. "Empirical Support for the Application of the Forgiveness Intervention Model to Postdivorce Coparenting." *Journal of Divorce & Remarriage* 50(1):38–54.

Bureau of Justice Statistics. 2011. "Intimate Partner Violence in the U.S.: Victim Characteristics." Available at http://bjs.ofp.usdoj.gov

Bureau of Labor Statistics. 2013a. *Women in the Labor Force: A Databook*. Report 1040. Available at www.bls.gov

Bureau of Labor Statistics. 2013b. *Employment Characteristics of Families in 2012*. Available at www.bls.gov

Carr, D., and K. W. Springer. 2010. "Advances in Families and Health Research in the 21st Century." *Journal of Marriage and Family* 72:743–761.

Carrington, Victoria. 2002. *New Times: New Families*. Dordrecht, the Netherlands: Kluwer Academic.

Carter, Lucy S. 2010. *Batterer Intervention: Doing the Work and Measuring the Progress*. Family Violence and Prevention Fund. Available at www.endabuse.org

Catalano, Shannon. 2012. *Intimate Partner Violence, 1993–2010*. Bureau of Justice Statistics. Available at www.ojp.usdoj.gov

Centers for Disease Control and Prevention. 2012. *Understanding Child Maltreatment*. Fact Sheet. Available at www.cdc.gov

Cherlin, Andrew J. 2009. *The Marriage-Go-Round: The State of Marriage and Family in America Today*. New York: Alfred A. Knopf.

Child Trends. 2012. Attitudes towards Spanking. Available at www.childtrendsdatabank.org

Coontz, Stephanie. 2005. *Marriage, a History*. New York: Penguin Books.

Coontz, Stephanie. 2004. "The World Historical Transformation of Marriage." *Journal of Marriage and Family* 66(4):974–979.

Coontz, Stephanie. 2000. "Marriage: Then and Now." *Phi Kappa Phi Journal* 80:10–15.

Coontz, Stephanie. 1997. *The Way We Really Are*. New York: Perseus.

Coontz, Stephanie. 1992. *The Way We Never Were: American Families and the Nostalgia Trap*. New York: Basic.

Cui, M., K. Ueno, M. Gordon, & F. D. Fincham. 2013. "The Continuation of Intimate Partner Violence from Adolescence to Young Adulthood." *Journal of Marriage and Family* 75(April):300–313.

Daniel, Elycia. 2005. "Sexual Abuse of Males." In *Sexual Assault: The Victims, the Perpetrators, and the Criminal Justice System*, Frances P. Reddington and Betsy Wright Kreisel, eds. (pp. 133–140). Durham, NC: Carolina Academic Press.

Davis, Lisa Selin. 2009. "Everything but the Ring." *Time* (May 25):57–58.

Decuzzi, A., D. Knox, and M. Zusman. 2004. "The Effect of Parental Divorce on Relationships with Parents and Romantic Partners of College Students." Roundtable Discussion, Southern Sociological Society, Atlanta, April 17.

DeGue, Sarah. 2009 (June). "Is Animal Cruelty a 'Red Flag' for Family Violence? Investigating Co-occurring Violence toward Children, Partners, and Pets." *Journal of Interpersonal Violence* 24(6):1033–1056.

DeMaris, Alfred, Laura A. Sanchez, and Kristi Krivickas. 2012. "Developmental Patterns in Marital Satisfaction: Another Look at Covenant Marriage." *Journal of Marriage and Family* 74(October):989–1004.

Demo, David H., Mark A. Fine, and Lawrence H. Ganong. 2000. "Divorce as a Family Stressor." In *Families and Change: Coping with Stressful Events and Transitions*, 2nd ed., P. C. McKenry and S. J. Price, eds. (pp. 279–302). Thousand Oaks, CA: Sage.

Dennison, R. P., and S. Koerner. 2008. "A Look at Hopes and Worries about Marriage: The Views of Adolescents Following a Parental Divorce." *Journal of Divorce & Remarriage* 48:91–107.

Doyle, Joseph. 2007. "Child Protection and Child Outcomes: Measuring the Effects of Foster Care." *American Economic Review* 97(5): 1583–1610.

Duncan, S., M. Phillips, S. Roseneil, J. Carter, and M. Stoilova. 2013 (April 22). "Living Apart Together: Uncoupling Intimacy and Co-residence." *Research Briefing*. Birkbeck, University of London. Available at www.bbk.ac.uk

Edin, Kathryn. 2000. "What Do Low-Income Single Mothers Say about Marriage?" *Social Problems* 47(1):112–133.

Emery, Robert E. 1999. "Postdivorce Family Life for Children: An Overview of Research and Some Implications for Policy." In *The Postdivorce Family: Children, Parenting, and Society*, R. A. Thompson and P. R. Amato, eds. (pp. 3–27). Thousand Oaks, CA: Sage.

Emery, Robert E., David Sbarra, and Tara Grover. 2005. "Divorce Mediation: Research and Reflections." *Family Court Review* 43(1):22–37.

Federal Bureau of Investigation. 2011. *Crime in the United States: 2011*. Available at www.fbi.gov

Fincham, F., M. Cui, M. Gordon, and K. Ueno. 2013. "What Comes before Why: Specifying the Phenomenon of Intimate Partner Violence." *Journal of Marriage and Family* 75(April):319–324.

Fincham, F. D., J. Hall, and S. R. H. Beach. 2006. "Forgiveness in Marriage: Current Status and Future Directions." *Family Relations* 55: 415–427.

Finkelhor, D., R. Ormrod, H. A. Turner, and S. L. Hamby. 2005. "The Victimization of Children and Youth: A Comprehensive National Survey." *Child Maltreatment* 10:5–25.

Fogle, Jean M. 2003. "Domestic Violence Hurts Dogs, Too." *Dog Fancy*, April, p. 12.

Follingstad, D. R., and M. Edmundson. 2010. "Is Psychological Abuse Reciprocal in Intimate Relationships? Data from a National Sample of American Adults." *Journal of Family Violence* 25:495–508.

Foubert, J. D., E. E. Godin, and J. L. Tatum. 2010. "In Their Own Words: Sophomore College Men Describe Attitude and Behavior Changes Resulting from a Rape Prevention Program 2 Years after Their Participation." *Journal of Interpersonal Violence* 25:2237–2257.

Fowler, K. A., and D. Westen. 2011. "Subtyping Male Perpetrators of Intimate Partner Violence." *Journal of Interpersonal Violence* 26(4):607–639.

Franklin, Emily. 2013. "How to Give the Dog a Home: Using Mediation to Solve Companion Animal Custody Disputes." *Pepperdine Dispute Resolution Law Journal* 12(2): Article 5. Available at http://law.pepperdine.edu

Gadalla, Tahany M. 2009. "Impact of Marital Dissolution on Men's and Women's Income: A Longitudinal Study." *Journal of Divorce & Remarriage* 50(1):55–65.

Gartrell, Nanette K., Henny M. W. Bos, and Naomi G. Goldberg. 2010. "Adolescents of the U.S. National Longitudinal Lesbian Family Study: Sexual Orientation, Sexual Behavior, and Sexual Risk Exposure." *Archives of Sexual Behavior*, online November 6.

Gelles, Richard J. 2000. "Violence, Abuse, and Neglect in Families." In *Families and Change: Coping with Stressful Events and Transitions*, 2nd ed., P. C. McKenry and S. J. Price, eds. (pp. 183–207). Thousand Oaks, CA: Sage.

Global Initiative to End All Corporal Punishment of Children. 2012. *Global Report 2010: Ending Legalised Violence against Children*. Available at www.endcorporalpunishment.org

Grogan-Kaylor, Andrew, and Melanie Otis. 2007. "The Predictors of Parental Use of Corporal Punishment." *Family Relations* 56:80–91.

Gromoske, Andrea N., and Kathryn Maguire-Jack. 2012. "Transactional and Cascading Relations between Early Spanking and Children's Social-Emotional Development." *Journal of Marriage and Family* 74:1054–1068.

Grych, John H. 2005. "Interparental Conflict as a Risk Factor for Child Maladjustment: Implications for the Development of Prevention Programs." *Family Court Review* 43(1):97–108.

Gustafsson, Hanna C., and Martha J. Cox. 2012. "Relations among Intimate Partner Violence, Maternal Depressive Symptoms, and Maternal Parenting Behaviors." *Journal of Marriage and Family* 74(October):1005–1020.

Hackstaff, Karla B. 2003. "Divorce Culture: A Quest for Relational Equality in Marriage." In *Family in Transition*, 12th ed., Arlene S. Skolnick and Jerome H. Skolnick, eds. (pp. 178–190). Boston: Allyn and Bacon.

Halligan, C., D. Knox, and J. Brinkley. 2013 (February 22). "TRAPPED: Technology as a Barrier to Leaving an Abusive Relationship." Poster, Southeastern Council on Family Relations, Birmingham, Alabama.

Halpern-Meekin, Sarah, Wendy D. Manning, Peggy C. Giordana, and Monica A. Longmore. 2013. *Journal of Marriage and Family* (February):2–12.

Hamilton, Brady E., Joyce A. Martin, and Stephanie Ventura. 2013. "Births: Preliminary Data for 2012." *National Vital Statistics Reports* 62(3). Available at www.cdc.gov/nchs

Hardesty, J. L., L. Khaw, M. D. Ridgway, C. Weber, and T. Miles. 2013 (May 13). "Coercive Control and Abused Women's Decisions about Their Pets When Seeking Shelter." *Journal of Interpersonal Violence* (May 13):1–24.

Hawkins, Alan J., Jason S. Carroll, William J. Doherty, and Brian Willoughby. 2004. "A Comprehensive Framework for Marriage Education." *Family Relations* 53(5):547–558.

Hewlett, Sylvia Ann, and Cornel West. 1998. *The War against Parents: What We Can Do for Beleaguered Moms and Dads*. Boston: Houghton Mifflin.

Hochschild, Arlie Russell. 1997. *The Time Bind: When Work Becomes Home and Home Becomes Work*. New York: Henry Holt.

Hochschild, Arlie Russell. 1989. *The Second Shift: Working Parents and the Revolution at Home*. New York: Viking.

Hymowitz, Kay, Jason S. Carroll, W. Bradford Wilcox, and Kelleen Kaye. 2013. *Knot Yet: The Benefits and Costs of Delayed Marriage in America*. The National Marriage Project, the National Campaign to Prevent Teen and Unplanned Pregnancy, and the Relate Institute. Available at nationalmarriageproject.org

Jackson, Shelly, Lynette Feder, David R. Forde, Robert C. Davis, Christopher D. Maxwell, and Bruce G. Taylor. 2003 (June). *Batterer Intervention Programs: Where Do We Go from Here?* U.S. Department of Justice. Available at www.usdoj.gov

Jalovaara, M. 2003. "The Joint Effects of Marriage Partners' Socioeconomic Positions on the Risk of Divorce." *Demography* 40:67–81.

Jasinski, J. L., L. M. Williams, and J. Siegel. 2000. "Childhood Physical and Sexual Abuse as Risk Factors for Heavy Drinking among African-American Women: A Prospective Study." *Child Abuse and Neglect* 24:1061–1071.

Jekielek, Susan M. 1998. "Parental Conflict, Marital Disruption, and Children's Emotional Well-Being." *Social Forces* 76:905–935.

Johnson, Michael P. 2001. "Patriarchal Terrorism and Common Couple Violence: Two Forms of Violence against Women." In *Men and Masculinity: A Text Reader*, T. F. Cohen, ed. (pp. 248–260). Belmont, CA: Wadsworth.

Johnson, Michael P., and Kathleen Ferraro. 2003. "Research on Domestic Violence in the 1990s: Making Distinctions." In *Family in Transition*, 12th ed., A. S. Skolnick and J. H. Skolnick, eds. (pp. 493–514). Boston: Allyn and Bacon.

Kalmijn, Matthijs, and Christiaan W. S. Monden. 2006. "Are the Negative Effects of Divorce on Well-Being Dependent on Marital Quality?" *Journal of Marriage and the Family* 68: 1197–1213.

Kamp Dush, Claire M. 2013. "Marital and Cohabitation Dissolution and Parental Depressive Symptoms in Fragile Families." *Journal of Marriage and Family* 75(February):91–109.

Kaufman, Joan, and Edward Zigler. 1992. "The Prevention of Child Maltreatment: Programming, Research, and Policy." In *Prevention of Child Maltreatment: Developmental and EcologicalPerspectives*, Diane J. Willis, E. Wayne Holden, and Mindy Rosenberg, eds. (pp. 269–295). New York: John Wiley.

Kitzmann, K. M., N. K. Gaylord, A. R. Holt, and E. D. Kenny. 2003. "Child Witnesses to Domestic Violence: A Meta-Analytic Review." *Journal of Clinical and Consulting Psychology* 71:339–352.

Knox, David (with Kermit Leggett). 1998. *The Divorced Dad's Survival Book: How to Stay Connected with Your Kids*. New York: Insight Books.

Koch, Wendy. 2009. "Fees Cut Down Private Adoptions." *USA Today*, April 27, 1A.

Lacey, K.K., D. G. Saunders, and L. Zhang. 2011. "A Comparison of Women of Color and Non-Hispanic White Women on Factors Related to Leaving a Violent Relationship." *Journal of Interpersonal Violence* 26:1036–1055.

LaFraniere, Sharon. 2005. "Entrenched Epidemic: Wife-Beatings in Africa." *New York Times*, August 11, pp. A1 and A8.

Lara, Adair, 2005. "One for the Price of Two: Some Couples Find Their Marriages Thrive When They Share Separate Quarters." *San Francisco Chronicle*, June 29. Available at www.sfgate.com

Laungani, P. 2005. "Changing Patterns of Family Life in India." In *Families in Global Perspective*, J. L. Roopnarine and U. P. Gielen, eds. (pp. 85–103). Boston: Pearson, Allyn and Bacon.

Levin, Irene. 2004. "Living Apart Together: A New Family Form." *Current Sociology* 52(2):223–240.

Lewin, Tamar. 2000. "Fears for Children's Well-Being Complicates a Debate over Marriage." *New York Times*, November 4. Available at www.nytimes.com

Lloyd, Sally A. 2000. "Intimate Violence: Paradoxes of Romance, Conflict, and Control." *National Forum* 80(4):19–22.

Lloyd, Sally A., and Beth C. Emery. 2000. *The Dark Side of Courtship: Physical and Sexual Aggression*. Thousand Oaks, CA: Sage.

Lofquist, D., T. Lugalia, M. O'Connell, and S. Feliz. 2012. "Households and Families: 2010." *2010 Census Briefs*. Available at www.census.gov

Mason, Mary Ann, Arlene Skolnick, and Stephen D. Sugarman. 2003. "Introduction." In *All Our Families*, 2nd ed., Mary Ann Mason, Arlene Skolnick, and Stephen D. Sugarman, eds. (pp. 1–13). New York: Oxford University Press.

Mental Health America. 2003. *Effective Discipline Techniques for Parents: Alternatives to Spanking*. Strengthening Families Fact Sheet. Available at www.nmha.org

Morin, Rich. 2011. (February 16). "The Public Renders a Split Verdict on Changes in Family Structure." Pew Research Center. Available at pewresearch.org

National Center for Injury Prevention and Control. 2012. *Understanding Intimate Partner Violence*. Available at www.cdc.gov

National Marriage Project and the Institute for American Values. 2012. *The State of Our Unions: Marriage in America 2012*. Available at www.stateofourunions.org

Nelson, B. S., and K. S. Wampler. 2000. "Systemic Effects of Trauma in Clinic Couples: An Exploratory Study of Secondary Trauma Resulting from Childhood Abuse." *Journal of Marriage and Family Counseling* 26:171–184.

Newport, Frank. 2012 (May 12). "Americans, Including Catholics, Say Birth Control is Morally OK." Gallup Organization. Available at www.gallup.com

Nock, Steven L. 1995. "Commitment and Dependency in Marriage." *Journal of Marriage and the Family* 57:503–514.

OECD. 2012. OECD Family Database. Available at www.oecd.org

Parker, K. 2011. "A Portrait of Stepfamilies." Pew Research Center. Available at www.pewsocialtrends.org

Parker, K., and W. Wang. 2013 (March 14). "Modern Parenthood." Pew Research Center. Available at www.pewsocialtrends.org

Parker, Marcie R., Edward Bergmark, Mark Attridge, and Jude Miller-Burke. 2000. "Domestic Violence and Its Effect on Children." *National Council on Family Relations Report* 45(4): F6–F7.

Pasley, Kay, and Carmelle Minton. 2001. "Generative Fathering after Divorce and Remarriage: Beyond the 'Disappearing Dad.'" In *Men and Masculinity: A Text Reader*, T. F. Cohen, ed. (pp. 239–248). Belmont CA: Wadsworth.

Pew Research Center. 2008. "Women Call the Shots at Home: Public Mixed on Gender Roles in Jobs." Available at http://pewresearch.org

Pilkauskas, Natasha V. 2012. "Three-Generation Family Households: Differences by Family Structure at Birth." *Journal of Marriage and Family* 74:931–943.

Planty, M., L. Langton, C. Krebs, M. Berzofsky, and H. Smiley-McDonald. 2013. "Female Victims of Sexual Violence, 1994–2010." U.S. Department of Justice, Bureau of Justice Statistics. Available at www.bjs.gov

Population Reference Bureau. 2011. *The World's Women and Girls 2011 Data Sheet.* Available at www.prb.org

Ricci, L., A. Giantris, P. Merriam, S. Hodge, and T. Doyle. 2003. "Abusive Head Trauma in Maine Infants: Medical, Child Protective, and Law Enforcement Analysis." *Child Abuse and Neglect* 27:271–283.

Rubin, D. M., C. W. Christian, L. T. Bilaniuk, K. A. Zaxyczny, and D. R. Durbin. 2003. "Occult Head Injury in High-Risk Abused Children." *Pediatrics* 111:1382–1386.

Russell, D. E. 1990. *Rape in Marriage.* Bloomington: Indiana University Press.

Scott, K. L., and D. A. Wolfe. 2000. "Change among Batterers: Examining Men's Success Stories." *Journal of Interpersonal Violence* 15:827–842.

Shepard, Melanie F., and James A. Campbell. 1992. "The Abusive Behavior Inventory: A Measure of Psychological and Physical Abuse." *Journal of Interpersonal Violence* 7(3):291–305.

Skinner, Jessica A. and Robin M. Kowalski. 2013. "Profiles of Sibling Bullying." *Journal of Interpersonal Violence* 28(8):1726–1736.

Smith, J. 2003. "Shaken Baby Syndrome." *Orthopaedic Nursing* 22:196–205.

Steimle, Brynn M., and Stephen F. Duncan. 2004. "Formative Evaluation of a Family Life Education Web Site." *Family Relations* 53(4):367–376.

Stone, R. D. 2004. *No Secrets, No Lies: How Black Families Can Heal from Sexual Abuse.* New York: Broadway Books.

Straus, Murray. 2010. "Prevalence, Societal Causes, and Trends in Corporal Punishment by Parents in World Perspective." *Law and Contemporary Problems* 73(1):1–30.

Straus, Murray. 2000. "Corporal Punishment and Primary Prevention of Physical Abuse." *Child Abuse and Neglect* 24:1109–1114.

Sullivan, Erin. 2010. "Abused Pasco Dog Taken in by Victim Advocate Now Pays It Forward." *St. Petersburg Times,* January 23. Available at www.tampabay.com

Swan, S. C., L. J. Gambone, J. E. Caldwell, T. P. Sullivan, and D. L Snow. 2008. "A Review of Research on Women's Use of Violence with Male Intimate Partners." *Violence and Victims* 23:301–315.

Sweeney, M. M. 2010. "Remarriage and Stepfamilies: Strategic Sites for Family Scholarship in the 21st Century." *Journal of Marriage and the Family* 72:667–684.

Swiss, Liam, and Celine Le Bourdais. 2009. "Father-Child Contact after Separation: The Influence of Living Arrangements." *Journal of Family Issues* 30(5):623–652.

Teaster, Pamela B., Tyler A. Dugar, Marta S. Mendiondo, Erin L. Abner, Kara A. Cecil, and Joanne M. Otto. 2006 (February). *The 2004 Survey of State Adult Protective Services: Abuse of Adults 60 Years of Age and Older.* National Center on Elder Abuse. Washington, DC.

Trinder, L. 2008. "Maternal Gate Closing and Gate Opening in Postdivorce Families." *Journal of Family Issues* 29:1298–1298.

Ulman, A. 2003. "Violence by Children against Mothers in Relation to Violence between Parents and Corporal Punishment by Parents." *Journal of Comparative Family Studies* 34:41–56.

Umberson, D., K. L. Anderson, K. Williams, and M. D. Chen. 2003. "Relationship Dynamics, Emotion State, and Domestic Violence: A Stress and Masculine Perspective." *Journal of Marriage and the Family* 65:233–247.

UNICEF. 2012. *Progress for Children: A Report Card on Adolescents.* Available at www.childinfo.org

UNICEF. 2010. *Child Disciplinary Practices at Home: Evidence from a Range of Low- and Middle-Income Countries.* Available at www.childinfo.org

United Nations Development Programme. 2009 (November 23). "Ending Violence against Women Helps Achieve Development Goals." Available at www.beta.undp.org

U.S. Census Bureau. 2012. *America's Families and Living Arrangements: 2012.* Available at www.census.gov

U.S. Department of Health and Human Services. Administration on Children, Youth, and Families. 2012. *Child Maltreatment 2011.* Washington, DC: U.S. Government Printing Office.

Walby, S. 2013. "Violence and Society: Introduction to an Emerging Field of Sociology." *Current Sociology* 61:95–111.

Walker, Alexis J. 2001. "Refracted Knowledge: Viewing Families through the Prism of Social Science." In *Understanding Families into the New Millennium: A Decade in Review,* Robert M. Milardo, ed. (pp. 52–65). Minneapolis, MN: National Council on Family Relations.

Wallerstein, Judith S. 2003. "Children of Divorce: A Society in Search of Policy." In *All Our Families,* 2nd ed., Mary Ann Mason, Arlene Skolnick, and Stephen D. Sugarman, eds. (pp. 66–95). New York: Oxford University Press.

Wang, Wendy, and Paul Taylor. 2011 (March 19). "For Millennials, Parenthood Trumps Marriage." Pew Research Center. Available at www.pewsocialtrends.org

Whiffen, V. E., J. M. Thompson, and J. A. Aube. 2000. "Mediators of the Link between Childhood Sexual Abuse and Adult Depressive Symptoms." *Journal of Interpersonal Violence* 15:1100–1120.

Whitehurst, Dorothy H., Stephen O'Keefe, and Robert A. Wilson. 2008. "Divorced and Separated Parents in Conflict: Results from a True Experiment Effect of a Court Mandated Parenting Education Program." *Journal of Divorce & Remarriage* 48(3/4):127–144.

Williams, K., and A. Dunne-Bryant. 2006. "Divorce and Adult Psychological Well-Being: Clarifying the Role of Gender and Child Age." *Journal of Marriage and the Family* 68: 1178–1196.

Yun, I., D. Ball, and H. Lim. 2011. "Disentangling the Relationship between Child Maltreatment and Violent Delinquency: Using a Nationally Representative Sample." *Journal of Interpersonal Violence* 26(1):88–110.

Zeitzen, Miriam K. 2008. *Polygamy: A Cross-Cultural Analysis.* Oxford: Berg.

Chapter 6

Acemoglu, Daron, and James A. Robinson. 2012. "The Problem with U.S. Inequality." *The Huffington Post Blog.* Available at www.huffingtonpost.com

Administration for Children and Families. 2002. *Early Head Start Benefits Children and Families.* U.S. Department of Health and Human Services. Available at www.acf.hhs.gov

Agazzi, Isoida. 2012 (November 26). "Fixing the 'Silent' Sanitation Crisis." Inter Press Service. Available at www.ipsnews.net

Alex-Assensoh, Yvette. 1995. "Myths about Race and the Underclass." *Urban Affairs Review* 31:3–19.

Anderson, Sarah, Scott Klinger, and Javier Rojo. 2013 (April 15). "Corporate Tax Dodgers: 10 Companies and Their Tax Loopholes." Washington, DC: Institute for Policy Studies.

Badgett, M. V. Lee, Laura E. Durso, and Alyssa Schneebaum. 2013. "New Patterns of Poverty in the Lesbian, Gay, and Bisexual Community." The Williams Institute. Available at williamsinstitute.law.ucla.edu

Bajak, Frank. 2010. "Chile-Haiti Earthquake Comparison: Chile was More Prepared." *Huffington Post,* February 27. Available at www.huffingtonpost.com

Bickel, G., M. Nord, C. Price, W. Hamilton, and J. Cook. 2000. *United States Department of Agriculture Guide to Measuring Household Food Security.* Alexandria, VA: U.S. Department of Agriculture, Food and Nutrition Service.

Bishaw, Alemayehu. 2013 (September). "Poverty: 2000 to 2012." *American Community Survey Briefs.* U.S. Census Bureau. Available at www.census.gov

Blumenthal, Susan. 2012 (March 12). "Debunking Myths about Food Stamps." SNAP to Health. Available at www.snaptohealth.org

Callahan, David, and J. Mijin Cha. 2013. *Stacked Deck: How the Dominance of Politics by the Affluent & Business Undermines Economic Mobility.* Demos. Available at www.demos.org

Chandy, Laurence. 2013 (May). "Counting the Poor." The Brookings Institution. Available at www.brookings.edu

Chandy, Laurence, Natasha Ledlie, and Veronika Penciakova. 2013 (April). *The Final Countdown: Prospects for Ending Extreme Poverty by 2030.* Policy Brief 2013–04. Washington, DC: The Brookings Institution.

Coleman-Jensen, Alisha, Mark Nord, and Margaret Andrews. 2012. *Household Food Security in the United States, 2011.* USDA Economic Research Service. Available at www.ers.usda.gov

Collins, Chuck. 2013 (May 28). "The Wealthy Kids Are All Right." *The American Prospect.* Available at prospect.org

Credit Suisse Research Institute. 2012 (October). *Global Wealth Databook 2012.* Available at www.usaagainstgreed.org/GlobalWealthDatabook2012.pdf

Davis, Kingsley, and Wilbert Moore. 1945. "Some Principles of Stratification." *American Sociological Review* 10:242–249.

DeNavas-Walt, Carmen, Bernadette D. Proctor, and Jessica C. Smith. 2013. *Income, Poverty, and Health Insurance in the United States: 2012.* U.S. Census Bureau, Current Population Reports P60-245. Washington, DC: U.S. Government Printing Office.

Dvorak, Petula. 2009. "Increase Seen in Attacks on Homeless." *Washington Post*, February 5, p. DZ01.

Epstein, William M. 2004. "Cleavage in American Attitudes toward Social Welfare." *Journal of Sociology and Social Welfare* 31(4):177–201.

FAO (Food and Agriculture Organization). 2012. *The State of Food Insecurity in the World.* Available at www.fao.org

Forster, Michael, and Marco Mira d'Ercole. 2005 (March 10). "Income Distribution and Poverty in OECD Countries in the Second Half of the 1990s." Organization for Economic Cooperation and Development. Available at www.oecd.org

Fry, Richard, and Paul Taylor. 2013. "A Rise in Wealth for the Wealthy: Declines for the Lower 93%." Pew Research Social & Demographic Trends. Available at www.pewsocialtrends.org

Gans, Herbert. 1972. "The Positive Functions of Poverty." *American Journal of Sociology* 78:275–289.

Garfinkel, Irwin. 2013. "The Welfare State: Myths & Measurement." *Spectrum* (Winter):6–7.

Giovanni, Thomas, and Roopal Patel. 2013. *Gideon at 50: Three Reforms to Revive the Right to Counsel.* Brennan Center for Justice. Available at www.brennancenter.org

Golden, Olivia. 2013 (May 9). "Poverty in America: How We Can Help Families." Urban Institute. Available at www.urban.org

Gould, Elise, Hilary Wething, Natalie Sabadish, and Nicholas Finio. 2013 (July 3). "What Families Need to Get By: The 2013 Update of EPI's Family Budget Calculator." Economic Policy Institute. Available at www.epi.org

Green, Autumn R. 2013. "Patchwork: Poor Women's Stories of Resewing the Shredded Safety Net." *Affilia* 28:51–64.

Grunwald, Michael. 2006. "The Housing Crisis Goes Suburban." *Washington Post*, August 27. Available at www.washingtonpost.com

Harell, Allison, Stuart Soroka, and Adam Mahon. 2008. "Is Welfare a Dirty Word?: Canadian Public Opinion on Social Assistance Policies." *Options Politiques* (September):53–56.

vanden Heuvel, Katrina. 2011. "Putting Poverty on the Agenda." *The Nation,* January 17. Available at www.thenation.com

Hoback, Alan, and Scott Anderson. 2007. "Proposed Method for Estimating Local Population of Precariously Housed." National Coalition for the Homeless. Available at www.nationalhomeless.org

Institute for Economics & Peace. 2013. *Pillars of Peace.* Available at economicsandpeace.org

Irvine, Leslie. 2013. "Animals as Lifechangers and Lifesavers: Pets in the Redemption Narratives of Homeless People." *Journal of Contemporary Ethnography* 42(1):3–36.

Katz, Michael B. 2013. *The Undeserving Poor: America's Enduring Confrontation with Poverty: Fully Updated and Revised.* New York: Oxford University Press.

Kondo, Naoki, Grace Sembajwe, Ichiro Kawachi, Rob M van Dam, S. V. Subramanian, and Zentaro Yamagata. 2009. "Income Inequality, Mortality, and Self- Rated Health: Meta-analysis of Multilevel Studies." *British Medical Journal* 339(7731):1178–1181.

Kraut, Karen, Scott Klinger, and Chuck Collins. 2000. *Choosing the High Road: Businesses That Pay a Living Wage and Prosper.* Boston: United for a Fair Economy.

Kroll, Luisa, and Kerry A. Dolan. 2013. "The World's Billionaires." *Forbes,* March 4. Available at www.forbes.com

Ku, Leighton, and Brian Bruen. 2013 (February 19). "The Use of Public Assistance Benefits by Citizens and Non-Citizen Immigrants in the United States." *CATO Working Paper.* Washington, DC: Cato Institute.

Ladd, Helen F. 2012. "Education and Poverty: Confronting the Evidence." *Journal of Policy Analysis and Management* 31(2):203–227.

Leigh, J. Paul. 2013 (March 6). "Raising the Minimum Wage Could Improve Public Health." The Economic Policy Institute Blog. Available at www.epi.org

Llobrera, Joseph, and Bob Zahradnik. 2004. *A HAND UP: How State Earned Income Tax Credits Helped Working Families Escape Poverty in 2004.* Center on Budget and Policy Priorities. Available at www.cbpp.org

Lowrey, Annie. 2013 (September 10). "The Rich Get Richer through the Recovery." *New York Times.* Available at http://economix.blogs.nytimes.com

Luker, Kristin. 1996. *Dubious Conceptions: The Politics of Teenage Pregnancy.* Cambridge, MA: Harvard University Press.

Massey, D. S. 1991. "American Apartheid: Segregation and the Making of the American Underclass." *American Journal of Sociology* 96:329–357.

Mayer, Susan E. 1997. *What Money Can't Buy: Family Income and Children's Life Chances.* Cambridge, MA: Harvard University Press.

McDonagh, Thomas. 2013. *Unfair, Unsustainable, and Under the Radar: How Corporations Use Global Investment Rules to Undermine a Sustainable Future.* The Democracy Center. Available at www.democracyctr.org

McKernan, Signe-Mary, Caroline Ratcliffe, C. Eugene Steuerle, and Sisi Zhang. 2013 (April). "Less Than Equal: Racial Disparities in Wealth." Urban Institute. Available at www.urban.org

McNamee, Stephen J., and Robert K. Miller, Jr. 2009. *The Meritocracy Myth,* 2nd edition. Lanham, MD: Rowman & Littlefield Publishing Group.

Milkman, Ruth, Penny Lewis, and Stephanie Luce. 2013. "The Genie's Out of the Bottle: Insiders' Perspectives on Occupy Wall Street." *The Sociological Quarterly* 54(2):194–198.

Mishel, Lawrence, and Nicholas Finio. 2013 (January 23). "Earnings of the Top 1.0 Percent Rebound Strongly in the Recovery." Economic Policy Institute. Available at www.epi.org

Mishel, Lawrence, and Natalie Sabadish. 2013 (June 26). "CEO Pay in 2012 Was Extraordinarily High Relative to Typical Workers and other High Earners." Economic Policy Institute. Available at www.epi.org

Narayan, Deepa. 2000. *Voices of the Poor: Can Anyone Hear Us?* New York: Oxford University Press.

National Alliance to End Homelessness. 2013. *The State of Homelessness in America.* Washington, DC: National Alliance to End Homelessness.

National Coalition for the Homeless. 2012. *Hate Crimes against the Homeless: The Brutality of Violence Unveiled.* Available at www.nationalhomeless.org

National Research Council and Institute of Medicine. 2013. *U.S. Health in International Perspective: Shorter Lives, Poorer Health.* Washington, DC: The National Academies Press.

Odede, Kennedy. 2010. "Slumdog Tourism." *The New York Times,* August 10, section A, p. 25. Available at www.nytimes.com

Office of Family Assistance. 2012. *Ninth Report to Congress.* Available at www.acf.hhs.gov

Pew Research Center. 2011 (July 26). "Wealth Gaps Rise to Record Highs between Whites, Blacks, Hispanics." Available at www.pewsocialtrends.org

Plumer, Brad. 2012. "Who Receives Government Benefits, in Six Charts." *The Washington Post,* September 18. Available at www.washingtonpost.com

Pugh, Tony. 2007. "U.S. Economy Leaving Record Numbers in Severe Poverty." *McClatchy Newspapers,* February 22. Available at www.mcclatchydc.com

Ramos, Alcida Rita, Rafael Guerreiro Osorio, and Jose Pimenta. 2009. "Indigenising Development." *Poverty in Focus* 17(May):pp. 3–5. International Policy Centre for Inclusive Growth.

Ratcliffe, Caroline, and Mary-Signe McKernan. 2010 (June 10). "Childhood Poverty Persistence: Facts and Consequences." Urban Institute. Available at www.urban.org

Rohde, David. 2012. "The Hideous Inequality Exposed by Hurricane Sandy." *The Atlantic,* October 12. Available at www.theatlantic.com

Roseland, Mark, and Lena Soots. 2007. "Strengthening Local Economies." In *2007 State of the World,* Linda Starke, ed. (152–169). New York: W. W. Norton & Co.

Saad, Lydia. 2013 (May 17). "Americans Say Family of Four Needs Nearly $60K to 'Get By.'" Gallup Poll. Available at www.gallup.com

Satterthwaite, David, and Gordon McGranahan. 2007. "Providing Clean Water and Sanitation." In *2007 State of the World: Our Urban Future*, L. Starke, ed. (pp. 26–45). New York: W. W. Norton & Company.

Seccombe, Karen. 2001. "Families in Poverty in the 1990s: Trends, Causes, Consequences, and Lessons Learned." In *Understanding Families into the New Millennium: A Decade in Review*, Robert M. Milardo, ed. (pp. 313–332). Minneapolis, MN: National Council on Family Relations.

Shierholz, Heidi. 2013 (June 11). "Unemployed Workers Still Far Outnumber Job Openings in Every Major Sector." The Economic Policy Institute. Available at www.epi.org

Sobolewski, Juliana M., and Paul R. Amato. 2005. "Economic Hardship in the Family of Origin and Children's Psychological Well-Being in Adulthood." *Journal of Marriage and Family* 67(1):141–156.

Stiglitz, Joseph E. 2013. "Student Debt and the Crushing of the American Dream." *New York Times*, May 12. Available at www.nytimes.com

Straus, Rebecca. 2013. "Schooling Ourselves in an Unequal America." *New York Times*, June 16. Available at www.nytimes.com

Susskind, Yifat. 2005 (May). *Ending Poverty, Promoting Development: MADRE Criticizes the United Nations Millennium Development Goals*. Available at www.madre.org

Talberth, John, Daphne Wysham, and Karen Dolan. 2013. "Closing the Inequality Divide: A Strategy for Fostering Genuine Progress in Maryland." Center for Sustainable Economy and Institute for Policy Studies. Available at www.sustainable-economy.org

Turner, Margery Austin, Susan J. Popkin, G. Thomas Kingsley, and Deborah Kaye. 2005 (April). *Distressed Public Housing: What It Costs to Do Nothing*. The Urban Institute. Available at www.urban.org

UNDP (United Nations Development Programme). 2013. *Human Development Report 2013*. Available at hdr.undp.org

UNDP 2010. *Human Development Report 2010*. Available at hdr.undp.org

UNDP. 2006. *Human Development Report 2006*. New York: Palgrave Macmillan.

UNDP. 1997. *Human Development Report 1997*. New York: Oxford University Press.

UN-Habitat. 2010. *State of the World's Cities 2010/2011*. Available at www.unhabitat.org

United for a Fair Economy. 2012. *Born on Third Base: What the Forbes 400 Really Says About Economic Equality & Opportunity in America*. Available at www.faireconomy.org

United Nations. 2005. *Report on the World Social Situation 2005*. New York: United Nations.

U.S. Census Bureau. 2013. *Poverty Thresholds for 2012*. Available at www.census.gov

U.S. Census Bureau. 2012a. *2011 American Community Survey*. Available at www.census.gov

U.S. Census Bureau, 2012b. *Current Population Survey, 2012 Annual Social and Economic Supplement*. Table POV26. Available at www.census.gov

U.S. Conference of Mayors. 2012. *Hunger and Homelessness Survey*. Available at usmayors.org

USDA Food and Nutrition Service. 2013. *Program Data, Supplemental Nutrition Assistance Program*. Available at www.fns.usda.gov

Wider Opportunities for Women. 2010. *The Basic Economic Security Tables for the United States*. Washington DC: Wider Opportunities for Women.

Wilson, William J. 1996. *When Work Disappears: The World of the New Urban Poor*. New York: Knopf.

Wilson, William J. 1987. *The Truly Disadvantaged: The Inner City, the Underclass, and Public Policy*. Chicago: University of Chicago Press.

World Bank. 2005. *Global Monitoring Report 2005*. Available at www.worldbank.org

World Health Organization. 2002. *The World Health Report 2002*. Available at www.who.int/pub/en

World Health Organization and UNICEF. 2013. *Progress on Sanitation and Drinking-Water 2013 Update*. Geneva: WHO Press.

Wright, Erik Olin, and Joel Rogers. 2011. *American Society: How It Really Works*. New York: W. W. Norton & Co.

Zedlewski, Sheila R. 2003. *Work and Barriers to Work among Welfare Recipients in 2002*. Urban Institute. Available at www.urban.org

Chapter 7

AFL-CIO. 2013. *Death on the Job: The Toll of Neglect*, 22nd ed. Available at www.aflcio.org

Austin, Colin. 2002. "The Struggle for Health in Times of Plenty." In *The Human Cost of Food: Farmworkers' Lives, Labor, and Advocacy*, C. D. Thompson Jr. and M. F. Wiggins, eds. (pp. 198–217). Austin: University of Texas Press.

Baily, Martin Neil, and Douglas J. Elliott. 2009 (June 15). "The U.S. Financial and Economic Crisis: Where Does It Stand and Where Do We Go from Here?" The Brookings Institution. Available at www.brookings.edu

Barsamian, David. 2012. "Capitalism and Its Discontents: Richard Wolff on What Went Wrong." *The Sun* 434:4–13.

Barstow, David, and Lowell Bergman. 2003. "Deaths on the Job, Slaps on the Wrist." *New York Times Online*, January 10. Available at www.nytimes.com

Bassi, Laurie J., and Jens Ludwig. 2000. "School-to-Work Programs in the United States: A Multi-Firm Case Study of Training, Benefits, and Costs." *Industrial and Labor Relations Review* 53(2):219–239.

Benjamin, Medea. 1998. *What's Fair About Fair Labor Association (FLA)?* Sweatshop Watch. Available at www.sweatshopwatch.org

Bonior, David. 2006. "Undermining Democracy: Worker Repression in the United States." *Multinational Monitor* 27(4). Available at www.essential.org/monitor

Brand, Jennie E., and Sarah A. Burgard. 2008. "Job Displacement and Social Participation over the Lifecourse: Findings for a Cohort of Joiners." *Social Forces* 87(1):211–242.

Bureau of Labor Statistics. 2013a. "The Employment Situation—June 2013." *Employment Situation Summary*. Available at www.bls.gov

Bureau of Labor Statistics. 2013b. "Census of Fatal Occupational Injuries." Available at www.bls.gov

Bureau of Labor Statistics. 2013c. *Union Members 2012*. Available at www.bls.gov

Butterworth, P., L. S. Leach, L. Strazdins, S. C. Olesen, B. Rodgers, and D. H. Broom. 2011. "The Psychosocial Quality of Work Determines Whether Employment Has Benefits for Mental Health: Results from a Longitudinal National Household Survey." *Occupational and Environmental Medicine*. Advance online publication. doi:10.1136/oem.2010.059030

Carney, Eliza Newlin. 2012 (October 12). "Rules of the Game: Workplace Intimidation Becomes Murky in Post-Citizens United Era." *Roll Call*. Available at www.rollcall.com

Cockburn, Andrew. 2003. "21st Century Slaves." *National Geographic*, September, pp. 2–11, 18–24.

Council of Economic Advisers. 2010 (March). *Work-Life Balance and the Economics of Workplace Flexibility*, Christina Romer, ed. Executive Office of the President. Available at www.whitehouse.gov

Davidson, Paul. 2012 (December 6). "More U.S. Service Jobs Heading Offshore." *USA Today*. Available at www.usatoday.com

Dorell, Oren. 2011. "Report Blames Massey for W. Va. Mine Explosion." *USA Today*, May 19. Available at www.usatoday.com

Ebeling, Richard M. 2009 (February 12). "Capitalism the Solution, Not Cause of the Current Economic Crisis." American Institute for Economic Research. Available at www.aier.org

The Editors. 2013 (May 12). "Your Future Will Be Manufactured on a 3-D Printer." Bloomberg. Available at http://bloomberg.com

Faux, Jeff. 2008 (February 29). "Overhauling NAFTA." Economic Policy Institute. Available at www.epi.org

FLA Watch. 2007a. *FLA Watch: Monitoring the Fair Labor Association*. Available at http://flawatch.usus.org/

FLA Watch. 2007b. *About FLA Watch*. Available at http://flawatch.usas.org/

Flounders, Sara. 2013 (February 4). "The Pentagon and Slave Labor in U.S. Prisons." Global Research. Available at www.globalresearch.ca

Frederick, James, and Nancy Lessin. 2000. "Blame the Worker: The Rise of Behavior-Based Safety Programs." *Multinational Monitor* 21(11). Available at www.essential.org/monitor

Galinsky, Ellen, Kerstin Aumann, James T. Bond. 2009. *Times Are Changing: Gender and Generation at Work and Home*. New York: Families and Work Institute.

Greenhouse, Steven. 2008. *The Big Squeeze*. New York: Alfred A. Knopf.

Hall, Charles A. S., and John W. Day, Jr. 2009. "Revising the Limits to Growth after Peak Oil." *American Scientist* (May–June):230–237.

Hardy, Quentin. 2013. "Global Slavery, by the Numbers." *Bits* (blog). *New York Times*, March 6. Available at http://bits.blogs.nytimes.com

Harris Poll. 2012 (December 5). "Few Spending Differences Year over Year." Available at www.harrisinteractive.com

Heymann, Jody. Alison Earle, and Jeffrey Hayes. 2007. *The Work, Family and Equity Index.* Montreal, QC: The Project on Global Working Families and The Institute for Health and Social Policy.

Huffstutter, P. J. 2009. "Struggling Cities Cancel Fourth of July Fireworks." *Los Angeles Times,* June 29. Available at www.latimes.com

Human Rights Watch. 2010. *Fields of Peril: Child Labor in US Agriculture.* Available at www.hrw.org

Human Rights Watch. 2009 (January). *The Employee Free Choice Act: A Human Rights Imperative.* Available at www.hrw.org

Human Rights Watch. 2007 (May). *Discounting Rights: Wal-Mart's Violation of US Workers' Right to Freedom of Association.* Volume 19, No. 2 (G). Available at www.hrw.org

ILO (International Labour Organization). 2013. "Child Labour." Available at www.ilo.org

ILO (International Labour Organization). 2011. *Global Employment Trends 2011: The Challenge of a Jobs Recovery.* Geneva: International Labour Office.

Institute for Global Labour & Human Rights. 2011 (March 23). "Triangle Returns: Young Women Continue to Die in Locked Sweatshops." Available at www.globallabourrights.org

International Trade Union Confederation. 2013. *Countries at Risk: 2013 Report on Violations of Trade Union Rights 2011.* Available at www.ituc-csi.org

Jensen, Derrick. 2002 (June). "The Disenchanted Kingdom: George Ritzer on the Disappearance of Authentic Culture." *The Sun,* pp. 38–53.

Kelly, Erin L., Phyllis Moen, and Eric Tranby. 2011. "Changing Workplaces to Reduce Work-Family Conflict: Schedule Control in a White Collar Organization." *American Sociological Review* 76(2):265–290.

Klerman, Jacob A., Kelly Daley, and Alyssa Pozniak. 2013. *Family and Medical Leave in 2012: Technical Report.* Prepared for the U.S. Department of Labor. Available at www.dol.gov

Lenski, Gerard, and J. Lenski. 1987. *Human Societies: An Introduction to Macrosociology,* 5th ed. New York: McGraw-Hill.

Leonard, Bill. 1996 (July). "From School to Work: Partnerships Smooth the Transition." *HR Magazine* (Society for Human Resource Management). Available at www.shrm.org

Lockard, C. Brett, and Michael Wolf. 2012. "Education and Training Outlook for Occupations, 2010–2020." Bureau of Labor Statistics. Available at www.bls.gov

Luhby, Tami. 2013. "Recent College Grads Face 36% 'Mal-Employment' Rate." *CNNMoney,* June 25. Available at http://money.cnn.com

MacEnulty, Pat. 2005 (September). "An Offer They Can't Refuse: John Perkins on His Former Life as an Economic Hit Man." *The Sun* 357:4–13.

Maher, Kris. 2011. "Mine Probe Faults Massey." *Wall Street Journal,* May 20. Available at http://online.wsj.com

Martinson, Karin, and Pamela Holcomb. 2007. *Innovative Employment Approaches and Programs for Low-Income Families.* Washington, DC: The Urban Institute.

Matos, Kenneth, and Ellen Galinsky. 2012. *2012 National Study of Employers.* Families and Work Institute. Available at familiesandwork.org

McGregor, Jena. 2013 (June 27). "New York Diners Relax. Paid Sick Leave is Now the Law." *Washington Post.* Available at www.washingtonpost.com

Mehta, Chirag, and Nik Theodore. 2005 (December). *Undermining the Right to Organize: Employer Behavior during Union Representation Campaigns.* American Rights at Work. Available at www.americanrightsatwork.org

Mendenhall, Ruby, Ariel Kalil, Laurel J. Spindel, and Cassandra M. D. Hart. 2008. "Job Loss at Mid-Life: Managers and Executives Face the 'New-Risk Economy.'" *Social Forces* 87(1):185–209.

Miers, Suzanne. 2003. *Slavery in the Twentieth Century: The Evolution of a Global Problem.* Walnut Creek, CA: AltaMira Press.

Miles, Kathleen. 2012 (December 14). "'Sweatshop Conditions' Found in LA Fashion District At Contractors for Urban Outfitters, Aldo, Forever 21." *Huffington Post.* Available at www.huffingtonpost.com

National Chicken Council. 2012. "Broiler Chicken Industry Key Facts." Available at www.nationalchickencouncil.org

National Labor Committee. 2007. "Senate Minority Leader Senator Harry Reid, Congressman Bernie Sanders, AFL-CIO and Others Endorse Anti-Sweatshop Bill." Available at www.nlcnet.org

Newport, Frank. 2012 (November 29). "Democrats, Republicans, Diverge on Capitalism, Federal Gov't." Gallup. Available at www.gallup.com

Newport, Frank. 2011 (March 31). "Americans' Top Job-Creation Idea: Stop Sending Work Overseas." Gallup. Available at www.gallup.com

"New OSHA Policy Relieves Employees." 1998. *Labor Relations Bulletin* no. 687, p. 8.

Parenti, Michael. 2007 (February 16). "Mystery: How Wealth Creates Poverty in the World." Common Dreams NewsCenter. Available at www.commondreams.org

Parker, Kim, and Wendy Wang. 2013 (March 14). *Modern Parenthood.* Pew Research Center. Available at www.pewresearch.org

Perkins, John. 2004. *Confessions of an Economic Hit Man.* San Francisco: Berrett-Koehler Publishers, Inc.

Pew Research Center for the People & the Press. 2012. *Trends in American Values: 1987-2012.* Available at www.people-press.org

Pew Research Center Global Attitudes Project. 2012. "Pervasive Gloom about the World Economy." Available at www.pewglobal.org

Public Citizen. 2013 (March 8). "U.S. Trade Deficit with Korea Soars to Highest Point in Record Under FTA." Eyes on Trade: Public Citizen's Blog on Globalization and Trade. Available at www.publiccitizen.org

Rampell, Ed. 2013 (April 16). "An Interview with Richard Wolff." Counterpunch. Available at www.counterpunch.org

Ritzer, George. 1995. *The McDonaldization of Society: An Investigation into the Changing Character of Contemporary Social Life.* Thousand Oaks, CA: Pine Forge Press.

Runyan, Carol W., Michael Schulman, Janet Dal Santo, Michael Bowling, Robert Agans, and Ta Myduc. 2007. "Work-Related Hazards and Workplace Safety of U.S. Adolescents Employed in the Retail and Service Sectors." *Pediatrics* 119(3):526–534.

Saad, Lydia. 2012 (November 12). "U.S. Workers Least Happy with Their Work Stress and Pay." Gallup Poll. Available at www.gallup.com

Schaeffer, Robert K. 2003. *Understanding Globalization: The Social Consequences of Political, Economic, and Environmental Change,* 2nd ed. Lanham, MD: Rowman & Littlefield.

Schieman, Scott, Melissa Milkie, and Paul Glavin. 2009. "When Work Interferes with Life: The Social Distribution of Work-Nonwork Interference and the Influence of Work-Related Demands and Resources." *American Sociological Review* 74:966–987.

Scott, Robert E., and David Ratner. 2005 (July 20). *NAFTA's Cautionary Tale.* Economic Policy Institute Briefing Paper 214. Available at www.epi.org

"Sex Trade Enslaves Millions of Women, Youth." 2003. *Popline* 25:6.

Shierholz, Heidi. 2013 (June 11). "Unemployed Workers Still Far Outnumber Job Openings in Every Major Sector." The Economic Policy Institute. Available at www.epi.org

Shierholz, Heidi. 2011 (May 11). "Continuing Dearth of Job Opportunities Leaves Many Workers Still Sidelined." Economic Policy Institute. Available at www.epi.org

Shierholz, Heidi, Natalie Sabadish, and Nicholas Finio. 2013. "The Class of 2013: Young College Graduates Still Face Dim Job Prospects." Briefing Paper #360. Economic Policy Institute. Available at www.epi.org

Shipler, David K. 2005. *The Working Poor.* New York: Vintage Books.

Skinner, E. Benjamin. 2008. *A Crime So Monstrous: Face-to-Face with Modern-Day Slavery.* New York: Free Press.

Southern Poverty Law Center and Alabama Appleseed. 2013. *Unsafe at These Speeds: Alabama's Poultry Industry and Its Disposable Workers.* Available at www.splc.org

Strully, Kate W. 2009. "Job Loss and Health in the U.S. Labor Market." *Demography* 46(2):221–247.

Students and Scholars Against Corporate Misbehavior. 2005 (August 12). "Looking for Mickey Mouse's Conscience: A Survey of the Working Conditions of Disney Factories in China." Available at www.nlcnet.org

SweatFree Communities. n.d. "Adopted Policies." Available at www.sweatfree.org

Tate, Deborah. 2007 (February 14). "U.S. Lawmakers Seek to Crack Down on Foreign Sweatshops." The National Labor Committee. Available at www.nlcnet.org

Thompson, Charles D., Jr. 2002. "Introduction." In *The Human Cost of Food: Farmworkers' Lives, Labor, and Advocacy,* C. D. Thompson, Jr., and M. F. Wiggins, eds. (pp. 2–19). Austin: University of Texas Press.

Turner, Anna, and John Irons. 2009 (July). "Mass Layoffs at Highest Level since at Least 1995." Economic Policy Institute. Available at www.epi.org

Uchitelle, Louis. 2006. *The Disposable American: Layoffs and Their Consequences.* New York: Knopf.

UNICEF. 2009. *Children and Conflict in a Changing World.* New York: UNICEF.

United Nations. 2005. *The Millennium Development Goals Report.* New York: United Nations.

Weiss, Tara. 2009. "Some New Grads Are Glad There Are No Jobs." *Forbes,* May 27. Available at www.forbes.com

Weissmann, Jordan. 2013 (February 26). "Here's Exactly How Many College Graduates Live Back at Home." *The Atlantic.* Available at www.theatlantic.com

White, D. Steven. 2012 (August 11). "The Top 175 Global Economic Enterprises." Available at http://dstevenwhite.com

Wolff, Richard D. 2013a. "Capitalism, Democracy, and Elections." *Democracy at Work* (blog), June 13. Available at www.democracyatwork.info

Wolff, Richard D. 2013b. "Alternatives to Capitalism." *Critical Sociology* 39(4):487–490.

Wright, Erik Olin. 2013. "Transforming Capitalism through Real Utopias." *American Sociological Review* 78(1):1–25.

Chapter 8

AASA (American Association of School Administrators). 2012 (February 26). "AASA Members Detailed Draconian Impact of Sequester Cuts." Available at www.aasa.org

AASA. 2009. *Bullying at School and Online.* Education.com. Available at www.education.com

Adams, Caralee. 2012. "Higher Education Costs and Borrowing Trends Start to Stabilize." *Education Week,* October 24. Available at http://blogs.edweek.org

Adams, Caralee. 2011. "Obama Calls Community Colleges 'Key to the Future.'" *Education Week,* October 5. Available at http://blogs.edweek.org

AFE (Alliance for Excellent Education). 2011. "The High Cost of High School Dropouts: What the Nation Pays for Inadequate High Schools." *Issue Brief* November:1–6.

ACSFA. (Advisory Committee on Student Financial Assistance). 2010 (June). *The Rising Price of Inequality.* Washington, DC. Available at www.ed.gov/acsfa

AACC (American Association of Community Colleges). 2013. *2013 Community College Fast Facts.* Available at www.aacc.nche.edu

AFT (American Federation of Teachers). 2009. *Building Minds, Minding Buildings: A Union's Roadmap to Green and Sustainable Schools.* Available at www.aft.org

ASCE (American Society of Civil Engineers). 2013. *2013 Report Card For America's Infrastructure.* Available at www.infrastructurereportcard.org

Allen, I. Elaine, and Jeff Seaman. 2011. *Going the Distance: Online Education in the United States, 2011.* Babson Survey Research Group. Available at www.babson.edu

Almond, Kyle. 2013. "Malala's Global Voice Stronger Than Ever." *CNN News,* June 17. Available at www.cnn.com

Amurao, Carla. 2013 "Fact Sheet: How Bad Is the School-to-Prison Pipeline?" *PBS.* Available at www.pbs.org

Associated Press. 2013. "U.S. Student Loan Bill Signed into Law in Rare Show of Bipartisan Compromise." *The Guardian,* August 9. Available at www.theguardian.com

Barton, Paul E. 2004. "Why Does the Gap Persist?" *Educational Leadership* 62(3):9–13.

Bausell, Carole Viongrad, and Elizabeth Klemick. 2007. "Tracking U.S. Trends." *Education Week* 26(30):42–44.

Blackwell, Brandon. 2013. "State Senator Aims to Give Tax Break to Homeschoolers." *The Cleveland Plain Dealer,* July 12. Available at www.cleveland.com

Boyd, Donald, Pamela Grossman, Hamilton Lankford, Susanna Loeb, and James Wyckoff. 2009. "Who Leaves? Teacher Attrition and Student Achievement." Working Paper No. 23. National Center for the Analysis of Longitudinal Data in Education Research.

Brenner, April D., and Sandra Graham. 2011. "Latino Adolescents Experiences of Discrimination across the First 2 Years of High School: Correlates and Influences on Educational Outcomes." *Child Development* 82(2):508–509.

BLS (Bureau of Labor Statistics). 2013. "Employment Projections." U.S. Department of Labor. Current Population Survey. Available at www.bls.gov

Bushaw, William J., and Shane Lopez. 2013. "Which Way Do We Go?" The 45th Annual Phi Delta Kappa/Gallup Poll of the Public's Attitudes toward the Public Schools. Available at http://pdkintl.org

Bushaw, William J., and Shane Lopez. 2012. "Public Education in the United States: A Nation Divided." The 44th Annual Phi Delta Kappa/Gallup Poll of the Public's Attitudes toward the Public Schools. Available at http://pdkintl.org

Carnevale, Anthony P., and Jeff Strohl. 2013 (July). *Separate and Unequal: How Higher Education Reinforces the Intergenerational Reproduction of White Racial Privilege.* Georgetown Public Policy Institute. Center on Education and the Workforce. Georgetown University.

Carter, J. Chelsea. 2013 (March 29). "Grand Jury Indicts 35 in Georgia School Cheating Scandal." Available at www.cnn.com

CDE (Center for Digital Education). 2012. *2012 Yearbook: A Research Report from the Center for Digital Education and Converge.* Retrieved from http://images.erepublic.com

CDF (Children's Defense Fund). 2012a. (July 22). *The State of America's Children 2012 Report.* Available at www.childrensdefense.org

CDF. 2012b. (November). *Portrait of Inequality 2012: Black Children in America.* Available at www.childrensdefense.org

Coleman, James S., J. E. Campbell, L. Hobson, J. McPartland, A. Mood, F. Weinfield, and R. York. 1966. *Equality of Educational Opportunity.* Washington, DC: U.S. Government Printing Office.

Corbett, Christianne, Catherine Hill, and Andresse St. Rose. 2008. *Where the Girls Are.* Washington, DC: American Association of University Women.

Cowen, Joshua M., David J. Fleming, John F. Witte, Patrick J. Wolf, and Brian Kisida. 2013. "School Vouchers and Student Attainment: Evidence from a State Mandated Study of Milwaukee's Parental Choice Program." *Policy Studies Journal* 44(1):147–168.

CREDO (Center for Research on Educational Outcomes). "Executive Summary." *National Charter School Study 2013.* Stanford, CA: Stanford University.

Davis, Michele R. 2013. "Education Industry Players Exert Public Policy Influence." From *When Public Mission Meets Private Opportunity.* Supplement to *Education Week* 32(29):S52–S16.

DeNeui, Daniel, and Tiffany Dodge. 2006. "Asynchronous Learning Networks and Student Outcomes." *Journal of Instructional Psychology* 33(4):256–259.

Dixon, Mark. 2013. *Public Education Finances: 2011.* May. U.S. Census Bureau. Available at www2.census.gov

Dobbs, Michael. 2005. "Youngest Students Most Likely to Be Expelled." *Washington Post,* May 16. Available at www.washingtonpost.com

Dalton, Peter, and Oscar Marcenaro-Gutierrez. 2011. "If You Pay Peanuts, Do You Get Monkeys? A Cross-Country Analysis Teacher Pay and Pupil Performance." *Economic Policy* 26(65):5–55.

Dylan Hockley Memorial web page. 2013. Available at www.dylanhockley.com

Eliot, Lise. 2010. *The Myth of PINK & BLUE Brains.* *Educational Leadership* 68(3):32–36.

Ewert, Stephanie. 2013 (January). "The Decline in Private School Enrollment." Social, Economic, and Housing Statistics Division. Working Paper. Number FY 12–117.

Fletcher, Robert S. 1943. *History of Oberlin College to the Civil War.* Oberlin, OH: Oberlin College Press.

Flexner, Eleanor. 1972. *Century of Struggle: The Women's Rights Movement in the United States.* New York: Atheneum.

Ferlazzo, Larry, and Katie Hull Sypnieski. 2012. "What to Do—and Not Do—for Growing Number of English Language Learners." *The Washington Post,* August 24. Available at www.washington post.com

Gardner, Walter. 2013. "Push Back on Standardized Testing." *Education Week,* January 14. Available at http://blogs.edweek.org

Goldenberg, Claude. 2008. "Teaching English Language Learners." *American Educator* (Summer): 8–11, 14–19, 22–23, 42–44.

Gordon Commission. 2013. *To Assess, To Teach, To Learn: A Vision for the Future of Assessment.* Final Report. Available at www.gordoncommission.org

Gupta, Sarita. 2013 (June 12). "Sallie Mae's Profits Soaring at the Expense of Our Nation's Students." Available at billmoyers.com

Greenhouse, Linda. 2007. "Supreme Court Votes to Limit the Use of Race in Integration Plans." *New York Times,* June 29. Available at www.nytimes.com

Haberman, Martin. 1991. "The Pedagogy of Poverty versus Good Teaching." *Phi Delta Kappan.* Available at www.det.nsw.edu.au

Hammer, Kate. 2012. "Global Rate of Adult Literacy: 84 Percent but 775 Million People Still Can't Read." *The Globe and Mail,* September 8. Available at www.theglobeandmail.com

Hanushek, Eric A., Steven G. Rivkin, and John J. Kain. 2005. "Teachers, Schools and Academic Achievement." *Econometrics* 73(2):417–458.

Harris, Philip, Bruce M. Smith and Joan Harris. 2011. *The Myths of Standardized Tests: Why They Don't Tell You What You Think They Do.* Lanham, Maryland: The Rowman and Littlefield Publishing Group, Inc.

Heckman, J., John Eric Humphries, Paul A. LaFontaine, and Pedro L. Rodríguez. 2012. "Taking the Easy Way Out: How the GED Testing Program Induces Students to Drop Out." *Journal of Labor Economics* 30(3):495–520.

Hightower, Amy M. 2013a (January 4). "States Show Spotty Progress on Education Gauges." Available at www.edweek.org

Hightower, Kyle. 2013b. "FAMU Hazing Case: 12 Charged with Manslaughter in Robert Champions Death." *The Huffington Post,* March 4. Available at www.huffingtonpost.com

Hopkinson, Natalie. 2011. "The McEducation of the Negro." *The Root,* January 3. Available at www.theroot.com

Horace Mann Educator Survey. 2013 (June). "2013 Horace Mann Educator Advisory Panel Survey." Race Mann Market Research. Available at http://horacemann.com

Hurdle, Jon. 2013. "Philadelphia Officials to Close 23 Schools." *New York Times,* March 7. Available at www.nytimes.com

Hurst, Marianne. 2005. "When It Comes to Bullying, There Are No Boundaries." *Education Week,* February 8. Available at www.edweek.org

Josephson Institute of Ethics. 2012. *2012 Report Card on the Ethics of American Youth.* Josephson Institute Center for Youth Ethics. Available at available at charactercounts.org

Kahlenberg, Richard D. 2013. "From All Walks of Life: New Hope for School Integration." *American Educator* (Winter):1–40.

Kahlenberg, Richard D. 2006. "A New Way of School Integration." *Issue Brief.* The Century Foundation. Retrieved from www.equaleducation.org

Kalet, Hank. 2013 "Arguing the Costs of Tuition Equality." New Jersey Spotlight, June 18. Available at www.njspotlight.com

Kanter, Rosabeth Moss. 1972. "The Organization Child: Experience Management in a Nursery School." *Sociology of Education* 45:186–211.

Kastberg, David, David Ferraro, Nita Lemanski, Stephen Roey, and Frank Jenkins. 2013. *Highlights from TIMSS 2011.* NCES 2013-009. Revised. Available at nces.ed.gov

Kohn, Alfie. 2011. "How Education Reform Traps Poor Children." *Education Week* 30(29):32–33.

Kozol, Jonathan. 1991. *Savage Inequalities: Children in America's Schools.* New York: Crown.

Kugler, Eileen Gale. 2013. "Understanding Our Diverse Students by Understanding Ourselves First." *Learning on the Edge,* July 22. Available at http://pdkintl.org

Lahey, Jessica. 2013. "The Benefits of Character Education." *The Atlantic,* May 6. Available at www.theatlantic.com

Leandro v. State, 488 S.E.2d 249 (N.C. 1997).

Lickona, Thomas, and Matthew Davidson. 2005. *A Report to the Nation: Smart and Good High Schools.* Available at www.cortland.edu

Losen, Daniel J., and Russell Skiba. 2010 (September). "Suspended Education." Southern Poverty Law Center. Available at www.spcenter.org

Loveless, Tom. 2013. "The Resurgence of Ability Grouping in Persistence of Tracking." Brown Center Report On American Education. Available at www.brookings.edu

Lu, Adrienne. 2013. "Parents Revolt against Failing Schools." *The Pew Charitable Trust,* July 1. Available at www.pewstates.org

Lubienski, Christopher, Janelle T. Scott. John Rogers, and Kevin G. Welner. 2012. "Missing the Target? The Parent Trigger as a Strategy for Parental Engagement and School Reform." *National Education Policy Center,* September 5. School of Education, University of Colorado at Boulder.

Lubienski, Sarah Theule, and Christopher Lubienski. 2006. "School Sector and Academic Achievement: A Multi-Level Analysis of NAEP Mathematics Data." *American Educational Research Journal* 43(4):651–698.

Lumina Foundation. 2013 (February 5). *Americans Call for Higher Education Redesign.* Available at luminafoundation.org

Lytle, Ryan. 2012. "Antioch University to Offer Online Course for Credit." *U.S. News,* November 21. Available at www.usnews.com

Manzo, Kathleen K. 2005. "College-Based High Schools Fill Growing Need." *Education Week,* May 25. Available at www.edweek.org

Martin, Michel. 2012. "Is Teach for America Failing?" NPR, June 11. Available at www.npr.org

Maxwell, Leslie A. 2013. "Head Start Gains Found to Wash Out by Third Grade." *Education Week,* January 9. Available at www.edweek.org

McDonnell, Sanford. 2009 (October 3). "America's Crisis of Character—And What to Do about It." Available at www.edweek.org

McKinsey & Company. 2009. *The Economic Impact of the Achievement Gap in America Schools.* Social Sector Office: McKinsey & Company. Available at www.mckinsey.com

Mead, Sara. 2006. "The Truth about Boys and Girls." *Education Sector,* June. Available at www.educationsector.org

Merton, Robert K. 1968. *Social Theory and Social Structure.* New York: Free Press.

MetLife. 2013. *MetLife Survey of the American Teacher: Past, Present and Future.* Available at www.metlife.com

MetLife. 2008. *MetLife Survey of the American Teacher: Challenges For School Leadership.* Available at www.metlife.com

Milkie, Melissa A., and Catherine H. Warner. 2011. "Classroom Learning Environments and Mental Health of First Grade Children." *Journal of Health and Social Behavior* 42:4–22.

Mikulecky, Marga. 2013. "Compulsory School Age Requirements." Education Commission of the United States. April. Available at www.ecs.org

Mitchell, Kenneth. 2010. "Taking Teacher Evaluation to Extremes." *Education Week* 30(15):1.

Molnar, Alex, Faith Boninger, Michael D. Harris, Ken Libby, and Joseph Fogarty. 2013 (April). *Health Threats Associated With Schoolhouse Commercialism.* Boulder, CO: National Education Policy Center.

Morse, Jodie. 2002. "Learning While Black." *Time,* May 27, pp. 50–52.

Moxley, Elle. 2013 (July 27). "With Georgia Out, What's Next for Common Core Testing Consortium PARCC." Available at stateimpact.npr.org

Muller, Chandra, and Katherine Schiller. 2000. "Leveling the Playing Field?" *Sociology of Education* 73:196–218.

Mullis, I. V. S., M. O. Martin, P. Foy, and A. Arora. 2012. *The TIMSS 2011 International Results In Mathematics.* Chestnut Hill, MA: International Study Center.

NAEP (National Assessment of Educational Progress). 2013. *Trends in Academic Progress.* NCES 2013-456. Available at http://nces.ed.gov/nationsreportcard

NBPTS (National Board for Professional Teaching Standards). 2012. "Promoting Student Learning, Growth Achievement." Available at www.nbpts.org

NCES (National Center for Educational Statistics). 2013a. *Digest of Education Statistics, 2012.* U.S. Department of Education. Available at nces.ed.gov

NCES. 2013b. *The Condition of Education, 2013b.* U.S. Department of Education. Available at http://nces.ed.gov

NCES. 2013c. *Indicators of School Crime and Safety: 2012.* U.S. Department of Education. Available at http://nces.ed.gov

NCES. 2013d. "Postsecondary Institutions and Cost of Attendance In 2012–13; Degrees and Other Awards Conferred, 2011–12; and Twelve-Month Enrollment, 2011–12." U.S. Department of Education. NCES 2013-289 rev. Available at http://nces.ed.gov

NCES. 2012. *Digest of Education Statistics, 2011.* U.S. Department of Education. Available at http://nces.ed.gov

NCES. 2011. *Digest of Education Statistics, 2010.* Available at nces.ed.gov

NCES. 2007. *Effectiveness of Reading and Mathematics Software Products.* Report to Congress. Washington, DC: U.S. Department of Education. NCES 2007–4005.

NCFOT (The National Center for Fair and Open Testing). 2012 (May 22). "What's Wrong with Standardized Tests?" Available at fairtest.org

NCSL (National Conference of State Legislatures). 2013a. "Parent Trigger Laws In The States." Available at www.ncsl.org

NCSL (National Conference of State Legislatures). 2013b. "School Vouchers." Available at www.ncsl.org

NPR (National Public Radio). 2013 (June 2). "Why Some Schools Want to Expel Suspensions." Available at www.npr.org

North Carolina Justice Center. 2012. "Leandro and the 2012 Court of Appeals Decision on NC Pre-K." *Education and Law Project.* Available at www.ncjustice.org

OECD (Organization for Economic Cooperation and Development). 2013a. *Education at a Glance 2012.* Available at www.oecd.org

OECD. 2013b. "Good Learning Strategies Reduce the Performance Gap between Advantaged and Disadvantaged Students?" *PISA in Focus.* Available at www.oecd.org

OECD. 2012. *Education at a Glance 2012.* Available at www.oecd.org

Office of Head Start. 2013. *Head Start Program Facts Fiscal Year 2012.* Available at http://eclkc.ohs.acf.hhs.gov

Orfield, Gary, John Kucsera, and Genevieve Siegel-Hawley. 2012 (September 19). "E Pluribus . . . Separation: Deepening Double Segregation for More Students." *The Civil Rights Project.* University of California at Los Angeles. Available at civilrightsproject.ucla.edu

Orfield, Gary, and Chungmei Lee. 2006. *Racial Transformation and the Changing Nature of Segregation.* Cambridge, MA: Harvard University, The Civil Rights Project.

PARCC (Partnership for Assessment of Readiness for College and Careers). 2013. "About PARCC." Available at www.parcconline.org

Perez-Pena, Richard. 2013. "College Enrollment Falls as Economy Recovers." *The New York Times,* July 25. Available at www.nytimes.com

Peskin, Melissa Fleschler, Susan R. Tortolero, and Christine M. Markham. 2006. "Bullying and Victimization among and Hispanic Adolescents." *Adolescence* 41(163):467–484.

Picciano, Anthony G., and Jeff Seaman. 2010 (August). *Class Connections: High School Reform in the Role of Online Learning.* Available at www.babson.edu

Poliakoff, Anne Rogers. 2006. "Closing the Gap: An Overview." *ASCD InfoBrief* 44(January):1–10. The Association for Supervision and Curricular Development.

Primary Sources. 2012. *Primary Sources: 2012 America's Teachers on the Teaching Profession.* Scholastic and the Bill and Melinda Gates Foundation. Available at www.scholastic.com

Ramey, Garey, and Valerie A. Ramey. 2010. "The Rug Rat Race." *Brookings Paper on Economic Activity* (Spring): 129 – 176.

Ravitch, Diane. 2010. *The Death and Life of the Great American School System: How Testing and Choice Are Undermining Education.* New York, NY: Basic Books.

Reardon, Sean F. 2013. "No Rich Child Left Behind." *The New York Times,* April 27. Available at www.nytimes.com

Reardon, Sean F., Allison Atteberry, Nicole Arshan, and Michal Kurlaend. 2009. *Effects of the California High School Exit Exam on Student Persistent, Achievement and Graduation.* Stanford: Institute for Research on Education Policy & Practice. Available at www.stanford.edu

Riehl, Carolyn. 2004. "Bridges to the Future: Contributions of Qualitative Research to the Sociology of Education." In *Schools and Society,* Jeanne Ballantine and Joan Spade, eds. (pp. 56–72). Belmont, CA: Thomson Wadsworth.

Rodriguez, Richard. 1990. "Searching for Roots in a Changing World." In *Social Problems Today,* James M. Henslin, ed. (pp. 202–213). Englewood Cliffs, NJ: Prentice-Hall.

Rooks, Noliew. 2012a. "Why It's Time to Get Rid of Standardized Tests." *Time,* October 11. Available at http://ideas.time.com

Rooks, Noliew. 2012b. "Why Do We Care More about Diversity on TV than in Our Schools?" *Time,* May 24. Available at http://ideas.time.com

Rosenthal, Robert, and Lenore Jacobson. 1968. *Pygmalion in the Classroom: Teacher Expectations and Pupils' Intellectual Development.* New York: Holt, Rinehart & Winston.

Rothstein, Richard, Helen F. Ladd, Diane Ravitch, Eva L. Baker, Paul E. Barton, Linda Darling-Hammond, Edward Haertel, Robert L. Linn, Richard J. Shavelson, and Lorrie A. Shepard. 2010 (August 29). *Problems with the Use of Student Test Scores to Evaluate Teachers.* Educational Policy Institute. Available at www.epi.org

Sadovnik, Alan. 2004. "Theories in the Sociology of Education." In *Schools and Society,* Jeanne Ballantine and Joan Spade, eds. (pp. 7–26). Belmont, CA: Thomson Wadsworth.

Save the Children. 2011. "An Uneducated Girl Is a Girl in Darkness." Available at www.savethechildren.org

Sawchuk, Stephen. 2011. "New Teacher Distribution Methods Hold Promise." *Education Week* 29(35):16–17.

SEIU (Service Employees International Union). 2013. *Falling Further Apart: Decaying Schools in New York City's Poorest Neighborhoods.* Available at www.seiu32bj.org

SBAC (Smarter Balanced Assessment Consortium). 2013. "Smarter Balanced Assessments." Available at www.smarterbalanced.org

Schott Report. 2012. *The Urgency of Now: The 2012 Schott 50 State Report on Public Education and Black Males.* Schott Foundation for Public Education. Available at http://blackboysreport.org

Segal, Tom. 2013. "The Impact of Investing in Education." *Education Week,* March 26. http://blogs.edweek.org

Semester Online. 2013. "Overview." Available at semesteronline.org

Shah, Nirvi. 2013 (January). "Discipline Policy Shift with Views on What Works." *Education Week.*

Shah, Nirvi and Michele McNeil. 2013. "Suspension, Expulsion Data Cast Some in Harsh Light." *Education Week* 32(16):12.

Smyth, Julie Carr. 2013. "Longer School Year: Will It Help or Hurt US Students?" *The Huffington Post,* January 13. Available at www.huffingtonpost.com

Stancill, Jane. 2011. "School Cuts Go before Judge." *News and Observer,* June 22. Available at www.newsandobservor.com

Stotsky, Sandra. 2009. "The Academic Quality of Teachers: A Civil Rights Issue." *Education Week,* June 26. Available at www.edweek.org

Tanner, C. K. 2008. "Explaining Relationships among Student Outcomes and the School's Physical Environment." *Journal of Advanced Academics* 19:444–471.

Toppo, Greg. 2013. "Growing Number of Educators Boycott Standardized Test." *USA Today,* February 1. Available at www.usatoday.com

TNTP. 2013. *The Irreplaceables: Understanding the Real Retention Crisis in America's Urban Schools.* Available at tntp.org

Tucker, Marc. 2013. "Teacher Quality: Who's On Which Side and Why." *Education Week,* November 2. Available at www.edweek.org

Tyler, John H., and Magnus Lofstrom. 2009. "Finishing High School: Alternative Pathways and Dropout Recovery." *The Future of Children* 19(1):78–102.

Tyre, Peg. 2008. *The Trouble with Boys: A Surprising Report Card on Our Sons, Their Problems at School, and What Parents and Educators Must Do.* New York: Crown Publishing Group.

Ujifusa, Andrew. 2013. "Tests Linked to Common Core in Critics' Cross Hairs." *Education Week* 32(37):1, 20.

UNESCO (United Nations Educational Scientific and Cultural Organization). 2012a. "Literacy for 2011." Available at http://www.uis.unesco.org

UNESCO. 2012b. *Youth and Skills: Putting Education to Work.* EFA Global Monitoring Report. Available www.unesco.org

U.S. Department of Education. 2012. "ESEA Flexibility." Available at www.ed.gov

Viadero, Debra. 2006. "Rags to Riches in U.S. Largely a Myth, Scholars Write." *Education Week,* October 25. Available at www.edweek.org

Viadero, Debra. 2005. "Study Sees Positive Effects of Teacher Certification." *Education Week,* April 27. Available at www.edweek.org

Vigdor, Jacob L., and Helen F. Ladd. 2010 (June). "Scaling the Digital Divide: Home Computer Technology and Student Achievement." The National Bureau of Economic Research, Working Paper No. 16078. Available at http://papers.nber.org/

Waggoner, Martha. 2009. "Judge: 'Academic Genocide' in Halifax Schools." *News & Observer,* March 19. Available at www.newsobserver.com

Wong, Jennifer S. 2009. *No Bullies Allowed: Understanding Peer Victimization, the Impacts on Delinquency, and the Effectiveness of Prevention Programs.* Rand Corporation. Available at www.rand.org

World Literacy Foundation. 2012. *The Economic and Social Cost of Illiteracy.* Available at www.worldliteracyfoundation.org/

Xu, Zeyu, Jane Hannaway, and Colin Taylor. 2007. *Making a Difference? The Effects of Teach for America in High School.* Working Paper No. 17. National Center for the Analysis of Longitudinal Data in Education Research.

Chapter 9

Alexander, Michelle. 2010. *The New Jim Crow: Mass Incarceration in the Age of Colorblindness.* New York: The New Press.

Allport, G. W. 1954. *The Nature of Prejudice.* Cambridge, MA: Addison-Wesley.

American Council on Education and American Association of University Professors. 2000. *Does Diversity Make a Difference? Three Research Studies on Diversity in College Classrooms.* Washington, DC: American Council on Education and American Association of University Professors.

Apfelbaum, Evan. 2011. "Prof. Evan Apfelbaum: A Blind Pursuit of Racial Colorblindness—Research has Implications for How Companies Manage Multicultural Teams." *MIT Sloan Experts.* MIT Sloan Management Blog. Available at mitsloanexperts.mit.edu

Armario, Christine. 2011. "Feds: All Kids, Legal or Not, Deserve K-12 Education." *Chron,* May 7. Available at www.chron.com

Associated Press. 2013. "Justice Department Adds Sikhs to Hate-Crimes List." *Washington Post*, August 5. Available at www.washingtonpost.com

Associated Press. 2011. "FBI Investigate Fatal Rundown of Black Miss. Man." *NPR*, August 17. Available at www.npr.org

Ayala, Elaine, and Ellen Huet. 2013. *San Francisco Chronicle*, February 4. Available at www.sfgate.com

Balko, Radley. 2009 (July 6). "The El Paso Miracle." Reasononline. Available at www.reason.com

Bauer, Mary, and Sarah Reynolds. 2009. *Under Siege: Life for Low-Income Latinos in the South.* Montgomery, AL: The Southern Poverty Law Center.

Beirich, Heidi. 2013 (Spring). "The Year in Nativism." *Intelligence Report* (149). Available at www.splcenter.org.

Beirich, Heidi. 2007 (Summer). "Getting Immigration Facts Straight." *Intelligence Report*. Available at www.splcenter.org

Bonilla-Silva, Eduardo. 2012. "The Invisible Weight of Whiteness: The Racial Grammar of Everyday Life in Contemporary America." *Ethnic and Racial Studies* 35(2):173–194.

Bonilla-Silva, Eduardo. 2003. *Racism without Racists: Color-Blind Racism and the Persistence of Racial Inequality.* Lanham, MD: Rowan and Littlefield.

Brace, C. Loring. 2005. *"Race" Is a Four-Letter Word.* New York: Oxford University Press.

Brooks, Roy. 2004. *Atonement and Forgiveness: A New Model for Black Reparations.* Berkeley: University of California Press.

Brown University Steering Committee on Slavery and Justice. 2007. *Slavery and Justice.* Providence, RI: Brown University.

Chisti, Muzaffar, and Faye Hipsman. 2013 (May 23). "As Congress Tackles Immigration Legislation, State Lawmakers Retreat from Strict Measures." Migration Policy Institute. Available at www.migrationinformation.org

CNN. 2009 (June 18). "Senate Approves Resolution Apology for Slavery." Available at www.cnn.com

Colford, Paul. 2013 (April 2). "'Illegal Immigrant' No More." *The Definitive Source* (AP Blog). Available at http://blog.ap.org

Conley, Dalton. 1999. *Being Black, Living in the Red: Race, Wealth, and Social Policy in America.* Berkeley: University of California Press.

Croll, Paul R. 2013. "Explanations for Racial Disadvantage and Racial Advantage: Beliefs about Both Sides of Inequality in America." *Ethnic and Racial Studies* 36(1):47–74.

"DOJ Study: Hate Crimes More Prevalent than Previously Known." 2013 (Fall). *Intelligence Report* 151:7.

Dudziak, Mary. 2000. *Cold War Civil Rights: Race and the Image of American Democracy.* Princeton, NJ: Princeton University Press.

Dwyer, Devin. 2011. "Opponents of Illegal Immigration Target Birthright Citizenship." ABC News, January 5. Available at www.abcnews.go.com

EEOC (Equal Employment Opportunity Commission). 2011 (June 22). "A. C. Widenhouse Sued by EEOC for Racial Harassment." Press Release. Available at www.eeoc.gov

Ennis, Sharon R., Merarys Rios-Vargas, and Nora G. Albert. 2011 (May). "The Hispanic Population: 2010." *2010 Census Briefs*. U.S. Census Bureau. Available at www.census.gov

Esposito, John L. 2011. "Getting It Right about Islam and American Muslims." *Huffington Post*, May 25. Available at www.huffingtonpost.com

FBI (Federal Bureau of Investigation). 2012. *Hate Crime Statistics 2011*. Available at www.fbi.gov

Follman, Mark. 2012. "Selling Trayvon Martin for Target Practice." *Mother Jones,* May 11. Available at www.motherjones.com

Frieden, Bonnie. 2013 (April 3). "'I Don't See Race': The Pitfalls of the Colorblind Mindset." *Washington University Political Review*. Available at www.wupr.org

Fry, Richard. 2009. "Sharp Growth in Suburban Minority Enrollment Yields Modest Gains in School Diversity." Pew Hispanic Center. Available at www.pewhispanic.org

Gaertner, Samuel L., and John F. Dovidio. 2000. *Reducing Intergroup Bias: The Common In-Group Identity Model.* Philadelphia: Taylor & Francis.

Gallup Organization. 2013. *Race Relations*. Available at www.gallup.com

Glaser, Jack, Jay Dixit, and Donald P. Green. 2002. "Studying Hate Crime with the Internet: What Makes Racists Advocate Racial Violence?" *Journal of Social Issues* 58(1):177–193.

Goldstein, Joseph. 1999. "Sunbeams." *The Sun* 277, January, p. 48.

Grieco, Elizabeth M., Yesenia D. Acosta, G. Patricia de la Cruz, Christine Gambino, Luke J. Larsen, Edward N. Trevalyan, and Nathan P. Walters. 2012. "The Foreign-Born Population in the United States: 2010." *American Community Survey Reports.* U.S. Census Bureau. Available at www.census.gov

Gurin, Patricia. 1999. "New Research on the Benefits of Diversity in College and Beyond: An Empirical Analysis." *Diversity Digest*, Spring, pp. 5–15.

Harwood, Stacy A., Margaret Browne Huntt, Ruby Mendenhall, and Jioni A. Lewis. 2012. "Racial Microaggressions in the Residence Halls: Experiences of Students of Color at a Predominantly White University." *Journal of Diversity in Higher Education* 5(3):159–173.

Healey, Joseph F. 1997. *Race, Ethnicity, and Gender in the United States: Inequality, Group Conflict, and Power.* Thousand Oaks, CA: Pine Forge Press.

"Hear and Now." 2000 (Fall). *Teaching Tolerance*, p. 5.

Higginbotham, Elizabeth, and Margaret L. Andersen. 2012. "The Social Construction of Race and Ethnicity." In *Race and Ethnicity in Society*, 3rd ed., E. Higginbotham and M. L. Andersen, eds. (pp. 3–6). Belmont, CA: Wadsworth, Cengage Learning.

Hodgkinson, Harold L. 1995. "What Should We Call People? Race, Class, and the Census for 2000." *Phi Delta Kappa*, October, pp. 173–179.

Hoefer, Michael, Nancy Rytina, and Bryan Baker. 2012. "Estimates of the Unauthorized Immigrant Population Residing in the United States: January 2011." U.S. Department of Homeland Security. Available at www.dhs.gov

hooks, bell. 2000. *Where We Stand: Class Matters.* New York: Routledge.

Humes, Karen R., Nicholas A. Jones, and Roberto R. Ramirez. 2011 (March). "Overview of Race and Hispanic Origin: 2010." *2010 Census Briefs*. Available at www.census.gov

Humphreys, Debra. 1999. "Diversity and the College Curriculum: How Colleges and Universities Are Preparing Students for a Changing World." *Diversity-Web*. Available at www.inform.umd.edu

Jensen, Derrick. 2001. "Saving the Indigenous Soul: An Interview with Martin Prechtel." *The Sun* 304(April):4–15.

Jones, Jeffrey M. 2013 (July 24). "In U.S., Most Reject Considering Race in College Admissions." Gallup Organization. Available at www.gallup.org

Kahlenberg, Richard D. 2013. "How to Fight Growing Economic and Racial Segregation in Higher Education." *The Chronicle of Higher Education* (August 7). Available at www.chronicle.com

King, Joyce E. 2000 (Fall). "A Moral Choice." *Teaching Tolerance* 18:14–15.

Knowles, Eric D., and Christopher K. Marshburn. 2012. "Understanding White Identity Politics Will Be Crucial to Diversity Science." *Psychological Inquiry* 21:134–139.

Kovac, Amy. 2009. "Transcript of Rev. Lowery's Inaugural Benediction." *Washington Post*, January 20. Available at http://voices.washingtonpost.com

Kozol, Jonathan. 1991. *Savage Inequalities: Children in America's Schools.* New York: Crown.

Kwok, Irene, and Yuzhou Wang. 2013. "Locate the Hate: Detecting Tweets against Blacks." Proceedings of the Twenty-Seventh AAAI Conference on Artificial Intelligence. Available at www.aaai.org

Leadership Council on Civil Rights Education Fund. 2009. *Confronting the New Faces of Hate: Hate Crimes in America.* Available at www.civilrights.org

Levin, Jack, and Jack McDevitt. 1995. "Landmark Study Reveals Hate Crimes Vary Significantly by Offender Motivation." *Klanwatch Intelligence Report*, August, pp. 7–9.

Ly, Laura. 2013 (March 4). "Oberlin Cancels Classes to Address Racial Incidents." CNN. Available at www.cnn.com

Marger, Martin N. 2012. *Race & Ethnic Relations: American and Global Perspectives*, 9th ed. Belmont CA: Wadsworth, Cengage Learning.

Maril, Robert Lee. 2011. *The Fence: National Security, Public Safety, and Illegal Immigration along the U.S.-Mexico Border.* Lubbock Texas: Texas Tech University Press.

Maril, Robert Lee. 2004. *Patrolling Chaos: The U.S. Border Patrol in Deep South Texas.* Lubbock, TX: Texas Tech University Press.

Martinez, Michael. 2013. "Florida Police Sergeant Fired for Having Trayvon Martin Shooting Targets." *CNN*, April 13. Available at http://edition.cnn.com

Massey, Douglas S., and Garvey Lundy. 2001. "Use of Black English and Racial Discrimination in Urban Housing Markets: New Methods and Findings." *Urban Affairs Review* 36(4):452–469.

McIntosh, Peggy. 1990 (Winter). "White Privilege: Unpacking the Invisible Knapsack." *Independent School* 49(2):31–35.

Morrison, Pat. 2002. "September 11: A Year Later—American Muslims Are Determined Not to Let Hostility Win." *National Catholic Reporter* 38(38):9–10.

Moser, Bob. 2004. "The Battle of 'Georgiafornia.'" *Intelligence Report* 116:40–50.

Mukhopadhyay, Carol C., Rosemary Henze, and Yolanda T. Moses. 2007. *How Real Is Race?* Lanham, MD: Rowman & Littlefield Education.

NCSL (National Conference of State Legislatures). 2013 (June). "Affirmative Action: An Overview." Available at www.ncsl.org

Orfield, Gary. 2001 (July). *Schools More Separate: Consequences of a Decade of Resegregation.* Cambridge, MA: Harvard University, Civil Rights Project.

Ossorio, Pilar, and Troy Duster. 2005. "Race and Genetics." *American Psychologist* 60(1): 115–128.

Pager, Devah. 2003. "The Mark of a Criminal Record." *American Journal of Sociology* 108(5):937–975.

Passel, Jeffrey S., and D'Vera Cohn. 2011. "Unauthorized Immigrant Population: National and State Trends." Pew Hispanic Research Center. Available at www.pewhispanic.org

Passel, Jeffrey S., and D'Vera Cohn. 2009. "A Portrait of Unauthorized Immigrants in the United States." Pew Hispanic Research Center. Available at www.pewhispanic.org

Passel, Jeffrey S., and Paul Taylor. 2009. "Who's Hispanic?" Pew Hispanic Research Center. Available at www.pewhispanic.org

Pew Hispanic Center. 2013 (January 29). *A Nation of Immigrants.* Available at www.pewhispanic.org

Pew Research Center. 2010 (January 12). "Blacks Upbeat about Black Progress, Prospects." Available at www.pewsocialtrends.org

Picca, Leslie, and Joe R. Feagin. 2007. *Two-Faced Racism.* New York: Routledge.

Pollin, Robert. 2011. "Economic Prospects: Can We Stop Blaming Immigrants?" *New Labor Forum* 20(1):86–89.

Saulny, Susan. 2011. "Census Data presents Rise in Multiracial Population of Youths." *New York Times,* March 24. Available at www.nytimes.com

Schiller, Bradley R. 2004. *The Economics of Poverty and Discrimination,* 9th ed. Upper Saddle River, NJ: Pearson Education.

Schmidt, Peter. 2004 (January 30). "New Pressure Put on Colleges to End Legacies in Admissions." *Chronicle of Higher Education* 50(21):A1.

Schuman, Howard, and Maria Krysan. 1999. "A Historical Note on Whites' Beliefs About Racial Inequality." *American Sociological Review* 64:847–855.

Shierholz, Heidi. 2010 (February 4). "Immigration and Wages—Methodological Advancements Confirm Modest Gains for Native Workers." EPI Briefing Paper #255. Available at www.epi.org

Shierholz, Heidi, Natalie Sabadish, and Nicholas Finio. 2013. "Young Graduates Still Face Dim Job Prospects." Briefing Paper #360. Economic Policy Institute. Available at www.epi.org

Shipler, David K. 1998. "Subtle vs. Overt Racism." *Washington Spectator* 24(6):1–3.

Sidanius, Jim, Shana Levin, Colette Van Laar, and David O. Sears. 2010. *The Diversity Challenge: Social Identity and Intergroup Relations on the College Campus.* New York: Russell Sage Foundation.

SPLC (Southern Poverty Law Center). 2013a. "Active U.S. Hate Groups." Available at www.splcenter.org

SPLC. 2013b. *Close to Slavery: Guestworker Programs in the United States (2013 edition).* Montgomery, AL: Southern Poverty Law Center.

SPLC. 2004 (Winter). "Neo-Nazi Label Woos Teens with Hate-Music Sampler." *Intelligence Report* 116:5.

Tanneeru, Manav. 2007 (May 11). "Asian-Americans' Diverse Voices Share Similar Stories." CNN.com. Available at www.cnn.com

Tavernise, Sabrina. 2011. "In Census, Young Americans Increasingly Diverse." *New York Times,* February 4. Available at www.nytimes.com

Teaching Tolerance. 2011 (Spring). "10 Myths about Immigration." Available at www.tolerance.org

Tolbert, Caroline J., and John A. Grummel. 2003. "Revisiting the Racial Threat Hypothesis: White Voter Support for California's Proposition 209." *State Politics and Policy Quarterly* 3(2):183–202, 215–216.

Turner, Margery Austin, Stephen L. Ross, George Galster, and John Yinger. 2002. *Discrimination in Metropolitan Housing Markets.* Washington, DC: Urban Institute.

Turner, Margery Austin, and Karina Fortuny. 2009. *Residential Segregation and Low-Income Working Families.* Washington, DC: Urban Institute.

Turn It Down. 2009 (May 3). "Social Networking: A Place for Hate?" Available at turnitdown.newcomm.org

Urban Dictionary. Available at www.urbandictionary.com

U.S. Census Bureau (2011; generated by C. Schacht, July 3). *2010 Census Data.* Available at http://2010.census.gov/2010census/data/index.php

U.S. Census Bureau. 2013. Current Population Survey. Table A-3. "Mean Earnings of Workers 18 Years and Over, by Educational Attainment, Race, and Hispanic Origin: 1975–2011." Available at www.census.gov

U.S. Citizenship and Immigration Services. 2011. *A Guide to Naturalization.* Available at www.uscis.gov

U.S. Customs and Border Protection. 2013. "U.S. Border Patrol Fiscal Year 2012 Statistics." Available at www.cbp.gov

U.S. Department of Labor. 2002. *Facts on Executive Order 11246 Affirmative Action.* Available at www.dol.gov

Wang, Wendy. 2012. "The Rise of Intermarriage." Pew Research Center. Available at www.pewresearch.org

Washington Post staff. 2013 (July 19). "President Obama's Remarks on Trayvon Martin (full transcript)." *The Washington Post.* Available at www.washingtonpost.com

Williams, Eddie N., and Milton D. Morris. 1993. "Racism and Our Future." In *Race in America: The Struggle for Equality,* Herbert Hill and James E. Jones Jr., eds. (pp. 417–424). Madison: University of Wisconsin Press.

Williams, Richard, Reynold Nesiba, and Eileen Diaz McConnell. 2005. "The Changing Face of Inequality in Home Mortgage Lending." *Social Problems* 52(2):181–208.

Winfrey, Oprah. 2009 (October). "Oprah Talks to Jay-Z." *O, The Oprah Magazine,* Available at www.oprah.com

Winter, Greg. 2003. "Schools Resegregate, Study Finds." *New York Times,* January 21. Available at www.nytimes.com

Wise, Tim. 2009. *Between Barack and a Hard Place: Racism and White Denial in the Age of Obama.* San Francisco: City Light Books.

Yeung, Jeffrey G., Lisa B. Spanierman, and Jocelyn Landrum-Brown. 2013. "'Being White in a Multicultural Society': Critical Whiteness Pedagogy in a Dialogue Course." *Journal of Diversity in Higher Education* 6(1):17–32.

Zinn, Howard. 1993. "Columbus and the Doctrine of Discovery." In *Systemic Crisis: Problems in Society, Politics, and World Order,* William D. Perdue, ed. (pp. 351–357). Fort Worth, TX: Harcourt Brace Jovanovich.

Chapter 10

AAUW. 2011. *Why So Few?* Available at www.aauw.org

Abernathy, Michael. 2003. *Male Bashing on TV.* Tolerance in the News. Available at www.tolerance.org

Abraham, Tamara, and Jennifer Madison. 2011. "Nailing a Trend: From Gwen Stefani to Jennifer Lopez, How J Crew Boss Took Her Lead from the Stars in Giving Son a Manicure." *The Daily Mail,* April 15. Available at www.dailymail.co.uk

ACLU (American Civil Liberties Union). 2013 (July 8). "Court Approves Settlement Reached in Challenge to West Virginia Single-Sex School Program Rooted in Stereotypes." Available at www.aclu.org

Adams, Jimi. 2007. "Stained Glass Makes the Ceiling Visible." *Gender and Society* 21(1):80–105.

Alvarado, Monsy. 2013. "Bergen County to Unveil Memorial to Victims of WWII Japanese 'Comfort Stations.'" *North Jersey News.* Available at northjersey.com

Amnesty International. 2013. "The International Violence against Women Act." June. Available at www.amnestyusa.org

Anderson, Margaret L. 1997. *Thinking about Women,* 4th ed. New York: Macmillan.

Arrindell, W. A., Sonja Van Well, Annemarie M. Kolk, Dick P. H. Barelds, Tian P. S. Oei, Pui Yi Lau, and the Cultural Clinical Psychology Study Group. 2013. "Higher Levels of Masculine Gender Role Stress in Masculine than in Feminine Nations: A 13-Nations Study." *Cross-Cultural Research* 47(1):51–67.

Askari, Sabrina F., Mirian Liss, Mindy J. Erchull, Samantha E. Staebell, and Sarah J. Axelson. 2010. "Men Want Equality, but Women Don't Expect It: Young Adults' Expectations for Participation in Household and Child Care Chores." *Psychology of Women Quarterly* 34(2): 243–252.

Badenhausen, Kurt. 2013. "Maria Sharapova Tops List of the World's Highest Paid Female Athletes." *Forbes,* August 5. Available at www.forbes.com

Bedi, Rahul. 2013 (February 27). "Indian Dowry Deaths on the Rise." Available at www.telegraph.co.uk

Begley, Sharon. 2000. "The Stereotype Trap." *Newsweek,* November 6, pp. 66–68.

Bertrand, Marianne, Claudia Goldin, and Lawrence F. Katz. 2009 (January). "Dynamics of the Gender Gap for Young Professionals in the Financial and Corporate Sectors." Working Paper. Available at www.economics.harvard.edu

Bittman, Michael, and Judy Wajcman. 2000. "The Rush Hour: The Character of Leisure Time and Gender Equity." *Social Forces* 79:165–189.

Blau, Francine D., and Lawrence M. Kahn. 2013. "Female Labor Supply: Why is the US Falling Behind?" January. NBER Working Paper No.18702. *The National Bureau of Economic Research.* Available at www.nber.org

BLS (Bureau of Labor Statistics). 2013. "Characteristics of Minimum Wage Workers: 2012." Labor Force Statistics from the Current Population Survey. Available at www.bls.gov

Bly, Robert. 1990. *Iron John: A Book about Men.* Boston: Addison-Wesley.

Bourin, Lenny, and Bill Blakemore. 2008. "More Men Take Traditionally Female Jobs." ABC World News, September 1. Available at http://abcnews.go.com

Brady, David, and Denise Kall. 2008. "Nearly Universal, but Somewhat Distinct: The Feminization of Poverty in Affluent Western Democracies, 1969–2000." *Social Science Research* 37:976–1,007.

Bryant, Christa . 2013. "In Israel, Women of the Wall Hit Raw Nerve over Religious Clout in State Life." *The Christian Science Monitor,* May 10. Available at www.csmonitor.com

CAWP (Center for American Women and Politics). 2013a. *Facts on Women Officeholders, Candidates, and Voters.* Available at www.cawp.rutgers.edu

CAWP. 2013b. *Women of Color in Elective Office 2013.* Available at www.cawp.rutgers.edu

Ceci, Stephen J., Wendy M. Williams, and Susan M. Barnett. 2009. "Women's Underrepresentation in Science: Socio-Cultural and Biological Considerations." *Psychological Bulletin* 135(2):218–261.

Clery Center. 2013 (March 7). "VAWA Reauthorization." Available at clerycenter.org

CNN News. 2013 (June 19). "Man's Book on 'Getting Awesome with Women' Hit by Critics as Guide for Rapists." Available at www.cnn.com

Cohen, Theodore. 2001. *Men and Masculinity.* Belmont, CA: Wadsworth.

Cohn, D'vera. 2011 (February 7). "India Offers Three Gender Options." Pew Research Center. Available at www.pewsocialtrends.org

Cook, Carolyn. 2009. "ERA Would End Women's Second-Class Citizenship: Only Three More States Are Needed to Declare Gender Bias Unconstitutional." *The Philadelphia Inquirer,* April 12. Available at www.philly.com

Corbett, Christianne, and Catherine Hill. 2012 (October 24). *Graduating to a Pay Gap: The Earnings of Women and Men One Year After College Graduation.* American Association of University Women. Available at www.aauw.org

Correll, Shelly J., Stephen Benard, and In Paik. 2007. "Getting a Job: Is There a Motherhood Penalty?" *American Journal of Sociology* 112(5):1,297–1,338.

Daily Mail. 2011. "Are These the Most PC Parents in the World? The Couple Raising a 'Genderless Baby'. . . to Protect His (or Her) Right to Choice." *The Daily Mail Online,* May 25. Available at www.dailymail.co.uk

Dallesasse, Starla L., and Annette S. Kluck. 2013. "Reality Television and the Muscular Male Ideal." *Body Image* 10:309–315.

Davis, James A., Tom W. Smith, and Peter V. Marsden. 2002. *General Social Surveys, 1972–2002: 2nd ICPSR Version.* Chicago: National Opinion Research Center.

Deen, Thalif. 2013. "New Push for U.S. to Ratify Major Women's Treaty." Available at www.ipsnews.net

Dunn, Marianne G, Aaron B. Rochlen, and Karen M. O'Brien. 2013. "Employee, Mother, and Partner: An Exploratory Investigation of Working Women With Stay-At-Home Fathers." *Journal of Career Development* 40(1):3–22.

Durkin, Erin, and Daniel Beekman. 2013 (May 7). "NRA Blasted for Endorsing Shooting Target that Looks Like a Woman and Bleeds." Available at www.nydailynews.com

EEOC (Equal Employment Opportunity Commission). 2013. *Charge Statistics: FY 1997 through FY 2012.* Available at www.eeoc.gov

EEOC. 2012 "Facts about Sexual Harassment." Available at www.eeoc.gov

Eisner, Manuel, and Lana Ghuneim. 2013. "Honor Killing Attitudes amongst Adolescents in Amman, Jordan." *Aggressive Behavior* 39(5):405–417.

ERA (Equal Rights Amendment). 2013. *The Equal Rights Amendment: Frequently Asked Questions.* Available at www.equalrightsamendment.org

Faludi, Susan. 2008. "Think the Gender War Is Over? Think Again." *The New York Times,* June 15. Available at www.nytimes.com

Faludi, Susan. 1991. *Backlash: The Undeclared War against American Women.* New York: Crown.

Fischer, Jocelyn, and Jeff Hayes. 2013 (August 13). "The Importance of Social Security in the Incomes of Older Americans." Institute for Women's Policy Research. Available www.iwpr.org

Fitzpatrick, Maureen, and Barbara McPherson. 2010. "Coloring within the Lines: Gender Stereotypes in Contemporary Coloring Books." *Sex Roles* 62(1/2):127–137.

Foerstel, Karen. 2008. "Women's Rights: Are Violence and Discrimination against Women Declining?" *Global Researcher* 2(5):115–147.

Gallagher, Sally K. 2004. "The Marginalization of Evangelical Feminism." *Sociology of Religion* 65:215–237.

Goffman, Erving. 1963. *Stigma.* Englewood Cliffs, NJ: Prentice Hall.

Goodstein, Laurie. 2009. "U.S. Nuns Facing Vatican Scrutiny." *The New York Times,* July 2. Available at www.nytimes.com

Goodstein, Laurie. 2012. "Vatican Reprimands a Group of U.S. Nuns and Plans Changes." *The New York Times,* April 18. Available at www.nytimes.com

Gorney, Cynthia. 2011. "Too Young to Wed." *National Geographic,* June. Available at ngm.nationalgeographic.com

Grant, Jaime M., Lisa A. Mottet, Justine Tants, Jack Harrison, Jody L. Herman, and Mara Ketsling. 2011. *Injustice at Every Turn: A Report of the National TransGender Discrimination Survey.* Washington, DC: National Center for Transgender Equality and the National Gay and Lesbian Task Force.

Gray, Melissa. 2013 (March 3). "Company Removes 'Rape' Shirt Listed on Amazon." Available at www.cnn.com

Grogan, Sarah. 2008. *Body Image: Understanding Body Dissatisfaction in Men, Women and Children.* New York: Routledge.

Gupta, Sanjay. 2003. "Why Men Die Young." *Time,* May 12, p. 84.

Guy, Mary Ellen, and Meredith A. Newman. 2004. "Women's Jobs, Men's Jobs: Sex Segregation and Emotional Labor." *Public Administration Review* 64:289–299.

Haines, Erin. 2006 (November 1). "Father Convicted in Genital Mutilation." Available at www.breitbart.com

Hamilton, Mykol C., David Anderson, Michelle Broaddus, and Kate Young. 2006. "Gender Stereotyping and Under-Representation of Female Characters in 200 Popular Children's Picture Books: A Twenty-First Century Update." *Sex Roles* 55:757–765.

Haub, Carl. 2011 (January 21). "A First for Census Taking: The Third Sex." Population Reference Bureau: *Behind the Numbers.* Available at prbblog.org

Hausmann, Ricardo, Laura D. Tyson, and Saadia Zahidi. 2012. *The Global Gender Gap Report 2012.* Geneva: World Economic Forum.

Halpern, Diane F., Lise Eliot, Rebecca S. Bigler, Richard A. Fabes, Laura D. Hanish, Janet Hyde, Lynn S. Liben, and Carol Lynn Martin. 2013. "Response—Single-Sex Education: Parameters Too Narrow." *Science* 13(January):166–168.

Hegewisch, Ariane, and Maxwell Matite. 2013 (April). *The Gender Wage Gap by Occupation.* Institute for Women's Policy Research. Available at www.iwpr.org

Hegewisch, Ariane, Claudia Williams, and Angela Edwards. 2012 (March). "The Gender Wage Gap: 2012." Institute for Women's Policy Research. Available at www.iwpr.org

Hersch, Joni. 2013. "Opting Out among Women with Elite Education." Vanderbilt Law and Economics Research Paper No. 13-05. Nashville, TN: Vanderbilt University.

Heyzer, Noeleen. 2003. "Enlisting African Women to Fight AIDS." *Washington Post,* July 8. Available at www.globalpolicy.org

Hines, Alice. 2012. "Walmart Sex Discrimination Claims Filed by 2000 Women." *The Huffington Post,* June 6. Available at www.huffingtonpost.com

Hochschild, Arlie. 1989. *The Second Shift.* London: Penguin.

Hoye, Sarah. 2013. "Girl, 11, Scores in Fight against Philadelphia Archdiocese to Play Football." CNN News, March 15. Available at www.cnn.com

HRC (Human Rights Campaign). 2012. "Sexual Orientation and Gender Identity: Terminology and Definitions" *Transgender FAQ.* Available at www.hrc.org

The Huffington Post. 2013 (June 11). "Employment Non-Discrimination Act 2013: The 'T' In LGBT Protections" (infographic). Available at www.huffingtonpost.com

Hvistendahl, Mara. 2011. *Unnatural Selection: Choosing Boys over Girls, and the Consequences of a World Full of Men.* Philadelphia: Public Affairs.

IASC (Inter–Agency Standing Committee). 2009. "IASC Policy Statement Gender Equality in Humanitarian Action." Available at http://www.humanitarianinfo.org

ILO (International Labour Organization). 2012a (December). *Global Employment Trends for Women.* Available at www.ilo.org

ILO. 2012b (June 25). "Gender Dimensions of the World of Work in a Globalized Economy." *Gender Rethinking Alternative Paths for Development.* Geneva: United Nations. Available at www.ilo.org

ILO. 2011. *Equality at Work: The Continuing Challenge.* Available at www.ilo.org

I-VAWA (International Violence Against Women Act). 2013 (June). "Issue Brief: The International Violence Against Women Act." Amnesty International. Available at www.amnestyusa.org

IPC (International Poverty Centre). 2008. *Poverty in Focus: Gender Equality.* No. 13. Available at www.undp-povertycentre.org

ITUC (International Trade Union. Commission). 2011 (March 8). *Decisions for Work: An Examination of the Factors Influencing Women's Decisions for Work.* Available www.ituc-csi.org

Italie, Leanne. 2013. "Breastfeeding Doll Wins Good Reviews But Raises Concerns." *The Chronicle Herald,* November 8. Available at thechronicleherald.ca

IWPR (Institute for Women's Policy Research). 2013 (April). "At Current Pace of Progress, Wage Gap for Women Expected to Close in 2057." *Quick Figures.* Available at www.iwpr.org

Jackson, Janna. 2010. "'Dangerous Presumptions': How Single-Sex Schooling Reifies False Notions of Sex, Gender, and Sexuality." *Gender and Education* 22(2):227–238.

Jackson, Robert, and Meredith A. Newman. 2004. "Sexual Harassment in the Federal Workplace Revisited: Influences on Sexual Harassment." *Public Administration Review* 64(6):705–717.

Jenkins, Colleen. 2013. "Couple Sues Over Adopted Sons Sex Assignment Surgery." Reuters News Service, May 14. Available at http://mobile.reuters.com

Jose, Paul, and Isobel Brown. 2009. "When Does the Gender Difference in Rumination Begin? Gender and Age Differences in the Use of Rumination by Adolescents." *Journal of Youth and Adolescence* 37:180–192.

Jourdan, A. 2013. "Female Bishops: Church of England Renews Pledge to Ordain Women." *The Huffington Post,* August 19. Available at www.huffingtonpost.com

Kaufman, David. 2009. "Introducing America's First Black, Female Rabbi." *Time,* June 6. Available at www.time.com

Kimmel, Michael. 2012 (December 19). "Masculinity, Mental Illness and Guns: A Lethal Equation?" *CNN News.* Available at www.cnn.com

Kimmel, Michael. 2011 (Winter). "Gay Bashing Is about Masculinity." *Voice Male.* Available at www.voicemalemagazine.org

Kravets, David. 2007 (February 6). *Court Says Wal-Mart Must Face Bias Trial.* Available at www.breitbart.com

Lawless, Jennifer L. and Richard L Fox. 2012 (January). *Men Rule: The Continued Underrepresentation of Women in US Politics.* Available at www.american.edu/spa

LCCREF (Leadership Conference on Civil Rights Education Fund). 2009. "Confronting the New Faces of Hate: Hate Crimes in America." Available at www.civilrights.org

Lepkowska, Dorothy. 2008. "Playing Fair?" *The Guardian,* December 16. Available at www.guardian.co.uk

Leslie, David W. 2007 (March). "The Reshaping of America's Academic Workforce." *Research Dialogue* 87. New York: TIAA-CREF Institute. Available at www.tiaa-crefinstitute.org

Levin, Diane E., and Jean Kilbourne. 2009. *So Sexy So Soon: The New Sexualized Childhood and What Parents Can Do to Protect Their Kids.* New York: Random House.

Lopez-Claros, Augusto, and Saadia Zahidi. 2005. *Women's Empowerment: Measuring the Global Gender Gap.* World Economic Forum.

Ludka, Alexandra. 2012 (August 20). "Augusta National Admits First Women Members, Condoleezza Rice and Darla Moore." Available at http://abcnews.go.com

Lutheran World Federation. 2009 (January 29). "Lutheran Woman Bishop Jeruma-Grinberga Succeeds Jagucki in Great Britain." Available at www.lutheranworld.org

Madera, Juan M., Michelle M. Hebl, and Randi C. Martin. 2009. "Gender and Letters of Recommendation for Academia: Agentic and Communal Differences." *Journal of Applied Psychology* 94(6):1,591–1,599.

M.C. v. Medical University of South Carolina. U.S. District Court. Filed May 14, 2013.

McGreevy, Patrick. 2013 (April 12). "California Transgender Students Given Access to Opposite-Sex Programs." Available at www.latimes.com

McGill. 2011 (July 12). "Two-Spirited People." Available at www.mcgill.ca

Media Matters. 2012. "61+ Women Rush Limbaugh Has Labeled 'Babe.'" *Media Matters,* April 17. Available at mediamatters.org

Media Matters. 2010. "Limbaugh: With Kagan Nomination, 'Obama Has Chosen Himself a Different Gender'—'She Is a Pure Academic Elitist Radical.'" *Media Matters TV,* May 10. Available at www.mediamatters.org

Media Matters. 2009. "Limbaugh Asserts 'Chicks… Have Chickified the News'; Again Refers to Female Reporter as Infobabe." Available at http://mediamatters.org

MediaSmarts. 2012. "Body Image – Girls" Available at mediasmarts.ca

Mehmood, Isha. 2009 (January 29). "Lilly Ledbetter Fair Pay Act Becomes Law." Available at www.civilrights.org

Misra, Joya, Jennifer Hickes Lundquist, Elissa Holmes, and Stephanie Agiomavritis. 2011. "The Ivory Ceiling of Service Work." *Academe* 97:22–26.

Messner, Michael A., and Jeffrey Montez de Oca. 2005. "The Male Consumer as Loser: Beer and Liquor Ads in Mega Sports Media Events." *Signs* 30:1,879–1,909.

MKP (ManKind Project). 2012. "About MKP-USA" The ManKind Project. Available at www.mkpusa.org

Models. 2011. "Andrej Pejic, Biography." Available at models.com/models/andrej-pejic

Moen, Phyllis, and Yan Yu. 2000. "Effective Work/Life Strategies: Working Couples, Working Conditions, Gender, and Life Quality." *Social Problems* 47:291–326.

Morin, Rich, and Paul Taylor. 2008 (September 15). *Revisiting the Mommy Wars: Politics, Gender and Parenthood.* Available at www.pewsocialtrends.org

Moskowitz, Clara. 2010. "When Teachers Highlight Gender, Kids Pickup Stereotypes." *Live Science,* November 16. Available at www.livescience.com

National Archives. 2010. "Pictures of Indians in the United States." *Native American Heritage.* Available at www.archives.gov

NCFM (National Coalition for Men). 2013. "Philosophy." Available at http://ncfm.org

NCFM. 2011. "About Us." Available at www.ncfm.org

NCWGE (National Coalition for Women and Girls in Education). 2009. "National Coalition for Women and Girls in Education Recommendations for the Obama Administration and the 111th Congress." Available at www.ncwge.org

NCES (National Center for Educational Statistics). 2013. *The Condition of Education, 2012.* U.S. Department of Education. Available at nces.ed.gov

NOMAS (National Organization for Men Against Sexism). 2013. "Statement of Principles." Available at www.nomas.org

NPR (National Public Radio). 2011. "Two Spirits: A Map of Gender—Diverse Cultures." Public Broadcasting System. Available at www.pbs.org

Nordberg, Jenny. 2010. "Afghan Boys Are Prized, So Girls Live the Part." *The New York Times,* September 20. Available at www.nytimes.com

Norman, Moss E. 2011. "Embodying the Double-Bind of Masculinity: Young Men and Discourses of Normalcy, Health, Heterosexuality, and Individualism." *Men and Masculinities* 14(4):430–449.

Nosek, B. A., F. L. Smyth, N. Sriram, N. M. Lindner, T. Devos, A. Ayala, Y. Bar-Anan, R. Bergh, H. Cai, K. Gonsalkorale, S. Kesebir, N. Maliszewski, F. Neto, E. Olli, J. Park, K. Schnabel, K. Shiomura, B. Tulbure, R. W. Wiers, M. Somogyi, N. Akrami, B. Ekehammar, M. Vianello, M. R. Banaji, and A. G. Greenwald. 2009. "National Differences in Gender-Science Stereotypes Predict National Sex Differences in Science and Math Achievement." *Proceedings of the National Academy of Sciences* 106:10,593–10,597.

Obama, Barack. 2009. "Remarks of President Barack Obama on the Lilly Ledbetter Fair Pay Restoration Act Bill Signing." January 29. Available at www.whitehouse.gov

Office of Civil Rights. 2012 (June 18). "Title IX and Sex Discrimination." Available at www.ed.gov

Padavic, Irene, and Barbara Reskin. 2002. *Men and Women at Work,* 2nd ed. Thousand Oaks, CA: Pine Forge Press.

Park, Hyunjoon, Jere R. Behrman, and Jaesung Choi. 2012. "Causal Effects of Single-Sex Schools on College Entrance Exams and College Attendance: Random Assignment in Seoul High Schools." *Demography* 50(2):447–469.

Parks, Janet B., and Mary Ann Roberton. 2001. "Inventory of Attitudes toward Sexist/ Nonsexist Language—General (IASNL-G): A Correction in Scoring Procedures." Erratum. *Sex Roles* 44(3/4):253.

Parks, Janet B., and Mary Ann Roberton. 2000. "Development and Validation of an Instrument to Measure Attitudes toward Sexist/Nonsexist Language." *Sex Roles* 42(5/6):415–438.

Pew Research Center. 2007 (July 12). "From 1997 to 2007 Fewer Mothers Prefer Full-time Work." Available at www.pewresearch.org

Polachek, Soloman W. 2006. "How the Life-Cycle Human Capital Model Explains Why the Gender Gap Narrowed." In *The Declining Significance of Race,* Francine D. Blau, Mary C. Brinton, and David B. Grusky, eds. (pp. 102–124). New York: Russell Sage.

Political Parity. 2013. *Why Women Do/Don't Run.* Available at www.politicalparity.org

Pollack, William. 2000. *Real Boys' Voices.* New York: Random House.

Quota Project. 2013. "Global Database of Quotas for Women." Available at www.quotaproject.org

Quist-Areton, Ofeibea. 2003. "Fighting Prejudice and Sexual Harassment of Girls in Schools." *All Africa,* June 12. Available at www.globalpolicy.org

Religious Tolerance. 2011. "Religious Sexism: When Faith Groups Started (and Two Stopped) Ordaining Women." Available at www.religioustolerance.org

Renzetti, Claire, and Daniel Curran. 2003. *Women, Men and Society.* Boston: Allyn and Bacon.

Reskin, Barbara, and Debra McBrier. 2000. "Why Not Ascription? Organizations' Employment of Male and Female Managers." *American Sociological Review* 65:210–233.

Ridgeway, Sicilia L. 2011. *Framed By Gender: How Gender Inequality Persists in the Modern World.* New York, New York: Oxford University Press.

Richey, Warren. 2013. "U.S. Supreme Court to Take up Michigan Affirmative Action Case." *Christian Science Monitor,* March 25. Available at www.csmonitor.com

Rose, Stephen J., and Heidi Hartman. 2008 (February). "Still a Man's Labor Market: The Long-Term Earnings Gap." IWPR# C366. New York: Institute for Women's Policy Research.

Rudnanski, Ryan. 2011 (April 10). "Augusta National Golf Club Won't Alter Tradition and Allow Women in Locker Room." Available at bleacherreport.com

Sadker, David, and Karen Zittleman. 2009. *Still Failing at Fairness: How Gender Bias Cheats Boys and Girls in Schools.* New York: Simon and Schuster.

Sanchez, Diana T., and Jennifer Crocker. 2005. "How Investment in Gender Ideals Affects Well-Being: The Role of External Contingencies of Self-Worth." *Psychology of Women Quarterly* 29:63–77.

Sayman, Donna M. 2007. "The Elimination of Sexism and Stereotyping in Occupational Education." *Journal of Men's Studies* 15(1):19–30.

Schneider, Daniel. 2012. "Gender Deviance and Household Work: The Role of Occupation." *American Journal of Sociology* 117(4): 1029–1072.

See Jane. 2013. PSA. *Geena Davis Institute on Gender in Media.* Available at seejane.org

Shierholz, Heidi. 2013 (June 12). "The Wrong Route to Inequality—Men's Declining Wages." Economic Policy Institute. Available at www.epi.org

Simister, John. 2013. "Is Men's Share of House Work Reduced by Gender Deviance Neutralization? Evidence from Seven Countries." *Journal of Comparative Family Studies* 44(3):311–325.

Simpson, Ruth. 2005. "Men in Non-Traditional Occupations: Career Entry, Career Orientation, and Experience of Role Strain." *Gender Work and Organization* 12(4):363–380.

Smith, Matt, and Kyung Lah. 2013 (August 16). "Hooters Blackballs San Diego Mayor." Available at www.cnn.com

Smith, Melanie. 2006. "Is Church Too Feminine for Men?" *The Decatur Daily,* July 1. Available at www.decaturdaily.com

Smith, Stacy L., Marc Choueiti, Ashley Prescott, and Catherine Pieper. 2013. "Gender Roles in Occupations: A Look at Character Attributes and Job-Related Aspirations in Film and Television." *Annenberg School for Communication and Journalism.* University of Southern California. Available at www.seejane.org

Snyder, Karrie Ann, and Adam Isaiah Green. 2008. "Revisiting the Glass Escalator: The Case of Gender Segregation in a Female Dominated Occupation." *Social Problems* 55(2):271–299.

SPLC (Southern Poverty Law Center). 2013 (May 14). "Groundbreaking SPLC Lawsuit Accuses South Carolina, Doctors and Hospitals of Unnecessary Surgery on Infant." Available at www.splcenter.org

Stancill, Jane. 2013. "UNC-CH Women Wage National Campaign against Sexual Assault." *The News and Observer,* June 1. Available at www.newsobserver.com

Stohr, Greg. 2011 (June 20). "Wal-Mart Million-Worker Bias Suit Thrown Out by High Court." Available at www.bloomberg.com

Strauss, Gary. 2013. "Pope: Door Closed on Women Priests." *USA Today,* July 29. Available at www.usatoday.com

Tenenbaum, Harriet R. 2009. "You'd Be Good at That: Gender Patterns in Parent-Child Talk about Courses." *Social Development* 18(2): 447–463.

Uggen, C., and A. Blackstone. 2004. "Sexual Harassment as a Gendered Expression of Power." *American Sociological Review* 69:64–92.

UN (United Nations). 2013. *The Millennium Development Goals Report 2013.* Available at www.un.org

UNESCO (United Nations Educational, Scientific and Cultural Organization). 2012 (September). "Adult and Youth Literacy." *Fact Sheet,* No. 20. Available at www.uis.unesco.org

UNESCO (United Nations Educational, Scientific and Cultural Organization). 2011. "EFA Global Monitoring Report." Available at www.unesco.org

UNICEF. 2011. *The State of the World's Children: 2010.* United Nations: United Nations Children's Fund. Available at www.unicef.org

UNICEF. 2007. *The State of the World's Children: 2007.* United Nations: United Nations Children's Fund. Available at www.unicef.org

UNIFEM (United Nations Development Fund for Women). 2007. "Harmful Traditional Practices." Available at www.unifem.org

USAID. 2013 (March 8). "Promoting Gender Equality through Health." Available at www.usaid.org

U.S. Census Bureau. 2013. *Statistical Abstract of the United States: 2012,* 130th ed. Washington, DC: U.S. Government Printing Office.

U.S. Census Bureau. 2009. *Statistical Abstract of the United States: 2008,* 128th ed. Washington, DC: U.S. Government Printing Office.

U.S. Department of State. 2012 (May 30). "The U.S. Response to Global Maternal Mortality: Saving Mothers, Giving Life." Available at www.state.gov

Vandello, Joseph A., Jennifer K. Bosson, Dov Cohen, Rochelle M. Burnaford, and Jonathan R. Weaver. 2008. "Precarious Manhood." *Journal of Personality and Social Psychology* 95 (6):1,325–1,339.

VAWA. (Violence Against Women Act). 2013. 113th Congress of the United States of America. Available at www.gpo.gov

Wang, Wendy, Kim Parker, and Paul Taylor. 2013 (May 29). *Breadwinner Moms.* Pew Research Social And Demographic Trends. Available at www.pewsocialtrends.org

Weeks, Linton. 2011. "The End of Gender?" National Public Radio, June 23. Available at www.npr.org

White, Alan, and Karl Witty. 2009. "Men's Under Use of Health Services—Finding Alternative Approaches." *Journal of Men's Health* 6(2):95–97.

Willer, Robb, Christabel L. Rogalin, Bridget Conlon, and Michael T. Wojnowicz. 2013. "Overdoing Gender: A Test of the Masculine Overcompensation Thesis." *American Journal of Sociology* 118(4): 980–1022.

WHO (World Health Organization). 2013 (February). "Female Genital Mutilation." Available at www.who.int

WHO (World Health Organization). 2011. Global Sector Strategy on HIV/AIDS 2011–2015. Available at http://whqlibdoc.who.ints

WHO. 2010. "Gender, Women and Primary Heal Care Renewal." Discussion Paper. Retrieved from http://wholibdoc.who.int

WHO. 2009. "Ten Facts about Women's Health." Available at www.who.int

WHO. 2008. *Eliminating Female Genital Mutilation: An Interagency Statement.* Available at http://whqlibdoc.who.int

Williams, Christine L. 2007. "The Glass Escalator: Hidden Advantages for Men in the 'Female' Occupations." In *Men's Lives*, 7th ed., Michael S. Kimmel and Michael Messner, eds. (pp. 242–255). Boston: Allyn and Bacon.

Williams, Joan. 2000. *Unbending Gender: Why Family and Work Conflict and What to Do About It.* Oxford: Oxford University Press.

Williams, Kristi, and Debra Umberson. 2004. "Marital Status, Marital Transitions, and Health: A Gendered Life Course Perspective." *Journal of Health and Social Behavior* 45:81–98.

Wood, Wendy and Alice H. Eagly. 2002. "A Cross-Cultural Analysis of the Behavior of Women and Men: Implications for the Origins of Sex Differences." *Psychological Bulletin* 128(5):699–727.

Yoder, P. Stanley, N. Abderrahim, and A. Zhuzhuni. 2004. *Female Genital Cutting in the Demographic and Health Surveys: A Critical and Comparative Analysis.* DHS Comparative Reports 7. Calverton, MD: ORC Macro.

Zakrzewski, Paul. 2005. "Daddy, What Did You Do in the Men's Movement?" *Boston Globe*, June 19. Available at www.bostonglobe.com

Chapter 11

AAA (American Anthropological Association). 2004. *Statement on Marriage and the Family*, February 26. Available at www.aaanet.org

Ackerman, Spencer. 2013. "Gay Military Couples Welcome Pentagon Decision to Extend Benefits." *The Guardian*, August 14. Available at www.theguardian.com

ACLU (American Civil Liberties Union). 2011 (July 22). "'Don't Ask, Don't Tell' Repeal Certified by President, Defense Secretary and Joint Chiefs Chairman." Available at www.aclu.org

Allen, J. L., and H. Messia. 2013. "Pope Francis on Gays: 'Who Am I to Judge?'" Available at www.religion.blogs.cnn.com

Allport, G. W. 1954. *The Nature of Prejudice.* Cambridge, MA: Addison-Wesley.

Amato, Paul R. 2013. "The Well-Being of Children with Gay and Lesbian Parents." *Social Science Research* 41(4):771–774.

Amato, Paul R. 2004. "Tension between Institutional and Individual Views of Marriage." *Journal of Marriage and Family* 66:959–965.

American College Health Association. 2012. "National College Health Assessment Reference Group Executive Summary." Available at www.acha-ncha.org

AMA (American Medical Association). 2011. "AMA Policy Regarding Sexual Orientation." *GLBT Advisory Committee.* Available at www.ama-assn.org

APA Task Force on Appropriate Therapeutic Responses to Sexual Orientation. (2009). *Report of the Task Force on Appropriate Therapeutic Responses to Sexual Orientation.* Washington, DC: American Psychological Association.

APA (American Psychological Association). 2008. "Answers to Your Questions: For a Better Understanding of Sexual Orientation and Homosexuality." Washington, DC: APA. Available at www.apa.org

AWAB (Association of Welcoming and Affirming Baptists). 2013. "Bonjour, Heloo, Bienvenue, Welcome." Available at www.awab.org

Badgett, M. V. Lee, Ilan H. Meyer, Gary J. Gates, Nan D. Hunter, Jennifer C. Pizer, and Brad Sears. 2011. (July 20). "Written Testimony of the Williams Institute, UCLA School of Law." Hearings on s. 598, *The Respect for Marriage Act: Assessing the Impact of DOMA on American Families.* Available at www.law.ucla.edu/williamsinstitute

Basu, M. 2012 (December 7). "Catholic Notre Dame Announces Services for Gay Students." Available at www.inamerica.blogs.cnn.com

Battista, J. 2013. "Ayanbadejo, Kluwe to Be Honored for Equality Efforts." *The New York Times*, March 29. Available at fifthdown.blogs.nytimes.com

Bayer, Ronald. 1987. *Homosexuality and American Psychiatry: The Politics of Diagnosis*, 2nd ed. Princeton, NJ: Princeton University Press.

Belkin, A., M. Ender, N. Frank, S. Furia, G. R. Lucas, G. Packard, T. S. Schultz, S. M. Samuels, and D. R. Segal. 2012. *One Year Out: An Assessment of DADT Repeal's Impact on Military Readiness.* Palm Center: The University of California, Los Angeles.

Bergman, K., R. J. Rubio, R. J. Green, and E. Padron. 2010. "Gay Men Who Become Fathers via Surrogacy: The Transition to Parenthood." *Journal of GLBT Family Studies* 6: 111–141.

Besen, Wayne. 2010. "Ex-Gay Group Should Repent, Not Revel." Huffington Post, June 17. Available at www.huffingtonpost.com

Bobbe, Judith. 2002. "Treatment with Lesbian Alcoholics: Healing Shame and Internalized Homophobia for Ongoing Sobriety." *Health and Social Work* 27(3):218–223.

Bonds-Raacke, Jennifer M., Elizabeth T. Cady, Rebecca Schlegel, Richard J. Harris, and Lindsey Firebaugh. 2007. "Can Ellen and Will Improve Attitudes toward Homosexuals?" *Journal of Homosexuality* 53:19–34.

Bos, H., N. Gartrell, and L. Van Gelderen. 2013. "Adolescents in Lesbian Families: *DSM*-Oriented Scale Scores and Stigmatization." *Journal of Gay and Lesbian Social Services* 25:121–140.

Brown, Devin. 2011. "Tracy Morgan's Gay Rant Riles Up Twitter and LGBT Community." Huffington Post, June 10. Available at www.huffingtonpost.com

Brown, Michael J., and Ernesto Henriquez. 2008. "Socio-Demographic Predictors of Attitudes towards Gays and Lesbians." *Individual Differences Research* 6:193–202.

Bruce-Jones, Eddie, and Lucas P. Itaborahy. 2011. "State-Sponsored Homophobia: A World Survey of Laws Criminalizing Same-Sex Sexual Acts between Consenting Adults." International Lesbian, Gay, Bisexual, Trans and Intersex Association (ILGA). Available at http://old.ilga.org

Burn, Shawn M. 2000. "Heterosexuals' Use of 'Fag' and 'Queer' to Deride One Another: A Contributor to Heterosexism and Stigma." *Journal of Homosexuality* 40:1–11.

Burn, Shawn M., Kelly Kadlec, and Ryan Rexer. 2005. "Effects of Subtle Heterosexism on Gays, Lesbians, and Bisexuals." *Journal of Homosexuality* 49:23–38.

Cahill, Sean. 2007. "The Coming GLBT Senior Boom." *The Gay and Lesbian Review*, January–February, pp. 19–21.

Caldwell, Alicia A. 2011. "Same-Sex Couples Denied Immigration Benefits by U.S." Huffington Post, March 30. Available at www.huffingtonpost.com

California Supreme Court. 2011. "In re Marriage Cases." Case No. 2147999. The Supreme Court of the State of California. Judicial Council Coordination Processing No. 4365. Available at http://outandequal.org/

Carney, Michael P. 2007 (September 5). "The Employment Non-Discrimination Act: Testimony by Officer Michael P. Carney." Springfield Massachusetts. Available at edlabor.house.gov

Casey, Bob. 2011. "Focus on the Family Stands Up for Bullying." Huffington Post, September 8. Available at www.huffingtonpost.com

Cianciotto, Jason, and Sean Cahill. 2007. "Anatomy of a Pseudo-Science." *The Gay and Lesbian Review*, July–August, pp. 22–24.

Ciarlante, Mitru, and Kim Fountain. 2010. "Why It Matters: Rethinking Victim Assistance for Lesbian, Gay, Bisexual, Transgender, and Queer Victims of Hate Violence and Intimate Partner Violence." National Center for Victims of Crime and the New York City Anti-Violence Project. Available at www.avp.org

CNN. 2009 (October 28). "Obama Signs Hate Crimes Bill into Law." Available at articles.cnn.com

Collins, J., and F. Lindz. 2013. "Why NBA Center Jason Collins is Coming Out Now." *Sports Illustrated*, May 6. Available at www.sportsillustrated.cnn.com

Condon, Stephanie. 2013. "Supreme Court Strikes Down Key Part of DOMA, Dismisses Prop 8 Case." *CBS News*, June 26. Available at www.cbs.com

Coogan, Steve. 2011. "Michael Strahan, Fiancé Film Same-Sex Marriage PSA." *USA Today*, June 12. Available at content.usatoday.com

DeSilver, Drew. 2013 (June 26). "How Many Same-Sex Marriages in the US? At Least 71,165, Probably More." *Pew Research Center.* Available at www.pewresearch.org

Dougherty, Jill. 2011. "U.N. Council Passes Gay Rights Resolution." CNN, June 17. Available at http://articles.cnn.com

Dudash, April. 2013. "Duke LGBT Center Receives New Name, Space for Fall Semester." *The Herald Sun*, July 13. Available at www.heraldsun.com

Durkheim, Emile. 1993 [1938]. "The Normal and the Pathological." In *Social Deviance*, Henry N. Pontell, ed. (pp. 33–63). Englewood Cliffs, NJ: Prentice-Hall. (Originally published in *The Rules of Sociological Method*.)

Eliason, Mickey. 2001. "Bi-Negativity: The Stigma Facing Bisexual Men." *Journal of Bisexuality* 1:136–154.

Equality Forum. 2013. "FORTUNE 500 Non-Discrimination Project." Available at www.equalityforum.com

Esselink, J. A. 2013 (June 27). "Respect for Marriage Act to Repeal DOMA Filed in Both the House and Senate." Available at http://thenewcivilrightsmovement.com

Esterberg, K. 1997. *Lesbian and Bisexual Identities: Constructing Communities, Constructing Selves.* Philadelphia: Temple University Press.

Exodus International. 2011. "Fact Sheet." Available at exodusinternational.org

Exodus International. 2013 (June 19). "Exodus International President to the Gay Community: 'We're Sorry.'" Available at exodusinternational.org

"FAIR Education Act." (2013). Available at www.faireducationact.com

Falomir-Pichastor, Juan Manuel, and Gabriel Mugny. 2009. "'I'm Not Gay . . . I'm a Real Man!': Heterosexual Men's Gender Self-Esteem and Sexual Prejudice." *Personality and Social Psychology Bulletin* 35:1233–1243.

FBI. 2012 (November). *Hate Crime Statistics, 2011.* US Department of Justice. Available at www.fbi.gov

FBI. 2011 (November). *Hate Crime Statistics, 2011.* US Department of Justice. Available at www.fbi.gov

Fidas, Deena. 2009. "At the Water Cooler." *Equality* (Spring):23, 29.

Fone, Byrne. 2000. *Homophobia: A History.* New York: Henry Holt.

Frank, Barney. 1997. "Foreword." In *Private Lives, Public Conflicts: Battles over Gay Rights in American Communities,* by J. W. Button, B. A. Rienzo, and K. D. Wald, eds. (pp. xi). Washington, DC: CQ Press.

Frank, David John, and Elizabeth H. McEneaney. 1999. "The Individualization of Society and the Liberalization of State Policies on Same-Sex Relations, 1984–1995." *Social Forces* 77(3): 911–944.

Gallup Organization. 2012. "Gay and Lesbian Rights." November 26–29. Available at www.gallup.com

Gardner, Lisa A., and Ryan M. Roemerman. 2011. "Iowa College Climate Survey: The Life Experiences of Lesbian, Gay, Bisexual, Transgender and Straight Allied (LGBTA) Students at Iowa's Colleges and Universities." Iowa Pride Network. Available at www.iowapridenetwork.org

Gates, Gary J. 2011. "How Many People are Lesbian, Gay, Bisexual, and Transgender?" Williams Institute. Available at www.law.ucla.edu

Gates, G. J. 2012 (October18). "Special Report: 3.4 percent of U.S. Adults Identify as LGBT." Available at www.gallup.com

Gay and Lesbian Victory Fund. 2013. "Out Officials." Available at www.victoryfund.org

Gilman, Stephen E., Susan D. Cochran, Vickie M. Mays, Michael Hughes, David Ostrow, and Ronald C. Kessler. 2001. "Risk of Psychiatric Disorders among Individuals Reporting Same-Sex Sexual Partners in the National Comorbidity Survey." *American Journal of Public Health* 91(6):933–939.

GLSEN and Harris Interactive (2012). Playgrounds and Prejudice: Elementary School Climate in the United States, A Survey of Students and Teachers. New York: GLSEN.

GLSEN. 2011a. "Damaging Language." Available at thinkb4youspeak.com

GLSEN. 2011b. "What We Do." Available at www.glsen.org

GLSEN. 2007. *Gay-Straight Alliances: Creating Safer Schools for LGBT Students and their*

Allies. GLSEN Research Brief. New York: Gay, Lesbian and Straight Education Network.

Gonzalez, Ivet. 2013. "Gay Parents in Cuba Demand Legal Right to Adopt." *Global Issues,* June 4. Available at www.globalissues.org

Goodstein, Laurie. 2010. "Lutherans Offer Warm Welcome to Gay Pastors." *The New York Times,* July 25. Available at www.nytimes.com

Goldbach, J. T., E. E. Tanner-Smith, M. Bagwell, and S. Dunlap. 2013. "Minority Stress and Substance Use in Sexual Minority Adolescents: A Meta-Analysis." *Prevention Science,* April 19.

Guadalupe, Krishna L., and Doman Lum. 2005. *Multidimensional Contextual Practice: Diversity and Transcendence.* Belmont, CA: Thomson Brooks/Cole, p. 11.

Guillory, Sean. 2013. "Repression and Gay Rights in Russia." *The Nation,* September 26. Available at www.thenation.com

Haider-Markel, Donald P., and Mark R. Joslyn. 2008. "Beliefs about the Origins of Homosexuality and Support for Gay Rights: An Empirical Test of Attribution Theory." *Public Opinion Quarterly* 72:291–310.

Hamilton, Julie. 2011. "Anti-Gay?! NARTH President Addresses Misperceptions about NARTH." Available at narth.com

Harper, Gary W., Nadine Jernewall, and Maria C. Zea. 2004. "Giving Voice to Emerging Science and Theory for Lesbian, Gay, and Bisexual People of Color." *Cultural Diversity and Ethnic Minority Psychology* 10:187–199.

Hatzenbuehler, M. L. 2011. "The Social Environment and Suicide Attempts in Lesbian, Gay, and Bisexual Youth." *Pediatrics* 127:896–904.

Herek, Gregory M. 2009. "Facts about Homosexuality and Child Molestation." Available at http://psychology.ucdavis.edu

Herek, Gregory M. 2004. "Beyond 'Homophobia': Thinking about Sexual Prejudice and Stigma in the Twenty-First Century." *Sexuality Research and Social Policy: A Journal of the NSRC* 1:6–24.

Herek, Gregory M. 2002. "Gender Gaps in Public Opinion about Lesbians and Gay Men." *Public Opinion Quarterly* 66:40–66.

Herek, Gregory M. 2000a. "The Psychology of Sexual Prejudice." *Current Directions in Psychological Science* 9:19–22.

Herek, Gregory M. 2000b. "Sexual Prejudice and Gender: Do Heterosexuals' Attitudes toward Lesbians and Gay Men Differ?" *Journal of Social Issues* 56:251–266.

Herek, Gregory M., and Linda D. Garnets. 2007. "Sexual Orientation and Mental Health." *Annual Review of Clinical Psychology* 3:353–375.

Herman, Joanne. 2009. *Transgender Explained for Those Who Are Not.* Bloomington: Author House.

Hewlett, Sylvia Ann, and Karen Sumberg. 2011. "For LGBT Workers, Being 'Out' Brings Advantages." *Harvard Business Review* 89(7/8): n/a.

Hicks, Stephen. 2006. "Maternal Men—Perverts and Deviants? Making Sense of Gay Men as Foster Carers and Adopters." *Family Studies* 2:93–114.

HRC (Human Rights Campaign). 2013a. "The Domestic Partnership Benefits and Obligations Act." Available at www.hrc.org

HRC. 2013b. *Every Child Deserves a Family Act.* Available at www.hrc.org

HRC. 2013c. "Parenting Laws: Joint Adoptions." Available at www.hrc.org

HRC. 2013d. "Parenting Laws: Second Parent Adoptions." Available at www.hrc.org

HRC. 2013e. "Safe Schools Improvement Act." Available at www.hrc.org

HRC. 2013f. "Statewide Employment Laws and Policies." Available at www.hrc.org

HRC. 2011a. "Faith Positions." Available at www.hrc.org

HRC. 2011b. "Professional Organizations on LGBT Parenting." Available at www.hrc.org

Huffington Post. 2013 (June 26). "27 Companies That Aren't Afraid to Support the Supreme Court's Gay Marriage Rulings." Available at www.huffingtonpost.com

Humphrey, Tom. 2011. (April 21). "'Don't Say Gay' Bill Clears Senate Panel." Available at www.knoxnews.com

ILGBTIA (International Lesbian, Gay, Bisexual, Trans, and Intersex Association). 2012 (May 17). "2012 ILGA State-Sponsored Homophobia Report: 40percent of UN Members Still Criminalize Same-Sex Sexual Acts." Available at www.ilga.org

It Gets Better Project. 2011. "What Is the It Gets Better Project?" Available at www.itgetsbetter.org

Joannides, Paul. (2011). *The Guide to Getting It On.* Waldport: Goofy Foot Press.

Johnson, Renee M., Jeremy D. Kidd, Erin C. Dunn, Jennifer G. Green, Heather L. Corliss, and Deborah Bowen. 2011. "Associations between Caregiver Support, Bullying, and Depressive Symptomatology among Sexual Minority and Heterosexual Girls: Results from the 2008 Boston Youth Survey." *Journal of School Violence* 10:185–200.

Jones, Jeffrey M. 2011. "Support for Legal Gay Relations Hits New High." Gallup Organization. Available at www.gallup.com/

Jones, J. M. 2013 (May 16). "More Americans See Gay, Lesbian Orientation as Birth Factor." Available at www.gallup.com/poll

Jones, J. M. 2012 (December 6). "Most in U.S. Say Gay/Lesbian Bias Is a Serious Problem." Available at www.gallup.com

Karimi, Faith. 2011. "Alleged Rape, Killing of Gay Rights Campaigner Sparks Call for Action." CNN. Available at www.cnn.com

Kastanis, Angeliki, and M. V. Lee Badgett. 2013 (March 6). "Estimating the Economic Boost of Marriage Equality in Minnesota." *Williams Institute.* Available at williamsinstitute.law.ucla.edu

Kennedy, K., and D. Temkin. 2013 (March 5). "The Time Is Now for a Federal Anti-Bullying Law." Available at www.huffingtonpost.com

Kimmel, Michael. 2011. "Gay Bashing Is about Masculinity." Winter. *Voice Male.* Available at www.voicemalemagazine.org

Kinsey, A. C., W. B. Pomeroy, and C. E. Martin. 1948. *Sexual Behavior in the Human Male.* Philadelphia: W. B. Saunders.

Kinsey, A. C., W. B. Pomeroy, C. E. Martin, and P. H. Gebhard. 1953. *Sexual Behavior in the Human Female.* Philadelphia: W. B. Saunders.

Kirkpatrick, R. C. 2000. "The Evolution of Human Sexual Behavior." *Current Anthropology* 41(3):385–414.

Kosciw, J. G., Greytak, E. A., Bartkiewicz, M. J., Boesen, M. J., & Palmer, N. A. (2012). *The 2011 National School Climate Survey: The experiences of lesbian, gay, bisexual and transgender youth in our nation's schools*. New York: GLSEN.

Kosciw, Joseph G., and Elizabeth M. Diaz. 2005. *The 2005 National School Climate Survey*. New York: GLSEN.

Lambda Legal. 2011a. "Health and Medical Organization Statements on Sexual Orientation, Gender Identity/Expression and 'Reparative Therapy.'" Available at data.lambdalegal.org

Lambda Legal. 2011b. "Preventing Censorship of LGBT Information in Public School Libraries." Available at data.lambdalegal.org

Liebelson, Dana. 2013. "Why Do So Many States Still Have Anti-Sodomy Laws?" *The Week*, April 8. Available at theweek.com

Lopez, Mark Hugo. 2013 (June 13). "Personal Milestones in the Coming-Out Experience." *Pew Research Social and Demographic Trends.* Available at www.pewsocialtrends.org

Louderback, L. A., and B. E. Whitley. 1997. "Perceived Erotic Value of Homosexuality and Sex-Role Attitudes as Mediators of Sex Differences in Heterosexual College Students' Attitudes toward Lesbians and Gay Men." *Journal of Sex Research* 34:175–182.

Loviglio, J. 2013. "ACLU to Pa. School: Allow Gay-Straight Alliance." *York Daily Record*, March 13. Available at www.ydr.com

Malmsheimer, T. 2013 (March 27). "California to Introduce More Gay-Themed Books into School Curriculum: Unsurprisingly, Backlash Ensues." Available at www.nydailynews.com

McGreevy, Patrick. 2011. "New State Law Requires Textbooks to Include Gays' Achievements." *The Los Angeles Times*, July 15. Available at articles.latimes.com

McKinley, James C. 2013. "Stars Align for a Gay Marriage Anthem." *The New York Times*, June 30. Available at www.nytimes.com

McKinley, Jesse. 2010. "Suicides Put Light on Pressures of Gay Teenagers." *The New York Times*, October 3. Available at www.nytimes.com

Mears, Bill. 2011. "Court Upholds Gay Judge's Ruling on Proposition 8." CNN, June 14. Available at articles.cnn.com

Meyer, Ilian. 2003. "Prejudice, Social Stress, and Mental Health in Lesbian, Gay, and Bisexual Populations: Conceptual Issues and Research Evidence." *Psychological Bulletin* 129:674–697.

Miller, Patti, McCrae A. Parker, Eileen Espejo, and Sarah Grossman-Swenson. 2002. *Fall Colors: Prime Time Diversity Report 2001–02*. Oakland CA: Children Now and the Media Program.

Mitz, B. 2013. "Milestone for Gay Athletes as Rogers Plays for Galaxy." *The New York Times*, May 27. Available at www.nytimes.com

Mohipp, C., and M. M. Morry. 2004. "Relationship of Symbolic Beliefs and Prior Contact to Heterosexuals' Attitudes toward Gay Men and Lesbian Women." *Canadian Journal of Behavioral Science* 36(1):36–44.

Morales, Lymari. 2010. (December 9). "In U.S., 67 percent Support Repealing 'Don't Ask, Don't Tell.'" Gallup Organization. Available at www.gallup.com

Morgan, E. M., M. G. Steiner, and E. M. Thompson. 2010. "Processes of Sexual Orientation Questioning among Heterosexual Men." *Men and Masculinities* 12:425–443.

Morgan, E. M. and E. M. Thompson. 2011. "Processes of Sexual Orientation Questioning among Heterosexual Women." *Journal of Sex Research* 48:16–28.

Morgan, G. 2013. "Brittney Griner, WNBA Draft Pick, Comes Out." *The Huffington Post*, April 8. Available at www.huffingtonpost.com/

Moulton, B. (2011). "HRC Supports 'Every Child Deserves a Family Act.'" Human Rights Campaign. Available at www.hrcbackstory.org

Mungin, Lateef. 2011. (May 27). "Expert: Use Gay Slurs Controversy to Tackle Homophobia in Sports." Available at www.cnn.com

Nash, Dannika. 2013 (April 7). "An Open Letter to the Church from My Generation." Available at dannikanash.com

NARTH (National Association for Research and Therapy of Homosexuality). 2011. "NARTH Mission Statement." Available at narth.com

NCAVP (National Coalition of Anti-Violence Programs). 2013. "National Report on Hate Violence against Lesbian, Gay, Bisexual, Transgender, Queer, and HIV-affected Communities." Media Release. Available at www.avp.org

National Gay and Lesbian Task Force. 2009 (July 1). "State Nondiscrimination Laws in the U.S." Available at www.thetaskforce.org

NPR (National Public Radio). 2013 (August 4). "Activists Fight Uganda's Anti-Gay Bill." Available at www.npr.org

Newcomb, Michael E., and Brian Mustanski. 2010. "Internalized Homophobia and Internalizing Mental Health Problems: A Meta-analytic Review." *Clinical Psychology Review* 30:1,019–1,029.

Newport, F. 2012a (December 17). "Americans Favor Rights for Gays, Lesbians to Inherit, Adopt." Available at www.gallup.com

Newport, F. 2012b (December 5). "Religion Big Factor for Americans Against Same-Sex Marriage." Available at www.gallup.com

Newport, F., and I. Himelfarb. 2013 (May 20). "In U.S., Record-High Say Gay, Lesbian Relations Morally OK." Available at www.gallup.com/

Oi, M. 2013 (April 22). "Is Singapore's Stance on Homosexuality Changing?" Available at www.bbc.co.uk

O'Keefe, Ed. 2010. "'Don't Ask, Don't Tell' is Repealed by Senate; Bill Awaits Obama's Signing." *The Washington Post*, December 19. Available at www.washingtonpost.com

Olson, Laura R., Cadge, Wendy, and Harrison, James T. 2006. "Religion and Public Opinion about Same-Sex Marriage." *Social Science Quarterly* 87:340–360.

Padilla, Yolanda C., Catherine Crisp, and Donna Lynn Rew. 2010. "Parental Acceptance and Illegal Drug Use among Gay, Lesbian, and Bisexual Adolescents: Results from a National Survey." *Social Work* 55:265–275.

Page, S. 2011. "Gay Candidates Gain Acceptance." *USA Today*, July 20. Available at www.usatoday.com

Patterson, Charlotte J. 2009. "Children of Lesbian and Gay Parents: Psychology, Law, and Policy." *American Psychologist* 64:727–736.

The Pew Forum. 2013a (February 8). "Gay Marriage around the World." Available at www.pewforum.org

The Pew Forum. 2013b (June 6). "Same-Sex Marriage State-By-State." *Religion in Public Life Project.* Available at features.pewforum.org

Pew Research Center. 2013a. "In Gay Marriage Debate, Both Supporters and Opponents See Legal Recognition as 'Inevitable'." June 6. Available at www.people-press.org

Pew Research Center. 2013b (March 20). "Growing Support for Same-Sex Marriage: Changed Minds and Changing Demographics." Available at www.people-press.org

Pew Research Center. 2013c (June 13). "A Survey of LGBT Americans: Attitudes, Experiences and Values in Changing Times." Available at www.pewsocialtrends.org

Pollack, William. 2000a. *Real Boys' Voices*. New York: Random House.

Pollack, William. 2000b. "The Columbine Syndrome." *National Forum* 80:39–42.

Price, Jammie, and Michael G. Dalecki. 1998. "The Social Basis of Homophobia: An Empirical Illustration." *Sociological Spectrum* 18:143–159.

Rapado, Donna, and Janie Campbell. 2011 (July 23). "New Anti-Bullying Rule in Effect for Miami-Dade Schools." NBC Miami. Available at www.nbcmiami.com

Regnerus, Mark, 2012. "How Different Are the Adult Children of Parents Who Have Same-Sex Relationships? Findings from the New Family Structures Study." *Social Science Research* 41(4): 752–770.

Röndahl, Gerd, and Sune Innala. 2008. "To Hide or Not to Hide, That Is the Question!" *Journal of Homosexuality* 52:211–233.

Rosky, Clifford, Christy Mallory, Jenni Smith, and M. V. Lee Badgett. 2011 (January). "Employment Discrimination against LGBT Utahns: Executive Summary." The Williams Institute. Available at www.law.ucla.edu

Rothblum, Esther D. 2000. "Sexual Orientation and Sex in Women's Lives: Conceptual and Methodological Issues." *Journal of Social Issues* 56:193–204.

Rust, P. C. R. 2002. "Bisexuality: The State of the Union." *Annual Review of Sex Research* 13:180–240.

Ryan, Caitlin, David Huebner, Rafael M. Diaz, and Jorge Sanchez. 2009. "Family Rejection as a Predictor of Negative Health Outcomes in White and Latino Lesbian, Gay, and Bisexual Young Adults." *Pediatrics* 123(1):346–352.

Saad, Lydia. 2013 (August 16). "In US, 52% Back Law to Legalize Gay Marriage in 50 States." Gallup politics. Available at www.gallup.com

Savage, D. 2013. "Gay Marriage Ruling: Supreme Court Finds DOMA Unconstitutional." *Los Angeles Times*, June 26. Available at www.latimes.com

Savage, Charlie, and Sheryl Gay Stolberg. 2011. "In Shift, U.S. Says Marriage Act Blocks Gay Rights." *The New York Times*, February 23, 2011. Available at www.nytimes.com

Savin-Williams, R. C. 2006. "Who's Gay? Does It Matter?" *Current Directions in Psychological Science* 15:40–44.

Savin-Williams, R.C., and Z. Vrangalova. 2013. "Mostly Heterosexual as a Distinct Sexual Orientation Group: A Systematic Review of the Empirical Evidence." *Developmental Review* 33:58–88.

Schiappa, E., P. B. Gregg, and D. E. Hewes. 2005. "The Parasocial Contact Hypothesis." *Communication Monographs* 72(1):92–115.

Schemo, Diana J. 2007. "Lessons on Homosexuality Move into the Classroom." *The New York Times*, August 15. Available at www.nytimes.com

Sears, Tim, and Christy Mallory. 2011 (July). "Evidence of Employment Discrimination on the Basis of Sexual Orientation in State and Local Government: Complaints Filed with State Enforcement Agencies 2003–2007." Williams Institute. Available at www.law.ucla.edu/

Shackelford, Todd K., and Avi Besser. 2007. "Predicting Attitudes toward Homosexuality: Insights from Personality Psychology." *Individual Differences Research* 5:106–114.

Shapiro, L. 2013. "UN Tackles Gay 'Conversion Therapy' for First Time." Huffington Post, February 1. Available at www.huffingtonpost.com

Shields, J. P., R. Cohen, J. R. Glassman, K. Whitaker, H. Franks, and I. Bertolini. 2012. "Estimating Population Size and Demographic Characteristics of Lesbian, Gay, Bisexual, and Transgender Youth in Middle School." *The Journal of Adolescent Health* 52:248–250.

Sieczkowski, C. 2013 (June 5). "Gay 11-Year-Old's Petition against Homophobic Politician Succeeds." Available at www.huffingtonpost.com/

Singer, P., and Belkin, A. 2012 (September 20). "A Year after DADT Repeal, No Harm Done." Available at www.cnn.com/

Smith, Tom W. 2011. "Cross–National Differences in Attitudes towards Homosexuality." April, GSS Cross-National Report No. 31. University of Chicago: National Opinion Research Center.

Stanley, Kim. 2009. *Resilience, Minority Stress, and Same Sex Populations: Toward a Fuller Picture*. ProQuest Dissertations and Theses. Available at www.proquest.com

Stone, Andrea. 2011. "Pentagon Discharged Hundreds of Service Members Under 'Don't Ask, Don't Tell' in Fiscal 2010: Report." Huffington Post, March 24. Available at www.huffingtonpost.com

Sue, Derald W. 2010 *Microaggressions in Everyday Life: Race, Gender, and Sexual Orientation*. Hoboken, NJ: John Wiley and Sons, Inc.: Hoboken, NJ.

Sullivan, A. 1997. "The Conservative Case." In *Same-Sex Marriage: Pro and Con*, A. Sullivan, ed. (pp. 146–154). New York: Vintage Books.

Summers, Bryce B. 2010. "Factor Structure and Validity of the Lesbian, Gay, and Bisexual Knowledge and Attitude Scale for Heterosexuals (LGB-KASH)." Proquest Dissertations and Theses. (UMI No. 3425038).

Szymanski, Dawn M., Susan Kashubeck-West, and Jill Meyer. 2008. "Internalized Heterosexism: Measurement, Psychosocial Correlates, and Research Directions." *The Counseling Psychologist* 36:525–574.

Tasker, Fiona, Helen Barrett, and Frederica De Simone. 2010. "'Coming Out Tales': Adults Sons and Daughters' Feeling about Their Gay Father's Identity." *Australian and New Zealand Journal of Family Therapy* 31(4):326–337.

Tobias, Sarah, and Sean Cahill. 2003. *School Lunches, the Wright Brothers, and Gay Families*. National Gay and Lesbian Task Force. Available at www.thetaskforce.org

Toomey, Russell B., Caitlin Ryan, Rafael M. Diaz, Noel A. Card, and Stephen T. Russell. 2010. "Gender Non-Conforming Lesbian, Gay, Bisexual, and Transgender Youth: School Victimization and Young Adult Psychosocial Adjustment." *Developmental Psychology* 46:1580–1589.

U.S. Census Bureau. 2010. "The Census: A Snapshot." Available at www.census.gov/2010census

U.S. Court of Appeals, District 3. 2010 (September 22). *In re: Matter of Adoption of X.X.G. and N.R.G. the State of Florida.* No. 3D08-3044. Available at www.3dca.flcourts.org

University of Alabama. 2011. "Safe Zone." Available at http://ua.edu

Varjas, Kris, Brian Dew, Megan Marshall, Emily Graybill, Anneliese Singh, Joel Meyers, and Lamar Birckbichler. 2008. "Bullying in Schools towards Sexual Minority Youth." *Journal of School Violence* 7:59–86.

Vrangalova, Z., and Savin-Williams, R. C. 2012. "Mostly Heterosexual and Mostly Gay/Lesbian: Evidence for New Sexual Orientation Identities." *Archives of Sexual Behavior* 41:85–101.

Vrangalova, Zhana, and Ritch C. Savin-Williams. 2010. "Correlates of Same-Sex Sexuality in Heterosexually Identified Young Adults." *Journal of Sex Research* 47: 92–102.

Wilcox, Clyde, and Robin Wolpert. 2000. "Gay Rights in the Public Sphere: Public Opinion on Gay and Lesbian Equality." In *The Politics of Gay Rights*, Craig A. Rimmerman, Kenneth D. Wald, and Clyde Wilcox, eds. (pp. 409–432). Chicago: University of Chicago Press.

Whitehead, Andrew L. 2010. "Sacred Rites and Civil Rights: Religion's Effect on Attitudes toward Same-Sex Unions and the Perceived Cause of Homosexuality." *Social Science Quarterly* 91:63–79.

Worthington, Roger L., Frank R. Dillon, and Ann M. Becker-Shutte. 2005. "Development, Reliability, and Validity of the Lesbian, Gay, and Bisexual Knowledge and Attitudes Scale for Heterosexuals (LGB-KASH)." *Journal of Counseling Psychology* 52:104–118.

Zweynert, Astrid. 2013 (June 25). "EU Foreign Affairs Ministers Adopt 'Groundbreaking' Global Policy to Protect Gay Rights." *Thompson Reuters Foundation*. Available at www.trust.org

Chapter 12

American Humane Association. 2011. "Pet Overpopulation." Available at www.americanhumane.org

ASPCA. 2011. "Position Statement on Mandatory Spay/Neuter Laws." Available at www.aspca.org

Aumann, Kerstin, Ellen Galinsky, Kelly Sakai, Melissa Brown, and James T. Bond. 2010. *The Elder Care Study: Everyday Realities and Wishes for Change*. Families and Work Institute. Available at www.familiesandwork.org

Barot, Sneha. 2011 (Spring). "Unsafe Abortion: The Missing Link in Global Efforts to Improve Maternal Health." *Guttmacher Policy Review* 14(2):24–28.

Bongaarts, John, and Susan Cotts Watkins. 1996. "Social Interactions and Contemporary Fertility Transitions." *Population and Development Review* 22(4):639–682.

Branigan, Tania. 2011. "China Considers Relaxing One-Child Policy." *The Guardian*, March 8. Available at www.guardian.co.uk

Brenoff, Ann. 2011 (May 26). "Foreclosures' Other Victims: Abandoned Pets." AOL Real Estate. Available at realestate.aol.com

Chou, Rita Jing-Ann. 2011. "Filial Piety by Contract? The Emergence, Implementation, and Implications of the 'Family Support Agreement' in China." *The Gerontologist* 51(1):3–16.

Cloutier-Fisher, Denise, Karen Kobayashi, and Andre Smith. 2011. "The Subjective Dimension of Social Isolation: A Qualitative Investigation of Older Adults' Experiences in Small Social Support Networks." *Journal of Aging Studies*, doi:10.1016/j.jaging.2011.03.012

Coate, Stephen, and Brian Knight. 2010. "Pet Overpopulation: An Economic Analysis." *The B.E. Journal of Economic Analysis & Policy* 10(1) (Advances), Article 106. Available at www.bepress.com

Dunn, Mark. 2008. "Darlington Will Have Australia's First Vertical Cemetery." *Herald Sun*, November 21. Available at www.heraldsun.com.au

Edwards, Kathryn A., Anna Turner, and Alexander Hertel-Fernandez. 2012. *A Young Person's Guide to Social Security*. Economic Policy Institute. Available at www.epi.org

Engelman, Robert. 2011 (July 18). "The World at 7 Billion: Can We Stop Growing Now?" *Yale Environment 360*. Available at e360.yale.edu

Farrell, Paul. 2009 (January 26). "Peak Oil? Global Warming? No, It's 'Boomsday'!" *MarketWatch*. Available at www.marketwatch.com

File, Thom, and Sarah Crissey. 2010 (May). "Voter Registration in the Election of November 2008." *Current Population Reports* P20-562. U.S. Census Bureau. Available at www.census.gov

Frost, Ashley E., and F. Nii-Amoo Dodoo. 2009. "Men Are Missing from African Family Planning." *Contexts: Understanding People in Their Social Worlds* 8(1):44–49.

Frynes-Clinton, Jane. 2011. "Problem of Grave Concern." *Courier Mail*, June 16. Available at www.couriermail.com.au

Greenfield, Beth. 2011. "The True Costs of Owning a Pet." *Forbes.com*, May 24. Available at www.forbes.com

Gullette, Margaret M. 2011. *Agewise: Fighting the New Ageism in America*. Chicago: University of Chicago Press.

Helman, Ruth, Nevin Adams, Craig Copeland, and Jack VanDerhei. 2013 (March). "2013 Retirement Confidence Survey: Perceived Savings Needs Outpace Reality for Many." EBRI Issue Brief, no. 384. Available at www.ebri.org

Here and Now. 2009 (October 18). "Japan's High-Tech Graveyard in the Sky." Public Radio International. Available at www.pri.org

Humane Society. 2010 (December 16). "Austin City Council Prohibits Retail Sales of Dogs and Cats." Available at www.humansociety.org

Humane Society. 2009 (November 23). "HSUS Pet Overpopulation Estimates." Available at www.humanesociety.org

Kidd, Andrew. 2009. "Shelters See Rise in Abandoned Pets as College Students' Year Ends." *Fox News,* May 9. Available at www.foxnews.com

Koch, Wendy. 2010 (May 25). "Curb Population Growth to Fight Climate Change?" *USAToday.* Available at http://content.usatoday.com

Kornadt, Anna E., and Klaus Rothermund. 2010. "Constructs of Aging: Assessing Evaluative Age Stereotypes in Different Life Domains." *Educational Gerontology* 36(6).

Kornblau, Melissa. 2009. "Social Security Systems around the World." *Today's Research on Aging* No. 15: all. Population Reference Bureau. Available at www.prb.org

Lahey, Johanna. 2008. "Age, Women, and Hiring: An Experimental Study." *Journal of Human Resources* 43:30–56.

Leonard, Matt. 2009. "The Kindest Cut." *Earth Island Journal.* Available at www.earthislandjournal.org

Livernash, Robert, and Eric Rodenburg. 1998. "Population Change, Resources, and the Environment." *Population Bulletin* 53(1):1–36.

Mesce, Deborah, and Donna Clifton. 2011. *Abortion: Facts and Figures 2011.* Washington DC: Population Reference Bureau.

Morrissey, Monique. 2011 (January 26). "Beyond 'Normal': Raising the Retirement Age is the Wrong Approach for Social Security." *EPI Briefing Paper #287.* Available at www.epi.org

Munnell, Alicia H., Anthony Webb, and Rebecca Cannon Fraenkel. 2013 (June). "The Impact of Interest Rates on the National Retirement Risk Index." Center for Retirement Research. Available at crr.bc.edu

National Council on Pet Population Study and Policy. 2009. "The Top Ten Reasons for Pet Relinquishment to Shelters in the United States." Available at petpopulation.org

Nelson, Todd D. 2011. "Ageism: The Strange Case of Prejudice against the Older You." In *Disability and Aging Discrimination,* R. L. Wiener and S. L. Willborn, eds. (p. 37). New York: Springer Science + Business Media.

Notkin, Melanie. 2013 (August 1). "The Truth about the Childless Life." *Huffington Post.* Available at www.huffingtonpost.com

Palmore, Erdman B. 2004. "Research Note: Ageism in Canada and the United States." *Journal of Cross-Cultural Gerontology* 19(1):41–46.

Parks, Kristin. 2005. "Choosing Childlessness: Weber's Typology of Action and Motives of the Voluntarily Childless." *Sociological Inquiry* 75(3):372–402.

Population Reference Bureau. 2010. *World Population Data Sheet.* Washington, DC: Population Reference Bureau. Available at prb.org

Population Reference Bureau. 2007. *World Population Data Sheet.* Washington, DC: Population Reference Bureau. Available at www.prb.org

Population Reference Bureau. 2004. "Transitions in World Population." *Population Bulletin* 59(1).

Ryerson, William N. 2011 (March 11). "Family Planning: Looking Beyond Access." *Science* 331:1,265.

Saad, Lydia. 2013 (May 23). "Three in Four U.S. Workers Plan to Work Past Retirement Age."

Gallup Organization. Available at www.gallup.com

Sandberg, Lisa. 2013 (September). "Inhumane: Nathan J. Winograd on Reforming Animal Shelters." *The Sun* 453:4–13.

Santora, Marc. 2010. "City Cemeteries Face Gridlock." *New York Times,* August 13. Available at www.nytimes.com

Schueller, Jane. 2005 (August). "Boys and Changing Gender Roles." YouthNet. *YouthLens* 16. Available at www.fhi.org

Scott, Laura. 2009. *Two Is Enough: A Couple's Guide to Living Childless by Choice.* Berkeley, CA: Seal Press.

Shikina, Rob. 2011. "Bill Mandates 'Fixing' of Cats, Dogs before Sale." *Star Advertiser,* April 17. Available at www.staradvertiser.com

Smith, Gar. 2009. "Planet Girth." *Earth Island Journal* 24(2):15.

Social Security Administration. 2013 (July). "Monthly Statistical Snapshot." Available at www.ssa.gov

Social Security Trustees. 2013. *The 2013 Annual Report of the Board of Trustees of the Federal Old Age and Survivors Insurance and Federal Disability Insurance Trust Funds.* Washington DC: U.S. Government Printing Office.

Sonfield, Adam. 2011 (Winter). "The Case for Insurance Coverage of Contraceptive Services and Supplies without Cost-Sharing." *Guttmacher Policy Review* 14(11):7–15.

SUNY College of Environmental Science and Forestry. 2009. "Worst Environmental Problem? Overpopulation, Experts Say." *ScienceDaily,* April 20. Available at www.sciencedaily.com

Szinovacz, Maximiliane E. 2011. "Introduction: The Aging Workforce: Challenges for Societies, Employers, and Older Workers." *Journal of Aging & Social Policy* 23(2):95–100.

United Nations. 2013. *World Population Prospects: The 2012 Revision.* Available at www.un.org/esa/population/publications/publications.htm

United Nations. 2012. *Population Ageing and Development 2012.* Available at www.un.org

United Nations Population Division. 2009. "What Would It Take to Accelerate Fertility Decline in the Least Developed Countries?" United Nations Population Division Policy Brief No. 2009/1. New York: United Nations.

Weeks, John R. 2012. *Population: An Introduction to Concepts and Issues,* 11th ed. Belmont, CA: Wadsworth, Cengage Learning.

Weiland, Katherine. 2005. *Breeding Insecurity: Global Security Implications of Rapid Population Growth.* Washington, DC: Population Institute.

Women's Studies Project. 2003. *Women's Voices, Women's Lives: The Impact of Family Planning.* Family Health International. Available at www.fhi.org

World Health Organization. 2013 (May). "Family Planning." Fact Sheet No. 351. Available at www.who.int

Chapter 13

American Lung Association. 2013. *State of the Air: 2013.* Available at lungaction.org

Aukema, J. E., B. Leung, K. Kovacs, C. Chivers, and K. O. Britton, et al. 2011. "Economic Impacts of Non-Native Forest Insects in the Continental United States." *PLoS ONE* 6(9):e24587. Available at www.plosone.org

Bechtel, Michael M., and Kenneth F. Scheve. 2013 (July 25; early online publication). "Mass Support for Global Climate Agreements Depends on Institutional Design." *PNAS.* Available at www.pnas.org

Beckel, Michael. 2010. "Congressmen Maintain Massive Portfolio of Oil and Gas Investments." *Open Secrets Blog,* August 27. Available at www.opensecrets.org

Beinecke, Frances. 2009 (Spring). "Debunking the Myth of Clean Coal." *OnEarth.* Available at www.onearth.org

Betts, Kellyn S. 2011. "Plastics and Food Sources: Dietary Intervention to Reduce BPA and DEHP." *Environmental Health Perspectives* 119(7): A306.

Blatt, Harvey. 2005. *America's Environmental Report Card: Are We Making the Grade?* Cambridge, MA: MIT Press.

Blunden, Jessica, and Derek S. Arndt. 2013. "State of the Climate in 2012." *Bulletin of the American Meteorological Society* 94(8):S1–S258.

Bogard, Paul. 2013 (August 19). "Bringing Back the Night: A Fight Against Light Pollution." *Yale Environment 360.* Available at e360.yale.edu

Bradshaw, Nancy. 2010 (March). *Fragrance-Free Policy Management Presentation.* Women's College Hospital, University of Toronto.

Brecher, Jeremy. 2011 (January 4). "Climate Protection Strategy: Beyond Business as Usual." Labor Network for Sustainability. Available at www.labor4sustainability.org

BP. 2011. *BP Statistical Review of World Energy.* Available at www.bp.com

Broder, John M. 2013 (February 13). "Keystone XL Protesters Seized at White House." *Green (New York Times* blog). Available at http://green.blogs.nytimes.com

Brody, Julia Gree, Kirsten B. Moysich, Olivier Humblet, Kathleen R. Attfield, Gregory P. Beehler, and Ruthann A. Rudel. 2007. "Environmental Pollutants and Breast Cancer." *Cancer* 109(S12):2667–2711.

Brown, Lester R. 2007. "Distillery Demand for Grain to Fuel Cars Vastly Understated: World May Be Facing Highest Grain Prices in History." *Earth Policy News,* January 4. Available at www.earthpolicy.org

Brown, Lester R., and Jennifer Mitchell. 1998. "Building a New Economy." In *State of the World 1998,* Lester R. Brown, Christopher Flavin, and Hilary French, eds. (pp. 168–187). New York: W. W. Norton.

Brulle, Robert J. 2009. "U.S. Environmental Movements." In *Twenty Lessons in Environmental Sociology,* Kenneth A. Gould and Tammy L. Lewis, eds. (pp. 211–227). New York: Oxford University Press.

Bruno, Kenny, and Joshua Karliner. 2002. *Earthsummit.biz: The Corporate Takeover of Sustainable Development.* CorpWatch and Food First Books. Available at www.corpwatch.org

Bullard, Robert D., Paul Mohai, Robin Saha, and Beverly Wright. 2007 (March). *Toxic Wastes and Race at Twenty 1987–2007.* Cleveland, Ohio: United Church of Christ.

Caress, Stanley M., and Anne C. Steinemann. 2004. "A National Population Study of the Prevalence of Multiple Chemical Sensitivity." *Archives of Environmental Health* 59(6):300–305.

Carlson, Scott. 2006. "In Search of the Sustainable Campus." *The Chronicle of Higher Education* LIII(9):A10–A12, A14.

Chafe, Zoe. 2006. "Weather-Related Disasters Affect Millions." In *Vital Signs*, L. Starke ed. (pp. 44–45). New York: W. W. Norton & Co.

Chafe, Zoe. 2005. "Bioinvasions." In *State of the World 2005*, L. Starke, ed. (pp. 60–61). New York: W. W. Norton.

Cheeseman, Gina-Marie. 2007. "Plastic Shopping Bags Being Banned." *The Online Journal*, June 27. Available at www.onlinejournal.com

Chepesiuk, Ron. 2009. "Missing the Dark: Health Effects of Light Pollution." *Environmental Health Perspectives* 117(1):A20–A27.

Cincotta, Richard P., and Robert Engelman. 2000. *Human Population and the Future of Biological Diversity*. Washington, DC: Population Action International.

Clarke, Tony. 2002. "Twilight of the Corporation." In *Social Problems, Annual Editions 02/03*, 30th ed., Kurt Finster-Busch, ed. (pp. 41–45). Guilford, CT: McGraw-Hill/Dushkin.

Clemmitt, Marcia. 2011. "Nuclear Power." *CQ Researcher* 21(22):all.

Cook, John, Dana Nuccitelli, Sarah A. Green, Mark Richardson, Barbel Winkler, Rob Painting, Robert Way, Peter Jacobs, and Andrew Skuce. 2013. "Quantifying the Consensus on Anthropogenic Global Warming in the Scientific Literature." *Environmental Research Letters* 8(2)1–7.

Cooper, Arnie. 2004. "Twenty-Eight Words that Could Change the World: Robert Hinkley's Plan to Tame Corporate Power." *The Sun* 345 (September):4–11.

Coyle, Kevin. 2005. *Environmental Literacy in America*. Washington, DC: The National Environmental Education and Training Foundation.

Dahl, Richard. 2013. "Cooling Concepts: Alternatives to Air Conditioning for a Warm World." *Environmental Health Perspectives* 121(1): A18–A125.

Denson, Bryan. 2000. "Shadowy Saboteurs." *IRE Journal* 23(May-June):12–14.

Edwards, Bob, and Adam Driscoll. 2009. "From Farms to Factories: The Environmental Consequences of Swine Industrialization in North Carolina." In *Twenty Lessons in Environmental Sociology*, Kenneth A. Gould and Tammy L. Lewis, eds. (pp. 153–175). New York: Oxford University Press.

Ehrlich, Paul R., and Anne H. Ehrlich. 2013. "Can a Collapse of Global Civilization Be Avoided?" *Proceedings of the Royal Society B* 280: 20122845. Available at http://dx.doi .org/10.1098/rspb.2012.2845

Elk, Mike. 2011 (September 5). "Which is More Likely to Rebuild the Labor Market: Environmental Allies or 6,000 Temp Jobs?" *Working in These Times*. Available at www .inthesetimes.com

Energy Information Administration. 2013 (September). *Monthly Energy Review*. Available at www.eia.gov

Environmental Working Group. 2005 (July 14). *Body Burden: The Pollution in Newborns*. Available at www.ewg.org

EPA (U.S. Environmental Protection Agency). 2013a. *National Priorities List (NPL)*. Available at www.epa.gov/superfund/sites

EPA. 2013b. *Municipal Solid Waste in the United States: 2011 Facts and Figures*. Available at www.epa.gov

EPA. 2004. *What You Need to Know about Mercury in Fish and Shellfish*. Available at www.epa.gov

Ewing, B., D. Moore, S. Goldfinger, A. Oursler, A. Reed, and M. Wackernagel. 2010. *The Ecological Footprint Atlas 2010*. Oakland: Global Footprint Network.

Fisher, Brandy E. 1999. "Focus: Most Unwanted." *Environmental Health Perspectives* 107(1). Available at http://ehpnet1.niehs.nih.gov/ docs/1999/107-1/focus-abs.html

Flavin, Chris, and Molly Hull Aeck. 2005 (September 15). "Cleaner, Greener, and Richer." Tom.Paine.com. Available at www.tompaine. com/articles/2005/09/15/cleaner_greener_and _richer.php

Food and Drug Administration. 2013. *Pesticide Residue Monitoring Program Results and Discussion FY 2009*. Available at www.fda.gov

Food & Water Watch and Network for New Energy Choices. 2007. *The Rush to Ethanol: Not All Biofuels Are Created Equal*. Available at www .newenergychoices.org

Foster, Joanna. 2011. "Impact of Gulf Spill's Underwater Dispersants Examined." *New York Times*, August 26. Available at nytimes.com

French, Hilary. 2000. *Vanishing Borders: Protecting the Planet in the Age of Globalization*. New York: W. W. Norton.

Gallup Organization. 2013. *Environment*. Available at www.gallup.com/poll

Gardner, Gary. 2005. "Forest Loss Continues." In *Vital Signs 2005*, Linda Starke, ed. (pp. 92–93). New York: W. W. Norton.

Global Footprint Network. 2013. "August 20th is Earth Overshoot Day." Press release. Available at www.footprintnetwork.org

Global Footprint Network. 2010. *2010 Annual Report*. Oakland, CA: Global Footprint Network. Available at www.footprintnetwork.org

Global Humanitarian Forum. 2009. *Human Impact Report: Climate Change—The Anatomy of a Silent Crisis*. Geneva: Global Humanitarian Forum.

Global Invasive Species Database. 2013. "Felis Catus." Available at www.issg.org

Goodstein, Laurie. 2005. "Evangelical Leaders Swing Influence behind Effort to Combat Global Warming." *New York Times*, March 10. Available at www.nytimes.com

Gottlieb, Roger S. 2003a. "Saving the World: Religion and Politics in the Environmental Movement." In *Liberating Faith*, Roger S. Gottlieb, ed. (pp. 491–512). Lanham, MD: Rowman & Littlefield.

Gottlieb, Roger S. 2003b. "This Sacred Earth: Religion and Environmentalism." In *Liberating Faith*, Roger S. Gottlieb, ed. (pp. 489–490). Lanham, MD: Rowman & Littlefield.

Greenpeace USA. 2013. *Dealing in Doubt: The Climate Denial Machine Vs. Climate Science*. Available at www.greenpeace.org

Griswold, Eliza. 2012. "How 'Silent Spring' Ignited the Environmental Movement." *New York Times*, September 21. Available at www .nytimes.com

Gunther, Marc. 2013 (September 3). "With Rooftop Solar on Rise, U.S. Utilities are Striking Back." *Environment 360*. Available at e360.yale.edu

Hager, Nicky, and Bob Burton. 2000. *Secrets and Lies: The Anatomy of an Anti-Environmental PR Campaign*. Monroe, ME: Common Courage Press.

Hefling, Kimberly. 2007. "Hearing Planned Today in Lejeune Water Case." *Marine Corps Times*, June 12. Available at www.marinecorpstimes .com

Hilgenkamp, Kathryn. 2005. *Environmental Health: Ecological Perspectives*. Sudbury, MA: Jones and Bartlett Publishers.

Horn, Steve. 2012 (March 21). "ALEC Climate Change Denial Model Bill Passes in Tennessee." *Desmogblog.com*. Available at www .Desmogblog.com

Hunter, Lori M. 2001. *The Environmental Implications of Population Dynamics*. Santa Monica, CA: Rand Corporation.

Intergovernmental Panel on Climate Change. 2013. *Climate Change 2013: The Physical Science Basis*. United Nations Environmental Programme and the World Meteorological Organization. Available at www.climatechange2013.org

Intergovernmental Panel on Climate Change. 2007. *Climate Change 2007: Impacts, Adaptation and Vulnerability*. United Nations Environmental Programme and the World Meteorological Organization. Available at www.ipcc.ch

The International Programme on the State of the Ocean. 2013. *The State of the Oceans 2013: Perils, Prognoses and Proposals*. Available at www .stateoftheocean.org

IUCN. 2013. *The IUCN Red List of Threatened Species, 2013.1*. Available at www.iucnredlist.org

Jamail, Dahr. 2011. "Fukushima Radiation Alarms Doctors." *Common Dreams*, August 20. Available at www.commondreams.org

Janofsky, Michael. 2005. "Pentagon Is Asking Congress to Loosen Environmental Laws." *New York Times*, May 11. Available at www.nytimes.com

Kaplan, Sheila, and Jim Morris. 2000. "Kids at Risk." *U.S. News and World Report*, June 19, pp. 47–53.

Kessler, Rebecca. 2013. "Sunset for Leaded Aviation Gasoline?" *Environmental Health Perspectives* 121(2):A54–A57.

Kiger, Patrick J. 2013 (August 7). "Fukushima's Radioactive Water Leak: What You Should Know." *National Geographic News*. Available at news.nationalgeographic.com

Kitasei, Saya. 2011. "Wind Power Growth Continues to Break Records Despite Recession." In *Vital Signs*, Linda Starke, ed. (pp. 26–28). Washington DC: Worldwatch Institute.

Knoell, Carly. 2007 (August 9). "Malaria: Climbing in Elevation as Temperature Rises." *Population Connection*. Available at www .populationconnection.org

Kumar, Supriya. 2013 (March 13). "The Looming Threat of Water Scarcity." *Vital Signs*. World Watch Institute. Available at vitalsigns. worldwatch.org

Lamm, Richard. 2006. "The Culture of Growth and the Culture of Limits." *Conservation Biology* 20(2):269–271.

Leahy, Stephen. 2009 (May 21). "Alien Species Eroding Ecosystems and Livelihoods." Interpress Service News Agency. Available at www.ipsnews.net

Leitzell, Katherine. 2011 (May 3). "When Will the Arctic Lose its Sea Ice?" National Snow and Ice Data Center. Available at nsidc.org

Little, Amanda Griscom. 2005. "Maathai on the Prize: An Interview with Nobel Peace Prize Winner Wangari Maathai." *Grist Magazine,* February 15. Available at www.grist.org

Lunden, Jennifer. 2013. "Exposed." *Orion Magazine* (September/October). Available at www.orionmagazine.com.

Malcolm, Jay R., Canran Liu, Ronald P. Neilson, Lara Hansen, and Lee Hannah. 2006. "Global Warming and Extinctions of Endemic Species from Biodiversity Hotspots." *Conservation Biology* 20(2):538–548.

Malewitz, Jim. 2013 (February 8). "Northeastern States Drastically Cut Emissions Cap." *Stateline* (The Daily News Service of The Pew Charitable Trust). Available at www.pewstates.org

McAllister, Lucy. 2013. "The Human and Environmental Effects of E-Waste." Population Reference Bureau. Available at www.prb.org

McCarthy, Michael. 2008. "Cleared: Jury Decides that Threat of Global Warming Justifies Breaking the Law." *The Independent*, September 11. Available at www.independent.co.uk

McCormick, James. 2011. "Nuclear Illinois Helped Shape Obama View of Energy in Dealings with Exelon." *Bloomberg,* March 23. Available at www.bloomberg.com

McDaniel, Carl N. 2005. *Wisdom for a Livable Planet.* San Antonio, TX: Trinity University Press.

McGinn, Anne Platt. 2000. "Endocrine Disrupters Raise Concern." In *Vital Signs 2000,* Lester R. Brown, Michael Renner, and Brian Halweil, eds. (pp. 130–131). New York: W. W. Norton.

McKibben, Bill. 2008 (October 24). "Meltdown: A Global Warming Travelogue." CNN.com. Available at www.cnn.com

McMichael, Anthony J., Kirk R. Smith, and Carlos F. Corvalan. 2000. "The Sustainability Transition: A New Challenge." *Bulletin of the World Health Organization* 78(9):1067.

Millennium Ecosystem Assessment. 2005. *Ecosystems and Human Well-Being: Synthesis.* Washington, DC: Island Press.

Miller, G. Tyler, Jr., and Scott E. Spoolman. 2009. *Living in the Environment,* 16th ed. Belmont, CA: Brooks/Cole, Cengage Learning.

Miranda, M. L., D. A. Hastings, J. E. Aldy, and W. H. Schlesinger. 2011. "The Environmental Justice Dimensions of Climate Change." *Environmental Justice* 4(1):17–25.

Mulrow, John, and Alexander Ochs (with Shakuntala Makhijani). 2011. "Glacial Melt and Ocean Warming Drive Sea Level Upward." In *Vital Signs,* Linda Starke, ed. (pp. 43–46). Washington DC: Worldwatch Institute.

Murtaugh, Paul, and Michael Schlax. 2009. "Reproduction and the Carbon Legacies of Individuals." *Global Environmental Change* 19:14–20.

Nader, Ralph. 2013 (October 14). "Why Atomic Energy Stinks Worse Than You Thought." *Counterpunch*. Available at www.counterpunch.org

National Commission on the BP Deepwater Horizon Oil Spill and Offshore Drilling. 2011. *Deepwater: The Gulf Oil Disaster and the Future of Offshore Drilling.* Available at www.gpo.gov/fdsys/pkg/GPO-OILCOMMISSION/contentdetail.html

National Geographic. 2007 (June 5). "Top Ten Tips to Fight Global Warming." *Green Guide.* Available at www.thegreenguide.com

NRDC (Natural Resources Defense Council). 2011 (April). *The BP Oil Disaster at One Year: A Straightforward Assessment of What We Know, What We Don't, and What Questions Need to Be Answered.* Available at www.nrdc.org

Normander, Bo. 2011 (February 23). "World's Forests Continue to Shrink." *Vital Signs.* World Watch Institute. Available at vitalsigns.worldwatch.org

"Nukes Rebuked." 2000. *Washington Spectator* 26(13):4.

OpenSecrets.org. 2013. "Top Interest Groups Giving to Members of Congress, 2012 Cycle." Available at www.opensecrets.org

Pearce, Fred. 2005. "Climate Warming as Siberia Melts." *New Scientist,* August 11. Available at www.NewScientist.com

Pew Center on the States. 2009. *State of the States 2009.* Available at www.stateline.org

Pew Research. 2013 (June 24). "Climate Change and Financial Instability Seen as Top Global Threats." Available at www.pewglobal.org

Pimentel, D., S. Cooperstein, H. Randell, D. Filiberto, S. Sorrentino, B. Kaye, C. Nicklin, J. Yagi, J. Brian, J. O'Hern, A. Habas, and C. Weinstein. 2007. "Ecology of Increasing Diseases: Population Growth and Environmental Degradation." *Human Ecology* 35(6):653–668.

Prah, Pamela M. 2007. "States Forge Ahead on Immigration, Global Warming." *Stateline,* July 30. Available at www.stateline.org

Price, Tom. 2006. "The New Environmentalism." *CQ Researcher* 16(42):987–1007.

Public Citizen. 2005 (February). *NAFTA Chapter 11 Investor-to-State Cases: Lessons for the Central America Free Trade Agreement. Public Citizens Global Trade Watch Publication E9014.* Available at www.citizen.org

Renner, Michael. 2004. "Moving toward a Less Consumptive Economy." In *State of the World 2004,* Linda Starke, ed. (pp. 96–119). New York: W. W. Norton.

Ridlington, Elizabeth, and John Rumpler. 2013. *Fracking by the Numbers: Key Impacts of Dirty Drilling at the State and National Level.* Environment America Research & Policy Center. Available at www.environmentamerica.org

Rifkin, Jeremy. 2004. *The European Dream: How Europe's Vision of the Future Is Quietly Eclipsing the American Dream.* New York: Tarcher/Penguin.

Rogers, Sherry A. 2002. *Detoxify or Die.* Sarasota, FL: Sand Key.

Saad, Lydia. 2013 (April 8). "Americans' Concerns about Global Warming on the Rise." Gallup, Inc. Available at www.gallup.com

Sawin, Janet. 2004. "Making Better Energy Choices." In *State of the World 2004,* Linda Starke, ed. (pp. 24–43). New York: W. W. Norton.

Scavia, Donald. 2011 (September 2). "Dead Zones in Gulf of Mexico and Other Waters Require a Tougher Approach: Donald Scavia." *Nola.com.* Available at www.nola.com

Schapiro, Mark. 2007. *Exposed: The Toxic Chemistry of Everyday Products and What's at Stake for American Power.* White River Junction, Vermont: Chelsea Green Publishing.

Schapiro, Mark. 2004. "New Power for 'Old Europe.'" *The Nation,* December 27, pp. 11–16.

Schulze, Karin. 2012 (December 23). "Plastic Chokes Oceans and Trashes Beaches." *ABC News.* Available at http://abcnews.go.com

Schwartz-Nobel, Loretta. 2007. *Poisoned Nation.* New York: St. Martin's Press.

Shapiro, Isaac, and John Irons. 2011. *Regulation, Employment, and the Economy.* EPI Briefing Paper #305. Washington, DC. Economic Policy Institute.

Shrank, Samuel. 2011. "Growth of Biofuel Production Slows." In *Vital Signs,* Linda Starke, ed. (pp. 16–18). Washington DC: Worldwatch Institute.

Sinks, Thomas. 2007 (June 12). *Statement by Thomas Sinks, PhD, Deputy Director, Agency for Toxic Substances and Disease Registry on ATSDR's Activities at U.S. Marine Corps Base Camp Lejeune before Committee on Energy and Commerce Subcommittee on Oversight and Investigations United States House of Representatives.* Available at www.hhs.gov

Staudinger, Michelle D., Nancy B. Grimm, Amanda Staudt, Shawn L. Carter, F. Stuart Chapin III, Peter Kareiva, Mary Ruckelshaus, Bruce A. Stein. 2012. *Impacts of Climate Change on Biodiversity, Ecosystems, and Ecosystem Services: Technical Input to the 2013 National Climate Assessment.* Available at assessment.globalchange.gov

Stoll, Michael. 2005 (September-October). "A Green Agenda for Cities." *E Magazine* 16(5). Available at www.emagazine.com

Sustainable Endowments Institute. 2011. *Greening the Bottom Line: The Trend toward Green Revolving Funds on Campus.* Cambridge MA: Sustainable Endowments Institute.

Swift, Anthony, Susan Casey-Lefkowitz, and Elizabeth Shope. 2011. *Tar Sands Pipelines Safety Risks.* Natural Resources Defense Council. Available at www.nrdc.org

Takada, Dr. Hideshige. 2013 (May 10). *Microplastics and the Threat to Our Seafood.* Ocean Health Index. Available at www.oceanhealthindex.org

TerraChoice Group Inc. 2009. *The Seven Sins of Greenwashing: Environmental Claims in Consumer Markets.* Available at http://sinsofgreenwashing.org

UNDP (United Nations Development Programme). 2013. *Human Development Report 2013.* Available at hdr.undp.org

United Nations Development Programme. 2007. *Human Development Report 2007/2008: Fighting Climate Change: Human Solidarity in a Divided World.* New York: Palgrave Macmillan.

UNEP. 2013. *UNEP Yearbook 2013: Emerging Issues in Our Global Environment.* Available at www.unep.org

UNEP. 2007. *GEO Yearbook 2007: An Overview of Our Changing Environment.* Available at www.unep.org

U.S. Department of Health and Human Services. 2011. *12th Report on Carcinogens.* Washington, DC: Public Health Service.

U.S. Geological Survey. 2007 (September 7). "Future Retreat of Arctic Ice Will Lower Polar Bear Populations and Limit Their Distribution." *USGS Newsroom.* Available at www.usgs.gov

Vedantam, Shankar. 2005a. "Nuclear Plants Not Keeping Track of Waste." *Washington Post,* April 19. Available at www.washingtonpost.com

Vedantam, Shankar. 2005b. "Storage Plan Approved for Nuclear Waste." *Washington Post,* September 10. Available at www.washingtonpost.com

Wald, Matthew L. 2013. "Ex-Regulator Says Reactors are Flawed." *New York Times,* April 8. Available at www.nytimes.com

Westerling, A. L., H. G. Hidalgo, D. R. Cayan, and T. W. Swetnam. 2006. "Warming and Earlier Spring Increase Western U.S. Forest Wildfire Activity." *Science* 313 (5789):940–943.

White House. 2012 (August 28). "Obama Administration Finalizes Historic 54.5 MPG Fuel Efficiency Standard." Office of the Press Secretary. Available at www.whitehouse.gov

Wire, Thomas. 2009 (August.) "Fewer Emitters, Lower Emissions, Less Cost." Optimum Population Trust. Available at www.optimumpopulation.org

Woodward, Colin. 2007. "Curbing Climate Change." *CQ Global Researcher* 1(2):27–50. Available at www.globalresearcher.com

World Nuclear Association. 2013. "Nuclear Basics." Available at world-nuclear.org

World Health Organization. 2012. "Indoor Air Pollution and Health." Fact Sheet No. 292. Available at www.who.int

World Resources Institute. 2000. *World Resources 2000–2001: People and Ecosystems—The Fraying Web of Life.* Washington, DC: World Resources Institute.

World Water Assessment Program. 2009. *World Water Development Report 3: Water in a Changing World.* Available at www.unesco.org

WWF (World Wildlife Federation). 2012. *Living Planet Report,* 9th ed. World Wildlife Fund, Zoological Society of London, Global Footprint Network, and the European Space Agency. Available at www.panda.org

Zelman, Joanna. 2011. "50 Million Environmental Refugees by 2020, Experts Predict." *Huffington Post,* February 22. Available at www.huffingtonpost.com

Chapter 14

AAEM (American Academy of Environmental Medicine). 2009 (May 8). "Genetically Modified Foods." Available at www.aaemonline.org

ACLU (American Civil Liberties Union). 2013 (June 13). "2013 Supreme Court Invalidates Breast and Ovarian Cancer Genes." Available at www.aclu.org

Anderson, Ross, Chris Barton, Rainer Bohme, Richard Clayton, Michel J. G. van Eeten,

Michael Levi, Tyler Moore, and Stefan Savage. 2012. "Measuring the Cost to Cybercrime." June 25–26. Workshop on the Economics of Information Security. Berlin, Germany.

Associated Press. 2013. "Protesters Across Globe Rally Against Monsanto." *The Denver Post,* May 25. Available at www.denverpost.com

Associated Press. 2010. "China's Internet Censorship." CBS News, January 11. Available at www.cbsnews.com

Atkinson, Robert D., and Daniel D. Castro. 2008 (October). "Digital Quality of Life: Understanding the Personal and Social Benefits of the Information Technology Revolution." The Information and Technology Foundation. Available at www.itif.org

Attewell, Paul, Belkis Suazo-Garcia, and Juan Battle. 2003. "Computers and Young Children: Social Benefit or Social Problem?" *Social Forces* 82:277–296.

Ball, James, Julian Borger, and Glenn Greenwood. 2013. "Revealed: How U.S. and U.K. Spy Agencies Defeat Internet Privacy and Security." *The Guardian.* September 5. Available www.guardian.com

Bell, Daniel. 1973. *The Coming of Post-Industrial Society: A Venture in Social Forecasting.* New York: Basic Books.

Beniger, James R. 1993. "The Control Revolution." In *Technology and the Future,* Albert H. Teich, ed. (pp. 40–65). New York: St. Martin's Press.

Bollier, David. 2009 (July 22). "Deadly Medical Monopolies." *On the Commons.* Available at onthecommons.org

Botelho, Greg. 2013. "Judge Orders New Samsung, Apple Faceoff; Strikes $450 Million in Damages." CNN, March 2. Available at www.cnn.com

Bowman, Lee. 2011. "Animal Testing: Biomedical Researchers Using Millions of Animals Yearly." Scripps Howard News Service, May 7.

Brynjolfsson, Erik, and Andrew McAfee. 2012. *Race against the Machine: How the Digital Revolution Is Accelerating Innovation, Driving Productivity, and Irreversibly Transforming Employment and the Economy.* Research Brief. January. The MIT Center for Digital Business. Available at ebusiness.mit.edu

Buchanan, Allen, Dan Brock, Norman Daniels, and Daniel Wikler. 2000. *From Chance to Choice: Genetics and Justice.* New York: Cambridge University Press.

Bumiller, Elizabeth, and Thom Shanker. 2012. "Panetta Warns of Dire Threat of Cyberattack on U.S." *The New York Times,* October 11. Available at www.nytimes.com

Bush, Corlann G. 1993. "Women and the Assessment of Technology." In *Technology and the Future,* Albert H. Teich, ed. (pp. 192–214). New York: St. Martin's Press.

Carr, Nicholas. 2010. *The Shallows: What the Internet Is Doing to Our Brains.* New York: W. W. Norton and Company, Inc.

Castro, Daniel. 2013. "Health IT 2013: A Renewed Focus on Efficiency and Effectiveness." *Electronic Health Reporter* (February 4). Available electronichealthreporter.com

Ceruzzi, Paul. 1993. "An Unforeseen Revolution." In *Technology and the Future,* Albert H. Teich, ed. (pp. 160–174). New York: St. Martin's Press.

Chan, Sewell. 2007. "New Scanners for Tracking City Workers." *New York Times,* January 23. Available at www.nytimes.com

Chen, Shirong. 2011. "China Tightens Internet Censorship Controls." British Broadcasting Corporation. Available at www.bbc.co.uk

Clarke, Adele E. 1990. "Controversy and the Development of Reproductive Sciences." *Social Problems* 37(1):18–37.

Cohen, S. P. 2002. "Can Pets Function as Family Members?" *Western Journal of Nursing Research* 24:621–638.

Colitt, Raymond, and Amaldo Galvao. 2013. "Rousseff Calls Off U.S. Visit Over NSA Surveillance." *Bloomberg News,* September 17. Available at www.bloomberg.com

Condon, Bernard, and Paul Wiseman. 2013. "Millions of Middle-Class Jobs Killed by Machines in Great Recessions Wake." *The Huffington Post,* January 23. Available at www.huffingtonpost.com

Crichton, Michael. 2007. "Patenting Life." *New York Times,* February 13. Available at www.nytimes.com

Dandekar, Pranav, Ashish Goel, and David T. Lee. 2013 (February 28). "Biased Assimilation, Homophily, and the Dynamics of Polarization." *Proceedings of the National Academy of Sciences of the United States of America.* Available at www.pnas.org

David-Ferdon, Corrine, and Marci Feldman Hertz. 2009. *Electronic Media and Youth Violence: A CDC Issue Brief for Researchers.* Atlanta, GA: Centers for Disease Control.

Davies, Michael J., Vivienne M. Moore, Kristyn J. Willson, Phillipa Van Essen, Kevin Priest, Heather Scott, Eric A. Hann, and Annabelle Chan. 2012. "Reproductive Technologies and the Risk of Birth Defects." *New England Journal of Medicine* 366:1803–1813.

Della Cava, Marco R. 2010. "Some Ditch Social Networks to Reclaim Time, Privacy." *USA Today,* February 10. Available at www.usatoday.com

DeNoon, Daniel. 2005 (March 8). *Study: Computer Design Flaws May Create Dangerous Hospital Errors.* Available at my.webmd.com

Department of Health and Human Services. 2009 (April 6). "The Genetic Information Nondiscrimination Act of 2008: Information for Researchers and Health Care Professionals." Available at www.genome.gov

Dewan, Shaila. 2010. "To Court Blacks, Foes of Abortion Make Racial Case." *The New York Times,* February 26. Available at www.nytimes.com

Dietrich, David R. 2013. "Avatars of Whiteness: Racial Expression in Video Game Characters." *Sociological Inquiry* 83(1):82–105.

Digest of Education Statistics. 2013. National Center for Education Statistics. Available at nces.ed.gov

Dignan, Larry. 2010. "Offshoring's Toll: IT Departments to Endure Jobless Recovery through 2014." *ZDNet,* November 18. Available at www.zdnet.com/

Duggan, Maeve, and Joanna Brenner. 2013 (February 14). "The Demographics of Social Media Users—2012." Pew Research Center. Available at www.pewresearch.org

Durkheim, Emile. 1973/1925. *Moral Education.* New York: Free Press.

Dutta, Soumitra, and Irene Mia. 2011. *The Global Information Technology Report 2010–2011.* World Economic Forum. Available at reports. weforum.org

Efrati, Amir. 2013. "Google to Find, Develop Wireless Networks in Emerging Markets." *The Wall Street Journal,* May 24. Available at online .wsj.com

Eibert, Mark D. 1998. "Clone Wars." *Reason* 30(2):52–54.

Eilperin, Juliet, and Rick Weiss. 2003. "House Votes to Prohibit All Human Cloning." *Washington Post,* February 28. Available at www.washingtonpost.com

Eisenberg, Anne. 2009. "Better Vision, with a Telescope Inside the Eye." *The New York Times,* July 19. Available at www.nytimes.com

ESA (Entertainment Software Association). 2013. "Essential Facts about the Computer and Video Game Industry." Available at www .theesa.com

ETC Group. 2003. *Contamination by Genetically Modified Maize in Mexico Much Worse Than Feared.* Available at www.etcgroup.org

Erickson, Jim. 2011 (June 9). "Banning Federal Funding for Human Embryonic Stem Cell Research Would Derail Related Work, U-M Researcher and Colleagues Conclude." U-M New Service. University of Michigan. Available at ns.umich.edu

FCC (Federal Communications Commission). 2011 (September 23). "Preserving the Open Internet: Final Rule." *Federal Register* 76(185): 59192–59235.

Ferdowsian, Hope. 2010 (November 7). "Animal Research: Why We Need Alternatives." *The Chronicle of Higher Education.* Available at chronicle.com

File, Thom. 2013. *Computer and Internet use in the United States.* May. U.S. Bureau of the Census. Publication No. 20-569. Washington, DC: U.S. Government Printing Office.

Fischer, Eric A. 2013 (June 20). *Federal Laws Relating to Cyber Security: Overview and Discussion of Proposed Revisions.* Congressional Research Service. Available at www .fas.org

Flecknell, Paul A. 2010. "Do Mice Have a Pain Face?" *Nature Methods* 7:437–438.

Fox, Susannah. 2013 (August 7). "51% of U.S. Adults Bank Online." Pew Internet and the American Life Project. Available at www .pewinternet.org

Fox, Susannah, and Maeve Duggan. 2013 (January 15). "Online Health 2013." Pew Internet and the American Life Project. Available at www .pewinternet.org

FTC (Federal Trade Commission). 2013 (March 1). "Top Complaint to the FTC? ID Theft, Again." Washington, DC. Available at www.consumer .ftc.gov

Gartner Research. 2013. "Gartner Says Declining Worldwide PC Shipments in Fourth Quarter of 2012 Signal Structural Shift in PC Market." Gartner Newsroom, January 14. Available at www.gartner.com

Genetics and Public Policy Center. 2010. "Frequently Asked Questions." Johns Hopkins University, Berman Institute of Bioethics. Washington, DC.

Genomics Law Report. 2011. "Myriad Gene Patent Litigation." A Publication of Robinson Bradshaw and Hinson. Available at www .genomicslawreport.com

Ghafour, Hamida. 2012. "Facebook Mix-Up Forced Iranian Woman to Flee from Her Life." *The Toronto Star,* October 11. Available at www .thestar.com

Goodman, Paul. 1993. "Can Technology Be Humane?" In *Technology and the Future,* Albert H. Teich, ed. (pp. 239–255). New York: St. Martin's Press.

Government Tracks. 2013. "Abortion." Available at www.govtrack.us

Greenhouse, Linda. 2007. "Justices Back Ban on Method of Abortion." *New York Times,* April 19. Available at www.nytimes.com

Greenwald, Glenn. 2013. "XKeyscore: NSA Tool Collects 'Nearly Everything a User Does on the Internet.'" *The Guardian,* July 31. Available at www.theguardian.com

Grossman, Lev. 2006. "Time's Person of the Year: You." *Time,* December 13. Available at www .time.com

Guttmacher Institute. 2013. (September 1). "State Policies in Brief: An Overview of Abortion Laws." Available at www.guttmacher.org

Hanson, Lawrence A. 2010 (November 7). "Animal Research: Groupthink in Both Camps." *The Chronicle of Higher Education.* Available at chronicle.com

Harmon, Amy. 2006. "That Wild Streak? Maybe It Runs in the Family." *New York Times,* June 15. Available at www.nytimes.com

Henry, Bill, and Roarke Pulcino. 2009. "Individual Differences and Study-Specific Characteristics Influencing Attitudes about the Use of Animals in Medical Research." *Society and Animals* 17:305–324.

Hinduja, Sameer, and Justin W. Patchin. 2013. "State Cyberbullying Laws." July. Cyberbullying Research Center. Available at www .cyberbullying.us

Holland, Earle. 2010 (November 7). "Animal Research: Activists' Wishful Thinking, Primitive Reasoning." *The Chronicle of Higher Education* Available at chronicle.com

Holt, Thomas J., and Max Kilger. 2012. "Examining Willingness to Attack Critical Infrastructure Online and Off-Line." *Crime and Delinquency* 58(5):798–822.

Honigman, Ryan. 2013. "100 Fascinating Social Media Statistics and Figures from 2012." The Huffington Post, November 29. Available at www.huffingtonpost.com

Houlahan, Brent. 2006. (September 18) "Telepresence Defined by Brent Houlahan with HSL's Thoughts and Analysis." Human Productivity Lab. Available at www.humanproductivitylab.com

Human Cloning Prohibition Act of 2007. U.S. House of Representatives, Washington, DC. Available at www.govtrack.us

Human Genome Project. 2007. *Medicine and the New Genetics.* Available at www.ornl.gov

ITIF (Information Technology and Innovation Foundation). 2012 (December). The 2012 *State New Economy Index.* Available at www2.itif.org

ITIF. 2011 (July). "The Atlantic Century II." European-American Business Council (February). Available at www.itif.org

ITIF. 2009. "Benchmarking EU & U.S. Innovation and Competitiveness." European-American Business Council (February). Available at http://www.itif.org

ICCVAM (Interagency Coordinating Committee on the Validation of Alternative Methods). 2011. "Since You Asked: Alternatives to Animal Testing." Available at www.niehs.nih.gov

IFR (International Federation of Robotics). 2013. "World Robotics 2012 Industrial Robots." Available at www.ifr.org

Internet Statistics. 2013. *Internet Usage Statistics.* Available at www.internetworldstats.com

Internet Statistics. 2011. "Top 20 Countries with the Highest Number of Internet Users." Available at www.internetworldstats.com

Jaschik, Scott. 2008. "If You Text in Class, This Prof Will Leave." *Inside Higher Ed,* April 2. Available at www.insidehighered.com

Jones, Cass. 2013. "Twitter Says 250,000 Accounts Have Been Hacked in Security Breach." *The Guardian,* February 1. Available at www .theguardian.com

Jordan, Tim, and Paul Taylor. 1998. "A Sociology of Hackers." *Sociological Review* 46(4):757–778.

Kahn, A. 1997. "Clone Mammals . . . Clone Man?" *Nature* 386:119.

Kaiser Family Foundation, 2010 (January 20). *Generation M2: Media in the Lives of 8- to 18-Year Olds.* Available at kff.org

Kaplan, Karen. 2009. "Corn Fortified with Vitamins Devised by Scientists." *Los Angeles Times,* April 29. Available at www.latimes.com

Kelly, David. 2013 (February 4). "Study Shows Facebook Unfriending Has Real Off-Line Consequences." *University Communications.* Available at www.ucdenver.edu

Kharfen, Michael. 2006. "1 of 3 and 1 in 6 Pre-Teens Are Victims of Cyber-Bullying." Available at www.fightcrime.org

Klotz, Joseph. 2004. *The Politics of Internet Communication.* Lanham, MD: Rowman and Littlefield.

Konrad, Alex. 2013. "After Security Breach Exposes 2.9 Million Adobe Users, How Safe Is Encrypted Credit Card Data?" *Forbes Magazine,* October 9. Available at www.forbes.com

Kuhn, Thomas. 1973. *The Structure of Scientific Revolutions.* Chicago: University of Chicago Press.

Langer, Gary. 2013. "Poll: Skepticism of Genetically Modified Foods." *ABC News,* June 19. Available at abcnews.go.com

Lawless, Jill. 2013. "Spread of DNA Databases Sparks Ethical Concerns." *Daily Chronicle,* July 12. Available at www.daily-chronicle.com

Legay, F. 2001 (January). "Genetics: Should Genetic Information Be Treated Separately?" *Virtual Mentor,* p. E5. Available at http://virtualmentor.ama-assn.org

Lemonick, Michael. 2006. "Are We Losing Our Edge?" *Time,* February 13, pp. 22–33.

Lemonick, Michael, and Dick Thompson. 1999. "Racing to Map Our DNA." *Time Daily* 153:1–6. Available at www.time.com

Lohr, Steve. 2011. "Carrots, Sticks and Digital Health Records." *The New York Times*, February 26. Available at www.nytimes.com

Lohr, Steve. 2009. "Wal-Mart Plans to Market Digital Health Records System." *The New York Times*, March 11. Available at www.nytimes.com

Lohr, Steve. 2008. "Health Care that Puts a Computer on the Team." *The New York Times*, December 27. Available at www.nytimes.com

Lunden, Ingrid. 2012. "Forrester: 760 Million Tablets in Use by 2016, Apple 'Clear Leader,' Frames Also Enter the Frame." TechCrunch, April 24. Available at techcrunch.com

Lynch, John, Jennifer Bevan, Paul Achter, Tim Harris, and Celeste M. Condit. 2008. "A Preliminary Study of How Multiple Exposures to Messages about Genetics Impact Lay Attitudes towards Racial and Genetic Discrimination." *New Genetics and Society* 27(1):43–56.

Madden, Mary, Amanda Lenhart, Maeve Duggan, Sandra Cortesi, and Urs Gasser. *Teens and Technology 2013*. March 13. Available at www.pewinternet.org

Malamud, Ofer, and Cristian Pop-Eleches. 2010. "Home Computer Use and the Development of Human Capital." National Bureau of Economic Research Working Paper 15814. Available at www.nber.org

MFA (Marketplace Fairness Act). 2013. "About." Available at www.marketplacefairness.org

Martinez, Michael. 2013. "After Ravages of Flesh Eating Bacteria, Aimee Copeland Uses New Bionic Hands." *CNN News*, May 20. Available at www.cnn.com

Mateescu, Oana. 2010. "Introduction: Life in the Web." *Journal of Comparative Research in Anthropology and Sociology* 1(2):1–21.

Mayer, Sue. 2002. "Are Gene Patents in the Public Interest?" *BIO-IT World*, November 12. Available at www.bio-itworld.com

Mayo Clinic. 2013. "Genetic Testing: Definition." Available at www.mayoclinic.com

McCollum, Sean. 2011. "Getting Past the 'Digital Divide.'" *Teaching Tolerance* 39. Available at www.tolerance.org

McCormick, S. J., and A. Richard. 1994. "Blastomere Separation." *Hastings Center Report*, March-April, pp. 14–16.

McDermott, John. 1993. "Technology: The Opiate of the Intellectuals." In *Technology and the Future*, Albert H. Teich, ed. (pp. 89–107). New York: St. Martin's Press.

McFarling, Usha L. 1998. "Bioethicists Warn Human Cloning Will Be Difficult to Stop." *Raleigh News and Observer*, November 18, p. A5.

Merton, Robert K. 1973. "The Normative Structure of Science." In *The Sociology of Science*, Robert K. Merton, ed. Chicago: University of Chicago Press.

Mesthene, Emmanuel G. 1993. "The Role of Technology in Society." In *Technology and the Future*, Albert H. Teich, ed. (pp. 73–88). New York: St. Martin's Press.

Mooney, Chris, and Sheril Kirshenbaum. 2009. *Unscientific America: How Scientific Literacy Threatens Our Future*. Philadelphia PA: Basic Books.

Murphie, Andrew, and John Potts. 2003. *Culture and Technology*. New York: Palgrave Macmillan.

NARAL. (National Abortion and Reproductive Rights Action League). 2013. *Mifepristine: The Impact of Abortion Politics on Women's Health in Scientific Research*. February 28. Available at www.prochoiceamerica.org

National Sleep Foundation. 2011 (March 7). "Annual Sleep in America Poll Exploring Connections with Communications Technology Use and Sleep." Available at www.sleepfoundation.org

Nature News. 2011 (February 23). "Animal Research: Battle Scars." *Nature* 470 (7335): 452–453.

NCSL (National Conference of State Legislatures). 2008. "Human Cloning Laws." Available at www.ncsl.org

Near, Christopher. 2013. "Selling Gender: Associations of Box Art Representation of Female Characters with Sales for Teen and Mature Rated Video Games." *Sex Roles* 68(3/4): 252–269.

Neuman, William, and Andrew Pollack. 2010. "Farmers Cope with Roundup Resistant Weeds." *The New York Times*, May 3. Available at www.nytimes.com

NIC (National Intelligence Council). 2003. "The Global Technology Revolution, Preface and Summary." Rand Corporation. Available at www.rand.org

NIEHS (National Institute of Environmental Health Sciences) 2009 (April 27). "Countries Unite to Reduce Animal Use in Product Toxicity Testing Worldwide." *Press Release*. Available at www.niehs.nih.gov

NIH (National Institute of Health). 2012. "The Promise of Stem Cells." U.S. Department of Health and Human Services. Available at stemcells.nih.gov

NSF (National Science Foundation). 2013. "Science and Engineering Indicators, 2012." Available at www.nsf.gov

Noonan, Mary C., and Jennifer L. Glass. 2012. "The Hard Truth about Telecommuting." *Monthly Labor Review* (June):38–45.

Oates, Thomas Patrick. 2009. "New Media in the Repackaging of NFL Fandom." *Sociology of Sport Journal* 26:31–49.

Occupy Monsanto. 2013. "March against Monsanto." Available at occupy-monsanto.com

Ogburn, William F. 1957. "Cultural Lag as Theory." *Sociology and Social Research* 41:167–174.

Olson, Parmy. 2013. "Rise of the Telepresence Robots." *Forbes*, June 27. Available at www.forbes.com

Olster, Marjorie. 2013. "GMO Foods: Key Points in the Genetically Modified Debate." *The Huffington Post*, August 2. Available at www.huffingtonpost.com

OpenNet. 2012 *OpenNet Initiative: Global Internet Filtering Map*. Available at map.opennet.net

Ophir, Eyal, Charles Nass, and Anthony D. Wagner. 2009. "Cognitive Control in Media Multitaskers." *Proceeds of the National Academy of Science* 106(37):15583–15587.

ORNL (Oak Ridge National Laboratory). 2011. "Medicine and the New Genetics." Office of Biological and Environmental Research. Available at www.ornl.gov

Padilla-Walker, Laura, Sarah M. Coyne, and Ashley M. Fraser. 2012. "Getting a High-Speed Family Connection: Associations between Family Media Use and Family Connection." *Family Relations* 61(July):426–440.

Park, Alice. 2009. "Stem-Cell Research: The Quest Resumes." *Time*, January 29. Available at www.time.com

PCAST (Presidents Council of Advisors on Science and Technology). 2012. *Report to the President and Congress on the Fourth Assessment of the National Nanotechnology Initiative*. Washington DC. April. Available at www.whitehouse.gov

Perlroth, Nicole. 2013. "Hackers in China Attacked the Times for Last Four Months." *The New York Times*, January 30. Available at www.nytimes.com

Pethokoukis, James M. 2004. "Meet Your New Co-Worker." *U.S. News and World Report*, March 7. www.usnews.com

Pew. 2013 (January 16). "*Roe v. Wade* at 40: Most Oppose Overturning Abortion Decision." Pew Research Center. Available at www.pewforum.org

Pew. 2009 (November 4). "Social Isolation and the New Technology." Available at www.pewresearch.org

Pew. 2007. *Trends in Political Values and Core Attitudes: 1987–2007*. Washington, DC: The Pew Research Center.

Pew Research Center. (April 22). 2013. "Public's Knowledge of Science and Technology." Available at www.people-press.org

Picard, Martin, and Doug M. Turnbull. 2013. "Linking the Metabolic State and Mitochondrial DNA in Chronic Disease, Health, and Aging." *Diabetes* 62 (March):672–678.

Pollack, Andrew. 2011. "U.S. Approves Genetically Modified Alfalfa." *The New York Times*, January 27. Available at www.nytimes.com

Postman, Neil. 1992. *Technopoly: The Surrender of Culture to Technology*. New York: Alfred A. Knopf.

Power, Emma. 2008. "Furry Families: Making a Human-Dog Family through Home." *Social and Cultural Geography* 9(5):535–555.

PCAT (President's Council of Advisors on Science and Technology). 2012 (February). *Report to the President—Engage to Excel: Producing 1 Million Additional College Graduates with Degrees in Science, Technology, Engineering, and Mathematics*. Executive Office of the President.

Presti, Ken. 2012 (December 21). "The Top 10 Security Breaches of 2012." Available at www.crn.com

Preston, Jennifer. 2011. "Facebook Officials Keep Quiet on Its Role in Revolts." *The New York Times*, February 14. Available at www.nytimes.com

Price, Tom 2008. "Science in America: Are We Falling Behind in Science and Technology?" *CQ Researcher* 18(2):24–48.

Rabino, Isaac. 1998. "The Biotech Future." *American Scientist* 86(2):110–112.

Research Animal Resources. 2003. "Ethics and Alternatives." University of Minnesota. Available at www.ahc.umn.edu

Reuters. 2013. "Spain Readies Hefty Jail Terms over Internet Piracy." *Reuters News Service*, September 20. Available at www.reuters.com

Richardson, Michelle, and Robin Greene. 2013. "NSA Legislation Since the Leaks Began." American Civil Liberties Union. Available at www.aclu.org

Robinson-Avila, Kevin. 2013. "Intel Launches Full Court Press into Mobile." *Albuquerque Journal of Business*, August 5. Available at www.abqjournal.com

Roe v. Wade. 1973. 410 U.S. 113.

Rovner, Julie. 2013. "Cloning, Stem Cells Long Mired in Legislative Gridlock." National Public Radio, May 16. Available at www.npr.org

Sedgh, Gilda, Susheela Singh, Iqbal H. Shah, Elizabeth Ahman, Stanley K. Henshaw, and Akinrinola Bankole. 2012. "Induced Abortion: Incidents and Trends Worldwide from 1995 to 2008." *The Lancet* 379 (9816):625–632.

Semantic Media. 2011. "Semantic Media—Smart Media for the Semantic Web." Available at semanticmedia.org

Sharkey, Joe. 2009. "A Meeting in New York? Can't We Videoconference?" *The New York Times*, May 12. Available at www.nytimes.com

Slade, Giles. 2012. *The Big Disconnect: The Story of Technology and Loneliness*. Amherst, New York: Prometheus Books.

Smith, Aaron. 2011 (March 17). "The Internet and Campaign 2010." Available at www.pewinternet.org

Spaeth, Matt. 2011. "What You Need to Know about GMOs." *Food Integrity Now*, March 8. Available at www.foodintegritynow.org

Stein, Rob, and Michaeleen Doucleff. 2013. "Scientists Clone Human Embryos to Make Stem Cells." National Public Radio, May 15. Available at www.npr.org

Stem Cell Research Advancement Act. 2013. H.R. 2433. Available at www.govtrack.us

Sterling, Toby. 2013. "EU Penalizes Microsoft $733m Breaking Browser Deal." *The Boston Globe*, March 7. Available at www.bostonglobe.com

Sticca, Fabio, and Sonja Perren. 2013. "Is Cyberbullying Worse than Traditional Bullying? Examining the Differential Roles of Media, Publicity, and Anonymity for the Perceived Severity of Bullying." *Journal of Youth and Adolescence* 42(5): 739–750.

Teitell, Beth. 2013. "Passing Notes in the Back of Coursera." National Public Radio. *Marketplace Commentary*, September 25. Available at www.marketplace.org

Thas, Angela, Chat Garcia Ramilo, and Cheekay Garcia Cinco. 2007. *Gender and ICT*. United Nations Development Programme. Bangkok, Thailand: Pacific Development Information Programme. Available at www.undp.org

The Economist. 2008 (December 18). "Pocket World in Figures." Available at www.economist.com

Toffler, Alvin. 1970. *Future Shock*. New York: Random House.

United Nations. 2012. "United Nations E-Government Survey 2012." Available at unpan1.un.org

UNEP (United Nations Environmental Programme). 2011 (May 12). "The International Treaty on Damage Resulting from Living Modified Organisms Receives Sixteen Signatures." Available at www.cbd.int

U.S. Census Bureau. 2013a. "Computer and Home Internet Use in the United States: Population Characteristics." *Document P20-569*. Available at www.census.gov

U.S. Census Bureau. 2013b. *Statistical Abstract of the United States, 2012*, 128 ed. Washington, DC: U.S. Government Printing Office.

U.S. Census Bureau. 2011 (May 26). "E-Stats." Available at www.census.gov

U.S. Citizenship and Immigration Services. 2013 (March 24). "H-1B Fiscal Year (FY) 2014 Season." Available at www.uscis.gov

U.S. Department of Justice. 2011 (April 22). "Joint Status Report on Microsoft's Compliance with the Final Judgment." *Document 927*. Available at www.justice.gov

U.S. Newswire. 2003 (October 2). *Feminists Condemn House Passage of Deceptive Abortion Ban, Urge Activists to March on Washington*. Available at www.usnewswire.com

Wait, Patience. 2012. "VA Computers Remain Unencrypted, Years after Breach." *Information Week*, October 19. Available at www.informationweek.com

Wakefield, Jane. 2010. "World Wakes Up to Digital Divide." *BBC New*, March 19. Available at news.bbc.co.uk

Weinberg, Alvin. 1966. "Can Technology Replace Social Engineering?" *University of Chicago Magazine* 59(October):6–10.

WEF (World Economic Forum). 2013. *The Global Information Technology Report 2013*. Available at www3.weforum.org

Welsh, Jonathan. 2009. "Late on a Car Loan? Meet the Disabler." *The Wall Street Journal*, March 25. Available at www.online.wsj.com

Welter, Cole H. 1997. "Technological Segregation: A Peek through the Looking Glass at the Rich and Poor in an Information Age." *Arts Education Policy Review* 99(2):1–6.

White, D. Steven, Angappa Gunasekaran, Timothy P. Shea, and Godwin C. Ariguzo. 2011. "Mapping the Global Digital Divide." *International Journal of Business Information Systems* 7:207–219.

White House. 2003 (November 5). *President Bush Signs Partial Birth Abortion Ban Act of 2003*. Available at www.whitehouse.gov

Whitlock. 2013. "Immersive Telepresence Solutions." Available at www.whitlock.com

Wikileaks. 2011. "Keep Us Strong." Available at www.wikileaks.org

Winner, Langdon. 1993. "Artifact/Ideas as Political Culture." In *Technology and the Future*, Albert H. Teich, ed. (pp. 283–294). New York: St. Martin's Press.

WIW (when I work). 2013. "The Easiest Way to Schedule and Communicate with Your Employees." Available at wheniwork.com

World Bank. 2009. *Global Economic Prospects: Technology Diffusion in the Developing World*. Washington DC: The World Bank

Whittaker, Zack. 2013. "CISPA Suffers Setback in Senate Citing Privacy Concerns." *C/Net News*, April 25. Available at news.cnet.com

WHO (World Health Organization). 2012. "Safe Abortion: Technical and Policy Guidance for Health Systems," 2nd ed. Available at apps.who.int

World Hunger Education Service. 2011. "2011 World Hunger and Poverty Facts and Statistics." Available at www.worldhunger.org

Zickuhr, Kathryn, and Aaron Smith. 2012 (April 13). *Digital Differences*. Pew Research Center's Internet and American Life. Available at pewinternet.org

Chapter 15

Alexander, Karen, and Mary E. Hawkesworth, eds. 2008. *War and Terror: Feminist Perspectives*. Chicago: University of Chicago Press.

Alvarez, Lizette. 2009. "Suicides of Soldiers Reach High of Nearly 3 Decades." *New York Times*, January 29. Available at www.nytimes.com

American Jewish Committee. 2011 (September 21). *Annual Survey of American Jewish Opinion*. Available at www.ajc.org

Animals at Arms. 2010. "Animals at Arms." *Army-Technology*, December 22. Available at www.army-technology.com

Arango, Tim, and Anne Bernard. 2013. "As Syrians Fight, Sectarian Strife Infects Mideast." *The New York Times*, June 1. Available at www.nytimes.com/

Arms Control Association. 2003 (May). *The Nuclear Proliferation Treaty at a Glance*. Available at www.armscontrol.org/factsheets

Associated Press. 2005. *Pentagon Tests Negative for Anthrax*. MSNBC, March 15. Available at www.msnbc.msn.com

Associated Press. 2008. *U.S. Officials: Scientist Was Anthrax Killer*. MSNBC, August 6. Available at www.msnbc.msn.com

Atomic Archive. 2011. *Arms Control Treaties*. Available at www.atomicarchive.com

Baker, Peter, Helene Cooper, and Mark Mazzetti. 2011. "Bin Laden is Dead, Obama Says." *The New York Times*, May 1. Available at www.nytimes.com

Baliunas, Sallie. 2004. "Anthrax Is a Serious Threat." In *Biological Warfare*, William Dudley, ed. (pp. 53–58). Farmington Hills, MA: Greenhaven Press.

Barkan, Steven, and Lynne Snowden. 2001. *Collective Violence*. Boston: Allyn and Bacon.

Barnes, Steve. 2004. "No Cameras in Bombing Trial." *New York Times*, January 29, p. 24.

Barstow, David. 2008. "One Man's Military-Industrial-Media Complex." *New York Times*, November 29. Available at www.nytimes.com

BBC. 2000. "UN Admits Rwanda Genocide Failure." *BBC News*, April 15. Available at news.bbc.co.uk

BBC. 2007. "Hamas Takes Full Control of Gaza." *BBC News*, June 15. Available at news.bbc.co.uk

Bercovitch, Jacob, ed. 2003. *Studies in International Mediation: Advances in Foreign Policy Analysis*. New York: Palgrave Macmillan.

Bergen, Peter. 2002. *Holy War, Inc.: Inside the Secret World of Osama bin Laden*. New York: Free Press.

Berrigan, Frida. 2009. "We Arm the World." *In These Times*, January 2. Available at www.inthesetimes.com

Berrigan, Frida, and William Hartung. 2005. *U.S. Weapons at War: Promoting Freedom or Fueling Conflict?* World Policy Institute Report. Available at www.worldpolicy.org/projects/arms/reports/wawjune2005.html

Bjorgo, Tore. 2003. *Root Causes of Terrorism.* Paper presented at the International Expert Meeting, June 9–11. Oslo, Norway: Norwegian Institute of International Affairs.

Bonner, Michael. 2006. *Jihad in Islamic History: Doctrines and Practice.* Princeton, New Jersey: Princeton University Press.

Borum, Randy. 2011. Radicalization into Violent Extremism II: A Review of Conceptual Models and Empirical Research. *Journal of Strategic Security* 4(4): 37–62. Available at http://scholarcommons.usf.edu/jss/vol4/iss4/3

Boustany, Nora. 2007. "Janjaweed Using Rape as 'Integral' Weapon in Darfur, Aid Group Says." *Washington Post*, July 3. Available at www.washingtonpost.com

Brauer, Jurgen. 2003. "On the Economics of Terrorism." *Phi Kappa Phi Forum*, Spring, pp. 38–41.

Braun, Stephen. 2011. "U.S. Defense Sales to Bahrain Rose before Crackdown." *ABC News*, June 11. Available at abcnews.go.com

Brinkley, Joel, and David E. Sanger. 2005. "North Koreans Agree to Resume Nuclear Talks." *New York Times*, July 10. Available at www.nytimes.com

Broad, William J., and David E. Sanger. 2009. "Report Says Iran Has Data to Make a Nuclear Bomb." *The New York Times*, October 3. Available at www.nytimes.com

Brooks, David. 2011. "Huntington's Clash Revisited." *The New York Times,* March 3. Available at www.nytimes.com

Brown, Michael E., Sean M. Lynn-Jones, and Steven E. Miller, eds. 1996. *Debating the Democratic Peace.* Cambridge, MA: MIT Press.

Bumiller, Elisabeth. 2011. "The Dogs of War: Beloved Comrades in Afghanistan." *The New York Times*, May 11. Available at www.nytimes.com

Buncombe, Andrew. 2009. "End of Sri Lanka's Civil War Brings Back Tourists." *The Independent*, August 16. Available at www.independent.co.uk

Burnham, Gilbert, Riyadh Lafta, Shannon Doocy, and Les Roberts. 2006. "Mortality after the 2003 Invasion of Iraq: A Cross-Sectional Cluster Sample Survey." *The Lancet*, October 11. Available at www.thelancet.com

Burns, Robert. 2011. "Panetta: U.S. within Reach of Defeating Al Qaeda." *The Washington Times*, July 9. Available at www.washingtontimes.com

Burrelli, David F. 2013. "Women in Combat: Issues for Congress." *Congressional Research Service* (May 9). Available at www.fas.org/sgp/crs/natsec/R42075.pdf

Carnegie Endowment for International Peace. 2009. *World Nuclear Arsenals 2009.* Available at www.carnegieendowment.org

Carneiro, Robert L. 1994. "War and Peace: Alternating Realities in Human History." In *Studying War: Anthropological Perspectives*, S. P. Reyna and R. E. Downs, eds. (pp. 3–27). Langhorne, PA: Gordon & Breach.

Carpenter, Dustin, Tova Fuller, and Les Roberts. 2013. "WikiLeaks and Iraq Body Count: The Sum of Parts May Not Add Up to the Whole—A Comparison of Two Tallies of Iraqi Civilian Deaths." *Prehospital and Disaster Medicine, 28,* 3(June):223–229.

Center for Arms Control and Non-Proliferation. 2013. "Fact Sheet: Global Nuclear Weapons Inventories in 2013." Available at www.armscontrolcenter.org.

Center for Defense Information. 2003. *Military Almanac.* Available at www.cdi.org

CENTCOM (U.S. Central Command). 2013 (July). "Contractor Support of U.S. Operations in the USCENTCOM Area of Responsibility to Include Iraq and Afghanistan." CENTCOM Quarterly Contractor Census Reports. Available at www.acq.osd.mil/log/PS/CENTCOM_reports.html

Clever, Molly, and David R. Segal. 2012. "After Conscription: The United States and the All-Volunteer Force." *Sicherheit und Frieden (Security and Peace) 30*(1): 9–18.

CNN. 2009 (December 31). "Charges Dismissed against Iraq Contractors." Available at www.cnn.com

CNN. 2003a (August 10). *Army Begins Chemical Weapons Burn.* Available at www.cnn.com

CNN. 2003b (February 26). *Poll: Muslims Call U.S. Ruthless, Arrogant.* Available at www.cnn.com

CNN. 2001 (January 31). *Libyan Bomber Sentenced to Life.* Available at www.europe.cnn.com

CNN/Opinion Research Corporation Poll. 2007 (June 22–24). *Iraq.* Available at www.pollingreport.com/iraq3.htm

Cohen, Ronald. 1986. "War and Peace Proneness in Pre- and Post-industrial States." In *Peace and War: Cross-Cultural Perspectives*, M. L. Foster and R. A. Rubinstein, eds. (pp. 253–267). New Brunswick, NJ: Transaction Books.

Conflict Research Consortium. 2003. *Mediation.* Available at www.colorado.edu/conflict/peace

Cowell, Alan, and A. G. Sulzberger. 2009. "Lockerbie Convict Returns to Jubilant Welcome." *The New York Times*, August 20. Available at www.nytimes.com

Dahl, Frederick. 2013. "North Korea Nuclear Test Still Shrouded in Mystery." *Reuters*, June 18. Available at www.reuters.com

Dalrymple, William. 2013 (June 25). "A Deadly Triangle: Afghanistan, Pakistan, & India." The Brookings Essay. Available at www.brookings.edu/series/the-brookings-essay

Dao, James. 2013. "Deployment Factors Are Not Related to Rise in Military Suicides, Study Says." *The New York Times*, August 6. Available at www.nytimes.com

Dao, James. 2009. "Veterans Affairs Faces Surge of Disability Claims." *New York Times*, July 12. Available at www.nytimes.com

Dareini, Ali Akbar. 2009. "Iran Missile Test: Ahmadinejad Says It's within Israel's Range." *Huffington Post*, May 20. Available at www.huffingtonpost.com

Davenport, Christian. 2005. "Guard's New Pitch: Fighting Words." *Washington Post*, April 28. Available at www.washingtonpost.com

DCAS (Defense Casualty Analysis System). 2013. "Conflict Casualties." U.S. Department of Defense. Available at www.dmdc.osd.mil

Deen, Thalif. 2000 (September 9). *Inequality Primary Cause of Wars, Says Annan.* Available at www.hartford-hwp.com/archives

Dixon, William J. 1994. "Democracy and the Peaceful Settlement of International Conflict." *American Political Science Review* 88(1):14–32.

Donnelly, John. 2011 (January 24). "More Troops Lost to Suicide." Available at www.congress.org

Environmental Media Services. 2002 (October 7). *Environmental Impacts of War.* Available at www.ems.org

Feder, Don. 2003. "Islamic Beliefs Led to the Attack on America." In *The Terrorist Attack on America*, Mary E. Williams, ed. (pp. 20–23). Farmington Hills, MA: Greenhaven Press.

Ferran, Lee, Brian Ross and James Gordon Meek. 2013. "Kenya Westgate Mall Attack: What is Al-Shabab?" September 23. ABC News. Available at www.abcnews.go.com

Frankel, Rebecca. 2011a. "War Dog: There's a Reason Why They Brought One to Get Osama Bin Laden." *Foreign Policy*, May 4. Available at www.foreignpolicy.com/articles

Frankel, Rebecca. 2011b. "War Dogs: The Legend of the Bin Laden Hunter Continues." *Foreign Policy*, May 12. Available at www.foreignpolicy.com

Frey, Josh. 2003 (August 12). "Anti-Swimmer Dolphins Ready to Defend Gulf." Available at www.navy.mil

Friedman, Brandon. 2013. "The Rise (and Fall) of the VA Backlog." *Time*, June 3. Available at nation.time.com

Funke, Odelia. 1994. "National Security and the Environment." In *Environmental Policy in the 1990s: Toward a New Agenda*, 2nd ed., Norman J. Vig and Michael E. Kraft, eds. (pp. 323–345). Washington, DC: Congressional Quarterly.

Gallup Poll. 2013 (April 25). "Terrorism in the United States." Available at www.gallup.com

Gamel, Kim, and Lee Keath. 2011. "Muammar Gaddafi Dead: Libya Dictator Maddened West, Captured, Killed in Sirte." *The Huffington Post*, October 20. Available at www.huffingtonpost.com

Gardner, Simon. 2007 (February 22). *Sri Lanka Says Sinks Rebel Boats on Truce Anniversary.* Available at www.reuters.com

Garrett, Laurie. 2001. "The Nightmare of Bioterrorism." *Foreign Affairs* 80:76.

Geitner, Paul. 2012. "U.S., Europe, and Japan Escalate Rare-Earth Dispute with China." *The New York Times*, June 27. Available at www.nytimes.com

German, Erik. 2011. "Flipper Goes to War." *The Daily*, July 18. Available at www.thedaily.com

Gettleman, Jeffrey. 2008. "Rape Victims' Words Help Jolt Congo into Change." *New York Times*, October 18. Available at www.nytimes.com

Gettleman, Jeffrey. 2009. "Symbol of Unhealed Congo—Male Rape Victims." *The New York Times*, August 4. Available at www.nytimes.com

Geneva Graduate Institute for International Studies. 2013. *Small Arms Survey, 2013.* New York: Oxford University Press.

Gioseffi, Daniela. 1993. "Introduction." In *On Prejudice: A Global Perspective*, Daniela Gioseffi, ed. (pp. xi–l). New York: Anchor Books, Doubleday.

Goodwin, Jeff. 2006. "A Theory of Categorical Terrorism." *Social Forces* 84(4):2027–2046.

Gordon, Michael R. 2013. "Iran is Said to Want Direct Talks with U.S. on Nuclear Program." *The New York Times*, July 26. Available at www.nytimes.com

Goure, Don. 2003. *First Casualties? NATO, the U.N.* MSNBC News, March 20. Available at www.msnbc.com/news

Greenburg, Jan Crawford, and Ariane de Vogue. 2008. "Supreme Court: Guantanamo Detainees Have Rights in Court." *ABC News,* June 12. Available at abcnews.go.com/

GreenKarat. 2007. *Mining for Gems.* Available at www.greenkarat.com

Greenwald, Glenn. 2013. "The Same Motive for Anti-US 'Terrorism' is Cited Over and Over." *The Guardian*, April 24. Available at theguardian.com

Haiken, Melanie. 2013. "Suicide Rate among Vets and Active Duty Military Jumps—Now 22 a Day." *Forbes*, February 5. Available at www.forbes.com

Healy, Jack, and Alissa J. Rubin. 2011. "U.S. Blames Pakistan-Based Group for Attack on Embassy in Kabul." *The New York Times,* September 14. Available at www.nytimes.com/

Hefling, Kimberly. 2011. "Former Marine, Advocate Kills Self after War Tour." 2011. *The Jacksonville Daily News*, April 15. Available at jdnews.com

Hewitt, J. Joseph, Jonathan Wilkenfeld, and Ted Robert Gurr. 2012. *Peace and Conflict 2012: Executive Summary.* College Park, MD: Center for International Development and Conflict Management.

Hicks, Josh. 2013. "Pentagon Extends Benefits to Same-Sex Military Spouses." *The Washington Post,* August 14. Available at www.washingtonpost.com

Hooks, Gregory, and Leonard E. Bloomquist. 1992. "The Legacy of World War II for Regional Growth and Decline: The Effects of Wartime Investments on U.S. Manufacturing, 1947–72." *Social Forces* 71(2):303–337.

Horwitz, Sari. 2013. "New Charges Brought against Former Blackwater Guards in Baghdad Shooting." *The Washington Post*, October 17. Available at www.washingtonpost.com

Howard, Michael. 2007. "Children of War: The Generation Traumatised by Violence in Iraq." *The Guardian*, February 6. Available at www.guardian.co.uk

Huntington, Samuel. 1996. *The Clash of Civilizations and the Remaking of World Order.* New York: Simon and Schuster.

IEP (Institute for Economics & Peace). 2013. The Economic Consequences of War on the U.S. Economy. Available at economicsandpeace.org

ICBL (International Campaign to Ban Landmines). 2013. *States not Party.* Available at www.icbl.org

Jeon, Arthur. 2011. "German Shepherd? Belgian Malinois? Navy SEAL Hero Dog Is Top Secret." *Global Animal*, May 5. Available at www.globalanimal.org

Johnston, David, and John Broder. 2007. "F.B.I. Says Guards Killed 14 Iraqis without Cause." *The New York Times*, November 14. Available at www.nytimes.com

Kaphle, Anup. 2013. "Timeline: Unrest in Syria." *The Washington Post,* June 13. Available at apps.washingtonpost.com/

Kemper, Bob. 2003. "Agency Wages Media Battle." *Chicago Tribune*, April 7. Available at www.chicagotribune.com

Klare, Michael. 2001. *Resource Wars: The New Landscape of Global Conflict.* New York: Metropolitan Books

Knickerbocker, Brad. 2002. "Return of the Military-Industrial Complex?" *Christian Science Monitor*, February 13. Available at www.csmonitor.com

Lamont, Beth. 2001. "The New Mandate for UN Peacekeeping." *The Humanist* 61:39–41.

Langley, Robert. n.d. "Chemical Sniffing Dogs Deployed along Borders." Available at http://usgovinfo.about.com

Laqueur, Walter. 2006. "The Terrorism to Come." In *Annual Editions 05–06*, Kurt Finsterbusch, ed. (pp. 169–176). Dubuque, IA: McGraw-Hill/Dushkin.

Larrabee, F. Stephen, Stuart E. Johnson, John Gordon, Peter A. Wilson, Caroline Baxter, Deborah Lai, and Calin Trenkov-Wermuth. *NATO and the Challenges of Austerity.* Santa Monica, CA: RAND Corporation, 2012. www.rand.org/pubs/monographs/MG1196

Larsen, Kaj. 2011. "Harnessing the Military Power of Animal Intelligence." CNN, July 31. Available at http://articles.cnn.com

LeardMann, Cynthia A. 2013. "Risk Factors Associated with Suicide in Current and Former U.S. Military Personnel." *Journal of the American Medical Association* 310(5):496–506.

Lederer, Edith. 2005 (May 20). "Annan Lays out Sweeping Changes to U.N." Associated Press. Available at www.apnews.com

Leinwand, Donna. 2003. "Sea Lions Called to Duty in Persian Gulf.*" USA Today*, February 16. Available at www.usatoday.com

Leland, Anne, and Mari-Jana Oboroceanu. 2010 (February 26). *American War and Military Operations Casualties: Lists and Statistics.* Washington, DC: Congressional Research Service. Available at www.fas.org

Levy, Clifford J., and Peter Baker. 2009. "U.S.-Russian Nuclear Agreement Is First Step in Broad Effort." *The Washington Post*, July 6. Available at www.washingtonpost.com

Levy, Jack S. 2001. "Theories of Interstate and Intrastate War: A Levels of Analysis Approach." In *Turbulent Peace: The Challenges of Managing International Conflict*, Chester A. Crocker, Fen Osler Hampson, and Pamela Aall, eds. (pp. 3–27). Washington, DC: U.S. Institute of Peace.

Li, Xigen, and Ralph Izard. 2003. "Media in a Crisis Situation Involving National Interest: A Content Analysis of Major U.S. Newspapers' and TV Networks' Coverage of the 9/11 Tragedy." *Newspaper Research Journal* 24:1–16.

Lindner, Andrew M. 2009. "Among the Troops: Seeing the Iraq War through Three Journalistic Vantage Points." *Social Problems* 56(1):21–48.

MacAskill, Ewen, Ed Pilkington, and Jon Watts. 2006. "Despair at UN over Selection of 'Faceless' Ban Ki-moon as General Secretary." *The Guardian*, October 7. Available at www.guardian.co.uk

Macfarquhar, Neil. 2010. "Change Will Not Come Easily to the Security Council." *The New York Times*, November 8. Available at www.nytimes.com

Marshall, Leon. 2007 (July 16). *Elephants "Learn" to Avoid Land Mines in War-Torn Angola.* National Geographic News. Available at www.nationalgeographic.com

Masood, Salman, and Ihsanullah Tipu Mehsud. 2013. "U.S. Drone Strike in Pakistan Kills at Least 16." *The New York Times,* July 3. Available at www.nytimes.com/

Matthews, Dylan. 2013. "Everything You Need to Know about the Drone Debate, in One FAQ." *The Washington Post*, March 8. Available at washingtonpost.com

Mazzetti, Mark, and Scott Shane. 2009a. "Interrogation Memos Detail Harsh Tactics by the C.I.A." *New York Times*, April 16. Available at www.nytimes.com

Mazzetti, Mark, and Scott Shane. 2009b. "C.I.A. Abuse Cases Detailed in Report on Detainees." *The New York Times*, August 24. Available at www.nytimes.com

McGirk, Tim. 2006. "Collateral Damage or Civilian Massacre in Haditha?" *Time*, March 19. Available at www.time.com

Military. 2007. *Tuition Assistance (TA) Program Overview.* Available at education.military.com

Military Professional Resources Inc. (MPRI) 2011. *InfoCenter: Brochures.* Available at www.mpri.com

Miller, Susan. 1993. "A Human Horror Story." *Newsweek*, December 27, p. 17.

Miller, T. Christian. 2007. "Private Contractors Outnumber U.S. Troops in Iraq." *Los Angeles Times*, July 4. Available at www.latimes.com

Millman, Jason. 2008. "Industry Applauds New Dual-Use Rule." *Hartford Business Journal*, September 29. Available at www.hartfordbusiness.com

Mittal, Dinesh, K.L. Drummond, D. Belvins, G. Curran, P. Corrigan, and G. Sullivan. 2013. "Stigma Associated with PTSD: Perceptions of Treatment Seeking Combat Veterans." *Psychiatric Rehabilitation Journal* 36,2(June):86–92.

Morgan, Jenny, and Alic Behrendt. 2009. *Silent Suffering: The Psychological Impact of War, HIV, and Other High-Risk Situations on Girls and Boys in West and Central Africa.* Woking: Plan. Available at plan-international.org

Montalván, Luis Carlos. 2011. *Until Tuesday: A Wounded Warrior and the Golden Retriever Who Saved Him.* New York: Hyperion.

Morales, Lymari. 2009 (April 6). *Americans See Newer Threats on Par with Ongoing Conflicts.* Available at www.gallup.com

Moroney, Jennifer D. P., Joe Hogler, Benjamin Bahney, Kim Cragin, David R. Howell, Charlotte Lynch, and Rebecca Zimmerman. 2009. "Building Partner Capacity to Combat Weapons of Mass Destruction." Rand National Defense Research Institute. Available at www.rand.org

Mueller, John. 2006. *Overblown: How Politicians and the Terrorism Industry Inflate National Security Threats, and Why We Believe Them.* New York: Free Press.

Myers, Steven Lee. 2009. "Women at Arms: Living and Fighting alongside Men, and Fitting In." *The New York Times*, August 16. Available at www.nytimes.com

Myers-Brown, Karen, Kathleen Walker, and Judith A. Myers-Walls. 2000. "Children's Reactions to International Conflict: A Cross-Cultural Analysis." Paper presented at the National Council of Family Relations, Minneapolis, November 20.

Nakamura, David, and Billy Kenber. 2013. "Obama Administration to Transfer Two Guantanamo Bay Detainees." *The Washington Post,* July 26. Available at www.washingtonpost.com

NASP (National Association of School Psychologists). 2003. *Children and Fear of War and Terrorism.* Available at www.nasponline.org

NCPSD (National Center for Posttraumatic Stress Disorder). 2007. *What Is Post Traumatic Stress Disorder?* Available at www.ncptsd.va.gov

National Counterterrorism Center. 2012 (March 12). *2011 Report on Terrorism.* Available at www.nctc.gov

National Priorities Project. 2011. *Cost of War: Trade-Offs.* Available at costofwar.com

NCVAS (National Center for Veterans Analysis and Statistics). 2013. "Expenditures," U.S. Department of Veterans' Affairs. Available at www.va.gov/vetdata/Expenditures.asp

New York Times. 2013. "A History of the Detainee Population." *The New York Times,* July 23. Available at projects.nytimes.com/Guantanamo/

Newport, Frank. 2011 (March 9). *Republicans and Democrats Disagree on Muslim Hearings.* Available at www.gallup.com

Nordland, Rod. 2012. "Risks of Afghan War Shift from Soldiers to Contractors." *The New York Times,* February 11. Available at nytimes.com

NPR (National Public Radio). 2009. "The Navy's Other Seals . . . Dolphins." National Public Radio, December 5. Available www.npr.org

Ochmanek, David, and Lowell H. Schwartz. 2008. "The Challenge of Nuclear-Armed Regional Adversaries." Rand Project Air Force. Available at www.rand.org

Office of the Coordinator for Counterterrorism. 2013 (July 30). "Foreign Terrorist Organizations." Available at www.state.gov/s/ct

Office of Management and Budget. 2013a. *Table S-5: Proposed Budget by Category.* Available at www.whitehouse.gov

Office of Management and Budget. 2013b. *Table 32-1: Policy Budget Authority and Outlays by Function, Category, and Program.* Available at www.whitehouse.gov/sites/default/files/omb/budget/fy2012/assets/32_1.pdf

Office of Weapons Removal and Abatement. 2013. *To Walk the Earth in Safety: The United States' Commitment to Humanitarian Mine Action and Conventional Weapons Destruction.* Available at www.state.gov

Oliver, Amy. 2011. "Mammals with a Porpoise . . . Meet the Dolphins and Sea Lions Who Go to War with the U.S. Navy." *The Daily Mail,* July 6. Available at www.dailymail.co.uk

Paul, Annie Murphy. 1998. "Psychology's Own Peace Corps." *Psychology Today* 31:56–60.

PBS NewsHour. 2013. "Bradley Manning Leaked Classified Documents to Spark 'Debate' on Foreign Policy." *PBS,* February 28. Available at www.pbs.org

Permanent Court of Arbitration. 2011 *About Us and Cases.* Available at www.pca-cpa.org

Pew Research Center. 2013 (June 12). *Sexual Assault in the Military Widely Seen as Important Issue, but No Agreement on Solution.* Available at www.pewresearch.org

Pew Research Center. 2011a (April 3). *A Nation of Flag Waivers.* Available at www.pewresearch.org

Pew Research Center. 2011b (May 17). *Obama's Challenge in the Muslim World: Arab Spring Fails to Improve U.S. Image.* Pet Global Attitudes Project. Available at www.pewresearch.org

Pew Research Center. 2009 (April 24). *Public Remains Divided Over Use of Torture.* Available at http://www.people-press.org

Pew Research Center for the People & the Press. 2010 (August 24). "Public Remains Conflicted over Islam." Available at www.people-press.org

Pilkington, Ed. 2013a."US Government Identifies Men on Guantanamo 'Indefinite Detainee' List." *The Guardian,* June 17. Available at www.theguardian.com

Pilkington, Ed. 2013b. "Bradley Manning Verdict: Cleared of 'Aiding the Enemy' but Guilty of Other Charges." *The Guardian,* July 30. Available at www.theguardian.com

Plumer, Brad. 2013. "The U.S. Gives Egypt $1.5 Billion a Year in Aid. Here's What It Does." *The Washington Post,* July 9. Available at www.washingtonpost.com

Porter, Bruce D. 1994. *War and the Rise of the State: The Military Foundations of Modern Politics.* New York: Free Press.

Portero, Ashley. 2013. "Women in Combat Units Could Help Reduce Sexual Assaults: US Joint Chiefs Chairman." *International Business Times,* January 25. Available at www.ibtimes.com

Powell, Bill, and Tim McGirk. 2005. "The Man Who Sold the Bomb." *Time,* February 14, pp. 22–31.

Price, Eluned. 2004. "The Served and Suffered for Us." November 1. The *Telegragh.* Available at www.telegraph.co.uk

Priest, Dana, and William M. Arkin. 2010. "Top Secret America: A Hidden World, Growing Beyond Control." *The Washington Post,* July 19. Available at www.washingtonpost.com

Puckett, Neal, and Haytham Faraj. 2009 (June 25). *Navy-Marine Corps Court Hears Appeal in Wuterich v U.S.* Retrieved from www.puckettfaraj.com

Rapoza, Kenneth. 2013. "Syria's Chemical Weapons Slowly Being Eradicated." Forbes, October 16. Available at www.forbes.com

Rashid, Ahmed. 2000. *Taliban: Militant Islam, Oil, and Fundamentalism in Central Asia.* New Haven, CT: Yale University Press.

Rasler, Karen, and William R. Thompson. 2005. *Puzzles of the Democratic Peace: Theory, Geopolitics, and the Transformation of World Politics.* New York: Palgrave Macmillan.

Renner, Michael. 2000. "Number of Wars on Upswing." In *Vital Signs: The Environmental Trends That Are Shaping Our Future,* Linda Starke, ed. (pp. 110–111). New York: W. W. Norton.

Renner, Michael. 1993. "Environmental Dimensions of Disarmament and Conversion." In *Real Security: Converting the Defense Economy and Building Peace,* Karl Cassady and Gregory A. Bischak, eds. (pp. 88–132). Albany: State University of New York Press.

Reuters. 2013. "U.S. Files Criminal Charges against Snowden over Leaks: Sources." *Reuters,* June 21. Available at www.reuters.com

Reuters. 2008. "Case Dropped against Officer Accused in Iraq Killings." *New York Times,* June 18. Available at www.nytimes.com

Romero, Anthony. 2003. "Civil Liberties Should Not Be Restricted during Wartime." In *The Terrorist Attack on America,* Mary Williams, ed. (pp. 27–34). Farmington Hills, MA: Greenhaven Press.

Rosenberg, Tina. 2000. "The Unbearable Memories of a U.N. Peacekeeper." *New York Times,* October 8, pp. 4, 14.

Roughton, Randy. 2011 (February 3). "Fallen Marine's Family Adopts His Best Friend." U.S. Air Force Official Website, Available at www.af.mil

Ryan, Jason, and Huma Khan. 2011. "In Reversal, Obama Orders Guantanamo Military Trial for 9/11 Mastermind Khaled Sheikh Mohamed." *ABC News,* April 4. Available at www.abcnews.go.com

Saad, Lydia. 2013 (April 26). *Post-Boston, Half in U.S. Anticipate More Terrorism Soon.* Available at poll.gallup.com/

Saad, Lydia. 2011 (May 4). *Majority in U.S. Say Bin Laden's Death Makes America Safer.* Available at www.gallup.com

Salem, Paul. 2007. "Dealing with Iran's Rapid Rise in Regional Influence." *The Japan Times,* February 22. Available at www.carnegieendowment.org

Salladay, Robert. 2003 (April 7). *Anti-War Patriots Find They Need to Reclaim Words, Symbols, Even U.S. Flag from Conservatives.* Available at www.commondreams.org

Sang-Hun, Choe, and Chris Buckley. 2013. "North Korean Leader Supports Resumption of Nuclear Talks, State Media Say." *The New York Times,* July 26. Available at www.nytimes.com/

Savage, Charlie. 2009. "Accused 9/11 Mastermind to Face Civilian Trial in N.Y." *New York Times,* November 13. Available at www.nytimes.com/

Savage, Charles, and Elisabeth Bumiller. 2012. "An Iraqi Massacre, a Light Sentence and a Question of Military Justice." *The New York Times,* January 27. Available at www.nytimes.com/

Save the Children. 2009. *Last in Line, Last in School 2009.* Available at www.savethechildren.org

Save the Children. 2007. *State of the World's Mothers: Saving the Lives of Children Under Five.* Available at www.savethechildren.org

Scheff, Thomas. 1994. *Bloody Revenge.* Boulder, CO: Westview Press.

Schmitt, Eric. 2005. "Pentagon Seeks to Shut Down Bases across Nation." *New York Times,* May 14. Available at www.nytimes.com

Schneier, Bruce. 2008. "America's Dilemma: Close Security Holes, or Exploit Them Ourselves." *Wired,* May 1. Available at www.wired.com

Schroeder, Matt. 2007. "The Illicit Arms Trade." Washington, DC: Federation of American Scientists. Available at www.fas.org

Schultz, George P., William J. Perry, Henry A. Kissinger, and Sam Nunn. 2007. "A World Free of Nuclear Weapons." *Wall Street Journal,* January 4. Available at www.wsj.com

Schwartz, Moshe, and Joyprada Swain. 2011. *Department of Defense Contractors in Afghanistan and Iraq: Background and Analysis.* Washington, DC: Congressional Research Service. Available at www.fas.org

Shanahan, John J. 1995. "Director's Letter." *Defense Monitor* 24(6):8.

Shanker, Thom, and David E. Sanger. 2009. "Pakistan is Rapidly Adding Nuclear Arms, U.S. Says." *New York Times*, May 17. Available at www.nytimes.com

Sheridan, Mary Beth, and William Branigin. 2010. "Senate Ratifies New U.S.-Russia Nuclear Weapons Treaty." *Washington Post*, December 22. Available at washingtonpost.com

Shrader, Katherine. 2005. "WMD Commission Releases Scathing Report." *Washington Post*, March 31. Available at www.washingtonpost.com

Silverstein, Ken. 2007. "Six Questions for Joost Hiltermann on Blowback from the Iraq-Iran War." *Harper's Magazine*, July 5. Available at www.harpers.org

Simon, Scott. 2003 (March 29). "Marine Mammals on Active Duty: Navy Uses Dolphins, Sea Lions to Patrol Waters in Persian Gulf." Available at www.npr.org

Skocpol, Theda. 1994. *Social Revolutions in the Modern World.* Cambridge, UK: Cambridge University Press.

Skocpol, Theda. 1992. *Protecting Soldiers and Mothers: The Political Origins of Social Policy in the United States.* Cambridge: The Belknap Press of Harvard University Press.

Slevin, Peter. 2009. "U.S. to Announce Transfer of Detainees to Ill. Prison." *Washington Post*, December 15. Available at www.washingtonpost.com

Smith, Craig. 2004. "Libya to Pay More to French in '89 Bombing." *New York Times*, January 9, p. 6.

Smith, Lamar. 2003. "Restricting Civil Liberties During Wartime Is Justifiable." In *The Terrorist Attack on America*, Mary Williams, ed. (pp. 23–26). Farmington Hills, MA: Greenhaven Press.

Smith-Spark, Laura, Jim Sciutto, and Elise Labbott. 2013. "Iran Nuclear Talks Start in Geneva amid 'Cautious Optimism.'" *CNN*, October 16. Available at www.cnn.com

Starr, J. R., and D. C. Stoll. 1989. *U.S. Foreign Policy on Water Resources in the Middle East.* Washington, DC: Center for Strategic and International Studies.

Stiglitz, Joseph E., and Linda J. Bilmes. 2010. "The True Cost of the Iraq War: $3 Trillion and Beyond." *The Washington Post*, September 5. Available at www.washingtonpost.com

Stiglitz, Joseph E., and Linda J. Bilmes. 2008. *The Three Trillion Dollar War: The True Cost of the Iraq Conflict.* New York, NY: W.W. Norton & Company.

Stimson Center. 2007. *Reducing Nuclear Dangers in South Asia.* Available at www.stimson.org

SIPRI (Stockholm International Peace Research Institute). 2013. *Recent Trends in Military Expenditure.* Available at www.sipri.org

Strobel, Warren, David Kaplan, Richard Newman, Kevin Whitelaw, and Thomas Grose. 2001. "A War in the Shadows." *U.S. News and World Report* 130:22.

Tanielian, Terri, Lisa H. Jaycox, Terry L . Schell, Grant N. Marshall, Audrey Burnam, Christine Eibner, Benjamin R. Karney, Lisa S. Meredith, Jeanne S. Ringe, Mary E. Vaiana, and the Invisible Wounds Study Team. 2008 "Invisible Wounds of War: Summary and Recommendations for Addressing Psychological and Cognitive Injuries." Available at www.rand.org

Tavernise, Sabrina. 2007. "U.S. Contractor Banned by Iraq over Shootings." *New York Times*, September 18. Available at www.nytimes.com

Taylor, Guy. 2013. "Foot-Draggers: U.S. and Russia Slow to Destroy Own Chemical Weapons amid Syria Smackdown." *Washington Times*, October 16. Available at www.washingtontimes.com

Tharoor, Ishaan. 2013. "Terrorist Attack at Nairobi Mall: Behind al-Shabaab's War with Kenya." *Time*, September 23. Available at www.world.time.com

Tilly, Charles. 1992. *Coercion, Capital and European States: AD 990-1992.* Cambridge, MA: Basil Blackwell.

Tremlatt, Giles. 2011. "Eta Declares Permanent Ceasefire." *The Guardian*, January 20. Available at www.guardian.co.uk

Treo. 2010. "Treo the Dog Awarded Animal VC." *The Telegraph*, February 6. Available at www.telegraph.co.uk

UCDP (Uppsala Conflict Data Program). 2013. *Ongoing Armed Conflicts.* Available at www.pcr.uu.se/research/ucdp/datasets/ucdp_prio_armed_conflict_dataset/

Uenuma, Francine. 2013 (August 23). "Number of Children Who Have Fled Reaches One Million." Save the Children Media Release. Available at www.savethechildren.org

UNMAO (United National Mine Action Office). 2011 (June). "UNMAO Regional Fact Sheet." *Southern Sudan.* Available at http://reliefweb.int/report/sudan/unmao-regional-fact-sheet-southern-sudan-updated-june-201

UNICEF (United Nations Children's Fund). 2011. *The State of the World's Children 2011: Adolescence—An Age of Opportunity.* Available at www.unicef.org/

United Nations. 2013. *United Nations Peacekeeping Operations.* Available at www.un.org/en/peacekeeping/resources/statistics/factsheet.shtml

United Nations. 2003. *Some Questions and Answers.* Available at www.unicef.org

United Nations. 1948. *Convention on the Prevention and Punishment of the Crime of Genocide.* Available at www.un.org

UNAUSA (United Nations Association of the United States of America). 2004. *Landmines Overview.* Available at www.unausa.org

UNHCR (United Nations High Commissioner for Refugees). 2013. *UNHCR Country Operations Profile: Iraq.* Available at www.unhcr.org/pages/49e486426.html

United Nations Security Council. 2011 (March 17). *Resolution 1973.S/RES/1973.* Available at www.un.org

U.S. Army Medical Command. 2007 (May). *Mental Health Advisory Team IV Findings.* Available at www.armymedicine.army.mil

U.S. Department of Defense. 2013. *Annual Report to Congress: Military and Security Developments Involving the People's Republic of China.* Available at www.defense.gov/pubs/2013_China_Report_FINAL.pdf

U.S. Department of Homeland Security. 2013. *FY 2012 Budget-in Brief.* Available at www.dhs.gov

U.S. Department of State. 2011 (February 14). The United States' Leadership in Conventional Weapons Destruction." *Fact Sheet: Bureau of Political-Military Affairs.* Available at www.state.gov

U.S. Department of State. 2007. *Small Arms/Light Weapons Destruction.* Available at www.state.gov

U.S. Navy. 2010. "Navy Marine Mammal Program Excels during Frontier Sentinel." Available at www.navy.mil/search/display.asp?story_id=53979

Vesely, Milan. 2001. "UN Peacekeepers: Warriors or Victims?" *African Business* 261:8–10.

Viner, Katharine. 2002. "Feminism as Imperialism." *The Guardian*, September 21. Available at www.guardian.co.uk

Wali, Sarah O., and Deena A. Sami. 2011. "Egyptian Police Using U.S.-Made Tear Gas against Demonstrators." *ABC News*, January 28. Available at abcnews.go.com

Walker, Peter. 2013. "The Bradley Manning Trial: What We Know from the Leaked WikiLeaks Documents." *The Guardian*, July 30. Available at www.theguardian.com

Washington Post. 2010. *Top Secret America: A Washington Post Investigation.* Available at www.washingtonpost.com

Waters, Rob. 2005. "The Psychic Costs of War." *Psychotherapy Networker*, March–April, pp. 1–3.

Watkins, Thomas. 2007. *Haditha Hearings Enter Fourth Day. Time*, May 11. Available at www.time.com.

Wax, Emily. 2003. "War Horror: Rape Ruining Women's Health." *Miami Herald*, November 3. Available at www.miami.com

Weimann, Gabriel. 2006. "Terror on the Internet: The New Arena, the New Challenges." Washington, DC: U.S. Institute of Peace Press.

Weisman, Jonathan. 2013. "Iran Talks Face Resistance in U.S. Congress." *The New York Times*, November 12. Available at www.nytimes.com

White House Statement. 2013. "Text of White House Statement on Chemical Weapons in Syria." *The New York Times*, June 13. Available at www.nytimes.com

Williams, Mary E., ed. 2003. *The Terrorist Attack on America.* Farmington Hills, MA: Greenhaven Press.

Wise, Lindsey. 2011. "Marine who Pushed Suicide Prevention Took Own Life." *Houston Chronicle*, April 9. Available at www.chron.com/news/

Woehrle, Lynne M., Patrick G. Coy, and Gregory M. Maney. 2008. *Contesting Patriotism: Culture, Power, and Strategy in the Peace Movement.* Lanham, Md.: Rowman & Littlefield Publishers, Inc.

Worsnip, Patrick. 2011. "South Sudan Admitted to U.N. as 193rd Member." *Reuters UK*, July 14. Available at uk.reuters.com

Zakaria, Fareed. 2000. "The New Twilight Struggle." *Newsweek*, October 12. Available at www.msnbc.com/news

Zaremba, John, Antonio Planas, and Laurel J. Sweet. 2013. "Dzhokhar Tsarnaev Pleads Not Guilty to Marathon Blasts." *The Boston Herald*, July 10. Available at www.bostonherald.com/

Zoroya, Gregg. 2006. "Lifesaving Knowledge, Innovation Emerge in War Clinic." *USA Today*, March 27. Available at www.usatoday.com

Name Index

Subject Index

Note: Page numbers in **boldface** type denote definitions. Page numbers in *italics* denote tables, figures, illustrations, and captions.